高等学校教材

INTRODUCTION TO ENVIRONMENTAL IMPACT ASSESSMENT

环境影响评价导论

（汉英双语）

（原著第四版）

(英) 约翰·格拉森 (John Glasson)
(英) 里基·泰里夫 (Riki Therivel) 著
(英) 安德鲁·查德威克 (Andrew Chadwick)
廖嘉玲 贾瑜玲 徐鹤 任伟 等译

Chinese English

化学工业出版社
·北京·

内容简介

《环境影响评价导论》由在世界环境影响评价领域享有极高声誉的三位专家 John Glasson、Riki Therivel 和 Andrew Chadwick 著写，书中很多新理念值得我国从事环境影响评价的相关人士借鉴，具有全面系统、结构清晰和深入浅出的特点。书中系统阐述了环境影响评价的原则和进程、实施程序、实践及远景；结合作者的实践经验，书中还介绍了典型环境影响评价的案例，比较了世界不同国家和地区的环境影响评价体系，讨论了战略环境评价发展。

《环境影响评价导论》可作为环境专业本科生、研究生的教材使用，还可作为环境评价、环境规划和环境管理技术人员的实践指导用书，并可供政府机构管理人员参考阅读。

Introduction to Environmental Impact Assessment, Fourth edition/by John Glasson, Riki Therivel, Andrew Chadwick

ISBN 9780415664707

北京市版权局著作权合同登记号：01-2015-101

图书在版编目（CIP）数据

环境影响评价导论：汉文、英文/(英) 约翰·格拉森 (John Glasson)，(英) 里基·泰里夫 (Riki Therivel)，(英) 安德鲁·查德威克 (Andrew Chadwick) 著；廖嘉玲等译. —北京：化学工业出版社，2021.10

书名原文：Introduction to Environmental Impact Assessment

高等学校教材

ISBN 978-7-122-38415-7

Ⅰ.①环… Ⅱ.①约…②里…③安…④廖… Ⅲ.①环境影响-评价-汉、英 Ⅳ.①X820.3

中国版本图书馆 CIP 数据核字（2021）第 018997 号

责任编辑：满悦芝　　文字编辑：丁海蓉　陈　雨

责任校对：王鹏飞　　装帧设计：张　辉

出版发行：化学工业出版社（北京市东城区青年湖南街 13 号　邮政编码 100011）

印　　装：大厂聚鑫印刷有限责任公司

787mm×1092mm　1/16　印张 51¾　字数 1268 千字　2021 年 10 月北京第 1 版第 1 次印刷

购书咨询：010-64518888　　售后服务：010-64518899

网　　址：http://www.cip.com.cn

凡购买本书，如有缺损质量问题，本社销售中心负责调换。

定　　价：**298.00 元**

译者的话

从环境影响评价第一次被《中华人民共和国环境保护法（试行）》提及，到2003年《中华人民共和国环境影响评价法》正式实施，在二十多年的曲折摸索过程中，环境影响评价从一个理念演变为一部法律，环境评价制度的内容不断得到丰富。伴随着环境影响评价制度的不断发展，环境影响评价作为环境科学专业的一门核心课程，受到了越来越多的大学和研究机构的重视，相应的环境影响评价教材也层出不穷。为了更为详细地了解国外的环境评价理论与实践，加快中国环境影响评价国际化，特此翻译了英国知名环境评价专家约翰·格拉森（John Glasson）等著写的《环境影响评价导论》（第四版），之所以选中本书：一是因为本书结合具体的案例研究，清晰易懂地讲述了环境影响评价的程序与过程；二是书中涉及的方法和技术对从事环境评价的人具有广泛的适用性。本书可以作为环境影响评价的入门双语教材，供环境科学、环境工程、环境管理专业的本科生或者相关专业的研究生使用。

本书的翻译工作由徐鹤教授总体负责，其中徐鹤教授负责了本书二分之一的翻译工作，廖嘉玲、贾瑜玲各负责了本书四分之一的翻译工作，徐鹤教授进行了全书的校核工作。其他参与翻译的人员有：王嘉炜、杨铁婷、刘婷、谷兆炜、任伟。刘政、王红磊对本书的翻译进行了指导。

本书得以翻译出版要感谢化学工业出版社的大力支持。

由于时间及水平有限，疏漏之处在所难免，希望专家、学者及广大读者批评指教。

译者

2021年9月

环境影响评价导论
第四版

《环境影响评价导论》为学生和从业者提供了一个清晰的环境评价主题概述。本书由具有环境影响评价业内广泛研究、培训和实践经验的三位作者撰写，涵盖了环境影响评价最新的法律法规、导则与良好实践。

本版本对案例研究部分进行拓展，相比之前的版本探索了更多的关键问题，与此同时，第四版还对如下基本信息进行更新：

- 环境评价的演变；
- 欧盟及英国环境评价程序的执行经验；
- 环境评价过程中的最佳实践；
- 世界各地环境评价制度比较；
- 战略环境评价/战略评价的立法和实践发展；
- 环境评价的未来展望。

虽然这本书的重点是英国和欧盟，但其中阐述的原则和技术在国际范围内适用。这本书以丰富多彩的图画和新颖的现代设计，为规划专业的本科生和研究生，以及学习环境管理和政策、环境科学、地理和建筑环境的学生提供环境影响评价的基本介绍。规划者、开发商、社会团体、政府及企业的决策者也将此书作为掌握环境影响评价主题的有效途径。

约翰·格拉森（John Glasson），牛津布鲁克斯大学环境规划荣誉教授，环境影响评价小组（IAU）和牛津可持续发展研究所创始理事。同时，他也是西澳大利亚科廷大学的客座教授。

里基·泰里夫（Riki Therivel），牛津布鲁克斯大学客座教授，环境影响评价小组（IAU）高级研究员，Levett-Therivel 可持续顾问成员。2010 年，Riki Therivel 和 John Glasson 被任命为英国基础设施规划委员会（IPC）委员。

安德鲁·查德威克（Andrew Chadwick），环境影响评价小组（IAU）高级研究员。

Introduction to Environmental Impact Assessment 4th edition

Introduction to Environmental Impact Assessment provides students and practitioners with a clearly structured overview of the subject, as well as critical analysis and support for further studies. Written by three authors with extensive research, training and practical experience in EIA, the book covers the latest EIA legislation, guidance and good practice.

Featuring an extended case studies section that explores more key issues than in previous editions, this 4th edition also updates essential information on:

- the evolving nature of EIA;
- experience of the implementation of the changing EU and UK EIA procedures;
- best practice in the EIA process;
- comparative EIA systems worldwide;
- development of SEA/SA legislation and practice; and
- prospects for the future of EIA.

Although the book's focus is on the UK and the EU, the principles and techniques it describes are applicable internationally. With colour images and a new modern design, the book provides an essential introduction to EIA for undergraduate and postgraduate students on planning courses, as well as those studying environmental management and policy, environmental sciences, geography and the built environment. Planners, developers, community groups and decision-makers in government and business will also welcome the book as an effective way to get to grips with this important and evolving subject that affects a wide range of development projects.

John Glasson is Emeritus Professor of Environmental Planning, Founding Director of the Impacts Assessment Unit (IAU) and of the Oxford Institute for Sustainable Development (OISD), at Oxford Brookes University. He is also Visiting Professor at Curtin University in Western Australia.

Riki Therivel is Visiting Professor at Oxford Brookes University, a Senior Research Associate in the IAU and partner in Levett-Therivel sustainability consultants. In 2010 both Riki Therivel and John Glasson were appointed Commissioners of the UK Infrastructure Planning Commission (IPC).

Andrew Chadwick is Senior Research Associate in the IAU.

第四版前言

和第一版相比，第四版的目标和范围是不变的。然而，正如在第一版前言中所指出的，环境影响评价在继续发展和完善，关于环境影响评价的所有讨论都会是未来进一步发展的重要部分。环境影响评价在世界范围内的传播正在逐步变得更全面。欧盟在《环境评价指令》方面已经积累了超过25年的实施经验，并且有10年的修订经验。在制订环境评价程序、加强已识别的弱点、扩大活动范围和评价成效方面，都引起了各界人士相当大的兴趣。为了反映这些变化，本修订版对评论注释部分进行了更新，引入了一些对学生和从业者都日益重要的问题。

本版在保留第一版结构的基础上，加上了第三版中的大部分材料，并将大的变化和补充设定为特定部分。在第一部分（有关原则及程序），重点介绍适应性环境评价的重要性与迅速增长的影响评价活动范围。对欧盟新修订的《环境评价指令》的实施情况进行了全面的讨论，其中包括新增成员国。第3章中列出了英国具体的2011年新法规和程序操作。在第二部分（讨论环境评价程序）中，大部分内容已更新，包括筛选和范围界定、替代方案、影响识别、预测、参与与陈述、减缓与加强、监测与审计。

根据对环境评价有效性和实际操作进行的重要评价的结果，我们已对第三部分（实践概况）做出重大修订。例如，第8章包含了许多关于环境评价中法律挑战含义的新材料。第9章包括一些新的实践案例研究。大多数案例研究都是基于英国项目级别的环境评价，但其中也讨论了两个关于战略环境评价的例子以及健康影响评估等新主题。虽然并不是所有选定的个案研究都是环境评价实践的最佳例子，但它们确实包括一些针对环境影响评价中特定问题的新颖和创新的方法，例如公众参与的新方法和累积效应的处理方法。我们提出要重视环境影响评价实践中的一些局限性。第10章（国际实践对比）也做了重大修订，反映了如非洲、中国和转型期国家日益增长的经验，并对如加拿大和澳大利亚已经建立好的环境评价制度做出审查。

为反映环境影响评估的一些变化前景，我们对第四部分（展望）也做了大幅修订。第11章讨论了战略环境评价（SEA）的必要性及其局限性。它回顾了美国、欧盟和联合国欧洲经委会以及中国的战略环境评价现状，讨论了有关欧盟《战略环境评价指令》在英国实施的细节。最后，以最近对《战略环境评价指令》有效性的研究结果作为总结。我们对第12章进行了大量修订与扩展。例如，它包括更多的累积影响、社会经济影响、健康影响评价、平等影响评价、适当性评价、新的弹性思考领域，在环境评价中对气候变化进行规划这一至关重要的主题，以及可能向更综合的评价的转变。本章最后讨论了环境管理制度和审计的并行和互补发展。这些议题共同构成了未来环境评价改进的行动清单。这一章特别引用了欧洲委员会、IAIA（国际影响评价协会）和IEMA（环境管理和评价协会）最近对环境影响评价实践的一些研究结果。

附录包括修订后的《环境影响评价指令》和《战略环境评价指令》的完整版本，修订后的影响评价小组（IAU）环境影响报告书审查包，以及全球关键的环境影响评价期刊和网站指南。

John Glasson
Riki Therivel
Andrew Chadwick
2011 年于牛津

Preface to the fourth edition

The aims and scope of this fourth edition are unchanged from those of the first edition. However, as noted in the preface to the first edition, EIA continues to evolve and adapt, and any commentary on the subject must be seen as part of a continuing discussion. The worldwide spread of EIA is becoming even more comprehensive. In the European Union there is now over 25 years' experience of the implementation of the pioneering EIA Directive, including 10 years' experience of the important 1999 amendments. There has been considerable interest in the development of the EIA process, in strengthening perceived areas of weakness, in extending the scope of activity and also in assessing effectiveness. Reflecting such changes, this fully revised edition updates the commentary by introducing and developing a number of issues that are seen as of growing importance to both the student and the practitioner of EIA.

The structure of the first edition has been retained, plus much of the material from the third edition, but considerable variations and additions have been made to specific sections. In Part 1 (on principles and procedures), the importance of an adaptive EIA, plus the burgeoning range of EA activity, are addressed further. In the EU context, the implementation of the amended EIA Directive is discussed more fully, including the divergent practice across the widening range of Member States. The specific new 2011 regulations and procedures operational in the UK are set out in Chapter 3. In Part 2 (discussion of the EIA process), most elements have been updated, including screening and scoping, alternatives, impact identification, prediction, participation and presentation, mitigation and enhancement, and monitoring and auditing.

We have made major changes to Part 3 (overview of practice), drawing on the findings of important reviews of EIA effectiveness and operation in practice. For example, Chapter 8 includes much new material on the implication of legal challenges in EIA. Chapter 9 includes some new practice case studies. Most of the case studies are UK-based and involve EIA at the individual project level, although two examples of SEA are also discussed, plus new topics such as health impact assessment. While it is not claimed that the selected case studies all represent best examples of EIA practice, they do include some novel and innovative approaches towards particular issues in EIA, such as new methods of public participation and the treatment of cumulative effects. They also draw attention to some of the limitations of the process in practice. Chapter 10 (Comparative practice) has also had a major revision, reflecting, for example, growing experience in African countries, China and countries in transition, and major reviews for some well-established EIA systems in, for example, Canada and Australia.

Part 4 of the book (Prospects) has also been substantially revised to reflect some of the changing prospects for EIA. Chapter 11 discusses the need for strategic environmental assessment (SEA) and some of its limitations. It reviews the status of SEA in the USA, European Union and UNECE, and China. It then discusses in more detail how the European SEA Directive is being implemented in the UK. It concludes with the results of recent research into the effectiveness of the SEA Directive. Chapter 12 has been extensively revised and extended. It includes, for example, more consideration of cumulative impacts, socio-economic impacts, health impact assessment, equalities impact assessment, appropriate assessment, the new area of resilience thinking, and the vitally important topic of planning for climate change in EIA, plus possible shifts towards more integrated assessment. The chapter concludes with a discussion of the parallel and complementary development of environmental management systems and audits. Together, these topics act as a kind of action list for future improvements to EIA. This chapter in particular, but also much else in the book, draws on some of the findings of recent reviews of EIA practice undertaken by, among others, the EC, the IAIA (International Association for Impact Assessment) and the IEMA (the Institute of Environmental Management and Assessment).

The Appendices include the full versions of the amended EIA Directive and the SEA Directive, a revised IAU EIS review package, and a guide to key EIA journals and websites worldwide.

John Glasson
Riki Therivel
Andrew Chadwick
Oxford 2011

第一版前言

过去几年，人们对环境问题产生了极大的兴趣。1987 年世界环境与发展委员会的报告（布伦特兰报告）提供了一个主要推动力；1992 年，里约高峰会议是对该推动力进行的一次加速尝试。有关环境问题和可持续发展的讨论大多是关于更好地管理与环境相协调的当前活动。然而，总是会有新的发展压力。若能在规划阶段避免或减轻未来发展对环境的潜在有害影响，那么好处何在？环境影响评价（EIA）预先评估计划活动对环境产生的影响，从而采取规避措施；预防优于治理。

1969 年，环境影响评价制度第一次在美国正式确立。随着《EC 指令》中关于环境影响评价的引入，环境影响评价已传播到世界各地，并在欧洲积极推行。该指令于 1988 年在英国实施。随后，环境评价活动迅速开展，目前，英国每年编写超过 300 份环境影响报告书（EISs）。环境评价是一种有效的方法。这也是许多从业者经验有限的领域。本书对环境影响评价的各个方面做了全面的介绍。它根据本科生和研究生的需求编撰，对规划人员、开发商和各利益组织等实践者也有一定价值。环境评价正处于快速发展的“学习曲线”上；本书也是该曲线上的一个点。

本书分为四个部分。第一部分介绍了环境评价的原则，概述了环境评价的发展、机构和立法背景。第二部分对环境评价程序展开逐步讨论和评价。第三部分是对目前的做法进行审查，主要是在英国和其他几个国家，并对选择的英国案例进行更加详细的介绍。第四部分考虑未来可能的发展。在 20 世纪 90 年代及以后，可能会有更多的环境评价“冰山”出现。本书最后对环境审计和战略环境评估方面的重要及相关发展进行了概述。

虽然这本书具有明显的英国特点，但它广泛借鉴了世界各地的环境评价经验，应该可以引起许多国家读者的兴趣。本书旨在突出最佳实践，并对方法提供充足见解以及参考文献支持，为从业者提供有价值的指导。有关影响评价特定主题领域（例如景观、空气质量、交通影响）的详细方法的信息，请参阅补充册《环境影响评价方法》（Morris 和 Therivel，1995，伦敦，伦敦大学学院出版社）。

John Glasson
Riki Therivel
Andrew Chadwick
牛津布鲁克斯大学

Preface to the first edition

There has been a remarkable and refreshing interest in environmental issues over the past few years. A major impetus was provided by the 1987 Report of the World Commission on the Environment and Development (the Brundtland Report); the Rio Summit in 1992 sought to accelerate the impetus. Much of the discussion on environmental issues and on sustainable development is about the better management of current activity in harmony with the environment. However, there will always be pressure for new development. How much better it would be to avoid or mitigate the potential harmful effects of future development on the environment at the planning stage. Environmental impact assessment (EIA) assesses the impacts of planned activity on the environment in advance, thereby allowing avoidance measures to be taken: prevention is better than cure.

Environmental impact assessment was first formally established in the USA in 1969. It has spread worldwide and received a significant boost in Europe with the introduction of an EC Directive on EIA in 1985. This was implemented in the UK in 1988. Subsequently there has been a rapid growth in EIA activity, and over 300 environmental impact statements (EISs) are now produced in the UK each year. EIA is an approach in good currency. It is also an area where many of the practitioners have limited experience. This text provides a comprehensive introduction to the various dimensions of EIA. It has been written with the requirements of both undergraduate and postgraduate students in mind. It should also be of considerable value to those in practice-planners, developers and various interest groups. EIA is on a rapid 'learning curve'; this text is offered as a point on the curve.

The book is structured into four parts. The first provides an introduction to the principles of EIA and an overview of its development and agency and legislative context. Part 2 provides a step-by-step discussion and critique of the EIA process. Part 3 examines current practice, broadly in the UK and in several other countries, and in more detail through selected UK case studies. Part 4 considers possible future developments. It is likely that much more of the EIA iceberg will become visible in the 1990s and beyond. An outline of important and associated developments in environmental auditing and in strategic environmental assessment concludes the text.

Although the book has a clear UK orientation, it does draw extensively on EIA experience worldwide, and it should be of interest to readers from many countries. The book seeks to highlight best practice and to offer enough insight to methods, and to supporting references, to provide valuable guidance to the practitioner. For information on detailed methods for assessment of impacts in particular topic areas (e. g. landscape, air quality, traffic impacts), the reader is referred to the complementary volume, *Methods of environmental impact assessment* (Morris and Therivel, 1995, London, UCL Press).

John Glasson

Riki Therivel

Andrew Chadwick

Oxford Brookes University

致　谢

我们感谢许多人的帮助，如果没有他们，这本书将不会出版。我们还要特别感谢我们家人的宽容和道义上的支持。还要感谢 Rob Woodward，感谢他制作了许多插图。此外，Taylor & Francis，副本编辑 Rosalind Davies 和编辑助理 Aimee Miles 也在把手稿转换成出版的文件过程中做出了重要的贡献。我们非常感谢我们的咨询顾问和研究经费资助机构，他们对牛津布鲁克斯大学（前牛津理工大学）规划学院的影响评估小组的工作给予了支持。我们特别希望记录英国政府部门（英国环境部，英国环境、运输和地区部，英国副首相办公室和英国社区与地方政府部）、欧共体环境委员会、经济和社会研究理事会（ESRC）、皇家鸟类保护协会（RSPB）、许多地方和地区主管部门，特别是英国能源工业的各个分支机构，为我们的环境评价研究和咨询工作提供了源动力和持续的正面支持。

牛津布鲁克斯大学的本科生和研究生对我们的想法进行了批判性检验。在此，特别感谢环境评价与管理学硕士课程的学生。非常感谢 Taylor & Francis 的工作人员对第四版的编辑和推广支持。我们得到规划、生物和分子科学学院同事的支持，以及环境评价学者、研究人员和顾问等更广泛的团体的支持，他们的支持使我们保持专注。非常感谢 Angus Morrison-Saunders 对我们的第三版建设性的评论，也非常感谢杨珊珊就中国环境影响评价的发展提出的建议。在这版书中，我们要特别感谢环境管理和评价协会的 Josh Fothergill 和英国社区与地方政府部的 Kim Chowns 提供环境管理和评价协会关于英国环境评价实践的 2011 年报告，以及英国社区与地方政府部新的 2011 年环境评价导则和指南。我们也非常感谢允许我们使用以下来源的材料的机构和个人：

英国自然保护主义者协会（卡通画：第二部分和第三部分）

RPS，Symonds/EDAW 和 Magnox Electric（图片 1.1）

环境评价审查（图 1.9）

环境数据服务（表 3.1 和表 3.2）

苏格兰政府（图 4.1 和图 4.2）

帕特森采石场（图 4.3）

南约克郡综合运输管理局（图 4.6）

苏格兰电力系统（图 4.8）

环境管理和评价协会（图 5.1 和图 12.6，表 8.5、表 12.4、表 12.6 和表 12.7）

法国电力公司能源，南安普顿每日回声，英国卫报（图 6.1）

大都会理事会（明尼阿波利斯/圣保罗），加拿大阿海珐资源，格里夫威格利，Evelop（图 6.2）

曼彻斯特大学环境影响评价中心（附录 4）

奥林匹克交付管理局（图 7.7）

高地和岛屿机关（图 9.3）

约翰威立国际出版公司（表 6.2）

基线环境咨询，西伯克利，加利福尼亚州（图 7.2）
英国环境部（表 6.3）
英国社区与地方政府部（表 3.5、表 3.6 和表 3.7，附录 2）
规划报（卡通图：第四部分）
山毛榉树出版社（图 7.8）
欧洲委员会（表 4.3、专栏 11.1、表 12.5）
西澳环境保护署（表 10.2、图 10.5）
西澳大利亚卫生部（图 12.2）
Scott Wilson（表 12.3）
多佛区议会（图 11.3）
副首相办公室（专栏 11.2）

Acknowledgements

Our grateful thanks are due to many people without whose help this book would not have been produced. We are particularly grateful for the tolerance and moral support of our families. Our thanks also go to Rob Woodward for his production of many of the illustrations. In addition, Louise Fox of Taylor and Francis, and copy-editor Rosalind Davies, and editorial assistant Aimee Miles have provided vital contributions in turning the manuscript into the innovative published document. We are very grateful to our consultancy clients and research sponsors, who have underpinned the work of the Impacts Assessment Unit in the School of Planning at Oxford Brookes University (formerly Oxford Polytechnic). In particular we wish to record the support of UK government departments (variously DoE, DETR, ODPM and DCLG), the EC Environment Directorate, the Economic and Social Research Council (ESRC), the Royal Society for the Protection of Birds (RSPB), many local and regional authorities, and especially the various branches of the UK energy industry that provided the original impetus to and continuing positive support for much of our EIA research and consultancy.

Our students at Oxford Brookes University on both undergraduate and postgraduate programmes have critically tested many of our ideas. In this respect we would like to acknowledge, in particular, the students on the MSc course in Environmental Assessment and Management. The editorial and presentation support for the fourth edition by the staff at Taylor and Francis is very gratefully acknowledged. We have benefited from the support of colleagues in the Schools of Planning and Biological and Molecular Sciences, and from the wider community of EIA academics, researchers and consultants, who have helped to keep us on our toes. We are grateful to Angus Morrison Saunders for some very useful pointers in his most constructive review of our third edition, and to Shanshan Yang for advice on the evolving approach to EIA in China. We owe particular thanks in this edition for the willingness of Josh Fothergill at IEMA, and Kim Chowns at DCLG, to provide advance copies of the IEMA 2011 Report on UK EIA practice, and the new 2011 DCLG EIA Regulations and Guidance. We are also grateful for permission to use material from the following sources:

British Association of Nature Conservationists (cartoons: Parts 2 and 3)

RPS, Symonds/EDAW and Magnox Electric (Plate 1. 1)

EIA Review (Figure 1. 9)

ENDS (Tables 3. 1 and 3. 2)

Scottish government (Figures 4. 1 and 4. 2)

Pattersons Quarries (Figure 4. 3)

South Yorkshire Integrated Transport Authority (Figure 4. 6)

Scottish Power Systems (Figure 4. 8)

IEMA (Figures 5. 1 and 12. 6, Tables 8. 5, 12. 4, 12. 6 and 12. 7)

EDF Energy, Southampton Daily Echo, Guardian Newspaper (Figure 6. 1)

Metropolitan Council (Minneapolis/St Paul), AREVA Resources Canada, Griff Wigley, Evelop (Figure 6. 2)

University of Manchester, EIA Centre (Appendix 4)

Olympic Delivery Authority (Figure 7. 7)

Highlands and Islands Enterprise (Figure 9. 3)

John Wiley & Sons (Table 6. 2)

Baseline Environmental Consulting, West Berkeley, California (Figure 7. 2)

UK Department of Environment (Table 6. 3)

UK Department of Communities and Local Government (Tables 3. 5, 3. 6 and 3. 7; Appendix 2)

Planning newspaper (cartoon: Part 4)

Beech Tree Publishing (Figure 7. 8)

European Commission (Table 4. 3, Box 11. 1, Table 12. 5)

West Australian Environmental Protection Agency (Table 10. 2, Figure 10. 5)

West Australian Department of Health (Figure 12. 2)

Scott Wilson (Table 12. 3)

Dover District Council (Figure 11. 3)

Office of the Deputy Prime Minister (Box 11. 2)

缩略语及术语

AA	适当性评价
ABI	英国年度业务调查
ADB	非洲开发银行
ADB	亚洲开发银行
AEE	环境影响的评价
AEP	环境专业人员协会
ANZECC	澳大利亚和新西兰环境和保护委员会
AONB	自然风光优美的区域
APC	空气污染控制
API	建议者资料信息评价
AQMA	空气质量管理区
BAA	机场有限公司（原英国机场管理局）
BANANA	完全没有建造任何东西
BG	保加利亚
BIO	生物情报服务
BME	黑色人种和少数族裔
BP	原英国石油公司
BPEO	最可行的环保方案
BS	英国标准
BWEA	英国风能协会
CAREC	中亚区域环境中心
CBA	成本效益分析
CC	郡议会
CCGT	联合循环燃气轮机
CCHP	冷却加热和电力结合
CCS	碳捕获和储存
CCW	威尔士乡村委员会
CE	绝对排斥
CEA	累积效应评价
CEAA	加拿大环境评估局
CEAM	累积效应评估与管理
CEARC	加拿大环境评价研究委员会
CEC	欧洲共同体委员会
CEGB	中央发电板
CEMP	建筑环境管理计划
CEPA	联邦环境保护署（澳洲）
CEQ	美国环境质量委员会
CEQA	加利福尼亚州环境质量法
CHP	热电联产
CIA	文化影响评价
CIE	社区影响评价
CISDL	国际可持续发展法律中心
CITES	《濒危物种贸易公约》
CO_2	二氧化碳
COWI	丹麦科威公司
CPO	强制采购订单
CPRE	英国乡村保护运动
CRM	条件排序法
CRS	美国国会研究服务
CRTN	道路交通噪声计算
CSR	企业社会责任
CVM	条件估值法
CY	塞浦路斯
CZ	捷克共和国
dB	分贝
dBA	加权分贝
DA	政务司司长下放行政权力（英国）
DBIS	英国商业、创新和技术部
DC	区议会
DCLG	英国社区与地方政府部
DECC	英国能源和气候变化部
DEFRA	英国环境、食品和农村事务部
DETR	英国环境、运输和地区部
DFID	英国国际发展部
DfT	英国交通部
DG	总局（CEC）

DMRB	道路和桥梁设计手册
DoE	英国环境部
DoEn	英国能源部
DoT	英国交通部
DTI	英国贸易和工业部
EA	环境评价
EA	英国环境局
EAGGF	欧洲农业指导与保障基金
EAP	环境行动计划
EBRD	欧洲复兴开发银行
EC	欧洲委员会
EcIA	生态影响评价
ECJ	欧洲法院
EDF	法国电力公司
EE	爱沙尼亚
EEA	欧洲环境局
EIA	环境影响评价
EIB	欧洲投资银行
EID	环境影响设计
EIR	环境影响报告
EIR	环境影响审查
EIS	环境影响报告书
EM&A	环境监察及环境监察及审核
EMAS	生态管理和审核计划
EMP	环境管理计划
EMS	环境管理系统
EN	英国自然委员会
ENDS	环境数据服务
EPA	英国环境保护法
EPA	美国环境保护法
EPA	美国环境保护署
EPA	西澳大利亚环境保护署
EPB	环境保护局（中国）
EPBCA	《环境保护和生物多样性保护法》（澳大利亚）
EPD	中国香港环境保护署
EqIA	平等影响评价
ERM	环境资源管理有限公司
ES	环境声明
ESRC	经济和社会研究理事会
ETSU	能源技术支援组
EU	欧盟
FEARO	联邦环境评价审查办公室
FEIS	最终环境影响报告书
FHWA	美国联邦公路管理局
FoE	地球之友
FONSI	无显著影响发现
G1；G2	1代；2代
GAM	目标实现矩阵
GHG	温室气体
GHK	咨询有限公司
GIS	地理信息系统
GNP	国民生产总值
GP	一般从业者
GPDO	一般允许开发令
GW	千兆瓦
ha	公顷
HEP	水力发电
HGV	重型货车
HIA	健康影响评价
HMG	女王政府
HMIP	女王政府污染监察局
HMSO	女王政府文书局
HPF	家庭生产函数
HPM	享乐成本估价法
HRA	生境管制评估
HSE	健康和安全执行局
HU	匈牙利
HWS	汉普郡废物服务中心
IA	影响评价
IAIA	国际影响评估协会
IAU	影响评价小组（牛津布鲁克斯）
IEA	环境评价研究所
IEMA	环境管理与评价研究所
IFI	国际资助机构
IIA	综合影响评估
IMD	复合剥夺指数
INEM	国际环境管理网络
IOCGP	国际组织间社会影响评价委员会

IPC	基础设施规划委员会
IPC	综合污染控制
IPCC	政府间气候变化专门委员会
IPHI	爱尔兰公共卫生研究所
ISO	国际标准化组织
IWM	废物管理研究所
JEAPM	环境评价政策和管理杂志
JNCC	联合自然保护委员会
KSEIA	韩国环境影响评价学会
kV	千伏特
L_{10}	监测期噪声水平不超过10%
LB	伦敦区
LCA	生命周期评价
LNG	液化天然气
LPA	地方规划局
LT	立陶宛
LTP	地方运输计划
LTP3	地方运输计划3
LULU	地方不可接受的土地用途
LV	拉脱维亚
MAFF	英国农业、林业和渔业部
MAUT	多属性效用理论
MBC	都市自治委员会
MCA	多准则评价
MCDA	多准则决策分析
MEA	环境评价手册
MMO	海事管理组织（英国）
MoD	英国国防部
MOEP	中国生态环境部
MT	马耳他
MW	兆瓦
NE	英国自然
NEPA	《美国国家环境政策法案》
NGC	国家电网公司
NGO	非政府组织
NHS	国民健康服务
NIMBY	不在我的后院（邻避）
NO_x	氮氧化物
NPDV	净现值
NPS	国家政策声明
NSIP	国家重点基础设施建设项目
NTS	非技术总结
ODA	奥林匹克交付管理局
ODPM	英国副首相办公室
OECD	经济合作与发展组织
OISD	牛津可持续发展研究所
OJ	欧共体官方刊物
OTP	营运计划
PADC	开发控制项目评估
PAS	规划咨询服务
PBS	平衡表
PEIR	计划环境影响报告
PEIS	计划环境影响声明
PER	公众环境审查（WA）
PIC	伙伴合租关系
PL	波兰
PM_{10}	直径小于10μm的颗粒物
PPG	规划政策指导
PPPs	政策、计划和规划
PPPP	政策、计划、规划或项目
PPS	规划政策声明
PWR	压水式反应堆
QBL	四倍底线
QOLA	生活质量评价
RA	恢复力联盟
RA	风险评价
RMA	资源管理法（新西兰）
RO	罗马尼亚
ROD	判定记录
RSPB	皇家鸟类保护协会
RTPI	皇家城市规划研究所
S106	第106节
SA	可持续性评价
SAC	特殊保护区域
SAIEA	南非环境评价研究所
SAVE	拯救英国的遗产
SD	可持续发展
SDD	苏格兰发展部
SEA	战略环境评价
SEERA	英格兰东南地区大会

S&EIA	社会经济和环境影响评价
SEPA	苏格兰环境保护署
SI	斯洛文尼亚
SIA	社会影响评估
SK	斯洛伐克
SNH	苏格兰自然遗产
SNIFFER	苏格兰和北爱尔兰环境研究论坛
SO_2	二氧化硫
SOER	环境报告状态
SoS	州秘书
SPA	特殊保护区
SSE	阻止斯坦斯特德扩张
SSSI	具有特殊科学价值的地点
TBL	三重底线
T&CP	城镇和乡村规划
TIA	交通影响评价
TRL	交通研究实验室
UKNEA	英国国家生态系统评价
UN	联合国
UNCED	联合国环境与发展会议
UNECE	联合国欧洲经济委员会
UNEP	联合国环境规划署
US	美国
USAID	美国国际开发署
VEC	已定价值的生态系统组成部分
VMP	访客管理计划
VROM	荷兰住房、空间规划与环境部
WA	西澳大利亚
WBCSD	世界可持续发展商业理事会
WHO	世界卫生组织
WID	美国国际开发署女性参与与发展项目
WTA	接受意愿
WTP	支付意愿

Abbreviations and acronyms

AA	Appropriate assessment
ABI	UK Annual Business Inquiry
ADB	African Development Bank
ADB	Asian Development Bank
AEE	Assessment of environmental effects
AEP	Association of Environmental Professionals
ANZECC	Australia and New Zealand Environment and Conservation Council
AONB	Area of Outstanding Natural Beauty
APC	Air pollution control
API	Assessment on Proponent Information (WA)
AQMA	Air quality management area
BAA	BAA Airports Limited (previously British Airports Authority)
BANANA	Build absolutely nothing anywhere near anything
BG	Bulgaria
BIO	Bio Intelligence Service S. A. S.
BME	Black and minority ethnic
BP	BP (previously British Petroleum)
BPEO	Best practicable environmental option
BS	British Standard
BWEA	British Wind Energy Association
CAREC	Regional Environmental Centre for Central Asia
CBA	Cost-benefit analysis
CC	County Council
CCGT	Combined-cycle gas turbine
CCHP	Combined cooling heat and power
CCS	Carbon capture and storage
CCW	Countryside Council for Wales
CE	Categorical exclusion
CEA	Cumulative effects assessment
CEAA	Canadian Environmental Assessment Agency
CEAM	Cumulative effects assessment and management
CEARC	Canadian Environmental Assessment Research Council
CEC	Commission of the European Communities
CEGB	Central Electricity Generating Board
CEMP	Construction environmental management plan
CEPA	Commonwealth Environmental Protection Agency (Australia)
CEQ	US Council on Environmental Quality
CEQA	California Environmental Quality Act
CHP	Combined heat and power
CIA	Cultural impact assessment
CIE	Community impact evaluation
CISDL	Centre for International Sustainable Development Law
CITES	Convention on Trade in Endangered Species
CO_2	Carbon dioxide
COWI	COWI A/S
CPO	Compulsory purchase order
CPRE	Campaign to Protect Rural England

CRM	Contingent ranking method
CRS	US Congressional Research Service
CRTN	Calculation of road traffic noise
CSR	Corporate social responsibility
CVM	Contingent valuation method
CY	Cyprus
CZ	Czech Republic
dB	Decibels
dBA	A-weighted decibels
DA	Devolved administration (in the UK)
DBIS	UK Department for Business, Innovation and Skills
DC	District Council
DCLG	UK Department for Communities and Local Government
DECC	UK Department of Energy and Climate Change
DEFRA	UK Department for Environment, Food and Rural Affairs
DETR	UK Department of Environment, Transport and the Regions
DFID	UK Department for International Development
DfT	UK Department for Transport
DG	Directorate General (CEC)
DMRB	Design manual for roads and bridges
DoE	UK Department of the Environment
DOEn	UK Department of Energy
DoT	UK Department of Transport
DTI	UK Department for Trade and Industry
EA	Environmental assessment
EA	UK Environment Agency
EAGGF	European Agricultural Guidance and Guarantee Fund
EAP	Environmental action plan
EBRD	European Bank for Reconstruction and Development
EC	European Commission
EcIA	Ecological impact assessment
ECJ	European Court of Justice
EDF	Electricité de France
EE	Estonia
EEA	European Environment Agency
EIA	Environmental impact assessment
EIB	European Investment Bank
EID	Environmental impact design
EIR	Environmental impact report
EIR	Environmental impact review
EIS	Environmental impact statement
EM&A	Environmental monitoring and audit
EMAS	Eco-Management and Audit Scheme
EMP	Environmental management plan
EMS	Environmental management system
EN	English Nature
ENDS	Environmental Data Services
EPA	UK Environmental Protection Act
EPA	US Environmental Protection Act
EPA	US Environmental Protection Agency
EPA	West Australian Environmental Protection Authority
EPB	Environmental Protection Bureau (China)
EPBCA	Environmental Protection and Biodiversity Conservation Act (Australia)
EPD	Hong Kong of China Environmental Protection Department
EqIA	Equality impact assessment

ERM	Environmental Resources Management Limited
ES	Environmental statement
ESRC	Economic and Social Research Council
ETSU	Energy Technology Support Unit
EU	European Union
FEARO	Federal Environmental Assessment Review Office
FEIS	Final environmental impact statement
FHWA	US Federal Highway Administration
FoE	Friends of the Earth
FONSI	Finding of no significant impact
G1; G2	Generation 1; Generation 2
GAM	Goals achievement matrix
GHG	Greenhouse gases
GHK	GHK Consulting Limited
GIS	Geographical information systems
GNP	Gross national product
GP	General practitioner
GPDO	General Permitted Development Order
GW	Gigawatt
ha	Hectare
HEP	Hydro-electric power
HGV	Heavy goods vehicle
HIA	Health impact assessment
HMG	Her Majesty's Government
HMIP	Her Majesty's Inspectorate of Pollution
HMSO	Her Majesty's Stationery Office
HPF	Household production function
HPM	Hedonic price methods
HRA	Habitats regulation assessment
HSE	Health and Safety Executive
HU	Hungary
HWS	Hampshire Waste Services
IA	Impact assessment
IAIA	International Association for Impact Assessment
IAU	Impacts Assessment Unit (Oxford Brookes)
IEA	Institute of Environmental Assessment
IEMA	Institute of Environmental Management and Assessment
IFI	International Funding Institution
IIA	Integrated impact assessment
IMD	Index of Multiple Deprivation
INEM	International Network for Environmental Management
IOCGP	Inter-organizational Committee on Guidelines and Principles for Social Impact Assessment
IPC	Infrastructure Planning Commission
IPC	Integrated pollution control
IPCC	Intergovernmental Panel on Climate Change
IPHI	Institute of Public Health in Ireland
ISO	International Organization for Standardization
IWM	Institute of Waste Management
JEAPM	Journal of Environmental Assessment Policy and Management
JNCC	Joint Nature Conservancy Council
KSEIA	Korean Society of Environmental Impact Assessment
kV	Kilovolt
L_{10}	Noise level exceeded for no more than 10 percent of a monitoring period

LB London Borough

LCA Life cycle assessment

LNG Liquified natural gas

LPA Local planning authority

LT Lithuania

LTP Local transport plan

LTP3 Third local transport plan

LULU Locally unacceptable land uses

LV Latvia

MAFF UK Ministry of Agriculture, Forestry and Fisheries

MAUT Multi-attribute utility theory

MBC Metropolitan Borough Council

MCA Multi-criteria assessment

MCDA Multi-criteria decision analysis

MEA Manual of Environmental Appraisal

MMO Marine Management Organization (UK)

MoD UK Ministry of Defence

MOEP Ministry of Environmental Protection (China)

MT Malta

MW Megawatt

NE Natural England

NEPA US National Environmental Policy Act

NGC National Grid Company

NGO Non-governmental organization

NHS National Health Service

NIMBY Not in my back yard

NO_x Nitrogen oxide

NPDV Net present day value

NPS National Policy Statement

NSIP Nationally significant infrastructure project

NTS Non-technical summary

ODA Olympic Delivery Authority

ODPM UK Office of the Deputy Prime Minister

OECD Organisation for Economic Co-operation and Development

OISD Oxford Institute for Sustainable Development

OJ Official Journal of the European Communities

OTP Operational Transport Programme

PADC Project Appraisal for Development Control

PAS Planning Advisory Service

PBS Planning balance sheet

PEIR Programme environmental impact report

PEIS Programmatic environmental impact statement

PER Public Environmental Review (WA)

PIC Partnerships in Care

PL Poland

PM_{10} Particulate matter of less than 10 microns in diameter

PPG Planning Policy Guidance

PPPs Policies, plans and programmes

PPPP Policy, plan, programme or project

PPS Planning policy statement

PWR Pressurized water reactor

QBL Quadruple bottom line

QOLA Quality of life assessment

RA Resilience Alliance

RA Risk assessment

RMA Resource Management Act (NZ)

RO Romania

ROD Record of decision

RSPB Royal Society for the Protection of Birds

RTPI Royal Town Planning Institute

S106 Section 106

SA Sustainability appraisal

SAC Special Area of Conservation

SAIEA	Southern African Institute for Environmental Assessment
SAVE	SAVE Britain's Heritage
SD	Sustainable development
SDD	Scottish Development Department
SEA	Strategic environmental assessment
SEERA	South East England Regional Assembly
S&EIA	Socio-economic and environmental impact assessment
SEPA	Scottish Environment Protection Agency
SI	Slovenia
SIA	Social impact assessment
SK	Slovakia
SNH	Scottish Natural Heritage
SNIFFER	Scotland and Northern Ireland Forum for Environmental Research
SO_2	Sulphur dioxide
SOER	State of the Environment Report
SoS	Secretary of State
SPA	Special Protection Area
SSE	Stop Stansted Expansion
SSSI	Site of Special Scientific Interest
TBL	Triple bottom line
T&CP	Town and country planning
TIA	Transport impact assessment
TRL	Transport Research Laboratory
UKNEA	UK National Ecosystem Assessment
UN	United Nations
UNCED	United Nations Conference on Environment and Development
UNECE	United Nations Economic Commission for Europe
UNEP	United Nations Environment Programme
US	United States
USAID	United States Agency for International Development
VEC	Valued ecosystem component
VMP	Visitor management plan
VROM	Netherlands Ministry of Housing, Spatial Planning and the Environment
WA	Western Australia
WBCSD	World Business Council for Sustainable Development
WHO	World Health Organization
WID	USAID Women in Development
WTA	Willingness to accept
WTP	Willingness to pay

目　录

第一部分

原则和程序

1
导言及原则

1.1 导言

过去的40年，人们对环境问题的关注日益增加，关注的焦点主要集中在可持续发展以及如何更好地处理发展与环境的协调问题上。与此同时，欧盟等一些国家或地区、国际机构开始制定新的法规，旨在协调环境与发展的关系。环境影响评价（EIA）就是一个重要的例子。美国早在40年前就建立了环境影响评价法规。欧洲共同体（EC）1985年的一项指令加快了环境影响评价在欧盟成员国及全球的应用。自从1988年英国引入环境影响评价制度以来，环境影响评价就逐渐成为规划实践的主要发展领域。在英国，最初预期的每年20份环境影响评价报告书（EIS）已迅速上升至几百份，而这还只是冰山一角，环境影响评价的应用范围仍在不断扩大和发展。

令人感到意外的是，环境影响评价在推行过程中遇到了多方面的强烈抵制，尤其是在英国。规划者们片面地认为他们在规划制定过程中已经做了相关的影响评价。很多开发商也认为环境影响评价既浪费金钱又浪费时间，并且制约发展，中央政府对此也不甚积极。有趣的是，英国最高立法机构（initial UK legislation）因为支持环境评价（EA），而忽略了一些对环境影响具有明显政治倾向的、负面的评论。目前，环境影响评价的应用范围还在不断扩展。本章主要介绍环境影响评价的程序及其目标、发展类型以及环境影响评价中的环境、影响及现存问题。

1.2 环境影响评价的本质

1.2.1 定义

环境影响评价的定义有多种。其中，我们通常引用 Munn 的广义定义（1979 年）：“识别和预测对环境、人类健康的影响以及立法提案、政策、规划、项目以及执行程序所带来的社会福利，并且就影响的信息进行说明与交流。”英国环境部（DoE）(1989 年）给出了一个狭义的定义：“‘环境评价’一词描述的是通过开发商以及其他渠道搜集有关项目环境影响的信息，规划部门对这些信息进行综合考虑后得出该项目是否可行的一种技术和过程。”联合国欧洲经济委员会（UNECE）(1991 年）做了一个简洁精炼的定义：“对规划活动（项目）所产生环境影响的评价”。欧盟环境影响评价指令要求对某些可能对环境产生严重影响的公共和私人项目在其获得建设批准之前做影响评价，并将此要求程序化（见附录 1)。国际影响评价协会（IAIA)(2009 年）定义环境影响评价为：“在重大决定和承诺做出之前，识别、预测、评价以及减轻所提出的开发建议在生物物理（biophysical)、社会及其他相关方面影响的过程。”此过程的重点目前正被广泛探索。

1.2.2 环境影响评价：程序

环境影响评价其实是预测开发活动可能带来的环境后果的一个系统的过程。与其他环境保护机制相比较，环境影响评价更强调预防为主。虽然规划者们基本上都考虑到了开发活动对环境的影响，但通常在环境影响评价所要求的系统性、全面性、学科交叉性上有所不足。该程序包括以下几个步骤，如图 1.1 所示。

下面将对该程序做一个简要描述，在第 4～7 章还将做进一步的详细讨论。虽然图 1.1 所示步骤是用线性图来表示的，但环境影响评价其实是一个循环过程，各步骤之间存在相互作用及反馈机制。还需指出的是，具体实践中的步骤有时会与图 1.1 中所描述的步骤有较大区别。例如，英国 EIA 法规体系中仍未要求跟踪监测。此外，程序中各步骤的顺序也不尽相同。

- 项目筛选将环境影响评价的应用范围缩小到那些可能产生重大环境影响的项目上。筛选可能部分地取决于国家现行的 EIA 法规体系。
- 评价因子要尽可能在评价的早期确定，从项目可能产生的全部影响以及提出的所有替代方案出发，确定关键的、重要的因子。
- 考察替代方案的目的是为了确保建议者能考虑其他可行的提议，包括可替代项目选址、规模、程序、布局、运行条件以及“零作为”方案（“no action”）等。
- 描述项目/开发活动包括阐述项目建设的目的和理由，了解项目的各种特征——即开发的阶段、选址以及过程等。
- 本底环境的描述要说明在项目不实施的情况下环境的现状及未来情况，要考虑自然条件及其他人类活动的作用对环境产生的影响。
- 主要影响识别的目的与前几个步骤相同，都是为了确保能识别出所有潜在显著的环境

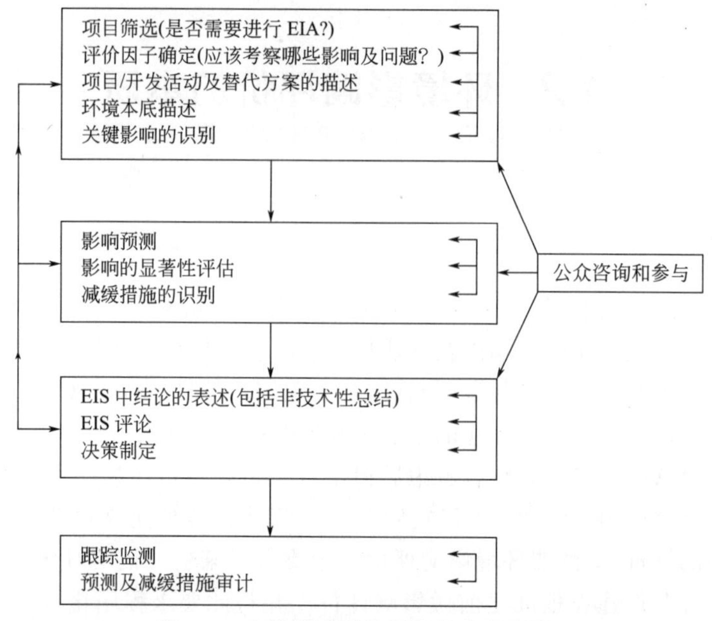

图 1.1 环境影响评价程序的重要步骤

注：环境影响评价应该是各个步骤相互作用的循环程序。例如：在整个程序中，公众参与在大多数阶段都是十分必要的。监测系统所选取的参数要与项目参数及环境本底数据联系起来。

影响（不利的和有利的），并且能在整个过程中考虑这些影响。

• 影响预测的目的是识别影响的大小与特征，比较项目/活动实施前后的环境变化。

• 显著性评估是评价所预测影响的相对重要性，以将评价重点放在主要的不利影响上。

• 减缓措施包括针对显著的不利影响所采取的避免、减少、修复/补救措施等。此外，应尽可能增强有利影响的发展。

• 公众咨询与公众参与旨在确保环境影响评价的质量、全面性和有效性，在决策过程中充分考虑公众意见。

• 环境影响报告书的表述是整个过程中至关重要的一环，如果做得不好，整个环境影响评价工作可能前功尽弃。

• 评审包括对环境影响报告书的质量进行系统评价，对决策过程起到重要的辅助作用。

• 项目决策包括相关机构将环境影响报告书（包括咨询反馈）与其他材料一起进行综合考虑。

• 跟踪监测是在项目实施决策后，记录各种开发活动的影响，它是项目有效管理的重要组成部分。

• 审计（auditing）在监测之后，它能够将实际结果与预测结果进行比较，并用来评价预测质量和减缓措施的有效性，是环境影响评价学习过程中至关重要的一步。

1.2.3 环境影响报告：文件编制

环境影响报告是对整个过程中各个步骤的信息进行整理，并在此基础上进行影响评价。预防优于治理，一份揭露不利影响的环境影响报告将为选择放弃还是适当修改的开发活动提供有价值的信息。不同的决策有不同的减缓措施，通过各种减缓措施可以有效地减少不利影响，所以往往会形成多种不同的决策。表 1.1 显示了项目环境影响报告所包含的内容。

表 1.1　项目环境影响评价报告内容实例

项目环境影响评价报告内容
非技术性总结
第一部分:导言、方法及关键问题
导言
方法
关键问题总结
第二部分:拟建项目背景
基本研究:需求、规划、替代方案及选址
选址描述,本底情况
对拟建项目的描述
建设方案,包括场址准备、建设、运行、撤销与恢复(如适用)
第三部分:环境影响评价——主题
土地利用
地质、地形和土壤
水文和水质
空气质量
气候变化
生态:陆生和水生生态
噪声和振动
社会-经济情况
交通
景观美学,视觉质量
历史环境
娱乐与舒适性
各影响间的相互关系
累积影响
残留影响汇总
第四部分:跟踪及管理
影响监测
影响管理

非技术性总结（non-technical summary）是报告书编制的重要组成部分。环境影响评价的过程可能很复杂，因此这样的总结有利于促进各参与者之间的交流。简介应反映环境影响评价过程的复杂性，需阐述例如谁是开发商、谁做的环境影响报告以及相关的法律框架等问题。同时，方法部分应提供阐述一些基本信息的机会（如采用了哪些方法、关键事项是如何鉴别的、咨询对象是谁、以何种方法进行的咨询、遇到了哪些困难、环境影响评价的局限是什么）。拟建项目的背景介绍涵盖环境影响评价程序的早期阶段，包括对项目基本情况的清晰描述（包括相关规划政策和规划等）。

环境影响报告的每个主题下，通常会有一些对现状、影响预测、减缓或增加的范围、遗留影响的讨论。表 1.1 所列为通用表格，但有一些主题仍未被提及，如气候变化以及累积影响（如适用）。结论部分虽经常在环境影响报告中被省略，也应包括监测和管理的项目后续跟踪事项。

环境影响评价和环境影响报告的实践因研究情况和国家的不同而各异，最佳实践在不断发展。联合国对一些国家进行的环境影响评价实践的早期研究提倡要在环境影响评价过程中以及报告书编制中进行有关改进（UNECE，1991），其中包括更加关注社会-经济影响、公众参与以及“决策后”行为（如监测）等。最近欧盟指令（CEC，2003）修正案的实施也提出了相似的要求，并提出了其他一些可能在 10 年后会出现的问题（第 2 章）。1996 年，Sadler 基于国际上关于环境影响评价有效性的研究，提出了 EIA 变化的范围（a wider agenda for change)(第 8 章和第 12 章)。

1.3　环境影响评价的目的

1.3.1　为决策提供帮助

环境影响评价可以为决策提供支持。对决策者（如地方权力机构）来说，在制定决策之前，环境影响评价可以系统地考察某个拟开发活动或替代方案的潜在环境影响。决策者在决策之前结合环境影响报告书和规划项目相关资料进行综合考虑。与费用-效益分析（CBA）等技术相比，通常环境影响评价的评价范围更广，但量化的内容可能较少。环境影响评价不能替代决策，但能帮助决策者认清拟开发行为的一些得失，从而做出更加合理、更加有组织的决策。环境影响评价是开发商、公共利益团体以及规划者之间交流的潜在基础，能促使开发行为与环境之间的利益均衡。

1.3.2 有助于开发行为规范化

开发商可能把环境影响评价看作是进行各种开发行为之前必须要跨越的障碍，他们认为在开发活动的批复过程中，进行EIA既浪费时间又浪费金钱。但实际上EIA能为他们带来巨大利益，因为EIA可以提供一个框架，从而将环境问题纳入项目选择和设计的考虑中。它有助于开发行为的规范化，即可以通过不断修改来减少或消除项目对环境的不利影响。在开发行为的早期，对可能造成的环境影响进行评价，可以实现环境友好发展；可以改善开发商、规划部门与当地社区之间的关系；可以更加顺利地得到规划部门的许可；有时也会像British Gas等一些开发商所讨论的那样，EIA的支出是值得的，因为它可能会产生额外的资金回报（Breakell和Glasson，1981）。O'Riordan（1990年）结合这些理念提出了一些重要的环境类专业词汇，如“绿色消费”“绿色资本”等。如果消费者对低环境影响物品的需求不断增长，加上清洁技术市场不断扩大，这样对开发商会产生很大的促进作用。对开发商来说，EIA可以成为潜在冲突的信号，明智的开发商可以通过环境影响评价获得“绿色收益”，用这种办法还可以消除或减轻不良的环境影响，减少当地居民的不满，避免昂贵的公共咨询费用。这可被看作是各大商家对企业社会责任（CSR）更广泛和更具现代意义的应用（Crane等，2008）。

1.3.3 利益相关者的咨询与参与工具

开发行为可能对环境产生广泛影响，从而影响到社会不同阶层与群体。许多不同级别的政府都越来越强调主要利益相关者的参与在项目规划和开发过程中的重要性，如奥胡斯公约（UNECE，2000）和欧盟公众参与指令（CEC，2003b）。环境影响评价可以成为联系社区和利益相关者的有用工具，帮助受拟开发行为影响的群体更好地了解和参与规划及开发过程。

1.3.4 实施可持续发展的手段

对于现有的不利于环境的开发行为应该加强管理，在极端情况下甚至予以取消，但其仍会在接下来的几十年里产生遗留环境问题。最好能在规划阶段就采取措施，预先减少不利环境影响或者避免某些特殊的开发行为，因为预防优于治理。这是美国和欧盟关于EIA法规首创的概念。例如，1985年欧盟EIA指令的序言指出：最好的环境政策是在源头防止污染物的产生，而非在过程中治理（CEC，1985）。这种理念理所当然确立了EIA在实施可持续发展的手段中的基础作用，虽然很多研究者认为这个作用还并未得到广泛认可（Jay等，2007）。

1.3.4.1 可持续发展的本质

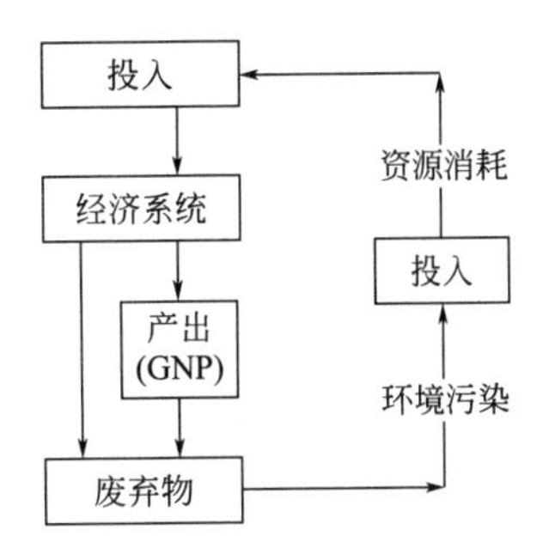

图 1.2 在环境背景下的经济发展过程
（改编自：Boulding，1966）

经济与社会的发展都必须考虑环境状况。Boulding（1996）生动地描述了“通量经济”与“宇宙飞船经济”的二分法（图 1.2）。发展经济的目的是提高国民生产总值（GNP），用更多的投入获得更多的产品和服务，但却埋下了自毁的祸根：高产出带来的不仅是产品与服务，同时还有更多的废弃物；不断增加的投入需要更多的资源，自然环境就成了资源的来源和废弃物的承载体。环境污染和资源耗竭总是伴随着经济的发展而产生。

地方政府和国际机构已经意识到，经济和社会的发展与自然环境是相互作用的，人类活动与生物界也会相互影响。为了更好地处理这种相互作用，人们做了很多尝试。然而 21 世纪的第一个 10 年末，欧洲环境署发布的《欧洲环境——2010 年现状与展望》（EEA，2010）报告指出：在良好发展的同时仍夹杂着遗留下来的根本性挑战，这些挑战可能会对环境质量造成非常严重的后果。例如，虽然温室气体排放削减，欧盟预计能在 2020 年实现减排 20%温室气体的目标，但 2008 年欧盟成员国仍排放了将近 50 亿吨当量的 CO_2。类似地，虽然欧洲的废弃物管理已由填埋稳步转化为再生利用和预防，但 2006 年欧盟 27 国产生的 30 亿吨废弃物中仍有一半被填埋处理。在自然和生物多样性方面，欧盟自然 2000 保护地网络已扩展到覆盖欧盟 18%的土地，但仍然没有达到 2010 年停止生物多样性损失的目标。欧洲的淡水资源也饱受缺水、洪涝灾害、物理变迁以及一系列污染物持续影响。环境空气质量和水质不容乐观，其对健康的影响普遍存在。与此同时，我们生活在一个相互关联的世界，欧洲的政策制定者面对的不仅是欧洲领土上复杂的系统相互作用，还有不断呈现的全球变化因素，这些因素也可能会影响欧洲的环境，并且其中很多因素不在欧洲的掌控范围之内。环境变化的趋势在发展中国家可能更为显著，因为这些地方的人口增长速度快且目前生活水平较低，这将对环境资源造成更大的压力。

联合国环境与发展署 1987 年的报告（通常称之为“布伦特兰报告”，它是以会议的女主席命名的）将可持续发展定义为“一种既满足当代人的需求又不破坏后代人满足其需求能力的发展”。可持续发展意味着我们交给后代的不仅是一个充满了公路、学校、历史建筑的“人造城市”和一个充满了知识与技术的“人类都市”，还是一个有着清洁空气、清洁水源、热带雨林、臭氧层和生物多样性的“自然/环境都市”。布伦特兰报告描述了可持续发展的几个重要特点：保持现有生活水平、保证对自然资源的持续利用、避免持续的环境破坏，同时表明我们应该依靠地球的自然产出来维持社会的发展，而不仅仅是消耗它的资源（DoE，1990）。此外，出于对环境和未来的担心，布伦特兰报告也强调了参与和公平，尤其是代内和代际的公平。这个定义比生态和自然环境的定义宽泛许多，它包含了代内和代际公平的社会组织，也强调了经济和文化方面的重要性，如防止两极分化、关注生活质量、关注社会福利伦理以及系统地组织利益相关者参与进来等。

不过，可持续发展很有可能成为一个宽泛并失去具体意义的术语，这是非常危险的。现在，各个领域已经有了很多对可持续发展的定义。1992 年，Holmberg 和 Sandbrook 已经发现了 70 多种关于可持续发展的定义。Redclift（1987）认为可持续发展是“一种代替思想的道德信念”，而 O'Riordan（1988）则认为可持续发展是“一种不易付诸实践的好想法”。

Skolimowski（1995）则认为可持续发展是：

一种抨击所有发展的激进想法，把发展看作与普通的商业行为一样平常。可持续发展的思想虽然宽泛、界限模糊，但却满足了所有人的要求，虽然有些激进但并不令人讨厌，这也是它最大的优点。

读者可以参考 Reid（1995）和 Kirkby 等（1995）来大体了解可持续发展的概念、讨论和反响。

随着时间的推移，“可持续性”已继承“可持续发展”一词的部分内涵（虽然二者仍可互为代名词），部分原因是“可持续发展”一词有被滥用之嫌（如政府寻求将可持续发展等同于可持续增长，公司寻求将可持续发展等同于可持续利润）。然而，尽管全球范围内对“可持续性/可持续发展”概念已认同，其范畴和本质还是具有争议和模糊性（Faber 等，2005）。“可持续性/可持续发展”的定义有很多，常用的一个是基于三重底线（TBL）的定义，其反映了环境、社会和经济因素在决策中的重要性，但也需超越这个定义来强调不同因素之间整合与协同的重要性（图 1.3）。然而对此协同作用的评估往往具有一定的挑战性。图 1.4 强调了在这个可持续性三元素定义之中存在的重要层次性。环境及其自然系统是任何可持续性概念的基础。我们无法离开地球的自然和物理系统提供的物质和服务而生存，包括可呼吸的空气、可饮用的水以及食物。生活在地球上，我们也需要社会系统提供社会正义、安全、文化认同和归属感。而离开运行良好的社会系统，经济体系就无法正常生产。

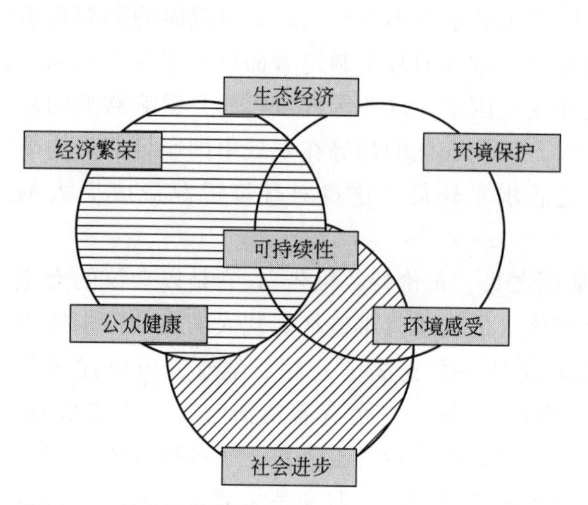

图 1.3 综合可持续发展的环境、社会、经济方面

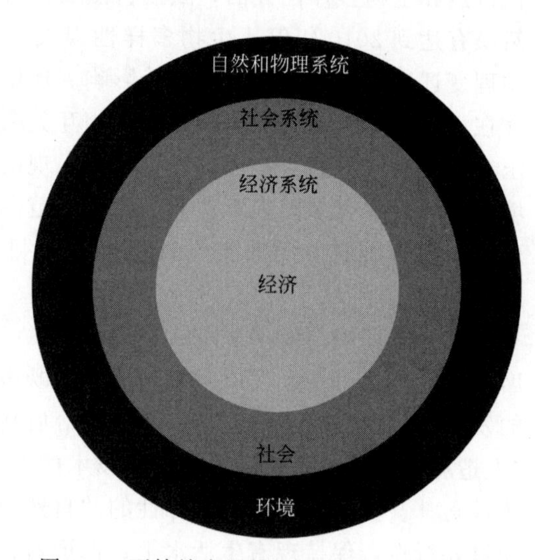

图 1.4 可持续发展维度的备选（层次）观点

1.3.4.2 可持续发展的制度对策

实现可持续发展目标的制度方面的对策，在以下几个层次有所要求。全球性环境问题，如臭氧层破坏、气候变化、热带雨林消失及生物多样性减少，需要全世界对其行为做出政治承诺。联合国环境与发展大会（UNCED）（1992 年里约会议）就是一个务实的会议，号召全世界共同行动起来，以处理这些全球性的问题。《21 世纪议程》，这部 800 多页的跨入 21 世纪国际社会的行动计划指明了各国应该为实现可持续发展做些什么，包含的主题有生物多样性、荒漠化、森林锐减、危险废物、生活污水、海洋及大气等环境问题。该议程中所包含的

115 个计划针对要做什么、要实现什么目标、要采取哪些行动、要采用什么方法等都一一进行了概述。《21 世纪议程》为实现消费、人口及地球生命支持能力之间的可持续平衡提供了一些政策和具体计划。遗憾的是，仅仅依靠国家、地方政府等来实现这些计划，并不具备法律约束力。

2002 年，约翰内斯堡全球峰会上重新强调了在一些环境问题上获得全球统一认识的困难性，不过会议也取得了一些积极的成果，如在水资源、医疗卫生（2015 年将不具备基本医疗卫生设施的人口控制在 12 亿左右）、贫困、健康、最低生活保障以及经济贸易全球化等方面取得的成果，不过其他问题就不是那么乐观了，通过法律手段来强制减少温室气体排放的《京都议定书》执行起来仍然十分困难。欧盟 2013 年在继《京都议定书》之后制定了一个雄心勃勃且具有法律约束力的全球气候条约，但 2009 年的哥本哈根气候大会并未向此目标迈进（Wilson 和 Piper，2010）。同样地，我们经常听到全球生物多样性和自然资源持续减少、许多国家在保障人权方面遇到挑战的消息。当然，以上所有的问题都因严峻的全球经济情况所带来的挑战和不确定性而变得更加复杂，进而严重阻碍了可持续发展的进程。

欧盟自 1972 年到 1992 年期间实施了 4 项有关环境的合作行动计划。这些计划的实施推动了各个领域专门法规的制定，包括废弃物管理、大气污染、自然保护和环境影响评价等。第五项计划，即“面向可持续发展”（1993～2000）是在实现统一欧洲市场的背景下建立的。统一欧洲市场关注的是清除各成员国间在金融、物资和技术上的壁垒，这些壁垒将导致经济发展发生重大变化，而且可能加重对环境的威胁。第五项计划认识到要综合各部门与环境保护相关的目标，包括制造、能源、运输和旅游等部门。欧盟在环境问题方面遵循“预防为主”的原则，就是说一定要采取预防措施，使得环境破坏必须在源头就予以整治，并且坚持污染者付费原则。尽管先前几项计划的执行几乎都依靠法律手段，第五项计划则提倡更大范围的融合，包括“市场手段”，如运用财政手段使环境成本内部化，以及“横向支持手段”，如改进本底值和统计数据、改进空间和部门的规划等。

第六项计划，即“我们的未来，我们的选择”（2001～2010）建立在近 10 年更广泛的方法基础上。它指出可持续发展包括社会、经济和自然环境等三方面，其焦点集中在四个优先考虑的问题上：解决气候变化、保护自然环境和生物多样性、减少环境污染对人体健康的影响、确保对资源和废弃物的可持续性管理。同时，它也意识到加强公民环境意识和改变日常行为习惯的重要性以及“绿色的土地利用和管理决策”的重要性。

由于环境影响评价的社区指令（the community directive on EIA）和对战略环境影响评价的建议，旨在确保规划的基础设施项目（规划）能够正确地实施，并且在实施过程中不会对环境造成破坏，这将确保决策者把环境因素更好地融入规划决策中（CEC，2001）。

欧盟委员会还未决定第七项计划的性质，包括关键性角色——气候变化是否包含在欧盟环境政策中或在相关组织里扮演更全面的角色。

在英国，《共同的遗产：英国环境战略》（DoE 等，1990）的出版成为该国第一部环境类的综合白皮书。报告包括对温室效应、城市与农村、污染控制、环保意识和环保组织的讨论。该书通篇强调应该由政府、企业和公众共同承担环境责任。政策范围包括立法、制定标准、规划和经济手段。最后，在 Pearce（1989）研究工作的基础上，还提出了包括税收、补贴、市场创新和激励机制等在内的方法。报告也指出，将 EIA 的方法纳入可持续发展的“工具箱”中。之后的英国政府报告，如《可持续发展：英国战略》（HMG，1994）指出了环境影响评价在可持续发展中的重要作用，并且提高了环境影响评价的地位。政府报告同时也提出了对可持续发展范围的扩展，包括社会、经济、环境等方面。这在《英国可持续发展战略：更高的生活质量》（DETR，1999a）一书中有所体现，它的四个目标是：

- 社会进步——认识到每个人的需要；
- 有效的环境保护；
- 自然资源的谨慎利用；
- 维持快速和稳定的经济增长和就业水平。

为了衡量所取得的成绩，英国政府建立了一系列可持续发展指标，包括15个关键指标（DETR，1999b）。它同时也要求在英联邦各个地区均要建立高水平的可持续发展框架（参见例如《东南部——更好的生活质量》，SEERA，2001）。

《规划政策宣言1》（PPS1，DCLG，2005）再次强调了对可持续发展的承诺："可持续发展是支撑规划的核心原则。简言之，可持续发展的中心是确保现在和未来子孙每一个人拥有更好的生活。"英国国家战略更新《确保未来：提供英国可持续发展战略》一书中再次修正了这个观点（DEFRA，2005），此书中英国政府引入了一套修订的指导原则，行动的优先次序以及20个关键标题指标，并强调其执行。指导原则如下：

- 在环境阈值范围下生活；
- 确保一个强大、健康和公正的社会；
- 实现可持续经济；
- 推进政府善治，有责任地使用合理的科学。

政府善治原则在可持续发展的其他三个支柱（环境、社会、经济）基础上新增了第四个重要支柱，在方法上从三大底线转到了四大底线。我们需要推进从中央政府到个人所有层面的善治，从而促进其他三个支柱的一体化。使环境影响评价再次成为此一体化过程的有用工具。

1.4 项目、环境与影响

1.4.1 大型项目的特点

如1.2节所述，环境影响评价与大范围的开发行为是息息相关的，包括政策、规划、方案和项目。这里主要对项目进行讨论，以反映出项目环境影响评价在实践中的重要作用，具有"更高层次"的战略环境影响评价将在第11章中做进一步探讨。环境影响评价覆盖项目的范围正在不断地扩大，这将在第3章和第4章中予以讨论。通常，项目环境影响评价多运用在大型项目上。但什么是大型项目呢，识别它们的标准又是什么？套用Lord Morley定义大象这种动物时的说法，即虽然很难确定，但是当你看到时很容易认出。与之相似的是，LULU（地方无法接受的土地利用）已经应用于美国的很多大型项目上，如能源、交通、制造业等，这能清楚地反映公众对这些开发行为所产生的消极影响的了解程度。大型项目的定义并不容易，但具有一些显著的特点（图片1.1和表1.2）。

(a) 伦敦国王十字车站——城市的重新发展

(b) 2012年伦敦奥林匹克建设项目选址

(c) 芬兰Olkiluoto核电站

(d) 厄勒海峡大桥连接瑞典和丹麦

(e) 丹麦的近海核电站

(f) 英国退役的欣克利角A核电站的ES

图片 1.1　一些主要项目的例子

[来源：Magnox Electric (2002)；RPS (2004)；Symonds/EDAW (2004)；Wikimedia]

表 1.2　大型项目的特征

大量的资金投入
占地广，雇工多(建设和/或运行阶段)
组织关系排列复杂
影响范围广(地理的和更多类型)
重大环境影响
需要特殊的实施程序
公共设施及公用事业；开采业和第一产业(包括农业)；服务业
带状开发、点状开发

绝大多数大型项目的投资额都很大。在英国，“大型项目”可分为两个部分，一部分是指如英吉利海峡隧道及其相关铁路、伦敦希斯罗机场第5航站楼、2012年伦敦奥运会工程、高速公路及其拓宽工程、核电站，天然气电站及可再生能源项目（如大型海上风电场和拟建的塞文河拦河坝）。另一部分则是指基础设施建设，如一小段公路、各种垃圾处理设施，虽然规模较小，但投资巨大。大型项目在建设期间会占用大量土地、雇佣大量工人，甚至在运行期也是如此。大型项目在各阶段都会产生一系列复杂的相互及内在组织行为。这些开发活动通常对环境产生大范围、长时间且非常显著的影响。显著性影响的确定是环境影响评价中非常重要的一个问题，它与开发活动的规模、选址的环境敏感性和不利影响的性质等有关。这将在后面的章节中做深入讨论。一石激起千层浪，大型项目的环境影响也是深远又广泛的。在很多方面，这些项目被视为是特殊的，需要专门的实施程序。在英国，这些程序包括公众咨询、需要由议会通过草案（例如海底隧道项目）并进行环境影响评价。2008规划法案（HMG，2008）中确定了一系列国家重要基础设施建设项目（NSIPs），这些项目将由基础设施规划委员会（IPC）(2012改名为英国规划督察的国家基础设施部）按照新程序来审查。NSIPs包括大型能源项目、交通项目（公路、铁路和港口）以及水利和废弃物设施。

大型项目也可按照开发行为的类型来划分，包括制造业和开采项目，如石化厂、钢厂、采矿和采石厂；服务项目，如休闲项目开发，郊区的购物中心、新的居住区以及教育和医疗设施；公共事业和基础设施，如电站、公路、水库、管网和水坝等。欧盟对点状、面状分布的基础设施进行了进一步区分，点状基础设施包括电站、桥梁、港口等，面状和线状基础设施包括输电网络、公路和运河等（CEC，1982）。

大型项目的规划和生命周期的开发包括不同的阶段。了解这些阶段是非常必要的，因为不同阶段受到的影响是不同的。图1.5给出了项目生命周期的主要阶段，各个阶段在时间上有所变化，其内部特点也不尽相同，但基本都遵循一般的程序。在环境影响评价中“决策前”（A、B阶段）和“决策后”（C、D、E阶段）存在重要的区别。如1.2节所述，环境影响评价过程经常缺少对项目的跟踪监测与审计。

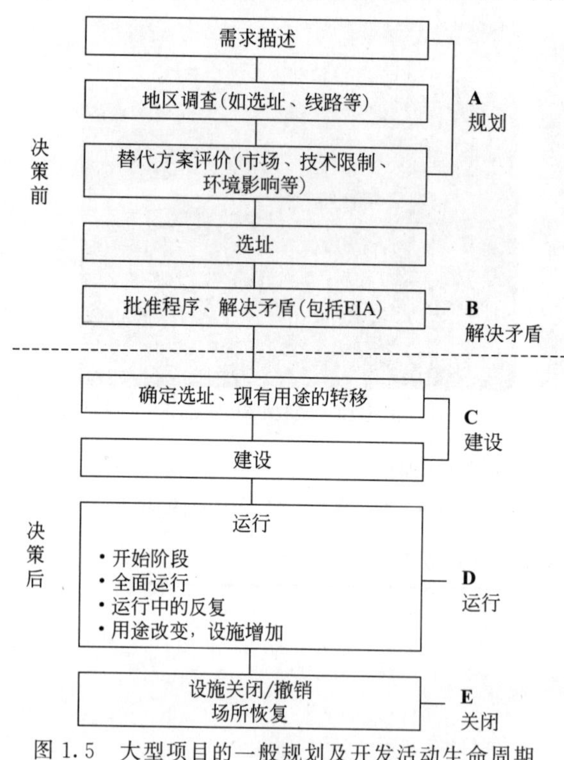

图1.5　大型项目的一般规划及开发活动生命周期（主要参考了主区域环境影响）

（节选自：Breese等，1965）

项目有几种启动方式：很多是因为市场需求（如度假村、市中心商业广场、天然气发电站、风电场等），有些则是必需的（如泰晤士河大坝），还有一些是具有明显象征意义的（如巴黎的巴士底歌剧院、奥赛美术馆、凯旋门等）。在许多国家，很多大型项目都是应公众要求而建设的，但随即就被私有化，这样就可以利用私人资金，其典型的项目有英国的North Midlands收费公路和海峡隧道，现在更多的是公共事业、能源、水和废弃物项目。最初的规划阶段 A 可能需要几年时间，从而得出详细的规划建议。在阶段 B，通常需要进行各种控制与调节程序，包括进行环境影响评价等。不同项目的建设阶段差异很大，有的项目可能要持续 10 年以上。除了开采业项目要比基础设施项目的运行期短外，很多大型项目毫无例外都具有很长的运行周期。环境影响并不是随着工厂关闭就可以忽略了，例如核电站的废料就会持续产生影响，图 1.6 描述了不同项目类型的生命周期各阶段的变化情况。

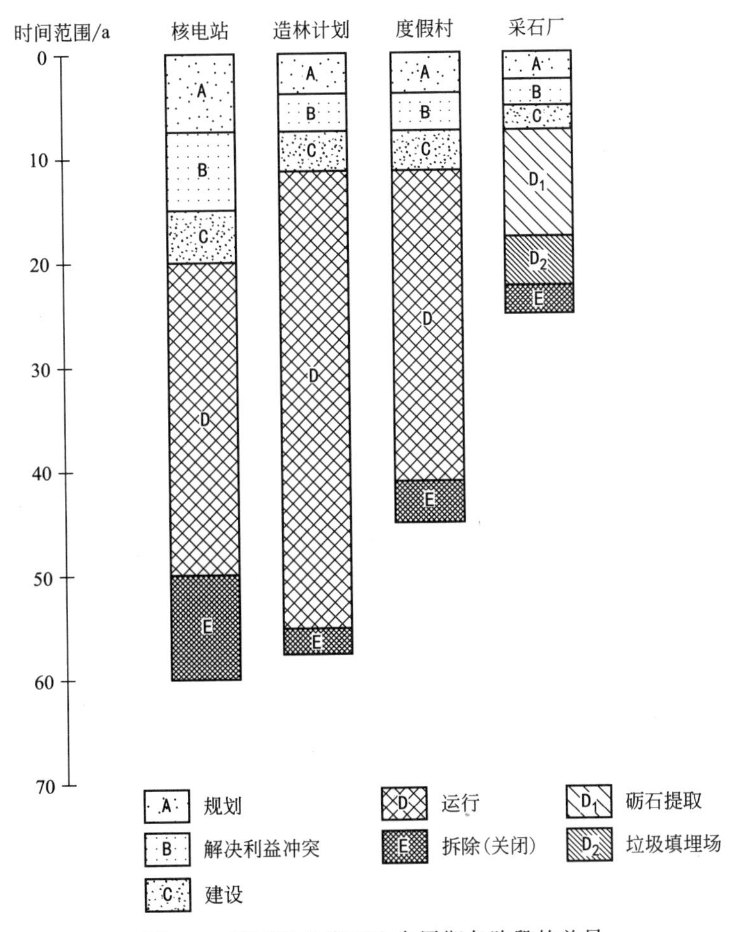

图 1.6 不同项目类型生命周期各阶段的差异

1.4.2 环境的属性

环境由几种要素构成，包括环境要素、空间范围和时间。狭义的环境要素主要指自然环境要素。例如，英国环境部曾把环境定义为易受污染影响的媒介，包括空气、水、土壤、植物、动物和人类、景观、城市、乡村保护区及建筑遗迹（DoE，1991）。DoE 所定义的环境要素清单见表 1.3。然而，如 1.2 节所述，环境还有很重要的经济属性和社会-文化属性，其中包括经济结构、劳动力市场、人口、住房、服务（教育、医疗、治安、防火等）、生活方式、价值观等，这些在表 1.3 中也有所体现。如 1.2.1 节中国际影响评价协会（IAIA）对环境影响评价的定义中所述，这种广义的环境定义与国际上的定义更为一致。类似的还有澳大利亚对环境的定义：从环境影响评价的目的出发，环境的含义综合了自然、生物、文化、经济和社会因素（ANZCEE，1991）。

表 1.3　环境组成要素

物理环境	
空气与大气	空气质量
水资源与水体	水的质量和数量
土壤与地质	等级、风险（如侵蚀、污染等）
动物群落与植物群落	鸟类、哺乳类、鱼类等；水生和陆生植物
人类	身体和精神健康、福利
景观	景观的特点与质量
文化遗产	保护区域；文物建筑；历史和考古学遗址；其他重要财产
气候	温度、降水、风等
能源	光、声、振动等
社会-经济环境	
人口统计学	人口结构和趋势
经济基础——直接	直接就业率；劳动力市场特点；本地和外来劳动力趋势
经济基础——间接	非基础性就业和服务业就业；劳动力供给和需求
住房；交通；娱乐	供应和需求特点
其他地方性服务	服务的供应和需求；健康；教育；警力等
社会-文化	生活方式、生活质量、社会问题；社会压力和冲突

来源：节选自 DoE 1991；DETR 2000；CEC 2003a。

我们也可以从不同的影响范围来对环境进行分析（图 1.7）。很多项目的空间效应影响着当地环境，虽然"当地"的影响范围会随着所考虑的环境角度和项目生命周期阶段的变化而变化，但是有些影响不仅仅局限于该地区。例如，交通噪声也许是地方性的问题，但是随着某个项目造成的交通流量增加会随之产生区域性的影响，如随之排放的 CO_2 就会导致全球温室效应加剧。环境也具有时间属性。在评价一个项目时都会需要环境本底数据。从地方到全国，互联网上可获取的数据量在迅猛增长（例如在英国，通过地方发展规划和国家统计数据，如环境食品和农村事务部发布的《环境统计摘要》，均可以提供一些相关数据）。在某些领域，这些数据可能被"量身定做"在环境质量报告和环境审计报告中（第 5、12 章将会提供详细的数据源指南）。对所有的数据而言，反映环境质量变化趋势的时间序列非常重要，因为环境本底数据在不断变化，因此在考察任何开发行为时都需要进行动态分析而不仅仅是静态分析。

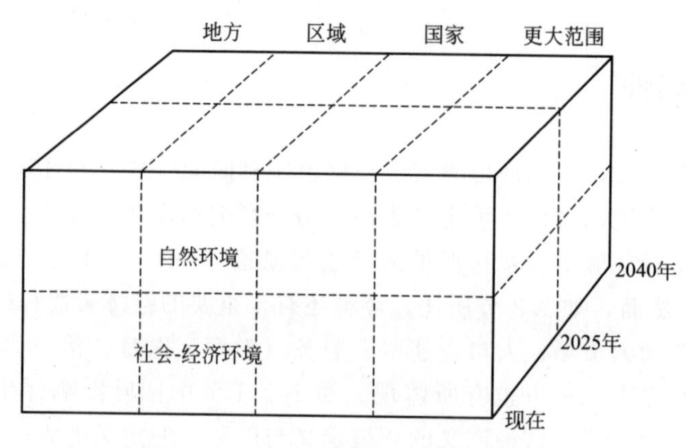

图 1.7　环境的要素：尺度和时间属性

1.4.3 影响的特性

所产生的环境影响是指与没有进行项目开发时相比，在时间和空间上环境参数的变化。这些参数可以是前面提到的任何一种类型的环境受体，如空气质量、水质量、噪声、地区失业及犯罪率等。图 1.8 对此进行了简单描述。

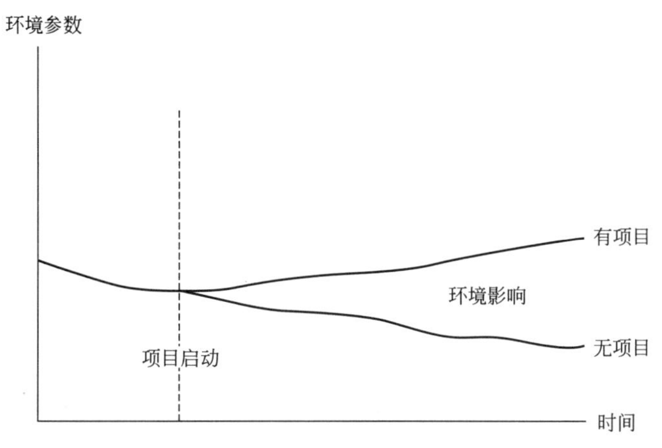

图 1.8 环境影响的特性

表 1.4 给出了一些在环境影响评价中可能会遇到的影响类型，自然的和社会-经济的影响都在其中得到体现，有利影响和不利影响通常相伴而生。例如，新的开发行为可能会产生有害废物，但是却可以为高失业率的地区创造就业机会。不过也并非完全如此，项目也可能会给环境带来好处。比如，被污染的或被弃置的土地得到了重新利用。同样，一个大型项目对当地群体产生的社会-经济影响包括对当地医疗卫生、房地产生压力，同时也增加了社会冲突和犯罪活动。项目的直接影响可能会导致之后的二次、间接的影响。如建在河流上的水库不仅占用了水域，而且可能会严重影响下游动植物的用水以及人类的活动，如渔业或航运等。直接/间接影响通常与长期/短期影响相关。对有些影响来说，长期和短期的区别也与项目处于建设阶段和运行阶段有关，比如，在建设阶段的有些影响（如改变土地利用方式）则可能是永久性影响。影响也具有空间属性。地方性影响与战略性影响的区别在于，后者的影响范围远远超过了地方的范围，这些影响经常是区域性的，有时对国家甚至全球造成显著影响。

表 1.4 影响类型

自然的和社会-经济的影响	可逆的和不可逆的影响
直接的和间接的影响	质的和量的影响
短期的和长期的影响	小范围和大范围的影响
地方性的和战略性的(包括地区、国家，甚至更大范围)影响	客观的和主观的影响
不利的和有利的影响	与其他开发行为相关的影响；累积影响

环境资源并不都具有可替代性，一旦破坏，可能就将永远失去。可逆影响与不可逆影响之间的区别是非常重要的，不可逆影响很难减缓，因而是环境影响评价中特别显著的影响。当然，也有可能在某些条件下替代、补偿或重建消失的资源，但是这种替代几乎都不理想。

资源的消失可能会在以后产生严重的影响，这就需要对其进行评估。有些影响可以被量化，有些则只能定性分析。但是不要忽视后者，同样也不能忽视一个拟开发行为产生的利益分配。对于不同的群体和区域来说，所受的影响是不一样的，一个项目的评价结果可能总体上产生了正面的影响，但有些群体和地区可能承受的更多的是不利影响，而其他地区则主要得到的是正面影响。客观影响和主观影响之间也存在差别，开发行为的主观影响可能在很大程度上依赖人们对拟建项目的态度。主观影响与客观影响预测是重要的信息源，应一起纳入考虑范围，社会建构不只是区别于现实的、单纯的认知或情感，而是社会形势决定了我们如何行事。此外，现实的社会建构是所有社会群体的特征，包括那些试图实施变革的机构以及受影响的社区。

最后，所有影响都应该与在“本底”情况（“do-nothing” situation）下和没有项目时所预测的环境状态做比较，也可以拓宽，与某地区替代方案的预期影响进行比较。有些项目现在和将来的其他开发行为可能会产生累积影响，如一个地区几个风力发电厂的协同影响，或是河口沿岸几个不同的大型开发项目（如港口、电站、钢铁厂、污水处理设施等）的累积影响。第 9 章和第 12 章将详细讨论累积影响的重要领域。

我们得出如下结论：“影响”（impact）与“效应”（effect）在 EIA 的著作及法规中有着广泛的应用，但两者之间是否可以相互转换或是否只用于表达特定的含义还不是十分清楚。美国《国家环境政策法案》（NEPA）的实施条例中明确规定“影响和效应在这些条例中的意思是一样的”，这个解释得到普遍认可，本书也采用这种说法。但很多学者也有不同的看法。如 Catlow 和 Thirlwall（1976）认为效应是指“由开发行为导致的直接或间接的物理和自然变化”，影响是指“由环境特征体现出来的可以被人们主观或客观评价的效应产生的结果”。而澳大利亚的研究报告（CEPA，1994）推翻了这一理论，认为“是由于影响才导致效应，这是逻辑先后的关系，而不是效应导致影响”。还有一些评论者引入了价值判断的概念来区分它们。Preston 和 Beford（1998）认为“影响不能体现价值判断”。Stakhiv（1998）也支持这个观点，他认为“对事实的科学评价是效应”，而“由分析家和公众对这些效应的相对重要性的评估则是影响”。该争论仍在继续。

1.5 不断变化的环境影响评价视角

1.5.1 适应性环境影响评价的重要性

对环境影响评价的讨论随时间、空间及参与方角度的不同而各异。从一个极简主义者的防御角度来看，一些开发商，可能还有一些政府部门，视环境影响评价如恶魔。行政练习可能只会给开发项目带来很小的（通常是表面上的）、可有可无的改变。对比之下，对“极端生态学家（deep ecologists）”或者“极端环保主义者（deep greens）”而言，环境影响评价并不能提供完全确定的拟建开发项目的环境后果，他们觉得在任何不确定或者有风险的情形下实施的项目都应被放弃。环境影响评价及其方法必将在弱可持续性和强可持续性上跨越这些观点。在当下，环境影响评价可以并且常常被看作是寻求开发项目和环境之间和谐关系的一个积极的过程。环境影响评价的本质和应用将随着相对价值和观点的变化而改变，环境影响评价必须去适应这些变化，就如同 O'Riordan（1990）在 20 多年前就非常乐观地指出的一样：

我们可以看到环境影响评价正慢慢脱离 20 世纪 70 年代主导的防御性工具的观点，而发展成为将在 20 世纪 90 年代盛行的、能够改善环境与社会的技术……如果我们不把环境影响评价看作是一项技术，而是在变化的环境政治和管理能力下不断改变的程序，那么我们可以想象其为一个在复杂环境社会中的敏感的环境价值晴雨表（a sensitive barometer of environmental values）。

环境影响评价必须不断适应快速变化的世界，因为这个世界存在着许多对可持续性支柱的严重挑战。气候变化目前被很多政府看作是 21 世纪最重要的挑战，需要专业性的解决措施，但这些措施在实施过程中收效甚微。最近几年，世界经济面临崩溃的边缘，严重的经济衰退既刺激了基础设施项目的投资，但也导致了削减赤字的激进措施，贫困和社会不公仍在继续并根深蒂固。但是，在强调影响评价系统变化的本质之前，我们需要首先考虑环境影响评价的理论背景。

1.5.2 环境影响评价的理论背景

环境影响评价必须在其理论背景之下被重新审视，特别是在决策理论的背景下（见 Lawrence，1997，2000；Bartlett 和 Kurian，1999；Weston，2000，2003）。环境影响评价是在 20 世纪 60 年代美国寻求理性决策方法的氛围之下诞生的（Caldwell，1988），关注点在于系统的程序、客观性、全面的方法、替代方案的考虑以及通常被视为主要是线性的方法。这个理性方法被假定为依赖于事实与逻辑完美的科学程序。在英国，这个理性方法在一些规划著作中有所体现，特别是 Faludi（1973）、McLoughlin（1969）以及 Friend 和 Jessop（1977）。

然而，其他关于环境影响评价理论背景的著作已经认识到环境影响评价程序的主观性本质的重要性。Kennedy（1988）把环境影响评价看作既是一门科学又是一门艺术，综合了政治和科学流程。更为精彩的是，Beattle（1995）在一篇文章中将环境影响评价称作“你已知道 EIA 的一切，但却不常承认”，再次强调了环境影响评价不是科学这一点，它们经常是在紧迫的时间和数据有差距的情况下做出来的，简化假设条件是这种情况下的规范做法。它们总是包含未检查和未解释的价值判断，也总是带有政治色彩。它们总是处理有争议的项目，并具有分配效应——即总是有赢家和输家。因此，当他们的作品被各方选择性使用时，环境影响评价的专业人员不应该对此感到吃惊或者沮丧。Leknes（2001）提到特别是在决策的后期，环境影响评价的结果很可能会给政治考虑让路。Weston（2003）提到 EIA 在科学、专家和理性方法方面的弱化。大型项目的决策信心往往被如核事故、化学泄漏、环境灾难以及项目的财务和时间的巨大超支等事件摧毁（Flyberg，2003）。公众越来越害怕他们很难掌控的变化所带来的后果，也越来越重视风险（见 Beck，1992，2008）。

然而，在决策理论的背景之下，这种对政治性、主观性以及价值判断的认识是体现在不同的行为和参与理论（behavioual/participative theories）之中的，并不是新的观点。例如，在 20 世纪 60 年代，Braybrooke 和 Lindblom（1963）将决策看作是不断优化调整的过程，此过程并不是全面、线性和有序的，更贴切的描述应为“摸索前进”。Lindblom（1980）通过“脱节渐进”的概念进一步发展了此观点，关注于满足通常是被政治定义的社会需求和目标。此时决策者已清晰认识到发现和面对两者的权衡（EIA 中的一个重要议题）的重要性。公众参与包括所有被影响团体之间的公开沟通过程。对众多受影响团体以及政府与公民之间的感知差异的认知激发了其他理论方法的产生，包括沟通和合作规划（Healey，1996，1997）。该方法是建于 Habermas（1984）、Forester（1989）以及其他一些前人的研究成果之上的，极大关注于建立共识、协调和沟通以及政府扮演的推进这些行动以处理不同利益相关者之间的利益冲突并促成合作的角色。批判者特别强调这种方法缺乏对社会中权力关系的考虑，尤其是私人开发商的角色——他们总是 EIA 中的支持者。

如今，可以认为环境影响评价当前的演变是趋于理性方法和行为方法之间的，两种方法的元素都有所体现。它包含了理性主义的重要思路，但同时也有众多参与方及决策点——政治、权力关系和专业判断通常走在前面。环境影响评价中涉及很多决策，比如，是否需要做 EIA（筛选）、EIA 的范围、考虑的替代方案、项目设计和再设计、减缓和改善措施的选择、项目生命周期中"重大决策后"阶段的实施和监测（Glasson，1999）。这与 Etzioni（1967）提倡的经典概念"混合审视"非常相符，都是运用评估的理性技术，并结合基于经验和价值基础上的更直觉的价值判断。Kaiser 等（1995）提出的理性适应方法也强调了决策中一系列步骤的重要性，既有以科学为基础的理性，也有让社区知晓的公众参与，中和对政策选项和期望结果的选择。

1.5.3 快速成长的影响评价（IA）家族中的一员——环境影响评价

过去的 40 年里，环境影响评价已加入了一个快速成长的评价工具家族。国际影响评价协会仅用评价影响（IA）这一通用词来概括所有的含义，但 Salder（1996）建议我们应将环境评价（EA）视为"包括具体项目的环境影响评价，政策、计划和方案（PPPs）的战略环境影响评价及涉及更多影响评价和规划的工具在内的通用程序"。不管这个家族的称谓如何，其成员在不断快速增加是不争的事实，其范围、规模和综合评价也在不断地扩展。如今，环境影响评价包括有 SIA、HIA、EqIA、TIA、SEA、SA、S&EIA、HRA/AA、EcIA、CIA 以及一系列相关技术如 RA、LCA、MCA、CBA 等。有些工具已被法规化，有些正被各行业的从业者推进，他们大多致力于区分和强调不同行业环评，由此形成了关注于不同主题的评价形式。Dalal-Clayton 和 Sadler（2004）观察到"首字母缩写（以及各种术语）现今造成了混乱的局面"。各种评价工具被简要地按其范围、规模和集成来概述如下，绝大多数将会在接下来的章节做进一步讨论。

1.5.3.1 范围

开发活动可能不只对物理环境造成影响，也会对社会和经济环境产生影响。特别是，就业机会、服务（如健康和教育）、社区结构、生活方式以及价值都可能受其影响。本书将社会-经济影响评价或社会影响评价（SIA）看作是 EIA 中不可分割的一部分。然而，有些国家却是（或曾经是）将之视为一份独立的程序，有时与 EIA 平行，读者应该意识到它的单独存在（Carley 和 Bustelo，1984；Finsterbusch，1985；IAIA，1994；Vanclay，2003）。一些领域明确用 S&SEA 表示社会-经济和环境影响评价。健康影响评价（HIA）从社会-经济影响评价演变而来，并在最近几年成了一个重要的发展领域，它关注于开发活动可能对其主要人口的健康产生的影响（IPHI，2009）。最新的领域是公平影响评价（EqIA），试图识别开发活动对不同社会群体的重要分布影响（如按不同性别、人种、年龄、残疾与否、性取向等，Downey，2005）。Vanclay 和 Bronstein（1995）等提到了几种其他相关定义，大部分是基于某些专业化焦点的，包括交通影响评价、人口统计评价影响、气候影响评价、性别影响评价、心理影响评价、噪声影响评价、经济影响评价以及累积影响评价等（Canter 和 Ross，2010）。

1.5.3.2 规模

战略环境评价（SEA）将评价规模从项目环境影响评价拓展到了战略层面的政策、计划和规划（PPPs）评价。开发活动可能是为了一个项目（如一个核电站）、一批项目［如一批压水式反应堆（PWR）核电站］，为了一个方案［如英国的城镇和乡村规划（T&CP）体系］，抑或是为了一项政策（如可再生能源的开发）。至今环境影响评价大多是被用于单个项目，这也是本书的首要关注点。对政策、计划和方案的环境影响评价也被称为战略环境影响评价（SEA），自 2004 年被欧盟（EU）引入后，已在全世界很多其他国家被采用（Therivel，2010；Therivel 和 Partidario，1996；Therivel 等，1992）。SEA 有助于更高、更早以及更具策略性层面的决策。理论上，EIA 应该按照先从政策的层面，再从规划和方案的层面，最后从项目层面的顺序来开展。SEA 的焦点主要是生物-物理方面的，与另一个相对较新的评价领域联系紧密，即生境调控评价/适宜性评价（habitats regulations assessment/appropriate assessment），欧盟要求对 Natura 2000 在生物多样性热点上、可能有重大影响的项目和规划都必须做此类评价。相比之下，SA 提供的战略评价的方法更为宽泛，包括了生物物理和社会-经济影响。英国要求 T&CP 体系下的规划影响评价须做 SA。在某些没有战略层面评价或规划的领域，项目层面的评价可以不同程度地参考 SEA 或 SA 的特点来采用战略的视角，巨型项目，如澳大利亚偏远地区的大型矿产开发项目，就是很好的例子。

1.5.3.3 集成

Hacking 和 Guthrie（2008）曾尝试提供一个合理的框架（图 1.9）来阐明各种评价工具在可持续发展规划和决策中的地位。除范围（被称为覆盖的全面性）和规模（重点和范围的战略性）以外，也包括技术和主体的集合。后者包括一整套技术以实现评价过程中的集成（如生物物理和社会经济影响之间的集成；Scrase 和 Sheate，2002）；这被 Lee（2002）称作"平行整合"。Petts（1999）提供了一个对包括如生命周期评价（LCA）、成本效益评价（CBA）、环境审计、多准则评价（MCA）和风险评价（RA）在内的一些技术的很好的综述。LCA 与 EIA 的区别在于它的重点不是某一个场址或设施，而是一个产品或系统及其造成的"从摇篮到坟墓"的环境影响（White 等，1995）。相比之下，CBA 关注的是开发项目的经济影响，从广阔而长远的角度来审视这些影响，并尽可能地将一个项目建议的成本和效益货币化。CBA 在 20 世纪 60 年代英国的大型交通项目中脱颖而出，之后又不断更新（Hanley 和 Splash，1993；Lichfield，1996）。环境审计是对设施运营和实践的环境表现进行系统的、周期性和记录性的评价，这个领域也见证了程序的发展，如国际标准 ISO 14001。

多准则决策评价（MCDA）通常是可量化的方法集合，可以用来帮助关键的利益相关者通过明确考虑多种准则来探索重要决策的替代方法（Belton 和 Stewart，2002），这个评价手段已被广泛使用。风险评价（RA）是另一个经常与 EIA 相联系的术语。例如英国

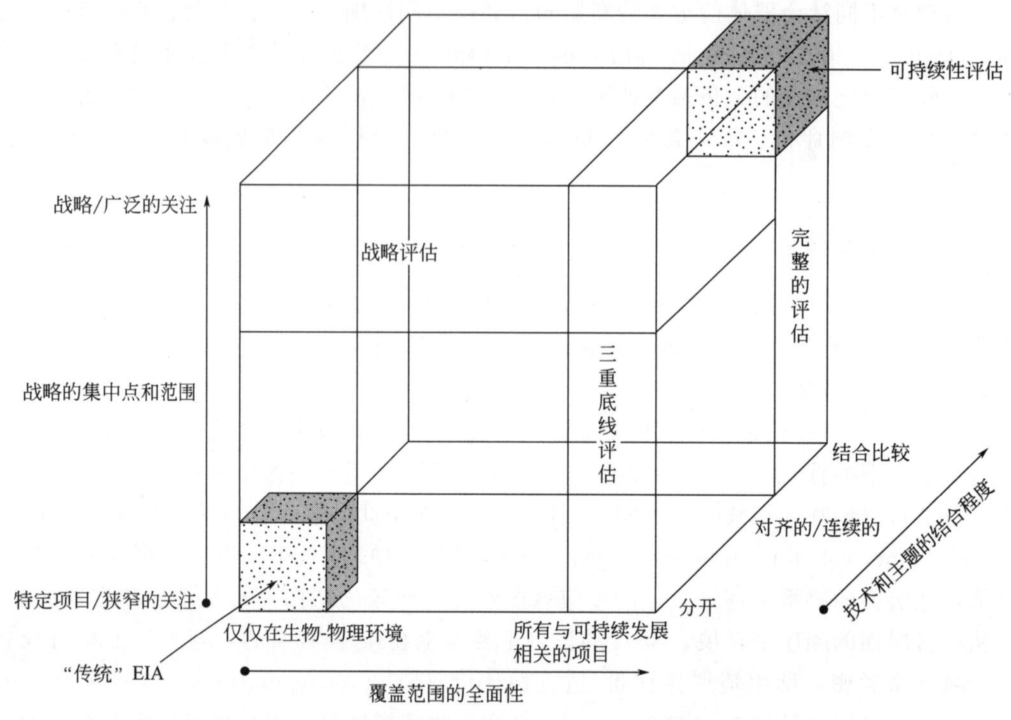

图 1.9　一种基于 SD 的评估工具的关系框架

（来源：Hacking and Guthrie，2008）

Flixborough 化工厂爆炸、美国 Three Mile Island 核电站事故以及乌克兰切尔诺贝利核电站事故等事件的响应，都使用了这一评价手段，RA 也发展成了分析各种类型开发项目风险的方法。Carlow（1997）回顾了环境风险评价和管理的发展领域，Flyberg（2003）也对风险评估实践进行了批判。虽然这些工具都偏重以技术为中心，它们仍可被看作是与 EIA 的互补，以寻求实现更综合的方法与决策。由此，第 5 章探索了 CBA 和 MCA 方法在 EIA 评估中的潜在作用，第 12 章进一步发展了综合评价的概念，并探索了环境审计和 LCA 相对于环境管理体系（EMSs）的作用。

本书对变化的视角、理论背景、相关工具和过程的简短讨论，强调了对持续再评价 EIA 的作用和其实施的需求以及适应性 EIA 的重要性。接下来几个章节，特别是在第 4 部分中，将进一步讨论。

1.6　目前环境影响评价所存在的问题

虽然 EIA 在美国已经有了 40 多年的历史，但其概念和实践在其他地方都是近些年才发展起来的。在很多国家，EIA 都得到了快速的发展，这其中也包括英国及其他欧盟成员国。很多发展的方向都是受欢迎的。Gibson（2002）提到一些世界范围内的 EIA 发展趋势，如过程提前了、更开放和更具参与性、更全面（不仅是生物物理环境）、更具强制性、监测更严密、运用更广泛（如在不同水平和层面上）、更综合、更加雄心勃勃（在可持续性目标方面）以及更加谨慎（认识到不确定性并加以防范）。然而这些进步也不是一成不变的，发展过程中也存在着一些问题。本章主要讲述目前 EIA 所存在的主要问题，并且将在后面的章节予以详细讨论。

1.6.1 评价方法的本质

如 1.2 节所述，EIA 的一些主要步骤（如审计和监测等）可能在很多研究中都被遗漏了。而且那些已被纳入的步骤也可能存在一些问题。影响预测中存在着很多概念与技术上的问题，例如前面已经提到过的确定环境本底情况时遇到的问题。把一个项目的空间范围和发展阶段都划分清楚也很困难，特别是对新技术项目而言。更为深入的概念性问题包括：如何确定在没有项目开发时的相关环境状况、认清多种现象相互作用的复杂性以及从整体出发进行利益权衡（如评价经济、社会、自然环境之间的利益平衡）。技术性问题包括：数据的缺失、将重点过多放在量化上、某些地区的指标比较单一等情况。有时，因果关系也可能存在延迟或不连续性，项目和政策之间也缺少关联。缺少对预测技术的审计也限制了对方法的有效性的反馈。不过，从简单的核查清单法、矩阵法到复杂的数学模型法和多准则方法，很多新方法正逐步被引入到影响预测中。需要注意的是，这些方法可能都不是中性的，方法越复杂，环境影响评价程序中的公众参与也越困难。

1.6.2 EIA 的质量和效率

环境影响报告（EIS）是一种直接输出程序的质量评价。很多环境影响报告可能都不能满足最低的标准。例如，Jones（1991）等基于英国 EIA 法规出版的环境影响报告的调查就揭露了许多问题。他们发现：

……1/3 的环境影响报告没有所要求的非技术性总结；1/4 的报告书中没有用来评价开发行为可能造成环境影响的数据；另外很多报告书都忽略了较为复杂的相互影响。

DoE（1996）认为虽然我们从实践中积累了一定经验，但是英国的很多报告书仍然无法令人满意（第 8 章将就此问题做进一步的讨论）。项目类型不同，报告书的质量也有所不同。而且在同一法律框架下，不同国家的报告书之间也存在差异。

环境影响报告书可能存在长篇大论、不够整合的风险，EIA 过程中的材料对于公众来说也很难看懂，这些结果都对 EIA 程序的效率提出了各种质疑。例如，“安全第一”的政策是否导致太多的项目接受 EIA 筛查？EIA 在范围界定阶段是否考虑了太多潜在影响？是否过于把重点放在了描述性的基础工作，而对事关重大的关键影响重视不够？环境影响报告书是否只是不同隔离的、专章的集合而非有机集合的文档？监测和审计的关键步骤是否很好融入了评估程序？对效率的考虑也可能违背对公平的考虑。

1.6.3 评价程序中的参与者的作用

环境影响评价程序中的参与者（开发商、受影响群体、一般公众和政府各级管理者）都会不同程度地参与进来，对结果产生的影响也不同。有些人认为英国等国家的环境影响评价多数是由开发商主导的。开发商及他们的咨询机构实施 EIA，编制 EIS，他们不太可能预测项目的开发会是一场环境灾难。尽管如此，开发商所关心的是提交环境影响报告书可能会耽误的时间以及产生的具体费用。根据早期的估计（Clark，1984；Hart，1984；Wathern，1988），EIA 的费用应该占到一般项目投资的 0.5%～2.0%。英国 DETR（1997）建议在新的环境影响评价法规下进行 EIA，较为合理的中间费用应在 35000 英镑左右，但大型项目的费用可大大高于这个数。最近的一份欧盟委托评估《欧盟环境影响评价指令》的研究表明，EIA 的费用占项目的投资比例从小项目的 1%到大项目的 0.1%不等（CEC，2006）。

各国的公众参与在环境影响评价中的程序和具体实践中均有所不同，有的非常复杂，有的则极其简单，甚至流于形式。其中的一个重要问题就是公众参与应该是在环境影响评价程序中的哪个阶段进行。在环境影响评价程序中，政府角色往往趋于谨慎，在这个快速发展的新领域里表现得缺乏经验和专业技能，并且过多的偏向于对资源的考虑。中央政府可能会对最佳实践方案提供有限的指导，并且可能会做出与之不一致的决策。地方政府可能会发现很难处理环境影响报告的范围和内容的复杂性，尤其是针对大型项目。

1.6.4 EIA 的有效性

虽然 EIA 体系在世界很多国家都已确立，但还是应该深刻反省其有效性，EIA 是否实现了如 1.3 节所述的目的？对如何评估 EIA 有效性也存在争议。对此，可以有不同的相关考虑。例如，狭义的程序/方法会把重心放在按照其国家要求的程序开展 EIA，广义的程序/方法可能会考虑 EIA 对增进关键利益相关者的环保意识和知识所做的贡献程度。这些方面在前面的小节（第 1.6.1～1.6.3 节）都有所讨论。然而，与 EIA 核心目的根本相关的是实质性方法。例如，狭义的实质性方法会关注 EIA 是否对规划决策的质量和开发活动的本质有直接影响；广义的实质性方法会关注 EIA 是否保持、恢复、提高了环境质量这一根本问题，是否促进了可持续发展。本书不同章节都对这些 EIA 有效性的问题进行了审视，特别是在第 8 章。

1.6.5 决策的延伸

大多数 EIS 都是针对某个特定项目的，因此很少有开发商主动去审计影响预测的质量以及监测影响，从而更好地对后续项目进行评价。EIA 在项目决策之前只是其过程中的一个部分，重要的是确保要求的减缓和改善措施都执行实施。在世界上的很多地区（如加利福尼亚、西澳大利亚、荷兰、中国香港等），影响的监测是法律强制执行的，并且必须在环境影响报告中包含监测程序。抓住这个周期性学习过程的机会也很重要，要尽可能全面地审计预测结果，检查预测的准确性。与环境管理程序的关系是另一个值得关注的重要领域，EIS 能有效指引项目实施的环境管理计划，但做得好的并不多。影响的监测等方法的延伸是当前在项目环境影响评价程序中出现的又一显著问题。

1.6.6 管理影响评价活动范围的扩大和复杂性

如 1.5 节所述，影响评价家族成长迅速，尤其是在最近几年。如何来管理其复杂性呢？例如，现代环境影响报告书内容的标准是什么？扩大环境影响评价考虑的方面，更全面地包括社会经济影响是一个强烈趋势。开发活动往往带来不利的生物物理影响（biophysical impacts）和有利的社会经济影响，这两者之间的权衡正是决策者的两难之处。影响扩展开来也可以包括其他一些类型的影响，但至今报道很少。EIS是否应该将社会、健康以及公平因素作为标准内容，还是将这些内容从EIS中分离开来？与此相类似的，哪些项目应该进行EIA？比如，大型项目需要强制使用EIA，在具体实施过程中又有哪些具体步骤需要强制使用EIA。如今很多国家都在建立相关的法律，但需要进行EIA的项目选择标准不太明确。

1.5节也提到，对PPPs的战略环境影响评价是项目环境影响评价范围的合理延伸。与项目评价相比，战略环境影响评价能更好地处理累积影响、替代方案和减缓措施。但不同规模的影响评价之间关系的本质是什么呢？战略层面上的规划和方案评价应该为具体项目评价提供有用的框架，以期减少工作量并趋向于更简洁和有效的EIAs。然而这种预期的层次关系可能理论胜于实践，从而导致了不必要和重复浪费的行为。

战略环境影响评价体系已经在加利福尼亚和荷兰得到了应用，在加拿大、德国、新西兰等国也有小范围的应用。之后的第五社区环境行动计划提到：要实现可持续发展的目标，对所有相关政策、规划和方案的潜在环境影响进行评价即使不是必要的，也是合理的（CEC，1992）。欧盟的有关战略环境影响评价法令已经出台，并于2004年开始施行（Therivel，2004和第12章）。

1.7 后续部分及章节的概述

本书共4部分。第一部分主要介绍了在人们对环境问题日益关注和制定相关法律法规的背景下，环境影响评价制度的建立及发展情况，尤其是英国和欧盟的一些情况。第1章简单介绍了环境影响评价概念及其主要原则。第2章主要讲述了1969年环境影响评价在美国NEPA的起源以及在英国的发展，之后介绍了欧共体85/337号指令和其之后的修订及发展。第3章论述了在英国T&CP和其他法规指导下，英国环境影响评价法规体系的详细情况。

第二部分介绍了环境影响评价程序中严格的逐步评价法，这也是本书的重点内容。第4章介绍了评价的早期开始阶段：建立管理框架；分析进行环境影响评价项目的类别；范围确定、替代方案选择、项目的描述、环境本底数据的收集及影响识别。第5章探讨了影响预测、显著性评价和不利影响的减缓措施等重要议题。这些方法勾勒出了影响预测操作的几大主要原则，并且附有实际案例加以说明。第6章详述了环境影响评价程序中的一个重要议题，即公众参与。本章也介绍了环境影响评价过程中的信息沟通、环境影响报告的表述以及环境影响评价的评审等。第7章介绍了项目决策之外的一些程序以及环境影响评价程序中监测、审计的重要性及其方法。

第三部分以实例说明环境影响评价程序。第8章对英国环境影响评价的实践进行了回顾，包括环境影响评价准备阶段的各种定性与定量分析。第9章介绍了环境影响评价在一些诸如能源、运输、废弃物管理及旅游等主要部门的实际应用情况，该章的特点是引用了大量英国及其他国家关于环境影响评价的新近典型案例，以说明环境影响评价的主要特征和问题。第10章将一些发达国家（如加拿大和澳大利亚）的经验与一些发展中国家及新兴经济体（如秘鲁、中国、贝宁和波兰）的经验进行了对比分析——提出了其他系统在实践中的优缺点，本章还讨论了一些国际机构（如欧洲复兴开发银行和世界银行等）在环境影响评价实践领域的重要作用。

第四部分着眼于未来，详细介绍了很多在 1.6 节中所提到的问题。第 11 章主要探讨了战略环境影响评价的需要及一些局限，回顾了战略环境影响评价在美国、欧盟、联合国欧洲经济委员会以及中国的现状，之后又详细讨论了欧洲战略环境影响评价指令在英国的实施情况。第 12 章重点讨论了项目环境影响评价有效性的改进以及项目环境影响评价的前景，这一章考虑了环境影响评价程序中不同参与者对变化的不同看法，紧接着也考虑了环境影响评价程序中一些重点领域和环境影响报告书本质的发展可能，结尾讨论了环境管理体系和审计的平行及互补发展。综合起来，这些主题将作为一种行动清单来促进环境影响评价未来的改善。本书附录中提供的某些法律和实践信息、网站和期刊的细节可能与正文内容不完全匹配。

问　题

以下问题贴合本章主要内容，帮助读者了解和明确 EIA 的主要原则和观点。

1. 回顾本章 EIA 的定义，选出你喜欢的，并说明理由。

2. EIA 中很多步骤在实践中难以实现，根据你的学习，选出难以实现的步骤并说明理由。

3. 选取本国最近的环评案例，对照书中内容，自己的理解是否有所提升？

4. 请找出两者的不同：（1）项目筛选和项目审查；（2）影响减缓和影响增加。

5. 反观 EIA 的目的，从自己的视角说明其重要性。

6. 结合表 1.2 的相关内容和你熟悉的两个环评案例，比较两个案例的区别并给出相应解释。

7. 结合问题 6，对环评案例进行生命周期评价。

8. 怎样理解 EIA 中多维度的环境手段？

9. EIA 中的影响是什么，怎样理解影响和效用之间的联系？

10. 对以下名词进行解释：（1）反作用；（2）累计影响；（3）分布影响（distributional impact）。

11. 为什么在环评时要采用适宜的方法？

12. 你觉得把 EIA 的过程认定为理性的、线性的、科学的过程是否合理？

13. EIA 和 SEA 的区别是什么？

14. 为什么要扩宽 EIA 的审查？

15. 怎么理解 EIA 的“beyond the decision”？

16. 如何衡量以下两个概念：（1）效率；（2）EIA 的有效性。

Part 1

Principles and procedures

1 Introduction and principles

1.1 Introduction

Over the last four decades there has been a remarkable growth of interest in environmental issues-in sustainability and the better management of development in harmony with the environment. Associated with this growth of interest has been the introduction of new legislation, emanating from national and international sources such as the European Commission, that seeks to influence the relationship between development and the environment. Environmental impact assessment (EIA) is an important example. EIA legislation was introduced in the USA over 40 years ago. A European Community (EC) directive in 1985 accelerated its application in EU Member States and it has spread worldwide. Since its introduction in the UK in 1988, it has been a major growth area for planning practice; the originally anticipated 20 environmental impact statements (EIS) per year in the UK has escalated to several hundreds, and this is only the tip of the iceberg. The scope of EIA continues to widen and grow.

It is therefore perhaps surprising that the introduction of EIA met with strong resistance from many quarters, particularly in the UK. Planners argued, with partial justification, that they were already making such assessments. Many developers saw it as yet another costly and timeconsuming constraint on development, and central government was also unenthusiastic. Interestingly, initial UK legislation referred to environmental assessment (EA), leaving out the apparently politically sensitive, negative-sounding reference to impacts. The scope of the subject continues to evolve. This chapter therefore introduces EIA as a process, the purposes of this process, types of development, environment and impacts, and current issues in EIA.

1.2 The nature of EIA

1.2.1 *Definitions*

Definitions of EIA abound. They range from the oft-quoted and broad definition of Munn (1979), which refers to the need 'to identify and predict the impact on the environment and on man's health and well-being of legislative proposals, policies, programmes, projects and operational procedures, and to interpret and communicate information about the impacts', to the narrow and early UK DoE (1989) operational definition:

> The term 'environmental assessment' describes a technique and a process by which information about the environmental effects of a project is collected, both by the developer and from other sources, and taken into account by the planning authority in forming their judgements on whether the development should go ahead.

UNECE (1991) had an altogether more succinct and pithy definition: 'an assessment of the impact of a planned activity on the environment'. The EU EIA Directive requires an assessment of the effects of certain public and private projects, which are likely to have significant effects on the environment, before development consent is granted; it is procedurally based (see Appendix 1). The EIA definition adopted by the International Association for Impact Assessment (IAIA 2009) is 'the process of identifying, predicting, evaluating and mitigating the biophysical, social and other relevant effects of proposed development proposals prior to major decisions being taken and commitments made'. This process emphasis is now explored further.

1.2.2 *EIA: a process*

In essence, EIA is *a process*, a systematic process that examines the environmental consequences of development actions, in advance. The emphasis, compared with many other mechanisms for environmental protection, is on prevention. Of course, planners have traditionally assessed the impacts of developments on the environment, but invariably not in the systematic, holistic and multidisciplinary way required by EIA. The process involves a number of steps, as outlined in Figure 1.1.

The steps are briefly described below, pending a much fuller discussion in Chapters 4—7. It should be noted at this stage that, although the steps are outlined in a linear fashion, EIA should be a cyclical activity, with feedback and interaction between the various steps. It

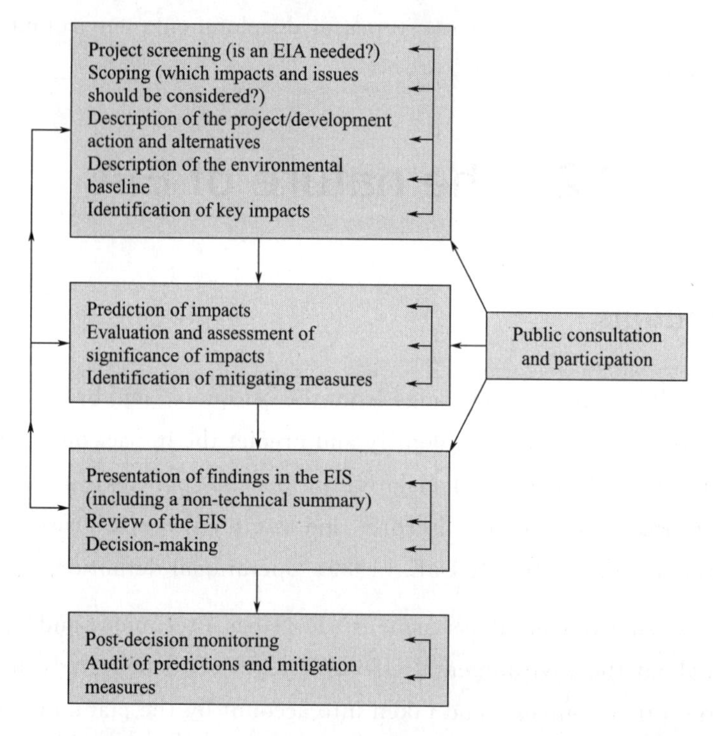

Figure 1. 1 Important steps in the EIA process

Note that EIA should be a cyclical process, with considerable interaction between the various steps. For example, public participation can be useful at most stages of the process; monitoring systems should relate to parameters established in the initial project and baseline descriptions.

should also be noted that practice can and does vary considerably from the process illustrated in Figure 1. 1. For example, UK EIA legislation still does not require post-decision monitoring. The order of the steps in the process may also vary.

- *Project screening* narrows the application of EIA to those projects that may have significant environmental impacts. Screening may be partly determined by the EIA regulations operating in a country at the time of assessment.
- *Scoping* seeks to identify at an early stage, from all of a project's possible impacts and from all the alternatives that could be addressed, those that are the crucial, significant issues.
- *The consideration of alternatives* seeks to ensure that the proponent has considered other feasible approaches, including alternative project locations, scales, processes, layouts, operating conditions and the 'no action' option.
- *The description of the project/development action* includes a clarification of the purpose and rationale of the project, and an understanding of its various characteristics—including stages of development, location and processes.
- *The description of the environmental baseline* includes the establishment of both the present and future state of the environment, in the absence of the project, taking into account changes resulting from natural events and from other human activities.
- *The identification of the main impacts* brings together the previous steps with the

aim of ensuring that all potentially significant environmental impacts (adverse and benef icial) are identified and taken into account in the process.

- *The prediction of impacts* aims to identify the magnitude and other dimensions of identified change in the environment with a project/action, by comparison with the situation without that project/action.
- *The evaluation and assessment of significance* assesses the relative significance of the pre-dicted impacts to allow a focus on the main adverse impacts.
- *Mitigation* involves the introduction of measures to avoid, reduce, remedy or compensate for any significant adverse impacts. In addition *enhancement* involves the development of beneficial impacts where possible.
- *Public consultation and participation* aim to ensure the quality, comprehensiveness and effectiveness of the EIA, and that the public's views are adequately taken into consideration in the decision-making process.
- *EIS presentation* is a vital step in the process. If done badly, much good work in the EIA may be negated.
- *Review* involves a systematic appraisal of the quality of the EIS, as a contribution to the decision-making process.
- *Decision-making* on the project involves a consideration by the relevant authority of the EIS (including consultation responses) together with other material considerations.
- *Post-decision monitoring* involves the recording of outcomes associated with development impacts, after a decision to proceed. It can contribute to effective project management.
- *Auditing* follows from monitoring. It can involve comparing actual outcomes with predicted outcomes, and can be used to assess the quality of predictions and the effectiveness of mitigation. It provides a vital step in the EIA learning process.

1.2.3 Environmental impact statements: the documentation

The EIS documents the information about and estimates of impacts derived from the various steps in the process. [1] Prevention is better than cure; an EIS revealing many significant unavoidable adverse impacts would provide valuable information that could contribute to the abandonment or substantial modification of a proposed development action. Where adverse impacts can be successfully reduced through mitigation measures, there may be a different decision. Table 1.1 provides an example of the content of an EIS for a project.

The *non-technical summary* is an important element in the documentation; EIA can be complex, and the summary can help to improve communication with the various parties involved. Reflecting the potential complexity of the process, an *introduction* should clarify, for example, who the developer is, who has produced the EIS, and the relevant legal framework. Also at the beginning, a *methodology section*, provides an opportunity to clarify some basic information (e.g. what methods have been used, how the key issues were identified, who was consulted and how, what difficulties have been encountered, and what are the limitations of the EIA). The *background to the proposed development* covers the early steps in the EIA process, including clear descriptions of a project, and baseline conditions (including relevant planning policies and plans).

Table 1.1 An EIS for a project-example of contents

Non-technical summary Part 1: Introduction, methods and key issues Introduction Methodology Summary of key issues	Hydrology and water quality Air quality Climate change Ecology: terrestrial and aquatic Noise and vibration
Part 2: Background to the proposed development Preliminary studies: need, planning, alternatives and site selection Site description, baseline conditions Description of proposed development Development programme, including site preparation, construction, operation, decommissioning and restoration (as appropriate)	Socio-economics Transport Landscape, visual quality Historic environment Recreation and amenity Interrelationships between effects Cumulative impacts Summary of residual impacts
Part 3: Environmental impact assessment-topic areas Land use Geology, topography and soils	Part 4: Follow-up and management Monitoring of impacts Management of impacts

Within each of the *topic areas* of an EIS there would normally be a discussion of existing conditions, predicted impacts, scope for mitigation and enhancement, and residual impacts. The list here is generic, and there are some topics that are still poorly covered, for example climate change and cumulative impacts (as appropriate). A concluding section, although often omitted from EISs, should cover key *follow-up issues*, including monitoring and management.

Environmental impact assessment and EIS practices vary from study to study, from country to country, and best practice is constantly evolving. An early UN study of EIA practice in several countries advocated changes in the process and documentation (UNECE 1991). These included giving a greater emphasis to the socio-economic dimension, to public participation and to 'after the decision' activity, such as monitoring. More recent reviews of the operation of the amended EC Directive (CEC 2003a, 2009) raised similar issues, and other emerging issues, a decade later (see Chapter 2). Sadler (1996) provided a wider agenda for change based on a major international study of the effectiveness of EIA, being updated in 2010-11 (see Chapters 8 and 12).

1.3 The purposes of EIA

1.3.1 An aid to decision-making

EIA is an aid to decision-making. For the decision-maker, for example a local authority, it provides a systematic examination of the environmental implications of a proposed action, and sometimes alternatives, before a decision is taken. The EIS can be considered by the decision-maker along with other documentation related to the planned activity. EIA is normally wider in scope and less quantitative than other techniques, such as cost-benefit analysis (CBA). It is not a substitute for decision-making, but it does help to clarify some of the trade-offs associated with a proposed development action, which should lead to more informed and structured decision-making. The EIA process has the potential, not always taken up, to be a basis for negotiation between the developer, public interest groups and the planning regulator. This can lead to an outcome that balances well the interests of the development action and the environment.

1.3.2 *An aid to the formulation of development actions*

Developers may see the EIA process as another set of hurdles to jump before they can proceed with their various activities; the process can be seen as yet another costly and time-consuming activity in the development consent process. However, EIA can be of great benefit to them, since it can provide a framework for considering location and design issues and environmental issues in parallel. It can be an aid to the formulation of development actions, indicating areas where a project can be modified to minimize or eliminate altogether its adverse impacts on the environment. The consideration of environmental impacts early in the planning life of a development can lead to more environmentally sensitive development; to improved relations between the developer, the planning authority and the local communities; to a smoother development consent process; and sometimes to a worthwhile financial return on the extra expenditure incurred. O'Riordan (1990) links such concepts of negotiation and redesign to the important environmental themes of 'green consumerism' and 'green capitalism'. The growing demand by consumers for goods that do no environmental damage, plus a growing market for clean technologies, is generating a response from developers. EIA can be the signal to the developer of potential conflict; wise developers may use the process to negotiate 'environmental gain' solutions, which may eliminate or offset negative environmental impacts, reduce local opposition and avoid costly public inquiries. This can be seen in the wider and contemporary context of corporate social responsibility (CSR) being increasingly practised by major businesses (Crane *et al*. 2008).

1.3.3 *A vehicle for stakeholder consultation and participation*

Development actions may have wide-ranging impacts on the environment, affecting many different groups in society. There is increasing emphasis by government at many levels on the importance of consultation and participation by key stakeholders in the planning and development of projects; see for example the 'Aarhus Convention' (UNECE 2000) and the EC Public Participation Directive (CEC 2003b). EIA can be a very useful vehicle for engaging with communities and stakeholders, helping those potentially affected by a proposed development to be much better informed and to be more fully involved in the planning and development process.

1.3.4 *An instrument for sustainable development*

Existing environmentally harmful developments have to be managed as best as they can. In extreme cases, they may be closed down, but they can still leave residual environmental problems for decades to come. It would be much better to mitigate the harmful effects in advance, at the planning stage, or in some cases avoid the particular development altogether. Prevention is better than cure. This is the theme of the pioneering US and EC legislation on EIA. For example, the preamble to the 1985 EC EIA Directive includes 'the best environmental policy consists in preventing the creation of pollution or nuisances at source, rather than subsequently trying to counteract their effects' (CEC 1985). This of course leads on to the fundamental role of EIA as an instrument for sustainable development-a role some writers have drawn attention to as one often more hidden than it should be when EIA effectiveness is being assessed (Jay *et al*. 2007).

The nature of sustainable development

Economic development and social development must be placed in their environmental contexts. The classical work by Boulding (1966) vividly portrays the dichotomy between the 'throughput economy' and the 'spaceship economy' (Figure 1.2). The economic goal of increased *gross national product* (GNP), using more inputs to produce more goods and services, contains the seeds of its own destruction. Increased output brings with it not only goods and services, but also more waste products. Increased inputs demand more resources. The natural environment is the 'sink' for the wastes and the 'source' for the resources. Environmental pollution and the depletion of resources are invariably the ancillaries to economic development.

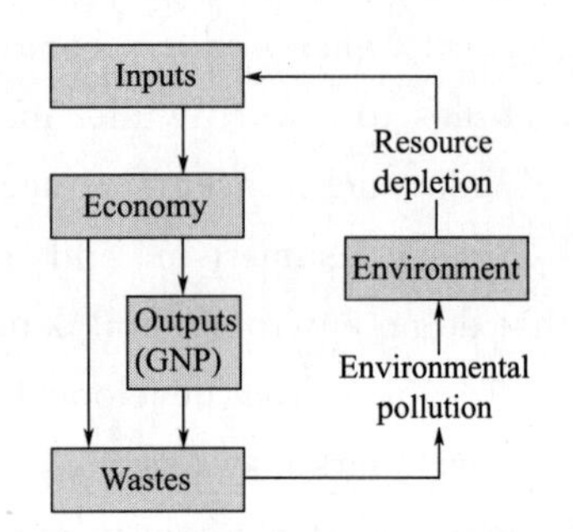

Figure 1.2 The economic development process in its environmental context (adapted from Boulding 1966)

The interaction of economic and social development with the natural environment and the reciprocal impacts between human actions and the biophysical world have been recognized by governments from local to international levels, and attempts have been made to manage the interaction better. However towards the end of the first decade of the twentieth-first century, the European Environment Agency report, *European Environment-State and Outlook* 2010 (EEA 2010), still showed some good progress mixed with remaining fundamental challenges, with potentially very serious consequences for the quality of the environment. For example, while greenhouse gas emissions have been cut and the EU is on track to reach a reduction target of 20 per cent by 2020, the Member States still produced close to 5 billion tonnes of CO_2 equivalent emissions in 2008. Similarly while Europe's waste management has shifted steadily from landfill to recycling and prevention, still half of the 3 billion tonnes of total waste generated in the EU-27 in 2006 was landfilled. In nature and biodiversity, Europe has expanded its Natura 2000 network of protected areas to cover 18 per cent of EU land, but missed its 2010 target to halt biodiversity loss. Europe's freshwaters are affected by water scarcity, droughts, floods, physical modifications and the continuing presence of a range of pollutants. Both ambient air and water quality remain inadequate and health impacts are widespread. We also live in an interconnected world. European policymakers aren't only contending with complex systematic interactions within Europe. There are also unfolding global drivers of change that are likely to affect Europe's environment, and many are beyond Europe's control. Some environmental trends are likely to be even more pronounced in developing countries, where, because population growth is greater and current living standards lower, there will be more pressure on environmental resources.

The 1987 Report of the UN World Commission on Environment and Development (usually referred to as the Brundtland Report, after its chairwoman) defined sustainable development as 'development which meets the needs of the present generation without compromising the ability of future generations to meet their own needs' (UN World Commission on Environment and Development 1987). Sustainable development means handing down to future generations not only 'manmade capital' (such as roads, schools and historic buildings) and 'human capital' (such as knowledge and skills), but also 'natural/environmental capital' (such as clean air, fresh water, rainforests, the ozone layer and biological diversity). The Brundtland Report identified the following chief characteristics of sustainable development: it maintains the quality of life, it maintains continuing access to natural resources and it avoids lasting environmental damage. It means living on the earth's income rather than eroding its capital (DoE *et al.* 1990). In addition to a concern for the environment and the future, Brundtland also emphasizes participation and equity, thus highlighting both inter- and intra-generational equity. This definition is much wider than ecology and the natural environment; it entails social organization of intra-and inter-generational equity. Importance is also assigned to economic and cultural aspects, such as preventing poverty and social exclusion, concern about the quality of life, attention to ethical aspects of human well being, and systematic organization of participation by all concerned stakeholders.

There is, however, a danger that 'sustainable development' becomes a weak catch-all phrase; there are already many alternative definitions. Holmberg and Sandbrook (1992) found over 70 definitions of sustainable development. Redclift (1987) saw it as 'moral convictions as a substitute for thought'; to O'Riordan (1988) it was 'a good idea which cannot sensibly be put into practice'. But to Skolimowski (1995), sustainable development

> ... struck a middle ground between more radical approaches which denounced all development, and the idea of development conceived as business as usual. The idea of sustainable development, although broad, loose and tinged with ambiguity around its

edges, turned out to be palatable to everybody. This may have been its greatest virtue. It is radical and yet not offensive.

Readers are referred to Reid (1995), Kirkby *et al.* (1995) and Faber *et al.* (2005) for an overview of the concept, responses and ongoing debate.

Over time, 'sustainability' has evolved as a partial successor to the term 'sustainable development' (although they can be seen as synonymous), partly because the latter has become somewhat ill used (for example, governments seeking to equate sustainable development with sustained growth, firms seeking to equate it with sustained profits). [2] How ever, despite the global acceptance of the 'sustainability/sustainable development' concept, its scope and nature are a somewhat contested and confused territory (Faber *et al.* 2005). There are numerous definitions, but a much-used one is that of the triple bottom line (TBL), reflecting the importance of environmental, social and economic factors in decision-making, although it is important to go beyond that to emphasize the importance of integration and synergies between factors (Figure 1.3); however the assessment of such synergies presents particular challenges. Figure 1.4 emphasizes that within this three-element definition of sustainability, there is an important hierarchy. The environment and its natural systems are the foundation to any concept of sustainability. We cannot survive without the 'goods and services' provided by Earth's natural and physical systems-breathable air, drinkable water and food. Living on Earth, we need social systems to provide social justice, security, cultural identity and a sense of place. Without a well-functioning social system, an economic system cannot be productive.

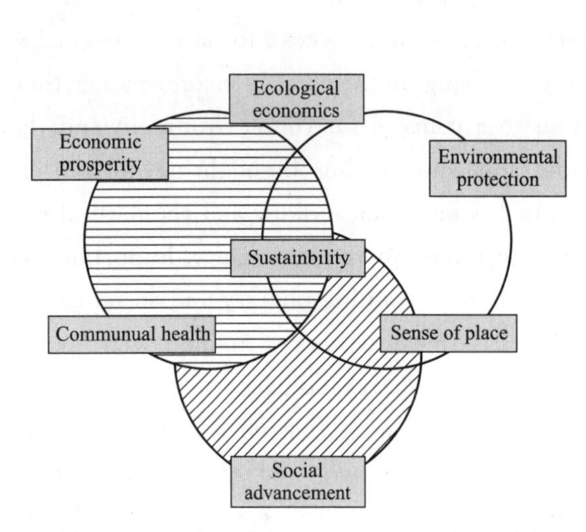

Figure 1.3 Integrating environmental, social and economic dimensions of sustainability

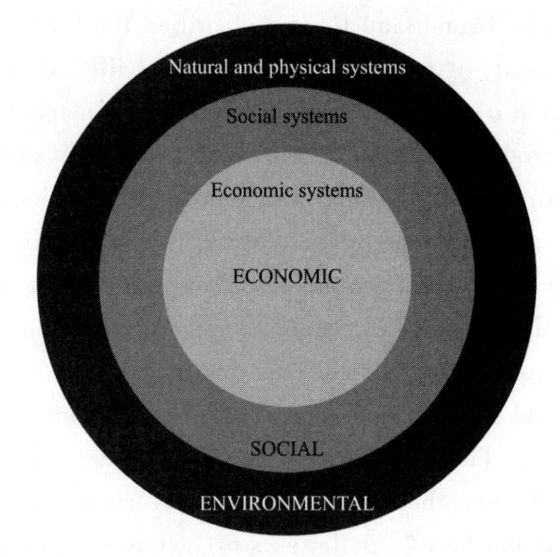

Figure 1.4 An alternative (hierarchical) perspective on the dimensions of sustainability

Institutional responses to sustainable development

Institutional responses to meet the goal of sustainable development are required at several levels. A *global response* is needed for issues of global concern, such as ozone-layer depletion, climate change, deforestation and biodiversity loss. The United Nations Conference on Environment and Development (UNCED) held in Rio de Janeiro in 1992 was an example not only of international concern, but also of the problems of securing concerted action to deal with such issues. Agenda 21, an 800-page action plan for the international community into the twenty first century, set out what nations should do to achieve sustainable development. It included topics such as biodiversity, desertification, deforestation, toxic wastes, sewage, oceans and the atmosphere. For each of its 115 programmes, the need for action, the objectives and targets to be achieved, the activities to be undertaken, and the means of implementation are all outlined. Agenda 21 offered policies and programmes to achieve a sustainable balance between consumption, population and Earth's life-supporting capacity. Unfortunately it was not legally binding, being dependent on national governments, local governments and others to implement most of the programmes.

The Johannesburg Earth Summit of 2002 reemphasized the difficulties of achieving international commitment on environmental issues. While there were some positive outcomes—for example, on water and sanitation (with a target to halve the number without basic sanitation—about 1.2 billion—by 2015), on poverty, health, sustainable consumption and on trade and globalization—many other outcomes were much less positive. Delivering the Kyoto Protocol on legally enforceable reductions of greenhouse gases continued to be difficult; the results of the 2009 Copenhagen climate conference fell short of the EU's goal of progress towards the finalization of an ambitious and legally binding global climate treaty to succeed the Kyoto Protocol in 2013 (Wilson and Piper 2010). Similarly, we hear regularly of the continuing loss of global biodiversity and of natural resources, and on the challenges of delivering human rights in many countries. All, of course, is now complicated further by the severe challenges and uncertainties of the serious global economic situation. Together, such problems severely hamper progress on sustainable development.

Within the EU, four Community Action Programmes on the Environment were implemented between 1972 and 1992. These gave rise to specific legislation on a wide range of topics, including waste management, the pollution of the atmosphere, the protection of nature and EIA. The Fifth Programme, 'Towards sustainability' (1993—2000), was set in the context of the completion of the Single European Market (CEC 1992). The latter, with its emphasis on major changes in economic development resulting from the removal of all remaining fiscal, material and technological barriers between Member States, could pose additional threats to the environment. The Fifth Programme recognized the need for the clear integration of performance targets—in relation to environmental protection—for several sectors, including manufacturing, energy, transport and tourism. EU policy on the environment would be based on the 'precautionary principle' that preventive action should be taken, that environmental damage should be rectified at source and that the polluter should pay. Whereas previous EU programmes relied almost exclusively on legislative instruments, the Fifth Programme advocated a broader mixture, including 'market-based instruments', such as the internalization of environmental costs through the application of fiscal measures, and 'horizontal, supporting instruments', such as improved baseline and statistical data and improved spatial and sectoral planning.

The Sixth Programme, *Our future, our choice* (2001—12), built on the broader approach introduced in the previous decade. It recognized that sustainable development has social and economic as well as physical environmental dimensions, although the focus is on four main priority issues: tackling climate change, protecting nature and biodiversity, reducing human health impacts from environmental pollution, and ensuring the sustainable management of natural resources and waste. It also recognized the importance of empowering citizens and changing behaviour, and of 'greening land-use planning and management decisions'.

> The Community directive on EIA and (the then) proposal on SEA, which aim to ensure that the environmental implications of planned infrastructure projects and planning are properly addressed, will also help ensure that the environmental considerations are better integrated into planning decisions. (CEC 2001)

The EC has not yet decided on the nature of a possible Seventh Programme, including the key role of climate change—either as within the EU environmental policy or as having a more overarching role in the Commission's organization.

In the UK, the publication of *This common inheritance: Britain's environmental strategy* (DoE *et al.* 1990) provided the country's first comprehensive White Paper on the environment. The report included a discussion of the greenhouse effect, town and country, pollution control, and awareness and organization with regard to environ mental issues. Throughout it emphasized that responsibility for our environment should be shared between the government, business and the public. The range of policy instruments advocated included legislation, standards, planning and economic measures. The last, building on work by Pearce *et al.* (1989), included charges, subsidies, market creation and enforcement incentives. The report also noted, cautiously, the recent addition of EIA to the 'toolbox' of instruments. Subsequent UK government reports, such as *Sustainable development: the UK strategy* (HMG 1994), recognized the role of EIA in contributing to sustainable development and raised the EIA profile among key user groups. The UK government reports also reflect the extension of the scope of sustainable development to include social, economic and environmental factors. This is reflected in the UK Strategy for Sustainable Development, *A better quality of life* (DETR 1999a), with its four objectives of:

- social progress which recognizes the needs of everyone;
- effective protection of the environment;
- prudent use of natural resources; and
- maintenance of high and stable levels of economic growth and employment.

To measure progress, the UK government published a set of sustainable development indicators, including a set of 15 key headline indicators (DETR 1999b). It also required a high-level sustainable development framework to be produced for each English region (see, for example, *A better quality of life in the South East*, SEERA, 2001).

Planning Policy Statement 1 (PPS1, DCLG 2005) reinforced the commitment to sustainable development. 'Sustainable development is the core principle underpinning planning. At the heart of sustainable development is the simple idea of ensuring a better quality of life for everyone, now and for future generations.' This was further reinforced and developed in an update of the national strategy, *Securing the future: delivering the UK sustainable development strategy* (DEFRA 2005), in which the UK government introduced a revised set of guiding principles, priorities for action and 20 key headline indicators, with a focus on delivery. The guiding principles are:

- living within environmental limits;
- ensuring a strong, healthy and just society;
- achieving a sustainable economy;
- promoting good governance;
- using sound science responsibly.

The good governance principle adds an important fourth pillar to the other three pillars (environmental, social and economic) of sustainable development, shifting from a triple to a quadruple bottom line (QBL) approach. Good governance, at all levels from central government to the individual, is needed to foster the integration of the three other pillars. Again, EIA can be a useful vehicle for such integration.

1.4 Projects, environment and impacts

1.4.1 *The nature of major projects*

As noted in Section 1.2, EIA is relevant to a broad spectrum of development actions, including policies, plans, programmes and projects. The focus here is on projects, reflecting the dominant role of project EIA in practice. The strategic environmental assessment (SEA) and sustainability appraisal (SA) of the 'upper tiers' of development actions are considered further in Chapter 11. The scope of projects covered by EIA is widening, and is discussed further in Chapters 3 and 4. Traditionally, project EIA has applied to major projects; but what are major projects, and what criteria can be used to identify them? One could take Lord Morley's approach to defining an elephant: it is difficult, but you easily recognize one when you see it. In a similar vein, the acronym LULU (locally unacceptable land uses) has been applied in the USA to many major projects, such as in energy, transport and manufacturing, clearly reflecting the public perception of the potential negative impacts associated with such developments. There is no easy definition, but it is possible to highlight some important characteristics (see Plate 1.1 and Table 1.2).

(a) Kings Cross, London - urban redevelopment

(b) Construction at London 2012 Olympics site

(c) Olkiluoto nuclear power plant, Finland

(d) The Oresund Bridge connecting Sweden and Denmark

(e) Danish offshore wind farm

(f) ES for decommissioning Hinkley Point A, UK

Plate 1.1 Some examples of major projects

Source: Magnox Electric (2002); RPS (2004); Symonds/EDAW (2004); Wikimedia.

Table 1.2 Characteristics of major projects

Substantial capital investment
Cover large areas; employ large numbers (construction and/or operation)
Complex array of organizational links
Wide-ranging impacts (geographical and by type)
Significant environmental impacts
Require special procedures
Infrastructure and utilities, extractive and primary (including agriculture); services
Band, point

Most large projects involve considerable investment. In the UK context, 'megaprojects' such as the Channel Tunnel and the associated Rail Link, London Heathrow Terminal 5, the Olympic 2012 project, motorways (and their widening), nuclear power stations, gas-fired power stations and renewable energy projects (such as major offshore wind farms and the proposed Severn Barrage) constitute one end of the spectrum. At the other end may be industrial estate developments, small stretches of road, and various waste-disposal facilities, with considerably smaller, but still substantial, price tags. Such projects often cover large areas and employ many workers, usually in construction, but also in operation for some projects. They also invariably generate a complex array of inter-and intra-organizational activity during the various stages of their lives. The developments may have wide-ranging, long-term and often very significant impacts on the environment. The definition of significance with regard to environmental effects is an important issue in EIA. It may relate, *inter alia*, to scale of development, to sensitivity of location and to the nature of adverse and beneficial effects; it will be discussed further in later chapters. Like a large stone thrown into a pond, a major project can create significant ripples, with impacts spreading far and wide. In many respects such projects tend to be regarded as exceptional, requiring special procedures. In the UK, these procedures have included public inquiries, hybrid bills that have to be passed through parliament (for example, for the Channel Tunnel) and EIA procedures. Under the 2008 Planning Act (HMG 2008), a special subset of nationally significant infrastructure projects (NSIPs) has been identified, with impacts to be examined by new procedures led by the Infrastructure Planning Commission (IPC) (to become the National Infrastructure Unit of the UK Planning Inspectorate in 2012). NSIPs include major energy projects, transport projects (road, rail and port), water and waste facilities.

Major projects can also be defined according to type of activity. In addition to the infrastructure and utilities, they also include manufacturing and extractive projects, such as petrochemical plants, steelworks, mines and quarries, and services projects, such as leisure developments, out-of-town shopping centres, new settlements and education and health facilities. An EC study adopted a further distinction between band and point infrastructures. Point infrastructure would include, for example, power stations, bridges and harbours; band or linear infrastructure would include electricity transmission lines, roads and canals (CEC 1982).

A major project also has a planning and development life cycle, including a variety of stages. It is important to recognize such stages because impacts can vary considerably between them. The main stages in a project's life cycle are outlined in Figure 1. 5. There may be variations in timing between stages, and internal variations within each stage, but there is a broadly common sequence of events. In EIA, an important distinction is between 'before the decision' (stages A and B) and 'after the decision' (stages C, D and E). As noted in Section 1. 2, the monitoring and auditing of the implementation of a project following approval are often absent from the EIA process.

Projects are initiated in several ways. Many are responses to market opportunities (e. g. a holiday village, a sub-regional shopping centre, a gas-fired power station; a wind farm); others may be seen

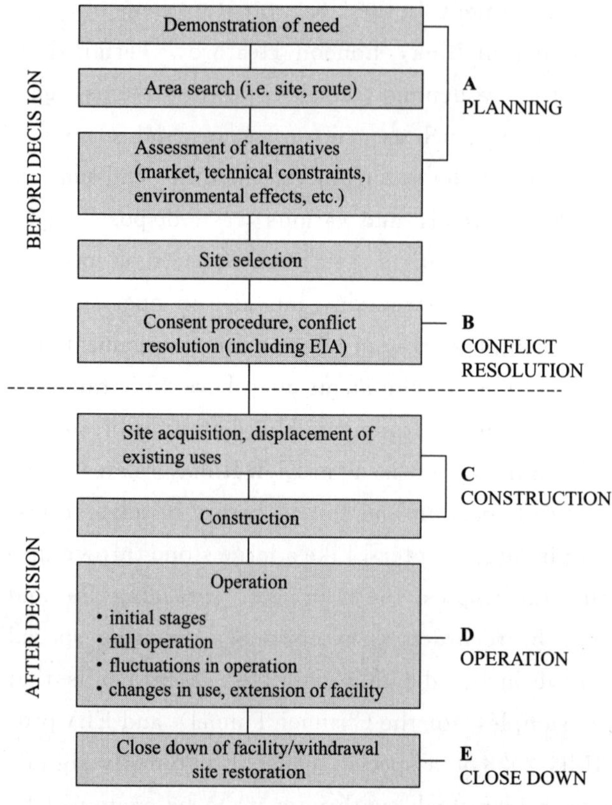

Figure 1. 5 Generalized planning and development life cycle for major projects (with particular reference to impact assessment on host area)

Source: Adapted from Breese *et al*. 1965

as necessities (e. g. the Thames Barrier); others may have an explicit prestige role (e. g. the programme of Grands Travaux in Paris including the Bastille Opera, Musée d'Orsay and Great Arch). Some major projects are public sector initiatives, but with the move towards privatization in many countries, there has been a move towards private sector funding, exemplified in the UK by such projects as the North Midlands Toll Road, the Channel Tunnel, and now most major utility energy, water and waste projects. The initial planning stage A may take several years, and lead to a specific proposal for a particular site. It is at stage B that the various control and regulatory procedures, including EIA, normally come into play. The construction stage can be particularly disruptive, and may last up to 10 years for some projects. Major projects invariably have long operational lives, although extractive projects can be short compared with infrastructure projects. The environmental impact of the eventual closedown/decommissioning of a facility should not be forgotten; for nuclear power facilities it is a major undertaking. Figure 1. 6 shows how the stages in the life cycles of different kinds of project may vary.

1. 4. 2 *Dimensions of the environment*

The environment can be structured in several ways, including components, scale/space and time. A narrow definition of environmental components would focus primarily on the biophysical environment. For example, the UK Department of the Environment (DoE) used the term to include all media susceptible to pollution, including: air, water and soil; flora, fauna and human beings; landscape, urban and rural conservation; and the built heritage (DoE 1991). The DoE checklist of environmental components is outlined in Table 1. 3. However, as already noted in Section 1. 2, the environment has important economic and socio-cultural dimensions. These include economic structure, labour markets, demography, housing, services (education, health, police, fire, etc.), lifestyles and values; and these are added to the checklist in Table 1. 3. This wider definition is more in line with international definitions, as noted by the IAIA definition of EIA in 1. 2. 1. Similarly, an

Australian definition notes, 'For the purposes of EIA, the meaning of environment incorporates physical, biological, cultural, economic and social factors' (ANZECC 1991).

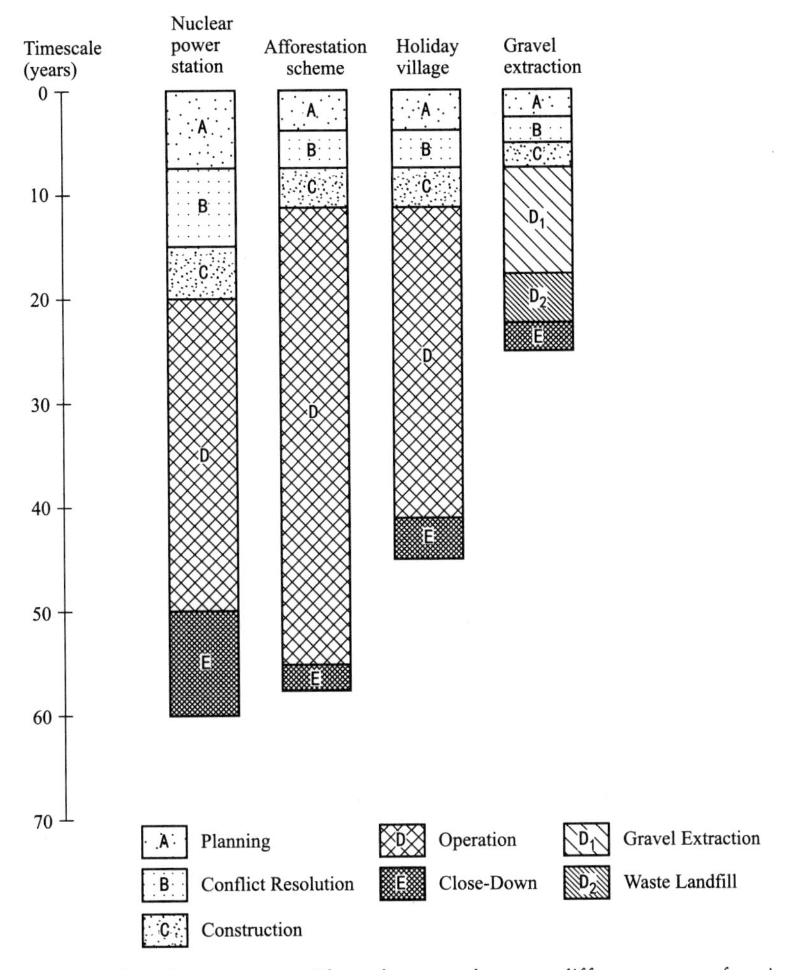

Figure 1. 6 Broad variations in life cycle stages between different types of project

Table 1. 3 Environmental components

Physical environment	
Air and atmosphere	Air quality
Water resources and water bodies	Water quality and quantity
Soil and geology	Classification, risks(e. g. erosion, contamination)
Flora and fauna	Birds, mammals, fish, etc. ; aquatic and terrestrial vegetation
Human beings	Physical and mental health and well-being
Landscape	Characteristics and quality of landscape
Cultural heritage	Conservation areas; built heritage; historic and archaeological sites; other material assets
Climate	Temperature, rainfall, wind, etc.
Energy	Light, noise, vibration, etc.
Socio-economic environment	
Demography	Population structure and trends
Economic base-direct	Direct employment; labour market characteristics; local and non-local trends
Economic base-indirect	Non-basic and services employment; labour supply and demand
Housing; transport; recreation	Supply and demand
Other local services	Supply and demand of services: health, education, police, etc.
Socio-cultural	Lifestyles, quality of life; social problems; community stress and conflict

Source: adapted from DoE 1991; DETR 2000; CEC 2003a

The environment can also be analysed at various scales (Figure 1.7). Many of the spatial impacts of projects affect the local environment, although the nature of 'local' may vary according to the aspect of environment under consideration and to the stage in a project's life. However, some impacts are more than local. Traffic noise, for example, may be a local issue, but changes in traffic flows caused by a project may have a regional impact, and the associated CO_2 pollution contributes to the global greenhouse problem. The environment also has a time dimension. Baseline data on the state of the environment are needed at the time a project is being considered. There has been a vast increase in data available on the Internet, from the local to the national level (e. g. in the UK via local authority development plans and national statistical sources, such as the e-Digest of Environment Statistics produced by the Department of Environment, Food and Rural Affairs). For some areas such data may be packaged in tailor-made state-of-the-environment reports and audits. See Chapters 5 and 12, for further guides to data sources. For all data it is important to have a time-series highlighting trends in environmental quality, as the environmental baseline is constantly changing, irrespective of any development under consideration, and requires a dynamic rather than a static analysis.

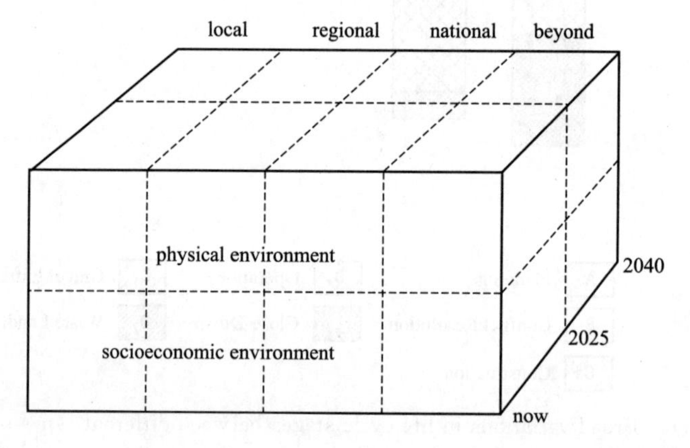

Figure 1.7 Environment: components, scale and time dimensions

1.4.3 *The nature of impacts*

The environmental impacts of a project are those resultant changes in environmental parameters, in space and time, compared with what would have happened had the project not been under taken. The parameters may be any of the type of environmental receptors noted previously: air quality, water quality, noise, levels of local unemployment and crime, for example. Figure 1.8 provides a simple illustration of the concept.

Table 1.4 provides a summary of some of the types of impact that may be encountered in EIA. The biophysical and socio-economic impacts have already been noted. These are sometimes seen as synonymous with adverse and beneficial, respectively. Thus, new developments may produce harmful wastes but also produce much needed jobs in areas of high unemployment. However, the correlation does not always apply. A project may bring physical

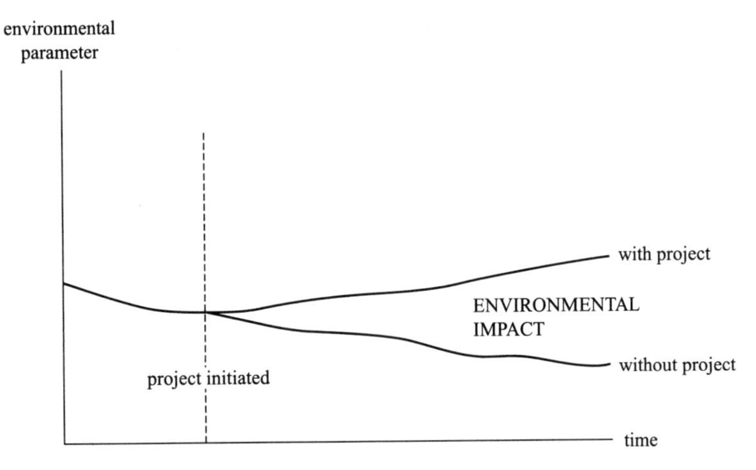

Figure 1.8 The nature of an environmental impact

benefits when, for example, previously polluted and derelict land is brought back into productive use; similarly, the socio-economic impacts of a major project on a community could include pressure on local health services and on the local housing market, and increases in community conflict and crime. Projects may also have immediate and direct impacts that give rise to secondary and indirect impacts later. A reservoir based on a river system not only takes land for the immediate body of water but also may have severe downstream implications for flora and fauna and for human activities such as fishing and sailing. The direct and indirect impacts may sometimes correlate with short-run and long-run impacts. For some impacts the distinction between short-run and long-run may also relate to the distinction between a project's construction and its operational stage; however, other construction-stage impacts, such as change in land use, are much more permanent. Impacts also have a spatial dimension. One distinction is between local and strategic, the latter covering impacts on areas beyond the immediate locality. These are often regional, but may sometimes be of national or even international significance.

Table 1.4 Types of impact

Physical and socio-economic
Direct and indirect
Short-run and long-run
Local and strategic(including regional,national and beyond)
Adverse and beneficial
Reversible and irreversible
Quantitative and qualitative
Distribution by group and/or area
Actual and perceived
Relative to other developments;cumulative

Environmental resources cannot always be replaced; once destroyed, some may be lost forever. The distinction between reversible and irreversible impacts is a very important one, and the irreversible impacts, not susceptible to mitigation, can constitute particularly significant impacts in an EIA. It may be possible to replace, compensate for or reconstruct a lost resource in some cases, but substitutions are rarely ideal. The loss of a resource may become more serious later, and valuations need to allow for this. Some impacts can be quantified, others are less tangible. The latter should not be ignored. Nor should the distributional impacts of a proposed development be ignored. Impacts do not fall evenly on affected parties and areas. Although a particular project may be assessed as bringing a general benefit, some groups and/or geographical areas may be receiving most of any adverse effects, the main benefits going to others elsewhere. There is also a distinction between actual and perceived impacts. Subjective perceptions of impacts may significantly influence the responses and decisions of people towards a proposed development. They constitute an important source of information, to be considered alongside more objective predictions of impacts.

> Social constructions are not mere perceptions or emotions, to be distinguished from reality; rather, how we view a social situation determines how we behave. Furthermore, social constructions of reality are characteristic of all social groups, including the agencies that are attempting to implement change as well as the communities that are affected. (IOCGP 2003)

Finally, all impacts should be compared with the 'do-nothing' situation, and the state of the environment predicted without the project. This can be widened to include comparisons with anticipated impacts from alternative development scenarios for an area. Some projects may also have cumulative impacts in combination with other development actions, current and future; for example, the impacts of several wind farms in an area, or the build-up of several major, but different, developments (e. g. port; power station; steel works; waste water facility) around an estuary. The important area of cumulative impacts is discussed further in Chapters 9 and 12.

We conclude on a semantic point: the words 'impact' and 'effect' are widely used in the literature and legislation on EIA, but it is not always clear whether they are interchangeable or should be used only for specifically different meanings. In the United States, the regulations for implementing the National Environmental Policy Act (NEPA) expressly state that 'effects and impacts as used in these regulations are synonymous'. This interpretation is widespread, and is adopted in this text. But there are other interpretations relating to timing and to value judgements. Catlow and Thirlwall (1976) make a distinction between effects that are 'the physical and natural changes resulting, directly or indirectly, from development' and impacts that are 'the consequences or end products of those effects represented by attributes of the environment on which we can place an objective or subjective value'. In contrast, an Australian study (CEPA 1994) reverses the arguments, claiming that 'there does seem to be greater logic in thinking of an impact resulting in an effect, rather than the other way round'. Other commentators have introduced the concept of value judgement into the differentiation. Preston and Bedford (1988) state that 'the use of the term "impacts" connotes a value judgement'. This view is supported by Stakhiv (1988), who sees a distinction between 'scientific assessment of facts (effects), and the evaluation of the relative importance of these effects by the analyst and the public (impacts) '. The debate continues!

1.5 Changing perspectives on EIA

1.5.1 *The importance of adaptive EIA*

The arguments for EIA vary in time, in space and according to the perspective of those involved. From a minimalist defensive perspective, some developers, and still possibly some parts of some governments, might see EIA as a necessary evil, an administrative exercise, something to be gone through that might result in some minor, often cosmetic, changes to a development that would probably have happened anyway. In contrast, for the 'deep ecologists' or 'deep greens', EIA cannot provide total certainty about the environmental consequences of development proposals; they feel that any projects carried out under uncertain or risky circumstances should be abandoned. EIA and its methods must straddle such perspectives on weak and strong sustainability. EIA can be, and now often is, seen as a positive process that seeks a harmonious relationship between development and the environment. The nature and use of EIA will change as relative values and perspectives also change. EIA must adapt, and as O'Riordan (1990) very positively noted over 20 years ago:

> One can see that EIA is moving away from being a defensive tool of the kind that dominated the 1970s to a potentially exciting environmental and social betterment technique that may well come to take over the 1990s ... If one sees EIA not so much as a technique, rather as a process that is constantly changing in the face of shifting environmental politics and managerial capabilities, one can visualize it as a sensitive barometer of environmental values in a complex environmental society. Long may EIA thrive.

EIA must continue to adapt in our rapidly changing world, a world where there are serious challenges to all the pillars of sustainability. Climate change is now recognized by many governments as the most important challenge of the twenty-first century, necessitating major initiatives—yet progress is sporadic. In recent years the world has also been on the edge of financial meltdown, and has endured serious economic recession, leading to stimulus investment, often through infrastructure projects, but also to drastic measures for deficit reduction. Poverty and social inequalities persist and are deep-seated. But before addressing the changing nature of the impact assessment family, we first consider EIA in its theoretical context.

1.5.2 *EIA in its theoretical context*

EIA must also be reassessed in its *theoretical context*, and in particular in the context of decision-making theory (see Lawrence 1997, 2000; Bartlett and Kurian, 1999; Weston

2000, 2003). EIA had its origins in a climate of a rational approach to decision-making in the USA in the 1960s (Caldwell 1988). The focus was on the systematic process, objectivity, a holistic approach, a consideration of alternatives and an approach often seen as primarily linear. This rational approach is assumed to rely on a scientific process in which facts and logic are pre-eminent. In the UK this rational approach was reflected in planning in the writings of, *inter alia*, Faludi (1973), McLoughlin (1969), and Friend and Jessop (1977).

However, other writings on the theoretical context of EIA have recognized the importance of the subjective nature of the EIA process. Kennedy (1988) identified EIA as both a 'science' and an 'art', combining political input and scientific process. More colourfully, Beattie (1995), in an article entitled 'Everything you already know about EIA, but don't often admit', reinforces the point that EIAs are not science; they are often produced under tight deadlines and data gaps, and simplifying assumptions are the norm under such conditions. They always contain unexamined and unexplained value judgements, and they will always be political. They invariably deal with controversial projects, and they have distributional effects—there are winners and losers. EIA professionals should therefore not be surprised, or dismayed, when their work is selectively used by various parties in the process. Leknes (2001) notes that it is particularly in the later stage of decision-making that the findings of EIA are likely to give way to political considerations. Weston (2003) notes the weakening of deference to science, experts and the rational approach. Confidence in decision-making for major projects is eroded by events such as nuclear accidents, chemical spills, numerous environmental disasters, and massive financial and time overruns of projects (Flyberg 2003). The public increasingly fear the consequences of change over which they have little control, and there is more emphasis on risk (see Beck 1992, 2008).

However, in the context of decision-making theory, this recognition of the political, the subjective and value judgement is reflected in a variety of behavioural/participative theories, and is not new. For example, in the 1960s Braybrooke and Lindblom (1963) saw decisions as incremental adjustments, with a process that is not comprehensive, linear and orderly, and is best characterized as 'muddling through'. Lindblom (1980) further developed his ideas through the concept of 'disjointed incrementalism', with a focus on meeting the needs and objectives of society, often politically defined. The importance of identifying and confronting trade-offs, a major issue in EIA, is clearly recognized. The participatory approach includes processes for open communication among all affected parties. The recognition of multiple parties and the perceived gap between government and citizens has stimulated other theoretical approaches, including communicative and collaborative planning (Healey 1996, 1997). This approach draws upon the work of Habermas (1984), Forester (1989) and others. Much attention is devoted to consensus-building, coordination and communication, and the role of government in promoting such actions as a means of dealing with conflicting stakeholder interests and achieving collaborative action. Critics of such an approach highlight in particular the lack of regard for power relationships within society, and especially the role of private sector developers—invariably the proponents in EIA.

It is probably now realistic to place the current evolution of EIA somewhere between the rational and behavioural approaches—reflecting elements of both. It does include important strands of rationalism, but there are many participants, and many decision points—and politics, power relationships and professional judgement are often to the fore. In EIA there are many decisions; for example, on whether EIA is needed at all (screening), the scope of the EIA, the alternatives under consideration, project design and redesign, the range of mitigation and enhancement measures, and implementation and monitoring during the 'post-key-decision' stages of the project life cycle (Glasson 1999). This tends to fit well with the classic concept of 'mixed scanning' advocated by Etzioni (1967), utilizing rational techniques of assessment, in combination with more intuitive value judgements, based upon experience and values. The rational-adaptive approach of Kaiser *et al.* (1995) also stresses the importance of a series of steps in decision-making, with both (scientific-based) rationality and (community-informed) participation, moderating the selection of policy options and desired outcomes.

1.5.3 *EIA in a rapidly growing Impact Assessment (IA) family*

Over the last 40 years, EIA has been joined by a growing family of assessment tools. The IAIA uses the generic term of impact assessment (IA) to encompass the semantic explosion; whereas Sadler (1996) suggested that we should view environmental assessment (EA) as 'the generic process that includes EIA of specific projects, SEA of PPPs, and their relationships to a larger set of impact assessment and planning-related tools'. Whatever the family name, there is little doubt that membership is increasing apace, with a focus on widening the *scope, scale and integration of assessment*. Impact assessment now includes, for example, SIA, HIA, EqIA, TIA, SEA, SA, S&EIA, HRA/AA, EcIA, CIA, plus a range of associated techniques such as RA, LCA, MCA, CBA and many more. Some of the tools have been led by legislation; others have been more driven by practitioners from various disciplines that have endeavoured to separate out and highlight the theme (s) of importance to their discipline, resulting in thematically focused forms of assessment. Dalal-Clayton and Sadler (2004) rightly observe that 'the alphabet soup of acronyms (and terms) currently makes for a confusing picture'. The various assessment tools are now briefly outlined in terms of scope, scale and integration; most are discussed much further in subsequent chapters.

Scope

Development actions may have impacts not only on the physical environment but also on the social and economic environment. Typically, employment opportunities, services (e. g. health, education), community structures, lifestyles and values may be affected. *Socio-economic impact assessment* or *social impact assessment* (SIA) is regarded in this book as an integral part of EIA. However, in some countries it is (or has been) regarded as a separate process, sometimes parallel to EIA, and the reader should be aware of its separate existence (Carley and Bustelo 1984; Finsterbusch 1985; IAIA 1994; Vanclay 2003). Some domains explicitly use *S&EIA* to denote *Socio-economic and environmental impact assessment*. *Health impact assessment* (*HIA*) has been a particularly important area of growth in recent years, evolving out of the socio-economic strand; its focus is on the effects that a development action may have on the health of its host population (IPHI 2009). A more recent area still is *equality impact assessment* (*EqIA*), which seeks to identify the important distributional impacts of development actions on various groups in society (e. g. by gender, race, age, disability, sexual orientation etc., Downey 2005). Vanclay and Bronstein (1995) and others note several other relevant definitions, based largely on particular foci of specialization and including, for example, transport impact assessment, demographic impact assessment, climate impact assessment, gender impact assessment, psychological impact assessment, noise impact assessment, economic impact assessment, and cumulative impacts assessment (Canter and Ross 2010).

Scale

Strategic environmental assessment (SEA) expands the scale of operation from the EIA of projects to a more strategic level of assessment of programmes, plans and policies (PPPs). Development actions may be for a project (e. g. a nuclear power station), for a programme (e. g. a number of pressurized water reactor (PWR) nuclear power stations), for a plan (e. g. in the town and country planning (T&CP) system in England) or for a policy (e. g. the development of renewable energy). EIA to date has generally been used for individual projects, and that role is the primary focus of this book. But EIA for programmes, plans and policies, otherwise known as SEA, has been introduced in the European Union (EU) since 2004 and is also used in many other countries worldwide (Therivel 2010; Therivel and Partidario 1996; Therivel *et al*. 1992). SEA informs a higher, earlier, more strategic tier of decision-making. In theory, EIA should be carried out in a tiered fashion first for policies, then for plans and programmes, and finally for projects. The focus of SEA has been primarily biophysical, and there are close links with another relatively new area of assessment, *habitats regulation assessment/appropriate assessment* (*HRA/AA*), which is required in the EU for projects and plans that may have significant impacts on key Natura 2000 sites of biodiversity. In contrast, a wider approach to strategic assessment, seeking to include biophysical and socio-economic impacts, is provided by *SA*. In England this is required for the assessment of the impacts of plans under the T&CP system. In some domains, where there is not a strategic level of assessment or planning, project-level assessment may adopt, to varying degrees, a strategic perspective, with features of either SEA or SA; good examples are provided by mega-projects, such as the major mineral development projects in the remote areas of Australia.

Integration

Hacking and Guthrie (2008) have sought to provide a relational framework (Figure 1. 9) to clarify the position of various assessment tools, in the context of planning and decision-making for sustainable development. In addition to scope (referred to as comprehensiveness of coverage) and scale (strategicness of the focus and scope), they also include integratedness of techniques and themes. The latter includes a package of techniques that seek to achieve integration in the assessment process (e. g. between biophysical and socio-economic impacts; Scrase and Sheate 2002); this was termed 'horizontal integration' by Lee (2002). Petts (1999) provides a good overview of some of the techniques that include, for example, *life cycle assessment* (LCA), *cost-benefit analysis* (CBA), *environmental auditing*, *multi-criteria assessment* (MCA) *and risk assessment* (RA). LCA differs from EIA in its focus not on a particular site or facility, but on a product or system and the cradle-to-grave environmental effects of that product or system (see White *et al*. 1995). In contrast, CBA focuses on the economic impacts of a development, but taking a wide and long view of those impacts. It involves as far as possible the monetization of all the costs and benefits of a proposal. It came to the fore in the UK in relation to major transport projects in the 1960s, but has subsequently enjoyed a new lease of life (see Hanley and Splash 1993; Lichfield 1996). Environmental auditing is the systematic, periodic and documented evaluation of the environmental performance of facility operations and practices, and this area has seen the development of procedures, such as the international standard ISO 14001.

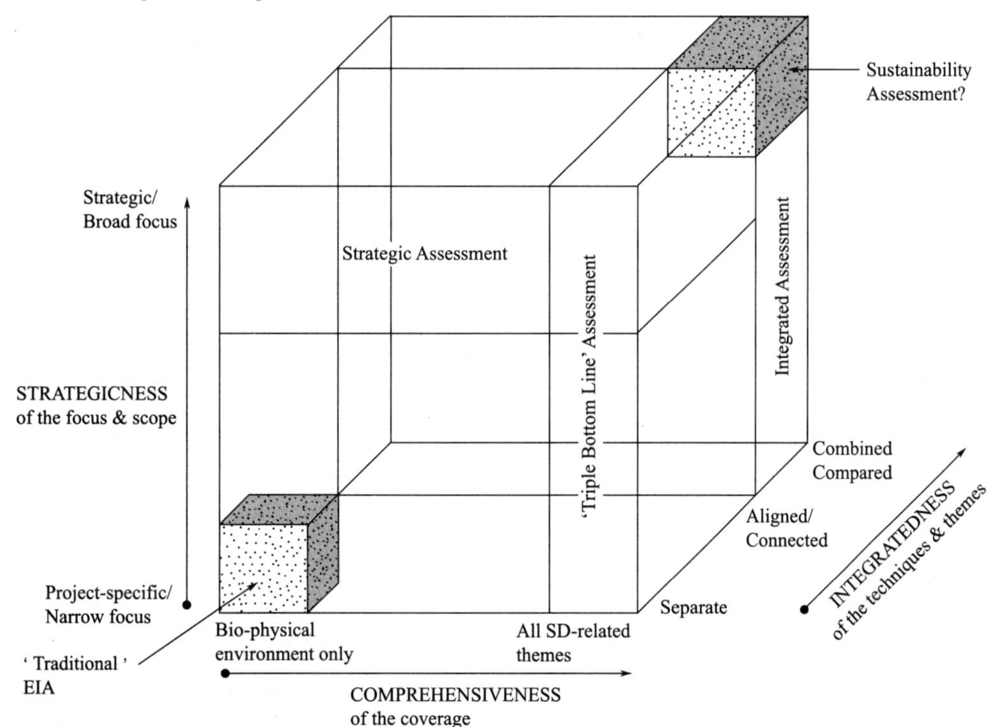

Figure 1. 9 A relational framework of SD-focused assessment tools

Source: Hacking and Guthrie 2008

Multi-criteria decision assessment (MCDA) covers a collection of approaches, often quantitative, that can be used to help key stakeholders to explore alternative approaches to important decisions by explicitly taking account of multiple criteria (Belton and Stewart 2002); it is quite widely used. *Risk assessment* is another term sometimes found associated with EIA. Partly in response to events such as the chemicals factory explosion at Flixborough (UK), and nuclear power station accidents at Three Mile Island (USA) and Chernobyl (Ukraine), RA developed as an approach to the analysis of risks associated with various types of development. Calow (1997) gives an overview of the growing area of environmental RA and management, and Flyberg (2003) provides a critique of risk assessment in practice. While these tools tend to be more technocentric, they can be seen as complementary to EIA, seeking to achieve a more integrated approach. Thus Chapter 5 explores the potential role of CBA and MCA approaches in EIA evaluation; Chapter 12 develops further the concept of integrated assessment, and explores the role of environmental auditing and LCA in relation to environmental management systems (EMSs).

This brief discussion on changing perspectives, on the theoretical context, associated tools and processes, emphasizes the need to continually reassess the role and operation of EIA and the importance of an adaptive EIA. This will be developed further in several chapters—especially in Part 4.

1.6 Current issues in EIA

Although EIA now has over 40 years of history in the USA, elsewhere the development of concepts and practice is more recent. Development is moving apace in many other countries, including the UK and the other EU Member States. There is much to welcome; Gibson (2002) noted some worldwide trends in EIA, such as that it is earlier in the process, more open and participative, more comprehensive (not just biophysical environment), more mandatory, more closely monitored, more widely applied (e. g. at various levels), more integrative, more ambitious (regarding sustainability objectives) and more humble (recognizing uncertainties, applying precaution). Yet such progress is variable, and has not been without its problems. A number of the current issues in EIA are highlighted here and will be discussed more fully in later chapters.

1.6.1 *The nature of methods of assessment*

As noted in Section 1.2, some of the main steps in the EIA process (e. g. auditing and monitoring) may be missing from many studies. There may also be problems with the steps that are included. The prediction of impacts raises various conceptual and technical problems. The problem of establishing the environmental baseline position has already been noted. It may also be difficult to establish the dimensions and development stages of a project clearly, particularly for new technology projects. Further conceptual problems include establishing what would have happened in the relevant environment without a project, clarifying the complexity of interactions of phenomena, and especially making trade-offs in an integrated way (i. e. assessing the trade-offs between economic apples, social oranges and physical bananas). Other technical problems relate to data availability and the tendency to focus on the quantitative, and often single, indicators in some areas. There may also be delays and gaps between cause and effect, and projects and policies may discontinue. The lack of auditing of predictive techniques limits the feedback on the effectiveness of methods. Nevertheless, innovative methods are being developed to predict and evaluate impacts, ranging from simple checklists and matrices to complex mathematical models and multi-criteria approaches. It should be noted however that these methods may not be neutral, in the sense that the more complex they are, the more difficult it becomes for the general public to participate in the EIA process.

1.6.2 The quality and efficiency of the EIA process

One assessment of quality is that of the immediate output of the process, the EIS. Many EISs may fail to meet even minimum standards. For example, an early survey by Jones *et al.* (1991) of the EISs published under UK EIA regulations highlighted some shortcomings. They found:

> ... that one-third of the EISs did not appear to contain the required non-technical summary, that, in a quarter of the cases, they were judged not to contain the data needed to assess the likely environmental effects of the development, and that in the great majority of cases, the more complex, interactive impacts were neglected.

The DoE (1996) later suggested that although there had been some learning from experience, many EISs in the UK were still unsatisfactory (see Chapter 8 for further and updated discussion). Quality may vary between types of project. It may also vary between countries supposedly operating under the same legislative framework.

EISs can run the risk of being voluminous, unintegrated, documents that can be difficult for most of the participants in the EIA process. Such outcomes raise various questions about the efficiency of the EIA process. For example, are 'safety first' policies resulting in too many projects being screened for EIA and the EIA scoping stage being too all embracing of potential impacts? Is there too much focus on over-descriptive baseline work and not enough focus on the key impacts that matter? Is the EIS still a set of segregated specialist chapters rather than a well-integrated document? Are the key steps of monitoring and auditing well enough built into the process? Considerations of efficiency, however, can also run counter to considerations of fairness in the process.

1.6.3 *The relative roles of participants in the process*

The various 'actors' in the EIA process—the developer, the affected parties, the general public and the regulators at various levels of government—have differential access to the process, and their influence on the outcome varies. Some would argue that in countries such as the UK the process is too developer-orientated. The developer or the developer's consultant carries out the EIA and prepares the EIS, and is unlikely to predict that the project will be an environmental disaster. Notwithstanding this, developers themselves are concerned about the potential delays associated with the requirement to submit an EIS. They are also concerned about cost. Details about costs are difficult to obtain. Early estimates (Clark 1984; Hart 1984; Wathern 1988) were of EIA costs of 0.5-2.0 per cent of a project's value. The UK DETR (1997) suggested £35, 000 as an appropriate median figure for the cost of undertaking an EIA under the EC regulations, but for major projects the monetary figure can be much higher than this. A more recent EU commissioned study evaluating the EIA Directive indicated that, as a share of the project costs, EIAs tend to range from an upper limit of 1 per cent for small projects to 0.1 per cent for larger projects (CEC 2006).

Procedures for and the practice of public participation in the EIA process vary between, and sometimes within, countries, from the very comprehensive to the very partial and largely cosmetic. An important issue is the stages in the EIA process to which the public have access. Government roles in the EIA process may be conditioned by caution at extending systems, by resource considerations and by limited experience and expertise for what in some domains is still a relatively new and developing area. A central government may offer only limited guidance on best practice, and make inconsistent decisions. A local government may find it difficult to handle the scope and complexity of the content of EISs, especially for major projects.

1.6.4 *The effectiveness of the EIA process*

While EIA systems are now well established in many countries of the world, there is considerable soul-searching about how effective it all is, whether EIA is achieving its purposes—as set out in Section 1.3? There is also considerable debate about *how we assess EIA effectiveness*. There can be various (interrelated) dimensions to this. For example, a procedural/narrow approach would focus on how well EIA is being carried out according to its own procedural requirements in the country of concern; a procedural/wider approach might consider the extent to which EIA is contributing to increased environmental awareness and learning among the array of key stakeholders. These dimensions are partly covered in the preceding sections (1.6.1-1.6.3). However, more fundamental, in relation to EIA core purposes, are substantive approaches. For example, a substantive/narrow approach would concentrate on whether EIA is having a direct impact on the quality of planning decisions and the nature of developments. A substantive/wider approach would focus on the fundamental question of whether EIA is maintaining, restoring, and enhancing environmental quality; is it contributing towards more sustainable development? These issues of EIA effectiveness are examined in various sections, and particularly in Chapter 8.

1.6.5 *Beyond the decision*

Many EISs are for one-off projects, and there may be little incentive for developers to audit the quality of the assessment predictions and to monitor impacts as an input to a better assessment for the next project. Yet EIA up to and no further than the decision on a project is a very partial exercise. It is important to ensure that the required mitigation and enhancement measures are implemented in practice. In some areas of the world (e.g. California, Western Australia, the Netherlands, and China Hong Kong of China to mention just a few), the monitoring of impacts is mandatory, and monitoring procedures must be included in an EIS. It is also important to take the opportunity for a cyclical learning process, auditing predicted outcomes as fully as possible—to check the accuracy of predictions. The relationship with environmental management processes is another vital area of concern; EISs can effectively lead to environmental management plans for project implementation—but, again, good practice is patchy. The extension of such approaches constitutes another significant current issue in the project-based EIA process.

1.6.6 *Managing the widening scope and complexity of IA activity*

As noted in Section 1.5, the IA family has grown apace, especially in recent years. How can this complexity be managed? For example, what should be the norm for the content of a contemporary EIS? There is a strong case for widening the dimensions of the environment under consideration to include socio-economic impacts more fully. The trade-off between the often adverse biophysical impacts of a development and the often beneficial socio-economic impacts can constitute the crucial dilemma for decision-makers. Coverage can also be widened to include other types of impacts only very partially covered to date. Should the EIS include social, health and equality elements as standard, or should these be separate activities, and documents? In a similar vein, which projects should have EIAs? For example, project EIA may be mandatory only for a limited set of major projects, but in practice many others may be included. Case law is now building up in many countries, but the criteria for the inclusion or exclusion of a project for EIA may not always be clear.

As also noted in Section 1.5, the SEA/SA of PPPs represents a logical extension of project assessment. SEA/SA can cope better with cumulative impacts, alternatives and mitigation measures than project assessment. But what is the nature of the relationship between the different scales of impact assessment? Strategic levels of assessment of plans and programmes should provide useful frameworks for the more site-specific project assessments, hopefully reducing workload and leading to more concise and effective EIAs. But the anticipated tiered relationship may be more in theory than practice, leading to unnecessary and wasteful duplication of activity.

1.7 An outline of subsequent parts and chapters

This book is in four parts. The first establishes the context of EIA in the growth of concern about environmental issues and in relevant legislation, with particular reference to the UK and the EU. Following from this first chapter, which provides an introduction to EIA and an overview of principles, Chapter 2 focuses on the origins of EIA under the US NEPA of 1969, on interim developments in the UK, and on the subsequent introduction of EC Directive 85/337 and subsequent amendments and developments. The details of the UK legislative framework for EIA, under T&CP and other legislation are discussed in Chapter 3.

Part 2 provides a rigorous step-by-step approach to the EIA process. This is the core of the text. Chapter 4 covers the early start-up stages, establishing a management framework, clarifying the type of developments for EIA, and outlining approaches to scoping, the consideration of alternatives, project description, establishing the baseline and identifying impacts. Chapter 5 explores the central issues of prediction, the assessment of significance and impact mitigation and enhancement. The approach draws out broad principles affecting prediction exercises, exemplified with reference to particular cases. Chapter 6 provides coverage of an important issue identified above: participation in the EIA process. Communication in the EIA process, EIS presentation and EIA review are also covered in this chapter. Chapter 7 takes the process beyond the decision on a project and examines the importance of, and approaches to, monitoring and auditing in the EIA process.

Part 3 exemplifies the process in practice. Chapter 8 provides an overview of UK practice to date, including quantitative and qualitative analyses of the EISs prepared. Chapter 9 provides a review of EIA practice in several key sectors, including energy, transport, waste management and tourism. A feature of the chapter is the provision of a set of case studies of recent and topical EIA studies from the UK and overseas, illustrating particular features of and issues in the EIA process. Chapter 10 draws on comparative experience from developed countries (e. g. Canada and Australia) and from a number of countries from the developing and emerging economies (Peru, China, Benin and Poland) —presented to highlight some of the strengths and weaknesses of other systems in practice. The important role of international agencies in EIA practice-such as the European Bank for Reconstruction and Development and the World Bank—is also discussed in this chapter.

Part 4 looks to the future; it illuminates many of the issues noted in Section 1. 6. The penultimate chapter discusses the need for SEA and some of its limitations. It reviews the status of SEA in the USA, European Union and UNECE, and China. It then discusses in more detail how the European SEA Directive is being implemented in the UK. Chapter 12, the final chapter, focuses on improving the effectiveness of, and the prospects for, project-based EIA. It considers the array of perspectives on change from the various participants in the EIA process, followed by a consideration of possible developments in some important areas of the EIA process and in the nature of EISs. The chapter concludes with a discussion of the parallel and complementary development of environmental management systems and audits. Together, these topics act as a kind of action list for future improvements to EIA. A set of appendices provide details of legislation and practice, and websites and journals not considered appropriate to the main text.

SOME QUESTIONS

The following questions are intended to help the reader focus on the important issues of this chapter, and to start building some understanding of the principles of EIA.

1 Revisit the definitions of EIA given in this chapter. Which one do you prefer and why?

2 Some steps in the EIA process have proved to be more difficult to implement than others. From your initial reading, identify which these might be and consider why they might have proved to be problematic.

3 Taking a few recent examples of environmental impact statements for projects in your country, review their structure and content against the outline information in this chapter. Do they raise any issues on structure and content?

4 What are the differences between (i) project screening and project scoping, and (ii) impact mitigation and impact enhancement?

5 Review the purposes for EIA, and assess their importance from your own perspective.

6 Apply the characteristics of major projects set out in Table 1. 2 to two major projects with which you are familiar. Are there any important variations between the applications? If so, can you explain why?

7 Similarly, for one of the projects identified in Q6, plot the likely stages in its life cycle-applying approximate timings as far as possible.

8 What do you understand by a multi-dimensional approach to the environment, in EIA?

9 What is an impact in EIA? Do you see any difference between impacts and effects?

10 What do you understand by (i) irreversible impacts, (ii) cumulative impacts and (iii) distributional impacts, in EIA?

11 Why should it be important to adopt an adaptive approach to EIA?

12 This question may be a little deep at this stage of your reading, but we will ask it all the same: do you think it is reasonable to consider the EIA process as a rational, linear scientific process?

13 What are the main differences between EIA and SEA?

14 What might be some of the reasons for the widening scope of EIA?

15 What do you understand by 'beyond the decision' in EIA?

16 How might we measure (i) the efficiency, and (ii) the effectiveness of EIA?

Notes

1 In some domains the EIS is referred to more simply as an ES; these terms are used interchangeably in this book.

2 Turner and Pearce (1992) and Pearce (1992) have drawn attention to alternative interpretations of maintaining the capital stock. A policy of conserving the whole capital stock (man-made, human and natural) is consistent with running down any part of it as long as there is substitutability between capital degradation in one area and investment in another. This can be interpreted as a 'weak sustainability' position. In contrast, a 'strong sustainability' position would argue that it is not acceptable to run down environmental assets, for several reasons: uncertainty (we do not know the full consequences for human beings), irreversibility (lost species cannot be replaced), life support (some ecological assets serve life-support functions) and loss aversion (people are highly averse to environmental losses). The 'strong sustainability' position has much to commend it, but institutional responses have varied.

References

ANZECC (Australia and New Zealand Environment and Conservation Council) 1991. *A national approach to EIA in Australia*. Canberra: ANZECC.

Bartlett, R. V. and Kurian, P. A. 1999. The theory of environmental impact assessment: implicit models of policy making. *Policy and Politics* 27 (4), 415-33.

Beattie, R. 1995. Everything you already know about EIA, but don't often admit. *Environmental Impact Assessment Review* 15.

Beck, U. 1992. *Risk society: towards a new modernity*. London: Sage.

Beck, U. 2008. *World at risk*. Cambridge: Polity Press.

Belton, V. and Stewart, T. J. 2002. *Multiple criteria decision analysis*. Boston/London: Kluwer Academic.

Boulding, K. 1966. The economics of the coming Spaceship Earth. In *Environmental quality to a growing economy*, H. Jarrett (ed), 3-14. Baltimore: Johns Hopkins University Press.

Braybooke, C., and Lindblom, D. 1963. *A strategy of decision*. New York: Free Press of Glencoe.

Breese, G. *et al*. 1965. *The impact of large installations on nearby urban areas*. Los Angeles: Sage.

Caldwell, L. 1988. Environmental impact analysis: origins, evolution and future directions. *Review of Policy Research*, 8 (1), 75-83.

Calow, P. (ed) 1997. *Handbook of environmental risk assessment and management*. Oxford: Blackwell Science.

Canter, L. and Ross, W. 2010. State of practice of cumulative effects assessment and management: the good, the bad and the ugly. *Impact Assessment and Project Appraisal* 28 (4), 261-8.

Carley, M. J. and Bustelo, E. S. 1984. *Social impact assessment and monitoring: a guide to the literature*. Boulder: Westview Press.

Catlow, J. and Thirlwall, C. G. 1976. *Environmental impact analysis*. London: DoE.

CEC (Commission of the European Communities) 1982. *The contribution of infrastructure to regional development*. Brussels: CEC.

CEC 1985. On the assessment of the effects of certain public and private projects on the environment. *Official Journal* L175, 5 July.

CEC 1992. *Towards sustainability: a European Community programme of policy and action in relation to the environment and sustainable development*, vol. 2, Brussels: CEC.

CEC 2001. *Our future, our choice. The sixth EU environment action programme 2001-10*. Brussels: CEC.

CEC 2003a. (Impacts Assessment Unit, Oxford Brookes University) Five years' report to the European Parliament and the Council on the application and effectiveness of the EIA Directive. Available on website of EC DG Environment: www. europa. eu. int/comm/environment/eia/home. htm.

CEC 2003b. *Directive 2003/35/EC of the European Parliament and of the Council of 26 May 2003 providing for public participation in respect of the drawing up of certain plans and programmes relating to the environment and amending with regard to public participation and access to justice Council Directives 85/337/EEC and 96/61/EC-statement by the Commission.* Brussels: CEC.

CEC 2006. Evaluation *of EU legislation-Directive 85/337EEC and associated amendments*. Carried out by GHK-Technopolis. Brussels: DG Enterprise and Industry.

CEC 2009. *Report to the Council, European Parliament, European Economic and Social Committee and the Committee of the Regions on the application and effectiveness of the EIA Directive*. Brussels: CEC.

CEPA (Commonwealth Environmental Protection Agency) 1994. *Assessment of cumulative impacts and strategic assessment in EIA*. Canberra: CEPA.

Clark, B. D. 1984. Environmental impact assessment (EIA): scope and objectives. In *Perspectives on environmental impact assessment*, B. D. Clark *et al*. (eds). Dordrecht: Reidel.

Crane, A., McWilliams, A., Matten, D., Moon, J. and Siegel, D. S (eds). 2008. *The Oxford handbook of corporate social responsibility*. Oxford: Oxford University Press.

Dalal-Clayton, B. and Sadler, B. 2004. *Sustainability appraisal: a review of international experience and practice, first draft of work in progress*. International Institute for Environment and Development. Available at: www. iied. org/Gov/spa/docs. html# pilot.

DCLG (Department for Communities and Local Government) 2005. Planning Policy Statement 1 (PPS1). London: DCLG.

DEFRA (Department for Environment, Food and Rural Affairs) 2005. *Securing the future: delivering the UK sustainable development strategy*. London: HM Government.

DETR (Department of Environment, Transport and the Regions) 1997. *Consultation paper: implementation of the EC Directive (97/11/EC) -determining the need for environmental assessment*. London: DETR.

DETR 1999a. *UK strategy for sustainable development: A better quality of life*. London: Stationery Office.

DETR 1999b. *Quality of life counts-Indicators for a strategy for sustainable development for the United Kingdom: a baseline assessment*. London: Stationery Office.

DETR 2000. *Environmental impact assessment: a guide to the procedures*. Tonbridge: Thomas Telford Publishing.

DoE (Department of the Environment) 1989. *Environmental assessment: a guide to the procedures*. London: HMSO.

DoE *et al*. 1990. *This common inheritance: Britain's environmental strategy* (Cmnd 1200). London: HMSO.

DoE 1991. *Policy appraisal and the environment*. London: HMSO.

DoE 1996. *Changes in the quality of environmental impact statements*. London: HMSO.

Downey, L. 2005. Assessing environmental inequality: how the conclusions we draw vary according to the definitions we employ. *Sociological Spectrum*, 25, 349-69.

EEA (European Environment Agency) 2010. *European environment: state and outlook 2010*. EEA: Copenhagen.

Etzioni, A. 1967. Mixed scanning: a 'third' approach to decision making. *Public Administration Review* 27 (5), 385-92.

Faber, N., Jorna, R. and van Engelen, J. 2005. A Study into the conceptual foundations of the notion of 'sustainability'. *Journal of Environmental Assessment Policy and Management*, 7 (1).

Faludi, A. (ed) 1973. *A reader in planning theory*. Oxford: Pergamon.

Finsterbusch, K. 1985. State of the art in social impact assessment. *Environment and Behaviour* 17, 192-221.

Flyberg, B. 2003. *Megaprojects and risk: on anatomy of risk*. Cambridge: Cambridge University Press.

Forester, J. 1989. *Planning in the face of power*. Berkeley, CA: University of California Press.

Friend, J. and Jessop, N. 1977. *Local government and strategic choice*, 2nd edn, Toronto: Pergamon Press.

Gibson, R. 2002. From Wreck Cove to Voisey's Bay: the evolution of federal environmental assessment in Canada. *Impact Assessment and Project Appraisal* 20 (3), 151-60.

Glasson, J. 1999. EIA-impact on decisions, Chapter 7 in *Handbook of environmental impact assessment*, J. Pettes (ed), vol. 1. Oxford: Blackwell Science. Habermas, J. 1984. *The theory of communicative action*, vol. 1: *Reason and the rationalisation of society*. London: Polity Press.

Hacking, T. and Guthrie, P. 2008. A framework for clarifying the meaning of triple bottom line, integrated and sustainability assessment. *Environmental Impact Assessment Review* 28, 73-89.

Hanley, N. D. and Splash, C. 1993. *Cost-benefit analysis and the environment*. Aldershot: Edward Elgar.

Hart, S. L. 1984. The costs of environmental review. In *Improving impact assessment*, S. L. Hart *et al.* (eds). Boulder, CO: Westview Press.

Healey, P. 1996. The communicative turn in planning theory and its implication for spatial strategy making. *Environment and Planning B: Planning and Design* 23, 217-34.

Healey, P. 1997. *Collaborative planning: shaping places in fragmented societies*. Basingstoke: Macmillan.

HMG, Secretary of State for the Environment 1994. *Sustainable development: the UK strategy*. London: HMSO.

HMG 2008. *Planning Act* 2008. London: Stationery Office.

Holmberg, J. and Sandbrook, R. 1992. Sustainable development: what is to be done? In *Policies for a small planet*, J. Holmberg (ed), 19-38. London: Earthscan.

IAIA (International Association for Impact Assessment) 1994. Guidelines and principles for social impact assessment. *Impact Assessment* 12 (2).

IAIA 2009. *What is impact assessment?* Fargo, ND: IAIA.

IOCGP (Inter-organizational Committee on Guidelines and Principles for Social Impact Assessment) 1994. *Guidelines and Principles for Social Impact Assessment*, 12 (Summer), 107-52.

IOCGP 2003. Principles and guidelines for social impact assessment in the USA, *Impact Assessment and Project Appraisal* 21 (3), 231-50.

IPHI (Institute of Public Health in Ireland) 2009. *Health impact assessment guidance*. Dublin and Belfast:

IPHI.

Jay, S., Jones, C., Slinn, P. and Wood, C. 2007.

Environmental impact assessment: retrospect and prospect. *Environmental Impact Assessment Review* 27, 287-300.

Jones, C., Lee, N., Wood, C. M. 1991. *UK environmental statements 1988-1990: an analysis*. Occasional Paper 29, Department of Planning and Landscape, University of Manchester.

Kaiser, E., Godshalk, D. and Chapin, S. 1995. *Urban land use planning*, 4th edn. Chicago: University of Illinois Press.

Kennedy, W. V. 1988. Environmental impact assessment in North America, Western Europe: what has worked where, how and why? *International Environmental Reporter* 11 (4), 257-62.

Kirkby, J., O'Keefe, P. and Timberlake, L. 1995. *The earthscan reader in sustainable development*. London: Earthscan.

Lawrence, D. 1997. The need for EIA theory building. *Environmental Impact Assessment Review* 17, 79-107.

Lawrence, D. 2000. Planning theories and environmental impact assessment. *Environmental Impact Assessment Review* 20, 607-25.

Lee, N. 2002. Integrated approaches to impact assessment: substance or make-believe? *Environmental Assessment Yearbook*. Lincoln: IEMA, 14-20.

Leknes, E. 2001. The role of EIA in the decision-making process. *Environmental Impact Assessment Review* 21, 309-03.

Lichfield, N. 1996. *Community impact evaluation*. London: UCL Press.

Lindblom, E. C. E. 1980. *The policy making process*, 2nd edn. Englewood Cliffs: Prentice Hall.

Magnox Electric 2002. *ES for Decommissioning of Hinkley Point A Nuclear Power Station*. Berkeley, UK: Magnox Electric.

McLoughlin, J. B. 1969. *Urban and regional planning—a systems approach*. London: Faber & Faber.

Munn, R. E. 1979. *Environmental impact assessment: principles and procedures*, 2nd edn, New York: Wiley.

O'Riordan, T. 1988. The politics of sustainability. In *Sustainable environmental management: principles and practice*, R. K. Turner (ed). London: Belhaven.

O'Riordan, T. 1990. EIA from the environmentalist's perspective. VIA 4, March, 13.

Pearce, D. W. 1992. *Towards sustainable development through environment assessment*. Working Paper PA92-11, Centre for Social and Economic Research in the Global Environment, University College London.

Pearce, D., Markandya, A. and Barbier, E. 1989. *Blueprint for a green economy*. London: Earthscan.

Petts, J. 1999. Environmental impact assessment versus other environmental management decision tools. In *Handbook of environmental impact assessment*, J. Petts (ed), vol. 1, Oxford: Blackwell Science.

Preston, D. and Bedford, B. 1988. Evaluating cumulative effects on wetland functions: a conceptual overview and generic framework. *Environmental Management* 12 (5).

Redclift, M. 1987. *Sustainable development: exploring the contradictions*. London: Methuen.

Reid, D. 1995. *Sustainable development: an introductory guide*. London: Earthscan.

RPS, 2004. *ES for Kings Cross Central*. RPS for Argent St George, London and Continental Railways& Excel.

Sadler, B. 1996. *Environmental assessment in a changing world: evaluating practice to improve performance*. International study on the effectiveness of environmental assessment. Ottawa: Canadian Environmental Assessment Agency.

Scrase, J. and Sheate, W. 2002. Integration and integrated approaches to assessment: what do they mean for the environment? *Journal of Environmental Policy and Planning* 4 (4), 276-94.

SEERA 2001. *A better quality of life in the south east*. Guildford: South East England Regional Assembly.

Skolimowski, P. 1995. Sustainable development-how meaningful? *Environmental Values* 4.

Stakhiv, E. 1988. An evaluation paradigm for cumulative impact analysis. *Environmental Management* 12 (5).

Symonds/EDAW 2004. *ES for Lower Lea Valley: Olympics and legacy planning application*. Symonds/EDAW for London Development Agency.

Therivel, R. 2010. *Strategic environmental assessment in action*, 2nd edn. London: Earthscan.

Therivel, R. and Partidario, M. R. 1996. *The practice of strategic environmental assessment*. London: Earthscan.

Therivel, R., Wilson, E., Thompson, S., Heaney, D. and Pritchard, D. 1992. *Strategic environmental assessment*. London: RSPB/Earthscan.

Turner, R. K. and Pearce, D. W. 1992. *Sustainable development: ethics and economics*. Working Paper PA92-09, Centre for Social and Economic Research in the Global Environment, University College London.

UNECE (United Nations Economic Commission for Europe) 1991. *Policies and systems of environmental impact assessment*. Geneva: United Nations.

UNECE 2000. Access to information, public participation and access to justice in environmental matters Geneva: United Nations.

UN World Commission on Environment and Development 1987. *Our common future*. Oxford: Oxford University Press.

Vanclay, F. 2003. International principles for social impact assessment. International Assessment and Project Appraisal 21 (1), 5-12.

Vanclay, F. and Bronstein, D. (eds) 1995. *Environment and social impact assessments*. London: Wiley.

Wathern, P. (ed) 1988. *Environmental impact assessment: theory and practice*. London: Unwin Hyman.

Weston, J. 2000. EIA, Decision-making theory and screening and scoping in UK practice. *Journal of Environmental Planning and Management* 43 (2), 185-203.

Weston, J. 2003. Is there a future for EIA? Response to Benson. *Impact Assessment and Project Appraisal* 21 (4), 278-80.

White, P. R., Franke, M. and Hindle, P. 1995. *Integrated solid waste management: a lifecycle inventory*. London: Chapman Hall.

Wilson, E. and Piper, J. 2010. *Spatial planning and climate change*. Abingdon: Routledge.

2 起源和发展

2.1 导言

环境影响评价制度最早于1969年在美国正式确立，此后以各种形式在其他国家不断推广和发展。在英国，环境影响评价最初是一个由地方规划部门和开发商共同实施的专门程序，主要针对石油和天然气相关的开发活动。1985年，欧洲共同体向所有成员国发布85/337号指令，对环境影响评价做了统一要求，极大地加快了英国环境影响评价制度的发展。然而，在该指令颁布10年后，各成员国执行的环境影响评价制度仍然五花八门，这有违"规范环评体系"的制定初衷。为了改善这种局面，欧盟又分别在1997年、2003年和2009年颁布了该体系的修正案。环境影响评价体系的特点（如强制性或自愿性、公众参与水平、需要进行环境影响评价的类别等）以及该体系在各国的实践都大相径庭，但环境影响评价理念的迅速传播及其作为政府环境管理的核心内容都显示了环境影响评价是一个非常有效的规划工具。

本章首先回顾环境影响评价体系在美国的发展历程，并简要介绍环境影响评价在世界范围内的发展现状（第10章将深入介绍一些国家的环境影响评价体系），然后深入介绍了英国和欧盟的环境影响评价制度，最后对其他欧盟成员国的几种环境影响评价体系进行了讨论。

2.2 美国《国家环境政策法案（NEPA）》及其法律体系

1969年美国《国家环境政策法案》（*The US National Environmental Policy Act*，NE-

PA）出台时，环境公害事件频频发生，环境问题正越来越受到社会的广泛关注，20 世纪 60 年代广为畅销的书籍《寂静的春天》（Carson，1962）和《人口爆炸》（Ehrlich，1968）便是生动的例子。《国家环境政策法案》是世界上第一个要求进行环境影响评价的法律。这种新的环境政策及其在后来发展中取得的经验和教训使其成为了其他环境影响评价体系的重要模板。自颁布之日起，《国家环境政策法案》已经推动了至少 25000 份全面或专项环境影响报告书（Environmental Impact Statements，EISs）的编制，影响了无数的决策，收集了丰富的环境信息。另一方面，《国家环境政策法案》也是独一无二的。由于该环境影响评价体系在解释和运作过程中带来了大量的诉讼纠纷，使得其他国家很难采用《国家环境政策法案》采用的形式和流程（procedures）。

本节将介绍《国家环境政策法案》的立法历程（即在立法生效之前的早期发展）、美国法院和美国环境质量委员会（the Council on Environmental Quality，CEQ）对《国家环境政策法案》的解释、《国家环境政策法案》规定的环境影响评价流程及环境影响评价未来可能的发展趋势等内容。更多信息可参阅 Anderson 等（1984）、Bear（1990）、Canter（1996）、美国环境质量委员会（1997a）、CRS（2006）、Greenberg（2012）、Mandelker（2000）、Orloff（1980）及美国环境质量委员会的年度报告等。

2.2.1 立法历程

《国家环境政策法案》在很大程度上是非常侥幸的，法院和立法机构原本打算通过修正案削弱它，结果反而使其强化。《国家环境政策法案》的立法历程之所以值得介绍，不仅是因为它解释了很多实践中遇到的异常情况，而且因为它触及了环境影响评价制度设计的一些关键问题。早在 20 世纪 60 年代，美国参议院和众议院就讨论过几个建立全国环境政策的议案。这些议案包括形成某种统一的环境政策、建立高级委员会来促进落实等内容。1969 年 2 月，参议院通过了 S1075 议案，提出了一项由联邦政府资助的生态研究计划，并成立了美国环境质量委员会（CEQ）。众议院也通过了一个相似的 HR6750 号议案，提议成立美国环境质量委员会，并就国家环境政策进行了简要陈述。参众两院的讨论之后集中在了以下几点：

- 需要对国家环境政策（即《国家环境政策法案》）进行声明。

- 议案陈述的“享受健康环境是每个公民都不可侵犯的基本权利”（该陈述将环境健康视为与言论自由同等重要的权利）在之后的 101（c）陈述中被削弱为“每个公民都应享有健康的环境”。

- 和《水质改善法案》（*Water Quality Improvement Act*）一样，议案也需要对强制执行作出规定，即联邦政府必须要对任何重大开发活动可能造成的环境影响进行详细报告，《国家环境政策法案》102（2）(c) 将其表述为“需要进行环境影响评价”。议案最初的意思是要得出“调查研究的结论”，要求将重点放在对环境保护的责任上，而不仅仅在“详细的报告”上。参议院弱化了这个提案，认为仅需要一个详细报告即可。因此，“详细评估”成了公众关注和争论的焦点；详细报告的公共有效性成了该法案实施之初的主要推动力量。《国家环境政策法案》于 1970 年正式生效。表 2.1 概括了其主要内容。

表 2.1 《国家环境政策法案》的要点

《国家环境政策法案》包括两个篇章。篇章Ⅰ制定了环境质量保护与恢复的国家政策。篇章Ⅱ建立了三方的环境质量委员会，负责评估环境计划和发展，并就这些问题向总统提出建议，同时要求总统每年向议会提交年度“环境质量报告”。篇章Ⅰ的规定在美国环境影响评价体系中起着决定性的作用，其内容概括如下：

第 101 条主要是原则要求。其中指出，联邦政府在“创造并维持人与自然的和谐相处，满足当代及后代美国公民的社会、经济及其他需求”上负有持续责任。为此，政府必须采取一切可能的手段，“在与国家政策及其他基本要求相一致的情况下”，通过各种联邦规划和方案最大限度地减少对环境的不利影响，保护和改善环境。最后，“每个公民都应享有健康的环境”，同时每个公民都有责任保护环境。

第 102 条主要是程序要求。联邦机构必须对其实施项目或开发活动的环境影响进行全面分析。102(1)条明确要求政府机构要按照《国家环境政策法案》来解释和管理其政策、规章及法律。102(2)条要求联邦机构：

- 应使用“系统的跨学科的方法”，确保社会、自然和环境等科学知识被应用于规划和决策。
- 识别和建立相关流程和方法，以便“在制定政策时，除了传统的经济和技术因素，也适当将尚无法量化的环境因素和环境价值纳入考虑”。
- 所有可能对人类环境产生显著影响的提案和其他政府活动的建议或报告书中，都应包括对以下内容的详细说明：
  - 拟议行动的环境影响；
  - 提案实施可能产生的不可避免的不利环境影响；
  - 拟议行动的替代方案；
  - 短期利用区域环境和维持提高长期生产力的关系；
  - 拟议行动的实施可能对提案中涉及的自然资源造成的任何不可挽回及不可弥补的后果。[强调]

第 103 条要求联邦机构评估其法规和流程是否与《国家环境政策法案》相一致，并提出可能的完善措施。

2.2.2 对《国家环境政策法案》的解释

由于《国家环境政策法案》的措辞较为笼统，需要尽快做出翔实的解释。为此，在《国家环境政策法案》要求下成立的环境质量委员会制定了解释该法案的指南（Guidelines）。然而，《国家环境政策法案》的权力大多来自早期的法院裁决，NEPA 很快就被环保主义者视为预防环境破坏的重要工具，以此为基础，20 世纪 70 年代初期美国产生了一系列影响深远的诉讼和判决。Orloff（1980）将这些诉讼主要分为了三大类：

（1）质疑那些不需要进行环境影响评价的政府决策。提出的质疑通常包括是否是重大项目、是否是联邦行为、是否是“开发活动”、是否会产生重大环境影响［参见《国家环境政策法案》102（2）(c)］。例如，当诉讼对象为涉及联邦投资的地方政府项目时，就需要考虑这个项目是否是联邦行为。

（2）质疑联邦机构的环境影响报告书的充分性。例如环境影响报告书是否充分论证了替代方案，是否涵盖了所有重大环境影响。由切萨皮克环境保护协会发起的诉讼案是早期一个比较著名的例子，该协会声称原子能委员会没有充分考虑其建设的核电站，特别是卡尔弗特崖（Calvert Cliffs）核电站对水质的影响，环境影响评价的充分性不足。原子能委员会认为，按照《国家环境政策法案》的要求，他们只需要考虑水质标准，而反对者则认为不应当只对达标与否进行评价。法院最终站在了反对者的一边。

（3）质疑机构决策的独立性，即环境影响报告书的内容是否能影响政府决定一个项目继续进行。另一个有影响的早期法院裁决明确了这个问题，即环境影响报告书对联邦立法的功能只能确保政府“重视”环境后果，不能代替政府做出裁决。

法院在早期的积极表现大大加强了美国环保运动的力量，许多项目都被停止或做出大幅调整。在很多案件中，诉讼对项目建设造成的拖延使其失去了经济可行性，这些项目选址后来成了国家公园或野生动物保护区（Turner，1988）。和早期决策相比，近期的决策明显削弱了环境保护的权重。早期的大量诉讼带来的时间和经济成本也为其他国家建立环境影响评价体系提供了经验。后文将会提到，许多国家在建立环境影响评价体系时都尽量避免涉及法律诉讼。

美国环境质量委员会还针对性地制定了很多指南来解释《国家环境政策法案》，如1970年的临时指南、1971年和1973年的指南等。总体来说，法院在进行诉讼裁决时都很好地执行了这些指南。不过这些指南也存在一些问题：不够详细，且被联邦机构自由裁量，不具有强制约束力。为了克服这些局限性，卡特总统于1997年签署了11992号行政令，授权美国环境质量委员会为推动《国家环境政策法案》发布强制性的条例（regulations）。新指南在1978年（CEQ，1978）正式发布，以期使NEPA程序对政策制定和公众更有用，减少书面工作和时间拖延，将重点放在真正的环境问题和替代方案上。

2.2.3 NEPA流程摘要

环境影响评价的程序（process）由《国家环境政策法案》确立，美国环境质量委员会发布条例对其做了进一步完善，图2.1对此做了总结。下面的引述内容摘自美国环境质量委员会条例（CEQ，1978）。

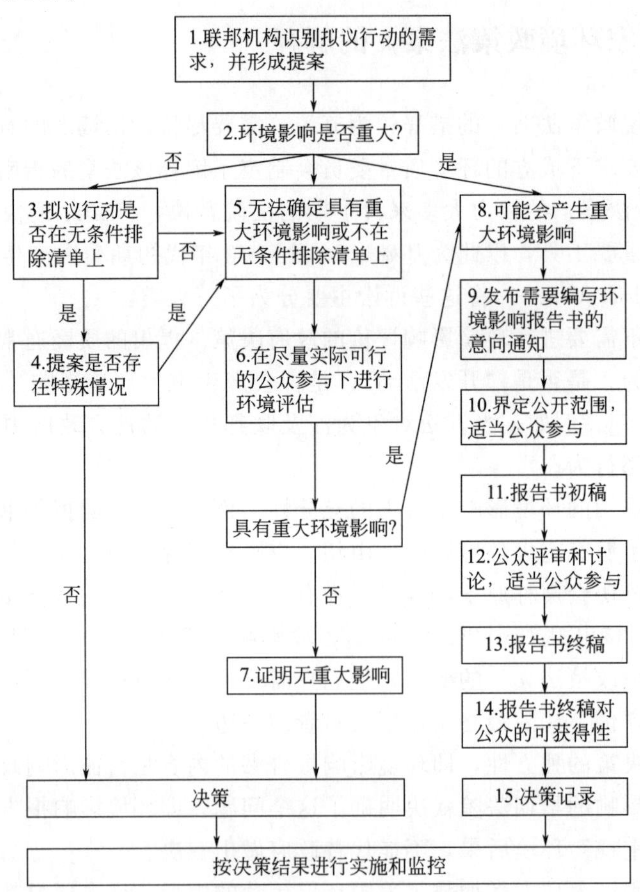

图2.1 《国家环境政策法案》确立的环境影响评价程序

（来源：CEQ，2007）

［环境影响评价程序］尽可能在政府机构制定或提交提案时开始……应提前准备好报告书，使其尽量为决策过程服务，而非为已经做出的决定进行合理性的证明或辩护。(1502.5 节)

要确立一个与环境影响评价过程相协调的“领导机构”。该领导机构首先需要通过一系列评估，确定提案是否要编制一份全面的报告书，抑或只需要得出一份“无重大影响”(finding of no significant impact，FONSI) 认定书。首先要评估联邦行为产生的重大环境影响是否具有可累积性。所有联邦机构都需要列出一个“无条件排除”(categorical exclusion) 清单来确认不会有上述影响。如果某个项目在这份清单上，则通常不需要进行环境影响评价。如果项目没有被无条件排除，就需要进行“环境评价”(environmental assessment) 以确定是否需要编制全面的环境影响报告书或无重大影响认定书。无重大影响认定书将被作为解释为什么项目不会产生重大环境影响的公开文件。

无重大影响认定书一旦完成，就需要进行公开讨论 (public discussion) 以决定是否通过。如果认为需要编制全面的环境影响评价报告书，领导机构就要发布一份“意向通知”(Notice of intent)，并开始确定评价范围。以明确环境影响评价将要评价哪些问题，排除那些非重要的问题，关注有重大影响的问题并寻找替代方案。领导机构会邀请该项目的支持者、受影响群体及其他感兴趣人群参与评价。

［替代方案］是环境影响报告书的核心……［它］应阐述清楚拟议行动的环境影响，对替代方案进行比选，以便认清问题，为决策提供明晰的支持。(1502.14 节)

如果在程序中识别出了合理的替代方案，联邦机构必须全面分析首选替代方案的直接、间接和累积效应。对《国家环境政策法案》来说，“效应”(effects) 和“影响”(impacts) 的含义是一样的，包括所有有利的和不利的生态、美学、历史、文化、经济、社会和健康影响。

接下来是准备环境影响报告书初稿，并交由相关机构和公众进行评审。评审意见需要在环境影响报告书的终稿中得到充分反映。环境影响报告书通常按照表 2.2 的格式进行编写，由于早期的环境影响报告书试图追求全面，往往内容烦琐，不易阅读。后来，CEQ 指南强调只需要针对重大问题准备相关报告文件：“环境影响报告书终稿一般应少于 150 页……用语应平实易懂……”。(1502.7～8 节)

表 2.2 《国家环境政策法案》给出的环境影响报告书的典型格式

(a)封面
- 相关责任部门名单
- 拟议行动的名称
- 部门联系人
- 说明报告书是初稿、终稿还是附录
- 报告书摘要
- 反馈意见的截止时间

(b)总论(通常不多于 15 页)
- 主要结论
- 尚有争议的部分
- 有待解决的问题

(c)目录
(d)拟议行动的目的和需求
(e)替代方案，包括拟议行动
(f)受影响的环境
(g)环境后果

续表

• 替代方案的环境影响，包括拟议行动 • 方案实施将产生的不可避免的不利环境影响 • 减缓措施及减缓措施实施后遗留的环境影响 • 短期利用区域环境和维持提高长期生产力的关系 • 拟议行动的实施可能对提案中涉及的自然资源造成的任何不可挽回及不可弥补的后果： —直接、间接和累积效应及其重大程度 —拟议行动和相关土地利用规划、政策和控制措施之间的潜在冲突 —替代方案的影响，包括拟议行动 —各替代方案及减缓措施的能源需求及保护效果 —对自然或可耗竭资源的需求，保护的各替代方案和减缓措施 —对城市质量、历史文化资源和建筑环境的影响 —不利影响的减缓措施 (h)监控计划和环境管理制度 (i)编制人员名单 (j)要递交报告书副本的机构名单等 (k)索引 (l)附录及相关支撑数据

在确定评价范围、形成报告书初稿和完成报告书终稿的各阶段，都要进行公众参与：

联邦机构应该（a）尽量让公众参与到 NEPA 流程的制定和实施过程中……（b）发布与 NEPA 有关的听证会和公众会议的公告，保证公众可以获取这些环境文件……（c）在适当时机……召开或举行听证会……（d）适当征求公众信息；（e）在流程允许范围内，向感兴趣公众解释信息和报告书的获取渠道……（f）编写环境影响报告书，接收反馈意见及其他符合信息自由法案要求的基本文件……（1506.6 节）

最后，做出拟议行动是否该被允许的决议：

联邦机构应按照流程确保做出的决策与相关政策和法案保持一致。这些流程包括但不限于：（d）执行 102（2）条的流程以实现 101 和 102（1）条的要求……（e）必须……考虑环境影响报告书中提供的替代方案。(1501.1 节)

当所有相关联邦机构都认为拟议行动不应该进行时，项目将被否定，并可尝试司法解决途径。当联邦机构认为拟议行动可以进行时，项目将获得批准（permission），但可能要适用特定条件（如接受监控、采取减缓措施等）。当联邦机构不同意时，美国环境质量委员会可以担任仲裁（1504 条）。在做出决策之前，“禁止采取提案中可能产生以下影响的任何行为：（1）对环境有不利影响的；（2）对合理替代方案的选择造成限制的……”（1506.1 节）

决策记录（The Record of Decision，ROD）是整个程序的最后一步。本部分需要陈述决策的内容；识别考虑的替代方案，包括环境优先的替代方案；讨论减缓计划，包括任何执行和监控的措施等。还需讨论是否找到避免或减缓环境损害的可行措施，如果没有找到，需要论述原因。该部分将是发布在联邦政府纪事或相关机构网站上的公开文件。

2.2.4 发展趋势

2.2.4.1 环境影响报告书的编制活动

《国家环境政策法案》实施的最初 10 年里，平均每年编制的环境影响报告书为 1500 份。后来，越来越多的拟议行动选择用协商的方式在“环境评估”的准备阶段解决问题，这引起了大量“采取减缓措施后无显著影响”的现象，从而大大减少了环境影响报告书的数量。1979年编制的报告书数量为1273份，到了1991年仅为456份，近年来稳定在每年500～550份（NEPA网站）。这种积极的发展趋势意味着环境影响在决策早期就被考虑，降低了编制环境影响报告书的成本。然而，程序精简使公众参与程度下降，也引起了一些问题。表2.3的总结表明，2008年提交环境影响报告书的联邦机构集中在农业部（主要是林业项目）和运输部（主要是道路建设）。从1979年到2008年，住房和城市发展部提交的环境影响报告书数量从170份下降到了0份！基于《国家环境政策法案》对联邦机构和部门提起的法律诉讼平均为每年100～150起，原告以公众利益团体和公民个人为主，常见控诉理由是“应该编写却没有编写环境影响报告书”以及“环境影响报告书不充分”。

表2.3　2008年编制的环境影响报告书数量

部门	编制数量	其中
农业部	128	林务局:124
商务部	36	国家海洋与大气管理局:36
国防部	79	工程兵部队:42;海军:24
环境事务部	36	联邦能源管理委员会:19
卫生及公共服务部	1	
国土安全部	8	美国海岸警卫队:6
住房及城市发展部	0	
内政部	128	美国土地管理局:48;国家公园管理局:25
司法部	2	
劳工部	0	
国务院	3	
运输部	104	联邦公路局:64
财政部	0	
退伍军人事务部	0	
独立机构	25	美国核管理委员会:14
总计	550	

资料来源：NEPA网站。

环境影响报告书的数量变化有其时代背景。虽然环境影响报告书非常重要，特别是在一些会产生巨大影响及涉及众多利益相关者的项目中，但环境影响报告书只是环境评估活动中的冰山一角。相比之下，根据美国环境质量委员会1997年的报告，联邦机构大约每年要产生50000份环境评估（CEQ，1997a）。要统计究竟有多少联邦活动涉及《国家环境政策法案》恐怕极为困难，但其中的一个联邦机构——联邦公路局（FHWA）对所有项目都进行了追踪。2001年，联邦公路局估计大约有3%的项目需要编制环境影响报告，7%的项目需要进行环境评估，90%的公路项目被划分在无条件排除清单里。

2.2.4.2　对体系的评审

1990年是《国家环境政策法案》实施的第20个年头，这一年美国就该法案召开了一系列会议，并向国会提交了《国家环境政策法案》修正案。由于提出在美国本土之外的联邦项目也应该进行环境影响评价（如联邦政府资助的在其他国家建设的项目），要求所有报告书将全球气候变化、臭氧层破坏、生物多样性减少及跨界污染等纳入考虑，该议案（HR1113）最后没有被通过。该修正案的争议在于：即使考虑项目对全球的影响具有必要性，但该议案在项目环评的层面不具有可行性。

环境影响评价的背景也是人们的关注焦点。环境影响评价只不过是宽泛的环境政策（《国家环境政策法案》）中的一部分，但《国家环境政策法案》102 (2)(c) 中关于程序的规定已经盖过了该法案的其他部分。有观点认为仅遵守这些流程要求是不够的，应当更加注重第 101 条对环境目标和政策所提出的要求。环境影响评价也必须遵守其他环境法规的要求。美国在 20 世纪 70 年代颁布并强化了很多环境类的专项法律，如《清洁水法案》和《清洁空气法案》等。这些方案在许多方面都超越了《国家环境政策法案》的实质性要求，可以作为环境政策程序要求的补充和支持。遵守这些法律就不一定要遵守《国家环境政策法案》。然而，遵守这些法律的相关许可程序已经成为评估项目影响的主要方式，在一些重大项目中，削弱《国家环境政策法案》的重要性有时会成为争论的焦点（Bear，1990）。

另一个关注焦点是环境影响评价的应用范围，特别是环境的社会属性。在黑色人种和少数族裔的长期抗争特别是对危险废物填埋场及焚烧炉分布不均衡的抗议下，美国环境质量委员会在环境保护法（EPA）许可范围内成立了一个工作小组，专门处理此类环境不公平问题（Hall，1994）。其结果是克林顿签发了“在执行联邦行动时为少数族裔居民和低收入居民实现环境正义”指令（白宫，1994）。根据这项指令，当《国家环境政策法案》要求进行环境影响分析时，各联邦机构必须分析联邦行动对人类健康、经济和社会等的影响，分析对象要涵盖少数族裔群体和低收入人群。如果可能的话，联邦机构应对同一社区内联邦行动产生的重大和不良环境影响采取减缓措施。此外，联邦机构必须提供公众参与 NEPA 程序的机会，通过向受影响社区的公众进行咨询和协商来识别潜在影响，并提出减缓措施，同时提高公众参与听证会和获取重要资料的机会。

其他一些问题，如《国家环境政策法案》的有效性、环境评估程序修订等仍处于争论中。Canter（1996）着重指出了《国家环境政策法案》需要继续完善的四个方面：

（1）联邦机构在发出环境影响报告书之前，应当在多大程度上识别和规划减缓措施；

（2）如何评估拟议开发活动的累积影响；

（3）开展“合理预测”（或最坏情况）分析的方法；

（4）预测结果的监控和审计。

1997 年，美国环境质量委员会对《国家环境政策法案》实施 25 年以来的有效性进行了评估，整体结论是《国家环境政策法案》取得了成功：

……它促使联邦机构认真审视其行动的环境后果，并将公众引入决策制定的程序中。在短短三页篇幅里，《国家环境政策法案》传达了保护和改善环境的国家共识，并促使各方凝聚到一起推动这一共同目标的实现。（CEQ，1997a）

但也有几个问题备受关注：过于注重文件编制，而不是提高决策；过于注重诉讼的证据文件；协商在环境影响评价程序中太过滞后；环境影响评价程序太长。有建议认为未来应增强跨部门协调，提高监控和适应性管理。还有人认为应探索汇聚多领域专业知识和信息的多学科综合方法，如专栏 2.1 所示。

专栏 2.1　综合提速案例——奥索卡山脉四车道公路

密苏里州的布兰森市是全国最火爆的娱乐中心之一，1991 年的六个月的旅游旺季里接纳游客超 370 万人次。高峰时期 Country Music Boulevard 大道每天车流量达 30000 辆，平均行驶速度在 10 英里[1]以下，造成了难以忍受的拥堵。1992 年 6 月上旬，密苏里州州长宣布布兰森地区的交通拥堵已陷入“经济紧急状态”，并宣布了一项针对 1.6 亿美元的奥索卡山脉四车道公路提案的规划和设计进程的快速追踪（fast-track）计划。对密苏里公路交通部门而言，在六个月内规划出一条全新的高速公路且不能影响到安全和环境过程（environmental process）完整性是一个巨大的挑战。有了快速追踪的意识和 NEPA 程序的指导，联邦机构成立了一个跨学科团队并定期会商，使人们在质量项目（quality project）的准备过程中可以兼顾敏感的奥索卡山脉生态系统的需求和增加区域交通承载力的需求。由于包含了所有的相关者和规则，各设计合理的替代方案及其相关影响都得到了讨论。

团队中的一些人曾经将《国家环境政策法案》的要求视为一种负担、障碍和需要克服的阻碍。但在公路取得成果后，这些人开始意识到《国家环境政策法案》能够帮助项目满足其目的和需求，同时保护环境和维持社区价值。最重要的是，在 NEPA 规划过程中形成的这种新的观念也正在其他的一些涉及地方、中央、联邦机构和咨询公司的项目中发挥作用。

资料来源：CEQ，1997a。

包括累积影响、环境影响报告书和环境管理体系（environmental management systems）的关系、整个体系的“精简”等在内的许多领域都已经取得了长足的进步。虽然最初的《国家环境政策法案》界定了累积影响的本质，但联邦机构已经意识到开发相关的流程和方法的必要性。然而，在 1997 年，美国环境质量委员会基于《国家环境政策法案》出台了一个关于考虑进行累积效应评估（无法律约束力）的指南。该 CEQ 手册介绍了衡量巧合效应（coincident effects）(包括有利的和不利的）的实践方法，用于对特定资源、生态系统、人类社区以及会触发评价程序的拟议项目和替代方案等进行评定（CEQ，1997b）。虽然这种“预测—减缓—实施”的模式得到了《国家环境政策法案》的认可，但实施环节非常薄弱。为了解决这一问题，促进环境管理体系实施，白宫颁布了一项行政命令（政令 13423 号，2007 年 1 月），要求所有联邦机构在各组织层次实施环境管理体系。美国环境质量委员会进一步出台了相关指导意见，要求根据环境管理体系对 NEPA 程序进行一致性调整（Aligning NEPA Processes with Environmental Management Systems）(CEQ，2007)。NEPA 程序的耗时性也正受到普遍持续的关注，越来越多的呼声要求进行“精简”。国会研究机构的一份报告指出主要有两类因素造成了 NEPA 程序的延迟：完成所需文件（主要是环境影响报告书）编制所耗费的时间，相关诉讼造成的时间拖延（CRS，2006)。这份报告进一步讨论了一系列改进措施，包括限制环境影响报告书的篇幅、建立司法审查限度、修订无条件排除和免责项目的清单等，其中一些措施已经得到了实施，另一些措施仍存在争议。

---

[1] 1 英里＝1.609 千米。

2.2.5 小型 NEPA 和加利福尼亚州的特例

除了《国家环境政策法案》外，美国许多州也建立了自己的环境影响评价体系。其中15个州以及华盛顿哥伦比亚特区和波多黎各的环境影响评价体系在很大程度上是对《国家环境政策法案》的模仿，被统称为“小型 NEPA”，要求对敏感地区的联邦行为（由联邦政府资助或许可的行为）和项目进行环境影响评价。其他州虽然没有专门的环境影响评价制度，但在《国家环境政策法案》的要求之外也有自己的环境影响评价要求。

加利福尼亚州的环评体系尤为特别，该体系是依据1973年的《加利福尼亚环境政策法案》(CEQA) 建立并逐渐修订完善的，被公认为是世界上最先进的环境影响评价体系之一。该法案不仅应用于政府活动，而且也应用于需要得到政府部门批准的私人活动。它不仅仅是一个程序，而且要求州政府和地方政府在环境影响评价（environmental impact reviews, EIRs）时采取切实可行的减缓措施的替代方案以保护环境。该法案还将评价范围从项目扩大到了更高层次的活动，1989年的修订案在法案中增加了强制减排、监控和报告的内容。加利福尼亚州政府每年都会就该环评体系出版指南，详细地解释《加利福尼亚环境政策法案》的规定和准则。2009年的最新版指南可参见环境专业协会（Association of Environmental Professionals，AEP）网站。2010年的修订版在法案规定的必须进行环境影响分析的清单中增加了温室气体排放量的内容。

2.3 环境影响评价在世界范围的推广

从《国家环境政策法案》制定以来，世界上许多国家都建立了各种形式的环境影响评价体系，最初是较发达国家（如加拿大1973年，澳大利亚1974年，联邦德国1975年，法国1976年等），继而扩展到欠发达国家。欧洲共同体于1985年发布的环境影响评价指令大大加快了20世纪80年代后期欧洲各国的环境影响评价的立法进程。1991年苏联解体后，新成立的诸国也大多在20世纪90年代中期进行了环境影响评价立法。非洲和南美洲也在20世纪90年代早期进行了大量的环境影响评价立法和指南出台工作。到1996年时，全球已经有100多个国家建立了环境影响评价体系。

这些环境影响评价体系存在较大差异。有些是强制性法规、法案或指令，当局通常要求在项目批准前提交一份充分的环境影响报告书。一些国家制定了环境影响评价导则，导则虽然不是强制性的，但通常要求管理部门承担一定责任。有些法律允许政府部门自行决定是否需要进行环境影响评价。有时环境影响评价会以一种特别的方式进行，通常是被作为项目资助机构（如世界银行、美国国际开发署等）的融资审批过程的一部分。然而，这些分类并不能说明环境影响评价的执行情况。例如，巴西、菲律宾的环境影响评价制度就没有得到很好的执行和实施（Glasson 和 Salvador，2000；Moreira，1988），而日本的指南则执行得比较彻底。

不同环境影响评价体系之间的另一个重要差异是环境影响评价项目的定义，有时是用定义的方式给出的（如美国的定义是“对人类环境质量有影响的重大联邦行动”），有时是以项目清单的方式给出的（如长度超过 10km 的公路建设项目）。大多数国家采用的是项目清单的方式，以尽可能避免因定义问题产生法律纠纷。另一个差异是需要进行环境影响评价的是政府项目（如《国家环境政策法案》）、私人项目，还是二者都要进行评价。最后，一些国际发展和基金组织也制定了环境影响评价导则，包括欧洲重建和发展银行（1992，2010）、英国海外开发署（1996）、英国国际发展部（2003）、联合国环境规划署（1997，2002）和世界银行（1991，1995，1997，1999，2002，2006）等。

2.4 环境影响评价制度在英国的发展

英国于 1988 年为环境影响评价正式立法，主要是把欧洲共同体 85/337 号指令及其修正案以法律的形式予以实施。如果没有来自欧盟委员会的推动，英国颁布环境影响评价法律的速度可能会比实际情况缓慢得多，因为英国政府认为已有的规划制度完全可以控制对环境不当的开发活动。当然，这并不是说英国在 1988 年以前没有任何环境影响评价体系，许多环境影响评价活动都是出于自愿或按地方政府要求开展的，英国还为这些评价活动制定了导则。

2.4.1 土地利用规划制度的局限性

1947 年起，英国就从法律上建立了土地利用规划制度（land-use planning system），要求地方规划管理部门（local planning authorities，LPAs）预测可能的发展压力，评估其影响程度，并合理调配土地。环境因素是评估时必须加以考虑的一个基本要素。由于大部分开发活动都需要得到规划许可，政府就通过不予许可来预防对环境有害的开发活动。这一制度积累了大量与拟议项目可能造成的后果有关的专业规划经验。

然而，从 20 世纪 60 年代中期以后，规划制度控制大型开发活动的影响的能力越来越小。由于开发活动的规模和复杂程度日益增加，其社会和物理环境影响愈发庞大，跨国开发活动也不断增加（如石油、天然气、化工等），对这些活动的影响进行预测和控制已经超过了开发控制体系（development control system）的能力。20 世纪 60 年代末，随着公众环保意识的增强，人们越来越关注法定规划控制（planning controls）和大型开发项目之间的关系。一个典型例子是拟建的伦敦第三机场。为了在英格兰东南部选出最合适的建设地点，人们成立了 Roskill 委员会对备选地点进行费用效益分析（CBA）。结果分析（HMSO，1971）把重点放在了社会-经济影响上而不是自然环境影响上，这使得人们意识到将费用效益分析扩展到环境影响评价上面临着许多困难，环境影响无法简单地用金钱进行衡量，需要采用其他的评价方法来实现社会经济和自然环境二者之间的平衡。

2.4.2 关于北海石油和天然气的环境影响评价计划

北海石油和天然气的发现促进了环境影响评价的进一步发展。开采这些资源必然就要在那些拥有秀丽景色及独特生活方式的地区（如设得兰群岛、奥克尼郡和高地地区）进行大规模建设开发。但这些地区的规划部门缺乏评价此类大型开发计划的影响的经验和资源，为此，苏格兰发展部（Scottish Development Department，SDD）向地方规划管理部门提供了一份技术建议书（technical advice note）(SDD，1974)。这份石油类开发项目影响评估（appraisal of the impact of oil-related development）指出要对此类开发活动及其他一些大型的、特殊的项目进行“严格的评价”，并建议地方规划管理部门在必要时委托相关机构对开发活动的影响进行研究。这是政府首次提出要对重大开发活动进行专项评价。20 世纪 70 年代早期开展的一些环境影响评价大部分是针对石油和天然气的开发项目。其中很多都是由苏格兰发展部和地方规划管理部门发起、环境咨询机构完成的，但有一些（如 Flotta 油库项目、Beatrice 油田项目等）是由开发商委托的。此外，早期的环境影响评价案例还有 Belvoir 山谷煤矿、Loch Lomond 抽水蓄能发电计划和各种高速公路和主干道的环境影响评价等。

1973 年，苏格兰事务部和英国环境事务部（DoE）委托阿伯丁（Aberdeen）大学建设项目开发控制评估小组（Project Appraisal for Development Control，PADC）开发一个系统化流程，用于帮助规划部门对大型工业开发活动的环境、经济和社会影响进行适当评估。建设项目开发控制评价小组提供了一份临时报告——大型工业项目评估指南（the assessment of major industrial applications—a manual）(Clark 等，1976)，该指南免费提供给英国所有的地方规划管理部门并“被中央政府推荐给规划部门、政府机构和开发商采用”。该流程在设计时尽量考虑了现有的规划框架，可用于评价各种项目（主要是私人项目）。1981 年该指南又出版了修订更新版（Clark 等，1981）。

1974 年，苏格兰和威尔士的环境事务秘书委托咨询公司研究“将影响分析的体系引入英国的诉求，该体系的适用情景、适用项目、如何纳入开发活动控制体系”等问题（Catlow 和 Thirwall，1976）。最终的研究报告在以下几个方面给出了建议：谁进行环境影响评价、谁支付评价费用、建立环境影响评价制度需要哪些立法改革以及其他类似的问题等。报告认为每年需要进行环境影响评价的项目数量为 25～50 个，包括公共项目和私人项目。Dobry 在一个关于开发活动控制体系的研究报告中对环境影响评价做了补充（Dobry，1975），主张地方规划管理部门应该要求开发商对特别重大的开发活动进行影响研究（impact studies）。该报告还列出了影响研究应该解决的主要议题，开发商应提供哪些信息等。虽然英国皇家环境污染委员会（Royal Commission on Environmental Pollution）认可了这份报告，但政府对它的反应褒贬不一。

2.4.3 环境部门的质疑

然而，整个环境事务部对环境影响评价的必要性、可操作性和成本仍然持怀疑态度。事实上，政府对环境影响评价制度的态度是从“起初的不情愿和消极”开始的（CPRE，1991）。针对 Catlow 和 Thirwall 的报告，环境事务部表示：“地方权力机构对该报告的考量不应当延误其正常的和新的规划流程，包括中央及地方政府对财力、人力的附加要求，在当前经济不景气的情况下，任何延迟都是不可接受的”（DoE，1977）。经过一年的考虑后，环境事务部对环境影响评价的态度才变得较为积极：

> 我们完全赞成……要确保对大规模开发活动可能产生的环境影响进行仔细评估……由 Thirwall/Catlow 提出的方法已经在许多［项目］得到应用……合理使用这种方法［应该］能提高对那些相对较少的、重大的拟议项目的掌控能力。(DoE，1978)

政府在1981年的PACD手册的序言里也强调要降低环境影响评价流程的成本："很重要的一点是，要根据拟议开发项目的具体情况和经济状况，选择性地使用本手册提供的方法"（Clark等，1981）。后面的章节里也要提到，正如欧洲共同体所承认的，政府有时仍会质疑环境影响评价的价值，质疑其扩大职权范围的必要性。但到20世纪80年代早期，英国针对项目环境影响开展了200多项研究（Petts和Hills，1982）。其中的许多项研究并不进行全面的环境影响评价，而只是关注某些方面的影响。不过，英国石油公司、英国天然气公司、中央电力局及国家煤业局等大型开发商越来越倾向于编制全面的环境影响报告。仅就英国天然气公司一家而言，该举措就在10年内为公司节约了3000万欧元（House of Lords，1981a）。

2.5 欧洲共同体85/337号指令

欧洲共同体85/337号指令的制定和实施极大地影响了英国和欧盟成员国的环境影响评价体系。英国中央政府在20世纪70年代中期以后就几乎停止了对英国环境影响评价体系的研究，而将重点转移到了如何确保欧盟环境影响评价体系满足英国对灵活性和自由裁量的需求上。欧盟成员国则希望指令能够反映他们自身更为严格的环评体系需求。指令实施后，欧盟各国的环境影响评价活动均显著增加。

2.5.1 立法历程

欧洲共同体建立统一的环境影响评价体系有两个主要原因。第一个原因是考虑到自然环境现状，欧洲共同体想通过该体系防止环境进一步恶化。欧洲共同体在1973年发出的首个环境行动纲领（Action Programme on the Environment）中（CEC，1973）提出要防止环境破坏："最好的环境政策应从源头上避免污染和损害的产生，而不是在后期试图消除其影响"，为此，"在所有技术规划和决策制定过程中，都应尽早考虑对环境的影响"。1977年、1983年、1987年、1992年和2001年的行动纲领都继续强调了这一点。土地利用规划被看作是将这些原则付诸实践的重要途径，而环境影响评价则被视为是将环境因素纳入规划过程的关键技术。

第二个原因是欧洲共同体想确保公平竞争不会被扭曲，如某成员国以环境理由否决的开发活动却在另一国得到许可，造成成员国间的不公平。换句话说，环境政策也是维护平等经济环境的必要条件。欧洲共同体的更深层动机包括希望鼓励各成员国更好地实践环境影响评价。此外，跨界环境污染问题（如欧洲的酸雨、水污染等）也是促使欧洲共同体制定统一行动框架的原因。

欧洲共同体委员会早在1975年就开始对环境影响评价进行研究。经过5年20多份草案的修改后，欧洲共同体向部长理事会提交了一份指令草案（CEC，1980），并散发给了所有成员国，试图协调各种冲突。这份草案借鉴了美国《国家环境政策法案》的经验，并结合欧洲的实际需求，目的是环境影响评价适用于所有可能产生重大环境影响的活动，同时又不失程序上的可操作性。最后，也是最有挑战性的一点是，它试图使环境影响评价具有足够的灵活性，以适应各成员国的需求和制度安排，但又要具有足够的一致性，以防止在环境影响评价的解释过程中出现歧义。其中，协调的不同项目类型、开发商的主要义务和环境影响评价的内容被认为尤其重要（Lee 和 Wood，1984；Tomlinson，1986）。

这两个因素使得指令草案具有一系列重要的特征。首先，项目要在经过充分的环境影响评价后才能获得规划许可；其次，地方规划管理部门和开发商必须共同提供拟议开发活动的环境影响的相关信息；再次，必须向对环境问题负责的法定机构咨询，在具有跨界效应时，还必须向其他成员国咨询；最后，必须要告知公众，并允许公众对项目开发的有关问题提出意见。

在英国该草案由上议院特别委员会进行了审查，并获得了广泛支持：

现行指令草案对各方面进行了很好的平衡：它提供了一个共同的管理实践框架（a framework of common administrative practices），允许各成员国在其各自体系的基础上进行有效的规划控制……它包含了足够的细节，以确保草案的意图得到实现……该指令在英国的适当实施并不会对规划流程产生过度的额外拖延和成本增加，因此并不会造成经济和其他方面的损失。（House of Lords，1981a）

然而，英国环境事务部副部长却对该指令草案提出了异议。他虽然认同需要进行环境影响评价，但却担心官僚主义机制的束缚会使得与草案相关的反对和诉讼被搁置拖延。英国皇家城市规划学会（RTPI）也对指令的几个草案发表过评论，学会总体上对草案持赞成态度，但也担心可能会使规划系统过于死板：

学会赞赏欧盟委员会对推广环境影响评价所做的努力，因为它相信适当实行环境影响评价能促进和提高某些拟议开发活动的决策质量。但由于实践经验有限，目前起草的这份指令会使流程过度形式化，难以提供足够的价值。因此，学会建议删除草案的第 4 条和附件。（House of Lords，1981a）

更概括地说，欧共体的这项法案实施得十分缓慢的原因是由于涉及的利益群体太广、缺乏公众对扩大城镇规划和环境保护整体流程的支持、各成员国不愿意为了适应其他国家的体制和法例对自己的规划制度和环境保护法规做出较大改动等（Williams，1988）。1982 年 3 月，在考虑各成员国提出的诸多意见之后，委员会对该法案进行了修订（CEC，1982）。由于丹麦政府担心那些被议会法案授权的项目会受影响，修正案的通过又被推迟。1985 年 3 月 7 日，欧洲外长会议通过了修正案；1985 年 6 月 27 日修正案被正式纳入指令（CEC，1985），1988 年 7 月 3 日正式生效。

之后，欧洲第五期行动计划《走向可持续》（*Towards sustainability*）（CEC，1992）强调了环境影响评价的重要性，特别是在帮助实现可持续发展方面的重要性，认为需要扩大环境影响评价的应用范围：

为了实现可持续发展的目标，应该对所有相关政策、规划和计划的环境影响进行评价，即使这一举措不是必需的，它至少也是唯一合乎逻辑的。在宏观决策中纳入环境影响评价不仅可以加强环境保护，促进资源管理优化，而且也能在进行新的开发活动时减少成员国之间不同的评价体系导致的国家和地区的竞争差异……

环境影响评价指令是动态发展的。它通过进行定期评审来改进流程，寻求应用的一致性。开展五年评审（CEC，1993）的结果是指令在 1997 年通过了修正案，2003 年和 2009 年又进行了小幅修订。附录 1 给出了修正指令的完整版本（截至 2011 年 6 月）。本章 2.5 节的剩余部分将对原版指令进行总结。2.6 节讨论指令 97/11 的重要修订。2.7 节重点对原版和 97/11 修正版指令的实施情况进行介绍，讨论取得的主要成果。2.8 节对其他评审和 2003 年及 2009 年的修正案进行总结，这些审查和修改在扩大后的欧盟 27 国的背景下也具有相当的重要性。

2.5.2 欧洲共同体 85/337 号指令流程概述

欧共体 85/337 号指令在关键问题上与《国家环境政策法案》存在较大差异。它要求公共部门和私人开发商都应实施环境影响评价，而《国家环境政策法案》只要求政府部门进行环境影响评价。它指定了一个需要进行环境影响评价的项目清单，而《国家环境政策法案》使用的是定义“重大联邦行动”。此外，它明确列出了环境影响评价中必须要评价的影响，而《国家环境政策法案》不做明确要求。最后，它对公众咨询的要求没有《国家环境政策法案》多。根据 1972 年的《欧洲共同体法案》的规定，欧洲共同体 85/337 号指令属于受控文件，为各成员国规定了环境影响评价的准则。各国在实施该指令时可以颁布自己的法规，并具有相当大的自主性。根据该指令的要求，需要进行环境影响评价的项目被分为两类，一类是强制执行（附件Ⅰ），另一类是自愿执行（附件Ⅱ）：

> 附件Ⅰ中的项目必须进行环境影响评价……附件Ⅱ中的项目成员国可根据下列方法自主决定是否要进行环境影响评价：(a) 逐案分析；(b) 根据成员国设定的阈值或标准确定……当采用［此方法］时，应考虑附件Ⅲ中给出的相关选择标准。（第 4 条）

表 2.4 总结了附件Ⅰ和附件Ⅱ中所列的项目。欧洲共同体（CEC，1995）还制定了指南来帮助成员国确定某个项目是否需要进行环境影响评价。

表 2.4 欧洲共同体 85/337 号指令给出的需要进行环境影响评价项目清单（修订）

附件Ⅰ（强制执行）
1. 原油提炼，煤/页岩气化和液化
2. 火电站和其他燃烧装置；核电站及其他核反应堆
3. 放射性废物的加工、储存装置
4. 炼铁、炼钢
5. 石棉开采、加工和运输
6. 集成化工装置
7. 高速公路、快速路、其他大型道路、铁路和机场建设
8. 贸易口岸和内陆水运涵道
9. 有毒和危险废物的焚烧、处置装置建设
10. 大型非危险废物的焚烧、处置装置建设
11. 大型地下水开采和回灌项目
12. 大型调水工程
13. 大型污水处理厂
14. 大型石油及天然气开采
15. 大型水坝及水库
16. 天然气、石油或化学品长距离输送管道
17. 大型家禽或生猪饲养场
18. 纸浆、木料及板材制造
19. 大型采石场或露天采矿场
20. 长距离高架输电线
21. 大型石油、石化或化工产品装置
22. 符合附件Ⅰ判断标准的其他调整、扩展项目
23. 碳储存场所
24. 碳捕集装置

续表

附件Ⅱ(自愿执行)
1. 农业、林业和水产养殖
2. 采掘业
3. 能源业
4. 金属生产及加工业
5. *矿*工业(未列入附件Ⅰ的项目)
6. 化学工业
7. 食品工业
8. 纺织、皮革、木材、造纸业
9. 橡胶工业
10. *基础设施项目*
11. *其他项目*
12. *旅游休闲*
13. 附件Ⅰ中的调整、扩展或临时测试项目

注：1997版修订的内容用斜体表示；第22～24条为2003版和2009版的修订内容。

同样地，指令在附件Ⅲ中给出了环境影响评价需要包含的信息（information），但这些信息只有在符合下述条件之一时才需要被提供：

(a) 成员国认为这些信息可能与同意流程（consent procedure）的特定阶段、特定项目的具体特点……及可能受到影响的环境特征相关；(b) 成员国认为有理由要求开发商在现有评价知识和方法的情况下对这些信息加以收集。(第5.1条)

表2.5总结了附件Ⅲ（修订后的附件Ⅳ）要求提供的信息。开发商所准备的环境影响报告书必须包含其所在成员国对指令的附件Ⅲ（修订后的附件Ⅳ）的解释里所指定的信息，并将其提交给“主管部门”。该环境影响报告书将被发给其他相关公共部门，并对公众公开：“成员国应采取必要措施确保该项目可能涉及的公众……有机会发表他们的意见”(第6.1条)。

表2.5 欧洲共同体85/337号指令规定环境影响评价应包含的信息（修订）①

附件Ⅲ(Ⅳ)
1. 项目描述
2. 选址(*简要概述对主要替代方案的研究,并说明最终选择的主要原因*)
3. 拟议项目可能产生重大影响的环境因素,包括人口、动物、土壤、水、空气气候因素、矿产、农业和建筑古迹、景观以及它们之间的相互影响
4. 拟议项目可能对环境产生的重大影响
5. 预防、减缓措施及可能补偿任何重大不利环境影响的措施
6. 非技术性总结
7. 搜集所需信息时可能遇到的任何困难

① 修订部分用斜体表示。

成员国还应确保：

• 按照［指令规定］向公众公开颁发许可证的所有要求及全部信息；

• 在项目启动前，给相关公众表达自己意见的机会。

成员国可自主决定这些信息和咨询的具体安排（第6.2条和第6.3条；也可见本书6.2节）。主管部门必须在同意流程中考虑环境影响报告书里陈述的内容、相关部门和公众的评论以及其他成员国的意见（如果适用）(第8条)。欧洲共同体委员会（1994）制定了一份清单帮助主管部门审查环境信息。最后，要将最终决定和所有附加条件告知公众(第9条)。

2.6 在欧盟 97/11 号指令基础上修订而来的欧洲共同体 85/337 号指令

85/337 号指令要求每五年进行一次评审，1993 年欧洲共同体发布了一份报告（CEC，1993）。虽然人们普遍满意“环境影响评价的基本内容”，但是仍然存在一些问题，没有完全包含某些项目、咨询和公众参与不足、缺乏替代方案的相关信息、缺少监控、成员国实施的一致性不足等。委员和成员国在对原版指令的评审过程中进行了大量的争论，指令几经修正，弱化了一些修改提议。该修正案最终于 1997 年 3 月完成，并在两年内予以实施。修订内容包括以下几项。

• 附件Ⅰ（强制执行）：新增了 12 类项目（如大坝和水库、输送管道、采石场、露天采矿场等）(表 2.4)。

• 附件Ⅱ（自愿执行）：新增了 8 类子项目（和 10 项其他项目），包括购物中心和停车场建设、旅游休闲（如建设房车营地和主题公园）等（表 2.4）。

• 新的附件Ⅲ列出了环境影响评价必须包括的内容：

—项目特征：规模，累积影响，自然资源利用，废弃物的产生、污染和扰民情况，事故风险等。

—项目选址：指定区域及其特点，原有及目前的土地利用情况。

—潜在影响的特征：地理范围，跨界影响，影响的大小及复杂程度，影响发生的可能性，影响的持续时间、频率和可逆性。

• 原附件Ⅲ修订为附件Ⅳ：内容上稍微有所改变。

• 其他修订的内容：

—第 2.3 条：所有跨界影响都必须要与其他成员国协商。

—第 4 条：当决定附件Ⅱ中的哪些项目需要环境影响评价时，成员国可采用阈值或标准法、逐案分析法或两种方法结合进行判断。

—第 5.2 条：当开发商提出时，主管部门应对环境影响报告书中应该包括的信息给出意见。成员国也可以要求主管部门在做出判定时不考虑开发商的要求。

—第 5.3 条：开发商提供的信息至少应包括主要替代方案的研究概述和确定最终选择的主要原因。

—第 7 条：要求受影响成员国和其他国家对跨界影响问题进行协商。

—第 9 条：主管部门必须让公众知道做出决策的主要原因和顾虑以及主要的减缓措施（CEC，1997a）。

附录 1 给出了包括以上修订的指令完整版，包含了强制执行的环境影响评价项目（附件Ⅰ）和自愿执行的环境影响评价项目（附件Ⅱ）。修正后的指令将替代方案变为了强制执行，更加强调咨询和参与的作用。今后欧盟各成员国可能会有更多的环境影响评价活动，在处理诸如替代方案、风险评估和累积影响等问题时也将面临许多挑战。

2.7 欧盟环境影响评价体系概述：同一体系下的不同实践

欧盟在制定环境政策方面一直非常积极，环境影响评价指令也被公认为是其卓越的环境成就之一（CEC，2001）。然而，有关（越来越多的）成员国之间指令实施的不一致的担忧一直并将继续存在（见 CEC，1993；CEC，2003；Glasson 和 Bellanger，2003）。这在某种程度上反映出 EC/EU 的本质只是为欧洲政策建立一个强制性框架，却将框架实施的“范围和方法”交由各成员国自行处理。此外，无论成员国的环境影响评价政策有多高的“法规一致性”，仍然会存在“实施一致性”的问题。法规的实施取决于各公共部门和私人部门的从业者，而这些人在文化背景和实现方法上本身就存在巨大的差异。

首先，各国对原版指令的实施时间就不一致。法国、荷兰和英国等国家准时执行了指令，其他国家（如比利时、葡萄牙）则没有。不难理解的是，另一些差异是来自各成员国在法律制度、管理和文化上的不同，一些差异概述如下。

- 各成员国对指令的实施情况存在很大差异。一些国家将该指令置于自然保护的广泛议程下（如法国、希腊、荷兰、葡萄牙）；一些国家将其置于规划体系下（如丹麦、爱尔兰、瑞典、英国）；也有一些国家制定了专门的环境影响评价法律（如比利时、意大利）；此外，比利时、德国和西班牙还要求在区域层面上也进行环境影响评价。
- 在大多数成员国，环境影响评价由开发商或开发商授权的咨询机构开展及支付费用。但在比利时的弗兰德斯，环境影响评价由环境主管部门授权的专家开展；在西班牙，环境影响评价由主管部门根据开发商得到的研究结果开展。
- 少数国家和地区建立了环境影响评价委员会。在荷兰，委员会负责协助确定评价范围、审查环境影响报告书的充分性以及搜集来自主管部门的监控信息。在弗兰德斯，委员会负责环境影响评价从业人员的资格评审、确定评价范围以及审查环境影响报告书与现行法律法规的符合情况。意大利也有这样的环境影响评价委员会。
- 在最简单的情况下，主管部门的职责就是决定一个项目是否应该继续（如在弗兰德斯、德国、英国），但在某些情况下，环境部长首先需要确定项目是否具有环境兼容性（如在丹麦、意大利、葡萄牙）。

2.7.1 对原版欧洲共同体 85/337 号指令的评审

对原版指令的第一个五年评审（CEC，1993）关注的就是各成员国在操作流程中的一些不一致问题（项目覆盖范围、替代方案、公众参与等）。第一个五年评审让一些成员国强化了他们的法规，以更全面地执行指令。1997 年的第二个五年评审（CEC，1997b）得到的主要结论如下：

- 环境影响评价是项目许可/授权制度的必需环节，尽管各操作流程有所差别（如程序步骤不同、和其他相关流程的关系不同）；
- 虽然所有成员国都要求对附件Ⅰ中规定的项目实施环境影响评价，但“对附件Ⅱ规定的项目的解释和操作流程却各不相同”；

• 对环境影响评价过程中的质量控制不足；

• 成员国没有对替代方案给予足够的重视；

• 公众参与和咨询还有待改进；

• 成员国抱怨指令中的一些关键项的定义模糊，或是没有定义。

1997 年的修正案试图消除剩余的几个差异。除了对附件Ⅰ和附件Ⅱ中的项目清单进行大幅扩展和修改以外，指令修正案（CEC，1997a）还强化了环境影响评价的程序基础。包括规定新的筛选程序、为附件Ⅱ中的项目规定了新的筛选标准。修正案还介绍了环境影响报告书的内容变化，包括开发商必须在考虑环境影响的前提下概述对主要替代方案的研究，并说明最终选择的主要原因。

2.7.2 对 97/11/EC 修正指令的评审

原版指令是在欧盟 97/11/EC 号指令（CEC，1997a）基础上修订的，第三次评审由欧盟委托牛津布鲁克斯大学（UK）的影响评估中心完成。这次评审提供了一份详细的总结，包括各成员国对指令（修正案）的实施情况以及进一步加强指令应用性和有效性的建议（CEC，2003）。实施中遇到的一些关键问题包括：

• 指令更新的速度过慢。许多成员国都超过了 1999 年的最后期限，奥地利、法国、比利时瓦隆和弗兰德斯地区在 2002 年底时仍未完成更新。而卢森堡甚至根本就没有进行更新。

• 用于判定附件Ⅱ中项目是否需要进行环境影响评价的阈值存在差异。附件Ⅰ对所有成员国都是强制执行的，一些国家还在附件Ⅰ中增加了一些项目类型。但直到指令修正案之前，成员国对附件Ⅱ的解释仍然存在着很大差异。有些国家（如荷兰）制定了需要进行环境影响评价的项目清单。随着修正案的出台，大部分成员国倾向于结合使用阈值法和逐案分析法对附件Ⅱ中的项目进行判定。然而，2003 年的评审结果表明各国执行的阈值仍然有很大不同。以造林项目为例，需要强制进行环境影响评价的造林面积阈值在丹麦是 $30hm^2$，而在葡萄牙是 $350hm^2$。类似地，瑞典规定风电场项目的涡轮机数量只要超过 3 个就要强制进行环境影响评价，而在西班牙则是 50 个。附件Ⅱ中这种对具体项目的规定差异仍将继续存在，特别是第 10（b）条中有关城市发展的项目。

• 各成员国开展环境影响评价的数量不同。一些国家缺乏充足的数据，增加了统计结果的复杂性。表 2.6 显示了各国每年开展环境影响评价的数量的巨大差异，从超过 7000 份（法国，主要原因是该国对财务标准的要求相对较低）到不足 20 份（奥地利）。这种差异是因为各国的经济发展水平不同，反映出各国设定的阈值不同。指令修正案似乎增加了一些成员国的环评项目数量。

表 2.6 欧盟 15 国（EU-15）开展环境影响评价的数量的变化

国家	1999 年之前(年平均)	1999 年之后(年平均)
奥地利	4	10～20
比利时布鲁塞尔	20	20
比利时弗兰德斯	无数据	增加了 20%
比利时瓦隆	无数据	有所增加
丹麦	28	100

续表

国家	1999年之前(年平均)	1999年之后(年平均)
芬兰	22	25
法国	6000～7000	>7000
德国	1000	有所增加
希腊	1600	1600
爱尔兰	140	178
意大利	37	无数据
卢森堡	20	20
荷兰	70	70
葡萄牙	87	92
西班牙	120	290
瑞典	1000	1000
英国	300	500

资料来源：CEC，2003。

• 虽然在确定评价范围、考虑替代方案等方面有所改善，但仍然存在一些问题。在修正案将确定评价范围作为一个正式的环境影响评价程序之前，只有少数国家将其作为独立的必须实施的步骤。修正案允许成员国自行决定是否将其作为一个强制性流程后，7个成员国将其纳入了流程中。而其他国家对这一问题的承诺却不尽相同。同样地，只有少数国家强制性地要求考虑拟议项目的替代方案，包括荷兰，其还要求分析各种情况下最具有环境可接受性的替代方案。指令修正案还要求开发商概述对主要替代方案的研究。2003年的评审结果表明，一些成员国已经将考虑替代方案作为其环境影响评价过程的核心内容，虽然它的覆盖范围还不够广——大多数国家对替代方案个数的要求仍然是零(或最少)。

• 环境影响评价过程中对公众咨询的要求不同。指令规定环境影响报告书应该在其被提交给主管部门后对公众公开，并且整个欧盟的公众都有机会对进行环境影响评价的项目发表意见。但是，各国的公众参与程度和对"相关公众"(the public concerned)的解释却大相径庭。在丹麦、荷兰和比利时瓦隆，公众咨询在确定环评范围时就进行。在荷兰和比利时弗兰德斯，公众听证会必须在环境影响报告书提交之后举行。在西班牙，公众咨询要在环境影响报告书提交之前进行。在奥地利，公众在环境影响评价的若干个阶段都可参与，其中公民团体和环境巡视员的地位尤其特殊。在环境影响评价立法时，《奥尔胡斯公约》(*the Aarhus Convention*)的引入也进一步提高了环评的公众参与水平(CEC，2001)。

• 环境影响评价/环境影响报告书中的一些关键内容不同，特别是与生物多样性、人类健康、风险和累积影响有关的内容。虽然指令并没有明确涉及生物多样性和人类健康影响，但这二者在环境影响评价中正变得越来越重要。荷兰和芬兰在生物多样性方面都有很好的实践，荷兰在健康影响评价上也有不错的案例。另一方面，指令修正案(附件Ⅲ)增加了风险和累积影响的内容。2003年的评审结果表明，虽然许多环境影响报告书都已经进行了风险评价，但大多数成员国仍然将风险视为独立于环境影响评价程序之外的因素，用其他的控制制度来处理。评审结果同时表明，越来越多的成员国开始重视累积影响，并采取了相应的措施（如法国、葡萄牙、芬兰、德国、瑞典和丹麦）。然而，成员国对累积影响的本质和属性仍然认识不足。

- 主管部门缺乏对项目实际影响的系统性监控。尽管人们普遍认识到这是该指令的致命缺陷，却强烈反对将监控列为强制性规定。因此，关于强制性系统性监控的成功案例很少（如荷兰）。荷兰法律规定主管部门制订一份评估计划，对报告书预测结果和实际影响进行比较。如果评估发现实际影响比预测的要糟，主管部门可责令采取额外的环保措施。在希腊，法律规定要对环境影响评价结果进行评审，将评审结果作为更新环境许可证流程的一部分。

总体来说，2003年的评审结果表明，修正指令的实施情况既有加强也有削弱。有很多成功的实践案例，而且修正案显著加强了环境影响评价的程序基础，并使环境影响评价促进了一些项目所在地区的和谐。但需要指出的是，各成员国对环境影响评价方法和环境影响评价的应用仍然存在巨大的差异，一些重大缺陷仍有待解决。评审结论中给了很多建议，包括建议各成员国尤其要加强对每年环境影响评价活动的记录；检查国家立法中与阈值、质量控制、累积影响有关的内容；更多地利用欧盟指南（如指南中关于筛选、确定范围和评审的内容）；加强环境影响评价培训等。2.8节将继续对扩大后的欧盟指令的评审和改善进行介绍。

2.8 欧盟27国的环境影响评价指令存在的问题、评审和改善

人们进一步评审了环境影响评价指令实施的有效性，并在2003年和2009年对指令进行了小幅修改。附录1给出了指令的综合版本（截至2011年6月），但下一轮协商后（2010年末）可能还会有进一步修改。对2003版指令修正案的最新一次评审着重审查了指令的实施情况，这次评审中有一个值得关注的内容，那就是2004年和2007年新加入欧盟的成员国对指令的实施。

2009年的评审对以下几个问题进行了全面而简要的审查：修正指令整体的实施情况，包括2003年新增的环境影响评价程序（主要指引入了奥尔胡斯公约中关于公众参与的要求）；指令的总体表现的变化情况；国家体系的地位；环境影响评价体系在旧成员国和新成员国的发展；和其他指令的协调性等。欧盟2003/35指令主要进行了以下几方面的修改：

- 对新增“公众”和“受影响公众”的定义（新第1.2条）；
- 过去国防项目可以免于进行环境影响评价，现在必须根据逐案分析的结果来确定是否排除（新第1.4条）；
- 增强公众咨询条款：在决策早期向公众提供详细的信息清单，制定合理的时间框架（新第6.3条）；
- 最终决策中应当包含公众参与过程的信息（第9.1条）；
- 关于公众参与评审流程的新规定[第10（a）条]；
- 变更和扩展了附件Ⅰ中的项目，修订了附件Ⅰ和附件Ⅱ中的某些项目。

2009 年版的修改幅度更小，主要是增加了一些新的项目——如碳捕获和储存装置等。总体来说，欧盟委员会对指令带来的效益给予了非常积极的评价：

> 效益主要体现在两个方面。第一，环境影响评价确保在决策过程中尽可能早地考虑环境因素；第二，通过公众参与，环境影响评价流程使得环境决策更透明、更持续、更能被社会认可。虽然环境影响评价的大部分效益都无法用货币进行衡量，但人们普遍认同进行环境影响评价的效益大于成本，现有的研究也证实了这一结论。(CEC，2009b)

环境影响评价在新增的 12 个成员国（保加利亚、塞浦路斯、捷克共和国、爱沙尼亚、匈牙利、拉脱维亚、立陶宛、马耳他、波兰、罗马尼亚、斯洛伐克和斯洛文尼亚）的实践和成果也值得关注。环境影响评价指令成为这些国家加入欧盟的要求的一部分，以确保这些国家的立法和欧盟共同体既定法体系（the EU Acquis）相协调，这些国家也为引入《奥尔胡斯公约》的要求做好了准备。从各种意义上说，这些新增国家在学习欧盟及旧成员国的环境影响评价流程和实践的发展历程时具有无可比拟的先天优势，可提供一些创造性的实践案例（参见第 10 章波兰的例子）。但同时也会遇到其他成员国在早期评审中遇到的一些问题。新成员国开展环境影响评价的数量日趋增加，但各成员国的具体情况有较大差异（虽然有些国家包含了筛选结果，但表 2.7 的数据仍不尽完善）。除了塞浦路斯和斯洛文尼亚以外，确定评价范围在所有新增成员国都是强制性的。所有新增成员国都要求由主管部门或专家委员会对环境影响评价文件进行评审，一些成员国还充分使用了欧盟指南中关于评审和筛选的内容。但仍存在如进行跨国环境影响评价、在环境影响评价中考虑人类健康保护的内容、为了达到环境影响评价的阈值要求而“分割”项目以及需要更新欧盟指南的累积影响等棘手问题。

表 2.7　新增成员国开展环境影响评价的数量

成员国	附件Ⅰ		附件Ⅱ		变化趋势
	2005	2006	2005	2006	
保加利亚	77	88	2212	2457(包括筛选阶段的项目)	增加
塞浦路斯		30		45	增加
捷克共和国		72		125	增加
爱沙尼亚		57		20	减少
匈牙利		70～90		370～400(包括筛选阶段的项目)	不变
拉脱维亚		40		5(该数字为完成环境影响评价的数字)	增加
立陶宛		12		838	不变
马耳他	4		6	增加	
波兰	不适用	不适用	不适用	不适用	不适用
罗马尼亚		179		643	增加
斯洛伐克	90	135	429	363	不变
斯洛文尼亚	不适用	不适用	不适用	不适用	不适用

资料来源：CEC，2009a。

尽管环境影响评价在实施20多年来取得了不错的进展，但仍然有一些悬而未决的问题，包括上文已经提到的——例如不同的筛选标准、跨界问题以及对质量控制的担忧。一个长久以来的严重问题是缺乏强制性的监控要求；另一个近期出现的问题是环境影响评价和战略环境影响评价指令的关系，二者在实践中含糊不清。此外，人们最近认识到需要尽快在环境影响评价的要求和指南中增加气候变化的内容，特别是对那些能源效率至关重要的能源和交通基础设施项目。2010年指令实施的下一轮协商里很有可能会对环境影响评价立法做出进一步修改。

2.9 总结

本章回顾了环境影响评价在世界范围内的发展，从最初在美国取得了出人意料的成功到如今在欧盟的发展。在实践中，环境影响评价具有非常丰富的内涵，从编写极为简单的专题报告到形成极其庞大而复杂的文件，从没有公众咨询到广泛的公众参与，从详细的定量预测到对未来发展趋势的宏观分析等。欧盟对环境影响评价取得的经验进行了总结，得出的结论是“总体来说，虽然环境影响评价的具体实践各不相同，但经验却未必不同，最近的举措，如出台修正案等似乎正在‘强化’法律框架，将可能促进各国环境影响评价的进一步融合”(Glasson和Bellanger，2003年)。全世界所有的环境影响评价体系都有一个共同的目的，那就是通过增加政策制定者对拟议项目的环境后果的认识，提高决策的水平。在过去40年里，环境影响评价已经成为项目规划的一个重要工具，它的应用将会更加广泛。第10章将进一步讨论世界各国的环境影响评价体系，第11章将进一步扩展到政策、规划和项目的战略环境影响评价。在下一章里，我们将着重讨论环境影响评价在英国的发展情况。

问　题

下列问题有助于读者理解本章重点，了解环境影响评价的起源和发展。

1. 为什么环境影响评价起源于《国家环境政策法案》，起源于20世纪60年代末的美国？

2. 根据《国家环境政策法案》的规定，简述无重大影响认定书和全面的环境影响报告书的编制流程有什么关键区别。

3. 如何解释过去10～15年里NEPA的环境影响报告书的编制活动的发展趋势？

4. 所有环境影响评价体系都在程序上存在一些问题。请列举近期NEPA的环境影响评价的实际操作中都出现了哪些问题。

5. 环境影响评价在各国和各大洲之间的传播模式是怎样的？

6. 为什么环境影响评价最初在英国推行时遇到了极大的阻力？

7. 促使欧洲共同体85/337号指令出台的关键因素是什么？

8. 欧盟环境影响评价指令的附件Ⅰ和附件Ⅱ中的项目有什么不同？

9. 欧盟97/11号修正案给环境影响评价指令带来了什么改变？

10. 请对2003年修正案的实施情况评审中着重强调的：(1) 成功实践，(2) 重大缺陷，举出几个实例。

11.哪些因素可以解释表2.6和表2.7中各国进行环境影响评价数量的差异?

12.什么是《奥尔胡斯公约》,它对环境影响评价指令产生了什么影响?

13.如何看待欧盟环境影响评价指令仍有一些难以解决的问题?

14.结合第2章的所学内容,现在可以充分证明欧盟27国建立了更一致的环境影响评价系统了吗?为什么?

2 Origins and development

2.1 Introduction

Environmental impact assessment was first formally established in the USA in 1969 and has since spread, in various forms, to most other countries. In the UK, EIA was initially an ad hoc procedure carried out by local planning authorities and developers, primarily for oil-and gas-related developments. A 1985 European Community directive on EIA (Directive 85/337) introduced broadly uniform requirements for EIA to all EU Member States and significantly affected the development of EIA in the UK. However, 10 years after the Directive was agreed, Member States were still carrying out widely diverse forms of EIA, contradicting the Directive's aim of 'levelling the playing field'. Amendments of 1997, 2003 and 2009 aimed to improve this situation. The nature of EIA systems (e. g. mandatory or discretionary, level of public participation, types of action requiring EIA) and their implementation in practice vary widely from country to country. However, the rapid spread of the concept of EIA and its central role in many countries' programmes of environmental protection attest to its universal validity as a proactive planning tool.

This chapter first discusses how the system of EIA evolved in the USA. The present status of EIA worldwide is then briefly reviewed (Chapter 10 will consider a number of countries' systems of EIA in greater depth). EIA in the UK and the EU are then discussed. Finally, we review the various systems of EIA in the EU Member States.

2.2 The National Environmental Policy Act and subsequent US systems

The US National Environmental Policy Act of 1969, also known as NEPA, grew out of increasing concern in the USA about widespread examples of environmental damage, vividly highlighted in the 1960s by the books *Silent Spring* (Carson 1962) and *The Population Bomb* (Ehrlich 1968). NEPA was the first legislation to require EIAs. Consequently it has become an important model for other EIA systems, both because it was a radically new form of environmental policy and because of the successes and failures of its subsequent development. Since its enactment, NEPA has resulted in the preparation of well over 25,000 full and partial EISs, which have influenced countless decisions and represent a powerful base of environmental information. On the other hand, NEPA is unique. Other countries have shied away from the form it takes and the procedures it sets out, not least because they are unwilling to face a situation like that in the USA, where there has been extensive litigation over the interpretation and workings of the EIA system.

This section covers NEPA's legislative history (i. e. the early development before it became law), the interpretation of NEPA by the courts and the Council on Environmental Quality (CEQ), the main EIA procedures arising from NEPA, and likely future developments. The reader is referred to Anderson *et al* (1984), Bear (1990), Canter (1996), CEQ (1997a), CRS (2006), Greenberg (2012), Mandelker (2000), Orloff (1980) and the annual reports of the CEQ for further information.

2. 2. 1 *Legislative history*

The National Environmental Policy Act is in many ways a fluke, strengthened by amendments that should have weakened it, and interpreted by the courts to have powers that were not originally intended. The legislative history of NEPA is interesting not only in itself but also because it explains many of the anomalies of its operation and touches on some of the major issues involved in designing an EIA system. Several proposals to establish a national environmental policy were discussed in the US Senate and House of Representatives in the early 1960s. All these proposals included some form of unified environmental policy and the establishment of a high-level committee to foster it. In February 1969, Bill S1075 was introduced in the Senate; it proposed a programme of federally funded ecological research and the establishment of a CEQ. A similar bill, HR6750, introduced in the House of Representatives, proposed the formation of a CEQ and a brief statement on national environmental policy. Subsequent discussions in both chambers of Congress focused on several points:

- The need for a declaration of national environmental policy (now Title I of NEPA).

• A proposed statement that 'each person has a fundamental and inalienable right to a healthful environment' (which would put environmental health on a par with, say, free speech). This was later weakened to the statement in § 101 (c) that 'each person should enjoy a healthful environment'.

• Action-forcing provisions similar to those then being proposed for the Water Quality Improvement Act, which would require federal officials to prepare a detailed statement concerning the probable environmental impacts of any major action; this was to evolve into NEPA's § 102 (2)(C), which requires EIA. The initial wording of the Bill had required a 'finding', which would have been subject to review by those responsible for environmental protection, rather than a 'detailed statement' subject to inter-agency review. The Senate had intended to weaken the Bill by requiring only a detailed statement. Instead, the 'detailed assessment' became the subject of external review and challenge; the public availability of the detailed statements became a major force shaping the law's implementation in its early years. NEPA became operational on 1 January 1970. Table 2. 1 summarizes its main points.

Table 2. 1 Main points of NEPA

NEPA consists of two titles. Title I establishes a national policy on the protection and restoration of environmental quality. Title II set up a three-member CEQ to review environmental programmes and progress, and to advise the president on these matters. It also requires the president to submit an annual 'Environmental Quality Report' to Congress. The provisions of Title I are the main determinants of EIA in the USA, and they are summarized here.

Section 101 contains requirements of a substantive nature. It states that the federal government has a continuing responsibility to 'create and maintain conditions under which man and nature can exist in productive harmony, and fulfil the social, economic and other requirements of present and future generations of Americans'. As such the government is to use all practicable means, 'consistent with other essential considerations of national policy', to minimize adverse environmental impact and to preserve and enhance the environment through federal plans and programmes. Finally, 'each person should enjoy a healthful environment', and citizens have a responsibility to preserve the environment.

Section 102 requirements are of a procedural nature. Federal agencies are required to make full analyses of all the environmental effects of implementing their programmes or actions. Section 102(1) directs agencies to interpret and administer policies, regulations and laws in accordance with the policies of NEPA. Section 102(2) requires federal agencies:

• To use 'a systematic and interdisciplinary approach' to ensure that social, natural and environmental sciences are used in planning and decision-making.

• To identify and develop procedures and methods so that 'presently unquantified environmental amenities and values may be given appropriate consideration in decision-making along with traditional economic and technical considerations'.

• To 'include in every recommendation or report on proposals for legislation and other *major Federal actions significantly affecting the quality of the human environment, a detailed statement* by the responsible official' on:

  • the environmental impact of the proposed action;
  • any adverse environmental effects that cannot be avoided should the proposal be implemented;
  • alternatives to the proposed action;
  • the relationship between local short-term uses of man's environment and the maintenance and enhancement of long-term productivity;
  • any irreversible and irretrievable commitments of resources that would be involved in the proposed action should it be implemented. [Emphasis added]

Section 103 requires federal agencies to review their regulations and procedures for adherence to NEPA, and to suggest any necessary remedial measures.

2.2.2 *An interpretation of NEPA*

The National Environmental Policy Act is a generally worded law that required substantial early interpretation. The CEQ, which was set up by NEPA, prepared guidelines to assist in the Act's interpretation. However, much of the strength of NEPA came from early court rulings. NEPA was immediately seen by environmental activists as a significant vehicle for preventing environmental harm, and the early 1970s saw a series of influential lawsuits and court decisions based on it. These lawsuits were of three broad types, as described by Orloff (1980):

1 Challenging an agency's decision not to prepare an EIA. This generally raised issues such as whether a project was major, federal, an 'action', or had significant environmental impacts (see NEPA § 102 (2)(C)). For instance, the issue of whether an action is federal came into question in some lawsuits concerning the federal funding of local government projects. [1]

2 Challenging the adequacy of an agency's EIS. This raised issues such as whether an EIS adequately addressed alternatives, and whether it covered the full range of significant environmental impacts. A famous early court case concerned the Chesapeake Environmental Protection Association's claim that the Atomic Energy Commission did not adequately consider the water quality impacts of its proposed nuclear power plants, particularly in the EIA for the Calvert Cliffs power plant. [2] The Commission argued that NEPA merely required the consideration of water quality standards; opponents argued that it required an assessment beyond mere compliance with standards. The courts sided with the opponents.

3 Challenging an agency's substantive decision, namely its decision to allow or not to allow a project to proceed in the light of the contents of its EIS. Another influential early court ruling laid down guidelines for the judicial review of agency decisions, noting that the court's only function was to ensure that the agency had taken a 'hard look' at environmental consequences, not to substitute its judgement for that of the agency. [3]

The early proactive role of the courts greatly strengthened the power of environmental movements and caused many projects to be stopped or substantially amended. In many cases the lawsuits delayed construction for long enough to make them economically unfeasible or to allow the areas where projects would have been sited to be designated as national parks or wildlife areas (Turner 1988). More recent decisions have been less clearly pro-environment than the earliest decisions. The flood of early lawsuits, with the delays and costs involved, was a lesson to other countries in how *not* to set up an EIA system. As will be shown later, many countries carefully distanced their EIA systems from the possibility of lawsuits.

The CEQ was also instrumental in establishing guidelines to interpret NEPA, producing interim guidelines in 1970, and guidelines in 1971 and 1973. Generally the courts adhered closely to these guidelines when making their rulings. However, the guidelines were problematic: they were not detailed enough, and were interpreted by the federal agencies as being discretionary rather than binding. To combat these limitations, President Carter issued Executive Order 11992 in 1977, giving the CEQ authority to set enforceable regulations for implementing NEPA. These were issued in 1978 (CEQ 1978) and sought to make the NEPA process more useful for decision-makers and the public, to reduce paperwork and delay and to emphasize real environmental issues and alternatives.

2.2.3 A summary of NEPA procedures

The process of EIA established by NEPA, and developed further in the CEQ regulations, is summarized in Figure 2.1. The following citations are from the CEQ regulations (CEQ 1978).

> [The EIA process begins] as close as possible to the time the agency is developing or is presented with a proposal... The statement shall be prepared early enough so that it can serve practically as an important contribution to the decision-making process and

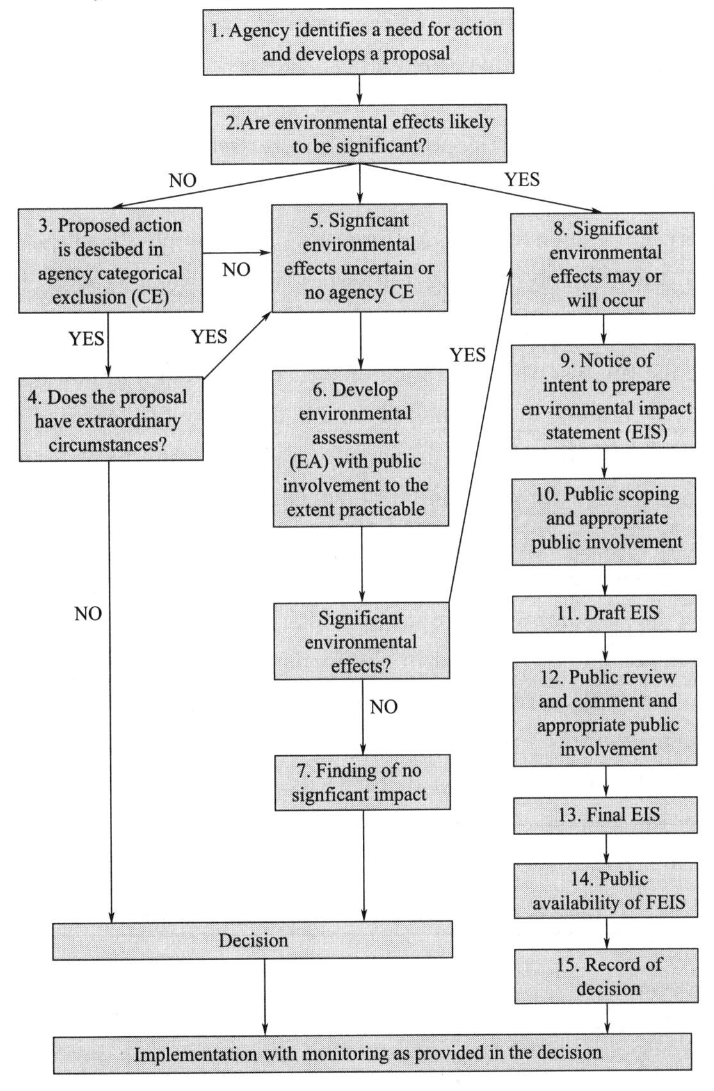

Figure 2.1 Process of EIA under NEPA

Source: CEQ 2007

will not be used to rationalize or justify decisions already made. (§1502.5)

A *'lead agency' is designated* that co-ordinates the EIA process. The lead agency first determines whether the proposal requires the preparation of a full EIS, no EIS at all, or a 'finding of no significant impact' (FONSI). This is done through a series of tests. A first test is whether a federal action is likely to individually or cumulatively have a significant environmental impact. All federal agencies have compiled lists of 'categorical exclusions' that are felt not to have such impacts. If an action is on such a list, then no further EIA action is generally needed. If an action is not categorically excluded, an 'environmental assessment' is carried out to determine whether a full EIS or a FONSI is needed. A FONSI is a public document that explains why the action is not expected to have a significant environmental impact.

If a FONSI is prepared, then a permit would usually be granted following public discussion. If a full EIS is found to be needed, the lead agency publishes a 'Notice of intent', and the *process of scoping begins*. The aim of the scoping exercise is to determine the issues to be addressed in the EIA: to eliminate insignificant issues, focus on those that are significant and identify alternatives to be addressed. The lead agency invites the participation of the proponent of the action, affected parties and other interested persons.

> [The alternatives] section is the heart of the environmental impact statement ... [It] should present the environmental impacts of the proposal and the alternatives in comparative form, thus sharply defining the issues and providing a clear basis for choice ... (§1502.14)

The agency must analyse the full range of direct, indirect and cumulative effects of the preferred alternative, if any, and of the reasonable alternatives identified in the process. For the purpose of NEPA, 'effects' and 'impacts' mean the same thing, and include ecological, aesthetic, historic, cultural, economic, social or health impacts, whether adverse or beneficial.

A draft EIS is then prepared, and is reviewed and commented on by the relevant agencies and the public. These comments are taken into account in the subsequent preparation of a final EIS. An EIS is normally presented in the format shown in Table 2.2. In an attempt to be comprehensive, early EISs tended to be so bulky as to be virtually unreadable. The CEQ guidelines consequently emphasize the need to concentrate only on important issues and to prepare readable documents: 'The text of final environmental impact statements shall normally be less than 150 pages ... Environmental impact statements shall be written in plain language...' (§1502.7-8).

Table 2.2 Typical format for an EIS under NEPA

(a) Cover sheet
- list of responsible agencies
- title of proposed action
- contact persons at agencies
- designation of EIS as draft, final or supplement
- abstract of EIS
- date by which comments must be received

Continued

(b) Summary (usually 15 pages or less)
- major conclusions
- areas of controversy
- issues to be resolved

(c) Table of contents
(d) Purpose of and need for action
(e) Alternatives, including proposed action
(f) Affected environment
(g) Environmental consequences
- environmental impacts of alternatives, including proposed action
- adverse environmental effects that cannot be avoided if proposal is implemented
- mitigation measures to be used and residual effects of mitigation
- relation between short-term uses of the environment and maintenance and enhancement of long-term productivity
- irreversible or irretrievable commitments of resources if proposal is implemented discussion of:

—direct, indirect and cumulative effects and their significance
—possible conflicts between proposed action and objectives of relevant land-use plans, policies and controls
—effects of alternatives, including proposed action
—energy requirements and conservation potential of various alternatives and mitigation measures
—natural or depletable resource requirements and conservation of various alternatives and mitigation measures
—effects on urban quality, historic and cultural resources, and built environment
—means to mitigate adverse impacts
(h) Monitoring arrangements, and environmental management system
(i) List of preparers
(j) List of agencies, etc. to which copies of EIS are sent
(k) Index
(l) Appendices, including supporting data

The public is involved in this process, both at the scoping stage and after publication of the draft and final EISs:

> Agencies shall: (a) Make diligent efforts to involve the public in preparing and implementing NEPA procedures … (b) Provide public notice of NEPA-related hearings, public meetings and the availability of environmental documents … (c) Hold or sponsor public hearings … whenever appropriate… (d) Solicit appropriate information from the public. (e) Explain in its procedures where interested persons can get information or status reports … (f) Make environmental impact statements, the comments received, and any underlying documents available to the public pursuant to the provisions of the Freedom of Information Act … (§ 1506. 6)

Finally, a decision is made about whether the proposed action should be permitted:

> Agencies shall adopt procedures to ensure that decisions are made in accordance with the policies and purposes of the Act. Such procedures shall include but not be limited to: (a) Implementing procedures under section 102 (2) to achieve the requirements of sections 101 and 102 (1) … (e) Requiring that … the decision-maker consider the alternatives described in the environmental impact statement. (§ 1505. 1)

Where all relevant agencies agree that the action should not go ahead, permission is denied, and a judicial resolution may be attempted. Where agencies agree that the action can proceed, permission is given, possibly subject to specified conditions (e. g. monitoring, mitigation). Where the relevant agencies disagree, the CEQ acts as arbiter (§ 1504). Until a decision is made, 'no action concerning the proposals shall be taken which could: (1) have an adverse environmental impact; or (2) limit the choice of reasonable alternatives ... ' (§ 1506. 1).

The Record of Decision (ROD) is the final step for agencies in the process. It states what the decision is; identifies the alternatives considered, including the environmentally preferred alternative; and discusses mitigation plans, including any enforcement and monitoring commitments. It also discusses if all practical means to avoid or minimize environmental harm have been adopted and, if not, why they were not. It is a publicly available document published in the Federal Register or on the website of the relevant agency.

2. 2. 4 *Recent trends*

EIS activity

During the first 10 years of NEPA's implementation, about 1,500 EISs were prepared annually. Subsequently, negotiated improvements to the environmental impacts of proposed actions have become increasingly common during the preparation of 'environmental assessments' (EA). This has led to many 'mitigated findings of no significant impact' (no perfect acronym exists for this), reducing the number of EISs prepared: whereas 1,273 EISs were prepared in 1979, only 456 were prepared in 1991 and the annual number has been approximately 500—550 in recent years (NEPA website). This trend can be viewed positively, since it means that environmental impacts are considered earlier in the decision-making process, and hence it reduces the costs of preparing EISs. However, the fact that this abbreviated process allows less public participation causes some concern. Table 2. 3 summarizes activity for 2008, indicating the predominance of EISs filed by the Department of Agriculture (primarily for forestry projects) and the Department of Transportation (primarily for road construction). Between 1979 and 2008, the number of EISs filed by the Department of Housing and Urban Development fell from 170 to 0! The number of legal cases filed against federal departments and agencies on the basis of NEPA has been on average 100—150 per annum. Plaintiffs are mainly public interest groups and individual citizens; common complaints are 'no EIS when one should have been prepared' and 'inadequate EIS'.

It is important to set this EIS activity in context. Important though they are, including some of the projects with the greatest impacts and highest stakeholder interest, the EISs represent only the tip of the iceberg of the wider EA activity. By comparison, in 1997 CEQ reported that federal agencies in total estimated that approximately 50,000 EAs were produced annually (CEQ 1997a). Determining the total number of federal actions subject to NEPA is difficult, but one agency, the Federal Highway Administration (FHWA), has tracked all projects. In 2001, FHWA estimated that 3 per cent of projects required an EIS, 7 per cent required an EA, and 90 per cent of all highway projects were classified as categorical exclusions.

Table 2.3 EISs filed under NEPA in 2008

Department	Number of EISs filed	Of which
Agriculture	128	Forest Service:124
Commerce	36	National Oceanic and Atmospheric Admin:36
Defense	79	Corp of Engineers:42;Navy:24
Energy	36	Federal Energy Regulatory Commission:19
Health and human services	1	
Homeland security	8	US Coast Guard:6
Housing and urban development	0	
Interior	128	Bureau of Land Management:48;National Parks Service:25
Justice	2	
Labor	0	
State	3	
Transportation	104	Federal Highway Admin:64
Treasury	0	
Veteran affairs	0	
Independent agency	25	Nuclear Regulatory Commission:14
TOTAL	550	

Source: NEPA website

System review

The National Environmental Policy Act's twentieth year of operation, 1990, was marked by a series of conferences on the Act and the presentation to Congress of a bill of NEPA amendments. Under the Bill (HR1113), which was not passed, federal actions that take place outside the USA (e.g. projects built in other countries with US federal assistance) would have been subject to EIA, and all EISs would have been required to consider global climatic change, the depletion of the ozone layer, the loss of biological diversity and transboundary pollution. This latter amendment was controversial: although the need to consider the global impacts of programmes was undisputed, it was felt to be infeasible at the level of project EIA.

The context of EIA has also become a matter of concern. EIA is only one part of a broader environmental policy (NEPA), but the procedural provisions set out in NEPA's § 102 (2)(C) have overshadowed the rest of the Act. It has been argued that mere compliance with these procedures is not enough, and that greater emphasis should be given to the environmental goals and policies stated in § 101. EIA must also be seen in the light of other environmental legislation. In the USA, many laws dealing with specific aspects of the environment were enacted or strengthened in the 1970s, including the Clean Water Act and the Clean Air Act. These laws have in many ways superseded NEPA's substantive requirements and have complemented and buttressed its procedural requirements. Compliance with these laws does not necessarily imply compliance with NEPA. However, the permit process associated with these other laws has become a primary method for evaluating project impacts, reducing NEPA's importance except for its occasional role as a focus of debate on major projects (Bear 1990).

The scope of EIA, and in particular the recognition of the social dimension of the environment, has been another matter of concern. After long campaigning by black and ethnic groups, particularly about inequalities in the distribution of hazardous waste landfills and incinerators, a working group was set up within the Environmental Protection Act (EPA) to make recommendations for dealing with environmental injustice (Hall 1994). The outcome was the Clinton 'Executive Order on Federal Actions to Address Environmental Justice in Minority Populations and Low-Income Populations' (White House 1994). Under this Order, each federal agency must analyse the environmental effects, including human health, economic and social effects, of federal actions, including effects on minority and low-income communities, when such analysis is required under NEPA. Mitigation measures, wherever feasible, should also address the significant and adverse environmental effects of federal actions on the same communities. In addition, each federal agency must provide opportunities for communities to contribute to the NEPA process, identifying potential effects and mitigation measures in consultation with affected communities and improving the accessibility of meetings and crucial documents.

Discussion of issues, NEPA effectiveness, and amendments to the EA process, has continued to date. Canter (1996) highlighted four areas for which NEPA requirements needed further elaboration:

1 how much an agency should identify and plan mitigation before issuing an EIS;
2 ways to assess the cumulative impacts of proposed developments;
3 ways to conduct 'reasonable foreseeability' (or worst-case) analyses; and
4 monitoring and auditing of impact predictions.

In 1997 CEQ carried out a review of the effectiveness of NEPA after 25 years. Overall the view was that NEPA had been a success:

> ...it had made agencies take a hard look at the environmental consequences of their actions, and it had brought the public into the agency decision-making process like no other statute. In a piece of legislation barely three pages long, NEPA gave a voice to the new national consensus to protect and improve the environment, and substance to the determination articulated by many to work together to achieve that goal. (CEQ 1997a)

But there was concern about several features: a focus on the documentation, rather than on the enhancement of decision making; an associated focus on litigation-proof documentation; consultation that was too late in the process; and an overlong process. Some recommendations for the future included better interagency co-ordination; and better monitoring and adaptive management. There was also a concern for a more integrated interdisciplinary place-based approach bringing together expertise and information from many fields, as illustrated in the example in Box 2. 1.

Box 2.1 Example of an integrated and accelerated approach—the Ozark Mountain Highroad

Branson, Missouri is one of the hottest entertainment centers in the country, receiving more than 3.7 million visitors during the six month tourist season in 1991. At peak times, 30,000 cars are jammed onto Country Music Boulevard each day, resulting in average speeds of 10 mph for much of the day and intolerable delays. In early June, 1992, the governor of Missouri declared the traffic congestion in the Branson area an "economic emergency" and announced a plan to fast-track the planning and design process of a proposed four-lane $160 million Ozark Mountain Highroad. The challenge to the Missouri Highway and Transportation Department was to plan a totally new highway in six months without compromising safety or the integrity of the environmental process. With the fast track in mind and the NEPA process in hand, an interdisciplinary team of agencies met on a regular and frequent basis. This resulted in the preparation of a quality project that integrated the needs of the environmentally sensitive Ozark Mountain Ecosystem with the need for increased recreational traffic in the area. With all the players and disciplines involved, every reasonable design alternative and associated impact was on the table for discussion.

There were those on the team who, in the past, had seen NEPA as a burden, a hindrance, and something to be overcome. But as a result of the Highroad experience, these same people came to realize that NEPA could help to shape projects in a way that met the project purpose and need while serving to protect the environment and preserve other community values. Most important, the new attitudes forged during the NEPA planning process have carried over into other projects that involve the same local, state, and federal agencies, and consulting firms.

Source: CEQ 1997a

There has been progress in a number of areas, including the consideration of: cumulative impacts, the relationship between the EIS and environmental management systems (EMS), and approaches to the 'streamlining' of the overall system. Although the original NEPA does define the nature of cumulative impacts, agencies have been left very much to their own devices to develop relevant procedures and methods. However, in 1997 (non-legally binding) guidance for the consideration of cumulative effects assessment under the NEPA was provided by the Council for Environmental Quality. The CEQ handbook presents practical methods for addressing coincident effects (adverse or beneficial) on specific resources, ecosystems, and human communities of all related activities, not just the proposed project or alternatives that initiate the assessment process (CEQ 1997b). Although NEPA affirms a 'predict-mitigate-implement' model, there can be major weaknesses in the implementation stage. The issuance of an Executive Order (Exec. Order 13423, January 2007), which directs all Federal agencies to implement EMSs at all organizational levels, helps to address this issue, providing a means to enhance EIS compliance. CEQ has again issued relevant guidance, on *Aligning NEPA Processes with Environmental Management*

Systems (CEQ 2007). There is also a general and ongoing concern about the time consumed by the NEPA processes, accompanied by calls for more 'streamlining'. A Congressional Research Service report identified two main categories of delay attributed to the NEPA process: those related to the time needed to complete the required documentation (primarily the EISs), and delays resulting from NEPA-related litigation (CRS 2006). The report discusses a range of responses, including limits on the length of EISs, establishing limits on judicial review, updating lists of Categorically Excluded and Exempt projects, and many others, some of which have been implemented while others are still subject to debate!

2.2.5 *Little NEPAs and the particular case of California*

Many state-level EIA systems have been established in the USA in addition to NEPA. Fifteen of the USA's states, plus the District of Columbia and Puerto Rico,[4] have their own EIA systems that, because they are largely modelled on the Federal NEPA, are collectively referred to as the 'little NEPAs'. They require EIA for state actions (actions that require state funding or permission) and/or projects in sensitive areas. Other states have no specific EIA regulations, but have EIA requirements in addition to those of NEPA.[5]

Of particular interest is the Californian system, established under the California Environmental Quality Act (CEQA) of 1973, and subsequent amendments. This is widely recognized as one of the most advanced EIA systems in the world. The legislation applies not only to government actions but also to the activities of private parties that require the approval of a government agency. It is not merely a procedural approach but one that requires state and local agencies to protect the environment by adopting feasible mitigation measures and alternatives in environmental impact reviews (EIRs). The legislation extends beyond projects to higher levels of actions, and an amendment in 1989 also added mandatory mitigation, monitoring and reporting requirements to CEQA. Guidance on the California system is provided in invaluable annual publications by the State of California, which sets out the CEQA Statutes and Guidelines in considerable detail. For the latest 2009 update of the Statute and Guidelines see the Association of Environmental Professionals, California website. A further amendment in 2010 added GHG emissions to the list of environmental impacts that must be analysed under CEQA.

2.3 The worldwide spread of EIA

Since the enactment of NEPA, EIA systems have been established in various forms throughout the world, beginning with more developed countries (e.g. Canada in 1973, Australia in 1974, West Germany in 1975, France in 1976) and later also in the less developed countries. The approval of a European Directive on EIA in 1985 stimulated the enactment of EIA legislation in many European countries in the late 1980s. The formation of new countries after the break-up of the Soviet Union in 1991 led to the enactment of EIA legislation in many of these countries in the early to mid-1990s. The early 1990s also saw a large growth in the number of EIA regulations and guidelines established in Africa and South America. By 1996, more than 100 countries had EIA systems (Sadler 1996).

These EIA systems vary greatly. Some are in the form of *mandatory regulations*, *acts* or *statutes*; these are generally enforced by the authorities requiring the preparation of an adequate EIS before permission is given for a project to proceed. In other cases, EIA *guidelines* have been established. These are not enforceable but generally impose obligations on the administering agency. Other legislation allows government officials to require EIAs to be prepared at their *discretion*. Elsewhere, EIAs are prepared in an ad hoc manner, often because they are required by funding bodies (e. g. the World Bank, USAID) as part of a funding approval process. However, these classifications are not necessarily indicative of how thoroughly EIA is carried out. For instance, the EIA regulations of Brazil and the Philippines are not well carried out or enforced in practice (Glasson and Salvador 2000; Moreira 1988), whereas Japan's guidelines are thoroughly implemented.

Another important distinction between types of EIA system is that sometimes the actions that require EIA are given as *a definition* (e. g. the USA's definition of 'major federal actions significantly affecting the quality of the human environment'), sometimes as *a list of projects* (e. g. roads of more than 10 km in length). Most countries use a list of projects, in part to avoid legal wrangling such as that surrounding NEPA's definition. Another distinction asks whether EIA is required for *government projects only* (as in NEPA), for *private projects only* or for both. Finally, some international development and funding agencies have set up EIA guidelines, including the European Bank for Reconstruction and Development (1992, 2010), UK Overseas Development Administration (1996), DFID (2003), UNEP (1997, 2002) and World Bank (1991, 1995, 1997, 1999, 2002, 2006).

2.4 Development in the UK

The UK has had formal legislation for EIA since 1988, in the form of several laws that implement European Community Directive 85/337/EEC (CEC 1985) and subsequent amendments. It is quite possible that without pressure from the European Commission such legislation would have been enacted much more slowly, since the UK government felt that its existing planning system more than adequately controlled environmentally unsuitable developments. However, this does not mean that the UK had no EIA system at all before 1988; many EIAs were prepared voluntarily or at the request of local authorities, and guidelines for EIA preparation were drawn up.

2.4.1 Limitations of the land-use planning system

The UK's statutory land-use planning system has since 1947 required local planning authorities (LPAs) to anticipate likely development pressures, assess their significance, and allocate land, as appropriate, to accommodate them. Environmental factors are a fundamental consideration in this assessment. Most developments require planning consent, so environmentally harmful developments can be prevented by its denial. This system resulted in the accumulation of considerable planning expertise concerning the likely consequences of development proposals.

After the mid-1960s, however, the planning system began to seem less effective at controlling the impacts of large developments. The increasing scale and complexity of developments, the consequent greater social and physical environmental impacts and the growing internationalization of developers (e. g. oil, gas and chemicals companies) all outstripped the capability of the development control system to predict and control the impacts of developments. In the late 1960s, public concern about environmental protection also grew considerably, and the relation between statutory planning controls and the development of large projects came under increasing scrutiny. This became particularly obvious in the case of the proposed third London Airport. The Roskill Commission was established to select the most suitable site for an airport in southeast England, with the mandate to prepare a cost-benefit analysis (CBA) of alternative sites. The resulting analysis (HMSO 1971) focused on socio-economic rather than physical environmental impacts; it led to an understanding of the difficulties of expanding CBA to impacts not easily measured in monetary terms, and to the realization that other assessment methods were needed to achieve a balance between socio-economic and physical environmental objectives.

2.4.2 North Sea oil-and gas-related EIA initiatives

The main impetus towards the further development of EIA, however, was the discovery of oil and gas in the North Sea. The extraction of these resources necessitated the construction of large developments in remote areas renowned for their scenic beauty and distinctive ways of life (e. g. the Shetlands, the Orkneys and the Highlands region). Planning authorities in these areas lacked the experience and resources needed to assess the impacts of such large developments. In response, the Scottish Development Department (SDD) issued a technical advice note to LPAs (SDD 1974). *Appraisal of the impact of oil-related development* noted that these developments and other large and unusual projects need 'rigorous appraisal', and suggested that LPAs should commission an impact study of the developments if needed. This was the first government recognition that major developments needed special appraisal. Some EIAs were carried out in the early 1970s, mostly for oil and gas developments. Many of these were sponsored by the SDD and LPAs, and were prepared by environmental consultants, but some (e. g. for the Flotta Oil Terminal and Beatrice Oilfield) were commissioned by the developers. Other early EIAs concerned a coal mine in the Vale of Belvoir, a pumped-storage electricity scheme at Loch Lomond, and various motorway and trunk road proposals (Clark and Turnbull 1984).

In 1973, the Scottish Office and DoE commissioned the University of Aberdeen's Project Appraisal for Development Control (PADC) team to develop a systematic procedure for planning authorities to make a balanced appraisal of the environmental, economic and social impacts of large industrial developments. PADC produced an interim report, *The assessment of major industrial applications—a manual* (Clark *et al.* 1976), which was issued free of charge to all LPAs in the UK and 'commended by central government for use by planning authorities, government agencies and developers'. The PADC procedure was designed to fit into the existing planning framework, and was used to assess a variety of (primarily private sector) projects. An extended and updated version of the manual was issued in 1981 (Clark *et al.* 1981).

In 1974, the Secretaries of State for the Environment, Scotland and Wales commissioned consultants to investigate the 'desirability of introducing a system of impact analysis in Great Britain, the circumstances in which a system should apply, the projects it should cover and the way in which it might be incorporated into the development control system' (Catlow and Thirwall 1976). The resulting report made recommendations about who should be responsible for preparing and paying for EIAs, what legislative changes would be needed to institute an EIA system, and similar issues. The report concluded that about 25—50 EIAs per year would be needed, for both public and private sector projects. EIA was given further support by the Dobry Report on the development control system (Dobry 1975), which advocated that LPAs should require developers to submit impact studies for particularly significant development proposals. The report outlined the main topics such a study should address, and the information that should be required from developers. Government reactions to the Dobry Report were mixed, although the influential Royal Commission on Environmental Pollution endorsed the report.

2.4.3 Department of the Environment scepticism

However, overall the DoE remained sceptical about the need, practicality and cost of EIA. In fact, the government's approach to EIA was described as being 'from the outset grudging and minimalist' (CPRE 1991). In response to the Catlow and Thirwall report, the DoE stated: 'Consideration of the report by local authorities should not be allowed to delay normal planning procedures and any new procedures involving additional calls on central or local government finance and manpower are unacceptable during the present period of economic restraint' (DoE 1977). A year later, after much deliberation, the DoE was slightly more positive:

> We fully endorse the desirability ... of ensuring careful evaluation of the possible effects of large developments on the environment ... The approach suggested by Thirwall/Catlow is already being adopted with many [projects] ... The sensible use of this

approach [should] improve the practice in handling these relatively few large and significant proposals. (DoE 1978)

The government's foreword to the PADC manual of 1981 also emphasized the need to minimize the costs of EIA procedures: 'It is important that the approach suggested in the report should be used selectively to fit the circumstances of the proposed development and with due economy' (Clark *et al*. 1981). As will be seen in later chapters, the government remained sceptical for some time about the value of EIA, and about extending its remit, as suggested by the EC. But by the early 1980s, more than 200 studies on the environmental impacts of projects in the UK had been prepared on an ad hoc basis (Petts and Hills 1982). Many of these studies were not full EIAs, but focused on only a few impacts. However, large developers such as British Petroleum, British Gas, the Central Electricity Generating Board and the National Coal Board were preparing a series of increasingly comprehensive statements. In the case of British Gas, these were shown to be a good investment, saving the company £30 million in 10 years (House of Lords 1981a).

2.5 EC Directive 85/337

The development and implementation of Directive 85/337 greatly influenced the EIA systems of the UK and other EU Member States. In the UK, central government research on a UK system of EIA virtually stopped after the mid-1970s, and attention focused instead on ensuring that any future Europe-wide system of EIA would fully incorporate the needs of the UK for flexibility and discretion. Other Member States were eager to ensure that the Directive reflected the requirements of their own more rigorous systems of EIA. Since the Directive's implementation, EIA activity in all the EU Member States has increased dramatically.

2.5.1 *Legislative history*

The EC had two main reasons for wanting to establish a uniform system of EIA in all its Member States. First, it was concerned about the state of the physical environment and eager to prevent further environmental deterioration. The EC's First Action Programme on the Environment of 1973 (CEC 1973) advocated the prevention of environmental harm: 'the best environmental policy consists of preventing the creation of pollution or nuisances at source, rather than subsequently trying to counteract their effects', and, to that end, 'effects on the environment should be taken into account at the earliest possible stage in all technical planning and decision-making processes'. Further Action Programmes of 1977, 1983, 1987, 1992 and 2001 have reinforced this emphasis. Land-use planning was seen as an important way of putting these principles into practice, and EIA was viewed as a crucial technique for incorporating environmental considerations into the planning process.

Second, the EC was concerned to ensure that no distortion of competition should arise through which one Member State could gain unfair advantage by permitting developments that, for environmental reasons, might be refused by another. In other words, it considered environmental policies necessary for the maintenance of a level economic playing field. Further motivation for EC action included a desire to encourage best practice across Member States. In addition, pollution problems transcend territorial boundaries (witness acid rain and river pollution in Europe), and the EC can contribute at least a sub-continental response framework.

The EC began to commission research on EIA in 1975. Five years later and after more than 20 drafts, the EC presented a draft directive to the Council of Ministers (CEC 1980); it was circulated throughout the Member States. The 1980 draft attempted to reconcile several conflicting needs. It sought to benefit from the US experience with NEPA, but to develop policies appropriate to European need. It also sought to make EIA applicable to all actions likely to have a significant environmental impact, but to ensure that procedures would be practicable. Finally, and perhaps most challenging, it sought to make EIA requirements flexible enough to adapt to the needs and institutional arrangements of the various Member States, but uniform enough to prevent problems arising from widely varying interpretations of the procedures. The harmonization of the types of project to be subject to EIA, the main obligations of the developers and the contents of the EIAs were considered particularly important (Lee and Wood 1984; Tomlinson 1986).

As a result, the draft directive incorporated a number of important features. First, planning permission for projects was to be granted only after an adequate EIA had been completed. Second, LPAs and developers were to cooperate in providing information on the environmental impacts of proposed developments. Third, statutory bodies responsible for environmental issues, and other Member States in cases of transfrontier effects, were to be consulted. Finally, the public was to be informed and allowed to comment on issues related to project development.

In the UK the draft directive was examined by the House of Lords Select Committee on the European Commission, where it received widespread support:

> The present draft Directive strikes the right kind of balance: it provides a framework of common administrative practices which will allow Member States with effective planning controls to continue with their system ... while containing enough detail to ensure that the intention of the draft cannot be evaded ... The Directive could be implemented in the United Kingdom in a way which would not lead to undue additions delay and costs in planning procedures and which need not therefore result in economic and other disadvantages. (House of Lords 1981a)

However, the Parliamentary Undersecretary of State at the DoE dissented. Although accepting the general need for EIA, he was concerned about the bureaucratic hurdles, delaying objections and litigation that would be associated with the proposed directive (House of

Lords 1981b). The UK Royal Town Planning Institute (RTPI) also commented on several drafts of the directive. Generally the RTPI favoured it, but was concerned that it might cause the planning system to become too rigid:

> The Institute welcomes the initiative taken by the European Commission to secure more widespread use of EIA as it believes that the appropriate use of EIA could both speed up and improve the quality of decisions on certain types of development proposals. However, it is seriously concerned that the proposed Directive, as presently drafted, would excessively codify and formalize procedures of which there is limited experience and therefore their benefits are not yet proven. Accordingly the Institute recommends the deletion of Article 4 and annexes of the draft. (House of Lords 1981a)

More generally, slow progress in the implementation of EC legislation was symptomatic of the wide range of interest groups involved, of the lack of public support for increasing the scope of town planning and environmental protection procedures, and of the unwillingness of Member States to adapt their widely varying planning systems and environmental protection legislation to those of other countries (Williams 1988). In March 1982, after considering the many views expressed by the Member States, the Commission published proposed amendments to the draft directive (CEC 1982). Approval was further delayed by the Danish government, which was concerned about projects authorized by Acts of Parliament. On 7 March 1985, the Council of Ministers agreed on the proposal; it was formally adopted as a directive on 27 June 1985 (CEC 1985) and became operational on 3 July 1988.

Subsequently, the EC's Fifth Action Programme, *Towards sustainability* (CEC 1992), stressed the importance of EIA, particularly in helping to achieve sustainable development, and the need to expand the remit of EIA:

> Given the goal of achieving sustainable development it seems only logical, if not essential, to apply an assessment of the environmental implications of all relevant policies, plans and programmes. The integration of environmental assessment within the macroplanning process would not only enhance the protection of the environment and encourage optimization of resource management but would also help to reduce those disparities in the international and interregional competition for new development projects which at present arise from disparities in assessment practices in the Member States ...

The EIA Directive can be seen as a work in progress. It has undergone regular reviews to improve procedures and to seek consistency of application. In response to a five-year review of the Directive (CEC 1993), amendments to the Directive were agreed in 1997. Further minor amendments followed in 2003 and 2009. Appendix 1 gives the complete consolidated version of the amended Directive as at June 2011. The rest of Section 2.5 now summarizes the original Directive. Section 2.6 discusses the important amendments in Directive 97/11. Section 2.7 considers in particular the findings of a major review of the implementation of the implementation of the amended Directive 97/11. This is followed in Section 2.8 with a summary of the more minor reviews and amendments in 2003, and in 2009; but importantly now set in the context of the enlarged EU-27.

2.5.2 *Summary of EC Directive 85/337 procedures*

The Directive differs in important respects from NEPA. It requires EIAs to be prepared by both public agencies and private developers, whereas NEPA applies only to federal agencies. It requires EIA for a specified list of projects, whereas NEPA uses the definition 'major federal actions'. It specifically lists the impacts that are to be addressed in an EIA, whereas NEPA does not. Finally, it includes fewer requirements for public consultation than does NEPA. Under the provisions of the European Communities Act of 1972, Directive 85/337 is the controlling document, laying down rules for EIA in Member States. Individual states enact their own regulations to implement the Directive and have considerable discretion. According to the Directive, EIA is required for two classes of project, one mandatory (Annex Ⅰ) and one discretionary (Annex Ⅱ):

> projects of the classes listed in Annex Ⅰ shall be made subject to an assessment … for projects listed in Annex Ⅱ, the Member States shall determine through: (a) a case-by-case examination; or (b) thresholds or criteria set by the Member State whether the project shall be made subject to an assessment … When [doing so], the relevant selection criteria set out in Annex Ⅲ shall be taken into account. (Article 4)

Table 2.4 summarizes the projects listed in Annexes Ⅰ and Ⅱ. The EC (CEC 1995) also published guidelines to help Member States determine whether a project requires EIA.

Table 2.4 Projects requiring EIA under EC Directive 85/337 (*as amended*) *

Annex Ⅰ (mandatory)	
1	Crude oil refineries, coal/shale gasification and liquefaction
2	Thermal power stations and other combustion installations; nuclear power stations and other nuclear reactors
3	Radioactive waste processing and/or storage installations
4	Cast-iron and steel smelting works
5	Asbestos extraction, processing or transformation
6	Integrated chemical installations
7	Construction of motorways, express roads, other large roads, railways, airports
8	Trading ports and inland waterways
9	Installations for incinerating, treating or disposing of toxic and dangerous wastes
10	*Large-scale installation for incinerating or treating non-hazardous waste*
11	*Large-scale groundwater abstraction or recharge schemes*
12	*Large-scale transfer of water resources*
13	*Large-scale waste water treatment plants*
14	*Large-scale extraction of petroleum and natural gas*
15	*Large dams and reservoirs*
16	*Long pipelines for gas, oil or chemicals*
17	*Large-scale poultry or pig-rearing installations*
18	*Pulp, timber or board manufacture*
19	*Large-scale quarries or open-cast mines*
20	*Long overhead electrical power lines*
21	*Large-scale installations for petroleum, petrochemical or chemical products*
22	*Any change or extension to an Annex I project that meets the thresholds***

Continued

23 *Carbon storage sites***
24 *Carbon capture installations***
Annex Ⅱ (discretionary)
1 Agriculture, silviculture and aquaculture
2 Extractive industry
3 Energy industry
4 Production and processing of metals
5 *Minerals industry* (projects not included in Annex Ⅰ)
6 Chemical industry
7 Food industry
8 Textile, leather, wood and paper industries
9 Rubber industry
10 *Infrastructure projects*
11 *Other projects*
12 *Tourism and leisure*
13 Modification, extension or temporary testing of Annex I projects

* 1997 amendments are shown in italic; ** are from later amendments in 2003 and 2009.

Similarly, the information required in an EIA is listed in Annex Ⅲ of the Directive, but must only be provided

> inasmuch as: (a) The Member States consider that the information is relevant to a given stage of the consent procedure and to the specific characteristics of a particular project … and of the environmental features likely to be affected; (b) The Member States consider that a developer may reasonably be required to compile this information having regard *inter alia* to current knowledge and methods of assessment. (Article 5. 1)

Table 2. 5 summarizes the information required by Annex Ⅲ (Annex Ⅳ, post-amendments). A developer is thus required to prepare an EIS that includes the information specified by the relevant Member State's interpretation of Annex Ⅲ (Annex Ⅳ, post-amendments) and to submit it to the 'competent authority'. This EIS is then circulated to other relevant public authorities and made publicly available: 'Member States shall take the measures necessary to ensure that the authorities likely to be concerned by the project … are given an opportunity to express their opinion' (Article 6. 1).

Table 2. 5 Information required in an EIA under EC Directive 85/337 (*as amended*)*

Annex Ⅲ (Ⅳ)
1 Description of the project
2 Where appropriate (*an outline of main alternatives studied and an indication of the main reasons for the final choice*)
3 Aspects of the environment likely to be significantly affected by the proposed project, including population, fauna, flora, soil, water, air climatic factors, material assets, architectural and archaeological heritage, landscape, and the interrelationship between them
4 Likely significant effects of the proposed project on the environment
5 Measures to prevent, reduce and where possible offset any significant adverse environmental effects
6 Non-technical summary
7 Any difficulties encountered in compiling the required information

* Amendment is shown in italics.

Member States are also to ensure that:

• any request for development consent and any information gathered pursuant to [the Directive's provisions] are made available to the public;

• the public concerned is given the opportunity to express an opinion before the project is initiated.

The detailed arrangements for such information and consultation shall be determined by the Member States (Articles 6. 2 and 6. 3) (see Section 6. 2 also). The competent authority must consider the information presented in an EIS, the comments of relevant authorities and the public, and the comments of other Member States (where applicable) in its consent procedure (Article 8). The CEC (1994) published a checklist to help competent authorities to review environmental information. It must then inform the public of the decision and any conditions attached to it (Article 9).

2.6 EC Directive 85/337, as amended by Directive 97/11/EC

Directive 85/337 included a requirement for a five-year review, and a report was published in 1993 (CEC 1993). While there was general satisfaction that the 'basics of the EIA are mostly in place', there was concern about the incomplete coverage of certain projects, insufficient consultation and public participation, the lack of information about alternatives, weak monitoring and the lack of consistency in Member States' implementation. The review process, as with the original Directive, generated considerable debate between the Commission and the Member States, and the amended Directive went through several versions, with some weakening of the proposed changes. The outcome, finalized in March 1997, and to be implemented within two years, included the following amendments:

• Annex Ⅰ (mandatory): the addition of 12 new classes of project (e. g. dams and reservoirs, pipelines, quarries and open-cast mining) (Table 2. 4).

• Annex Ⅱ (discretionary): the addition of 8 new sub-classes of project (plus extension to 10 others), including shopping and car parks, and particularly tourism and leisure (e. g. caravan sites and theme parks) (Table 2. 4).

• New Annex Ⅲ lists matters that must be considered in EIA, including:

—Characteristics of projects: size, cumultive impacts, the use of natural resources, the production of waste, pollution and nuisance, the risk of accidents.

—Location of projects: designated areas and their characteristics, existing and previous land uses.

—Characteristics of the potential impacts: geographical extent, transfrontier effects, the magnitude and complexity of impacts, the probability of impact, the duration, frequency and reversibility of impacts.

- Change of previous Annex III to Annex IV: small changes in content.
- Other changes:

—Article 2. 3: There is no exemption from consultation with other Member States on transboundary effects.

—Article 4: When deciding which Annex II projects will require EIA, Member States can use thresholds, case by case or a combination of the two.

—Article 5. 2: A developer may request an opinion about the information to be supplied in an environmental statement (ES), and a competent authority must provide that information. Member States may require authorities to give an opinion irrespective of the request from the developer.

—Article 5. 3: The minimum information provided by the developer *must include* an outline of the main alternatives studied and an indication of the main reasons for the final choice between alternatives.

—Article 7: This requires consultation with affected Member States, and other countries, about transboundary effects. [6]

—Article 9: A competent authority must make public the *main* reasons and considerations on which decisions are based, together with a description of the *main* mitigation measures (CEC 1997a).

A consolidated version of the full Directive, as amended by these changes, is included in Appendix 1. There are more projects subject to mandatory EIA (Annex I) and discretionary EIA (Annex II). Alternatives also became mandatory, and there is more emphasis on consultation and participation. The likely implication is more EIA activity in the EU Member States, which also have to face up to some challenging issues when dealing with topics such as alternatives, risk assessment (RA) and cumulative impacts.

2.7 An overview of EIA systems in the EU: divergent practice in a converging system?

The EU has been active in the field of environmental policy, and the EIA Directive is widely regarded as one of its more significant environmental achievements (see CEC 2001). However, there has been, and continues to be, concern about the inconsistency of application across the (increasing number of) Member States (see CEC 1993, CEC 2003, Glasson and Bellanger 2003). This partly reflects the nature of EC/EU directives, which seek to establish a mandatory framework for European policies while leaving the 'scope and method' of implementation to each Member State. In addition, whatever the degree of 'legal harmonization' of Member State EIA policies, there is also the issue of 'practical harmonization'. Implementation depends on practitioners from public and private sectors, who invariably have their own national cultures and approaches.

An early inconsistency was in the timing of implementation of the original Directive. Some countries, including France, the Netherlands and the UK, implemented the Directive relatively on time; others (e. g. Belgium, Portugal) did not. Other differences, understandably, reflected variations in legal systems, governance and culture between the Member States, and several of these differences are outlined below.

- The legal implementation of the Directive by the Member States differed considerably. For some, the regulations come under the broad remit of nature conservation (e. g. France, Greece, the Netherlands, Portugal); for some they come under the planning system (e. g. Denmark, Ireland, Sweden, the UK); in others specific EIA legislation was enacted (e. g. Belgium, Italy). In addition, in Belgium, and to an extent in Germany and Spain, the responsibility for EIA was devolved to the regional level.
- In most Member States, EIAs are carried out and paid for by the developers or consultants commissioned by them. However, in Flanders (Belgium) EIAs are carried out by experts approved by the authority responsible for environmental matters, and in Spain the competent authority carries out an EIA based on studies carried out by the developer.
- In a few countries, or national regions, EIA commissions have been established. In the Netherlands the commission assists in the scoping process, reviews the adequacy of an EIS and receives monitoring information from the competent authority. In Flanders, it reviews the qualifications of the people carrying out an EIA, determines its scope and reviews an EIS for compliance with legal requirements. Italy also has an EIA commission.
- The decision to proceed with a project is, in the simplest case, the responsibility of the competent authority (e. g. in Flanders, Germany, the UK). However, in some cases the minister responsible for the environment must first decide whether a project is environmentally compatible (e. g. in Denmark, Italy, Portugal).

2. 7. 1 Reviews of the original Directive 85/337/EC

The first five-year review of the original Directive (CEC 1993) expressed concern about a range of inconsistencies in the operational procedures across the Member States (project coverage, alternatives, public participation, etc.). As a result, several Member States strengthened their regulations to achieve a fuller implementation. A second five-year review in 1997 (CEC 1997b) had the following key findings:

- EIA is a regular feature of project licensing/authorization systems, yet wide variation exists in relation to those procedures (e. g. different procedural steps, relationships with other relevant procedures);
- While all Member States had made provision for the EIA of the projects listed in Annex I, there were different interpretations and procedures for Annex II projects;
- quality control over the EIA process is deficient;
- Member States did not give enough attention to the consideration of alternatives;
- improvements had been made on public participation and consultation; and

• Member States, themselves, complained about the ambiguity and lack of definitions of several key terms in the Directive.

The amendments of 1997 sought to reduce further several of the remaining differences. In addition to the substantial extensions and modification to the list of projects in Annex I and Annex II, the amended Directive (CEC 1997a) also strengthened the procedural base of the EIA Directive. This included a provision for new screening arrangements, including new screening criteria (in Annex III) for Annex II projects. It also introduced EIS content changes, including an obligation on developers to include an outline of the main alternatives studied, and an indication of the main reasons for their choices, taking into account environmental effects. The amended Directive also enables a developer, if it so wishes, to ask a competent authority for formal advice on the scope of the information that should be included in a particular EIS. Member States, if they so wish, can require competent authorities to give an opinion on the scope of any new proposed EIS, whether the developer has requested one or not. The amended Directive also strengthens consultation and publicity, obliging competent authorities to take into account the results of consultations with the public and the reasons and considerations on which the decision on a project proposal has been based.

2.7.2 Review of the amended Directive 97/11/EC

A third review of the original Directive, as amended by Directive 97/11/EC (CEC 1997a), undertaken for the EC by the Impacts Assessment Unit at Oxford Brookes University (UK), provided a detailed overview of the implementation of the Directive (as amended) by Member States, and recommendations for further enhancement of application and effectiveness (CEC 2003). Some of the key implementation issues identified included:

• Further delays in the transposition of the Directive. Many Member States missed the 1999 deadline, and by the end of 2002 transposition was still incomplete for Austria, France, and Walloon and Flanders regions of Belgium. There was a complete lack of transposition for Luxembourg.

• Variations in thresholds used to specify EIA for Annex II projects. In all Member States, EIA is mandatory for Annex I projects, and some countries have added in additional Annex I categories. Until the amendments to the Directive, Member States differed considerably in their interpretation of which Annex II projects required EIA. In some (e. g. the Netherlands), a compiled list specified projects requiring EIA. Subsequent to the amendments, most of the Member States appeared to make use of a combination of both thresholds and a case-by-case approach for Annex II projects. However, the 2003 review revealed that there were still major variations in the nature of the thresholds used. For example, with afforestation projects the area of planting that triggered mandatory EIA ranged from 30 ha in Denmark to 350 ha in Portugal. Similarly, three turbines would trigger mandatory EIA for a wind farm in Sweden, compared with 50 in Spain. Considerable variations also continue to exist in the detailed specification of which projects were covered by some Annex II categories, with 10 (b) (urban development) being particularly problematic.

• Considerable variation in the number of EIAs being carried out in Member States. Documentation is complicated by inadequate data in some countries, but Table 2.6 shows the continuing great variation in annual output from over 7, 000 (in France, where a relatively low financial criterion is a key trigger) to fewer than 20 (in Austria). While some of the variation may be explained by the relative economic conditions within countries, it also relates to the variations in levels at which thresholds have been set. The amendments to the Directive do seem to be bringing more projects into the EIA process in some Member States.

Table 2.6 Change in the amount of EIA activity in EU-15 Member States

Country	Pre-1999(average p. a.)	Post-1999(average p. a.)
Austria	4	10-20
Belgium-Brussels	20	20
Belgium-Flanders	No data	20% increase
Belgium-Walloon	No data	est. increase
Denmark	28	100
Finland	22	25
France	6,000—7,000	7,000+
Germany	1,000	est. increase
Greece	1,600	1,600
Ireland	140	178
Italy	37	No data
Luxembourg	20	20
Netherlands	70	70
Portugal	87	92
Spain	120	290
Sweden	1,000	1,000
UK	300	500

Source: CEC 2003

• Some improvement, but still issues in relation to the scoping stage, and consideration of alternatives. Until the amendments made it a more formal stage of the EIA process, scoping was carried out as a discrete and mandatory step in only a few countries. The amended Directive allows Member States to make this a mandatory procedure if they so wish; seven of the Member States have such procedures in place. Commitment to scoping in the other Member States is more variable. Similarly, the consideration of alternatives to a proposed project was mandatory in only a very few countries, including the Netherlands, which also required an analysis of the most environmentally acceptable alternatives in each case. The amended Directive required developers to include an outline of the main alternatives studied. The 2003 review showed that in some Member States the consideration of alternatives is a central focus of the EIA process; elsewhere the coverage is less adequate—although the majority of countries do now require assessment of the zero ('do minimum') alternative.

- Variations in nature of public consultation required in the EIA process. The Directive requires an EIS to be made available after it is handed to the competent authority, and throughout the EU the public is given an opportunity to comment on the projects that are subject to EIA. However, the extent of public involvement and the interpretation of 'the public concerned' varied from quite narrow to wide. In Denmark, the Netherlands and Wallonia, the public is consulted during the scoping process. In the Netherlands and Flanders, a public hearing must be held after the EIS is submitted. In Spain, the public must be consulted before the EIS is submitted. In Austria, the public can participate at several stages of an EIA, and citizens' groups and the Ombudsman for the Environment have special status. The transposition of the Aarhus Convention into EIA legislation provided an opportunity for further improvements in public participation in EIA (CEC 2001).

- Variations in some key elements of EIA/EIS content, relating in particular to biodiversity, human health, risk and cumulative impacts. While the EIA Directive does not make explicit reference to biodiversity and to health impacts, both can be seen as of increasing importance for EIA. There are some examples of good practice in the Netherlands and Finland for biodiversity, and in the Netherlands again for health impact assessment. On the other hand, the amended Directive (Annex III) includes risk and cumulative impacts. The 2003 review showed that although RAs appear in many EISs, for most Member States risk was seen as separate from the EIA process and handled by other control regimes. The review also showed a growing awareness of cumulative impacts, with measures put in place in many Member States (e. g. France, Portugal, Finland, Germany, Sweden and Denmark) to address them. However, it would seem that Member States are still grappling with the nature and dimensions of cumulative impacts.

- Lack of systematic monitoring of a project's actual impacts by the competent authority. Despite widespread concern about this Achilles' heel in the EIA Directive, there was considerable resistance to the inclusion of a requirement for mandatory monitoring. As such, there are very few good examples (e. g. the Netherlands) of a mandatory and systematic approach. Dutch legislation requires the competent authority to draw up an evaluation programme, which compares actual outcomes with those predicted in the EIS. If the evaluation shows effects worse than predicted, the competent authority may order extra environmental measures. In Greece, legislation provides for a review of the EIA outcome as part of the renewal procedure for an environmental permit.

Overall, the 2003 review showed that there were both strengths and weaknesses in the operation of the Directive, as amended. There are many examples of good practice, and the amendments have provided a significant strengthening of the procedural base of EIA, and have brought more harmonization in some areas—for example on the projects subject to

EIA. Yet, as noted here, there is still a wide disparity in both the approach and the application of EIA in the Member States, and significant weaknesses remain to be addressed. The review concluded with a number of recommendations. These included advice to Member States to, *inter alia*, better record on an annual basis the nature of EIA activity; check national legislation with regard to aspects such as thresholds, quality control, cumulative impacts; make more use of EC guidance (e. g. on screening, scoping and review); and improve training provision for EIA. Section 2. 8 continues the ongoing story of review and refinement of the Directive in the enlarged EU.

2.8 Continuing issues, review and refinement of the EIA Directive in EU-27

There have been further reviews of the application and effectiveness of the EIA Directive, and as a result some limited changes were made to the Directive in 2003 and in 2009. Appendix 1 provides a consolidated version of the Directive (as at June 2011), but there may be further changes following another round of consultation (in late 2010) on implementation. A particular interesting feature of recent reviews, especially on the operation of the 2003 amended Directive, is the nature of implementation by the new Member States that joined the EU in 2004 and 2007 (CEC 2009a).

The 2009 review had a wide brief, examining: the application of the amended Directive as a whole, including the additional 2003 EIA procedures (which focused particularly on the integration of the requirements of the Aarhus Convention on public participation into the Directive), general trends in the performance of the Directive, the status of national systems, developments in EIA systems in the old Member States and the new Member States, and the relationships with other Directives. The main changes introduced by Directive 2003/35/EC were:

- definition of the 'public' and 'the public concerned' (new Article 1. 2);
- option to include provisions in law exempting national defence projects from EIA now only allowed on a case-by-case basis (new Article 1. 4);
- strengthened public consultation provision: early in the decision-making procedure, detailed list of information to be provided, reasonable time frames (new Articles 6. 2 and 6. 3);
- information on the public participation process within the information provided on the final decision (Article 9. 1);
- new provisions on public access to a review procedure (Article 10 (a)); and
- changes or extensions of Annex I projects and other modifications of Annex I projects and modifications of Annex II projects.

The 2009 changes were much more limited and focused on the addition of new projects-for example carbon capture and storage installations. Overall, the European Commission was very positive about the benefits of the Directive:

> Two major benefits have been identified. Firstly, the EIA ensures that environmental considerations are taken into account as early as possible in the decision-making process. Secondly, by involving the public, the EIA procedure ensures more transparency in environmental decision-making and, consequently, social acceptance. Even if most benefits of the EIA cannot be expressed in monetary terms, there is widespread agreement, confirmed by the studies available, that the benefits of carrying out an EIA outweigh the costs of preparing an EIA. (CEC 2009b)

The experience and performance of the 12 new Member States (Bulgaria, Cyprus, Czech Republic, Estonia, Hungary, Latvia, Lithuania, Malta, Poland, Romania, Slovakia and Slovenia) is of particular interest. In these countries, the EIA Directive had been transposed as part of the accession requirements to ensure harmonization of the national legislation with the EU *Acquis*, and these Member States were ready to incorporate the requirements of the Aarhus Convention. In many respects, the new Member States have the advantage of learning from the evolving EIA procedures and practice of the EU and its old Member States, and can provide some examples of innovative practice (see example of Poland in Chapter 10). But they have also encountered some of the issues raised by the other states in earlier reviews. The number of EIAs is increasing in many of the new Member States, but there is considerable variation between the states (although the data in Table 2.7 are distorted by the inclusion of total screening decisions in some cases). Scoping is mandatory in all the states, with the exception of Cyprus and Slovenia. Quality review of the EIA documentation, either by the competent authority, or an expert committee, is a legal requirement in all the states, and several make good use of EU guidance on topics such as review and screening. But there are concerns about carrying out EIA in a transboundary context, consideration of human health protection in EIA, the 'salami slicing' of projects to fit under EIA thresholds, and the need for up-dated EU guidance on several issues, including how to address the thorny issue of the cumulative impacts of projects.

Table 2.7 Number of EIAs carried out in the new EU Member States

Member State	Annex I		Annex II		Tendencies in EIAs carried out
	2005	2006	2005	2006	
Bulgaria	77	88	2,212	2,457(incl. screening decisions)	Increase
Cyprus		30		45	Increase
Czech Republic		72		125	Increase
Estonia		57		20	Decrease

Continued

Member State	Annex I		Annex II		Tendencies in EIAs carried out
	2005	2006	2005	2006	
Hungary		70—90		370—400(incl. screening procedures)	Static
Latvia		40		5(the number indicates finished EIAs)	Increase
Lithuania		12		838	Static
Malta	4		6	Increase	
Poland	n/a	n/a	n/a	n/a	n/a
Romania		179		643	Increase
Slovakia	90	135	429	363	Static
Slovenia	n/a	n/a	n/a	n/a	n/a

Source: CEC 2009a

Notwithstanding good progress over 20 years of implementation, the EU has ongoing concerns about some stubborn issues, including those already mentioned—such as variations in screening, transboundary problems, and concerns about quality control. Another serious and longstanding issue is the lack of a mandatory monitoring requirement; a more recent issue includes the relationship between the EIA and SEA Directives, and the very limited hierarchical tiering in practice. There is also recognition of the urgent need to improve requirements and guidance on covering climate change issues in EIA, especially for energy and transport infrastructure projects, and those for which energy efficiency is a key issue. Further changes to EIA legislation are likely following the 2010 round of consultation on implementation of the Directive.

2.9 Summary

This chapter has reviewed the development of EIA worldwide, from its unexpectedly successful beginnings in the USA to recent developments in the EU. In practice, EIA ranges from the production of very simple ad hoc reports to the production of extremely bulky and complex documents, from wide-ranging to non-existent consultation with the public, from detailed quantitative predictions to broad statements about likely future trends. In the EU, reviews of EIA experience show that 'overall, although practice is divergent, it may not be diverging, and recent actions such as the amended Directive appear to be "hardening up" the regulatory framework and may encourage more convergence' (Glasson and Bellanger 2003). All these systems worldwide have the broad aim of improving decision-making by raising decision-makers' awareness of a proposed action's environmental consequences. Over the past 40 years, EIA has become an important tool in project planning, and its applications are likely to expand further. Chapter 10 provides further discussion of EIA systems internationally and Chapter 11 discusses the widening of scope to strategic environmental assessment of policies, plans and programmes. The next chapter focuses on EIA in the UK context.

SOME QUESTIONS

The following questions are intended to help the reader focus on the important issues of this chapter, and to start building some understanding of the origins and development of EIA.

1 Why do you think EIA had its origins, in NEPA, in the USA in the late 1960s?

2 What are the key differences, under NEPA, of the processes for a FONSI and for a full EIS?

3 How do you explain the recent trends in EISs filed under NEPA over the last 10-15 years?

4 As for all EIA systems, there are concerns about procedural issues. Note some of the recent concerns about the operation of EIA under NEPA.

5 Is there any clear pattern to the spread of EIA across countries and continents?

6 Why was there strong initial resistance to the introduction of EIA in the UK?

7 What were the key drivers behind the introduction of the EC EIA Directive 85/337?

8 What is the difference between Annex I and Annex II projects under the EC EIA Directive?

9 What were the main changes introduced to the EIA Directive under the 97/11/EC amendments?

10 Identify several examples of (i) good practice and (ii) significant weaknesses highlighted by the 2003 review of the implementation of the amended EIA Directive.

11 What factors might explain the variations in Member States' EIA activity illustrated in Tables 2.6 and 2.7?

12 What is the Aarhus Convention, and what are its implications for the EIA Directive?

13 What do you see as some of the stubborn issues still to be resolved in the EC EIA Directive?

14 From what you have covered in Chapter 2, are there now grounds for saying that there is clear evidence of a more consistent system of EIA across the 27 EU Member States?

Notes

1 For example, *Ely v. Velds*, 451 F. 2d 1130, 4th Cir. 1971; *Carolina Action v. Simon*, 522 F. 2d 295, 4th Cir. 1975.

2 *Calvert Cliff's Coordinating Committee, Inc. v. United States Atomic Energy Commission* 449 F. 2d 1109, DC Cir. 1971.

3 *Natural Resources Defense Council*, *Inc. v. Morton*, 458 F. 2d 827, DC Cir. 1972.

4 California, Connecticut, Georgia, Hawaii, Indiana, Maryland, Massachusetts, Minnesota, Montana, New York, North Carolina, South Dakota, Virginia, Washington and Wisconsin, plus the District of Columbia and Puerto Rica.

5 Arizona, Arkansas, Delaware, Florida, Louisiana, Michigan, New Jersey, North Dakota, Oregon, Pennsylvania, Rhode Island and Utah.

6 Amendments to Articles 7 and 9 were influenced by the requirements of the Espoo Convention on EIA in a Transboundary Context, signed by 29 countries and the EU in 1991. This widened and strengthened the requirements for consultation with Member States where a significant transboundary impact is identified. The Convention deals with both projects and impacts that cross boundaries and is not limited to a consideration of projects that are in close proximity to a boundary.

References

Anderson, F. R., Mandelker, D. R. and Tarlock, A. D. 1984. *Environmental protection: law and policy*. Boston: Little, Brown.

Association of Environmental Professionals (AEP) 2009. *California Environmental Quality Act: statute and guidelines*. Palm Desert, CA: AEP.

Bear, D. 1990. EIA in the USA after twenty years of NEPA. *EIA Newsletter* 4, EIA Centre, University of Manchester.

Canter, L. W. 1996. *Environmental impact assessment*, 2nd edn. London: McGraw-Hill.

Carson, R. 1962. *Silent Spring*. Boston: Houston Mifflin Company.

Catlow, J. and Thirwall, C. G. 1976. *Environmental impact analysis* (DoE Research Report 11). London: HMSO.

CEC (Commission of the European Communities) 1973. First action programme on the environment. *Official Journal* C112, 20 December.

CEC 1980. Draft directive concerning the assessment of the environmental effects of certain public and private projects. COM (80), 313 final. *Official Journal* C169, 9 July.

CEC 1982. Proposal to amend the proposal for a Council directive concerning the environmental effects of certain public and private projects. COM (82), 158 final. *Official Journal* C110, 1 May.

CEC 1985. On the assessment of the effects of certain public and private projects on the environment. *Official Journal* L175, 5 July.

CEC 1992. *Towards sustainability*. Brussels: CEC. CEC 1993. *Report from the Commission of the Implementation of Directive* 85/337/*EEC on the assessment of the effects of certain public and private projects on the environment*. COM (93), 28 final. Brussels: CEC.

CEC 1994. *Environmental impact assessment review checklist*. Brussels: EC Directorate-General XI.

CEC 1995. *Environmental impact assessment: guidance on screening DG XI*. Brussels: CEC.

CEC 1997a. Council Directive 97/11/EC of 3 March 1997 amending Directive 85/337/EEC on the assessment of certain public and private projects on the environment. *Official Journal* L73/5, 3 March.

CEC 1997b. *Report from the Commission of the Implementation of Directive 85/337/EEC on the assessment of the effects of certain public and private projects on the environment*. Brussels: CEC.

CEC 2001. *Proposal for a directive of the European Parliament and of the Council providing for public participation in respect of the drawing up of certain plans and programmes relating to the environment and amending council directives 85/337/EEC and 96/61/EC*. COM (2000), 839 final, 18 January 2001. Brussels: DG Environment, EC.

CEC 2003. (Impacts Assessment Unit, Oxford Brookes University). *Five years' report to the European Parliament and the Council on the application and effectiveness of the EIA Directive*. Available on the website of DG Environment, EC: www. europa. eu. int/comm/environment/ eia/home. htm.

CEC 2009a. *Study concerning the report on the application and effectiveness of the EIA Directive: Final Report*. Brussels: DG Env.

CEC 2009b. *Report from the Commission to the Council, the European Parliament, the European Economic and Social Committee and the Committee of the Regions on the application and effectiveness of the EIA Directive (Directive 85/337/EEC, as amended by Directives 97/11/EC and 2003/35/EC)*. Brussels: CEC.

CEQ (Council on Environmental Quality) 1978. National Environmental Policy Act. Implementation of procedural provisions: final regulations. *Federal Register* 43 (230), 55977-6007, 29 November.

CEQ 1997a. *The National Environmental Policy Act: a study of its effectiveness after 25 years*. Washington, DC: US Government Printing Office.

CEQ 1997b. *Considering cumulative effects-under the NEPA*. Washington, DC: US Government Printing Office.

CEQ 2007. *A citizens' guide to the NEPA*. Washington, DC: US Government Printing Office.

CEQ 2007. *Aligning NEPA processes with environmental management systems*. Washington, DC: US Government Printing Office.

Clark, B. D. and Turnbull, R. G. H. 1984. Proposals for environmental impact assessment procedures in the UK. In *Planning and ecology*, R. D. Roberts and T. M. Roberts (eds), 135-44. London: Chapman & Hall.

Clark, B. D., Chapman, K., Bisset, R. and Wathern, P. 1976. *Assessment of major industrial applications: a manual* (DoE Research Report 13). London: HMSO.

Clark, B. D., Chapman, K., Bisset, R., Wathern, P. and Barrett, M. 1981. *A manual for the assessment of major industrial proposals*. London: HMSO.

CPRE 1991. *The environmental assessment directive: five years on*. London: Council for the Protection of Rural England.

CRS (Congressional Research Services) 2006. *The National Environmental Policy Act-streamlining NEPA*. Washington, DC: Library of Congress.

DFID (Department for International Development) 2003.

Environment guide. London: DFID. Available at: www. eldis. org/vfile/upload/1/document/0708/ DOC12943. pdf.

Dobry, G. 1975. *Review of the development control system: final report*. London: HMSO.

DoE (Department of the Environment) 1977. Press Notice 68. London: Department of the Environment.

DoE 1978. Press Notice 488. London: Department of the Environment.

Ehrlich, P. 1968. *The population bomb*. New York: Ballantine.

European Bank for Reconstruction and Development 1992. *Environmental procedures*. London: European Bank for Reconstruction and Development.

European Bank for Reconstruction and Development 2010. *Environmental and social procedures*. Available at: www. ebrd. com/downloads/ about/sustainability/esprocs10. pdf.

Glasson, J. and Bellanger, C. 2003. Divergent practice in a converging system? The case of EIA in France and the UK. *Environmental Impact Assessment Review* 23, 605-24.

Glasson, J. and Salvador, N. N. B. 2000. EIA in Brazil: a procedures-practice gap. A comparative study with reference to the EU, and especially the UK. *Environmental Impact Assessment Review* 20, 191-225.

Greenberg, M. 2012. *The environmental impact statement after two generations*. Abingdon: Routledge.

Hall, E. 1994. *The environment versus people? A study of the treatment of social effects in EIA* (MSc dissertation). Oxford: Oxford Brookes University.

HMSO (Her Majesty's Stationery Office) 1971. *Report of the Roskill Commission on the Third London Airport*. London: HMSO.

House of Lords 1981a. *Environmental assessment of projects*. Select Committee on the European Communities, 11th Report, Session 1980-81. London: HMSO.

House of Lords 1981b. *Parliamentary debates (Hansard) official report, session 1980-81*, 30 April, 1311-47. London: HMSO.

Lee, N. and Wood, C. M. 1984. Environmental impact assessment procedures within the European Economic Community. In *Planning and ecology*, R. D. Roberts and T. M. Roberts (eds), 128-34. London: Chapman & Hall.

Mandelker, D. R. 2000. *NEPA law and litigation*, 2nd edn. St Paul, MN: West Publishing.

Moreira, I. V. 1988. EIA in Latin America. In *Environmental impact assessment: theory and practice*, P. Wathern (ed), 239-53. London: Unwin Hyman.

NEPA (National Environmental Policy Act) 1970. 42 USC 4321-4347, 1 January, as amended.

O'Riordan, T. and Sewell, W. R. D. (eds) 1981. *Project appraisal and policy review*. Chichester: Wiley.

Orloff, N. 1980. *The National Environmental Policy Act: cases and materials*. Washington, DC: Bureau of National Affairs.

Overseas Development Administration 1996. *Manual of environmental appraisal: revised and updated*. London: ODA.

Petts, J. and Hills, P. 1982. *Environmental assessment in the UK*. Nottingham: Institute of Planning Studies, University of Nottingham.

Sadler, B. 1996. *Environmental assessment in a changing world: evaluating practice to improve performance*, International study on the effectiveness of environmental assessment. Ottawa: Canadian Environmental Assessment Agency.

SDD (Scottish Development Department) 1974. *Appraisal of the impact of oil-related development*, DP/TAN/16. Edinburgh: SDA.

Tomlinson, P. 1986. Environmental assessment in the UK: implementation of the EEC Directive. *Town Planning Review* 57 (4), 458-86.

Turner, T. 1988. The legal eagles. *Amicus Journal* (winter), 25-37.

UNEP (United Nations Environment Programme) 1997. *Environmental impact assessment training resource manual*. Stevenage: SMI Distribution.

UNEP 2002. *UNEP Environmental Impact Assessment Training Resource Manual*, 2nd edn. Available at www. unep. ch/etu/publications/EIAMan _ 2edition _ toc. htm.

White House 1994. *Memorandum from President Clinton to all heads of all departments and agencies on an executive order on federal actions to address environmental injustice in minority populations and low income populations*. Washington, DC: White House.

Williams, R. H. 1988. The environmental assessment directive of the European Community. In *The role of environmental assessment in the planning process*, M. Clark and J. Herington (eds), 74-87. London: Mansell.

World Bank 1991. *Environmental assessment sourcebook*. Washington, DC: World Bank, www. worldbank. org.

World Bank 1995. *Environmental assessment: challenges and good practice*. Washington, DC: World Bank.

World Bank 1997. *The impact of environmental assessment: A Review of World Bank Experience*. World Bank Technical Paper no. 363. Washington, DC: World Bank.

World Bank 1999. *Environmental assessment*, BP 4. 01. Washington, DC: World Bank.

World Bank 2002. Environmental impact assessment systems in Europe and Central Asia Countries. Available at: www. worldbank. org/eca/environment.

World Bank 2006. Environmental impact assessment regulations and strategic environmental assessment requirements: practices and lessons learned in East and Southeast Asia. Environment and social development safeguard dissemination note no. 2. Available at: www. vle. worldbank. org/bnpp/files/ TF055249EnvironmentalImpact. pdf.

3

英国的环境影响评价机构及立法背景

3.1 导言

本章对英国 EIA 实施的法律框架进行了探讨。首先简要地介绍了规划、开发过程以及相关 EIA 的主要参与者。之后，回顾了相关的法律及其适用的项目类型，然后介绍了最新修订的城乡法规体系 2011 中所要求的 EIA 程序，这些是最“一般”的 EIA 法规，可以应用于大多数项目，并且为其他法规的制定提供了参考。最后是总结。这些法律在应用中的主要成果和局限性将在第 8 章进行讨论，可供读者参考。

3.2 主要参与者

3.2.1 概述

任何提案开发都具有基本的结构特点：利益、战略和远景。但是，无论何种开发活动，如公路、电站、水库或者森林等，规划和开发过程的参与者都可能分为四个主要团体，包括：

(1) 开发商；

(2) 直接或间接地受到开发活动影响的或对开发活动感兴趣的群体；

(3) 政府和管理机构；

(4) 对以上三方的相互关系感兴趣的调解者（顾问、建议者和拥护者）(图 3.1)。

介绍 EIA 参与者的范围，是了解英国 EIA 法律框架的重要的第一步。

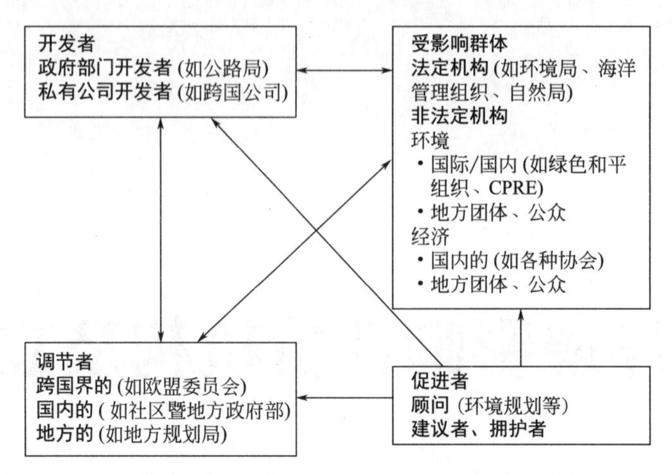

图 3.1　环境影响评价、规划、开发过程的主要参与者

3.2.2　开发商

在英国，政府和私有企业的项目都需要进行 EIA，但是也存在一些特例，如国防部和皇家委员会的一些项目就不需要。政府的开发项目由中央政府部门（如交通部）负责，也可以由地方当局以及其他主管部门（如环境局和公路局）负责。有些由国有企业（如前面提到的英国铁路公司和核工业公司）负责，1980 年以后，快速的私有化进程使很多国有企业变为私人所有。有些部门，如主要的能源公司（英国天然气公司、国家电网和法国电力集团）和地区水利部门等，他们掌控某行业领域的多数项目及持续的项目计划，他们的经验会对发展和完善 EIA 程序大有益处。其他许多私有公司（通常指跨国公司），也会开发一系列的项目。然而，对很多开发商来说，项目通常是一次性行为，而且，他们对 EIA 程序及相关的规划和开发过程都不太熟悉，需要迅速进行了解，同时他们也希望能够获得好的建议。

3.2.3　受影响群体

有很多群体都直接或间接地受到开发活动的影响。在图 3.1 中，按照他们的职责或影响力（如法定的和咨询性质的）、实施层次（如国际的、国内的、地方的）或侧重点（如环境、经济）的不同将他们进行了大致的分类。绿色和平组织、地球之友、英格兰乡村保护委员会（CPRE）、英国皇家鸟类保护协会（RSPB）等环境组织随着公众对环境问题的关注而不断发展。例如，英国皇家鸟类保护协会的成员从 1970 年的 10 万发展到 1997 年的 100 多万，并且至今仍保持如此之高的人数。可持续交通组织是一家鼓励人们使用自行车代替私家车的慈善机构，其会员从 1993 年的 4000 人发展到 1996 年的 2 万人，目前至少拥有 6 万人。这些组织虽然经常受到各种因素的限制，却有着相当大的“道德力量”。开发商对受影响群体的利益调解通常被认为是项目“合法化”的重要步骤。和开发商一样，一些环境团体（尤其是国家层次的团体）对此都有长期而持续的职责。一些地方组织也具有持续的职责，并且积累了很多有价值的地方环境认知。其他团体（通常是地方性的）的生命周期短，只与某个特定的项目有关。最后一类作为地方性压力团体，他们可以快速涌现出来以反对开发活动。这些组织有时被称为 NIMBY（“别在我的后院”），他们的目的通常包括维护财产的价值和现存的生活方式，并力争将必须开发的项目转移到其他地方。另外，这些组织中还有一些被称为 BANANA（“别在任何地方建设任何东西”）。

在 EIA 程序中，法定咨询机构是一个非常重要的团体。规划部门在对一个需要进行 EIA 的重点项目做出决定时，必须向这些机构进行咨询。英格兰的法定咨询机构包括英格兰自然署（NE）、海洋管理组织（MMO）、环境局（部分开发项目）以及拟建项目所在地的地方议会。其他被咨询者通常包括地方高速公路局和地方郡县考古学家等。如上所述，非法定团体（如英国皇家鸟类保护协会和一般公众）也都可以就环境问题提供有价值的信息。

3.2.4 调解者

各级政府通常在调解和管理上述各团体间的关系中扮演着重要的角色。如第 2 章所述，欧盟委员会已经采纳了有关 EIA 程序方面的指令（CEC，1985 及修订指令）。英国政府也通过颁布一系列的法规和导则将该指令付诸实施（3.3 节）。目前（2012）主要的调解部门是英国社区与地方政府事务部（以前是 ODPM、DTLR、DETR、DoE），它通过伦敦总部履行其职责。EIA 程序中尤其重要的是地方权力机构，特别是相关的规划部门（地方规划管理局 LPA）。这包括地区、郡和自治市镇的部门。这些部门就像过滤器一样，开发商提出的计划通常需要通过这些部门的审核。此外，地方规划管理局也为其他机构提供参与开发过程的机会。

3.2.5 促进者

最后一个团体在 EIA 程序中发挥着重要作用，包括参与规划、开发和 EIA 程序的各类顾问、建议者和拥护者。他们通常受聘于开发商，有时也受聘于地方组织、环境组织及对规划提案有反对意见的组织。调解机构也偶尔会聘请他们协助对环评程序进行监督。

环境与规划咨询机构完成了大多数的 EIA 工作，这与一些小型专业咨询机构的支持密不可分，这些小型咨询机构精于考古、噪声、健康和社会经济影响等领域的评价。环境咨询服务的数量已经大幅上升（图 3.2），自 20 世纪 80 年代中期以来，增加量超过了 400%，据估计，2008 年客户在咨询服务上的花费约为 15 亿英镑（E4NDS，2009）。导致咨询业务增长的主要因素有：英国于 1990 年实施的《环境保护法案》（EPA）、EIA 法规、英国环境管理系统带来的持续增长的商业利益（如 BS77509 ISO 14001 的出台）以及欧盟有关生态审计、战略环境评价（SEA）等法律法规和水框架指令。表 3.1 和表 3.2 列出了在环境咨询机构需要的工作技能和主要领域。环境影响评价和规划是最需要的工作技能，这也被 1/4 参加过环境数据服务（ENDS）数据调查的人提起。我们将在第 8 章对最近咨询活动的特点进行深入讨论。

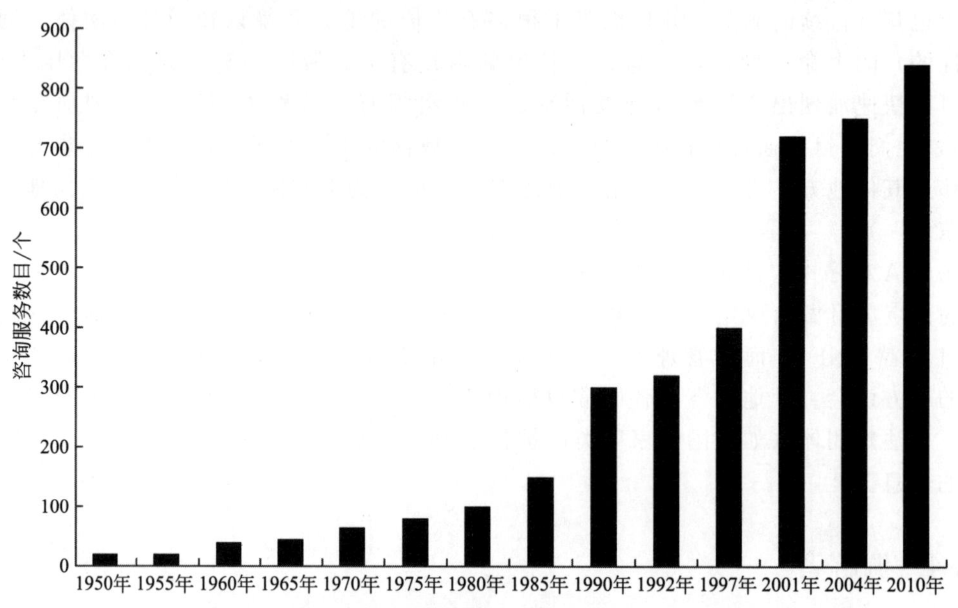

图 3.2 英国环境咨询服务数量的增加（1950～2010 年）
（资料来源：根据 ENDS 1993、1997、2001 以及网站）

表 3.1 在环境咨询机构需要的工作技能（2008 年）

专长	占比/%
废物管理	62
能源管理	53
可持续发展	52
环境影响评价和规划	50
环境管理和审计	49
水和废物管理	46
GHG 和碳管理	44
污染阻止和控制(IPPC)	44
健康安全以及环境管理	42
污染土地治理和修复	40
可再生能源和清洁能源	40
空气污染和控制	38
企业政策、CSR 以及交流	37
生态和自然保护	35
水文和水文地质	30
公共事务和相关者交流	29
危险品管理	28
声音	22
工艺学	20

资料来源：根据 ENDS，2009。

表 3.2 环境咨询机构主要领域（2009）

活动	占比/%
环境管理	44
废物管理/回收	35
环境保护/立法	29
健康安全以及环境管理	26
可持续发展	24
审计/认证	23
环境教育/培训	23
环境影响评价（包括 EIA、SEA 和 SA）以及规划	23
碳管理	19
能源	19
气候变暖以及 GHG 管理	18
污染控制	18
风险管理	18
企业政策、CSR 以及交流	16

资料来源：根据 ENDS，2010。

3.2.6 机构间的相互影响

这里所列的各种机构都代表不同的利益和目的，成为一个复杂的序列，不同的组合对开发活动的影响也各不相同。该序列具有几个维度，并且每个维度内部又存在很多矛盾。例如，地方和国家层次上观点的矛盾，利润最大化和环境保护之间的矛盾，短期和长期愿景之间的矛盾，组织与个人之间的矛盾。各部门之间也存在千丝万缕的联系，有些联系是受法令约束的，有些是顾问性质的；有些是合同形式的，有些是规章形式的。EIA 法规和导则提供了一套程序，下面将对其进行概述。

3.3 环境影响评价法规：概况

在英国，欧洲共同体 85/337 号指令的实施涉及 1972 年《欧洲共同体法案》2.2 条中 40 多个不同的二级法规。这些法规可以阐述英国是如何实施 EIA 的。不同的法规应用于不同的项目，如规划系统的项目、其他授权系统的项目，以及属于非授权系统但仍需进行 EIA 的项目。英格兰、威尔士独立政府、苏格兰、北爱尔兰所采用的法规也各不相同。从 1999 年（最近的法规在 2011 年）以来的修订法规促进了修订过的欧盟指令的实施，同时对相关的法规进行了调整。然而，如表 3.3 所示，为了确保满足指令的所有要求，还需制定很多其他法规。这些法规也被引入到一些政府及其他组织的环境影响评价导则作为补充（表 3.3）。此外，1991 年的《规划与补偿法案》也允许政府要求指令之外的项目进行环境影响评价。

表 3.3　英国的主要环境影响评价法规及其实施日期

有关城乡规划系项目的英国法规
英格兰
城乡规划(环境影响评价)法案 2011
城乡规划(一般允许开发项目)指令 1995
威尔士
正在评审
苏格兰
苏格兰城乡规划(环境影响评价)法案 2011
北爱尔兰
正在评审
英国环境评价法规中需要提交代替征询意见的法案
农业
环境影响评价(农业)法案(英格兰)(第 2 号)2006
造林
环境影响评价(林业)法案(英格兰和威尔士)1999
环境影响评价(林业)法案(北爱尔兰)1999
环境影响评价(林业)法案(苏格兰)1999
基础设施/主要项目
基础设施规划(环境影响评价)法案 2009
地面排水系统的完善
环境影响评价(地面排水改进工程)法案 1999
地面排水(环境影响评价)法案(苏格兰)1999
排水系统(环境评价)法案(北爱尔兰)2001
渔业农场
环境影响评价(海洋渔业)法案 1999
环境影响评价(海洋渔业)法案(北爱尔兰)2006
干路和高速公路
高速公路(环境影响评估)法案 1999
公路(环境影响评价)法案(北爱尔兰)1999
铁路、轻轨、运河以及涉及航行权的相关工程
交通和工程(环境影响评价)规则 2000
港口和海港以及海洋捕捞
环境影响评价和栖息地法案(从海洋捕捞或提取矿产)2007
海港工程(环境影响评价)法案 1999
海港工程(环境影响评估)法案(北爱尔兰)1990
发电站、高架输电线以及远距离石油和天然气运输通道
电力工程(环境影响评估)法案(英格兰和威尔士)2000
电力工程(环境影响评价)法案(苏格兰)2000
管道工程(环境影响评价)法案 2000
核反应堆退役(环境影响评价)法案 1999
公用天然气运输管网工程(环境影响评价)法案 1999
海洋石油生产及运输管网工程(环境影响评估)法案 1999
水资源
水资源(环境影响评价)法案 2003

与美国EIA系统相比，欧盟的EIA指令不仅适用于政府项目，也适用于私人企业项目。开发商进行环境影响评价，要将环评结果提交给规划权力机构。在英格兰，欧盟指令附件Ⅰ和附件Ⅱ中所列的绝大多数开发项目都被划入了规划系统的审议范围，并且应遵守2011年制定的《城乡规划（环境影响评价）法案》（T&CP法案），继承了1988年和1999年制定的《城乡规划（环境影响评价）法案》。随着时间的推移，各种附加法规和修正案修正了《城乡规划（环境影响评价）法案》的不足，并扩大其应用范围，例如：

- 扩充和明确需要进行EIA的原始项目清单（例如，高速公路服务区域、风力发电场、最新加入的碳捕获与封存技术）；
- 审批时需要EIA的项目（如土地开垦、废水处理项目以及位于简化规划区域的项目）；
- 对（原有）规划实施公告上诉成功，进而要求进行EIA的项目；
- 即使法定清单中没有列出，但负责相关领域的内阁大臣仍有权直接要求特殊的项目进行环评；
- 与欧盟和英国法规系统保持一致性，例如关于项目筛选和文本化的程序。

除上述项目外，欧盟指令中的其他类型项目需要单独的法律，因为它们不受规划系统的约束。对于各种交通项目，地方高速公路和机场的开发是在T&CP法规的指导下由地方规划（高速公路）部门进行管理。由交通部提议和管理的高速公路及主干道项目遵循1999年的《高速公路（环境影响评估）法案》。港口类项目由交通部管理，遵循《海港口工程（环境影响评价）法案》。新建铁路和轨道交通工程在2000年的《交通和工程（环境影响评估）法案》要求下进行环评。

低于50MW的能源项目由地方当局按照T&CP法规进行管理。50MW及以上的大多数输电线缆和管道项目（在苏格兰、英格兰和威尔士）由能源和气候变化部（DECC）依照多条2000年的电力及管道工程（EIA）法案进行管理。

新建地面排水工程，包括防洪和海岸防护工程，需要获得规划许可并遵循T&CP法规。排水工程的改善工作由环境机构和其他的排水部门负责实施，在环境、食品和农业部（DEFRA）的调控下，该工程需依照《环境影响评价法规（地面排水改进工程）》来进行EIA。林业项目需要依照1999年的《环境影响评价法案（林业）》开展EIA。在英格兰、威尔士及苏格兰，距海岸线两千米以内的渔业活动需要获得皇家遗产委员会（the Crown Estates Commission）的租赁许可，而不是规划许可，此类开发活动需要依照1991年的《环境评价法案（海洋渔业）》的要求开展EIA。

在苏格兰、威尔士和北爱尔兰，越来越多的项目被当地的法律囊括其中。例如，最近修订的《苏格兰城乡规划法案（EIA）》2011，建立了与英格兰法规体系平行的法规系统。这里也为繁忙的读者提供了一些法规的简要版本（表3.4）。威尔士也制定了独立的《城乡规划法案》。

表3.4 英国政府环境影响评价导则

DoE 1991。环境评价和规划监控。伦敦：HMSO
DoE 1994。规划项目环境信息评估：一个实用性较强的导则。伦敦：HMSO
DoE 1995。需要进行环境评价的规划项目的环境报告准备。伦敦：HMSO
DoE 1996。规划项目环境报告质量的提高。伦敦：HMSO
环境局 1996。项目界定手册。伦敦：HMSO
DETR 1997。环境报告中的减缓措施。伦敦：HMSO
苏格兰执行发展部 1999。规划建议第58条。爱丁堡：SEDD

续表

威尔士国家立法机构 1999。通报 11/99 环境影响评价。加的夫：国家立法机构
规划服务(北爱尔兰)。开发控制建议第 10 条 1999。Belfast：NI 规划服务
DETR 2000。环境影响评价；程序导则。伦敦：DCLG
DCLG 2006。通告和导则实用性范本。伦敦：DCLG
DfT 2007b。交通评价导则。伦敦：DfT
DfT 2011。桥梁设计手册导论。伦敦：DfT
苏格兰政府 2011a。通告 3/2011：城乡规划环境影响评价(苏格兰)法案 2011。爱丁堡：苏格兰政府(附带简读本)
DCLG 2011。英格兰环境影响评价法案 2011 导则。伦敦：DCLG

第 8 章将要讲到，在所有需要进行 EIA 的项目中，60%～70%的 EIA 需要遵循《城乡规划法案》，大约 10%的 EIA 需要遵循《苏格兰环境影响评价法案》和《高速公路环境影响评价法案》，项目几乎都是英格兰和威尔士的地面排水、电力、管网工程和林业项目以及北爱尔兰与规划相关的开发项目。

如此大范围 EIA 法规的制定弥补了很多早期法规体系的问题（例如，CPRE，1991；Fortlage，1990）。但仍存在一些问题，特别是这一套法规系统的复杂性。首先是对“项目”的定义模糊不清。例如电力生产及传输，从 EIA 的目的出发，电站建设和电能传输线通常被看作是独立的项目，但事实上它们之间有着密切的联系（Sheate，1995，9.2 节）。又例如公路建设被分成几个独立的部分来进行规划和评价，事实上它们也是不可能完全独立的。

3.4 《城乡规划（环境影响评价）法案》2011 [原《城乡规划（环境影响评价）法案》1999 和《城乡规划（环境影响评估）法案》1988]

在英格兰，T&CP 法案体系使欧盟指令落实到那些需要获得规划许可的项目上（威尔士目前有独立的法规）。该法案体系是英国履行欧盟指令的主要方式，另外还有其他的 EIA 法案对 T&CP 法案体系未包括的项目进行了补充。本节介绍了 T&CP 法案下的 EIA 程序。图 3.3 对这些程序进行了总结，图中的粗体字母部分将在下面进行解释。3.5 节将讨论由 T&CP 法案衍生的其他主要的 EIA 法案。3.6 节将对欧盟指令（修订）所带来的变化进行评论。

最早的 T&CP 法案在 1988 年 7 月出台。不久后相应的指南出台，环境事务部 15/88 号通告（威尔士事务部 23/88 号通告）的主要目的是给地方规划管理局提供指导，《环境评价：程序指南》（DoE，1989）主要是面向开发商及顾问的。针对 EIA 有效执行与评审的《环境事务部导则》修订版分别于 1994 年、1995 年（DoE，1994a、b，1995）和 1997 年（DETR）出版。新的 EIA 通告（DETR，1999；Scottish Executive Development Department 1999；NAFW，1999）伴随着 1999 年新的 T&CP 法案而出台，这些通告对法案进行了全面指导。新的指南《环境影响评价：程序指南》（DETR，2000）也随之出版。《英格兰环境影响评价法案指南》（2011）以及苏格兰相应的文件为这个领域提供了最新的资讯。这些通告和政府的其他导则具有很强的可读性。不过，其中只有法规是强制实施的，指南解释和建议是非强制性的。

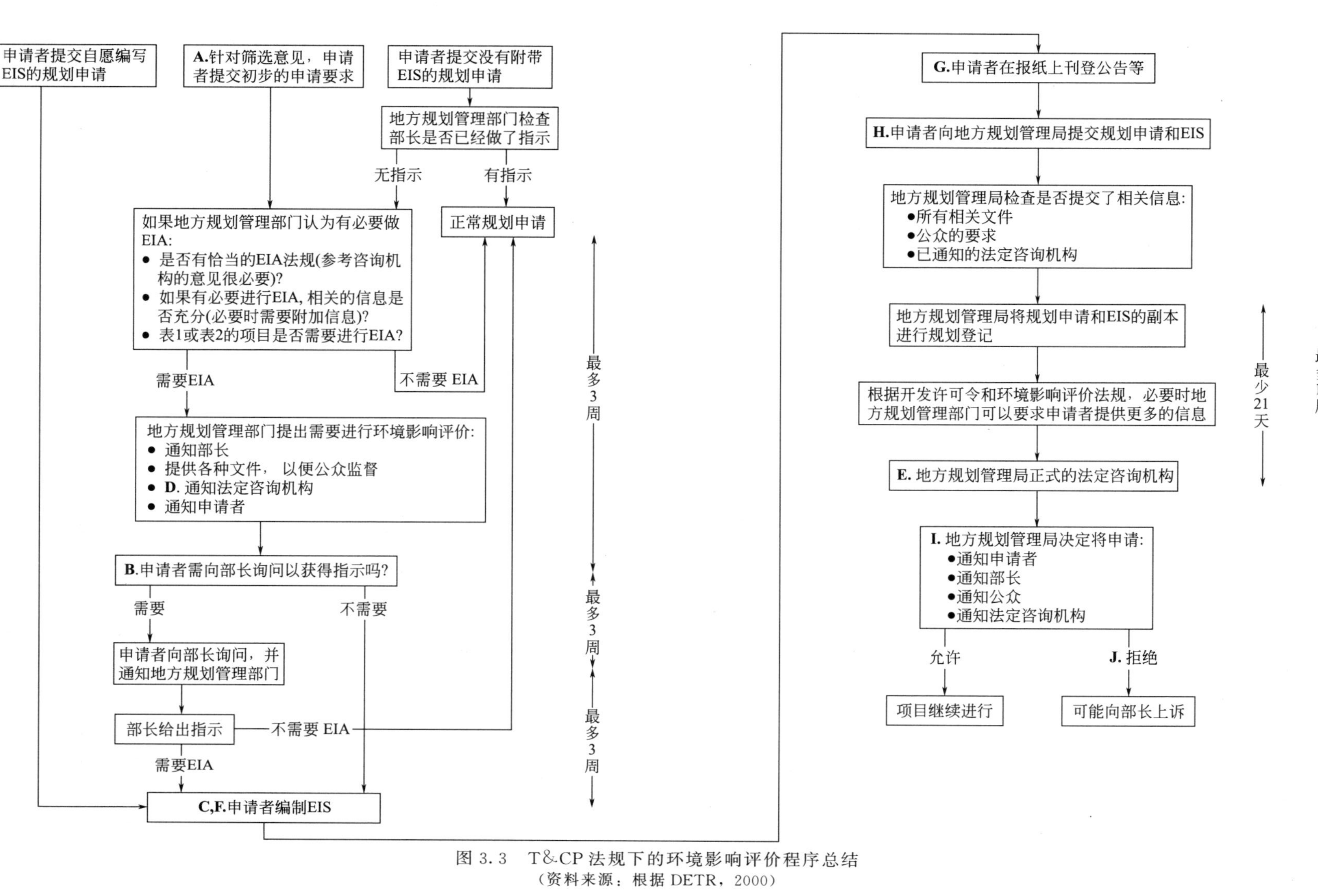

图 3.3　T&CP 法规下的环境影响评价程序总结
（资料来源：根据 DETR，2000）

3.4.1 哪些项目需要进行环境影响评价

T&CP 法案将需要进行 EIA 的项目分成两大类，详见清单 1 和清单 2。修订后的表与修订欧盟指令的附件Ⅰ和附件Ⅱ也很相似，表 2.4 对此进行了概述，详细叙述见清单 1❶。清单 1 与附件Ⅰ只是在用词上有微小的变化，另外加上附件Ⅰ的 1.20，即关于长途高空电力输送电缆的规定——2000 年《电力工程（环境影响评估）法案》（S1，1927）。另外，英国还建立了关于规定核电站以及反应堆淘汰的 EIA 法案。清单 2 也只对附件Ⅱ做了很小的修改，如附件 2.10（b）增加的范围包括体育场馆、休闲中心及多功能电影院。此外，对高速服务区也进行了单独分类（p），并且对其他的少量分类进行了重组。清单 2.12 还增添了对高尔夫球场及相关开发活动的分类（f）。清单 1 的项目都需要进行环境影响评价。清单 2 只有那些“可能会产生显著环境影响”的项目需要进行环境影响评价。

项目的环境影响显著性由一系列清单 2（附件 2）中的标准和条件组成以及法规中清单 3（表 3.5）的选择条件。清单 3 中的三个标准如下（DETR，2000；DCLG，2011a）：

① 该开发活动产生的影响是否超出了当地承载力（例如在自然尺度方面）；

② 拟建项目是否具有显著的环境敏感性或选址区域的环境脆弱性（例如国家公园或具有特殊科学研究价值的区域）；

③ 开发活动是否会产生异常复杂的、不利于环境的潜在影响（例如污染物排放）。

表 3.5 清单 2 发展规划筛选标准

1. 发展规划的特点
关于规划的特点必须特别考虑到以下几个方面：
(a)规划的面积；
(b)与其他规划的兼容性；
(c)使用自然资源；
(d)废物的产生；
(e)污染物；
(f)在使用特殊或者科技时，事故的风险。
2. 发展规划的选址
必须考虑到关于规划发展对环境敏感地区在以下几方面的影响。
(a)目前的土地使用；
(b)这个地区自然资源的数量和质量以及可再生能力；
(c)自然资源的吸收能力，尤其是以下几个方面需要特别注意：
i. 湿地；
ii. 海岸地区；
iii. 山脉和森林；
iv. 自然保护区以及公园；

❶ 目前有些争议，例如，在清单 1 里面发电量等于或者超过 300 万瓦的发电站，虽然它们全部在《电力工程（环境影响评价）法案》的范围里面，并且包含所有“特殊道路”；法案范围包括在地方规划局裁决的特殊道路。

续表

v. 欧盟 2009/147/EC 指令中确立的野生鸟类保护区，92/43/EEC 指令确立的自然栖息地[①]和野生动植物保护区[②]；

vi. 欧盟法案中提到的环境质量标准日益下降的地区；

vii. 人口密集区；

viii. 历史文化或者文化古迹地区。

3. 潜在影响的特点

规划的潜在重要影响必须考虑到在第一段和第二段中设置的标准，特别是：

(a)现存的影响(受影响群体的地理位置和面积)；

(b)跨界自然影响；

(c)影响的尺度和复杂性；

(d)影响的可能性；

(e)影响的持续时间、频率以及是否可恢复。

资料来源：DCLG，2011a。

① O. J. no. L 20，26. 1. 2010，p. 7。

② O. J. no. L 206，22. 7. 1992，p. 7。

第三类设计旨在确定前两类的相互作用（例如在发展规划和它的环境之间）是否会产生潜在的环境影响。一个项目如果满足清单 2 中规定的条件则需要进行 EIA。(a) 满足标准或者超过清单 2 中第 2 列的临界值（附录 2）；或者（b）全部或部分项目位于法规定义的敏感区内（表 3.5）。该地区环境敏感度越高，项目的潜在环境影响可能就越大，则越可能需要进行 EIA。

英国法规中的筛选环节引发过很多问题，包括很多著名的法律案例，这些案例将会在第 8 章 8.6 节中进行讨论。案例包括项目的扩建、多阶段的咨询以及在筛选环节中拟建项目减缓措施的考虑。

2011 年的法规对引起的很多问题做出了回应。

① 对清单 2.13 进行了修改，清单 2 中的标准适用于修改和扩建后的整体规划，而非拟修改和扩建的部分。例如，一个工厂扩建 0.2hm^2，因为施工面积在 0.5hm^2 以下可能不需要进行 EIA。目前这个项目扩建必须考虑原有规划，如果原有规划是 0.4hm^2，那么项目总面积是 0.6hm^2，则需要进行 EIA。

② 2011 年法规也包括欧盟法院（ECJ）关于筛选程序的主要决议，被称为梅勒案例，要求在筛选程序中的消极意见对公众可见。

③ 欧盟法院（ECJ）在多阶段的咨询决议中，2011 年法规解决了在项目初期取得许可但没有进行筛选，然而之后申请者被要求进行更改的缺陷。

④ 两个清单中包括一个新的类别，即碳获取和储存（CCS）。

2011 年法规对 1999 年规定中相关准则的适用标准进行了简化。在清单 2 中显示，这些规定大多是空间性的（例如规划区域超过 0.5hm^2；新的建筑面积超过 1000m^2）。目前的一个有趣的问题（受到媒体日益关注）是关于对越来越多的停运或者拆除项目的案例进行筛选。欧盟法院在 2011 年决议项目拆除在 EIA 指令范围里面，项目拆除可能会有潜在的严重的环境影响（例如历史建筑的拆解），当地规划机关必须发布是否需要进行 EIA 的意见。同理，相关申请人当想拆解一些建筑结构的时候，必须咨询环境筛选意见，考虑是否有严重的环境影响和是否需要 EIA。鉴于上述情况以及欧盟和英国的法律法规，2011 年法规也必须在五年之内进行更深层次的审查。

A. 根据 T&CP 法案，开发商可以决定需要进行环境影响评价的项目，或者对未要求环境影响评价的项目自愿进行环境影响评价。如果开发商不能确定，可以向地方规划管理局（LPA）询问（“筛选意见”）寻求意见。为此，开发商必须向地方规划管理局提供详细的报告，包括开发项目的地点、拟开发项目的总体描述以及指出可能产生的环境影响。地方规划管理局必须在接到报告后的 3 周内做出决定。地方规划管理局也可以要求开发商提供更为详细的信息，但这些必须在三周的决策期限内完成。

如果地方规划管理局认为不需要进行环境影响评价，申请就可以被视为正常的规划申请。反之，如果地方规划管理局认为需要进行环境影响评价，那么就必须要解释进行环境影响评价的原因，并且要将开发商提供的信息和最终决定公布于众。在所有的情况下，地方规划管理局必须对项目是否需要进行环境影响评价做出完整和精确的解释。如果地方规划管理局接到的规划申请并没有经过环境影响评价，但又觉得有必要进行时，地方规划管理局必须在 3 周内通知开发商，并要解释需要进行环境影响评价的原因。然后，开发商有三周的时间告知地方规划管理局他们是否打算准备环境影响评价报告，或者是选择向部长上诉；如果开发商什么都没有做，那么规划申请就自动废止了。

B. 如果开发商对地方规划管理局决定需要进行环境影响评价的意见有异议，开发商可以求助部长来裁决[1]。部长也必须在 3 周内给出决定。如果部长也要求进行环境影响评价，也需提供解释，并刊登在《规划与环境法》期刊上。如果不需要进行环境影响评价则不需解释说明。如果开发商没有要求部长进行裁决，即使地方规划管理局决定不需要进行 EIA 时，部长也可以依据来自其他团体或机关的信息，决定是否需要进行环评。

3.4.2 环境影响评价的内容

表 3.6 对 T&CP 法案中的清单 4 做了说明，列出了环境影响评价应包含的信息。按照指令第 5 条的标准，清单 4 阐释了环境影响评价指令附件Ⅳ的要求，如下。

成员国需采取必要的措施以确保开发商准确恰当地提供附件Ⅳ中指定的信息：

（a）成员国认为与批准程序的特定阶段、特殊项目的特定特性、项目类型和环境特点相关的信息可能会受到影响；

（b）成员国认为应要求开发商用新的知识和评价方法编写报告。

表 3.6 T&CP 法规 2011 要求的环境影响报告应该包含的内容——清单 4

根据第 2 条第 1 款法规的定义，“环境报告”是指：
(a)包括清单 4 中第一部分所要求的信息，因为理论上需要用它们来评价建设项目的环境影响，并且要求申请者在写报告时特别注意采用最新的评价技术和方法。
(b)至少包含清单 4 中第二部分所要求的信息。
第一部分
1. 对项目的介绍，尤其应该包括：
(a)整个项目的自然属性的描述以及在其建设和运行阶段需要利用的土地；
(b)对生产过程的主要特点的描述，例如，所利用资源的种类和数量；

[1] 在 2011 年以后废除了关于由相关政府部分对区域间被关注的项目做出决定的规定。第 8 章将会详细讨论这点，因为两个或者多个政府部门有时候会针对类似的案例做出不同决定。

(c)通过类型和数量，估计项目在建设和运行阶段可能产生的残余物和废弃物(水、空气和土壤污染物、噪声、振动、光、热、辐射等)。
2.在充分考虑环境影响的前提下，给出申请者或者是反对者所研究的主要代替方案的概要及其选择的主要原因。
3.指出建设项目可能引起的显著环境影响，尤其是对人口、动植物、土壤、水、大气、气候因素、物质资产(包括建筑和考古学遗产)、风景的影响以及以上因素之间的相互作用。
4.指出项目对环境可能造成的显著影响，这应该包括直接影响以及任何间接影响，二次影响，累积影响，短期、中期和长期的影响，长期和暂时性影响，积极和消极影响，这些是由上而下的活动产生的：
(a)现有的开发项目；
(b)对自然资源的利用；
(c)污染物排放、产生的臭气以及废弃物排放，申请者对评价环境影响的预测方法的陈述。
5.描述预防、降低以及补偿任何重大的消极环境影响所采取的措施。
6.对本部分第一条到第五条提供的信息进行非技术性总结。
7.指出申请者在汇总所需信息时可能遇到的各种困难(技术缺陷或者实际知识的缺乏)。
第二部分
1.对建设项目的介绍，包括项目的选址、设计和规模。
2.给出为了避免、减少以及可能补偿消极影响等所采取的措施。
3.识别和评价建设项目可能产生的主要环境影响所需要的数据。
4.在充分考虑环境影响的前提下，给出申请者或反对者所研究的主要替代方案的概要及其选择的主要原因。
5.对于这部分要求的第一条到第四条信息的非技术性总结。

资料来源：DCLG，2011a。

在清单4中，原来附件Ⅳ的信息被分为了两个部分。环境影响评价必须包含第二部分列出的信息，并且还要包括在第一部分里那些被认为是“评价项目影响的合理要求，并且开发商在编写报告时应予以合理考虑”的信息。这个区别非常重要：第8章就此进行了探讨，目前的环境影响报告大多数没能很好地反映第一部分列出的信息，虽然这部分包含了一些重要的内容，如被考察的替代方案和开发活动产生的预期废弃物及污染物排放。此外，在指南的附录5(DETR，2000)中，DETR给出了一份很长的环境影响评价需要包括的项目清单，这份清单虽然只是指导性的，但有助于确保把所有较为显著的环境影响都考虑进去(表3.7)。

表3.7　环境报告应考虑的问题核查清单

该核查清单给出了在环境报告准备阶段所需要的一些信息。当然这里的每个内容并不一定适用于所有的项目。
要分别考虑建设项目在建设、试运行以及真正运行阶段的环境影响。当一个项目的运行期有限时，也应分别考虑项目解体或者土地恢复的影响。
第一部分　项目的相关信息
1.1项目的目的和自然属性，包括对提议的通道和交通布置的详细信息以及员工的数量及其来源。
1.2土地利用要求和项目其他的自然特征：
• 建设阶段；
• 运行阶段；
• 运行终止以后(如果适当的话)。
1.3项目生产过程和运行方面的特征：
• 原材料、能源以及所消耗的其他资源的种类和数量；
• 残余物与污染物的种类、数量、组成以及浓度，包括：水污染物；大气污染物；噪声；振动；光污染；热污染；辐射；地面和土壤中堆放的废弃物/残渣；其他。
1.4在适当条件下，考虑主要的替代选址和生产过程以及最终选择的原因。

续表

第二部分　有关项目选址及其环境的信息

自然特征

2.1 人口密度和数量。

2.2 动植物(包括它们的栖息地和种类)尤其是受保护物种及其栖息地。

2.3 土壤:农业特性、地质和地形概况。

2.4 水:蓄水层、河道、海岸线,还包括各种现存污染物的种类、数量、组成及其浓度。

2.5 空气:气候因素、空气质量等。

2.6 建筑和历史遗迹、考古遗址及特征以及其他物质资产。

2.7 景观和地形。

2.8 娱乐用途。

2.9 各种其他相关的环境特征。

政策框架

2.10 如果可行,这部分信息还应该包括所有相关的法定区域,例如国家自然保护区、具有特定科研价值的地点、国家公园、著名自然风景区、有历史价值的海岸、地区公园、乡村公园以及指定的绿化带、地方自然保护区、受林业保护法规制约的地区、水资源保护区、自然保留地、受保护的建筑、历史悠久的纪念碑和特定的有考古价值的区域。另外还要参考相关的国家政策[包括规划政策导则(PPG)条款]和区域、地方计划和政策(包括已经通过的或正在制订的计划)。

2.11 还要参考国际公约,例如欧盟"野生鸟类"或"栖息地"指令、《生物多样性公约》和《拉姆萨尔公约》。

第三部分　影响评价

包括项目的直接影响和间接影响,二次影响,累积影响,短期、中期和长期影响,永久性和暂时性影响,积极和消极影响。

对人类、建筑和人造资本的影响

3.1 项目导致的人口数变化以及由此产生的环境影响。

3.2 项目对周边地区和景观的视觉影响。

3.3 在项目正常运行阶段产生的废弃物等级及其影响。

3.4 项目的噪声等级和影响。

3.5 项目对当地道路和交通的影响。

3.6 项目对建筑物、古建筑和历史遗迹、考古方面和其他人工制品的影响,例如,通过污染物排放、视觉污染、振动等途径造成的影响。

对动植物和地质的影响

3.7 栖息地、动植物物种的消失和破坏。

3.8 地质、古生物和地形特征的消失和破坏。

3.9 其他的生态影响。

对土地的影响

3.10 项目的自然影响,例如当地地形的改变、对土壤运动稳定性的影响、土壤侵蚀等。

3.11 化学废物和堆放废弃物对该地土壤和周围土壤的影响。

3.12 对土地利用/资源的影响:

- 被占农田的质量和数量;
- 矿产资源枯竭;

续表

• 其他可替代的土地，包括“土地闲置”方案；

• 对周围土地利用(包括用于农业)的影响；

• 废弃物处置。

对水的影响

3.13 项目对区域排水系统的影响。

3.14 其他水文特征的变化，例如地上水位、河道以及地下水的流动等。

3.15 对海岸或河口水文的影响。

3.16 污染物、废弃物等对水质的影响。

对于空气和气候的影响

3.17 化学排放物的数量和浓度以及它们的环境影响。

3.18 颗粒物。

3.19 刺激性物质。

3.20 其他气候影响。

与项目相关的其他间接影响和二次影响

3.21 与项目相关的交通(公路、铁路、航空、水运)所产生的影响。

3.22 与项目相关的资源、水、能源或者其他资源的开采和使用所产生的影响。

3.23 与本项目相关的其他项目产生的影响，例如新的道路、下水管道、住房、电线、管道、电信等的建设。

3.24 该项目与其他现存的或提议的项目共同产生的影响。

3.25 由上述单独的直接影响相互作用造成的二次影响。

第四部分　减缓措施

4.1 当已经确认显著的不利影响应该提出避免、降低或者补偿这些影响的减缓措施。例如：

(a)选址规划。

(b)技术措施，例如：

• 程序选择；

• 回收；

• 污染控制和预防；

• 密封(例如将储存容器加封)。

(c)美学和生态措施，例如：

• 假山；

• 设计、上色等，

• 景观设计；

• 植树；

• 保护特殊栖息地或建造替代栖息地的措施；

• 记录考古地点；

• 保护古建筑和历史遗址的措施。

4.2　评价可能有效的减缓措施。

续表

第五部分　风险和开发危险
5.1 环境影响评价指令或者最后实施的法规中未包括的事故风险。然而，如果所提议的项目需要使用那些在事故中对环境（包括人类）造成危害的材料，那么环境报告中就应该包括需要采取的预防措施，以保证一旦发生事故不会造成严重的影响。如果条件允许，还应该适当参考健康和安全法。
5.2 对于有害物质的保存和使用已经有现行的法规，并且对各种计划实施中涉及有害物质的设施，健康与安全管理局（HSM）都会向当地规划部门提供风险评价需要的专家意见。
5.3 但如果可能的话，最好将项目的事故风险和一般环境影响进行综合考虑，发展和规划部门都应该谨记这一点。

资料来源：DETR，2000。

C. 1999 年修订指令实施前，英国对正式的“范围确定”阶段一直没有强制的法律要求，在此阶段，地方规划管理局、开发商以及其他相关团体需要就环境影响评价内容达成共识。尽管指南（DoE，1989）早就强调了对环境影响评价的范围进行早期咨询以及较早达成共识的好处，但事实上，在正式的环境影响评价和规划申请提交之前，在开发商与其他受影响的群体之间从未要求过任何形式的咨询。1999 年以及之后的法规规定，开发商在真正的规划申请之前需要向地方规划管理局申请对环境影响报告所包含的信息进行正式的“范围评价”。这可以使开发商了解地方规划管理局对预期重要影响所持的观点。开发商在提出请求的同时还必须提供与筛选评价相同的信息，也可以与筛选评价同时提供。地方规划管理局必须咨询某些团体（参见 D），并且必须在 5 周内完成筛选评价。在征得开发商同意的前提下，时间也可以延长，如果地方规划管理局和开发商在环境影响评价报告的内容上存在分歧，这里没有规定可以向部长求助。但如果地方规划管理局在规定时间内仍然无法完成筛选评价，开发商就可以请部长（或者议会）在 5 周内做出筛选指导，并且也可以向某些团体进行咨询。在开发商和地方规划管理局的讨论中，核查清单（DETR，2000）起到了有效的辅助作用（表 3.7）。

3.4.3　法定的及其他咨询机构

如 3.2 节所述，在 T&CP 法案的要求下，环境影响评价程序中有一系列法定咨询机构。这些机构主要参与到环境影响评价的两个阶段，另外也可能参与到“范围确定”阶段。

D. 首先，当地方规划管理局决定某个项目需要进行环境影响评价时，它必须通知这些法定咨询机构。目前，英格兰的相关咨询机构包括相关规划部门、英格兰自然署、环境局和最近成立的海洋管理组织（主要针对项目对海洋产生的潜在影响）。如果开发商需要咨询而且支付的费用较为合理，这些咨询机构应该向开发商提供其所掌握的所有相关的环境信息。例如，英国自然署可以提供有关地区生态的信息。这些信息不包括任何机密和空白信息。

E. 其次，一旦提交了环境影响评价报告，地方规划管理局或者开发商必须免费向这些法定咨询机构提供环境影响报告的副本。在收到环境影响评价报告后的两周内，咨询机构可以向地方规划管理局提出意见。当地方规划管理局在决定是否授予开发许可时，必须考虑咨询机构的意见。开发商在编制环境影响报告时，也可以与其他的咨询机构及公众进行联系。政府指南中指出，这些机构可能在相关领域有着相当丰富的经验，或者可能关注与项目相关的某些重要环境问题。尽管开发商没有任何义务与这些机构联系，但是政府导则仍再三强调早期全面的咨询可能带来的好处。

3.4.4 实施环境影响评价：编制环境影响报告

F. 对于环境影响评价中应该应用哪些技术和方法，政府没有给出正式的指导，仅指明它们会因开发活动、环境团体与可利用信息的不同而不同，而且与之相关的影响预测往往也是不确定的。

3.4.5 提交环境影响报告和规划申请以及公众咨询

G. 当环境影响报告完成后，开发商必须在当地报纸上或者当地公告栏上发布通告。这些通告必须满足 1995 年《城乡规划法案》（一般允许的开发项目）的要求，即应该向公众提供环境影响报告的副本以便于公众监督；尽可能给出一个可以获得环境影响报告副本的当地地址，并且标明编制环境影响报告的费用。在公告发出的 20 天内，公众可向地方规划管理局提出书面陈述，但必须在地方规划管理局收到规划申请的 21 天之内。

H. 环境影响评价报告需至少公布 21 天，之后开发商可向地方规划管理局提交计划申请、环境影响报告副本[1]以及已经向公众发出了必要公告的证明。地方规划管理局必须将环境影响评价报告交给法定咨询机构，并邀请它们在指定时间内给出书面意见（至少在接到环境影响评价报告两周内），同时也要向部长递交一份报告，并且将此环境影响报告进行规划登记。在做出决策之前，必须确定是否还需要与项目相关的其他信息，如果需要，应向开发商索取。但并不能因为这样就耽搁时间，决策仍必须在规定的时间内完成。

3.4.6 计划决策

I. 在对规划申请做出决策之前，地方规划管理局必须在收到规划申请的 3 周内收集公众的书面意见，并且在收到环境影响报告两周内收集法定咨询机构的书面意见。从收到规划申请到最终决策至少要 14 天的时间。与那些必须在 8 周内做出决定的常规规划申请相比，附带环境影响报告的规划申请至少需要 16 周才能做出决定。如果 16 周后，地方规划管理局还没做出最后决定，申请者可以向部长寻求最终决策。即使地方规划管理局觉得附带的环境影响评价报告不够充分，也不能由此断定规划申请无效，只能在 16 周内要求提供更多的信息。

在决策过程中，地方规划管理局必须考虑环境影响评价报告，公众和咨询机构的意见以及其他重要因素。环境信息和其他重要因素一样，也只是地方规划管理局所考虑因素的一部分。实质上，决策仍然是一种政策性行为，但必须确保了解该项目的环境影响。地方规划管理局可以有条件或无条件地批准或拒绝申请。对地方规划局来讲，考虑环境评价报告的减缓措施是授予许可的重要前提和附加条件；也可通过强制规划条款获得许可证。针对修订欧盟指令所带来的变化，地方规划管理局除了必须按照正常的要求通知申请者外，还应该告知部长，并在当地媒体上刊登公告，给出最后决策以及制定决策的主要依据，并提出主要的减缓措施。

---

[1] 然而开发商没有提供足够的文件：这个包括为所有法定咨询机构提供足够的文件，为地方规划局提供一份文件，为部长提供几份文件。

J. 对于一个正常的规划申请，如果遭到地方规划管理局的拒绝，开发商可以求助于部长，而部长可以在决策之前索要更多的信息。

3.5 其他环境影响评价法规

依据迄今为止大量的环境影响报告，本节还对其他的环境影响评价法规程序进行了总结，下面将根据使用频率的高低依次给出这些法规，第七项可能在最近排名中上升很快。

(1)《环境影响评价法案（苏格兰）》2011；

(2)《高速公路（环境影响评估）法案》1999；

(3)《环境影响评价（地面排水改进工程）法案》1999；

(4)《电力工程（环境影响评估）法案》2000；

(5)《管道工程（环境影响评价）法案》2000；

(6)《环境评价法案（林业）》1999；

(7)《基础设施规划（环境影响评价）法案》2009。

3.5.1 《环境影响评价法案（苏格兰）》（包含苏格兰城乡规划法案） 2011

《环境影响评价法案（苏格兰）》与《英格兰环境影响评价法案》十分相似。该指令不仅可用于需要获得规划许可的项目，还可用于一些地面排水系统和主干道等建设项目。苏格兰法规有其独立的导则（表 3.2）。对某些项目而言，如核电站废弃项目，也包含在苏格兰法规之中，并且该法规在英国的其他地区也有应用。在北爱尔兰，指令适用于那些根据 1999 年（目前有修改）《规划（环境影响评价）法案（北爱尔兰）》需要获得规划许可的项目。同理，《环境影响评价法案（苏格兰）》也根据英国的法案同步进行了修改。对 2011 年法案的注释如下：

因考虑到欧盟法院和国内的案例决议以及欧盟指令和国内的法律体系，现存的 1999 版法案被反复修改（11 年中修改了 12 次）。致使 1999 版法案更加复杂并且很难采用。苏格兰议会法规委员会为了让法规更易理解而要求整理原有法规。

在苏格兰的法规中，对清单 2 中项目的筛选引入具体问题具体分析的筛选原则：规划的特点、规划的选址以及潜在影响。鼓励地方规划管理局使用包括苏格兰政府（规划）环境影响评价网站中的调查表。

3.5.2 《高速公路（环境影响评估）法案》 1999

《高速公路（环境影响评估）法案》适用于交通部（DoT）拟建的高速公路以及主干道建设项目。该法案是在 1980 年《高速公路法案》所设定程序的基础上通过的，它规定在编制某些新的高速公路或者对现有高速公路的大规模改造的草案时，应该要求交通部部长对拟建路线编制环境影响报告。由部长确定该项目隶属于指令的附件Ⅰ还是附件Ⅱ，以及是否需要环境影响评价。对于新建高速公路和其他四车道以上（含四车道）的公路以及某些公路的改造，都需要强制进行环境影响评价。其规定环境影响报告必须包括：

- 对制订的方案及选址的描述；
- 对不利环境影响的减缓措施的描述；
- 充足的数据，用以识别和评价项目可能产生的主要环境影响；
- 非技术性总结。

1993年之前，《高速公路（MI）法案》中的要求在DoT的AD18/88标准（DoT，1989）和《环境评价手册》（DoT，1983）中均做了详细的说明。为了应对激烈的反对意见[1]，尤其是来自主干道评价咨询委员会（SACTRA）的反对意见（1992），法规被1993年推出的《公路及桥梁设计手册》（DMRB）第Ⅱ卷和《环境评价》（DoT，1993）所取代。该手册建议环境影响评价分为3个阶段，并且针对如何实施环境影响评价给出了大量的详细建议。《公路及桥梁设计手册》（DfT，2011）提供了公路项目导则的最新进展。在交通部交通分析指南网站上还有大量包括《交通评估导则》等的关于交通评估的有用信息。

3.5.3 《环境影响评价（地面排水改进工程）法案》1999

《环境影响评价（地面排水改进工程）法案》适用于除公共卫生排水系统之外的大部分的英格兰和威尔士的河道。如果某个排水单位（包括作为排水机构的地方当局）认为其拟改建工程提案有可能产生显著的环境影响，那么就必须至少在当地两份报纸上对其拟建工程进行说明，并且表明是否打算编制环境影响报告。如果该单位不打算编制环境影响报告，公众可以在28日内就该提案可能产生的任何环境影响提出意见。如果没有提出意见，该单位就可以在没有环境影响报告的情况下实施拟建工程。如果提出了意见，而该单位仍想在没有环境影响报告的情况下继续实施该项目，DEFRA（威尔士议会）可以在部长级会议上就此问题进行裁决。

这些法规对环境影响报告内容的要求与T&CP法规的要求实际上是完全相同的。当环境影响评价报告完成后，排水单位必须在当地两家报纸上刊发公告，并向英国自然署、环境局（EA）、海洋保护组织及其他相关单位递交环境影响报告副本，且将副本价位控制在公众能接受的范围内。公众必须在28日内提出意见，并确保排水单位在做出决策时对这些意见给予充分的考虑。只有当所有异议都消除时，项目才可继续进行，否则将交由主管部长进行裁决。总之，这些法规相对于T&CP法规而言要薄弱，因为它们更侧重于提议的通过，除非存在异议，否则将对环境组织的咨询要求降到最小。

3.5.4 《电力工程（环境影响评估）法案》2000和《核反应堆退役（环境影响评价）法案》1999

依照1989年《电力法案》第36、37节的要求，超过50MW电站的新建和扩建以及高压缆线的铺设都需要得到相关部长（目前是能源与气候变暖部）的批准。《电力工程（环境影响评估）法案》2000是这些条款下申请程序的一部分。需要进行环境影响评价的项目有：

---

[1] 批评的声音铿锵有力。通告断定“……个别的高速公路项目对气候的严重影响有限，在大多数情况下，不可能对土地和水源造成严重影响”，这是对私有公路建设的累计影响非常有趣的观点。

• 指令附件Ⅰ中所涉及的热电及核电站建设（例如发电量在300MW或以上的热电站以及发电量在50MW以上的核电站）；

• 电压在200kV以上的高架线或距离在15km以上的输电线路的铺设工程。

另外，法规也要求对附件Ⅰ中没有涉及的电站类型以及电压在132kV以上的所有高压输电线路铺设等工程进行环境影响评价。低于50MW的电站可以由规划法规以及T&CP（EIA）法规来决定是否进行环境影响评价。《电力工程（环境影响评估）法案》允许开发商向部长提出书面请求，以决定是否需要进行环境影响评价。部长在做出决策之前还必须要咨询地方规划管理局。当开发商宣布已完成环境影响报告之后，部长必须通知地方规划管理局或者相关地区的重要委员会、乡镇办公室、英国自然署、环保局、海洋保护组织有关电站的信息，以便于它们向申请者提供相关信息。环境影响报告所要求的内容与T&CP法案中列出的内容几乎相同。

1999年，法规中还添加了有关核电站退役方面的内容。设备拆除及整体退役需要健康安全局（HSE）批准。申请者必须向健康安全局提供环境影响报告。该法规也适用于那些现有的且可能对环境产生重大影响的设备拆除或整体废弃的项目。

3.5.5 《管道工程（环境影响评价）法案》2000、《公共天然气输送管网工程（环境影响评价）法案》1999及《海洋石油生产及运输管网工程（环境影响评估）法案》

这一系列有关管网铺设工程的法案不仅反映出此类开发活动的增多，也反映出英国环境影响评价法规体系的不断完善。关于海岸管网工程和公用天然气输送管网工程的法律仅应用于英格兰、威尔士和苏格兰；《海洋石油生产及运输管网工程（环境影响评价）法案》将应用于整个英国。油气管道的直径在800mm以上且距离在40km以上的铺设项目均隶属于指令的附件Ⅰ；如果有一项指标低于该标准，则隶属于附件Ⅱ。对于后者，10km左右的管网工程可以由规划条例来决定。纳入以上管道法规的其他项目，通常由相关部门的部长通过联合审批和授权程序以及根据各种标准和阈值来进行裁决。例如，指令附件Ⅱ中滨海天然气管网如果设计操作压力超过7个标准大气压[1]，或者它们完全或部分穿过某个敏感区域（如国家公园），那么就可能需要进行环境影响评价。

3.5.6 《环境影响评价（林业）法案（英格兰和威尔士）》1999

根据最初的环境影响评价指令和相关的英国法规，林业环境影响评价仅限于项目申请者以绿化为目的并希望获得林业部门的同意或者提供贷款的项目。很多林业管理局主管的项目没有开展环境评价，导致了该局没有开展环境影响评价的项目遭到了一致的批评（例如英格兰乡村保护委员会的批评，1991）。之后，修订指令和其他相关的英国法规都就此问题进行了改进。

植树造林和森林开发项目均隶属于指令附件Ⅱ。根据以上法规，任何想进行可能造成显著环境影响的林业项目的团体和个人，在项目开发前，都必须取得林业局的审批同意，并且

[1] 1个标准大气压=101.325kPa。

申请者需要准备环境影响报告。法规中包括的项目有：植树造林（开辟新的林地）、伐木（将林地转作他用）、铺设森林公路和筑路所需原材料的开采。如果项目的面积在 $1hm^2$ 以下或者造林项目在 $5hm^2$ 以下，除非是处于敏感区，否则均可认为其对环境不会产生显著影响。考虑到地点和项目的不确定性，林业局需要逐案考察申请。如果申请者不同意林业局的决定，他们可以向相关的部长申请寻求帮助。林业法所要求的环境影响评价的内容与T&CP法案的要求基本一致。

3.5.7 《基础设施规划（环境影响评价）法案》2009

3.5.7.1 内容

在1.4.1节中提到，在2008年规划法中（HMG，2008），一组国家级重要基础建设项目（NSIPs）被列出来，这些项目影响的测评将由一个新机构——基础设施委员会（IPC）负责（该机构于2012年与英国规划检测委员会合并）。NSIPs包括主要的能源项目、交通项目（道路、铁路和港口）、水和废品设施等。其中的很多项目在前面介绍的法案中曾经提到（特别是高速公路、电力和管道）。IPC等检测范围包括在《基础设施规划（环境影响评价）法案》2009提到的重大环境影响。环评法规加强了对某些NSIPs拟建项目进行环境影响评价的程序要求。例如，NSIPs项目中的核电站是要求进行环境影响评价的，但是别的项目，例如风力农场，只有在考虑到其项目本质规模或者位置会造成严重环境影响情况下才要求进行环境影响评价。在环境影响评价法规中IPC的作用包括：

- “筛选”拟建项目来决定其是否需要进行环境影响评价；
- 拟定评价“范围”，建议申请者在环评报告中应该提供哪些信息及向其他咨询单位寻求哪些意见；
- 协助准备环境影响评价报告，提醒相关咨询机构有义务提供相关信息和告知申请者；
- 在给出决定前，评估在环评报告中的环境信息以及环境影响提议；
- 发布IPC“筛选”和“范围”的意见；
- 发布对包含环评报告的申请的决定。

如果不考虑环境信息，IPC可能不会让任何项目通过（环境影响评价报告或者/以及任何进一步关于项目环境影响的信息）。它可以以环境信息不足为理由不接受申请，或者不接受一个需要进行环境影响评价但是没有提供环境影响评价报告的项目。

3.5.7.2 筛选

一个拟建项目可能在如下情况下需要进行环境影响评价：如果申请者通知IPC有意向提交一份环境影响评价报告；部长认为该项目需要进行环境影响评价；或者拟建项目在环境影响评价法规清单1中。很多NSIPs项目都有可能在清单1中，它们是需要进行强制环评的。另外，一些在清单2中的项目，如果进行环境影响评价，则要考虑是否与《环境影响评价法案》清单3中“选择条件”有冲突。拟建项目与2011年T&CP法案（环境影响评价）有所不同，有些项目需要进行环评而2011年法案的条件和阈值却不要求。其他的项目中，如果申请者没有提交环评报告的意向，他们必须要求IPC出具一份关于该拟建项目的筛选意见。理想的情况下申请者需要提供该项目特点的信息、项目的地理位置和潜在影响的特点。筛选意见的要求提出以后，IPC必须在21天以内出具一个筛选意见。

3.5.7.3 范围

在项目申请许可以前，申请者有机会向IPC申请要求提供一份正式“范围意见”，该意见会被包括在环评报告中。当提出申请时，申请者必须提供一个最低要求：一份翔实的用地规划，关于项目本质和目的的一份简单说明和潜在环境影响以及其他申请者希望提供的意见和建议。然而目前的现实是，申请者向IPC要求正式范围意见时，同时自己提供一份范围报告。理想的情况下这份报告需包括：

- 关于项目或者拟定规划的描述；
- 关于项目地点设置以及周围环境的介绍；
- 罗列关于备选方案以及如何达成意见一致的方法；
- 二手资料研究及尽可能开展基线调查的结果；
- 预测影响的方法以及认定标准框架；
- 对减缓措施和残余影响的考虑；
- 范围中部分关键问题的考虑；
- 环境影响评价报告的提纲。

采用一个范围意见以前，IPC有义务去咨询环境影响法规中的咨询机构。这些机构有28天时间去回复。IPC必须在收到范围要求的42天以内给出范围意见，目前为止，这些意见趋向于更加丰富而相关。最后，IPC所有的环境影响评价项目都可以在其网站上找到[1]。

3.6 修正法规的概括总结

最初的（和修正的）欧盟环境影响评价指令在英国的实施，是通过一系列法律将开发商、受影响群体、调解者以及咨询者以各种方式联系起来。其中，T&CP环境影响评价法案是最主要的。其他法规所包含的项目并不属于英国及威尔士规划系统，如高速公路及主干道建设、电站、管网、地面排水工程、林业项目以及苏格兰、北爱尔兰的开发项目。

原有的英国法规存在许多缺陷，所涉及的项目包括：环境影响评价的程序、筛选方法、范围确定、替代方案选择、环境影响报告书中的强制和自选内容、公众咨询及其他。欧盟97/11/EC指令以及后续修正案力求解决在英国和其他成员国已经产生的这些问题。

为针对欧盟指令和国内法律的修改以及欧盟和国内法院的决议，英国法规曾经被反复修改过多次。新的英国法规与欧盟修订指令中现行的4个附件的要求非常接近。这使得英国环境影响评价体系所涉及的项目范围更广，增加了需要强制进行环境影响评价的项目种类，并且促进了环境影响评价活动的增加和大量环境影响报告的编制。筛选程序改进后，环评的范围确定和替代方案的选择也都有了更清晰的框架。2011年以后，T&CP环境影响评价法案修改后，法规虽然已经逐步合理化，但仍非常复杂。环境影响评价导则在英国环境影响评价系统中具有重要的地位，政府刊物，尤其是相关通告和程序指导手册，推动着这一法律体系的发展。

---

[1] 基础设施建设委员会的实践活动发展迅速，目前主要包含了很多规划申请前的活动，包括环境影响评价。《共同声明（SOCG）》在开发商、地方规划局以及地方规划局颁布的当地专用影响报告（LIRs）之间起到了重要作用。

问题

以下一些问题致力于帮助读者注意到本章所述的一些重要问题以及帮助建立对英国环境影响评价相关的法律体系和相关部门的理解。

1. 在一个你熟悉的项目中，找出不同的主要参与者，罗列出他们各自利益中可能引起与彼此冲突的方面。

2. 在下面三个项目中，《英格兰环境影响法案》和调节者的作用分别是什么？

(a) 清单2中高尔夫球场项目；(b) 高速公路项目；(c) 核电站的拆除项目。

3. 在环境影响评价程序中法定咨询机构的作用是什么？

4. 在T&CP环境影响评价法案中，一个项目如果被认为“有潜在的重要环境影响”则需要进行环评。那么决定重要影响的条件和阈值是什么？

5. 罗列出2011年T&CP环境影响评价法案中的新变化以及造成这些变化的原因。

6. 社会经济基础设施的变化以及新科技的发展导致被列入环评体系中的项目越来越多。请你预测在下一次2016年法案重新评审中可能会有哪些变化？

7. 目前英国废弃和拆除部分项目的环境影响评价只应用于核电站和反应堆。检测如果把类似范围扩展到所有项目中的假设。

8. 什么是“筛选意见”？谁提供此意见？

9. 指出《基础设施规划（环境影响评价）法案》2009中有何创新特点？

3

UK agency and legislative context

3.1 Introduction

This chapter discusses the legislative framework within which EIA is carried out in the UK. It begins with an outline of the principal actors involved in EIA and in the associated planning and development process. It follows with an overview first of relevant regulations and the types of project to which they apply, and then of the EIA procedures required by the recently revised Town and Country Planning (T&CP) regulations 2011. These can be considered the 'generic' EIA regulations that apply to most projects and provide a model for the other EIA regulations. The latter are then summarized. Readers should refer to Chapter 8 for a discussion of the main effects and limitations of the application of these regulations.

3.2 The principal actors

3.2.1 *An overview*

Any proposed major development has an underlying configuration of interests, strategies and perspectives. But whatever the development, be it a motorway, a power station, a reservoir or a forest, it is possible to divide those involved in the planning and development process broadly into four main groups. These are:

- the developers;
- those directly or indirectly affected by or having an interest in the development;
- the government and regulatory agencies; and
- various intermediaries (consultants, advocates, advisers) with an interest in the interaction between the developer, the affected parties and the regulators (Figure 3.1).

An introduction to the range of 'actors' involved is an important first step in understanding the UK legislative framework for EIA.

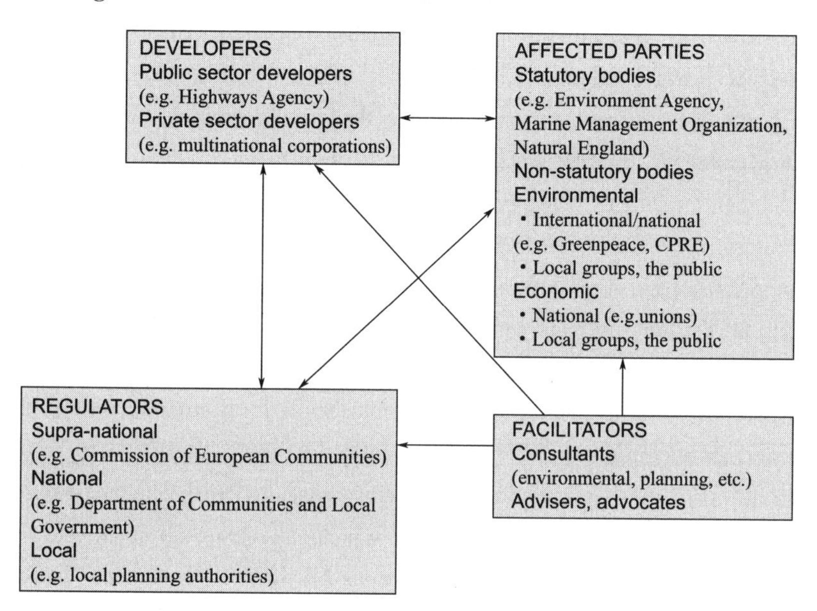

Figure 3.1 Principal actors in the EIA and planning and development processes

3.2.2 Developers

In the UK, EIA applies to projects in both the public and private sectors, although there are notable exemptions, including Ministry of Defence developments and those of the Crown Commission. Public sector developments are sponsored by central government departments such as the Department for Transport (DfT), by local authorities and by statutory bodies, such as the Environment Agency and the Highways Agency. Some were also sponsored by nationalized industries (such as the former British Rail and the nuclear industry), but the rapid privatization programme since the 1980s has transferred most former nationalized industries to the private sector. Some, such as the major energy companies (British Gas, National Grid and EDF) and the regional water authorities, have major and continuing programmes of projects, where it may be possible to develop and refine EIA procedures, learning from experience. Many other private-sector companies, often of multinational form, may also produce a stream of projects. However, for many developers, a major project may be a one-off or 'once in a lifetime' activity. For them, the EIA process, and the associated planning and development process, may be much less familiar, requiring quick learning and, it is to be hoped, the provision of some good advice.

3.2.3 Affected parties

Those parties directly or indirectly affected by such developments are many. In Figure 3.1 they have been broadly categorized, according to their role or degree of power (e.g. statutory, advisory), level of operation (e.g. international, national, local) or emphasis (e.g. environmental, economic). The growth in environmental groups, such as Greenpeace, Friends of the Earth, the Campaign to Protect Rural England (CPRE) and the Royal Society for the Protection of Birds (RSPB), is of particular note and is partly associated with the growing public interest in environmental issues. For instance, membership of the RSPB grew from 100,000 in 1970 to over a million in 1997, and has maintained this high level since. Membership of Sustrans, a charity that promotes car-free cycle routes, rose from 4,000 in 1993 to 20,000 in 1996; CPRE has over 60,000 members. Such groups, although often limited in resources, may have considerable 'moral weight'. The accommodation of their interests by a developer is often viewed as an important step in the 'legitimization' of a project. Like the developers, some environmental groups, especially at the national level, may have a long-term, continuing role. Some local amenity groups also may have a continuing role and an accumulation of valuable knowledge about the local environment. Others, usually at the local level, may have a short life, being associated with one particular project. In this latter category we can place local pressure groups, which can spring up quickly to oppose developments. Such groups have sometimes been referred to as NIMBY ('not in my back yard'), and their aims often include the maintenance of property values and existing lifestyles, and the diversion of any necessary development elsewhere. Another colourful relation of this group is BANANA ('build absolutely nothing anywhere near anything').

Statutory consultees are an important group in the EIA process. The planning authority must consult such bodies before making a decision on a major project requiring an EIA. Statutory consultees in England include Natural England (NE), the Marine Management Organization (MMO), the Environment Agency (for certain developments), and the principal local council for the area in which the project is proposed. Other consultees often involved include the local highway authority and the county archaeologist. As noted above, non-statutory bodies, such as the RSPB and the general public, may provide additional valuable information on environmental issues.

3.2.4 Regulators

The government, at various levels, will normally have a significant role in regulating and managing the relationship between the groups previously outlined. As discussed in Chapter 2, the European Commission has adopted a Directive on EIA procedures (CEC 1985 and amendments). The UK government has subsequently implemented these through an array of regulations and guidance (see Section 3. 3). The principal department involved currently is (2012) the Department for Communities and Local Government (DCLG; formerly ODPM, DTLR, DETR and DoE!) through its London headquarters. Of particular importance in the EIA process is the local authority, and especially the relevant local planning authority (LPA). This may involve district, county and unitary authorities. Such authorities act as filters through which schemes proposed by developers usually have to pass. In addition, the LPA often opens the door for other agencies to become involved in the development process.

3. 2. 5 *Facilitators*

A final group, but one of particular significance in the EIA process, includes the various consultants, advocates and advisers who participate in the EIA and the planning and development processes. Such agents are often employed by developers; occasionally they may be employed by local groups, environmental groups and others to help to mount opposition to a proposal. They may also be employed by regulatory bodies to help them in their examination process.

Environmental and planning consultancies carry out most of the EIA work, supported by smaller consultancies specializing in such issues as archaeology, noise, health and socio-economic impacts. There has been a massive growth in the number of environmental consultancies in the UK (Figure 3. 2). The numbers have increased by over 400 per cent since the mid-1980s, and it has been estimated that clients in the year 2008 were spending approximately 1. 5 billion on their services (ENDS 2009). Major factors underpinning the consultancy growth included the advent of the UK Environmental Protection Act (EPA) in 1990,

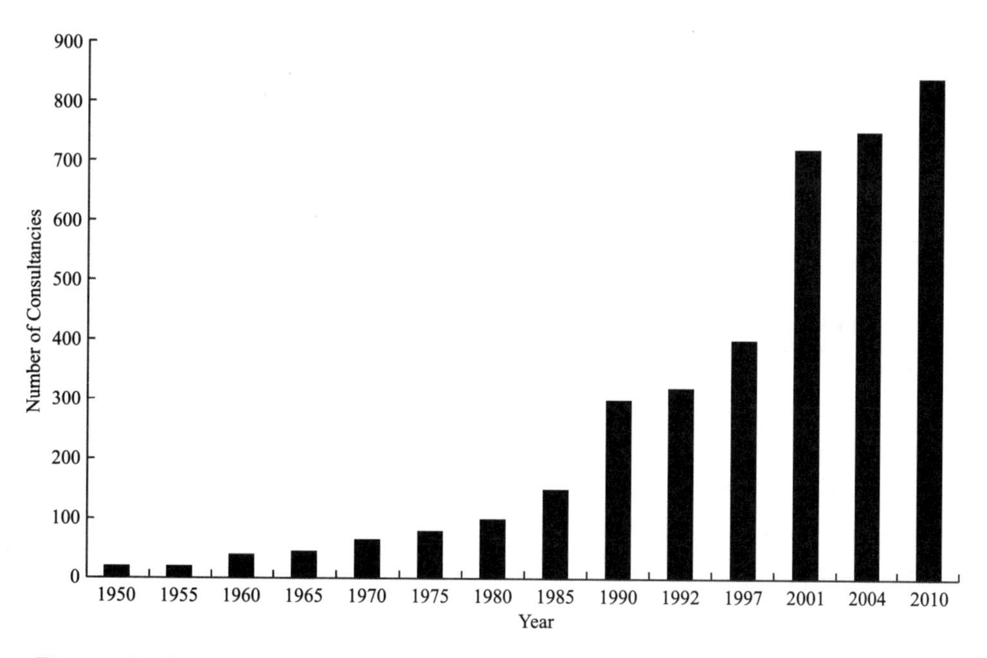

Figure 3. 2 Increase in the number of environmental consultancies in the UK (1950—2010)

Source: Based on ENDS 1993, 1997, 2001 and website

EIA regulations, the growing UK business interest in environmental management systems (e. g. BS 7750, ISO 14001), and a whole raft of EC regulations including on SEA, eco-auditing, and the Water Framework Directive. Tables 3. 1 and 3. 2 provide a summary of skills in demand and the main work areas of work for UK environmental consultancies. EIA and planning is a specialism in considerable demand, and the area was also mentioned by about a quarter of all respondents to the Environmental Data Services (ENDS) survey as a major area of activity. Further characteristics of recent consultancy activity are discussed in Chapter 8.

Table 3. 1 Skills in demand by UK environmental consultancies (2008)

Specialism	%
Waste management	62
Energy management	53
Sustainable development	52
EIA and planning	50
Environmental management and auditing	49
Water and waste management	46
GHG and carbon management	44
Pollution prevention and control(IPPC)	44
Health, safety and environmental management	42
Contaminated land and remediation	40
Renewable energy and clean energy	40
Air pollution and control	38
Corporate policy, CSR and communications	37
Ecology and nature conservation	35
Hydrology and hydrogeology	30
Public affairs and stakeholder communication	29
Hazard risk management	28
Acoustics	22
Process engineering	20

Source: ENDS 2009

Table 3. 2 Environmental consultants' main professional activities (2009)

Activity	%
Environmental management	44
Waste management/recycling	35
Environmental protection/regulation	29
Health, safety and environmental management	26
Sustainable development	24
Auditing/verification	23
Environmental education/training	23
Environmental impact assessment(inc. EIA, SEA and SA) and planning	23

Continued

Activity	%
Carbon management	19
Energy	19
Climate change and GHG management	18
Pollution control	18
Risk assessment	18
Corporate policy, CSR and communications	16

Source: ENDS 2010

3.2.6 *Agency interaction*

The various agencies outlined here represent a complex array of interests and aims, any combination of which may come into play for a particular development. This array has several dimensions, and within each there may be a range of often conflicting views. For example, there may be conflict between local and national views, between the interests of profit maximization and those of environmental conservation, between short-term and long-term perspectives and between corporate bodies and individuals. The agencies are also linked in various ways. Some links are statutory, others advisory. Some are contractual, others regulatory. The EIA regulations and guidance provide a set of procedures linking the various actors discussed, and these are now outlined.

3.3 EIA regulations: an overview

In the UK, EC Directive 85/337 was implemented through over 40 different secondary regulations under Section 2. 2 of the European Communities Act 1972. The large number of regulations was symptomatic of how EIA has been implemented in the UK. Different regulations apply to projects covered by the planning system, projects covered by other authorization systems and projects not covered by any authorization system but still requiring EIA. Different regulations apply to England, and the Devolved Administrations (DA) of Wales, Scotland and Northern Ireland. The introduction of various revisions to the regulations from 1999 onwards (the most recent being those of 2011), to implement the amended EC Directive, provided opportunities for some tidying up of the list, but as Table 3. 3 shows, there are still many regulations to ensure that all of the Directive's requirements are met. The regulations are supplemented by an array of EIA guidance from government and other bodies (Table 3. 4). In addition, the Planning and Compensation Act 1991 allows the government to require EIA for other projects that fall outside the Directive.

Table 3.3 Key UK EIA regulations and dates of implementation

UK regulations for projects subject to the Town and Country Planning system
England
Town and Country Planning(EIA)Regulations 2011
Town and Country Planning(General Permitted Development)Order 1995
Wales
Under review
Scotland
Town and Country Planning(EIA)(Scotland)Regulations 2011
Northern Ireland
Under review
UK EIA regulations for projects subject to alternative consent systems
Agriculture
Environmental Impact Assessment(Agriculture)(England)(no. 2)Regulations 2006
Afforestation
Environmental Impact Assessment(Forestry)(England and Wales)Regulations 1999
Environmental Impact Assessment(Forestry)Regulations(Northern Ireland)1999
Environmental Impact Assessment(Forestry)Regulations(Scotland)1999
Infrastructure/major projects
Infrastructure Planning(EIA)Regulations 2009
Land drainage improvements
Environmental Impact Assessment(Land Drainage Improvement Works)Regulations 1999
Land Drainage(EIA)(Scotland)Regulations 1999
Drainage(EA)Regulations(Northern Ireland)2001
Fish farming
Environmental Impact Assessment(Fish Farming in Marine Waters)Regulations 1999
Environmental Impact Assessment(Fish Farming in Marine Waters)Regulations(Northern Ireland)2006
Trunk roads and motorways
Highways(AEE)Regulations 1999
Roads(EIA)Regulations(Northern Ireland)1999
Railways,tramways,inland waterways and works interfering with navigation rights
Transport and Works(AEEs)2000
Ports and harbours,and marine dredging
Environmental impact assessment and Natural Habitats(Extraction of Minerals by Marine Dredging)(England and Northern Ireland)Regulations 2007
Harbour Works(EIA)Regulations 1999
Harbour Works(AEE)Regulations(Northern Ireland)1990
Power stations,overhead power lines and long-distance oil and gas pipeline
Electricity Works(AEE)(England and Wales)Regulations 2000
Electricity Works(EIA)(Scotland)Regulations 2000
Pipeline Works(EIA)Regulations 2000
The Nuclear Reactors(EIA for Decommissioning)Regulations 1999
The Gas Transporter Pipeline Works(EIA)Regulations 1999
Offshore Petroleum Production and Pipelines(AEE)Regulations 1999
Water Resources
The Water Resources(EIA)(England and Wales)Regulations 2003

Table 3.4 UK government EIA procedural guidance

DoE 1991. Monitoring environmental assessment and planning. London: HMSO
DoE 1994. Evaluation of environmental information for planning projects: a good practice guide. London: HMSO
DoE 1995. Preparation of environmental statements for planning projects that require environmental assessment. London: HMSO
DoE 1996. Changes in the quality of environmental statements for planning projects. London: HMSO
Environment Agency 1996. A scoping handbook for projects. London: HMSO
DETR 1997. Mitigation measures in environmental statements. London: HMSO
Scottish Executive Development Department 1999b. Planning advice note 58. Edinburgh: SEDD
National Assembly for Wales 1999. Circular 11/99 Environmental impact assessment. Cardiff: National Assembly
Planning Service (Northern Ireland) *Development control advice note* 10 1999. Belfast: NI Planning Service.
DETR 2000. Environmental impact assessment: a guide to the procedures. London: DCLG
DCLG 2006. Circular and guide to good practice. London: DCLG
DfT 2007b. Guidance on transport assessment (GTA). London: DfT
DfT 2011. Design manual for roads and bridges. London: DfT
Scottish Government 2011a. Circular 3/2011: The Town and Country Planning Environmental Impact Assessment (Scotland) Regulations 2011. Edinburgh: Scottish government. (Also available as EasyRead Guide)
DCLG 2011. Guidance on the Environmental Impact Assessment (EIA) Regulations 2011 for England. London: DCLG

In contrast to the US system of EIA, the EC EIA Directive applies to both public and private sector development. The developer carries out the EIA, and the resulting EIS must be handed in with the application for authorization. In England, most of the developments listed in Annexes I and II of the Directive fall under the remit of the planning system, and are thus covered by the Town and Country Planning (EIA) Regulations 2011 (the T&CP Regulations), previously the Town and Country Planning (Assessment of Environmental Effects (AEE)) Regulations 1988 and 1999. Over time various incremental additions and amendments have been made to the T&CP Regulations to plug loopholes and extend the remit of the regulations, for instance:

- to expand and clarify the original list of projects for which EIA is required (e. g. to include motorway service areas and wind farms, and more recently to add carbon capture and storage projects);
- to require EIA for projects that would otherwise be permitted (e. g. land reclamation, waste water treatment works, projects in Simplified Planning Zones);
- to require EIA for projects resulting from a successful appeal against a planning enforcement notice;
- to allow the relevant Secretary of State (SoS) to direct that a particular development should be subject to EIA even if it is not listed in the regulations; and
- to be consistent with various EC and UK legal rulings, for instance about screening processes and documentation.

Other types of projects listed in the EC Directive require separate legislation, since they are not governed by the planning system. Of the various *transport* projects, local highway developments and airports are dealt with under the T&CP Regulations by the local planning (highways) authority, but motorways and trunk roads proposed and regulated by the Department for Transport (DfT) fall under the Highways (Assessment of environmental effects) Regulations 1999. Applications for harbours are regulated by the DfT under the various Harbour Works (EIA) Regulations. New railways and tramways require EIA under the Transport and Works (AEE) procedure 2000.

Energy projects producing less than 50 MW are regulated by the local authority under the T&CP Regulations. Those of 50 MW or over, most electricity power lines, and pipelines (in Scotland as well as in England and Wales) are controlled by the Department of Energy and Climate Change (DECC) under the various Electricity and Pipeline Works (EIA) Regulations 2000.

New *land drainage* works, including flood defence and coastal defence works, require planning permission and are thus covered by the T&CP Regulations. Improvements to drainage works carried out by the Environment Agency and other drainage bodies require EIA through the EIA (Land Drainage Improvement Works) Regulations, which are regulated by the Department for Environment, Food and Rural Affairs (DEFRA). *Forestry* projects require EIA under the EIA (Forestry) Regulations 1999. *Marine fish farming* within 2 km of the coast of England, Wales or Scotland requires a lease from the Crown Estates Commission, but not planning permission. For these developments, EIA is required under the EA (Fish Farming in Marine Waters) Regulations 1999.

Many other developments in Scotland, Wales and Northern Ireland are increasingly being covered by country-specific legislation. For example the T&CP (EIA) (Scotland) Regulations 2011 have also been revised recently, and provide an interesting parallel system with the regulations for England. They also include a very useful *User Guide/EasyRead* short version for the busy reader (see Table 3. 4). Wales is also developing its own separate T&CP (EIA) Regulations.

As will be discussed in Chapter 8, about 60—70 per cent of all the EIAs prepared in the UK fall under the T&CP Regulations, about 10 per cent fall under each of the EIA (Scotland) Regulations and the Highways (EIA) Regulations; almost all the rest involve land drainage, electricity and pipeline works, forestry projects in England and Wales and planning-related developments in Northern Ireland.

The enactment of this wide range of EIA regulations has made many of the early concerns regarding procedural loopholes (e. g. CPRE 1991, Fortlage 1990) obsolete. However, several issues still remain-not least the sheer complexity of this array of regulation. First is the ambiguity inherent in the term project. An example of this is the EIA procedures for electricity generation and transmission, in which a power station and the transmission lines to and from it are seen as separate projects for the purposes of EIA, despite the fact that they are inextricably linked (Sheate 1995; see also Section 9. 2). Another example is the division of road construction into several separate projects for planning and EIA purposes even though none of them would be independently viable.

3.4 The Town and Country Planning (EIA) Regulations 2011 (previously the Town and Country Planning (EIA) Regulations 1999 and the Town and Country Planning (AEE) Regulations 1988)

The T&CP Regulations implement the EC Directive for those projects that require planning permission in England (Wales now has separate regulations). They are the central form in which the Directive is implemented in the UK; the other UK EIA regulations were established to cover projects that are not covered by the T&CP Regulations. As a result, the T&CP Regulations are the main focus of discussions on EIA procedures and effectiveness. This section presents the procedures of the T&CP Regulations. Figure 3. 3 summarizes these procedures; the letters in the figure correspond to the letters in bold preceding the explanatory paragraphs below. Section 3. 5 considers other main EIA regulations as variations of the T&CP Regulations and Section 3. 6 comments on the changes following from the amended EC Directive.

The original T&CP Regulations were issued in July 1988. Guidance on the Regulations followed soon after; DoE Circular 15/88 (Welsh Office Circular 23/88) was aimed primarily at local planning authorities; a guidebook entitled *Environmental assessment: a guide to the procedures* (DoE 1989), was aimed more at developers and their advisers. Further DoE guidance on good practice in carrying out and reviewing EIAS was published in 1994 and 1995 (DoE 1994a, b, 1995) and in 1997 (DETR). The revised 1999 T&CP Regulations were accompanied by new circulars on EIA (DETR 1999; Scottish Executive Development Department 1999, NAFW 1999), which give comprehensive guidance on the Regulations. A valuable guidebook, *Environmental impact assessment: a guide to the procedures* (DETR 2000) was also issued. *Guidance on the Environmental Impact Assessment (EIA) Regulations* 2011 *for England* (DCLG 2011b), and the parallel document for Scotland, now provide the latest versions in this stream of very useful documents. The circulars and other government guidance are strongly recommended reading. However, only the regulations are mandatory: the guidance interprets and advises, but cannot be enforced.

3. 4. 1 *Which projects require EIA?*

The T&CP Regulations require EIAs to be carried out for two broad categories of project, given in Schedules 1 and 2. These schedules correspond very closely to Annexes I and II in the amended EC Directive, as outlined in Table 2. 4 and detailed in Appendix 1. [1] Schedule 1 has very minor wording changes from Annex I, plus the switch of Annex I, 1. 20,

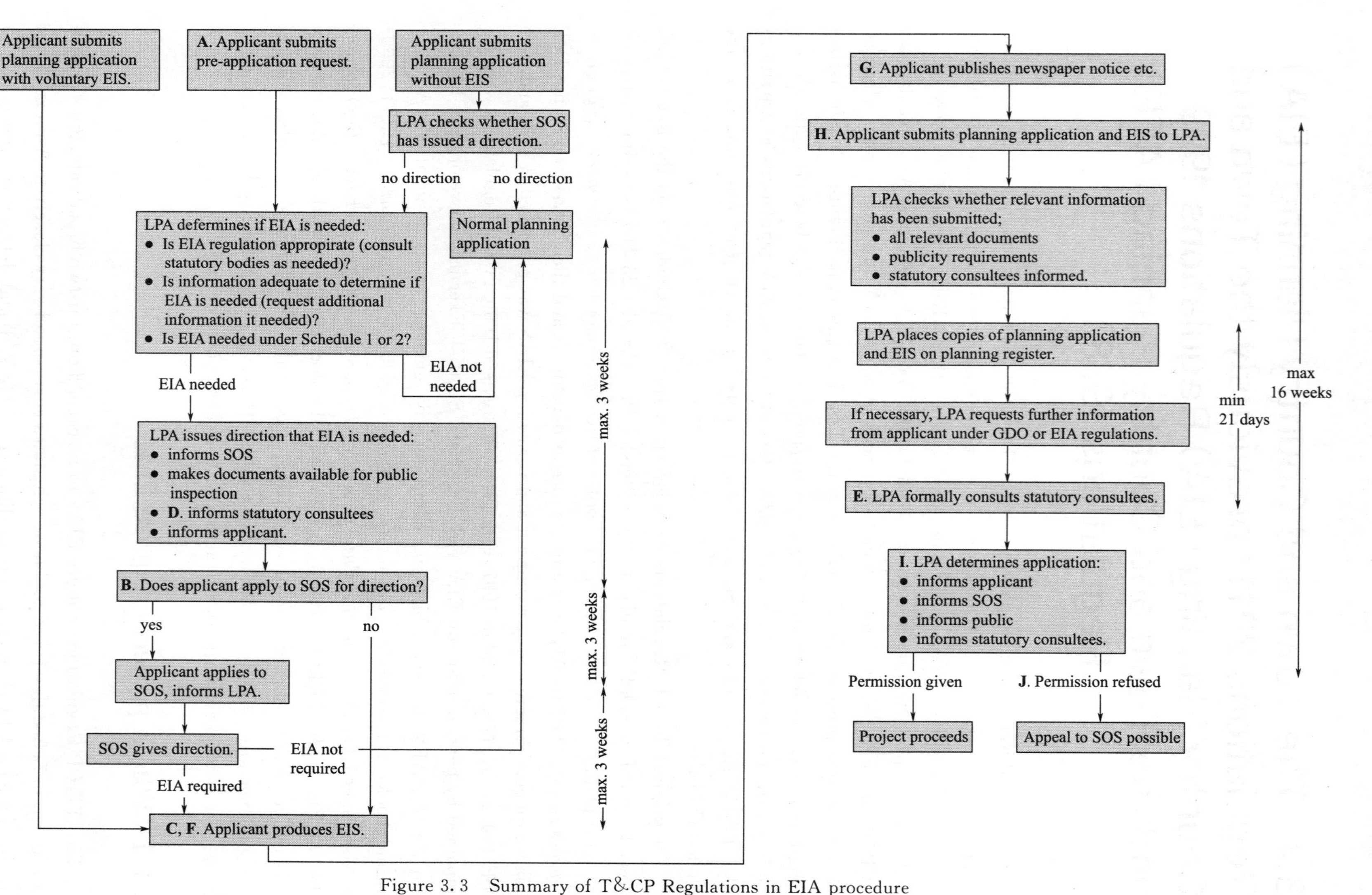

Figure 3.3 Summary of T&CP Regulations in EIA procedure

Source：Based on DETR 2000

long overhead electrical power lines, to the Electricity Works (AEE) Regulations 2000 (S1 1927). The decommissioning of nuclear power stations and reactors is also covered by separate EIA regulations in the UK. Schedule 2 has only very minor modifications from Annex II; primarily in 2. 10 where (b) also includes sports stadiums, leisure centres and multiplex cinemas. Also, there is a separate category (p) for motorway service areas, and a few other categories are split or relocated. Schedule 2. 12 also includes an additional category (f) for golf courses and associated developments. For Schedule 1 projects, EIA is required in every case. A Schedule 2 project requires EIA if it is deemed 'likely to give rise to significant environmental effects'.

The 'significance' of a project's environmental effects is determined on the basis of a set of applicable thresholds and criteria for Schedule 2 development (see Appendix 2), and the selection criteria in Schedule 3 of the Regulations (see Table 3. 5). Three categories of criteria are listed in Schedule 3 (DETR 2000; DCLG 2011a):

Table 3. 5 Selection criteria for screening Schedule 2 development

1 Characteristics of development

The characteristics of development must be considered having regard, in particular, to:

(a) the size of the development;

(b) the cumulation with other development;

(c) the use of natural resources;

(d) the production of waste;

(e) pollution and nuisances;

(f) the risk of accidents, having regard in particular to substances or technologies used.

2 Location of development

The environmental sensitivity of geographical areas likely to be affected by development must be considered, having regard, in particular, to:

(a) the existing land use;

(b) the relative abundance, quality and regenerative capacity of natural resources in the area;

(c) the absorption capacity of the natural environment, paying particular attention to the following areas

(i) wetlands;

(ii) coastal zones;

(iii) mountain and forest areas;

(iv) nature reserves and parks;

(v) areas designated by Member States pursuant to Council Directive 2009/147/EC on the conservation of wild birds(1) and Council Directive 92/43/EEC on the conservation of natural habitats and of wild fauna and flora(2);

(vi) areas in which the environmental quality standards laid down in EU legislation have already been exceeded;

(vii) densely populated areas;

(viii) landscapes of historical, cultural or archaeological significance.

3 Characteristics of the potential impact

The potential significant effects of development must be considered in relation to criteria set out under paragraphs 1 and 2 above, and having regard, in particular, to:

(a) the extent of the impact (geographical area and size of the affected population);

(b) the transfrontier nature of the impact;

(c) the magnitude and complexity of the impact;

(d) the probability of the impact;

(e) the duration, frequency and reversibility of the impact.

Source: DCLG 2011a

Notes: (1) O. J. no. L 20, 26. 1. 2010, p. 7. (2) O. J. no. L 206, 22. 7. 1992, p. 7.

• whether it is a development of more than local importance [for example, in terms of physical scale];

• whether the development is proposed for a particularly environmentally sensitive or vulnerable location [for example, a national park or a site of special scientific interest]; and

• whether the development is likely to have unusually complex and potentially hazardous environmental effects [for example, in terms of the discharge of pollutants].

The third category is designed to help in determining whether the interactions between the first two categories (i. e. between a development and its environment) are likely to be significant. A project constitutes Schedule 2 development for EIA when: (a) it meets criteria or exceeds thresholds listed in the second column of the Schedule 2 table (see Appendix 2); or (b) is located in, or partly in, a 'sensitive area', as defined in the regulations (see Table 3. 5). The more environmentally sensitive the location, the more likely it is that the effects will be significant and require EIA.

Screening criteria have raised many issues over the life of the UK EIA regulations, including giving rise to several celebrated legal cases—many of which are covered in Section 8. 6, Chapter 8. These include, for instance, cases about dealing with extensions to projects, multi-stage consents, and the extent to which proposed project mitigation measures can be taken into account in screening decisions.

The 2011 Regulations have responded on many of the legal issues raised.

• Schedule 2. 13 has been amended so that the thresholds in Schedule 2 apply to the development as a whole once changed or extended, and not just to the change or extension. For example, the construction of a 0. 2 ha extension to an industrial estate may not have previously qualified for EIA by virtue of falling below the relevant 0. 5 ha threshold. Now, this extension must be considered with the original development. If the latter was 0. 4 ha, this would take the combined development over the EIA threshold size of 0. 6ha.

• The 2011 Regulations also include amendments in relation to a European Court of Justice (ECJ) preliminary ruling on screening decisions, known as the Mellor case, which now requires reasons for a negative EIA screening decision to be made available on request.

• Further to another ECJ ruling on multi-stage consents, the 2011 Regulations provide for the limitation to the requirement for subsequent applications to be subject to screening in those cases where the development is likely to have significant effects on the environment, which were not identified at the time the initial planning permission was granted.

• New categories for carbon capture and storage (CCS) have been included in both Schedules.

The 2011 Regulations also include the simplification of the thresholds from the more comprehensive applicable *and* indicative thresholds and criteria introduced under the 1999 Regulations, to applicable thresholds and criteria only. As can be seen in Appendix 2, these are largely spatially based ('the area of the development exceeds 0. 5 ha'; 'the area of new floorspace exceeds 1000 square metres'). A current interesting issue (at the time of going to press) concerns a widening of screening actions to more cases of project demolition/decommissioning. The European Court ruled in 2011 that demolition work comes within the scope of the EIA Directive; the effect is that where demolition works are likely to have significant environmental effects (for instance demolition of a listed building), the LPA must issue a screening opinion on whether EIA is required. Similarly, applicants who intend to demolish a structure must consider whether this may have significant environmental effects and require EIA. If this is the case the applicant must ask for a screening opinion. Finally, reflecting the continuing momentum in the evolving nature of developments, and EU and domestic legislation, there is a commitment to a further review of the 2011 Regulations within five years of their introduction!

A. A developer may decide that a project requires EIA under the T&CP Regulations, or may want to carry out an EIA even if it is not required. If the developer is uncertain, the LPA can be asked for an opinion ('screening opinion') on whether an EIA is needed. To do this the developer must provide the LPA with a plan showing the development site, a description of the proposed development and an indication of its possible environmental impacts. The LPA must then make a decision within three weeks. The LPA can ask for more information from the developer, but this does not extend the three week decision-making period.

If the LPA decides that no EIA is needed, the application is processed as a normal planning application. If instead the LPA decides that an EIA is needed, it must explain why, and make both the developer's information and the decision publicly available. In all cases the LPA must explain clearly and precisely the full reasons for its conclusion whether EIA is required or not. If the LPA receives a planning application without an EIA when it feels that it is needed, the LPA must notify the developer within three weeks, explaining why an EIA is needed. The developer then has three weeks in which to notify the LPA of the intention either to prepare an EIS or to appeal to the Secretary of State (SoS); if the developer does not do so, the planning application is refused.

B. If the LPA decides that an EIA is needed but the developer disagrees, the developer can refer the matter to the SoS for a ruling.[2] The SoS must give a decision within three weeks. If the SoS decides that an EIA is needed, an explanation is needed; it is published in the *Journal of Planning and Environment Law*. No explanation is needed if no EIA is required. The SoS may make a decision if a developer has not requested an opinion, and may rule, usually as a result of information made available by other bodies, that an EIA is needed where the LPA has decided that it is not needed.

3.4.2 The contents of the EIA

Schedule 4 of the T&CP Regulations, which is shown in Table 3.6, lists the information that should be included in an EIA. Schedule 4 interprets the requirements of the EIA Directive Annex IV according to the criteria set out in Article 5 of the Directive, namely:

Member States shall adopt the necessary measures to ensure that the developer supplies in an appropriate form the information specified in Annex IV in as much as:

(a) the Member States consider that the information is relevant to a given stage of the consent procedure and to the specific characteristics of a particular project or type of project and of the environmental features likely to be affected;

(b) the Member States consider that a developer may reasonably be required to compile this information having regard *inter alia* to current knowledge and methods of assessment.

Table 3.6 Content of EIS required by the T&CP Regulations (2011) -Schedule 4

Under the definition in Regulation 2.1, 'environmental statement' means a statement:

(a) that includes such of the information referred to in Part 1 of Schedule 4 as is reasonably required to assess the environmental effects of the development and which the applicant can, having regard in particular to current knowledge and methods of assessment, reasonably be required to compile, but

(b) that includes at least the information referred to in Part 2 of Schedule 4.

Part 1

1 Description of the development, including, in particular:

(a) a description of the physical characteristics of the whole development and the land-use requirements during the construction and operational phases;

(b) a description of the main characteristics of the production processes, for instance, nature and quantity of the materials used;

(c) an estimate, by type and quantity, of expected residues and emissions (water, air and soil pollution, noise, vibration, light, heat, radiation, etc.) resulting from the operation of the proposed development.

2 An outline of the main alternatives studied by the applicant or appellant and an indication of the main reasons for the choice made, taking into account the environmental effects.

3 A description of the aspects of the environment likely to be significantly affected by the development, including, in particular, population, fauna, flora, soil, water, air, climatic factors, material assets, including the architectural and archaeological heritage, landscape and the inter-relationship between the above factors.

4 A description of the likely significant effects of the development on the environment, which should cover the direct effects and any indirect, secondary, cumulative, short, medium and long-term, permanent and temporary, positive and negative effects of the development, resulting from:

(a) the existence of the development;

(b) the use of natural resources;

(c) the emission of pollutants, the creation of nuisances and the elimination of waste, and the description by the applicant of the forecasting methods used to assess the effects on the environment.

5 A description of the measures envisaged to prevent, reduce and, where possible, offset any significant adverse effects on the environment.

6 A non-technical summary of the information provided under paragraphs 1 to 5 of this Part.

7 An indication of any difficulties (technical deficiencies or lack of know-how) encountered by the applicant in compiling the required information.

Part 2

1 A description of the development comprising information on the site, design and size of the development.

2 A description of the measures envisaged in order to avoid, reduce and, if possible, remedy significant adverse effects.

3 The data required to identify and assess the main effects which the development is likely to have on the environment.

4 An outline of the main alternatives studied by the applicant or appellant and an indication of the main reasons for his choice, taking into account the environmental effects.

5 A non-technical summary of the information provided under paragraphs 1 to 4 of this Part.

Source: DCLG 2011a

In Schedule 4, the information required in Annex IV has been interpreted to fall into two parts. The EIS must contain the information specified in Part 2, and such relevant information in Part 1 'as is reasonably required to assess the effects of the project and which the developer can reasonably be required to compile'. This distinction is important: as will be seen in Chapter 8, the EISs prepared to date have generally been weaker on Part 1 information, although this includes such important matters as the alternatives that were considered and the expected wastes or emissions from the development. In addition, in Appendix 5 of the guidebook (DETR 2000), there is a longer checklist of matters that may be considered for inclusion in an EIA: this list is for guidance only, but it helps to ensure that all the possible significant effects of the development are considered (Table 3.7).

Table 3.7 Checklist of matters to be considered for inclusion in an environmental statement

This checklist is intended as a guide to the subjects that need to be considered in the course of preparing an environmental statement. It is unlikely that all the items will be relevant to any one project.

The environmental effects of a development during its construction and commissioning phases should be considered separately from the effects arising while it is operational. Where the operational life of a development is expected to be limited, the effects of decommissioning or reinstating the land should also be considered separately.

Section 1: *Information describing the project*

1.1 Purpose and physical characteristics of the project, including details of proposed access and transport arrangements, and of numbers to be employed and where they will come from.

1.2 Land-use requirements and other physical features of the project:

- during construction;
- when operational;
- after use has ceased (where appropriate).

1.3 Production processes and operational features of the project:

- type and quantities of raw materials, energy and other resources consumed;
- residues and emissions by type, quantity, composition and strength including:

discharges to water; emissions to air; noise; vibration; light; heat; radiation; deposits/residues to land and soil; others.

1.4 Main alternative sites and processes considered, where appropriate, and reasons for final choice.

Section 2: *Information describing the site and its environment*

Physical features

2.1 Population-proximity and numbers.

2.2 Flora and fauna (including both habitats and species)-in particular, protected species and their habitats.

2.3 Soil: agricultural quality, geology and geomorphology.

2.4 Water: aquifers, watercourses, shoreline, including the type, quantity, composition and strength of any existing discharges.

2.5 Air: climatic factors, air quality, etc.

2.6 Architectural and historic heritage, archaeological sites and features, and other material assets.

2.7 Landscape and topography.

2.8 Recreational uses.

2.9 Any other relevant environmental features.

The policy framework

2.10 Where applicable, the information considered under this section should include all relevant statutory designations such as national nature reserves, sites of special scientific interest, national parks, areas of outstanding natural beauty, heritage coasts, regional parks, country parks and designated green belt, local nature reserves, areas affected by tree preservation orders, water protection zones, conservation areas, listed buildings, scheduled ancient monuments and designated areas of archaeological importance. It should also include references to relevant national policies (including Planning Policy Guidance (PPG) notes) and to regional and local plans and policies (including approved or emerging development plans).

2.11 Reference should also be made to international designations, e.g. those under the EC 'Wild Birds' or 'Habitats' Directives, the Biodiversity Convention and the Ramsar Convention.

Section 3: *Assessment of effects*

Including direct and indirect, secondary, cumulative, short-, medium-and long-term, permanent and temporary, positive and negative effects of the project.

Effects on human beings, buildings and man-made features

Continued

3.1 Change in population arising from the development, and consequential environment effects.

3.2 Visual effects of the development on the surrounding area and landscape.

3.3 Levels and effects of emissions from the development during normal operation.

3.4 Levels and effects of noise from the development.

3.5 Effects of the development on local roads and transport.

3.6 Effects of the development on buildings, the architectural and historic heritage, archaeological features and other human artefacts,

e. g. through pollutants, visual intrusion, vibration.

Effects on flora, fauna and geology

3.7 Loss of, and damage to, habitats and plant and animal species.

3.8 Loss of, and damage to, geological, palaeontological and physiographic features.

3.9 Other ecological consequences.

Effects on land

3.10 Physical effects of the development, e. g. change in local topography, effect of earth-moving on stability, soil erosion, etc.

3.11 Effects of chemical emissions and deposits on soil of site and surrounding land.

3.12 Land-use/resource effects:

- quality and quantity of agricultural land to be taken;
- sterilization of mineral resources;
- other alternative uses of the site, including the 'do nothing' option;
- effect on surrounding land uses including agriculture;
- waste disposal.

Effects on water

3.13 Effects of development on drainage pattern in the area.

3.14 Changes to other hydrographic characteristics, e. g. groundwater level, watercourses, flow of underground water.

3.15 Effects on coastal or estuarine hydrology.

3.16 Effects of pollutants, waste, etc. on water quality.

Effects on air and climate

3.17 Level and concentration of chemical emissions and their environmental effects.

3.18 Particulate matter.

3.19 Offensive odours.

3.20 Any other climatic effects.

Other indirect and secondary effects associated with the project

3.21 Effects from traffic (road, rail, air, water) related to the development.

3.22 Effects arising from the extraction and consumption of materials, water, energy or other resources by the development.

3.23 Effects of other development associated with the project, e. g. new roads, sewers, housing, power lines, pipelines, telecommunications, etc.

3.24 Effects of association of the development with other existing or proposed development.

3.25 Secondary effects resulting from the interaction of separate direct effects listed above.

Section 4: *Mitigating measures*

4.1 Where significant adverse effects are identified, a description of the measures to be taken to avoid, reduce or remedy those effects, e. g. :

(a) site planning;

(b) technical measures, e. g. :

- process selection;
- recycling;
- pollution control and treatment;
- containment (e. g. bunding of storage vessels).

(c) aesthetic and ecological measures, e. g. :

- mounding;
- design, colour, etc.;
- landscaping;
- tree plantings;
- measures to preserve particular habitats or create alternative habitats;
- recording of archaeological sites;
- measures to safeguard historic buildings or sites.

4.2 Assessment of the likely effectiveness of mitigating measures.

Section 5: *Risk of accidents and hazardous development*

5.1 Risk of accidents as such is not covered in the EIA Directive or, consequently, in the implementing Regulations. However, when the proposed development involves materials that could be harmful to the environment (including people) in the event of an accident, the environmental statement should include an indication of the preventive measures that will be adopted so that such an occurrence is not likely to have a significant effect. This could, where appropriate, include reference to compliance with Health and Safety legislation.

5.2 There are separate arrangements in force relating to the keeping or use of hazardous substances and the HSE provides local planning authorities with expert advice about risk assessment on any planning application involving a hazardous installation.

5.3 Nevertheless, it is desirable that, wherever possible, the risk of accidents and the general environmental effects of developments should be considered together, and developers and planning authorities should bear this in mind.

Source: DETR 2000

C. Until the implementation of the amended Directive in 1999, there was no mandatory requirement in the UK for a formal 'scoping' stage at which the LPA, the developer and other interested parties could agree on what would be included in the EIA. Indeed, there was no requirement for any kind of consultation between the developer and other bodies before the submission of the formal EIA and planning application, although guidance (DoE 1989) did stress the benefits of early consultation and early agreement on the scope of the EIA. The 1999 and subsequent Regulations enable a developer to ask the LPA for a formal 'scoping opinion' on the information to be included in an EIS—in advance of the actual planning application. This allows a developer to be clear on LPA views on the anticipated key significant effects. The request must be accompanied by the same information provided for a screening opinion, and may be made at the same time as for the screening opinion. The LPA must consult certain bodies (see D), and must produce the scoping opinion within five weeks. The time period may be extended if the developer agrees. There is no provision for appeal to the SoS if the LPA and developer disagree on the content of an EIS. But if the LPA fails to produce a scoping opinion within the required timescale, the developer may apply to the SoS (or Assembly) for a scoping direction, also to be produced within five weeks, and also to be subject to consultation with certain bodies. The checklist (DETR 2000) provides a useful aid to developer—LPA discussions (see Table 3.7).

3.4.3 Statutory and other consultees

Under the T&CP Regulations, a number of statutory consultees are involved in the EIA process, as noted in Section 3.2. These bodies are involved at two stages of an EIA, in addition to possible involvement in the scoping stage.

D. First, when an LPA determines that an EIA is required, it must inform the statutory consultees of this. Current consultation bodies in England include the relevant planning authority, Natural England, the Environment Agency and more recently the Marine Management Organization (primarily in relation to projects with potential marine impacts). The consultees in turn must make available to the developer, if so requested and at a reasonable charge, any relevant environmental information in their possession. For example, Natural England might provide information about the ecology of the area. This does not include any confidential information or information that the consultees do not already have in their possession.

E. Second, once the EIS has been submitted, the LPA or developer must send a free copy to each of the statutory consultees. The consultees may make representations about the EIS to the LPA for at least two weeks after they receive the EIS. The LPA must take account of these representations when deciding whether to grant planning permission. The developer may also contact other consultees and the general public while preparing the EIS. The government guidance explains that these bodies may have particular expertise in the subject or may highlight important environmental issues that could affect the project. Although the developer is under no obligation to contact any of these groups, again the government guidance stresses the benefits of early and thorough consultation.

3.4.4 *Carrying out the EIA: preparing the EIS*

F. The government gives no formal guidance about what techniques and methodologies should be used in EIA, noting only that they will vary depending on the proposed development, the receiving environment and the information available, and that predictions of effects will often have some uncertainty attached to them.

3.4.5 *Submitting the EIS and planning application: public consultation*

G. When the EIS has been completed, the developer must publish a notice in a local newspaper and post notices at the site. These notices must fulfil the requirements of Article 8 of the Town and Country Planning Act (General Permitted Development) Order 1995 (GPDO), state that a copy of the EIS is available for public inspection, give a local address where copies may be obtained and state the cost of the EIS, if any. The public can make written representations to the LPA for at least 20 days after the publication of the notice, but within 21 days of the LPA's receipt of the planning application.

H. After the EIS has been publicly available for at least 21 days, the developer submits to the LPA the planning application, copies of the EIS,[3] and certification that the required public notices have been published and posted. The LPA must then send copies of the EIS to the statutory consultees, inviting written comments within a specified time (at least two weeks from receipt of the EIS), forward another copy to the SoS and place the EIS on the planning register. It must also decide whether any additional information about the project is needed before a decision can be made, and, if so, obtain it from the developer. The clock does not stop in this case: a decision must still be taken within the appropriate time.

3.4.6 *Planning decision*

I. Before making a decision about the planning application, the LPA must collect written representations from the public within three weeks of the receipt of the planning application, and from the statutory consultees at least two weeks from their receipt of the EIS. It must wait at least 14 days after receiving the planning application before making a decision. In contrast to normal planning applications, which must be decided within eight weeks, those accompanied by an EIS must be decided within 16 weeks. If the LPA has not made a decision after 16 weeks, the applicant can appeal to the SoS for a decision. The LPA cannot consider a planning application invalid because the accompanying EIS is felt to be inadequate: it can only ask for further information within the 16-week period.

In making its decision, the LPA must consider the EIS and any comments from the public and statutory consultees, as well as other material considerations. The environmental information is only part of the information that the LPA considers, along with other material considerations. The decision is essentially still a political one, but it comes with the assurance that the project's environmental implications are understood. The LPA may grant or refuse permission, with or without conditions. It is important for LPAs to consider how mitigation measures proposed in an ES are likely to be secured; they may be included in conditions attached to a planning permission; they can also be secured through enforceable planning obligations. Further to the changes resulting from the EC Directive the LPA must, in addition to the normal requirements to notify the applicant, notify the SoS and publish a notice in the local press, giving the decision, the main reasons on which the decision was based, together with a description of the main mitigation measures.

J. If an LPA refuses planning permission, the developer may appeal to the SoS, as for a normal planning application. The SoS may request further information before making a decision.

3.5 Other EIA regulations

This section summarizes the procedures of the other EIA regulations under which a large number of EISs have been prepared to date. We discuss the regulations in approximate descending order of frequency of application to date, although category 7 is likely to rise quickly up the order soon.

1 The Town and Country Planning Environmental Impact Assessment (Scotland) Regulations 2011.

2 Highways (AEE) Regulations 1999.

3 Environmental Impact Assessment (Land Drainage Improvement Works) Regulations 1999.

4 Electricity Works (AEE) Regulations 2000.

5 Pipeline Works (Environmental Impact Assessment) Regulations 2000.

6 Environmental Assessment (Forestry) Regulations 1999.

7 Infrastructure Planning (Environmental Impact Assessment) Regulations 2009.

3.5.1 *EIA (Scotland) Regulations: including The Town and Country Planning (EIA) (Scotland) Regulations 2011*

The EIA (Scotland) Regulations are broadly similar to those for England. They implement the Directive for projects that are subject to planning permission, but also cover some land drainage and trunk road projects. There is separate guidance on the Scottish Regulations (see Table 3.2). For some projects (for example, for the decommissioning of nuclear power stations), Scotland is included in regulations that also apply to other parts of the UK. In Northern Ireland, the Directive is implemented for projects subject to planning permission by the Planning (EIA) Regulations (Northern Ireland) 1999 (currently being revised).

The T&CP (EIA) (Scotland) Regulations were amended roughly in line with the English Regulations and for similar reasons. The commentary on the 2011 Regulations notes that:

> The existing 1999 Regulations had been repeatedly amended (12 times in 11 years) to take into account case law from domestic and European Courts, and changes to the Directive and/or domestic legislation. This made the 1999 Regulations increasingly complex and difficult to follow. The Scottish Parliament's subordinate legislation committee called for the Regulations to be consolidated in order to make them more accessible. (Scottish Government 2011b)

Under the Scottish Regulations, screening for Schedule 2 projects is on a case by case taking into account the key selection criteria: characteristics of the development, location of the development, and characteristics of the potential impacts. LPAs are encouraged to use checklists, including the one on the Scottish government's (Planning) EIA web page.

3.5.2 *Highways (AEE) Regulations 1999*

The Highways (AEE) Regulations apply to motorways and trunk roads proposed by the Department of Transport (DoT). The regulations are approved under procedures set out in the Highways Act 1980, which require the SoS for Transport to publish an EIS for the proposed route when draft orders for certain new highways, or major improvements to existing highways, are published. The SoS determines whether the proposed project comes under Annex I or Annex II of the Directive, and whether an EIA is needed. EIA is mandatory for projects to construct new motorways and certain other roads, including those with four or more lanes, and for certain road improvements. The regulations require an EIS to contain:

- a description of the published scheme and its site;
- a description of measures proposed to mitigate adverse environmental effects;
- sufficient data to identify and assess the main effects that the scheme is likely to have on the environment; and
- a non-technical summary.

Before 1993, the requirements of the Highways (AEE) Regulations were further elaborated in DoT standard AD 18/88 (DoT 1989) and the *Manual of Environmental Appraisal* (DoT 1983). In response to strong criticism,[4] particularly by the SACTRA (1992), these were superseded in 1993 by the *Design manual for roads and bridges* (DMRB), vol. II: *Environmental assessment* (DoT 1993). The manual proposed a three-stage EIA process and gave extensive, detailed advice on how these EIAs should be carried out. DMRB (DfT 2011) provides the latest evolution of the guidance for road projects. Other very useful transport assessment documentation can be found on the DfT WebTAG Transport Analysis Guidance website, including Guidance on Transport Assessment (GTA) (DfT 2007a).

3.5.3 Environmental Impact Assessment (Land Drainage Improvement Works) Regulations 1999

The EIA (Land Drainage Improvement Works) Regulations apply to almost all watercourses in England and Wales except public-health sewers. If a drainage body (including a local authority acting as a drainage body) determines that its proposed improvement actions are likely to have a significant environmental effect, it must publish a description of the proposed actions in two local newspapers and indicate whether it intends to prepare an EIS. If it does not intend to prepare one, the public can make representations within 28 days concerning any possible environmental impacts of the proposal; if no representations are made, the drainage body can proceed without an EIS. If representations are made, but the drainage body still wants to proceed without an EIS, DEFRA (National Assembly in Wales) gives a decision on the issue at ministerial level.

The contents required of the EIS under these regulations are virtually identical to those under the T&CP Regulations. When the EIS is complete, the drainage body must publish a notice in two local newspapers, send copies to NE, the Environment Agency (EA), the MMO and any other relevant bodies and make copies of the EIS available at a reasonable charge. Representations must be made within 28 days and are considered by the drainage body in making its decision. If all objections are then withdrawn, the works can proceed; otherwise the minister gives a decision. Overall, these regulations are considerably weaker than the T&CP Regulations because of their weighting in favour of consent, unless objections are raised, and their minimal requirements for consultation with environmental organizations.

3.5.4 *Electricity Works (AEE) Regulations 2000; Nuclear Reactors (EIA for Decommissioning) Regulations 1999*

The construction or extension of power stations exceeding 50 MW, and the installation of overhead power lines, requires consent from the relevant SoS (currently Department of Energy and Climate Change) under Sections 36 and 37 of the Electricity Act 1989. The Electricity Works (EIA) Regulations 2000 is part of the procedure for applications under these provisions. EIA is required for:

- all thermal and nuclear power stations that fall under Annex 1 of the Directive (i. e. thermal power stations of 300 MW or more, and nuclear power stations of at least 50 MW); and
- construction of overhead power lines of 220 kV or more and over 15 km in length.

The regulations also require proposed power stations not covered by Annex I, and all overhead power lines of at least 132 kV, to be screened for EIA. Power stations of less than 50 MW are approved under the planning legislation, through the T&CP (EIA) Regulations. The Electricity Works Regulations allow a developer to make a written request to the SoS to decide whether an EIA is needed. The SoS must consult with the LPA before making a decision. When a developer gives notice that an EIS is being prepared, the SoS must notify the LPA or the principal council for the relevant area, NE, the EA and the MMO, in the case of a power station, so that they can provide relevant information to the applicant. The contents required of the EIS are almost identical to those listed in the T&CP Regulations.

The regulations on decommissioning of nuclear power stations were added in 1999. Dismantling and decommissioning require the consent of the Health and Safety Executive (HSE). A licensee who applies for consent must provide the HSE with an EIS. The regulations apply also to changes to existing dismantling or decommissioning projects that may have significant effects on the environment.

3.5.5 *The Pipeline Works (EIA) Regulations 2000; The Gas Transporter Pipeline Works (EIA) Regulations 1999; Offshore Petroleum Production and Pipelines (AEE) Regulations*

The evolving array of regulations relating to pipelines reflects not only the growing importance of such development, but also the continuation of the fragmented UK approach to

EIA legislation. The onshore Pipeline Works, and the Gas Transporter Pipeline Works, Regulations apply to England, Wales and Scotland; the Offshore Petroleum Production and Pipelines Regulations apply to the whole of the UK. Oil and gas pipelines with a diameter of more than 800 mm and longer than 40 km come within Annex I of the Directive; those that fall below either of these thresholds are in Annex Ⅱ. For the latter, pipelines 10 miles long or less are approved under the planning legislation. The rest fall under the above pipeline regulations, normally with determination by the relevant SoS in relation to associated consent and authorization procedures and to various criteria and thresholds. For example, on-shore gas pipeline works in Annex Ⅱ of the Directive may be subject to EIA if they have a design operating pressure exceeding 7 bar gauge or they either wholly or in part cross a sensitive area (e. g. national park).

3.5.6 Environmental Impact Assessment (Forestry) (England and Wales) Regulations 1999

Under the original EIA Directive and associated UK regulations, forestry EIAs were limited to those projects where applicants wished to apply for a grant or loan, for afforestation purposes, from the Forestry Agency. The lack of EIA requirements for other forestry projects, the perceived vested interest of the then Forestry Agency as a promoter of forestry and the lack of EIA requirements for the Agency's own projects have all been criticized (e. g. by the CPRE; see CPRE 1991). The amended Directive and associated UK legislation have subsequently brought about some changes.

Afforestation and deforestation come under Annex II of the Directive. Under the above Regulations, anyone who proposes to carry out a forestry project that is likely to have significant effects on the environment must apply for consent from the Forestry Commission before starting work. Those who apply for consent will be required to produce an EIS. The Regulations include: afforestation (creating new woodlands); deforestation (conversion of woodland to another use) and constructing forest roads and quarrying material to construct forest roads. Where projects are below 5 ha (afforestation) and 1 ha (others), they may be deemed unlikely to have significant effects on the environment, unless they are in sensitive areas. Given the variability of sites and projects, the Forestry Commission considers applications on a case-by-case basis. An applicant who disagrees with the Forestry Commission's opinion may apply to the relevant SoS for a direction. The contents required of an EIA under the Forestry Regulations are almost identical to those required under the T&CP Regulations.

3.5.7 Infrastructure Planning (EIA) Regulations 2009

Context

As noted in Section 1.4.1, under the 2008 Planning Act (HMG 2008), a special subset of nationally significant infrastructure projects (NSIPs) has been identified, with impacts to be examined by new procedures led by the Infrastructure Planning Commission (IPC) (to be merged with the UK Planning Inspectorate in 2012). NSIPs include major energy projects, transport projects (road, rail and port), water and waste facilities—many of which were formally covered by some of the previously discussed regulations (especially highways, electricity, and pipelines). The IPC examination involves the consideration of environmental impacts, as relevant, under the Infrastructure Planning (EIA) Regulations 2009. The EIA Regulations impose procedural requirements for carrying out EIA on certain NSIP proposals. For example, EIA is always required for NSIPs such as nuclear power stations, but others, such as wind farms, only require EIA if they are likely to have significant effects on the environment by virtue of their nature, size or location. The role of the IPC under the EIA regulations includes:

- 'screening' proposals to determine whether they are EIA developments;
- 'scoping' proposals to advise the applicant what information should be provided within the environmental statement—involving seeking views from consultation bodies;
- facilitating the preparation of ESs by notifying consultation bodies about their duty to provide information and informing the applicant;
- evaluating environmental information in the ES and any representations made about the environmental effects before making a decision;
- publicizing the IPC's screening and scoping opinions; and
- publicizing any decision in relation to an application that has been accompanied by an ES (IPC 2010).

The IPC may not grant development consent unless it has first taken account of the 'environmental information' (ES and/or any further information about the environmental effects of the development). It will not accept an application if the supporting ES is considered inadequate, or it is deemed to be an EIA development but is not accompanied by an ES.

Screening

A proposal will be an EIA development if: the applicant notifies the IPC that it intends to submit an ES; the IPC adopts a screening opinion to the effect that the proposal is an EIA development; the SoS directs that it is an EIA development; or the proposal falls within Schedule 1 of the EIA Regulations. Many NSIPs are likely to fall within Schedule 1, for which EIA is mandatory. Others that fall within Schedule 2 must be considered for EIA against 'selection criteria' specified in Schedule 3 of the EIA Regulations; this differs from proposals under the T&CP (EIA) Regulations 2011, which are considered for EIA against applicable thresholds and criteria. For others, and where the applicant has not notified an intention to submit an ES, they must request that the IPC adopts a screening opinion in respect of the proposed development. Ideally the applicant should provide information on the characteristics of the development, the location of the development and characteristics of the potential impacts. Following the submission of a screening request, the IPC must issue its screening opinion within 21 days.

Scoping

Before making an application for a development consent order, the applicant has the opportunity to ask the IPC for a formal 'scoping opinion' on the information to be included in the ES. When making the request, the applicant must provide as a minimum: a plan sufficient to identify the land involved; a brief description of the nature and purpose of the development and of its possible effects on the environment; and such other information or representations as the applicant may wish to provide. However it is common practice for applicants to provide a scoping report as part of their formal request for a scoping opinion. Ideally this should include:

- a description of the scheme or proposal;
- interpretation of the site settings and surroundings;
- outline of alternatives and methods used in reaching a preferred opinion;
- results of desktop and baseline studies where available;
- methods to predict impacts and the significance criteria framework;
- mitigation and residual impacts to be considered;
- key topics covered as part of the scoping exercise; and
- an outline of the structure of the proposed ES.

Before adopting a scoping opinion the IPC has a duty to consult the 'consultation bodies', as defined in the EIA Regulation. These bodies have 28 days to respond. The IPC must adopt a scoping opinion within 42 days of receiving a scoping request; to date, these scoping opinions have tended to be long and detailed. Finally, all IPC project EIAs are made available on the IPC website.[5]

3.6 Summary and conclusions on changing legislation

The original (and amended) EC EIA Directive has been implemented in the UK through an array of regulations that link those involved-developers, affected parties, regulators and facilitators—in a variety of ways. The T&CP (EIA) Regulations are central. Other regulations cover projects that do not fall under the English and Welsh planning systems, such as motorways and trunk roads, power stations, pipelines, land drainage works, forestry projects, and development projects in Scotland and Northern Ireland.

The original UK Regulations had a number of weaknesses, relating to the range of projects included in the ambit of the EIA procedures, approaches to screening, scoping, consideration of alternatives, mandatory and discretionary EIS content, public consultation and others. Directive 97/11/EC, and subsequent amendments, has sought to address some of these issues that had arisen in the UK and other Member States.

UK regulations have been amended many times, partly in response to changes to the EC Directive and to domestic legislation, and partly also to case law from domestic and European Courts. The evolving UK Regulations replicate closely the four annexes of the amended EC Directive. This has brought a wider array of projects into the UK EIA system, has increased the number of mandatory categories and has led to some growth in EIA activity and EIS output. Screening procedures have been developed, and the consideration of scoping and alternatives now has a higher profile. There has been considerable rationalization of legislation, most recently in the 2011 consolidation of the T&CP (EIA) Regulations, but the array is still wide and complex. EIA guidance is a particular strength of the UK system, and government publications, especially the relevant circulars and guide to the procedures, help to navigate the legislative array.

SOME QUESTIONS

The following questions are intended to help the reader focus on the important issues of this chapter, and to start building some understanding of the UK agency and legislative context of EIA.

1 For a project with which you are familiar, identify the various sets of principal actors, and outline the potential areas of conflict between their interests with regard to the project.

2 How does EIA legislation, and the relevant key regulators, vary in England between (a) a Schedule 2 golf course project; (b) a motorway development; and (c) the decommissioning of a nuclear power station?

3 What is the role of statutory consultees in the EIA process?

4 Under the T&CP (EIA) Regulations, a Schedule 2 project requires EIA if it is deemed 'likely to give rise to significant environmental effects'. What criteria and thresholds are used to determine that significance?

5 Outline some of the changes introduced in the T&CP (EIA) Regulations for England (2011). What were the key drivers behind the changes?

6 Over time the projects covered by EIA have grown in response to both new technology and changes in social and economic infrastructure. What new additions might you anticipate for inclusion in the next review of legislation by 2016?

7 Currently a separate EIA for decommissioning or dismantling of a project only applies to nuclear power stations and reactors in the UK. Examine the case for extending this requirement to other projects.

8 What is a 'screening opinion', and who provides it?

9 Identify any particularly innovative features associated with the Infrastructure Planning (EIA) Regulations 2009.

Notes

1 There are some discrepancies. For instance, power stations of 300 MW or more are included in Schedule 1, although they actually fall under the Electricity Works (AEE) Regulations, and all 'special roads' are included, although the regulations should apply to special roads under local authority jurisdiction.

2 Decisions were actually made by the relevant government office in the region concerned (or the Assembly)—subsequently disbanded in 2011. As will be discussed in Chapter 8, this led to some discrepancies where two or more offices made different decisions on very similar projects.

3 This includes enough copies for all the statutory consultees to whom the developer has not already sent copies, one copy for the LPA and several for the Secretary of State.

4 The criticism was well deserved. The circular's assertion that '... individual highway schemes do not have a significant effect on climatic factors and, in most cases, are unlikely to have significant effects on soil or water' is particularly interesting in view of the cumulative impact of private transport on air quality.

5 IPC practice is developing fast and includes much more focus on pre-application activities, including the EIA process. There is also an important role for a Statement of Common Ground (SOCG) between the developer and the LPAs, and the production of separate Local Impact Reports (LIRs) by the relevant LPAs.

References

CEC (Commission of the European Communities) 1985. On the assessment of the effects of certain public and private projects on the environment. *Official Journal* L175, 5 July.

CPRE (Council for the Protection of Rural England) 1991. *The environmental assessment directive: five years on*. London: CPRE.

DCLG 2011a Town and Country (T&CP) (Environmental impact assessment (EIA)) regulations 2011. London: DCLG.

DCLG 2011b *Guidance on the Environmental Impact Assessment (EIA) Regulations 2011 for England*. London: DCLG.

DETR 1997. *Mitigation measures in environmental assessment*. London: HMSO.

DETR 1999. *Circular 02/99. Environmental impact assessment*. London: HMSO.

DETR 2000. *Environmental impact assessment: a guide to the procedures*. Tonbridge, UK: Thomas Telford Publishing.

DfT (Department for Transport) 2007a. *WebTAG: Transport Analysis Guidance*. (www.webtag.org.uk). London: DfT.

DfT 2007b. *Guidance on transport assessment* (GTA). London: DfT.

DfT 2011. *Design manual for roads and bridges* (DMRB). London: DfT.

DoE (Department of the Environment) 1989. *Environmental assessment: a guide to the procedures*. London: HMSO.

DoE 1991. *Monitoring environmental assessment and planning*. London: HMSO.

DoE 1994a. *Evaluation of environmental information for planning projects: a good practice guide*. London: HMSO.

DoE 1994b. *Good practice on the evaluation of environmental information for planning projects: research report*. London: HMSO.

DoE 1995. *Preparation of environmental statements for planning projects that require environmental assessment: a good practice guide*. London: HMSO.

DoE 1996. *Changes in the quality of environmental statements for planning projects*. London: HMSO.

DoT (Department of Transport) 1983. *Manual of environmental appraisal*. London: HMSO.

DoT 1989. *Departmental standard HD 18/88: environmental assessment under the EC Directive 85/337*. London: Department of Transport.

DoT 1993. *Design manual for roads and bridges, vol. 11: Environmental assessment*. London: HMSO.

ENDS 1993. *Directory of environmental consultants 1993/94*. London: Environmental Data Services.

ENDS 1997. *Directory of environmental consultants 1997/98*. London: Environmental Data Services.

ENDS 2001. *Environmental consultancy directory*, 8th edn. London: Environmental Data Services.

ENDS 2003. *Directory of environmental consultants* 2003/2004. London: Environmental Data Services.

ENDS 2009. *Salary and careers survey*. London: Environmental Data Services.

ENDS 2010. *Survey of environmental professionals*. London: Environmental Data Services.

Environment Agency 1996. *A scoping handbook for projects*. London: HMSO.

Fortlage, C. 1990. *Environmental assessment: a practical guide*. Aldershot: Gower.

HMG (Her Majesty's Government) 2008. *Planning Act* 2008. London: The Stationery Office.

IPC 2010. *Identifying the right environmental impacts: advice note 7-EIA, screening and scoping*. Bristol: IPC.

NAFW (National Assembly for Wales) 1999. *Circular 11/99 Environmental impact assessment*. Cardiff: National Assembly.

Planning Service (Northern Ireland). *Development control advice note 10* 1999. Belfast: NI Planning Service.

SACTRA (Standing Advisory Committee on Trunk Road Assessment) 1992. *Assessing the environmental impact of road schemes*. London: HMSO.

Scottish Executive Development Department 1999a. *Circular 1999 15/99. The environmental impact assessment regulations* 1999. Edinburgh: SEDD.

Scottish Executive Development Department 1999b. *Planning Advice Note 58*. Edinburgh: SEDD. Scottish Government 2011a. Town and Country Planning (Environmental impact assessment) (Scotland) Regulations 2011 Edinburgh: Scottish Government.

Scottish Government 2011b. EasyRead version-Town and Country Planning (Environmental impact assessment) (Scotland) Regulations 2011 Edinburgh: Scottish Government.

Sheate, W. R. 1995. Electricity generation and transmission: a case study of problematic EIA implementation in the UK. *Environmental Policy and Practice* 5 (1), 17-25.

第二部分

程序

（本插图和第三部分的开头插图由 Neil Bennett 创作，转载自 Browers，J.（1990），《环境经济学：环保主义者对皮尔斯报告的回应》，英国自然保护主义者协会，摄政街 69 号，惠灵顿，特尔福德，什罗浦郡 TE1 1PE）

4 EIA程序开始：早期阶段

4.1 导言

本章是讨论怎样实施环境影响评价（EIA）的四章内容中的第一章。重点是讨论英国法律体系所要求的EIA程序和最佳实践理念。虽然第4～7章都在试图提出一种符合逻辑、可以按部就班实施EIA程序的方法，但没有哪种方法是十全十美的。每种EIA程序方法都是在一定制度背景下形成的，而各国的制度不尽相同（见第10章）。正如我们之前提到的，即使在同一个国家（如英国），针对不同的项目也可能会使用不同的法规（见第9章）。在具体案例中，各步骤实施顺序可能不同，有些步骤也可能被跳过。实施程序不应该是线性的，而应该是可循环的，以便后面阶段产生的信息能反馈到前面的阶段。

第4章讨论EIA程序实施的早期阶段。具体包括为EIA建立管理程序、明确是否需要进行EIA"筛选"以及是否需要概括EIA的评价范围（"范围确定"），对此可以向第3章列出的主要参与者进行咨询。此外，EIA的早期阶段也应该对项目的可替代方案进行初步研究。本章对环境本底数据、开发行为的参数（包括相关的政策立场）、现在和将来环境状况的参数进行了研究。本章讨论的重点内容为影响的识别，影响识别在EIA实施的早期阶段非常重要，后续阶段也可能会用到本章讨论的影响识别的方法，反映出EIA各阶段的可循环性以及相互作用的特点。当然，第5章讨论的一些预测、评估、信息交流、减缓措施等方面的方法和第6章讨论的公众参与方法也可以应用到评价的早期阶段。本章我们先从EIA程序的管理谈起。

4.2 EIA 程序的管理

环境影响评价是一个强调管理的过程。它常常用来对大型项目（有时很难界定）进行评价，这些项目产生的影响通常存在较大争议，而且影响范围较大。因此，EIA 程序的有效管理就显得尤为重要。本节主要讨论程序管理中几个互相关联的内容：EIA 团队、EIA 程序类型以及 EIA 成本和资源。

4.2.1 EIA 团队

EIA 程序中总是包括多学科交叉的方法。美国早期的法律强烈推荐以下方法：

应用交叉学科的方法编制环境影响报告，能够确保自然科学、社会科学和环境设计艺术相结合。报告中涉及的学科应该符合范围确定步骤中所确定的范围和得到的结论。（CEQ，1978. par. 1502. 6）

该方法不但能反映 EIA 研究从自然到社会经济的正常范围，还可以给实施程序带来很多益处，例如能够对复杂问题的不同观点和意见提出良好的建议。

EIS 的编制团队可以是项目倡议者、外聘顾问专家、助理顾问以及其他专家中的一个或几个。对某些项目，团队规模从几人到十几个人不等，这个团队成员所具备的能力应该是可以相互补充的：技术上要能够满足这项任务的需求；同时也需要一个较好的管理团队。以 3 人小组为例，他们的专业可以涵盖物理/化学、生物/生态和文化/社会经济领域，其成员可能包括一个环境工程师、一个生态学家和一个规划者，其中至少有一个成员具有 EIA 和管理的经验或培训经历。不过，小组的成员需要在确定评价范围之后才能最终确定。

很多 EIA 团队在“核心/焦点”管理团队与相关专家组之间区分得很清楚，这反映了一个事实，即没有一个组织能够涵盖一个大型项目 EIS 编制所需的全部信息。还有一些评论家（Weaver 等，1996）也在宣传这种方法的优点。例如，一个以 Weaver 等为核心组成的 5 人项目组对南非一个大型露天采矿项目进行研究，其成员包括一个项目管理者、两个资深作家、一个编辑顾问和一个文字处理员，该团队管理着 EIA 实施程序的各种信息，梳理了至少 60 条学术性和非学术性意见，制定了多种多样的公众参与及交流方案。

显然，项目管理者在团队中起关键作用。除了人事和团队管理技巧，管理者还应该对所考察项目的类型有一个正确的评价，掌握相关程序和 EIA 方面的知识，并具备识别重要问题和选择更具价值的专家意见的能力等。项目管理者的关键作用包括：

- 挑选一个合适的项目团队，处理工作人员变动等特殊人事问题；
- 管理专家意见；
- 在程序的实施中，与相关人士进行沟通，包括各个利益相关者、公众，并决定其参与的技术方式和后续的内容；
- 制订团队计划，保证 EIA 实施的高效性，增强团队合作，使之能应对预期之外的情况；
- 确保 EIA 程序对重点问题的关注，与最终目标保持一致；
- 在编写各种报告书时，协调好团队的内部意见（IEMA，2011；Lawrence，2003；Morrison-Saunders and Bailey，2009；Petts and Eduljee，1994）。

4.2.2 EIA 程序类型

EIA 中包含许多对相关项目的优缺点以及造成的影响持有不同看法的参与者。所以跨学科的团队共事时，互补性、相似性及协调性显得尤为重要。合作是其中最重要的内容：协调各种结论和数据（例如，团队成员可以按照统一的地图比例和统一的版式开展各自的工作），并将协调好的结论和数据汇总到中心数据库。

对大多数评价团队来说，他们最薄弱的环节之一是：所有的顾问必须随时了解其他人所做的工作，避免产生漏洞、异常现象和矛盾，因为这些很可能成为反对者或媒体攻击的对象（Forrtlage，1990）。

此外，不同的项目（包括 EIA）有不同的技术要求、参与方法、关注点和突发事件的解决方案。一个有争议的项目环评在人口密集地区要远远复杂于一般的或人口稀少的地区；数据完备的项目环评不同于一个崭新的项目环评；居民或生态环境已受影响地区的环评不同于一般地区的项目环评。

Lawrence（2003）指出：

- 严格的环评程序适用于可以通过科学分析做出决策的情况，例如，环境或资源信息可以被科学地分析出来的情况。
- 合理的环评程序适用于利益相关者可自由参与、观点不过度分化、可以自由提出建议的情况。
- 简化的环评程序适用于资源有限的极端情形，并且变化可能会以增量的形式改变现状。
- 民主的环评程序适用于倡议者愿意将他们的决策权反映给代表们，反过来，这些代表也要有时间、精力与其他团体一道参与计划和决策的过程。
- 协作的环评程序类似于民主程序，区别是利益相关者可直接参与过程且有能力参与。
- 道德环评程序适用于公平正义占主导地位时，利益相关者愿意认同和调解这些伦理冲突的情况。
- 适应性的环评程序主要针对风险、健康隐患突出且需要考虑知识的局限性和不确定性的混乱复杂的情况。

4.2.3 EIA 成本和资源

EIA 团队组成和程序类型在影响资源的同时，也受资源的约束。EIA 成本多耗费在环境现状研究和 EIS 编写阶段，目前这笔开销由开发者承担，但是主管部门、法律顾问和公众在参与 EIA 的过程中也产生了成本。

EC（欧洲委员会）研究表明，有多年 EIA 经验的国家的 EIA 成本为项目成本的 0.1%～1%，虽然实际成本会随项目成本的增加而增加，但是它在总成本中所占的比例是下降的。对于小型项目，EIA 成本大约为项目成本的 1%，对于大型项目，该比例约为 0.1%（EC，2006；COWI，2009；Oosterhuis，2007）。英国的 EIA 情形与该情况一致，将在第 8 章中进一步讨论。世界银行（1999）《环境评价资料》指出 EIA 成本从几千到几百万美元不等，很少有超过项目成本 1%的情况。

法国的一项研究（BIO，2006）指出，对于线型工程来说 EIA 成本将更加高昂，如道路、电力、核能和工业活动等项目；涉及海洋环境、大型企业或行政部门的项目需要进行健康影响评价。世界银行（1999）指出如果当地人员参与EIA的大部分工作，则可以降低其成本。

EIA程序通常持续6～18个月，有时会根据需要适当延长时间。例如，项目开发者中途更改项目提案或政府要求变动的时候。准确划定范围，找到关键问题，可以节省EIA的时间。相反，数据的缺失以及相关信息的补充会延长EIA的时间（EC，2006）。

4.3 项目筛选——需要进行EIA吗？

需要进行EIA的项目可能很多，但其中多数项目并没有重大或显著的环境影响。筛选机制的建立试图将重点放在那些具有潜在显著不利环境影响或影响尚不清楚的项目上，那些影响较小或没有影响的项目则被“筛出”评价范围，按照正常的规划许可执行管理程序即可，不需要其他评价，这在一定程度上可以避免浪费额外的时间和资金。

筛选在某种程度上是由一个国家当时的评价制度决定的。第3章中指出欧盟（包括英国）被“筛选进”评价范围的项目（附录1）是由于它们的规模和潜在的环境影响（例如原油提炼厂、较大规模的热电厂、特殊的道路建设等），该类项目需要进行全面的评价。还有一些项目（附录2）的筛选结果不是很明确，其筛选结果可能出现两种情况，一种需要进行全面评价，一种则不需要。筛选过程通过相关标准，根据项目规模、选址的敏感性以及对不利环境影响的预测得出结论。在这样的案例中，工作导则、指导性标准以及可能产生显著环境影响的阈值都非常重要（3.4节）。

在美国加利福尼亚州，“如果有确凿的证据，根据权力机构之前的全部记录，表明一个项目可能对环境产生重大影响”，就需要一份环境影响报告的草稿。如果根据整个记录，有大量证据表明该项目将造成重大影响，则需要一份完整的环境影响评价报告表明对环境质量、生境和物种或历史文物的影响；对长期环境目标有负面影响；累积有相当大的影响；或对人类造成重大不利影响（加利福尼亚州，2010年）。这些构成了“包含列表”方法。此外，可能有一个“排除列表”，如在加利福尼亚州和加拿大使用，它确定了那些不需要进行环境影响评估的项目类别，因为经验表明它们的不利影响并不显著。

一些EIA程序包含最初的环境影响评价研究框架，以检查可能的环境影响及影响的显著性。《加利福尼亚环境质量法案》规定，项目支持者可以提交“否决声明”，即声明该项目只有很小的环境影响，而不需要进行全面的EIA。该声明必须通过最初的研究给予证实，这种研究通常为简单的核查表法，通过对可能的环境影响回答“是”“可能”“否”来进行，如果回答主要是“否”，那么对“是”和“可能”就可以忽略了，然后这个项目就不需要进行全面的EIA。在加拿大和澳大利亚，“筛选”程序也开展得很好（见第10章）。

总的来说，主要的“筛选”方法有两种。“阈值法”的应用包括将项目进行分类，然后给不同类型项目设置不同的阈值。阈值的设置或与开发规模有关（如项目占地$20hm^2$及$20hm^2$以上），或与预期的项目影响有关（如某一个工厂每年废弃物的产生量减少5万吨或5万吨以上），或与项目所处位置有关（如指定的风景区）。表 3.3 列出了英国的指示性阈值。

逐案分析法涉及对项目特点的评估，当提出进行“筛选”时，可以借助导则和标准的核查表。表 4.1 对这两种方法的一些优缺点进行了总结。欧盟（2001 年）公布的一份指南有助于逐案分析法的实施。此外，还有很多交叉的方法，例如阈值法与逐案分析法相结合的方法。图 4.1 对英国所采取的阈值法系统进行了简要说明，其范围从附件Ⅰ的法定阈值到附件Ⅱ的指示性阈值，最后再到那些通常不需要进行阈值法的 EIA 项目。

表 4.1 阈值法和逐案分析法的优缺点

方法	优点	缺点
阈值法	使用简单； 使用快捷，更加明确； 区域之间能保持协调； 区域内的决策保持协调； 项目类型间保持协调	将强制性的、不变的规则应用于多变的环境（除非是多级的）； 很少有常识或明智判断； 可能与附近相关区域不协调； 很难设定，而且一旦设定就很难改变； 导致低于阈值的项目数量增多
逐案分析法	常识和明智的判断； 灵活——可以将项目和环境的变化相结合；	可能会很复杂且不明确； 可能会花费大量时间和金钱； 由于政治和经济利益而容易被决策者滥用； 决策者可能做出不合理的判断； 可能受惯例的影响而动摇，并且因此失去灵活性

DCLG（2011）EIA 导则对如何筛选给出了详细指导，并提供了法案参考（见 3.4 节和 8.6 节）。

阈值法和逐案分析法都不能给出一致的、可信的结论，并且不同国家利用阈值法得到的结果不同。表 4.2 列出了 2006～2008 年，欧盟成员国执行 EIA 的情况。此表显示，不同国家由于法律体系的不同，人均 EIA 执行情况会有 50 以上的差别（EIA 指令）。这一差别是由不同政体、发展水平、环境敏感度和文化价值导致的。例如，低（高）水平的 EIA 活动不是导致问题的必要条件（IEMA，2011）。

表 4.2 2006～2008 年欧盟成员国家 EIA 数量

成员国	平均 EIA/年（2006～2008 年）	平均 EIA/百万人（2006～2008 年）	成员国	平均 EIA/年（2006～2008 年）	平均 EIA/百万人（2006～2008 年）
奥地利	23	3	拉脱维亚	11	5
比利时	183	17	立陶宛	142	42
保加利亚	249	33	卢森堡	70	149
塞浦路斯	117	136	马耳他	10	24
捷克共和国	96	9	荷兰	123	7
丹麦	125	22	波兰	4000	105
爱沙尼亚	80	61	葡萄牙	323	29
芬兰	38	7	罗马尼亚	596	27
法国	3867	59	斯洛伐克	670	124
德国	1000	12	斯洛文尼亚	108	54
希腊	425	38	西班牙	1054	23
匈牙利	152	15	瑞典	288	31
爱尔兰	197	44	英国	334	5
意大利	1548	26			

资料来源：GHK，2010。

一系列欧洲法院条例中都提及如何进行 EIA 筛选（EU，2010），其中明确不论 EIA 中用什么评价方法都不能违反 EIA 指令中的目标，即对环境产生影响的项目都应进行环境影响评价，特别是：

- EIA 指令对筛选这一程序的解读为“范围广泛并且目的深入的”。项目不能被简单地筛选出来，因为同 EIA 指令中的实施条例不完全相符［Kraaiieveld（荷兰堤坝）C-72/95］。
- 敏感地区的小项目也有可能对环境产生很大的影响（C-392/96）。
- 可能对环境产生影响的翻新、改建、拆除项目也应进行 EIA（C-142/07，C-2/07，C-50/09）。
- EIA 指令的目的不是将需要进行 EIA 的大项目变成不需要进行 EIA 的小项目，相反，EIA 必须考虑各个项目的累积影响（C-392/96，C-142/07，C-205/08）。
- 虽然 EIA 指令允许成员国对于特殊项目不用进行 EIA，但是对于特殊项目的限制却是苛刻的。特殊项目是指紧急的、特殊用途的项目，不能提前实施的项目，或者不满足 EIA 指令的项目（C-435/97，C-287/98）。
- 虽然 EIA 的筛选过程不需要提供解释原因的说明，但是权力部门有责任提供决策的相关信息（C87/02，C121/03，C-75/08）。

关于筛选过程的英国法案在 8.6 节进行集中讨论。

4.4 范围确定——需要考虑哪些影响和问题？

EIA 的范围即所提出的环境影响及问题的范围。“范围确定”的过程就是从项目所有可能的影响和所有替代方案中选出具有显著影响的决策过程。通过确定可能产生影响的最初范围，以识别潜在的、显著的影响以及那些不显著的和显著性尚不明确的影响。更加深入地研究不同项目分类中的各种影响，确定没有显著影响的项目；还没有明确分类的项目则归入到其他具有潜在的、显著的影响分类中。通过范围的确定能提炼出 EIA 程序中持续的、最显著的影响。事实证明，恰当的“范围确定”是高质量完成 EIS 的关键因素（IEMA，2011）。

“范围确定”通常通过开发商、主管部门和其他相关机构之间的讨论来完成，更为理想的讨论还应包括公众参与。开发商与其他利益团体间的协商和意见征求通常为第一步，在 EIA 中这是非常重要的一步，因为这可以使 EIA 团队的有限资源得到合理配置，以产生最好的效果，并且可以防止各相关团队之间产生误解。“范围确定”也能够识别之后应该监控的问题，虽然它在 EIA 中很重要，但目前在英国还不属于法律强制的步骤。在准备编写 EIS 之前，开发商可以咨询主管部门及法定咨询机构，这一步也很重要，但在英国早期的 EIA 中，大约只有一半的案例实施了这一步。到目前为止，缺乏前期咨询是 EIA 无法充分发挥效应的主要局限之一。政府强烈建议进行这些咨询，自从 1999 年欧盟指令修订以来，开发商现在可以向地方规划管理局（LPA）提出要求，以获得正式的“范围确定意见”。

“范围确定”应该从可能受项目影响的个体、团体、地方权力部门及法定咨询机构的认识开始，有效的实践会使他们形成一个工作组，和/或者与开发商交流意见。4.8 节中讨论了影响识别的技术方法，这些技术方法可以用来组织讨论和建议所要考虑的重点问题。其他问题包括：

- 有特殊价值的环境特征；
- 受影响团体特别关注的影响；
- 预测和评估不同影响的方法；
- 考虑影响的范围；[1]
- 考虑广泛的替代方案。

信息收集和协商的过程应该是对主要问题、所造成的影响及替代方案的确定以及解释为什么其他问题未被重要考虑。并且对于每个影响，应确定其时间和空间的测量范围。

专栏 4.1 提供了一份筛选报告的摘录，这份报告是根据开发商的要求由当地规划部门编写的。报告表明上述各点都在实践中得到处理。

专栏 4.1　国家公园采石场环评的范围确定

社会经济影响：社会经济影响中应该考虑土地占用对公众的影响，也包括对场地安全性的考虑。同时也应该考虑发展对普通老百姓的权利所造成的损害。尽管在 EIA 的法定范围外，申请人会考虑对采石场就业市场和产品市场的基本评价是否有利于相关规划的实施。

考古学：目前存在的记录没有提供与考古有关的影响，即使如此，EIA 的这部分也应该由熟悉地下条件及性质的考古学家进行步行调查和矿井测试，从而提供评估(例如，泥煤情况和底土深度等)，任何被认为有必要的深入调查都应作为 EIA 的一部分实施。

水文地质影响：厂址涉及蓄水层，距离水资源保护区 450m，距离得文港大约 200m。EIA 评价中应考虑对这些区域的影响。关于环保局提供的更多建议请见《项目 EIA 筛选导则》(2002)(第二部分——露天采矿和采石业务)。

生态和生物多样性影响：当局要求进行第一阶段栖息地调查和专家调查，之后识别其必要性。评价要求包括必要的避免、减缓措施。

累积影响和替代方案：除了以上所述，EIA 中也应该包括对累积影响和替代方案的考虑。当局建议评价工作应该涉及其项目可能对国家公园造成的潜在影响。

根据《英格兰和威尔士城乡规划法案》1999 中的第 19 条规定：采用这种范围的意见并不排除需要从矿产规划部门获取额外的信息。

资料来源：达特穆尔国家公园，2010

欧洲法院规定，EIA 中不仅应考虑直接影响，还应考虑项目的整体影响。在列日机场（C-2/07）的案例中，规定 EIA 不仅需要提出针对机场设施建设的修正方案，还应提高机场的利用效率。同样，跨区域的影响也应考虑到 EIA 范围内（C-205/08）。

虽然筛选在 EIA 程序中是一个重要的步骤，但在英国却不是法律要求的。从 1999 年开始，一些开发商编写筛选报告作为更优操作，如果开发商需要，则 LPA 必须在 EIS 中提供正式的“筛选意见”。关于英国筛选实践的研究显示，大约一半的 EIA 项目开发商要求包含正式筛选过程，超过 18%的项目中开发商要求包含一般的筛选过程。在大约 2/3 的案例中，

---

[1] 评价范围是指将覆盖的区域和已经覆盖的区域。Joao（2002）表明，后者（迄今为止被广泛忽视的问题）可能会因为规模的不同而导致不同的决策。

开发商都会要求进行筛选过程，大多数筛选意见由 LPA 通过专家咨询的方法提出。2/3 的反馈表明，通过筛选提升报告的专注度、增加讨论和减少后期调查内容的方法提升了 EIS 的质量（DCLG，2006）。相反，IEMA（2011）提出很多筛选意见过于苛刻，并没有对确定主要问题的影响范围提供帮助，同时也没有避免当地政府部门或者法律顾问在 EIS 的准备过程中索要更多的项目相关信息。

其他国家（如加拿大、荷兰）也已经有正式的“范围确定”阶段。该阶段有时在公众咨询之后，开发商与主管部门及独立的 EIA 调查机构需要就 EIA 内容达成一致。欧盟（2001）已经公布了一份“范围确定”核查表。Carroll 和 Turpin 也在 2009 年提供了一份与一系列类型发展项目相结合的可能出现的潜在问题的列表。此外，政府和其他管理机构还能对特殊类型的项目所涉及的影响进行指导，例如英国环境事务部（2002）和新南威尔士政府（1996）合作。

4.5 替代方案的考虑

4.5.1 法律要求

美国环境质量委员会（CEQ，1978）称替代方案的讨论是“EIA 的心脏”：环境影响评价中替代方案的提出情况将会决定之后的决策程序。进行替代方案讨论能够确保开发商既能 考虑项目的其他方案，同时还要考虑防止环境破坏的方法。替代方案的讨论能鼓励分析者们将分析重点放在所选择方案与替代方案之间的差别上。它允许没有直接参与决策制定过程的人们评估拟建项目的各个方面，并参与决策的制定。此外，它还为主管部门的决策提供了一个参考框架，而不仅仅是为开发活动辩护。最后，如果在项目的建设或运行阶段中出现不可预见的困难，对替代方案进行重新审视有助于得出快速、经济的解决方案。

美国环保者要求联邦机构分析“拟议行动的替代方案”。这一要求的执行一直是一系列法律挑战的主题，主要是因为联邦机构没有考虑到一整套合理的替代办法，或者说他们有不当的建造目的和需要，所以选择面太窄了。一般来说，法院已经决定支持联邦机构，接受他们在环境影响评估中排除看似合理的替代方案的理由（Smith，2007）。

原始的欧洲共同体 85/337 号指令中明确规定了在 EIA 中应该考虑替代方案，同时应遵从第 5 条的要求（如果是相关的信息，那么要求开发商收集这些信息就是合理的）。附件Ⅲ要求“开发商在对主要替代方案进行研究和对所选方案的主要原因进行说明时都要适当考虑环境影响”。在英国，该要求是自愿执行的，致使替代方案的考虑成为 EIS 质量最薄弱的环节之一（Barker 和 Wood，1999；Eastman，1997；Jones 等，1991）。修订后的欧盟 97/11（CEC，1997）号指令的最大变化就是加强了对替代方案的要求：现在的 EIS 要求包括“开发商在对主要替代方案进行研究说明和对所选方案的主要原因进行说明时要适当考虑环境影响”。政府的指南（DCLG 2011）如下：

考虑替代方法的发展，包括工艺或设计的备选，或分期建造，ES 应包括主要的概要以及作出选择的主要原因。虽然该指令和条例没有明确要求申请人研究替代品，也不定义“替代品"，但某些发展的性质及其所在地可能使考虑替代场址成为一种审议材料……在这种情况下，ES 必须记录对其他地点的考虑。

4.5.2 替代方案的类型

在项目规划期间，需要决定拟建项目的类型、规模、选址和实施程序等。由于经济、技术和制度的原因，开发商将会否决提出的大多数可能的替代方案。EIA 的作用是可以确保将环境标准纳入项目的早期计划中。在开发项目的类型、规模及选址达成一致之前，需要在规划过程中尽早地全面考虑替代方案。此外，还应该考虑替代方案的众多类型："不作为"（noaction）的选择、可替代的选址、项目的替代规模、可替代的程序或装备、可替代的选址规划、可替代的实施条件与可替代的处理环境影响的方式。针对后面几项，我们将在 5.4 节中加以讨论。

如果一个项目不再继续进行，那么这（"不作为"选择）通常与所涉及的环境条件有关。其实，对"不作为"选择的考虑与对项目进行必要的讨论差不多，即考虑项目的效益是否超过其成本，在一些国家（如美国）要求对此予以考虑，但在英国的 EIS 中却很少提及。[1]

可替代选址的考虑是项目规划程序的一个基本组成部分。在一些案例中，项目的选址在不同程度上会受到约束，例如，采石厂只能建在石头充足的地方；风力发电场则要求所处地要达到一定的风速。在很多案例中，好的选址可以实现经济、规划和环境的最优化。对于工业项目，其经济效益可能很重要，如土地价值、基础设施的可用性、资源到市场的距离以及劳动力供应等。对于公路项目，工程标准会对公路的规划产生极大影响。然而，在所有这些案例中，都应该考虑把项目建在环境功能强的区域，或远离指定区域与环境敏感区域。

项目规划也应该考虑开发活动的不同规模。在一些案例中，开发规模是可以调整的。例如，垃圾处理厂的规模可以随填埋场的空间、其他选址的可用性、附近居民区或环境敏感地点而变化。风力发电场的涡轮机数量也可以改变。在其他一些案例中，需要开发商决定是否构建完整的工作单元。例如，压水反应堆核电站的核反应堆是一个巨大、独立的结构，其规模很难缩小；管道或桥梁要实现应有功能，其部件也是不能缩小的。

程序和装置的替代是指用不同的方法实现相同的目标。例如，1500MW 的电力可以由一个联合循环燃气轮机电站产生，也可以由潮汐发电、燃烧垃圾发电以及数量众多的风轮机电站产生。用于干法或湿法的砾石可以直接开采，也可以重复利用；废弃物也可以重复利用、焚烧或者填埋。

当项目选址、开发规模和程序都确定后，不同的选址规划仍然会产生不同的影响。例如，有噪声的工厂应该尽量远离居民区；电厂的冷却塔可以减少数量、增加高度（减少对土地资源的占用），或者增加数量、降低高度（降低视觉影响）。建筑选址要考虑减少显著影响或视觉影响。同样，运行条件的变化也可以使得影响最小化。图 4.1 展示了不同的桥梁设计。例如，相同的噪声级在夜晚通常比白天显得影响更大，因此要尽量避免夜间施工。为项目制定专门交通路线可以使对居民的干扰最小化。建筑施工可在环境影响最小的时候进行，如要考虑鸟类的迁徙和筑巢。这些"替代"行为与减缓措施相似。

---

[1] 在美国，应该考虑不采取措施的后果；考虑机构职权范围以外的替代方案，并且考虑为降低影响，而达成一部分目标的后果。

单塔形　　H形

钻石形

A形

图 4.1　设计案例：Forth Replacement Crossing 设计考虑

（资料来源：苏格兰政府，Jacobs 和 Arup，2009）

由图 4.1 可知，早期的替代方案可以有不同的层次，尽早确定的替代方案会为之后确定的替代方案打基础。在此案例中，先确定桥梁建设地点，之后再确定建设方案。另一个案例是新建风力发电厂的选址问题。首先考虑对涡轮机的设计，然后才是每个涡轮机的具体位置如何选择。

“替代方案”必须是合理的：首先应该是合法的、技术可行的。根据特定开发商的实际考虑，替代方案的类型也可能变化。矿产开采公司投资一块土地，以期能开采沙石，而没有考虑将其用于风力发电。在这样的案例中，“合理”意味着选择其他地方开采沙石，或者重新拟订开发规模和程序。基本上，替代方案应该让主管部门明白规划项目（非其他项目）选址在这个地方（而不是其他地方）的原因。

另一方面，从美国的情况来看（EIS 由政府部门编制），Steinemann（2001）认为如果替代方案不能实现项目的既定目标，那么就很容易被否决，替代方案反映的不只是某个机构的目标，还应该反映社会目标。她还建议：

目前的顺序——建议方案、设定目标和要求、形成替代方案、分析替代方案等，需要重新修改。否则建议方案会偏向那些用于分析的替代方案。政府部门应该在提出开发计划前研究一些更环保的方法，然后再确立目标和需求说明，以确保危害较小的替代方案不被淘汰，也不会过于支持拟提议的方案。政府部门也应该谨慎，以避免在早期只关注单一的“问题”

和“解决方法”。

Smith（2007）在美国案例中得出以下结论：

• 项目建议者应解释他们作出选择的原因，特别是如何选择在环评中详细研究的备选方案范围；

• 如果有人提出一个他们认为“合理”的替代方案，项目的倡议者应仔细考虑，而环评则应提供一个合理的解释，说明该方案被否决的原因；

• 如果拟议的行动不太可能产生重大影响，那么开发“最低限度行动”替代方案的其他替代方案是没有意义的。

这些非常合理的经验法则在其他情况下也有适用。

4.5.3 替代方案说明与比较

替代方案的成本因不同的团体和环境要素而不同。通过与当地居民、法律顾问和特殊利益团体进行讨论，可以很快淘汰一些替代方案。然而，一个让各方都接受的替代方案几乎是不可能存在的。EIS应该保留合理且切实可行的替代方案的信息，以便与公众讨论并最终决策。替代方案的比较和说明方法包括简单的、定性的描述以及相当详细的定量模型。

本章后面谈到的许多影响识别方法也有助于对替代方案的比较。叠图法是比较不同选址影响的一种定性分析方法。核查表法和简单矩阵法也可以用于对不同替代方案的比较，这可能是直观地提出替代方案影响的最有效方法。其他一些用于影响识别的技术——阈值核查表法、权重矩阵法和环境评估体系（EES），都可以用于对替代方案的比较。通过对环境要素赋权，即根据对每种环境要素的影响，对每一个替代方案赋予权重系数（定量的），然后乘以权重系数得到一个权重影响，再将这些权重影响相加得出每一个替代方案的总分。得出的分数可以相互进行比较，以识别最优的替代方案。除阈值核查表法外，其他方法不能用于讨论中的替代方案的确切说明，并且不能明确不同替代方案的影响。

4.6 对项目/开发活动的理解

4.6.1 了解项目的特征

初期可能会认为对拟议开发项目的描述是EIA程序中较直观简单的步骤之一。但是，项目具有很多特征，并且相关信息可能也很有限。因此，项目在这一开始阶段可能就存在很多困难。项目的重要特征必须阐述清楚，包括项目的目标、生命周期、自然现状、程序、政策背景及相关政策等。

2011年EIA条例要求环境现状调查包括以下内容：

• 对建设项目的描述，特别是对开发项目的自然特性的描述以及在建设和运行阶段对土地利用要求的描述；

• 生产过程主要特征的描述，例如，所用原材料的特性和数量；

• 拟开发项目运行所产生的废物及排放物（水、空气、土壤污染，噪声，振动，光，热，放射等）类型和数量的预测。

在评价开发活动的环境影响时，可以合理地要求申请者提供这些信息，并且特别重视在报告书编制中合理地要求应用当前的评价知识和方法（DCLG，2011）。

项目的目标和基本原则能对项目描述提供帮助。例如，这可以将特定项目放在一个较大的背景下——主干公路的缺失部分、发展计划中的电站、在主要人口增长的地区设立新的居住区等。目标的讨论可能包括项目类型、项目选址、开发时序等基本原则。讨论目标和原则的同时也可以更新有关规划和设计的背景资料。

如 1.4 节所提到的，所有项目都有一个生命活动周期，所以描述项目时应该明确所考察项目的生命周期的不同阶段以及相对应的持续时间。最简单的描述通常包含建设和运行阶段的识别以及与之相关的各种活动。较精简的描述可能还包括规划和设计、项目授权、扩建、拆除和建设处复原等阶段。此外，还应该明确在生命周期的不同阶段项目开发规模的变化，这主要参考项目的投入、产出、规模、员工人数等。

项目选址和自然现状应该在早期阶段就加以明确。包括拟议开发与其他开发活动（如住宅区域、工作地点和娱乐场所）和行政区域在相关地图上的大致位置。而具体位置应该体现在更大比例的地图上，它能够对土地使用区域和项目元素的主要部署（例如预留区域、主要工厂、废物回收区域、该项目选址的交通等）加以说明。项目选址的规划也会随生命周期的不同阶段而变化，这对形成规划的最终结果非常重要（图 4.2）。任何与拟建项目相关的项目和开发活动（如与项目选址的交通、管道和各种传输线等）都应加以识别和描述，这是项目的组成部分，但也可以分离出来单独讨论（例如，某地区堰坝的建设会影响另一地区的一系列开发活动）。对项目自然现状的描述通常可以通过可视化的三维视图予以展现，包括采用图片合成等方法来展现项目的运行情况。在评价一些开发活动造成的影响时，选址和自然现状的清晰表述非常重要，如土地利用性质变化、对其他基础设施的物理干扰、开发活动造成的隔绝性影响（例如农业自留地、村庄）、视觉干扰及景观变化等。

了解一个项目还包括对整个项目程序的了解。程序的性质不尽相同，根据工业项目、服务项目、基础设施项目而异，但大多都可以看作是输入流，它们经过某一过程转变为输出流。输入和输出的特点、来源和目标以及其经历的时间都应该加以识别。图 4.3 显示了固废管理体系和物质流，其中包括了废物分类和能源流动过程。系统的识别应该既考虑自然特征，又考虑社会经济特征。虽然很多社会经济特征都是伴随着自然特征而形成的，但应该对它们之间的相互作用有明确的认识。

自然特征包括：

- 项目选址的土地利用和自然性状改变（如空旷地、等级），可能会随项目生命周期的不同阶段而变化；
- 整个程序的操作流程（通常用流程图来表示）；
- 所利用资源的类型和数量（如用水、矿物、能源）；
- 交通要求（输入和输出）；
- 废弃物的产生，包括对水污染物、大气污染物、颗粒物污染物、固体污染物、噪声污染、电磁污染、热、光、辐射等的种类、数量和浓度（或强度）的预测；
- 潜在的事故、危险和紧急情况；
- 废弃物的控制、治理和处置程序以及事故的控制和处理程序，监控和监督系统。

社会-经济特征包括：

- 项目的劳动力需求——包括规模、持续时间、来源、特定技能种类与培训；
- 提供给员工的住房、交通、健康及其他服务；
- 当地商业组织及其他商业组织所要求的直接服务；
- 从项目到更广区域的资金流动（来自雇工和次级承包者）；
- 社会活动流（服务需求、团体参与、团体冲突）。

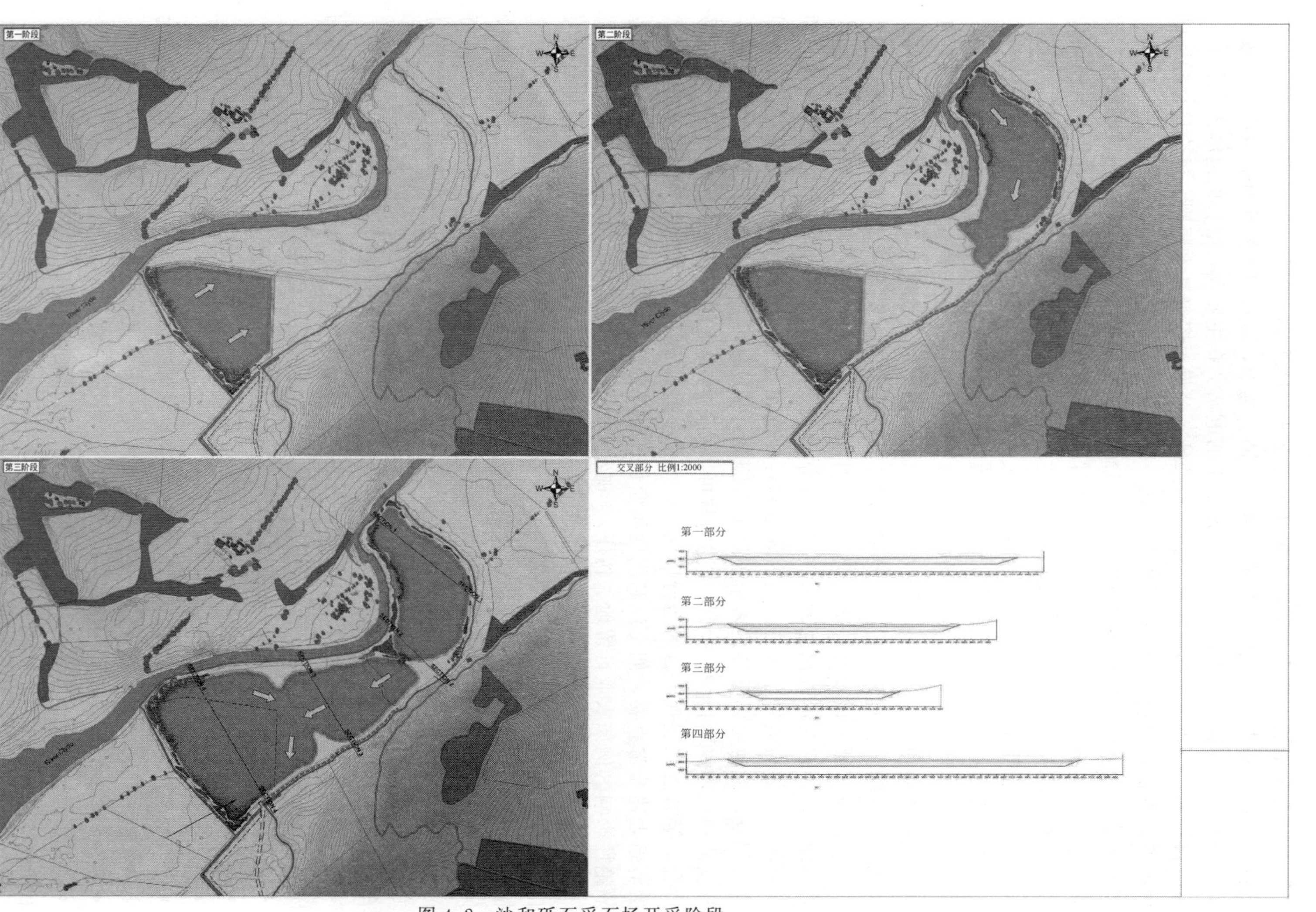

图 4.2 沙和砾石采石场开采阶段
（资料来源：帕特森采石场/SLR，2009）

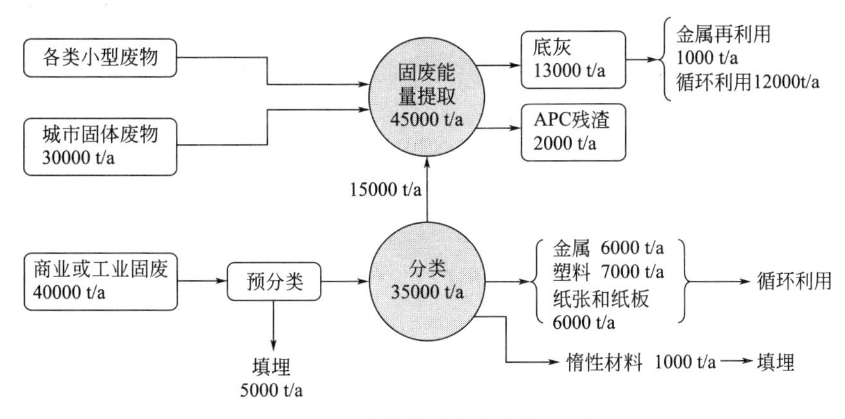

图 4.3 固废管理体系和物质流

显然，项目的物质/生态和社会经济方面是相互作用的。关于生态系统服务的研究（如UKNEA 2011）表明生态系统以产品（如食品、纤维、燃料）和服务（如授粉、碳固定）的形式提供社会价值。反过来，培训或人们的旅行行为等社会干预措施可以改善或削弱生态系统提供服务的能力。

项目也可能与政策相关，这在项目选址规划和项目运行程序流程图中体现的并不明显，但却对最终结果有重要的影响。例如，轮班工作制所产生的交通噪声可能会对附近的居民产生显著的影响。旅馆、乡间宿营区或度假村的建设会对当地房地产市场和社区产生显著的内在影响。提供本地或异地培训将会极大地影响本地和外来劳动人员的融入以及社会-经济影响的平衡。

各个项目应该有其规划政策背景。在英国，《地方发展框架》对主要的地方政策背景分别进行了简单但详细的叙述。对项目选址的描述必须考虑区域的土地利用指标和开发限制，这种限制一般都暗含一些指标。当然，最重要的是与各种环境指标相关的项目选址（如著名自然风景区、具特殊科学价值景点、绿化带及地方和国家自然保护区等）。同时，还要考虑区域空间战略和国家规划指南。在英国，这些由副首相办公室所发布的一系列重要的规划政策报告（NPS）提供。

4.6.2 数据来源与说明

初始数据主要来源于开发商的最初资料。在理想情况下，开发商应该对拟建项目的特点，如规划和运行流程等相当了解，从而可以借鉴以前的经验。然而，在最初设计阶段，项目的选址规划图和流程图都还只是一个临时的框架。即使在理想情况下，分析人员和开发商也需要反复研究以找出项目的特征。分析人员可以参照其他类似项目的 EIS 补充信息，但是这些通常没有被检测，所以对它们的预测的正确性是未经检验的（第 7 章、第 8 章）。分析人员也可以参考一些与环境影响评价相关的文献（书籍和期刊）、指导方针、手册和统计数据，包括 NPS、CEC（1993）、Morris 和 Therivel（2009）、Rodriguez-Bachiller 和 Glasson（2004）及联合国大学（United Nations University，2007）。可以对类似的项目进行现场调查，可以从对此类项目有经验的咨询机构那里获得建议。

随着项目设计和评价程序的开展，这在一定程度上是对早期 EIA 结果的相应检验。所以开发商必须提供与项目特征相关的更为详细的信息。对潜在的、显著的影响进行识别可能会优化项目的规划和流程。

各种信息和插图应明确确定项目各阶段之间的主要差异。项目的生命周期可以用线性柱状图表示，某些特殊阶段可以进行详细说明，因为这些阶段很可能会产生显著影响，这些阶段通常是大型项目建设阶段需要调查的。从大的项目定位到特殊项目选址，规划项目位置和自然现状在图纸上都有清晰的说明。按照所涉及的数据来源和问题，通过航拍照片、图片合成、可视化叠加等提供补充说明。

项目的数据可以有多种呈现方式。图 4.4 以劳动力需求图为例，需求量是根据规划和运行阶段的绝对人数和技能分类来确定的。此外，比较复杂的流程图还应该指明每个操作的类型、频率（正常的、批量的、可歇的或紧急的）和周期（分钟或小时每天或周）。同时也应说明周期性和材料变数，应包括污染负荷最高的时间段等。

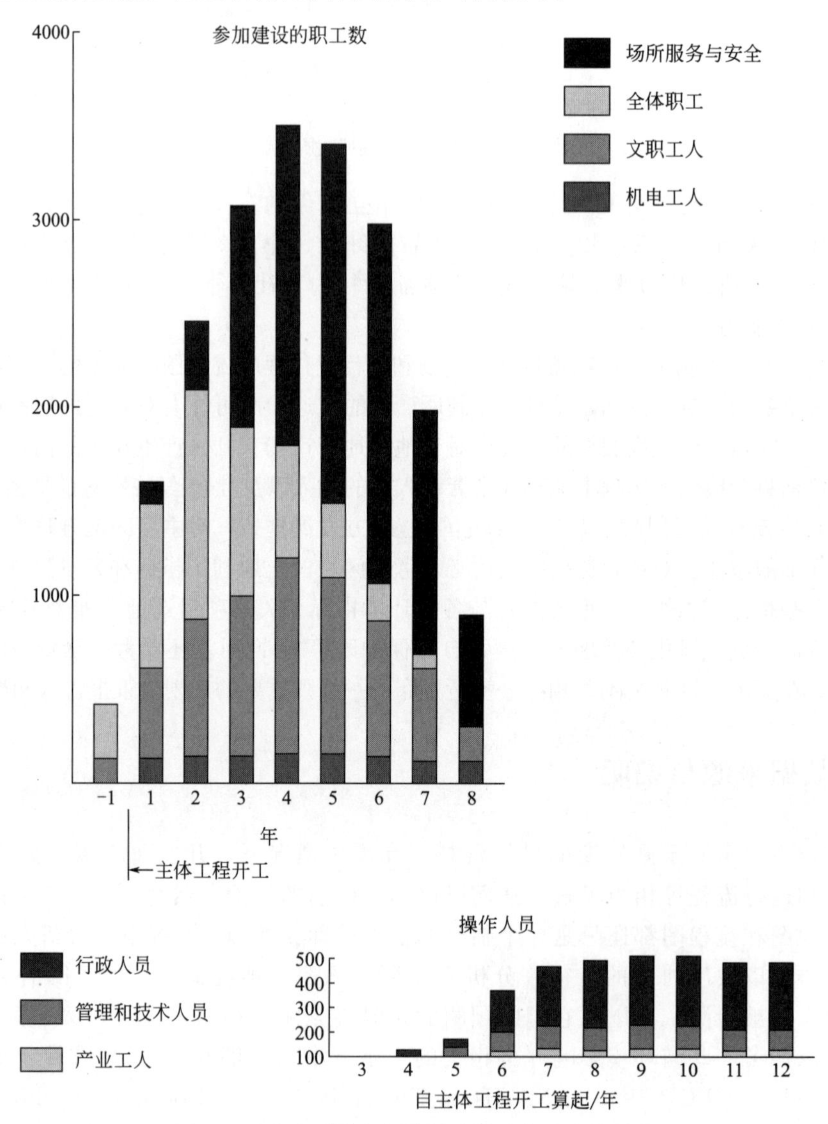

图 4.4　项目不同生命周期阶段对劳动力的需求

遗憾的是，现实与理想通常相差很远，开发商提供的信息很有限，特别是对于新建项目。场地布局图和流程柱状图描述的只是最初设计阶段大概的、临时的情况，项目的实施阶段可能会有重大调整（Frost，1994），对于规划的应用也只是大框架的应用。欧洲法院的条例规定，连贯的 SEA 程序应该包括原则和一系列符合原则的实施决定，之后由 EIS 评价项目的环境影响，并为决策提供支持。只有对实施决定进行评价后，这些影响才能被识别出来（C-201/02，C-508/03，C-290/03，C-2/07）。

关于这个案例的其他信息在 COWI（2009）中。即使之前已经对相关项目参数进行过说明，仍然需要考虑 EIS 分析和开发商之间的关系，以便优化项目。

4.7 环境背景的建立

4.7.1 总则

环境背景的建立包括环境现状和未来可能的状况，在拟建项目还没有运行的情况下，考虑自然活动和人类活动所引起的环境变化。例如，湖中某种鱼类的数量可能在湖边拟建工业项目运行之前就已经减少了。图 1.7 表明在建立环境背景时，需要考虑不同时间、环境组成和空间跨度等因素。预测未来某一时段的环境状况应该与拟开发项目的生命周期对照，这种对照可能会持续几十年。环境组成要素包括生物自然环境和社会经济环境。空间跨度范围以当地为主，但是也可以参考更广的区域，考虑更多的环境要素。

最初的环境背景研究可能范围很广，大部分的观点都认为这是一种资源浪费。背景研究应尽量将重点放在拟建项目可能产生的直接或间接的重大环境影响上：

在环境描述中应提供对发展的描述，在步骤 4 中强调了对重大环境影响的描述，环境描述不应超过其所需要的详细程度，对环境没有影响的或有轻微影响的应进行简述（DCLG 2011）。

选择的合理性能在一定程度上解释评价因子的筛选过程。虽然筛选过程通常都考虑了各种环境要素，但了解它们之间的相互作用和功能关系也非常重要。例如，植物会受空气质量和水质的影响，而动物又会受植物的影响。这种了解有利于预测。如同环境影响评价程序需要很多步骤一样，建立环境背景也不是一蹴而就的。它是一个从宽泛转向详细、集中的过渡过程。新的潜在影响识别可能会引入新的环境调查要素，选择有效的减缓影响方法能缩小调查的区域。

4.7.2 数据的来源和说明

环境数据的质量和可靠性差别很大，在影响评价中会对这些数据的使用产生影响。Fortlage（1990）用以下几点进行了说明：

- “硬”数据来源可靠，便于核对并且短期内不随时间变化，如地质学、地形学及基础设施的实况调查；
- “中级”数据同样可靠，但却不能完全确定，如水环境质量、土地价值、植被条件和交通量等，因为它们是变量；
- “软”数据是指主观选择或具有社会价值的数据，如调查意见、景观视觉享受和相应的享受人数等，这些结果取决于人们的态度和公众的意识。

在英国，较为重要的数据来源于地区气象台和地方权威监测机构。它们常常能提供与自然环境、社会环境和经济环境相关的有用数据，这些数据能及时更新，并且在互联网上的共享程度也越来越大。在英国，地方性数据可以利用中央政府所公布的数据来进行补充，因为中央政府的数据来源广泛（包括人口普查、街区统计、生活质量状况、地区发展趋势、环境统计分类、交通统计等），并且越来越多的数据来源于欧盟。但是，很多有用的信息都没有公开或只是“半公开”，且只在组织内部使用。在英国 EIA 制度下，法定咨询机构（如乡村机构、英国自然协会、英国遗产和环境局）有义务向开发商提供他们所拥有的信息，这些信息可能有助于环境报告书的编写（DCLG，2011）。相关英国和国际数据见第 5 章（5.3 节）。

当然，有很多地方性的及其他层次的非法定咨询机构也可以提供有价值的信息。地方史志、自然保护和博物学家学会等也拥有丰富的信息，如有关地方动植物、道路通行权和历史遗迹等信息。国家性研究团体，如英国皇家鸟类保护协会（RSPB）和林业部等，都可以提供详细的专业知识。在 EIA 的早期阶段，向地方服务组织（amenity group）进行咨询不仅能帮助获取数据，而且有利于明确关键环境问题所需的基础数据。

从现有数据来源获取的数据都应该充分利用，但是这些数据与拟建项目所需的环境本底数据之间总是存在一定差距。环境监测和调查也十分必要。调查和监测还存在一些问题，它们不可避免地会受到时间和经费的限制，因此必须有选择地进行。然而，这种选择必须确保监测和调查的时间跨度与任务的进行步调协调一致。例如，某一环境特征（如动植物的很多种类）可能需要一年或更长的时间来进行调查，这些调查需要考虑到季节变化或迁徙模式。调查通常采用抽样的方法，在抽样过程中，应该明确涉及抽样误差的范围和含义。

在 EIS 中，背景研究可以由多种方式来表示。这些方式通常包括对研究范围内的自然和社会经济环境进行简单的概述，接下来进行项目描述，在后面的影响研究章节中将重点详述（如空气质量、地质概况、就业等）。或者是在较早阶段进行一系列更全面的详细研究，从而为后面的影响研究章节提供参考。互联网和地理信息系统（GIS）的不断深入运用对环境数据的获取和表述来说是一个很好的创新。GIS 在计算机中建立数据库，包括存储不同空间图层参数，从而可以简单快捷地显示、合并和分析不同的地图（Rodriguez-Bachiller，2000）。

分析人员也应该谨慎，在定性分析的基础上也要注重定量分析，因为二者在建立环境背景时都做出了积极的贡献。最后还应该注意所有数据都会受某些不确定因素的影响，在环境影响预测时需要识别这些因素（见第 5 章）。

4.8 影响识别

4.8.1 目标和方法

影响识别通过结合项目特征、环境背景特征来进行分析，以确保在 EIA 过程中能识别和考虑所有潜在的显著性环境影响（负面或正面的）。在对众多的影响识别方法进行选择时，分析人员需要考虑具体分析目标，这些目标的主要原则是：

- 确保与法规一致；
- 提供影响的所有覆盖范围，包括社会、经济和自然环境；
- 区别积极的和消极的、大的和小的、长期的和短期的、可消除的和永久性的影响；
- 与识别直接影响一样，需要识别次要的、间接的和累积的影响；

- 区分显著/重大影响与不显著/非重大影响；
- 允许对替代开发方案进行比较；
- 在该区域承载力范围内考虑各种影响；
- 综合定性和定量分析信息；
- 运行简单且经济；
- 客观公正地给出分析结论；
- 利用好 EIS 中的总结和影响说明。

最简单的影响识别方法是使用影响清单，以确保所列影响具有全面性。最复杂的包括使用能相互作用的计算机程序、网络来说明能量流动，为不同影响分配权重。这一部分列出了相关方法，从需要依照条例的最简单的核查表法到复杂的方法。本节意在提出这些方法的适用范围，这些也是开发商、咨询和研究机构都愿意深入研究和实践的方法。这些方法分为以下几类：

- 核查表法；
- 矩阵法；
- 定量法；
- 网络法；
- 图层叠加法。

在英国，用来识别和总结影响的方法就是简单的核查表法，或者是专家咨询法。这可能与英国 EIA 指令的高度灵活性和谨慎态度有关。在英国，人们通常不愿使环境影响评价程序过于复杂，或者尝试那些较为复杂的方法。由于这个原因，虽然指令的早期版本包含了更复杂的方法（如 Sorensenn 和 Moss，1973），本版本却进行了调整。

这里所讨论的方法主要与影响识别相关，但是很多方法在 EIA 程序的其他阶段也有相当多的应用，例如在影响预测、评估、讨论、减缓措施、说明、监测和审计等阶段的应用。因此，本书第 4～7 章之间存在很大的关联性，这与在实际工作中不同阶段的相互作用是一样的。

4.8.1.1 核查表法

核查表法基于一系列可能会受开发活动影响的特殊的自然、社会和经济因素。简单的核查表法只能帮助识别影响，确保影响不被忽略。核查表中通常不包括与项目活动相关的直接影响。不过，它具有易于使用的优点。表 3.7（DETR，2000）就是一个简单核查表法的例子。

问卷调查核查表法以一系列问题的回答为基础，一些问题可能涉及间接影响和可能的减缓措施。也可以对估计的影响进行等级分类，包括从极度不利到非常有利的影响。表 4.3 对欧盟（2001b）的一部分问卷调查核查表法进行了说明。

表 4.3 问卷调查核查表法节选

编号	在确定范围时应考虑的问题	是/否	项目环境的哪些特征会受到影响以及如何影响	影响是否明显，为什么
7	如果建设项目的污染物排放到地面、下水道、地表水、地下水、沿海或者海洋，是否会导致土壤或水体污染			
7.1	是否存在危险或者有毒物质的接触、储藏、使用或溢出			

续表

编号	在确定范围时应考虑的问题	是/否	项目环境的哪些特征会受到影响以及如何影响	影响是否明显，为什么
7.2	是否存在废水或者其他污染物向水体或土壤中的排放（无论是处理过的还是未经处理的）			
7.3	排放到大气、土壤或者水体中的污染物是否沉积			
7.4	是否还存在其他污染源			
7.5	这些污染源排放的污染物在环境中的累积会不会存在风险			

资料来源：EC，2001b。

4.8.1.2 矩阵法

在 EIA 的影响识别中最常用到的方法是矩阵法。简单的矩阵只是二维的图表，一轴表示环境要素，另一轴表示开发活动。从某种程度上说，矩阵法是在了解开发项目不同要素的（如绘图、运行、拆除、建筑、连接道路等）不同影响基础上对核查表法的拓展。可能对环境要素产生影响的活动通过在相应的框内画“×”来表示。表 4.4 是一个简单矩阵法的例子。三维矩阵目前也有所发展，第三维是指经济和社会制度，如确定 EIA 程序所需数据的提供机构，并且明确目前那些缺乏了解的领域。

表 4.4 简单矩阵法节选

环境要素	项目活动				
	建设		运行		
	公共设施	居民和商业建筑	居民建筑	商业建筑	公园和开放空间
土壤和地质	√	√			
植物	√	√			√
动物	√	√			√
空气质量				√	
水质量	√	√	√		
人口密度			√	√	
就业		√		√	
交通	√	√	√	√	
住房			√		
社区结构		√	√		√

量值矩阵法已经不只是影响识别，而是根据影响的大小、重要性或时间范围（如短期、中期或长期）进行描述。表 4.5 就是量值矩阵法的一个例子。

分配影响矩阵法代表矩阵法可能的另一种发展方向，它旨在识别建设项目潜在影响中的获益方和受损方。这是非常有用的信息，而在其他矩阵法中鲜有体现，实际上在 EIS 中也很少见。影响随着空间的变化而变化，如城市和农村。空间的变化可以用线性项目来表示，如轻轨系统。同一个项目对不同社会群体也会产生不同的影响（如一个居住区拟建方案对于已经拥有房屋的退休老人、寻求能够支付的住房和进入住房供给市场的年轻人以及有孩子的人的影响都是不同的）（图 5.8）。

表 4.5　量值矩阵法节选

环境要素	项目活动				
	建设		运行		
	公共设施	居民和商业建筑	居民建筑	商业建筑	公园和开放空间
土壤和地质	•	•			
植物	•	●			○
动物	•	•			o
空气质量				•	
水质量	○	•	•		
人口密度			o	o	
就业		○		○	
交通	•	•	•	●	
住房			○		
社区结构		•	○		o

注：• 为小的负面影响；o 为小的正面影响；●为大的负面影响；○为大的正面影响。

通过利用数据进行矩阵分析，人们也可以尝试对这些数字进行计算，得出项目开发活动影响的综合值，然后与其他开发进行比较。不过由于矩阵法无法反映不同影响分配相对重要性的权重，因此这种想法目前还不能实现。权重矩阵法是为了解决上述问题而发展起来的。它可以给环境要素（有时也给项目要素）分配权重，以此来反映它们各自的重要程度。首先评价项目（或项目要素）对环境要素的影响，然后赋予相应的权重，最后通过影响与权重相乘而得到该项目的赋权影响。权重矩阵可以反映环境要素与标准或阈值的接近程度。项目要素可能会产生长期或不可逆的影响（Odum 等，1975），或者产生的长期影响要大于短期影响（Stover，1972）。

表 4.6 所示的是一个小的权重矩阵，对项目选址的 3 个替代方案进行了比较。每个相关的环境要素都分配了一个权重（a）：空气质量占了总环境要素的 21%。每个项目对每个环境要素的影响大小（c）被分成 0～10 个量级来评价，然后乘以（a）就得到了一个赋权影响（a×c）。例如，选址 A 对空气质量的影响是 3，乘以 21 即得出赋权影响为 63。对每个选址来说，赋权影响可以相加，从而得出项目总的赋权影响。在该例中，选址 B 的权重影响最小，对环境的危害也最小。

表 4.6　权重矩阵：可替代的项目选址

环境要素	(a)	可替代的项目选址					
		选址 A		选址 B		选址 C	
		(c)	(a×c)	(c)	(a×c)	(c)	(a×c)
空气质量	21	3	63	5	105	3	63
水质量	42	6	252	2	84	5	210
噪声	9	5	45	7	63	9	81
生态环境	28	5	140	4	112	3	84
总计	100		500		364		438

注：(a)＝环境要素的相对权重（总分为 100）。
(c)＝位于特定选址的项目对环境要素的影响（0～100）。

矩阵法（或更广义来说，多指标分析法）的优势是可以定量地比较各选择的优势，从而更有利于决策者的取舍，保证评价与结果的一致性。然而，该方法过多地依赖于权重值和划定的影响等级，可接受度依赖于假设的建立（特别是权重的分配）。不同人所得结果可能会因假设的不同而有所差异，第 5 章将对其进行深入讨论。

更普遍地说，核查表法和矩阵法将环境看作是由离散单元组成的：影响只与特定的参数有关，没有描述环境成分之间的复杂关系。当影响减少到用符号或数字表示时，大量信息就会丢失。清单和矩阵没有具体说明发生影响的可能性；其评分系统天生具有主观性，容易产生偏见，也不能揭示事态发展的间接影响。然而，它们是（相对）快速和简单的，可以用来呈现发现范围以及确定可能产生的影响。

4.8.1.3 网络法

网络法明确地认识到环境系统是由各种复杂的、相互联系的关系网组成，并且尝试着再现该关系网。环境识别过程中运用网络法，包括通过模型中环境参数的改变来识别开发活动所带来的一系列影响。

英国的许多地方性交通规划中对该技术有较为简单的描述。因果链分析（因果关系图）由规划者起草，以识别某一行为：如对公路和连接处的维护、翻新与改善（图 4.5），如何导致社会、经济和环境条件的变化，同时也要明确能够产生积极影响的先决条件，避免出现问题。网络法没有确定环境要素或变化范围之间相互影响的大小或重要性，确定这些需要大量的环境知识。该方法最大的优点是可以追踪拟开发活动产生的较为显著的影响。

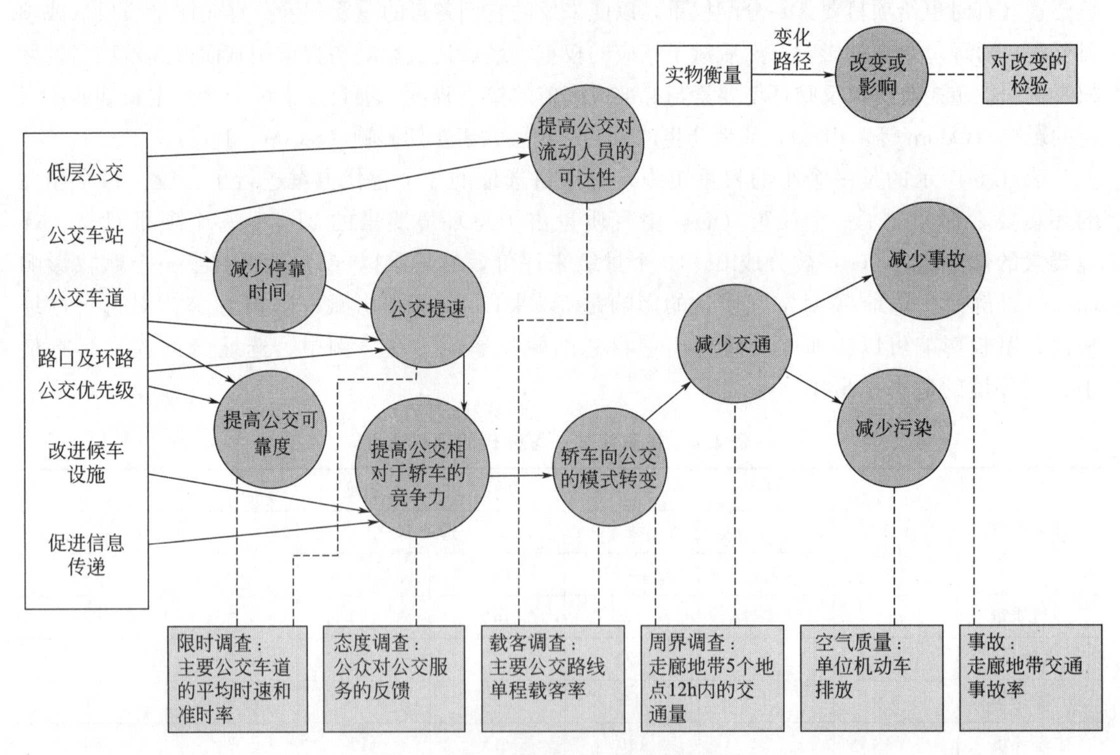

图 4.5　公交走廊改进的因果链分析

（资料来源：改编自南约克郡综合运输管理局，2001）

4.8.1.4 图层叠加（或约束）法

20世纪60年代在美国《国家环境政策法案》颁布之前，图层叠加法（McHarg，1968）就在环境规划中有所应用。一系列透明的图片被用来识别、预测、分析相对重要性和扩散影响。首先准备一张工作草图，以表示项目选址区域的总范围，然后为环境要素准备连续的透明图层，专家认为这可能会受到项目（如农业、林地、噪声）的影响。项目对环境特征的影响程度用阴影度来表示，阴影越深表示影响越大。项目的综合影响通过叠加图层和标记相应的总阴影度来获取。没有阴影的区域说明项目开发不会对其造成明显的影响。图4.6展示了该方法的原理。

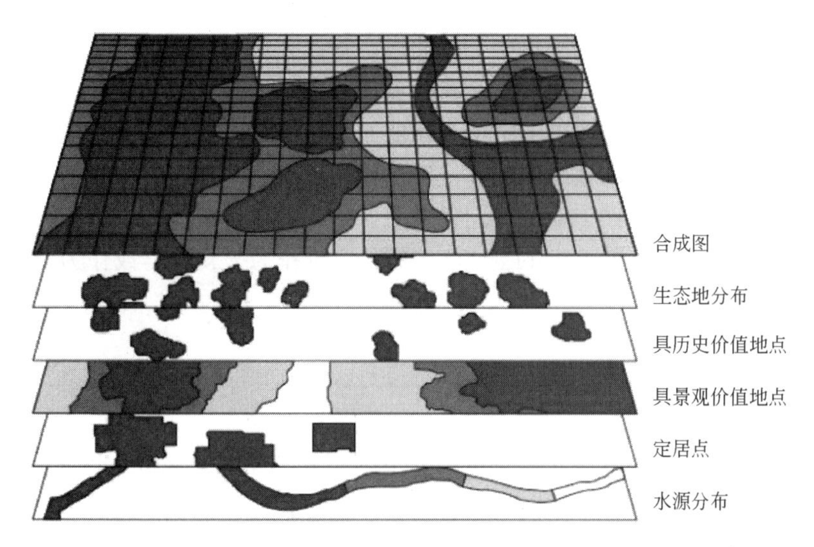

图4.6 图层叠加示例

当然，同样的过程也可以通过运用GIS和赋权来完成：这使灵敏度分析得以实现，从而判断对假设影响重要性的改变是否会影响决策。

图层叠加法简单易懂，因此很受欢迎。它能很好地显示影响的空间分布。此外，图层叠加法还有利于做出对环境影响较小的决策。图层叠加法在识别开发的最佳路线，例如输电缆线和公路等，同时在比较替代方案、评价较大区域性开发方面非常有用。不过，该方法也有一定的局限性，因为它没有考虑诸如影响的可能性、次级影响或可恢复影响与不可恢复影响之间的区别等情况，而且它还需要有明确的边界划分（如森林和田地的边界），因此它不能真实地反映各种情况。此外，在使用图层叠加法之前还需要研究者去识别可能的影响。

图层叠加法还可以被应用于显示影响分布中（例如谁是获利者、谁是受损者），以及是否将会不成比例地失去传统弱势群体。

4.8.2 生活质量评价

生活质量评价法（QOLA）是由乡村局等联合提出来的（2001），是综合不同部门环境管理方法的手段。QOLA的重点不是事物，而是受到拟建项目影响的各种利益群体。它首先假设所有的事物（林地、历史建筑）都很重要，因为人们从中受益（如视觉享受、娱乐、吸收二氧化碳等），相反，对这些事物的管理应该以它们所提供的利益最优化为目标。这种对福利的强调后来被纳入生态系统服务方法（例如 DEFRA2007）。

QOLA 包括 6 个步骤（A～F）。明确评价目的（A），描述拟建项目的位置（B），评价选址对可持续性（例如对当代及后代的发展）的利/弊（C）。之后该技术提出了以下问题（D）：

- 各项利弊的重要性，它们的对象是谁？它们产生的原因？
- 以目前的趋势，利弊是否会增加？
- 什么可以代替这些益处？

这些问题的答案得出了一系列管理线索（E），其中考虑设计"购物清单"，任何对该选址的开发、管理都应达到"购物清单"上的要求，而且要知道是如何达到的以及它们的相对重要性。最后，提出对这些益处的监测方案（F）。因此，在决策方案通过之前，需要明确地保证实现各种利益，最后需要得出结论，说明开发活动放在哪里最不合适。此外，还可以为在指定地点（以及对更大区域的管理）进行的任何开发活动（例如用于规划条件或者减缓措施）建立一个管理框架（如 106 条规定的职责、规划状况等）。QOLA 法已经用来确定环境影响评价的范围（特别是对 Bristol 港口），而且可以作为公众参与或作为对项目选址的专家意见综合分析的媒介。

4.8.3 小结

表 4.7 总结了本节所讨论的几种主要影响识别方法的优点。考虑到很多影响识别技术的复杂性，英国的环境影响报告中采用了核查表法、简单矩阵法和网络法，或综合以上几种方法。在选择影响识别方法时需要注意：它们是没有政治色彩的，并且方法越灵活，交流讨论和有效地参与就会越难（第 6 章将做详细讨论）。通常越简单的方法越易使用，并且在 EIS 中的信息表述会更加通畅和有效，但是显著影响的覆盖范围、间接影响或者替代方案却会非常有限或者根本没有。综合了这些方面的较为复杂的模型却不具有时效性。

表 4.7　影响识别方法的比较

标准
1. 符合法规；
2. 全面覆盖(经济、社会和环境影响)；
3. 积极或消极的影响，可恢复或不可恢复的影响；
4. 次级影响、间接影响、累积影响；
5. 显著与不显著的影响；
6. 替代方案比较；
7. 承载力的比较；
8. 利用定性和定量的信息；
9. 便于使用；
10. 客观的、一致的；
11. 总结 EIS 中的影响

续表

方法	1	2	3	4	5	6	7	8	9	10	11
核查表法	√	√						√	√	√	√
矩阵法											
• 简单	√	√						√	√	√	√
• 量级	√	√	√					√	√	√	√
• 权重	√	√			√	√		√	√		√
网络法	√			√		√		√		√	
图形叠加法		√	√		√	√		√	√	√	√

4.9 总结

EIA 的早期阶段以几个相互联系的步骤为代表，主要包括：决定是否需要进行 EIA（筛选），咨询各团体以将最初的重点放在主要的影响上（范围确定）以及对项目可能的替代方案的概述，包括替代选址、规模和程序。范围确定和对替代方案的考虑可以极大地提高 EIA 程序的质量。在程序的早期阶段，专家希望了解项目的特征及可能受影响区域的环境背景情况。项目在其生命周期的各个阶段具有多重属性（例如目的、自然环境、程序、政策等）。对环境背景的调查也具有多重属性。对项目和受影响的环境来说，获取相关数据将面临不小的挑战。

前面讨论的大部分内容都包含在影响识别中。从简单的核查表法和矩阵法到复杂的计算机模型和网络法，通常都有影响识别方法的应用。在英国，运用的正式影响识别方法一般都是较为简单的类型。和环境影响的预测、评价、信息交流和减缓措施有关的方法将在下一章进行讨论。

问　题

以下问题旨在帮助读者更好地明确本章讨论的重点问题。

1. 假设开发商计划在一个区域建设一个风电场（或者其他项目），在 EIS 过程中需要考虑以下哪些因素？

- 在你的团队中你希望邀请哪方面的专家？
- 在 EIA 范围确定阶段，需要收集哪些信息？
- 有哪些相关的替代方案？这些替代方案如何分级？
- 在项目影响识别阶段，需要采用哪些技术？为什么？

2. 表 4.2 表明不同欧盟成员国的 EISs 的数量差别很大，在筛选阶段有哪些问题会影响 EIA 的进行？

3. 是否所有的 EIS 都应考虑“不作为”替代方案？包括不同地点、不同规模、不同过程、不同设计。为什么？

4. 在 EIS 中都应呈现哪些信息，可能有哪些有用的附加信息？

5. 在本章的图表中，你认为哪两个或者哪三个对环评的帮助最大？

6. 在 4.8.1 部分对过去一些复杂的影响识别方法进行了修正，这其中的原因是什么？

Part 2

Process

This illustration and the illustration opening Part 3 are by Neil Bennett, reproduced from Bowers, J. (1990), *Economics of the environment: the conservationist's response to the Pearce Report*, British Association of Nature Conservationists, 69 Regent Street, Wellington, Telford, Shropshire TE1 1PE.

4 Starting up: early stages

4.1 Introduction

This is the first of four chapters that discuss how an EIA is carried out. The focus throughout is on both the procedures required by UK legislation and the ideal of best practice. Although Chapters 4—7 seek to provide a logical step-by-step approach through the EIA process, there is no one exclusive approach. Every EIA process is set within an institutional context, and the context will vary from country to country (see Chapter 10). As already noted, even in one country, the UK, there may be a variety of regulations for different projects (see also Chapter 9). The various steps in the process can be taken in different sequences. Some may be completely missing in certain cases. The process should also not just be linear but build in cycles, with feedback from later stages to the earlier ones.

Chapter 4 covers the early stages of the EIA process. These include setting up a management process for the EIA activity, clarifying whether an EIA is required at all ('screening') and an outline of the extent of the EIA ('scoping'), which may involve consultation between several of the key actors outlined in Chapter 3. Early stages of EIA should also include an exploration of possible alternative approaches for a project. Baseline studies, setting out the parameters of the development action (including associated policy positions) and the present and future state of the environment involved, are also included in Chapter 4. However, the main section in the chapter is devoted to impact identification. This is important in the early stages of the process, but, reflecting the cyclical, interactive nature of the process, some of the impact identification methods discussed here may also be used in the later stages. Conversely, some of the prediction, evaluation, communication and mitigation approaches discussed in Chapter 5 can be used in the early stages, as can the participation approaches outlined in Chapter 6. The discussion in this chapter starts, however, with a brief introduction to the management of the EIA process.

4.2 Managing the EIA process

Environmental impact assessment is a management-intensive process. EIAs often deal with major (and sometimes poorly defined) projects, with many wide-ranging and often controversial impacts. It is important that the EIA process is well managed. This section discusses some of the interrelated aspects of such management: the EIA team, the style of the EIA process, and costs and resourcing.

4.2.1 *The EIA team*

The EIA process invariably involves an *inter-disciplinary team* approach. Early US legislation strongly advocated such an approach:

> Environmental impact statements shall be prepared using an interdisciplinary approach which will ensure the integrated use of the natural and social sciences and the environmental design arts. The disciplines of the preparers shall be appropriate to the scope and issues identified in the scoping approach. (CEQ 1978, par. 1502.6)

Such an interdisciplinary approach not only reflects the normal scope of EIA studies, from the biophysical to the socio-economic, but also brings to the process the advantages of multiple viewpoints and perspectives on the complex issues involved.

The team producing the EIS may be one, or a combination, of proponent in-house, lead external consultant, external sub-consultants and individual specialists. The team size can range from just a few people to more than a dozen members for larger projects. The team's skills should be complementary: technically for the skills needed to complete the task, and personally for those in the core management team. A small team of three, for instance, could cover the areas of physical/chemical, biological/ecological and cultural/socio-economic, with a membership that might include, for example, an environmental engineer, an ecologist and a planner, at least one member having training or experience in EIA and management. Additional input could be required from experts in ecology, archaeology, air quality, traffic and other specialist fields. However, the finalization of a team's membership may be possible only after an initial scoping exercise has been undertaken.

Many EIA teams make a clear distinction between a 'core/focal' management team and associated specialists, often reflecting the fact that no one organization can cover all the inputs needed in the production of an EIS for a major project. Some commentators (see Weaver *et al*. 1996) promote the virtues of this approach. On a study for a major open-cast mining project in South Africa, Weaver *et al*. had a core project team of five people: a project manager, two senior authors, an editorial consultant and a word processor. This team managed the inputs into the EIA process, co-ordinated over 60 scientific and non-scientific contributors, and organized various public participation and liaison programmes.

The *team project manager* has a pivotal role. In addition to personnel and team management skills, the manager should have a broad appreciation of the project type under consideration, knowledge of the relevant processes and impacts subject to EIA, the ability to identify important issues and preferably a substantial area of expertise. The project manager must:

- select an appropriate project team, and deal with typical personnel issues including staff turn-over;
- manage specialist inputs;
- liaise with the project proponent, various stakeholders and the public, including choosing the participation techniques to use, and subsequent follow-through;
- keep the EIA team on schedule, make sure that the EIA is carried out efficiently, and adapt the team's work to unanticipated events;
- ensure that the EIA process focuses on key issues, and is fit for purpose, internally coherent and robust; and
- co-ordinate the contributions of the team in the various documentary outputs. (IEMA 2011; Lawrence 2003; Morrison-Saunders and Bailey 2009; Petts & Eduljee 1994)

4.2.2 The style of the EIA process

EIAs can involve many participants with very different perspectives on the relative merits and impacts of projects. In interdisciplinary team work, co-ordination is particularly important: findings and data should be co-ordinated (e.g. they should work to agreed map scales, spatial and temporal boundaries, mitigation measures, and EIS chapter formats) and should be fed into a central source.

This is one of the weakest aspects of most assessment teams; all consultants must be aware, and stay aware, of others' work, in order to avoid lacunae, anomalies and contradictions that will be the delight of opposing counsel and the media (Fortlage 1990).

Beyond this, different projects—and their EIAs—will require different scientific techniques, methods of participation, focuses and ways of responding to unanticipated events. A large, controversial project proposed near a population centre will require a different style of EIA process than a less controversial project, or one in an uninhabited area. An EIA carried out where much detailed data already exist will be different from one in a remote and unstudied location. An EIA for a project that affects groups of people that have traditionally been deprived, or environmental components already subject to cumulative impacts, will be different from one for a homogeneous, wealthy population or a robust environment.

Lawrence (2003) suggests that:

- A *rigorous* EIA process is more appropriate where scientific analysis can contribute significantly to decision-making: for instance, where the environment can be scientifically analysed, and where the resources for such an analysis exist.
- A *rational* EIA process is appropriate for situations where stakeholders can engage in the process in a free and 'reasonable' manner, where views are not overly polarised, and where well-defined options and proposals can be put forward.
- A *streamlined* EIA process is appropriate in a polarised situation where resources are limited, relatively little data exists, and changes are likely to take the form of incremental adjustments to the status quo.
- A *democratic* EIA process works best when pro-ponents are willing to delegate their decision-making authority to representatives, who in turn have the time, energy and resources to participate in planning and decision-making processes with other parties.
- A *collaborative* EIA process is like a democratic process, but with stakeholders being directly engaged in the process, and having the resources to do so.
- An *ethical* EIA process is required when issues of fairness, equity and justice predominate, and when the stakeholders are willing to identify and reconcile these ethical conflicts.
- An *adaptive* EIA process is appropriate for turbulent and complex situations where risk, uncertainty and health predominate, and where the EIA needs to take into account knowledge limits and uncertainty-related concerns.

4.2.3 EIA costs and resources

The EIA team and the style of the EIA process will affect, and be affected by, resources. Most of the cost of EIA is incurred in carrying out environmental studies and writing the EIS, and is borne by the developer. However, the planning authority, statutory consultees and the public will also incur costs in reviewing the EIA and commenting on it.

European Commission research showed that, for countries with several years of EIA experience, the costs of carrying out EIAs tend to range from 0.1 per cent to 1 per cent of the capital cost of the project. Although the actual cost of an EIA tended to rise with the capital cost of the project, it fell as a percentage of the total cost of the project. For smaller projects, EIA costs were typically closer to 1 per cent of the project cost, whereas for larger projects they were typically closer to 0.1 per cent (EC 2006; COWI 2009; Oosterhuis 2007). This is broadly consistent with EIA costs in the UK, which are discussed at Chapter 8. The World Bank (1999) *Environmental assessment sourcebook* also states that EIA costs typically range from a few thousand to over a million dollars, and that they rarely exceed 1 per cent of the total capital cost of the project.

A French study (BIO 2006) suggested that EIAs are generally more expensive for linear projects such as roads and electricity lines, nuclear and industrial activities, projects where health impact assessments are required, projects related to the marine environment, and large companies or administrations. The World Bank (1999) notes that EIA costs can be reduced if local personnel are used to do most of the work.

The sources above suggest that a typical EIA will take 6—18 months to carry out. However, the time required for the full EIA process can be significantly extended, for instance if the developer proposes modifications to the project, or if there are changes in government. Good scoping can reduce the time needed for EIA by ensuring that the EIA process focuses on key issues and is carried out efficiently. In contrast, a main cause of delay is where the EIA does not provide adequate or relevant data, leading to the need for supplementary information (EC 2006).

4.3 Project screening: is an EIA needed?

The number of projects that could be subject to EIA is potentially very large. Yet many projects have no substantial or significant environmental impact. The screening stage seeks to focus on those projects with potentially significant adverse environmental impacts or whose impacts are not fully known. Those with few or no impacts are 'screened out' and allowed to proceed to the normal planning permission and administrative processes without any additional assessment or additional loss of time and expense.

Screening can be partly determined by the EIA regulations operating in a country at the time of an assessment. Chapter 3 indicated that in the EC, including the UK, there are some projects (Annex/Schedule 1) that will always be 'screened in' for full assessment, by virtue of their scale and potential environmental impacts (for example a crude oil refinery, a sizeable thermal power station, a special road). There are many other projects (Annex/Schedule 2) for which the screening decision is less clear. Here two examples of a particular project may be screened in different ways (one 'in' for full assessment, one 'out') by virtue of the project scale, the sensitivity of the proposed location and/or the expectation of adverse environmental impacts. In such cases it is important to have guidelines, indicative criteria or thresholds on conditions considered likely to give rise to significant environmental impacts (see Section 3. 4).

In California, a draft environmental impact report is required 'if there is substantial evidence, in light of the whole record before a lead agency, that a project may have a significant effect on the environment'. A full environmental impact report is required where there is substantial evidence, in light of the whole record, that the project would cause significant impacts on environmental quality, habitats and species, or historical artefacts; negatively affect long-term environmental goals; have 'cumulatively considerable' effects; or would cause substantial adverse effects on human beings (State of California 2010). These constitute 'inclusion list' approaches. In addition, there may be an 'exclusion list', as used in California and Canada, which identifies those categories of project for which an EIA is not required because experience has shown that their adverse effects are not significant.

Some EIA procedures require an initial outline EIA study to check on likely environmental impacts and their significance. Under the California Environmental Quality Act a 'negative declaration' can be produced by the project proponent, claiming that the project has minimal significant effects and does not require a full EIA. The declaration must be substantiated by an initial study, which is usually a simple checklist against which environmental impacts must be ticked as *yes*, *maybe* or *no*. If the responses are primarily *no*, and most of the *yes* and *maybe* responses can be mitigated, then the project may be screened out from a full EIA.

In general there are two main approaches to screening. The use of *thresholds* involves placing projects in categories and setting thresholds for each project type. These may relate, for example, to project characteristics (e. g. 20 hectares and over), anticipated project impacts (e. g. 50,000 tonnes or more of waste per annum to be taken from a site) or project location (e. g. a designated landscape area). Appendix 2 shows the applicable thresholds used in the UK.

A *case-by-case* approach involves the appraisal of the characteristics of projects, as they are submitted for screening, against a checklist of guidelines and criteria. Some of the advantages and disadvantages of these two approaches are summarized in Table 4. 1. The EC (2001a) has published guidance to help in such case-by-case screening processes. There are also many hybrid approaches with, for example, indicative thresholds used in combination with a flexible case-by-case approach.

Table 4. 1 Thresholds versus case-by-case approach to screening: advantages and disadvantages

Advantages	Disadvantages
Thresholds	
Simple to use	Place arbitrary, inflexible rules on a variable environment (unless tiered)
Quick to use; more certainty	Less room for common sense or good judgement
Consistent between locations	May be or become inconsistent, depending on neighbouring receivers and developments
Consistent between decisions within locations	Difficult to set and, once set, difficult to change
Consistent between project types	Lead to a proliferation of projects lying just below the thresholds
Case by case	
Allows common sense and good judgement	Likely to be complex and ambiguous
Flexible—can incorporate variety in project and environment	Likely to be slow and costly
	Open to abuse by decision-makers because of political or financial interests
	Open to poor judgement of decision-makers
	Likely to be swayed by precedent and therefore lose flexibility

The DCLG (2011) EIA guidance gives detailed guidance on how screening should be carried out for English development projects, with extensive reference to case law (see Sections 3. 4 and 8. 6).

Neither the threshold nor the case-by-case approach gives wholly consistent results, and the thresholds used by different countries can vary widely. Table 4. 2 takes another view of EIAs carried out in European Union Member States (here for the period 2006—08). It shows that the number of EIAs prepared per head of population varies by a factor of 50+between countries with a consistent legislative basis, namely the EIA Directive; this further reinforces the earlier (Chapter 2) theme of divergent practice in a converging system. These differences can be attributed to factors such as different consenting regimes, levels of development, environmental sensitivity and cultural values. As such, lower (or indeed higher) levels of EIA activity do not necessarily indicate a problem (IEMA 2011).

Table 4. 2 Number of EIAs carried out in European Union Member States, 2006—08

Member State	Average number of EIAs/yr(2006—08)	Average number of EIAs/yr per million population (2006—08)	Member State	Average number of EIAs/yr(2006—08)	Average number of EIAs/yr per million population (2006—08)
Austria	23	3	Latvia	11	5
Belgium	183	17	Lithuania	142	42
Bulgaria	249	33	Luxembourg	70	149
Cyprus	117	136	Malta	10	24
Czech Rep.	96	9	Netherlands	123	7
Denmark	125	22	Poland	4,000	105
Estonia	80	61	Portugal	323	29
Finland	38	7	Romania	596	27
France	3,867	59	Slovak Rep.	670	124
Germany	1,000	12	Slovenia	108	54
Greece	425	38	Spain	1,054	23
Hungary	152	15	Sweden	288	31
Ireland	197	44	United Kingdom	334	5
Italy	1,548	26			

Source: Adapted from GHK 2010

A series of European Court of Justice rulings have provided further guidance on how EIA screening should be carried out (EU 2010). They have clarified that, whatever method is used by a Member State to determine whether a project requires EIA or not, the method must not undermine the objective of the EIA Directive, which is that no project likely to have significant environmental effects should be exempt from assessment. In particular:

• The EIA Directive should be interpreted as having a 'wide scope and broad purpose' with respect to screening. A project should not be screened out simply because that type of project is not directly referred to in the EIA Directive or implementing regulations (Kraaijeveld ('Dutch Dykes') case C-72/95).

• Even small-scale projects can have significant effects on the environment if they are in a sensitive location (C-392/96).

- EIA is required for refurbishment, improvement and demolition projects that are likely to have significant effects on the environment (C-142/07, C-2/07, C-50/09).
- The purpose of the EIA Directive cannot be circumvented by splitting larger projects that would require EIA into smaller projects that would not. Similarly, EIAs must consider the cumulative effects of several projects where, individually, these might not have significant environmental effects (C-392/96, C-142/07, C-205/08).
- Although the EIA Directive allows Member States to exempt 'exceptional case' projects from EIA, the interpretation of 'exceptional cases' should be narrow. Possible rules for qualifying as an 'exceptional case' are that there is an urgent and substantial need for the project; inability to undertake the project earlier; and/or inability to meet the full requirements of the EIA Directive (C-435/97, C-287/98).
- Although a decision to screen a project out of EIA does not require a formal statement explaining the reasons for doing so, planning authorities have a duty to provide further information on the reasons for the decision if an interested person subsequently requests (C87/02, C121/03, C-75/08).

UK court cases regarding screening are discussed at Section 8. 6.

4.4 Scoping: which impacts and issues to consider?

The scope of an EIA is the impacts, issues and alternatives it addresses. The process of scoping is that of deciding, from all of a project's possible impacts and from all the alternatives that could be addressed, which are the significant ones. Effective scoping can help to save time and money, shorten the length of EISs, and reduce the need for developers to provide further environmental information after a planning application has been submitted (IEMA 2011).

An initial scoping of possible impacts may identify those impacts thought to be potentially significant, those thought to be not significant and those whose significance is unclear. Further study should examine impacts in the various categories in more depth. Those confirmed by such a study to be not significant are eliminated; those in the uncertain category are added to the initial category of other potentially significant impacts. This refining of focus onto the most significant impacts continues throughout the EIA process.

Scoping is generally carried out in discussions between the developer, the LPA, other relevant agencies and, ideally, the public. It is often the first stage of negotiations and consultation between a developer and other interested parties. It is an important step in EIA because it enables the limited resources of the EIA team to be allocated to best effect, and prevents misunderstanding between the parties concerned about the information required in an

EIS. Scoping can also identify issues that should later be monitored.

Scoping should begin with the identification of individuals, communities, local authorities and statutory consultees likely to be affected by the project; good practice would be to bring them together in a working group and/or meetings with the developer. One or more of the impact identification techniques discussed in Section 4. 8 can be used to structure a discussion and suggest important issues to consider. Other issues could include:

- environmental attributes that are particularly valued;
- impacts considered to be of particular concern to the affected parties;
- the methodology that should be used to predict and evaluate different impacts;
- the scale at which those impacts should be considered;[1] and
- broad alternatives that might be considered.

The result of this process of information collection and negotiation should be the identification of the chief issues, impacts and alternatives, an explanation of why other issues are not considered significant, and, for each key impact, a defined temporal and spatial boundary within which it will be measured.

Box 4. 1 shows excerpts from an admirably brief scoping report prepared by a local planning authority in response to a developer's request for such a report. It illustrates how many of the above points are dealt with in practice.

Box 4. 1 Extracts from a scoping opinion for extension of a quarry in a National Park

Socio economic impact: This subject area should include a consideration of access and public amenity issues in relation to the loss of land open to public access, as well as a consideration of the impact of the security/safety of the site. It should include a consideration of the impact of the development on the grazing rights of commoners. Although perhaps outside the statutory scope of the EIA the applicant may like to consider whether a basic assessment of the role of Yennadon Quarry in the local employment/product market may support any accompanying planning application.

Archaeology: Existing records do not provide a lot of evidence of in terms of archaeological features already identified in this area. As such this section of the EIA should include a walkover survey and a test pit to provide an assessment by a qualified archaeologist of the nature of below ground conditions (e. g. presence of peat, depth of subsoil). Any further survey work identified as necessary should then also be undertaken as part of the EIA.

Geology and hydrogeology: The site is identified as being on an Aquifer of Intermediate Vulnerability; it is approximately 450m from the inner water Source Protection Zone 1 and approximately 200m from the Devonport Leat. The EIA should consider the impact of the development on these features. Further advice from the Environment Agency is available in their publication *Scoping guidelines on the EIA of projects* (2002) (part D2-opencast mining and quarrying operations).

Ecological impacts and biodiversity: This Authority would require assessment to comprise a Phase 1 habitat survey, as well as any specialist surveys then identified as necessary. The assessment should include a consideration of avoidance, mitigation and compensation measures as necessary.

Cumulative impacts and an assessment of alternatives: Further to the above topic areas it is essential that the EIA includes a consideration of cumulative impacts and demonstrates a consideration of alternatives. The Authority would advise that the assessment should refer each subject to its potential impact upon the special qualities of the National Park and the purposes of National Park designation.

The adoption of this scoping opinion does not preclude the Mineral Planning Authority from requesting additional information following submission by the applicant, under Regulation 19 of the Town and Country Planning (EIA) (England and Wales) Regulations 1999.

Source: Dartmoor National Park Authority 2010

The European Court of Justice has ruled that an EIA should consider the overall effects of a project, not just its direct effects. In the case of Liege Airport (C-2/07), it ruled that the EIA should consider not only the proposed modifications to the airport's infrastructure but also the increased airport activity that these modifications would permit. Similarly, transboundary impacts are within the scope of EIA (C-205/08).

Although scoping is an important step in the EIA process, it is not legally required in the UK. Some developers produce a scoping report as a matter of good practice, and since 1999 LPAs must provide a formal 'scoping opinion' on the information to be included in an EIS when a developer requests one. A study on UK scoping practice showed that developers formally requested scoping opinions from LPAs for half of the EIA projects examined, with less formal scoping discussions occurring in a further 18 per cent of cases. In almost two-thirds of cases, developers provided scoping reports to accompany their requests for a scoping opinion. Most of the scoping opinions were prepared in-house by the LPA, and they consulted with other statutory bodies in nearly three-quarters of cases. Two-thirds of the respondents felt that scoping improved the quality of the EIS subsequently submitted, by improving the report's focus, bringing a wider range of concerns to the discussion, and reducing the need to request further information at later stages (DCLG 2006). In contrast, IEMA (2011) suggests that many scoping opinions are overly exigent, do not help to scope out issues, and do not prevent local authorities or statutory consultees from requesting further information after an EIS is prepared.

Other countries (e. g. Canada and The Netherlands) have a formal scoping stage, in which the developer agrees with the competent authority or an independent EIA commission, sometimes after public consultation, on the subjects the EIA will cover. The EC (2001b) has published a detailed scoping checklist; Carroll and Turpin (2009) provide lists of potential issues associated with a range of development projects; and the Environment Agency (2002) and Government of New South Wales (1996) are examples of organizations that have developed EIA guidance for particular types of projects.

4.5 The consideration of alternatives

4.5.1 *Regulatory requirements*

The US Council on Environmental Quality (CEQ 1978) calls the discussion of alternatives 'the heart of the environmental impact statement': how an EIA addresses alternatives will determine its relation to the subsequent decision-making process. A discussion of alternatives ensures that the developer has considered both other approaches to the project and the means of preventing environmental damage. It encourages analysts to focus on the *differences* between real choices. It can allow people who were not directly involved in the decision-making process to evaluate various aspects of a proposed project and understand how decisions were arrived at. It also provides a framework for the competent authority's decision, rather than merely a justification for a particular action. Finally, if unforeseen difficulties arise during the construction or operation of a project, a re-examination of these alternatives may help to provide rapid and cost-effective solutions.

The US NEPA requires federal agencies to analyse 'alternatives to the proposed action'. The implementation of this requirement has been the subject of a range of legal challenges, mostly on the basis that federal agencies had not considered a full range of reasonable alternatives, or that they had improperly constructed the purpose and need for their projects so that the resulting alternatives were too narrow. Generally the courts have ruled in favour of the federal agencies, accepting their reasons for eliminating seemingly 'reasonable' alternatives from analysis in the EIA (Smith 2007).

The original EC Directive 85/337 stated that alternative proposals should be considered in an EIA if the information was relevant and if the developer could reasonably be required to compile this information. Annex Ⅲ required 'where appropriate, an outline of the main alternatives studied by the developer and an indication of the main reasons for this choice, taking into account the environmental effects'. In the UK, this requirement was interpreted as being discretionary, and in the 1990s the consideration of alternatives was one of the weakest aspects of EIS quality (Barker and Wood 1999; Eastman 1997; Jones *et al*. 1991). One of the main changes resulting from the 1997 amendments to the EIA Directive (CEC 1997) was a strengthening of the requirements on alternatives: EISs are now required to include 'an outline of the main alternatives studied by the developer and an indication of the main reasons for the developer's choice, taking into account the environmental effects'. Government guidance (DCLG 2011) is that:

> Where alternative approaches to development have been considered, including alternative choices of process or design, or phasing of construction, the ES should in-

clude an outline of the main ones, and the main reasons for the choice made. Although the Directive and the Regulations do not expressly require the applicant to study alternatives, and do not define 'alternatives', the nature of certain developments and their location may make the consideration of alternative sites a material consideration ... In such cases, the ES must record this consideration of alternative sites.

4.5.2 *Types of alternative*

During the course of project planning, many decisions are made concerning the type and scale of the project proposed, its location and the processes involved. Most of the possible alternatives that arise will be rejected by the developer on economic, technical or regulatory grounds. The role of EIA is to ensure that environmental criteria are also considered at these early stages, and to document the results of this decision-making stage. A thorough consideration of alternatives begins early in the planning process, before the type and scale of development and its location have been agreed on. A number of broad types of alternative can be considered: the 'no action' option, alternative locations, alternative scales of the project, alternative processes or equipment, alternative site layouts, alternative operating conditions and alternative ways of dealing with environmental impacts. We shall discuss the last of these in Section 5.4.

The '*no action*' or '*business as usual*' option refers to environmental conditions if a project were not to go ahead. In essence, consideration of the 'no action' option is equivalent to a discussion of the need for the project: do the benefits of the project outweigh its costs? This option must be considered in some countries (e.g. the USA), but is rarely discussed in UK EISs.[2]

The consideration of alternative *locations* is an essential component of the project planning process. In some cases, a project's location is constrained in varying degrees: for instance, gravel extraction can take place only in areas with sufficient gravel deposits, and wind farms require locations with sufficient wind speed. In other cases, the best location can be chosen to maximize, for example, economic, planning or environmental considerations. For industrial projects, for instance, economic criteria such as land values, the availability of infrastructure, the distance from sources and markets, and the labour supply are likely to be important. For road projects, engineering criteria strongly influence the alignment. In all these cases, however, siting the project in 'environmentally robust' areas, or away from designated or environmentally sensitive areas, should be considered.

The consideration of different *scales* of development is also integral to project planning. In some cases, a project's scale will be flexible. For instance, the size of a waste disposal site can be changed, depending, for example, on the demand for land-fill space, the availability of other sites and the presence of nearby residences or environmentally sensitive sites. The number of turbines on a wind farm could vary widely. In other cases, the developer will need to decide whether an entire unit should be built or not. For instance, the reactor building of a nuclear power station is a large discrete structure that cannot easily be scaled down. Pipelines or bridges, to be functional, cannot be broken down into smaller sections.

Alternative *processes and equipment* involve the possibility of achieving the same objective by a different method. For instance, 1500 MW of electricity can be generated by one combined-cycle gas turbine power station, by a tidal barrage, by several waste-burning power stations or by hundreds of wind turbines. Gravel can be directly extracted or recycled, using wet or dry processes. Waste may be recycled, incinerated or put in a landfill.

Once the location, scale and processes of a development have been decided upon, different *project/site layouts and designs* can still have different impacts. For instance, noisy plants can be sited near or away from residences. Power station cooling towers can be few and tall (using less land) or many and short (causing less visual impact). Buildings can be sited either prominently or to minimize their visual impact. Figure 4.1 shows different bridge designs. Similarly, *operating conditions* can be changed to minimize impacts. For instance, a level of noise at night is usually more annoying than the same level during the day, so night-time work could be avoided. Establishing designated routes for project-related traffic can help to minimize disturbance to local residents. Construction can take place at times of the year that minimize environmental impacts on, for example, migratory and nesting birds. These kinds of 'alternatives' act like mitigation measures.

Figure 4.1 Example of alternative designs: designs considered for the Forth Replacement Crossing

Source: Scottish Government, Jacobs, and Arup 2009

As can be seen from Figure 4. 1, alternatives can be tiered, with decisions made at an earlier stage (or higher tier) acting as the basis for the consideration of lower-tier decisions. In the case of the Forth Bridge Replacement, the bridge route was chosen first, and then various bridge designs for that route were considered. Another example of tiering is the choice of a location for a new wind farm, then the choice of the preferred turbine design, and then the detailed siting of individual turbines.

Alternatives must be reasonable: they should not include ideas that are not technically possible, or illegal. The type of alternatives that can realistically be considered by a given developer will also vary. A mineral extraction company that has put a deposit on a parcel of land in the hope of extracting sand and gravel from it will not consider the option of using it for wind power generation: 'reasonable' in such a case would be other sites for sand and gravel extraction, or other scales or processes. Essentially, alternatives should allow the competent authority to understand why this project, and not some other, is being proposed in this location and not some other.

On the other hand, from a US context (where EISs are prepared by government agencies) Steinemann (2001) argues that alternatives that do not meet a narrow definition of project objectives tend to be too easily rejected, and that alternatives should reflect social, not just agency, goals. She also suggests that

> the current sequence-propose action, define purpose and need, develop alternatives, then analyze alternatives—needs to be revised. Otherwise the proposed action can bias the set of alternatives for the analysis. Agencies should explore more environmentally sound approaches before proposing an action. Then, agencies should construct a purpose and need statement that would not summarily exclude less damaging alternatives, nor unduly favour the proposed action. Agencies should also be careful not to adhere to a single 'problem' and 'solution' early on.

Smith (2007) concludes, in a US context, that:

- project proponents should explain the reasons for the choices they make, particularly in terms of how they selected the range of alternatives that is studied in detail in the EIA;
- if someone proposes an alternative that they feel is 'reasonable', the project proponent should carefully consider this, and the EIA should provide a well-reasoned explanation for why it is dismissed;
- where a proposed action is unlikely to have significant impacts, then developing additional alternatives to the 'do minimum' alternative does not make sense.

These very reasonable rules of thumb also make sense in other contexts.

4. 5. 3 The presentation and comparison of alternatives

The impacts and costs of alternatives vary for different groups of people and for different environmental components. Discussions with local residents, statutory consultees and special interest groups may rapidly eliminate some alternatives fromconsideration and suggest others. However, it is unlikely that one alternative will emerge as being most acceptable to all the parties concerned. The EIS should distil information about a reasonable number of realistic alternatives into a format that will facilitate public discussion and, finally, decision-making. Methods for comparing and presenting alternatives span the range from simple, non-quantitative descriptions to quite detailed, quantitative modelling.

Many of the impact identification methods discussed later in this chapter can also help to compare alternatives. Overlay maps compare the impacts of various locations in a non-quantitative manner. Checklists or less complex matrices can be applied to various alternatives and compared; this may be the most effective way to present the impacts of alternatives visually. Weighted or multi-criteria matrices assign quantitative importance weightings to environmental components, rating each alternative (quantitatively) according to its impact on each environmental component, multi-plying the ratings by their weightings to obtain a weighted impact, and aggregating these weighted impacts to obtain a total score for each alternative. These scores can be compared with each other to identify preferable alternatives.

4.6 Understanding the project/development action

4.6.1 *Understanding the dimensions of the project*

At first glance, the description of a proposed development would appear to be one of the more straightforward steps in the EIA process. However, projects have many dimensions, and relevant information may be limited. As a consequence, this early step may pose challenges. Crucial dimensions to be clarified include the purpose of the project, its life cycle, physical presence, process (es), policy context and associated policies.

The 2011 EIA Regulations require an environmental statement to include

- a description of the development, including in particular a description of the physical characteristics of the whole development and the land-use requirements during the construction and operational phases;
- a description of the main characteristics of the production processes, for instance, nature and quantity of the materials used; and
- an estimate, by type and quantity, of expected residues and emissions (water, air and soil pollution, noise, vibration, light, heat, radiation, etc.) resulting from the operation of the proposed development.

Such information must be given where it 'is reasonably required to assess the environmental effects of the development and which the applicant can, having regard in particular to current knowledge and methods of assessment, reasonably be required to compile' (DCLG 2011).

An outline of *the purpose and rationale* of a project provides a useful introduction to the project description. This may, for example, set the project in a wider context—the missing section of road, a power station in a programme of developments, a new housing project in an area of regeneration or major population growth. A discussion of purpose may include the rationale for the particular type of project, for the choice of the project's location and for the timing of the development. It may also provide background information on planning and design activities to date.

As we noted in Section 1.4, all projects have a *life cycle of activities*, and a project description should clarify the various stages in the project's life cycle, and their relative duration. A minimum description would usually involve the identification of construction and operational stages and associated activities. Further refinement might include planning and design (including consultation), project commissioning, expansion, closedown and site rehabilitation stages. The size of the development at various stages in its life cycle should also be specified. This can include reference to physical size, inputs, outputs, and the number of people to be employed.

The *location and physical presence* of a project should also be clarified at an early stage. This should include its general location on a base map in relation to other activities (e.g. housing, employment sites and recreational areas) and to administrative boundaries. A more detailed site layout of the proposed development, again on a large-scale base map, should illustrate the land area and the main disposition of the elements of the project (e.g. storage areas, main processing plant, waste-collection areas, transport connections to the site). Where the site layout may change substantially between different stages in the life cycle, it is valuable to have a sequence of anticipated layouts or phases of working (see Figure 4.2 for an example). Any associated projects and activities (e.g. transport connections to the site; pipes and transmission lines from the site) should also be identified and described, as should elements of a project that, although integral, may be detached from the main site (e.g. the construction of a barrage in one area may involve opening up a major quarry development in another area). A description of the physical presence of a project is invariably improved by a three-dimensional visual image, which may include a photo-montage of what the site layout may look like at, for example, full operation. A clear presentation of location and physical presence is important for an assessment of change in land uses, any physical disruption to other infrastructures, severance of activities (e.g. agricultural holdings, villages) and visual intrusion and landscape changes.

Understanding a project also involves an understanding of the *processes* integral to it. The nature of processes varies between industrial, service and infrastructure projects, but many can be described as a flow of inputs through a process and their transformation into outputs. Where relevant, a process diagram for the different activities associated with a project should accompany the location and site-layout maps. This could identify the nature, origins and destinations of the inputs and outputs, and the timescale over which they are expected. This may be presented in the form of a simplified pictorial diagram or in a block flow

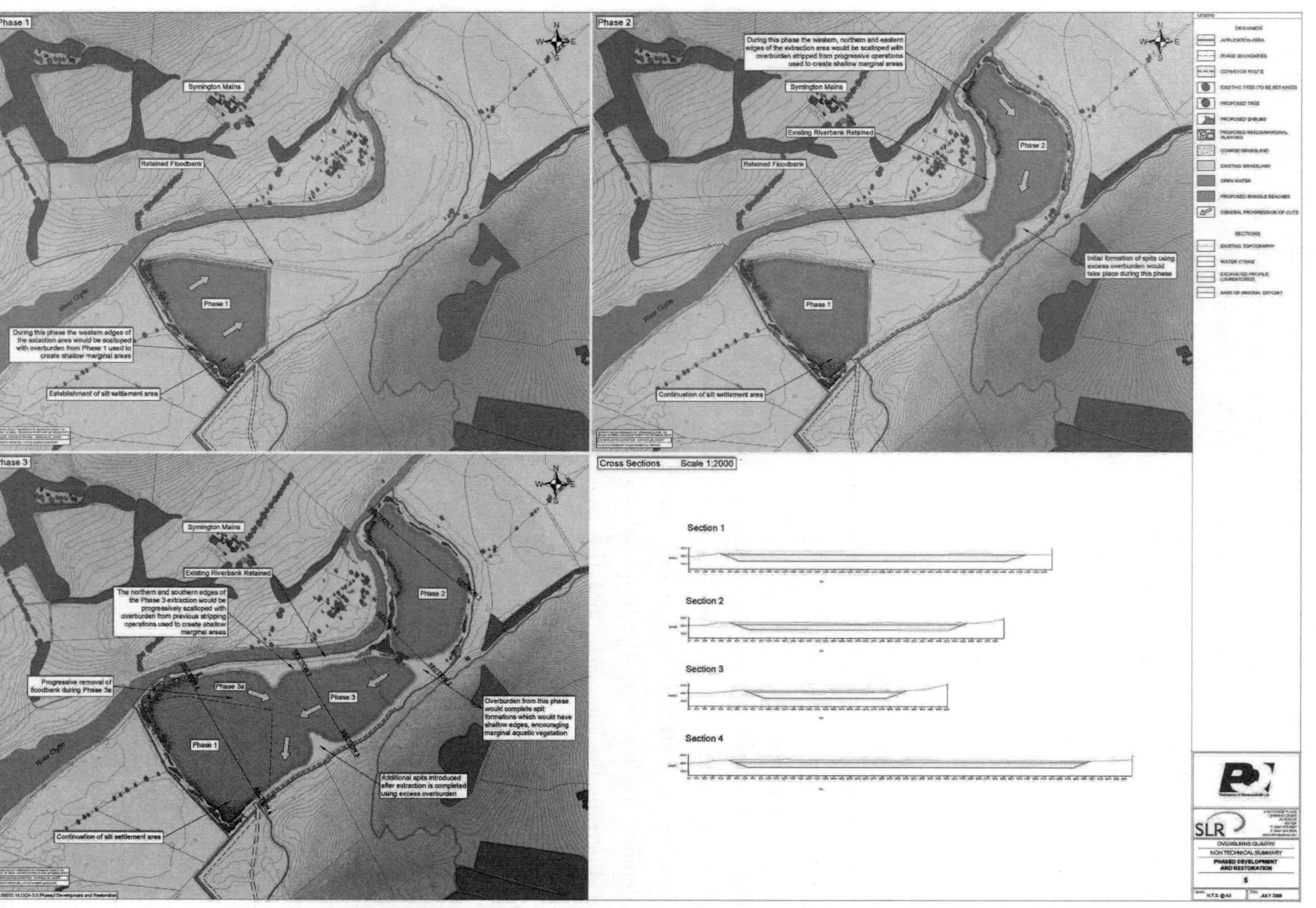

Figure 4.2 Three extraction phases for a sand and gravel quarry

Source: Pattersons Quarries/SLR 2009

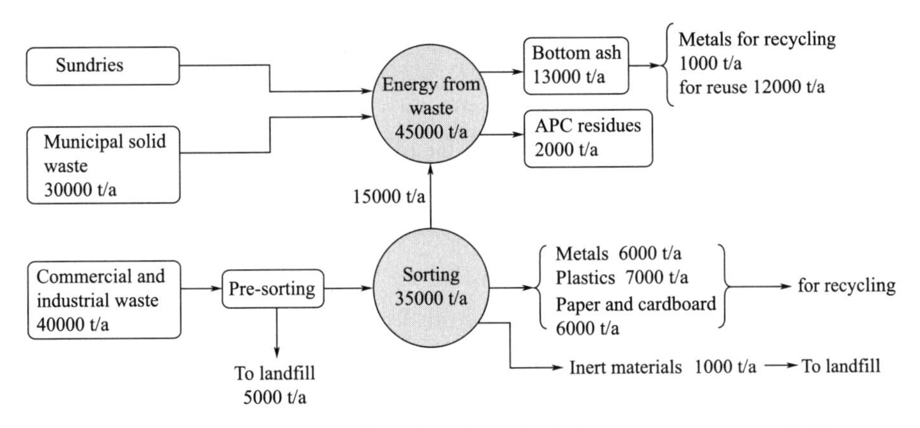

Figure 4. 3 Materials flow chart for a hypothetical waste management facility

chart. Figure 4. 3 shows an example of a materials flow chart for a waste management facility comprising waste sorting and energy from waste production; it outlines the types of inputs and outputs, their expected quantities, and where the outputs will end up. A comprehensive flow chart of a production process should include the types, quantities and locations of resource inputs, intermediate and final product outputs, and wastes generated by the total process.

Physical characteristics of the project may include:

- the land take and physical transformation of a site (e. g. clearing, grading), which may vary between different stages of a project's life cycle;
- the total operation of the process involved (usually illustrated with a process-flow diagram);
- the types and quantities of resources used (e. g. water abstraction, minerals, energy);
- transport requirements (of inputs and out-puts);
- the generation of wastes, including estimates of types, quantity and strength of aqueous wastes, gaseous and particulate emissions, solid wastes, noise and vibration, heat and light, radiation, etc. ;
- the potential for accidents, hazards and emergencies; and
- processes for the containment, treatment and disposal of wastes and for the containment and handling of accidents; monitoring and surveillance systems.

Socio-economic characteristics may include:

- the labour requirements of a project-including size, duration, sources, particular skills categories and training;
- the provision or otherwise of housing, transport, health and other services for the work-force;
- the direct services required from local businesses or other commercial organizations;
- the flow of expenditure from the project into the wider community (from the employees and subcontracting); and
- the flow of social activities (service demands, community participation, community conflict).

Clearly the physical/ecological and socio-economic dimensions of projects interact with each other. Research on ecosystem services (e. g. UKNEA 2011) is clarifying the value to society of ecosystems in the form of products (e. g. food, fibre, fuel) and services (e. g. pollination, carbon fixing). In turn, social interventions such as training or people's travel behaviour can improve or reduce the delivery of ecosystem services.

The projects may also have *associated policies*, not obvious from site layouts and process-flow diagrams, but which are nevertheless significant for subsequent impacts. For example, shift-working will have implications for transport and noise that may be very significant for nearby residents. The use of a construction site hostel, camp or village can significantly internalize impacts on the local housing market and on the local community. The provision of on-or off-site training can greatly affect the mixture of local and non-local labour and the balance of socio-economic effects.

Projects should be seen in their *planning policy context*. In the UK, the main local policy context is outlined and detailed in Local Development Frameworks. The description of location must pay regard to land-use designations and development constraints that may be implicit in some of the designations. Of particular importance is a project's location in relation to various environmental designations (e. g. areas of outstanding natural beauty, Special Areas of Conservation and Special Protection Areas, heritage designations such as listed buildings). Attention should also be given to national planning guidance, provided in the UK by Planning Policy Guidance and Statements and National Policy Statements (NPSs).

4.6.2 Sources and presentation of data

The initial brief from the developer provides the starting point. Ideally, the developer would have detailed knowledge of the proposed project's characteristics, likely layout and production processes, and be able to draw on previous experience. An analyst can supplement information with reference to other EISs for similar projects; although these are normally not monitored so the correctness of their predictions is untested (see Chapters 7 and 8). The analyst may also draw on EIA literature (books and journals), guidelines, manuals and statistical sources, including NPSs, CEC (1993), Morris and Therivel (2009), Rodriguez-Bachiller with Glasson (2004), and United Nations University (2007). Site visits can be made to comparable projects, and advice can be gained from consultants with experience of the type of project under consideration.

As the project design and assessment process develop—in part in response to early EIA findings—so the developer will have to provide more detailed information on the characteristics specific to the project. The identification of sources of potential significant impacts may lead to changes in layout and process.

Data about the project can be presented in different ways. The life cycle of a project can be illustrated on a linear bar chart. Particular stages may be identified in more detail where the impacts are considered of particular significance; this is often the case for the construction stage of major projects. Location and physical presence are best illustrated on a map base, with varying scales to move from the broad location to the specific site layout. This may be supplemented by aerial photographs, photo-montages and visual mock-ups according to the resources and issues involved.

The various information and illustrations should clearly identify the main variations between a project's stages. Figure 4. 4 illustrates a labour-requirements diagram that identifies the widely differing requirements, in absolute numbers and in skill categories, of the construction and operational stages. More sophisticated flow diagrams could indicate the type, frequency (normal, batch, intermittent or emergency) and duration (minutes or hours per day or week) of each operation. Seasonal and material variations, including time periods of peak pollution loads, can also be documented.

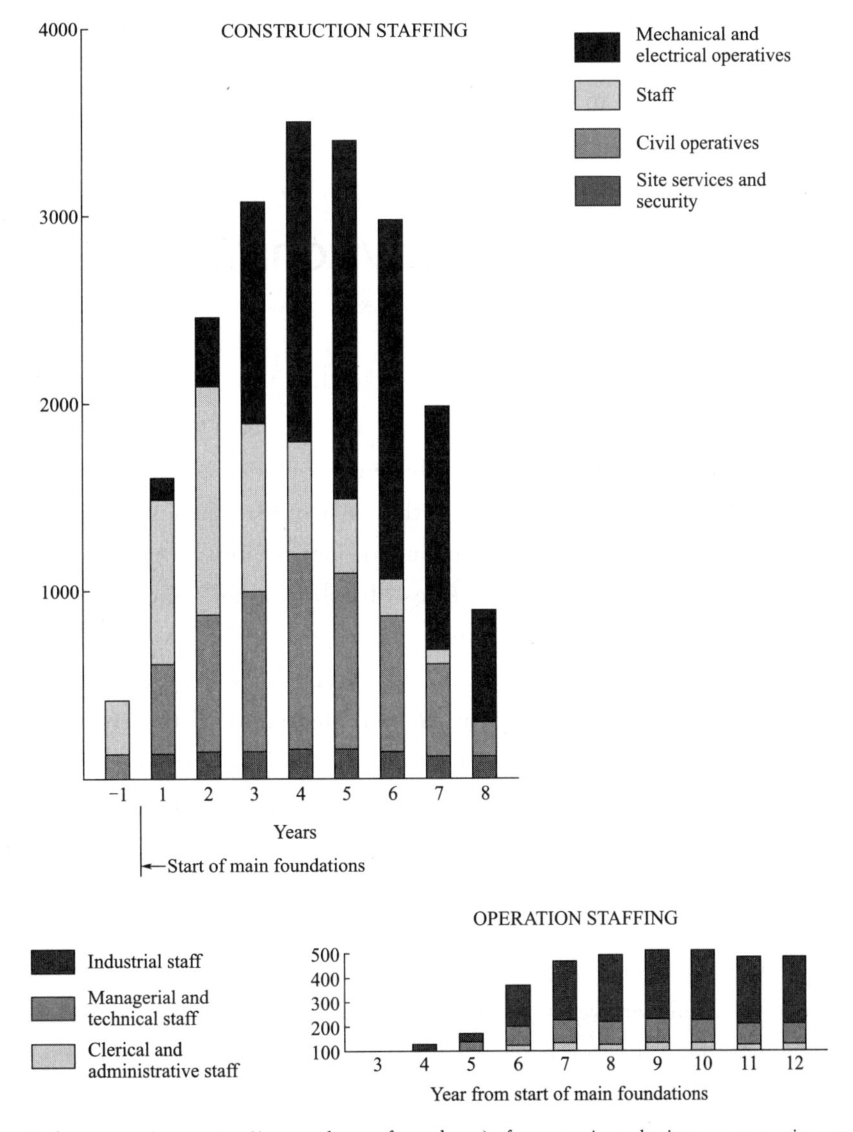

Figure 4. 4 Labour requirements (in numbers of workers) for a project during construction and operation

Unfortunately, in some cases the situation may be far from ideal, with inadequate information provided by the developer—especially in the case of new types of projects. Site layout diagrams and process-flow charts may be only in outline, provisional form at the initial design stage, and the project may change significantly during the planning stage (Frost 1994). The planning application may be an outline application only. A series of European Court of Justice rulings has concluded that, where a consent procedure involves a principal decision and a subsequent implementing decision that cannot go beyond the parameters set by the principal decision, then the EIS must assess the project's environmental effects in time to inform for the principal decision. Only those effects that are not identifiable until the implementing decision should be assessed to inform the later decision (C-201/02, C-508/03, C-290/03; and C2/07).

Further information on these cases is provided in COWI (2009). Even where the project's parameters are well understood early on, and the planning application is a detailed application, there will still need to be considerable interaction between the EIS analyst and the developer to refine the project's characteristics.

4.7 Establishing the environmental baseline

4.7.1 *General considerations*

The establishment of an environmental baseline includes both the present and likely future state of the environment, assuming that a proposed project is not undertaken, taking into account changes resulting from natural events and from other human activities. For example, the population of a species of fish in a lake may already be declining before the proposed introduction of an industrial project on the lake shore. Figure 1.7 illustrated the various time, component and scale dimensions of the environment, and all these dimensions need to be considered in the establishment of the environmental baseline. The period for the prediction of the future state of the environment should be comparable with the life of the proposed development; this may mean predicting for several decades. Components include both the biophysical and socio-economic environment. Spatial coverage may focus on the local, but refer to the wider region and beyond for some environmental elements.

Initial baseline studies may cover a wide range of environmental, social and economic variables, but comprehensive overviews can be wasteful of resources. The studies should focus as quickly as possible on those aspects of the environment that may be significantly affected by the project, either directly or indirectly:

> While every environmental statement (ES) should provide a full factual description of the development, the emphasis of Schedule 4 is on the 'main' or 'sig-

nificant' environmental effects to which a development is likely to give rise. The ES should not be any longer than is necessary to properly assess those effects. Impacts that have little or no significance for the particular development in question will need only very brief treatment to indicate that their possible relevance has been considered (DCLG 2011).

The rationale for the choice of focus should be explained as part of the documentation of the scoping process. Although the studies would normally consider the various environmental elements separately, it is also important to understand the interaction between them and the functional relationships involved; for instance, flora will be affected by air and water quality, and fauna will be affected by flora. This will facilitate prediction. As with most aspects of the EIA process, establishing the baseline is not a 'one-off' activity. Studies will move from broad-brush to more detailed and focused approaches. The identification of new potential impacts may open up new elements of the environment for investigation; the identification of effective measures for mitigating impacts may curtail certain areas of investigation.

4.7.2 Sources and presentation of data

The quality and reliability of environmental data vary a great deal, and this can influence the use of such data in the assessment of impacts. Fortlage (1990) clarifies this in the following useful classification:

- 'hard' data from reliable sources which can be verified and which are not subject to shortterm change, such as geological records and physical surveys of topography and infrastructure;
- 'intermediate' data which are reliable but not capable of absolute proof, such as water quality, land values, vegetation condition and traffic counts, which have variable values;
- 'soft' data which are a matter of opinion or social values, such as opinion surveys, visual enjoyment of landscape and numbers of people using amenities, where the responses depend on human attitudes and the climate of public feeling.

Important UK data sources are the *Census of Population*, *Measuring Progress*, *Environmental Accounts*, *Transport Statistics*, public health observatories and other national and regional Internet data sources, many of which are compiled for the local and ward level in *Neighbourhood statistics*. Local authority monitoring units can also provide very useful data on the physical, social and economic environment. Additional data is unpublished or 'semi-published' and internal to various organizations. In the UK, consultation bodies (e.g. Natural England, Environment Agency and Marine Management Organization) must make environmental information available to any person who requests it, particularly applicants preparing EISs. However, they do not have to undertake research or obtain information that they do not already have (DCLG 2011). See also Chapter 5 (Section 5.3), for other UK and international data sources.

There are of course many other organizations, at local and other levels, which may be able to provide valuable information. Local history, conservation and naturalist societies may have a wealth of information on, for example, local flora and fauna, rights of way and archaeological sites. National bodies such as the RSPB and the Forestry Commission may have particular knowledge and expertise to offer. Consultation with local amenity groups at an early stage in the EIA process can help not only with data but also with the identification of those key environmental issues for which data should be collected.

Even where every use has been made of data from existing sources, there will invariably be gaps in the required environmental baseline data for the project under consideration. Environmental monitoring and surveys may be necessary. Surveys and monitoring raise a number of issues; they are inevitably constrained by budgets and time, and must be selective. However, such selectivity must ensure that the length of time over which monitoring and surveys are undertaken is appropriate to the task in hand. For example, for certain environmental features (e. g. many types of flora and fauna) a survey period of 12 months or more may be needed to take account of seasonal variations or migratory patterns. Sampling procedures will often be used for surveys; the extent and implications of the sampling error involved should be clearly established.

Baseline studies can be presented in the EIS in a variety of ways. A brief overview of the biophysical and socio-economic environments for the area of study may be followed by the project description, with the detailed focused studies in subsequent impact chapters (e. g. air quality, geology, employment), or a more comprehensive set of detailed studies at an early stage providing a point of reference for future and often briefer impact chapters. Maps are typically—but not necessarily—prepared using geographical information systems (GIS). GIS can also be used for more complex analytical functions, for instance map overlays and intersections, buffering around given features, multi-factor map algebra, and visibility analysis derived from terrain modelling (Rodriguez-Bachiller 2000).

The analyst should also be wary of the seductive attraction of quantitative data at the expense of qualitative data; each type has a valuable role in establishing baseline conditions. Finally, it should be remembered that all data sources suffer from some uncertainty, and this needs to be explicitly recognized in the prediction of environmental effects (see Chapter 5).

4.8 Impact identification

4. 8. 1 *Aims and methods*

Impact identification brings together project characteristics and baseline environmental characteristics with the aim of ensuring that all potentially significant environmental impacts (adverse or favourable) are identified and taken into account in the EIA process. When choosing among the existing wide range of impact identification methods, the analyst needs to consider more specific aims, some of which conflict:

- to ensure compliance with regulations;
- to provide a comprehensive coverage of a full range of impacts, including social, economic and physical;
- to distinguish between positive and negative, large and small, long-term and short-term, reversible and irreversible impacts;
- to identify secondary, indirect and cumulative impacts, as well as direct impacts;
- to distinguish between significant and insignificant impacts;
- to allow a comparison of alternative development proposals;
- to consider impacts within the constraints of an area's carrying capacity;
- to incorporate qualitative as well as quantitative information;
- to be easy and economical to use;
- to be unbiased and to give consistent results; and
- to be of use in summarizing and presenting impacts in the EIS.

The simplest impact identification methods involve the use of lists of impacts to ensure that none has been forgotten. The most complex include the use of interactive computer programmes, networks, or the use of weightings to denote impact significance. This section presents a range of these methods, from the simplest checklists needed for compliance with regulations to more complex approaches that developers, consultants and academics who aim to further 'best practice' may wish to investigate further. The methods are divided into the following categories:

- checklists;
- matrices;
- quantitative methods;
- networks; and
- overlay maps.

In the UK, simple checklists and consultation with the local planning authority and statutory consultees are the most widely used impact identification methods. This focus on simple approaches may be attributable to the high degree of flexibility and discretion in the UK's implementation of the EIA Directive, a general unwillingness to make the EIA process over-complex, or disillusionment with some of the early complex approaches. For this reason, although earlier editions of this book covered the more complex impact identification methods (e. g. Sorensen and Moss 1973), this edition does not.

The discussion of the methods here relates primarily to impact identification, but most of the approaches can also be useful in other stages of the EIA process-in impact prediction, evaluation, communication, mitigation and enhancement, presentation, monitoring and auditing. As such, there is considerable interaction between Chapters 4—7, paralleling the interaction in practice between these various stages.

Checklists

Most checklists are based on a list of special biophysical, social and economic factors which may be affected by a development. The simple checklist can help only to identify impacts and ensure that impacts are not overlooked. Checklists do not usually include direct cause-effect links to project activities. Nevertheless, they have the advantage of being easy to use. Table 3. 7 (DETR 2000) is an example of a simple checklist.

Questionnaire checklists are based on a set of questions to be answered. Some of the questions may concern indirect impacts and possible mitigation measures. They may also provide a scale for classifying estimated impacts, from highly adverse to highly beneficial. Table 4. 3 shows part of the EC's (2001b) questionnaire checklist.

Table 4. 3 Part of a questionnaire checklist

No. Questions to be considered in scoping	Yes/No/?	Which characteristics of the project environment could be affected and how?	Is the effect likely to be significant? Why?
7 Will the project lead to risks of contamination of land or water from releases of pollutants onto the ground or into sewers, surface waters, ground water, coastal waters or the sea?			
7. 1 From handling, storage, use or spillage of hazardous or toxic materials?			
7. 2 From discharge of sewage or other effluents (whether treated or untreated) to water or the land?			
7. 3 By deposition of pollutants emitted to air, onto the land or into water?			
7. 4 From any other sources?			
7. 5 Is there a risk of long-term build-up of pollutants in the environment from these sources?			

Source: EC 2001b

Matrices

Simple matrices are merely two-dimensional charts showing environmental components on one axis and development actions on the other. They are, essentially, expansions of checklists that acknowledge the fact that various components of a development project (e. g. construction, operation, decommissioning, buildings, and access road) have different impacts. The action likely to have an impact on an environmental component is identified by placing a tick or cross in the appropriate cell. Table 4. 4 shows an example of a *simple matrix*.

Table 4. 4 Example of a simple matrix

	Project action				
	Construction		Operation		
	Utilities	Residential and commercial buildings	Residential buildings	Commercial buildings	Parks and open spaces
Soil and geology	√	√			
Flora	√	√			√
Fauna	√	√			√
Air quality				√	
Water quality	√	√	√		
Population density			√	√	
Employment		√		√	
Traffic	√	√	√	√	
Housing			√		
Community structure		√	√		√

Magnitude matrices go beyond the mere identification of impacts by describing them according to their magnitude, importance and/or time frame (e. g. short, medium or long-term). Table 4. 5 is an example of a magnitude matrix that represents whether the impact is positive or negative with either a red or green circle (red/green/amber 'traffic light' type colours are often used in EIA), and the magnitude of the impact by the depth of colour.

Table 4. 5 Example of a magnitude matrix

	Project action				
	Construction		Operation		
	Utilities	Residential and commercial buildings	Residential buildings	Commercial buildings	Parks and open spaces
Soil and geology					
Flora					
Fauna					
Air quality					
Water quality					
Population density					
Employment					
Traffic					
Housing					
Community structure					

large positive impact　small positive impact　neutral impact　small negative impact　large negative impact

Distributional impact matrices aim to broadly identify who might lose and who might gain from the potential impacts of a development. This is useful information, which is rarely included in the matrix approach, and indeed is often missing from EISs. Impacts can have varying spatial impacts—varying, for example, between urban and rural areas. Spatial variations may be particularly marked for a linear project, such as a road or rail line. A project can also have different impacts on different groups in society (for example the impacts of a proposed new settlement on old people, retired with their own houses, and young people, perhaps with children, seeking affordable housing and a way into the housing market; see Figure 5. 8).

Where matrices use numerical values to describe impacts, people may attempt to add these values to produce a composite value for the development's impacts and compare this with that for other developments; this should not be done unless each of the impacts is broadly as important as the others, otherwise these differences in impact importance will not be reflected in the outcome. *Weighted matrices* try to deal with this issue by assigning weightings to different impacts—and sometimes to different project components—to reflect their relative importance. The impact of the project (component) on the environmental component is then assessed and multiplied by the appropriate weighting (s), to obtain a total for the project. Weightings could reflect environmental components that are close to legal standards or thresholds; project components that will have long-term and irreversible impacts (Odum *et al*. 1975); or future long-term impacts that could be given more weight than short-term impacts (Stover 1972).

Table 4. 6 shows a small weighted matrix that compares three alternative project sites. Each environmental component is assigned an importance weighting (a), relative to other environmental components: in the example, air quality is weighted 21 per cent of the total environmental components. The magnitude (c) of the impact of each project on each environmental component is then assessed on a scale 0—10, and multiplied by (a) to obtain a weighted impact (a × c): for instance, site A has an impact of 3 out of 10 on air quality, which is multiplied by 21 to give the weighted impact, 63. For each site, the weighted impacts can then be added up to give a project total. The site with the lowest total, in this case site B, is the least environmentally harmful.

Table 4. 6 A weighted matrix: alternative project sites

Environmental component	(a)	Alternative sites					
		Site A		Site B		Site C	
		(c)	(a×c)	(c)	(a×c)	(c)	(a×c)
Air quality	21	3	63	5	105	3	63
Water quality	42	6	252	2	84	5	210
Noise	9	5	45	7	63	9	81
Ecosystem	28	5	140	4	112	3	84
Total	100		500		364		438

(a) = relative weighting of environmental component (total 100)

(c) = impact of project at particular site on environmental component (0-10)

The attraction of weighted matrices (or, more generally, multi-criteria analysis) lies in their ability to 'substantiate' numerically that a particular course of action is better than others. This may save decision-makers considerable work, and it ensures consistency in assessment and results. However, these approaches, and the 'answers' they lead to, depend heavily on the weightings and impact scales. They effectively take decisions away from decision-makers; they are difficult for lay people to understand; and their acceptability depends on the assumptions (especially the weighting schemes, built into them). People carrying out assessments may manipulate—or be perceived to manipulate—results by changing assumptions. The main problems implicit in such weighting approaches are considered further in Chapter 5.

More generally, checklists and matrices treat the environment as if it consisted of discrete units: impacts are related only to particular parameters, and the complex relationship between environmental components is not described. Much information is lost when impacts are reduced to symbols or numbers. Checklists and matrices do not specify the probability of an impact occurring, their scoring systems are inherently subjective and open to bias, and they cannot reveal indirect effects of developments. However, they are (relatively) quick and simple, and they can be used to present scoping findings as well as identifying possible impacts.

Networks or causal chain analyses

Network methods explicitly recognize that environmental systems consist of a complex web of relationships, and try to reproduce that web. Impact identification using networks involves following the effects of development through changes in the environmental parameters in the model.

A simple network method is used in the development of many UK Local Transport Plans. Network diagrams are drawn by planners to identify how one action—say, changes to carriageways, junctions or public transport provision (Figure 4. 5)—leads to changes in social, economic and environmental conditions. They also identify what preconditions are needed to achieve a positive outcome, and problems to avoid. Network methods do not establish the magnitude or significance of interrelationships between environmental components, or the extent of change. They can require considerable knowledge of the environment. Their main advantage is their ability to trace the higherorder impacts of proposed developments.

Overlay (or constraints) maps

Overlay maps have been used in environmental planning since the 1960s (McHarg 1968) before the NEPA was enacted. A series of overlays—originally in the form of transparencies, now more typically in the form of GIS layers—is used to identify, predict, assign relative significance to and communicate impacts. A base map is prepared, showing the general area within which the project may be located. Successive overlay maps are then prepared for the environmental components that are likely to be affected by the project

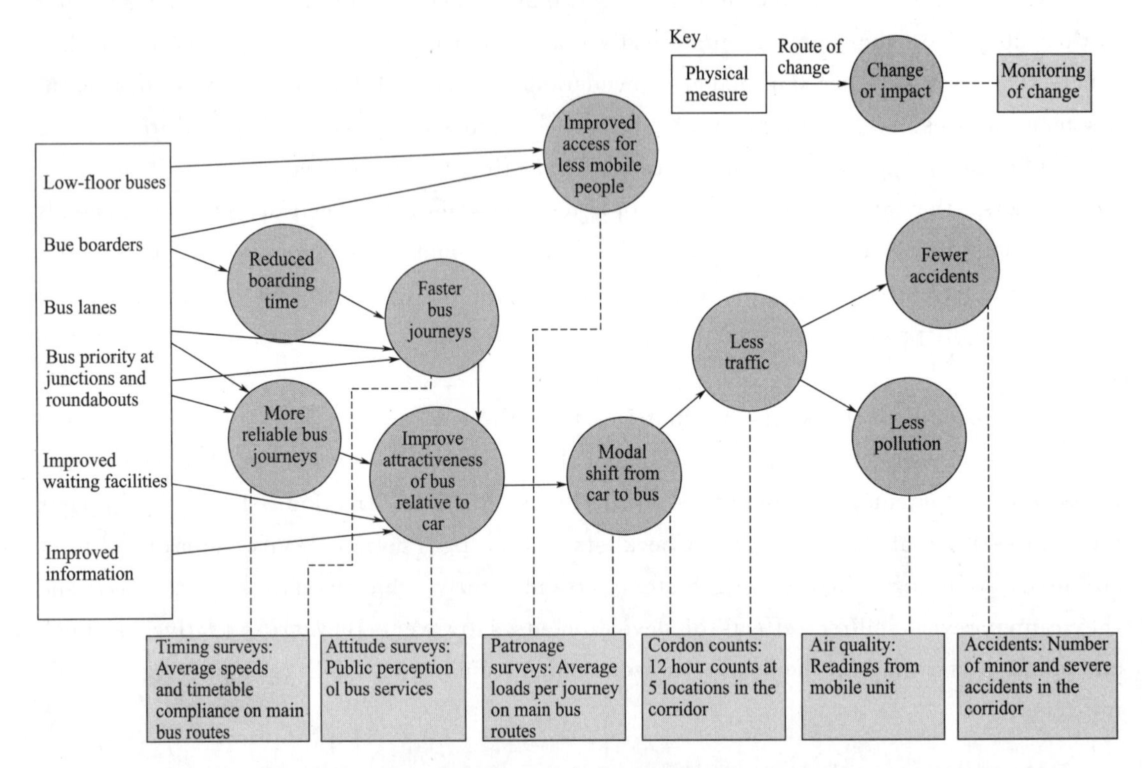

Figure 4. 5 Causal chain analysis for bus corridor improvements

Source: Adapted from South Yorkshire Integrated Transport Authority 2001

(e. g. agriculture, woodland, noise). The composite impact of the project is found by superimposing the overlay maps and noting the relative intensity of the total shading. Areas with little or no shading are those where a development project would not have a significant impact. Figure 4. 6 shows the principle of overlay maps.

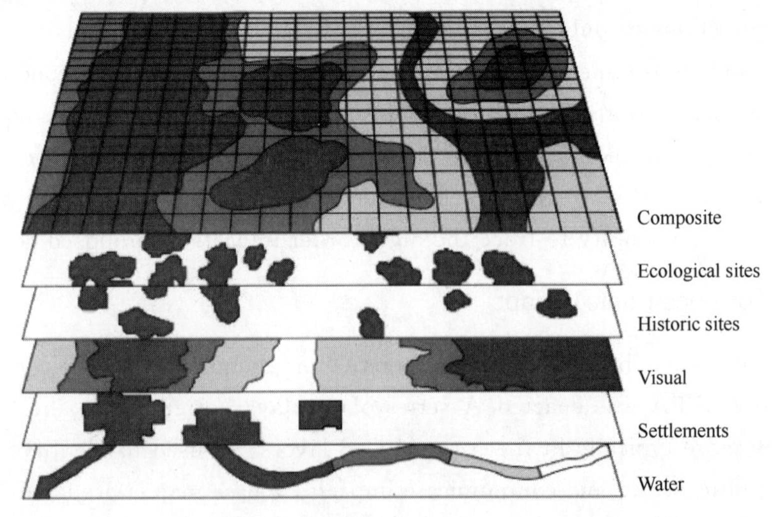

Figure 4. 6 Overlay map: general principle

GIS can be used to assign different importance weightings to the impacts: this enables a sensitivity analysis to be carried out, to see whether changing assumptions about impact importance would alter the decision.

Overlay maps are easy to use and understand and are popular. They are an excellent way of showing the spatial distribution of impacts. They also lead intrinsically to a low-impact decision. The overlay maps method is particularly useful for identifying optimum corridors for developments such as electricity lines and roads, for comparisons between alternatives, and for assessing large regional developments. However, the method does not consider factors such as the likelihood of an impact, indirect impacts or the difference between reversible and irreversible impacts. It requires the clear classification of often indeterminate boundaries (such as between forest and field), and so may not be a true representation of conditions on the ground. It relies on the user to identify likely impacts before it can be used.

Maps can also be used to show distributional impacts—who wins, who loses, and whether groups that are traditionally disadvantaged will lose out disproportionately.

4.8.2 Quality of life assessment

The quality of life assessment (QOLA) (or quality of life capital) approach was developed jointly by the Countryside Agency *et al.* (2001) as a way of integrating the different agencies' approaches to environmental management. QOLA focuses not on the *things* but on the *benefits* that would be affected by a development proposal. It starts with the assumption that things (e.g. woodlands, historical buildings) are important because of the benefits that they provide to people (e.g. visual amenity, recreation, CO_2 fixing), and conversely that management of those things should aim to optimize the benefits that they provide. This emphasis on benefits has subsequently been incorporated in the ecosystem services approach (e.g. DEFRA 2007).

Quality of life assessment involves six steps (a—f). Having identified the purpose of the assessment (a) and described the proposed development site (b), the benefits/disbenefits that the site offers to sustainability (i.e. to present and future generations) are identified (c). The technique then asks the following questions (d):

- How important is each of these benefits or disbenefits, to whom, and why?
- On current trends, will there be enough of each of them?
- What (if anything) could substitute for the benefits?

The answers to these questions lead to a series of management implications (e), which allow a 'shopping list' to be devised of things that any development/management on that site should achieve, how they could be achieved, and their relative importance. Finally, monitoring of these benefits is proposed (f). Thus the process concludes by clearly stipulating the benefits that the development would have to provide before it was considered acceptable and, as a corollary, indicates where development would not be appropriate. It can be used to set a management framework (e.g. for planning conditions or mitigation measures) for any development on a given site (and also for management of larger areas). The QOLA approach can be used as a vehicle for public participation and/or the integration of different experts' analyses of a site.

4.8.3 Summary

Table 4.7 summarizes the respective advantages of the main impact identification methods discussed in this section. Impact identification methods need to be chosen with care: they are not politically neutral, and the more sophisticated the method becomes, often the more difficult becomes clear communication and effective participation (see Chapter 6 for more discussion). The simpler methods are generally easier to use, more consistent and more effective in presenting information in the EIS, but their coverage of impact significance, indirect impacts or alternatives is either very limited or non-existent. The more complex models incorporate these aspects, but at the cost of immediacy.

Table 4.7 Comparison of impact identification methods

Criterion	
1 Compliance with regulations	7 Compare against carrying capacity
2 Comprehensive coverage (social, economic and physical impacts)	8 Uses qualitative and quantitative information
3 Positive vs. negative, reversible vs. irreversible impacts, etc.	9 Easy to use
4 Secondary, indirect, cumulative impacts	10 Unbiased, consistent
5 Significant vs. insignificant impacts	11 Summarizes impacts for use in EIS
6 Compare alternative options	

	1	2	3	4	5	6	7	8	9	10	11
Checklists	√	√						√	√	√	√
Matrices											
• simple	√	√						√	√	√	√
• magnitude	√	√	√					√	√	√	√
• weighted	√	√			√	√		√	√		√
Network	√			√		√		√		√	
Overlay maps		√	√		√	√		√	√	√	√

4.9 Summary

The early stages of the EIA process are typified by several interacting steps. These include deciding whether an EIA is needed at all (screening), consulting with the various parties involved to help focus on the chief impacts (scoping), and an outline of possible alternative approaches to the project, including alternative locations, scales and processes. Scoping and the consideration of alternatives can greatly improve the quality of the process. Early in the process an analyst will also wish to understand the nature of the project concerned, and the environmental baseline conditions in the likely affected area. Projects have several dimensions (e. g. physical presence, processes and policies) over several stages in their life cycles; a consideration of the environmental baseline also involves several dimensions. For both projects and the affected environment, obtaining relevant data may present challenges.

Impact identification includes most of the activities already discussed. It usually involves the use of impact identification methods, ranging from simple checklists and matrices to more complex networks and maps. The methods discussed herehave relevance also to the prediction, assessment, communication and mitigation of environmental impacts, which are discussed in the following chapters.

SOME QUESTIONS

The following questions are intended to help the reader focus on the key issues of this chapter.

1 Assume that a developer is proposing to build a wind farm (or another project of your choice) in an area that you know well, and for which you have been asked to manage the EIS for the project.

- What kind of experts would you want on your team?
- What information about the project would you need to know before you could carry out the EIA scoping stage?
- What types of project alternatives might be relevant? Would there be tiers of alternatives?
- What technique would you use to identify the impacts of the project, and why?

2 Table 4. 2 shows that the number of EISs prepared in different European Union Member States varies widely. What issues associated with the screening process might account for these differences?

3 Should all EISs consider the 'no action' alternative? Different locations? Different scales? Different processes? Different designs? Why or why not?

4 What minimum level of information about the project should be presented in the EIS? What additional information could be useful?

5 Of the different figures and tables presented in this chapter, which two or three would you find most helpful when trying to understand a project and its impacts?

6 Section 4. 8. 1 suggests that quite complex impact identification methods have been devised in the past but not used much in practice. What might be the reason for this?

Notes

1 This refers both to the spatial extent that will be covered and to the scale at which it is covered. João (2002) suggests that the latter—which has been broadly ignored as an issue to date—could be crucial enough to lead to different decisions depending on the scale chosen.

2 In the US, 'agencies should: consider the option of doing nothing; consider alternatives outside the remit of the agency; and consider achieving only a part of their objectives in order to reduce impact'.

References

Barker, A. and Wood, C. 1999. An evaluation of EIA system performance in eight EU countries. *Environmental Impact Assessment Review* 19, 387-404.

BIO 2006. Cost and benefits of the implementation of the EIA Directive in France, Unpublished document.

Carroll, B. and Turpin, T. 2009. *Environmental impact assessment handbook*. London: Thomas Telford.

CEC (Commission of the European Communities) 1993. *Environmental manual: user's guide; sectoral environment assessment sourcebook*. Brussels: CEC DG VIII.

CEC 1997. Council Directive 97/11/EC amending Directive 85/337/EEC on the assessment of certain public and private projects on the environment. *Official Journal* L73/5, 3 March.

CEQ (Council on Environmental Quality) 1978. National Environmental Policy Act. Implementation of procedural provision: final regulations. *Federal Register* 43 (230), 5977-6007, 29 November.

Countryside Agency, English Nature, Environment Agency, English Heritage 2001. *Quality of Life Capital: What matters and why*. Available at: www. qualityoflifecapital. org. uk.

COWI 2009. Study concerning the report on the application and effectiveness of the EIA Directive. Final report to the European Commission, DG ENV. Kongens Lyngby, Denmark.

Dartmoor National Park Authority 2010. Town and Country Environmental Impact Assessment (England and Wales), Regulations 1999 (Part IV) Scoping Opinion (12 August 2010). Available at: www. dartmoornpa. gov. uk/pl-2010-06-10 _ yennadon _ eia _ scoping _ opinion. pdf.

DCLG (Department for Communities and Local Government) 2006. *Evidence review of scoping in environmental impact assessment*. London: DCLG.

DCLG (Department for Communities and Local Government) 2011. *Guidance on the environmental impact assessment (EIA) regulations 2011 for England*. London: DCLG.

DEFRA (Department for Environment, Food and Rural Affairs) 2007. *An introductory guide to valuing ecosystem services*. Available at: www. defra. gov. uk/ environment/policy/natural-environ/documents/eco-valuing. pdf.

DETR (Department of the Environment, Transport and the Regions) 2000. *Environmental impact assessment: a guide to the procedure*. Available at: www. communities. gov. uk/publications/planningandbuilding/environmentalimpactassessment.

Eastman, C. 1997. *The treatment of alternatives in the environmental assessment process* (MSc dissertation). Oxford: Oxford Brookes University.

EC (European Commission) 2001a. *Screening checklist*. Brussels: EC.

EC 2001b. *Scoping checklist*. Brussels: EC.

EC 2006. Evaluation of EU legislation-Directive 85/337/EEC (Environmental Impact Assessment, EIA) and associated amendments. Brussels: DG Enterprise and Industry.

Environment Agency 2002. *Environmental impact assessment: scoping guidelines for the environmental impact assessment of projects*. Reading: Environment Agency.

EU (European Union) 2010. *Environmental Impact Assessment of Projects: Rulings of the Court of Justice*. Brussels: EU.

Fortlage, C. 1990. *Environmental assessment: a practical guide*. Aldershot: Gower.

Frost, R. 1994. *Planning beyond environmental statements* (MSc dissertation). Oxford: Oxford Brookes University, School of Planning.

GHK 2010. *Collection of information and data to support the impact assessment study of the review of the EIA Directive*. London: GHK.

Government of New South Wales 1996. *EIS guidelines*. Sydney: Department of Urban Affairs and Planning.

IEMA (Institute of Environmental Management and Assessment) 2011. *The State of Environmental Impact Assessment Practice in the UK*. Grantham: IEMA.

João, E. 2002. How scale affects environmental impact assessment. *Environmental Impact Assessment Review* 22, 289-310.

Jones, C. E., Lee, N. and Wood, C. 1991. *UK environmental statements 1988-1990: an analysis*. Occasional Paper 29, Department of Town and Country Planning, University of Manchester.

King County International Airport 2004. NEPA environmental assessment, SEPA Environmental impact statement for proposed master plan improvements at King County Inernational Airport (Boeing Field), Seattle. Available at: www. your. kingcounty. gov/airport/plan/EIS-EA _ 2-23-04. pdf.

Lawrence, D. P. 2003. *Environmental Impact Assessment: practical solutions to recurrent problems*. Hoboken: Wiley-Interscience.

McHarg, I. 1968. *A comprehensive route selection method*. Highway Research Record 246. Washington, DC: Highway Research Board.

Morris, P. and Therivel, R. (eds) 2009. *Methods of environmental impact assessment*, 3rd edn. London: Routledge.

Morrison-Saunders, A. and Bailey, M. 2009. Appraising the role of relationships between regulators and consultants for effective EIA. *Environmental Impact Assessment Review* 29, 284-94.

Odum, E. P., Zieman, J. C., Shugart, H. H., Ike, A. and Champlin, J. R. 1975. In *Environmental impact assessment*, M. Blisset (ed). Austin: University of Texas Press.

Oosterhuis, F. 2007. Costs and benefits of the EIA Directive: Final report for DB Environment under specific agreement no. 07010401/2006/447175/FRA/G1. Available at: www. ec. europa. eu/ environment/eia/pdf/Costs%20and%20benefits%20of%20the%20EIA%20Directive. pdf.

Pattersons Quarries/SLR 2009. Planning application for proposed sand and gravel quarry at Overburns Farm, Lamington, Non-technical summary. Available at: www. patersonsquarries. co. uk/images/downloads/ Overburns _ Quarry _ NTS-SM. pdf.

Petts, J. and Eduljee, G. 1994. Integration of monitoring, auditing and environmental assessment: waste facility issues. *Project Appraisal* 9 (4), 231-41.

Rodriguez-Bachiller, A. 2000. Geographical information systems and expert systems for impact assessment, Parts I and II. *Journal of Environmental Assessment Policy and Management* 2 (3), 369-448.

Rodriguez-Bachiller, A. with J. Glasson 2004. *Expert systems and geographical information systems for impact assessment*. London: Taylor and Francis.

Scottish Government, Jacobs, and Arup 2009, Forth Replacement Crossing environmental statement. Edinburgh: Scottish Government.

Smith, M. D. 2007. A review of recent NEPA alternatives analysis case law, *Environmental Impact Assessment Review* 27, 126-40.

Sorensen, J. C. and Moss, M. L. 1973. *Procedures and programmes to assist in the environmental impact statement process*. Berkeley, CA: Institute of Urban and Regional Development, University of California.

South Yorkshire Integrated Transport Authority 2001. South Yorkshire Local Transport Plan 2001-06. Available at: www. southyorks. gov. uk/ index. asp? id=3086.

State of California 2010. Title 14 California Code of Regulations section 15000 *et seq*. Sacramento, CA: State of California.

Steinemann, A. 2001. Improving alternatives for environmental impact assessment. *Environmental Impact Assessment Review* 21, 3-21.

Stover, L. V. 1972. *Environmental impact assessment: a procedure*. Pottstown, PA: Sanders & Thomas.

UKNEA (National Ecosystem Assessment) 2011.

Understanding nature's value to society. Available at: www. uknea. unep-wcmc. org.

United Nations University (2007). Environmental Impact Assessment Open Educational Resource. Available at: www. eia. unu. edu.

Weaver, A. B., Greyling, T., Van Wilyer, B. W. and Kruger, F. J. 1996. Managing the ETA process. Logistics and team management of a large environmental impact assessment-proposed dune mine at St. Lucia, South Africa. *Environmental Impact Assessment Review* 16, 103-13.

World Bank (1999) Environmental Assessment Sourcebook 1999. Available at: www. siteresources. worldbank. org/INTSAFEPOL/1142947-111649557 9739/20507372/Chapter1The Environmental Review Process. pdf.

5

影响预测、评价、减缓及改善措施

5.1 导言

本章着重介绍了影响预测、评价、减缓及改善措施的核心步骤。虽然我们已经意识到EIA不是一个线性过程，但影响预测、评价、减缓及改善措施却是EIA程序的核心。事实上，EIA的全过程都与预测密切相关。一个项目，包括它的替代方案，需要在最早期的阶段就对其预测进行计划和设计，并且将始终贯穿于方案的减缓、改善、监测和审计过程。然而，尽管预测在EIA中处于核心地位，可是由于其研究多为描述性的内容，导致在很多研究中还没有得到足够的重视。在EIA中，预测通常被看作是一个不确定过程，在研究中往往忽视了那些定义清晰的模型。一些模型即使得以应用，也不是很详细，而且其局限性很少被讨论。5.2节讨论了预测范围（预测什么）、预测方法和模型（如何预测）以及实际应用中模型固有的局限性（不确定性）。本章也对一些有用的国际及英国的预测资料进行了简要总结。

评价紧随预测之后，并且包括了显著性影响的评价。评价的方法变化多样，从简单到复杂，从直观到抽象，从定性到定量，从正式到非正式。费用效益分析理论（CBA）、货币估价技术、多标准/多属性方法及计分和权重系统，都为评价提供了有效的方法。本章也讨论了越来越受重视的影响改善措施这一方面，也就是项目可能产生的重要的有益影响的方面。

5.2 预测

5.2.1 预测范围（预测什么）

预测的目的是为了确定某一项目或行为实施后与其实施前相比，可能引起的环境变化的程度和范围。预测也可以为重大影响评价提供基础，对此，我们将在5.3节中加以讨论。

在英国，相关法律的要求是预测范围识别的出发点之一（表3.6，第1、2部分）。如表5.1所示，《环境影响评价：程序指南》（DETR，2000）对预测范围进行了详细说明，但此表对于社会-经济影响的评价还很有限。表1.3提供了可能会受到拟建项目影响的环境范围和环境受体。

预测包含了对该环境受体指标潜在变化的识别。变化的范围将确定所考虑的相关项目的影响类别。如果某种特定的环境指标（如空气中的SO_2水平）显示了某一区域中某个问题的加剧与某项目或行为（如发电站）无关，那么这就可以作为该项目的特定指标的预测本底值。不过需要将这些指标细化并加以详细描述，进而提供可测得的相关变量。例如，经济影响可以逐步描述为：

直接雇佣→当地雇佣→当地技术雇佣。

表5.1 英国法律对影响评价的概述

影响评价(包括项目引起的直接的和间接的，二次的，累积的，短期，中期和长期的，永久的和暂时的，积极的和消极的影响)
人类、建筑物和人造景观的影响
1.开发引起的人口数量变化以及继而产生的环境影响。
2.开发对周围地区和景观的视觉影响。
3.正常运行中排放物的水平和影响。
4.开发产生的噪声水平和影响。
5.开发对当地道路和交通的影响。
6.开发对建筑物、建筑和历史遗产、考古遗迹以及其他人类活动的影响，例如污染物排放、视觉侵扰、振动带来的影响。
动、植物和地质的影响
7.动、植物物种及其栖息地的破坏和消失。
8.地质特征、古生物学特征和地形学特征的破坏和消失。
9.其他的生态影响。
土壤环境影响
10.开发对自然环境的影响，如当地地形变化、地层移动稳定性的影响、土壤腐蚀等。
11.化学物质对周边地域土壤的排放和沉积。
12.土地利用/资源的影响：
(1)占用耕地的质量和数量；
(2)矿产资源的消亡；
(3)土地使用方式的改变，包括"土地闲置"方案；
(4)对周围土地利用(包括农业用地)的影响；
(5)废弃物处置。
水环境影响
13.开发商对区域排水模式的影响。
14.其他水文特征的变化，例如地下水水位、河道、地下水水流的变化。
15.海岸或河口水文的影响。

16. 污染物、废弃物等对水质的影响。
大气和气候的影响
17. 化学物质排放的水平与浓度以及对环境的影响。
18. 特殊物质。
19. 臭气。
20. 其他气候影响。
与项目有关的其他间接和二次影响
21. 与交通开发(公路、铁路、机场和水运)有关的影响。
22. 开发过程中原料、水、能源及其他资源的开采和消耗产生的影响。
23. 与项目有关的其他方面的开发产生的影响,例如新的道路、下水道、住宅供电线路、管道、电视通信等。
24. 其他现有开发或拟议开发产生的综合影响。
25. 上述各自直接影响的交互作用产生的二次影响。

资料来源：DETR，2000。

通过这种方式，一系列与政策相关的重大影响指标就可以不断地得以完善。

对可能性等级（如大小）的预测与影响显著性（如对决策的重要性）的预测进行区分通常是十分重要的，这是因为大小与显著性并不总是一致的。例如，在健康的环境中，某种污染物大量增加所产生的结果可能仍在可接受的标准范围之内，然而，在敏感性环境中，污染物极少量的增加就可能会超过其可接受的标准（图 5.1），这也突出了客观和主观方法的区别。尽管很难，但对某种影响程度的预测应该是一种客观的实践，而对重要性的判定通常是更为主观的应用，因为它通常含有价值判断。

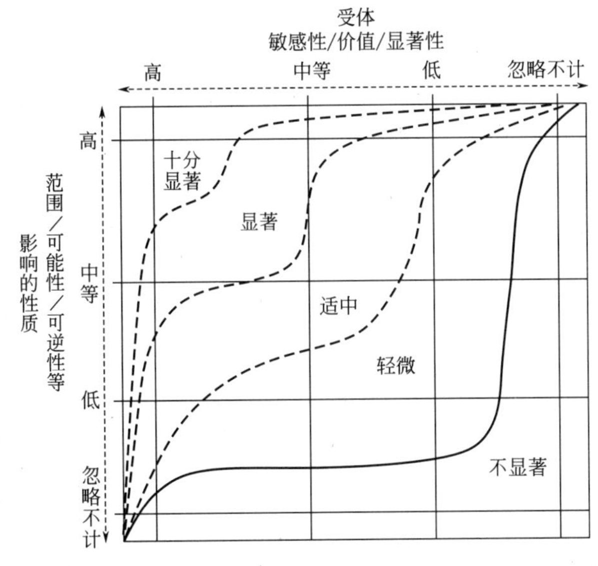

图 5.1　通用的重要性矩阵示例（以“影响的程度”与“影响的敏感性”为例）

（资料来源：IEMA，2011）

如表 1.4 所示，预测还应该确定出直接影响和间接影响（这里可以用简单的因果关系图表示）、影响的地理范围（如本地、区域和国家）、有利影响还是不利影响以及影响的持续时间等。另外，对于项目生命周期的预测（包括它的建设、运行和其他阶段），分析人员还应注意影响的变化速率。慢速变化造成的影响也许比快速变化造成的影响更容易被接受，在偏远或不发达地区进行旅游项目开发就是快速变化造成破坏性影响的具体例子。项目可能具有非线性过程、因果滞后的特征，因而要预测其影响的间歇性。影响的可逆性及其他一些特性，如持久性、累积性及协同性也应该被预测。累积影响（或叠加影响）是所有影响共同作用的结果，虽然独立看来，每一个的影响都很小，但是将其综合起来，久而久之，其影响是非常大的。这种累积的影响很难预测，在EIA研究中往往很少涉及或者是完全忽略（第12章）。

预测的另一个方面是量度单位以及定性影响与定量影响的区别。与其他指标相比，一些指标更容易定量（例如，对比与某一项目有关的社会压力的变化，饮用水水质的变化可能更易定量）。如果可能，预测的影响应该有明确的单位，以便为评估和协调提供基础。这允许评估的预测影响可以按不同地区、国家和国际标准进行对比。预测还应该包括对某一影响所产生的可能性的估算，这提出了不确定性这一重要问题。

5.2.2 预测的方法和模型（如何预测）

预测影响的可行方法有很多。在20世纪80年代早期，环境资源有限公司为荷兰政府开展的一项研究表明，在荷兰和北美的140项EIA研究中，共识别出150种不同的预测方法（VROM，1984）。不过，对预测来说，没有一种方法是万能的。

万物运行方式的概念模型是所有预测的基础。模型的复杂性从直观判断到对环境过程本质的假设而不断变化着……然而环境与模型的假设条件并不一致，这将阻碍评价者接受现成的公式。（Munn，1979）

预测模型有很多类型，而且它们之间并非相互排斥。关于评价范围，所有方法对于影像的覆盖面都有一定的局限性，有些相较其他而言评价范围要完整一些。有些方法可以根据项目的类型（如单一的影响评价）或影响的类型（如广泛的经济影响）分类。一些是外推法，还有一些方法则更加标准化。对于外推法，预测的制定要与历史和现状数据相一致。外推法包括趋势分析法（考虑项目造成的变化并进行修正，推断现在的趋势）、情景分析法（基于一系列假设对未来情况的常识预测）、类推法（把别处的经验应用于当前研究）、直接预测法［如使用德尔菲法（Delphi法，也称专家判断法）获取某项目影响的一致性］等（Green等，1989；Briedenhann与Butts，2006）。标准化的方法是先确定项目的预期目标，并在其环境背景下，评价该项目的预期目标是否能够实现。例如，一个主要项目的建设阶段要求的社会-经济目标可能是雇佣50%的本地劳动力，而这一目标的实现还要受到项目以及相关雇佣政策（如进行培训）的限制。通过对各种情景进行筛选进而确定一种最有可能达到所要求目标的方案。预测方法可以根据其形式进行分类，以下是几种典型的模型。

5.2.2.1 数学和计算机模型

数学模型试图通过使用数学函数来表示环境行为。它们通常以科学定律、统计分析或者两者的结合为依据，以计算机为基础。基础函数既可以是简单直接的输入输出关系，也可以是一系列表达相互关系的、更加复杂的动态数学模型。数学模型可以是空间上的集合（如预测某一人群生存率的模型或者某个特定区域经济增长的模型），或者更多的基于某一区域的特定地点，预测整个研究区域的净变化。后者即单一影响模型，利用引力模型原理预测零星支出的分布，如Harris、Lowry、Cripps等的土地综合利用规划区域模型就提供了较全面的案例（Bracken，2008）。数学模型也可以分为确定模型和随机模型。确定模型以某种固定的关系为依据，如引力模型。而随机模型是以概率为基础的，它“通过对一些特定事件在既定区域和给定时间段内发生的统计学概率分析，以显示某一特定事件发生的可信度”（Loewenstein，1996）。

能够预测特定影响的数学模型有很多，参考各类 EIS（尤其是来自美国的）和著作（如 Bregman 和 Mackenthun，1992；Hansen 和 Jorgensen，1991；Rau 和 Wooten，1980；Sutter，1993；US Environmental Protection Agency，1993；Westman，1985），都能说明数学模型的广泛适用性。例如，Kristensen 等（1990）列出了 21 种关于湖泊中磷富集的数学模型。在社会-经济影响预测中，常常用到一个确定性数学模型——乘数模型（Lewis，1988），其具体示例见图 5.2。对地方、区域或国家经济系统进行投资，使得该经济系统产生的收益增加为原始投入资金的几倍。对基本模型校正，可以使其用于预测项目周期的各个阶段产生的各方面的收益和雇佣的影响（Glasson 等，1988）。乘数模型中投入-产出成分的细化（按工业类型）为经济影响的预测提供了一种独特且完善的方法，不过这需要大量的数据支持，而且有一定的局限性。

统计模型利用回归分析或成分分析等统计学技术来描述数据之间的关系、检验假设或者外推数据。例如，在污染监测研究中，可以使用基于水流速率和向下游的流经距离来确定污染物浓度的方法。比较污染点和控制点的情况，以决定不同监测数据的重要性。利用外推法可以得到监测数据之外的情况（如从污染物的高毒性情况外推出低毒性情况），或者是从易得数据外推出不易得的数据（如从鼠类毒性试验外推人类的反应）。

$$Y_r = \frac{1}{1-(1-s)(1-t-u)(1-m)} J$$

此公式中：

Y_r=区域内（r）收益（Y）水平的变化，单位英镑；

J=初始投入的收益（倍数）；

t=直接税收和国家保险资助的额外收益比例；

s=剩余收益比例（也就是未用于当地的花费）；

u=由于当地收入与就业的增加造成的转移支付（失业救济）的减少；

m=用于购买进口消费品的额外收益比例。

图 5.2　简化的用于预测当地经济影响的乘数模型

5.2.2.2　物理/建筑模型与实验方法

物理模型、图像模型或建筑模型可以再现项目与环境要素间相互作用的例证模型或比例模型。例如，一个比例模型（或计算机图表算法）可以预测开发对景观或建筑环境的影响。图片合成可以用来表现“受体”区域内项目所在地的环境现状，这些图像和与之叠加在一起的项目图像可以加深影响的直观印象。这类图像可以是项目模型的照片，也可能是一些简单的线形轮廓，或者是一些更加复杂的 3D 图像。若条件允许，可以使用更加复杂的展示方法，这种方法可以包括电脑动画图像，能从多种角度展现拟建区域的拟建项目的情况。

现场考察和实验室试验方法使用现有的数据（常常会通过专门的调查进行补充）来预测项目对受体的影响。现场考察在无约束的条件下实施，通常与影响预测的范围大致相同，如野外池塘中杀虫剂的试验。实验室试验的运行成本通常较低，但外推到自然系统状态下的效果可能不佳，如污染物对树苗生长影响的实验。

5.2.2.3 专家判断法和类比模拟模型

EIA中所有的预测方法都会用到专家判断。专家判断可以利用其他一些预测方法，如数学模型、因果网络图或者流程图，这些将在后面部分进行讨论。专家判断有时也可以使用类比模拟模型——以模拟情况为基础进行预测。它包括将拟议开发的影响与现存相似的开发进行类比；将某一地点与其他相似地点的环境状况进行对比；将未知的环境影响（如风轮机对无线电接收）与已知的环境影响（如其他形式的开发对无线电接收）进行对比。类比模拟模型能够通过现场调查、文献检索或相似项目的监测而得到发展。现如今，网络为潜在的比较项目提供了大量的信息。

5.2.2.4 其他预测方法

第4章中所讨论的各种影响识别方法对影响预测也是有参考价值的。例如，叠图法也可以用来预测空间影响。

5.2.2.5 预测方法选择

预测方法的性质与选择根据我们所研究的影响而变化，Rodriguez-Bachiller和Glasson（2004）确立了以下类型。

- 实体模拟影响：以数学模拟模型为主要影响预测方法。其应用范围包括大气和噪声影响。大气污染影响预测以“高斯扩散模型”方法为主，高斯扩散模型模拟开发项目含污染物的形状（Elsom，2009）。
- 虚体模拟影响：利用数学模型对虚拟对象进行影响预测。该类主要包括对陆地生态和景观影响的预测。陆地生态影响预测往往依赖于野外各种动植物的样本调查，这种预测中专家对所需样本的预测选择是非常重要的（Morris和Thurling，2001）。景观评价中专家的预测也非常重要，不过，简单的图片合成和GIS的利用对影响的预测也有一定的帮助（Knight，2009；Wood，2000）。图5.3为景观评价核心步骤。
- 混合模拟影响：在该类影响预测中模拟模型需要由技术含量较低的方法进行补充（有时会被代替）。交通影响预测中往往大量使用此类模型，但通常需要输入调查的样本值。社会-经济影响预测可以使用简单的流程图和专门用于经济影响预测的数学模型（见图5.2和图5.3），不过它们倾向于大量调查和专家判断，这在社会影响预测中表现的比较突出。

在选择预测方法时，评价人员应当考虑可利用资源对于所评价的任务是否适合（Lee，1987）。所选择的方法能够从可利用的资源（包括时间、数据、相关技术）中提取出我们所需要的东西（如各阶段中一定区域内的影响范围）吗？另外，在选择标准时还应该考虑标准的复现性（允许分析者采取不同方法）、一致性（同一方法运用于不同的项目，对其预测进行比较）和适应性。很多情况下，合适的方法不止一种。例如，在Rau和Wooten（1980）的近165页的课题中记录了用于大气质量影响预测的多种方法。表5.2提供了一些关于最初污染物扩散的预测方法的概述，污染物扩散加上大气相互作用，会降低大气质量，对人类产生不利影响。

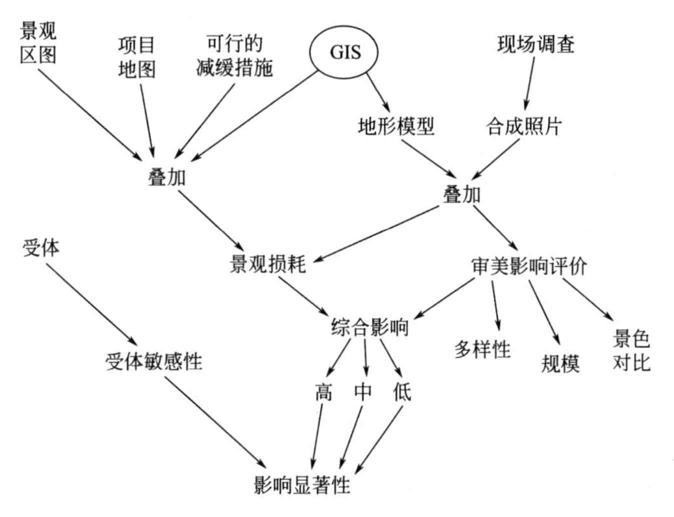

图 5.3　景观评价核心步骤

（资料来源：Rordriguez-Bachiller 和 Glasson，2004）

表 5.2　预测污染物扩散所使用方法的实例

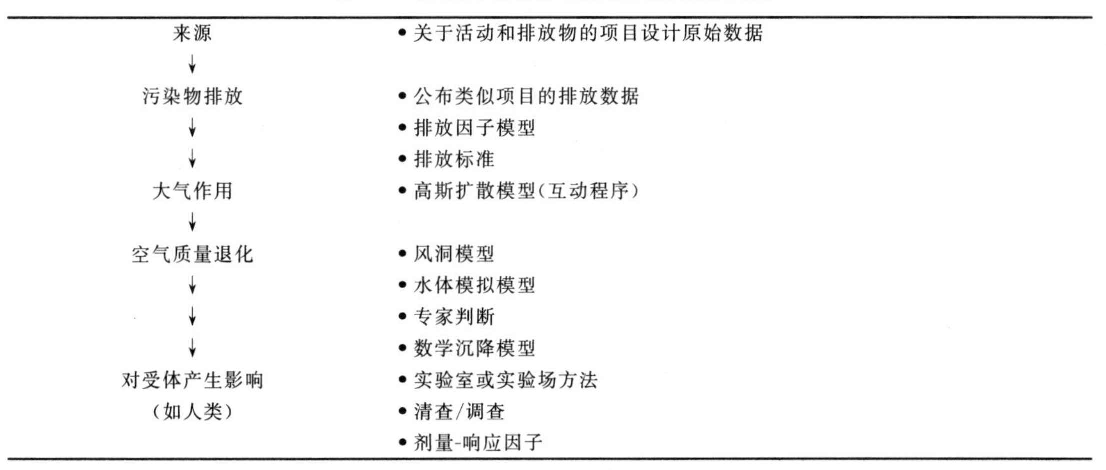

来源	• 关于活动和排放物的项目设计原始数据
↓	
污染物排放	• 公布类似项目的排放数据
↓	• 排放因子模型
↓	• 排放标准
大气作用	• 高斯扩散模型(互动程序)
↓	
空气质量退化	• 风洞模型
↓	• 水体模拟模型
↓	• 专家判断
↓	• 数学沉降模型
对受体产生影响	• 实验室或实验场方法
（如人类）	• 清查/调查
	• 剂量-响应因子

资料来源：VROM，1984；Rau 和 Wooten，1980。

在实践中，使用一些非正规的预测方法已经成为趋势，尤其是专家判断法（VROM，1984）。即使使用那些正规方法，也会使其应用趋向于简化，如直观影响评价中图片合成的应用或水体质量评价中的简单稀释和恒定扩散模型。然而，简化的方法也有一定的适用性，特别是在 EIA 的初期阶段，需要精密度不高但十分简便的方法。Lee（1987）的说明如下。

（1）一个专家就可能提供一个简洁的、定性的观点。

（2）要求专家证实他的观点：

① 口头或数学形式描述他所考虑的影响之间的关系；

② 简单说明支持其观点的经验性论据。

（3）如（2）所述，不包括其他专家也支持该观点。

（4）如（3）所述，不包括要求专家就支持理由、资格等达成共识。

（5）如（4）所述，不包括要求专家使用广泛认同的技术（如基于“Delphi”技术，Golden 等，1979）达成共识。

深入研究复杂方法是非常耗时耗材的，特别是很多复杂的方法因为被限定在某些特定的环境组成和物理过程中，因此只能用于一些相对类似的项目中。尽管我们强调简化的非正式方法，数学模拟模型在预测阶段仍有其应用范围。尤其是当按国家要求处理大量简单计算、影响过程依赖于时间、某些评价关系只能依据统计学概率来定义时。

5.2.2.6 环境影响评价预测中的因果关系网

在前面一些方法的叙述中提及过预测中使用的一种重要方法：因果关系。然而在 EIA 程序中，这种方法即使有也很少被表述出来。有一个好的案例可以说明运用因果网络（展示元素之间因果关系的图表）作为工具可以很容易地将原因与结果联系在一起，并且清楚地表述出来（Perdicoulis 和 Glasson，2006，2009）。因果关系网特殊的标识是元素间相互关系的图标表示以及这些相互关系的因果归因。这个关系网可以用节点与链路组成的抽象图表来表示。网络逻辑和因果逻辑都与 EIA 过程密切相关。抽象图假设：(a) 环境的独立元素与项目之间存在着联系（网络逻辑）；(b) 当一个元素受到影响时，与之关联的元素也将会受到影响（因果逻辑）(CEC，1999)。因果关系网的一个重要的优势是它们通过相互作用的顺序推演出不同层面受到影响的能力，这一事实给了它们“顺序图表”的别名（Canter，1996)。其两个缺点是：很难处理好时间与空间关系；存在复杂性不断增加的潜在风险(CEC，1999)。如果变得太过复杂的话，它们将会以一种特殊的方式被简化或者是整个被忽略掉（Goldvarg 和 Johnson-Laird，2001)。

这里举出了两个因果关系网的例子。这个有向图，或者称为定向图（图 5.4)，可能是因果关系网最简单的形式。元素都是以节点来表示，通过定向连接（单向箭头）联系在一起，而且这些元素都直接标有选配的附加信息。正负号分别表示协同变化（+）及相反变化（—)。

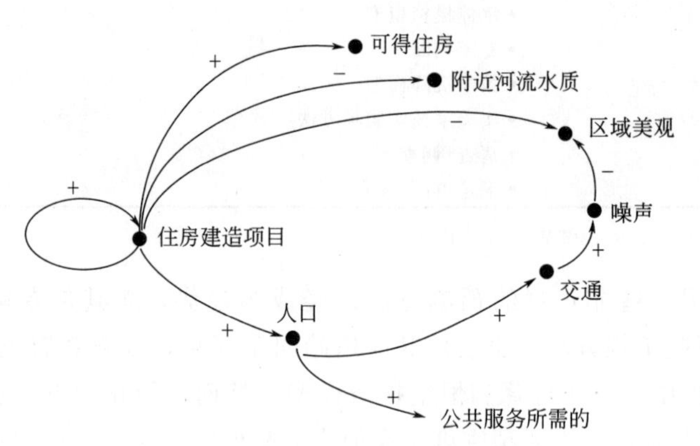

图 5.4 流程图或定向图示例

（资料来源：Canter，1996）

因果流程图是定向图，它们的元素在文章中可以以各种形状来呈现，大部分是矩形。原因关系以单向箭头来表示，通常不会含有定量的信息。总体来说，通常都是图形化的，比较精细，与简单的图表相比，可能在信息的丰富度上要稍微差一些。在 EIA 中，因果流程图通常都是被用于发展项目的影响识别与预测（CEC，1999；Glasson，2009)。图 5.5 提供了一个简单的发电站开发带来的当地社会-经济影响预测的流程图表。这个模型中，关键的决定性因素是项目具体的劳动力需求、当地经济条件、相关部门及开发者对相关问题（如培训）的政策以及当地招募及旅游的津贴。本地招募的比率是决定后续影响的关键性因素。

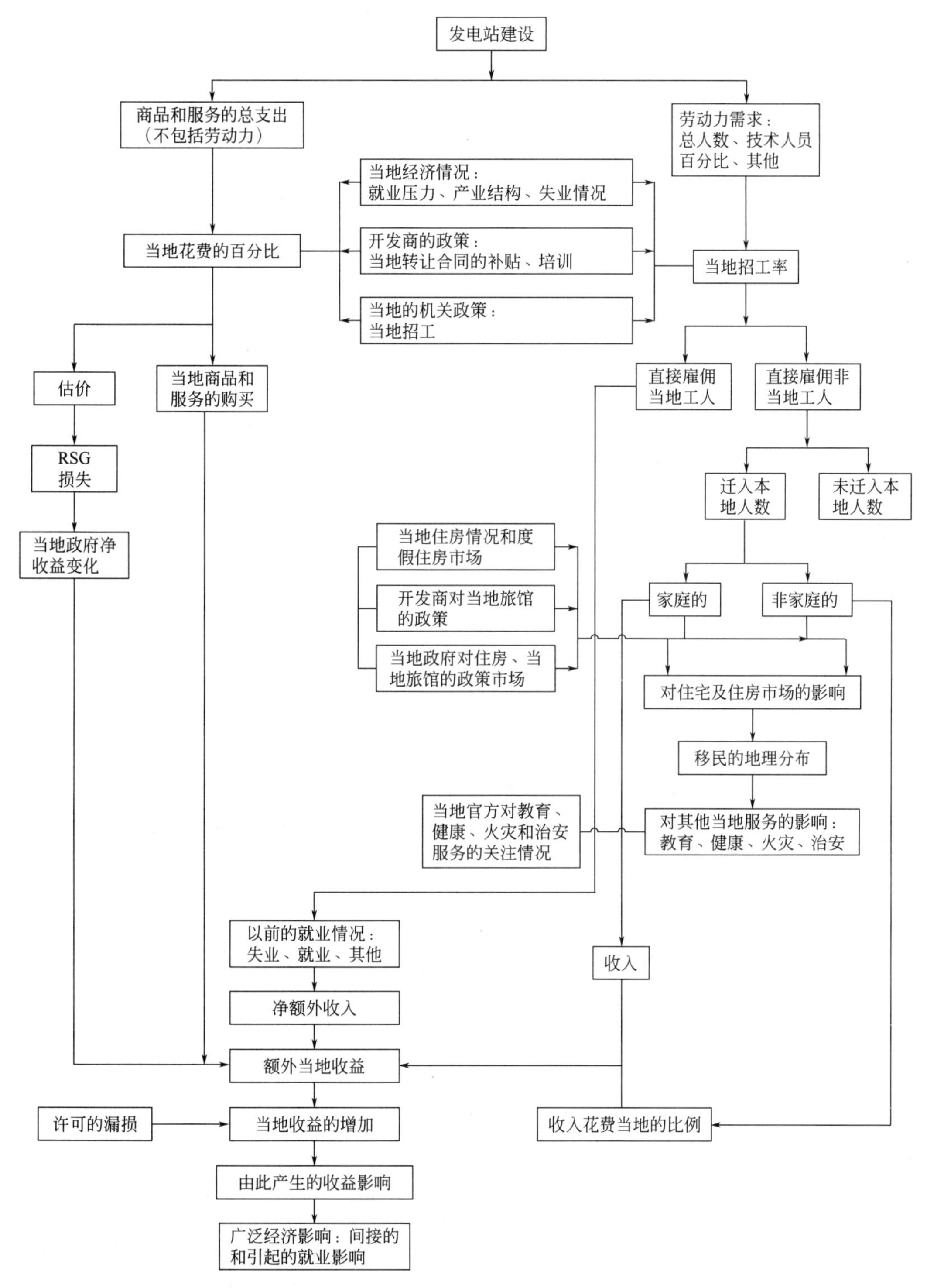

图 5.5　发电站提案对当地社会-经济影响预测的因果流程图

（资料来源：Glasson 等，1987）

5.2.3 不确定性

在环境影响报告中，环境影响的预测往往比在实际中看来更加确定。这反映出人们不关注可信度或者不愿接受不确定性。所有的预测都有不确定性，这种不确定性应该在 EIA 过程中得到认可（Beattic，1995；De Jongh，1988）。修正后的 EIA 指令（CEC，1997）及后来的英国法规（DETR，1999）都指出在考虑项目的潜在影响特征时都必须包括"影响概率"。从总体上说，在 EIA 程序中有许多不确定性的因素。在 Friend 和 Jessop（1977）与 Friend 和 Hickling（1987）有关战略选择的代表性著作中，他们将不确定性划分为三类：自然、社会和经济环境的不确定性（UE），导向价值的不确定性（UV）和相关决策的不确定性（UR）(图 5.6)。这些不确定性都会影响预测的准确性，但是 EIA 研究的重点主要是环境的不确定性。这些不确定性主要包括：项目和环境背景状况不准确或者信息不全，项目生命周期内一个或多个阶段产生的不可预测的变化，方法和模型的过度简单化处理及产生的误差。对社会-经济情况的预测是非常困难的，因为一些潜在的社会价值在一个项目周期（如 30～40 年）内可能会发生很大的变化。

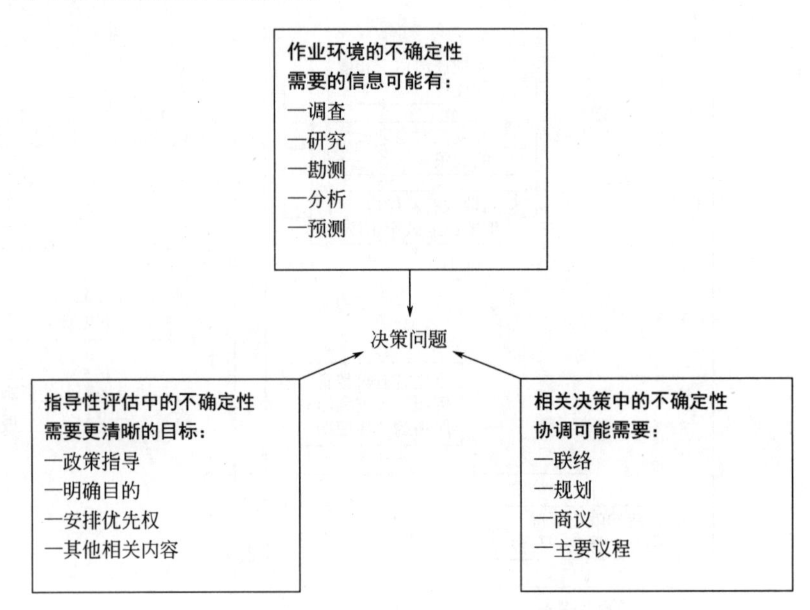

图 5.6 决策中的不确定性的类型

（资料来源：Friend 和 Hickling，1987）

EIA 中不确定性的预测可以通过多种方式来处理。应该明确设定基本的预测条件，说明预测结果的可信度和发生的概率，并指出真实结果在预测范围内的发生概率。例如，研究新工业项目产生的噪声，其 95%的置信区间是 65～70dB（A），这意味着仅 5%的噪声值不在此范围内。Tomlinson（1989）引起人们对预测中的可能性和置信度足够重视。可能性和置信度这一对孪生因素的表达方式常常是一致的。例如，在预测"石油重大泄漏产生的重大生态影响中"，其可能性和置信度都很高。然而，也可能发生这样的情况：基于低置信水平所预测的小概率事件，这可能比高置信水平的大概率事件更重要，因为低置信水平可能会在减缓措施范围之外，因而其重要性易被大家忽略。监测措施可能会对该种情况产生有效响应。对于项目的某一特定阶段，显示其影响的峰值或平均值也是很有用的，其往往与主体工程的建设阶段有关。

灵敏度分析可以用来评价变量间关系的一致性。如果输入 A 和输出 B 之间的关系是不论 A 怎么变化 B 几乎不变，则不需要更多的信息。然而，当影响变化明显时，就可能需要更多的信息。当然，对预测准确度最好的检验是决策后项目的执行结果。对于已建项目而言，这么做就太迟了，不过对于未来项目而言还是很有用的。相反，对相似项目的监测可以为正在进行的项目提供有用的信息。Holling（1978）认为“EIA 成果的核心是如何在不确定的情况下做出决策”，他推荐了一种适应性的 EIA 策略，即在项目的整个周期内对 EIA 进行定期跟踪评价。这种运用了“预测、监测及管理”方法的适应性评价是一种有价值的且灵敏度较高的方法，可以用于处理影响评价预测的内在问题。一种方法是要求将不确定性报告作为 EIA 程序的一个步骤，这类报告能将项目中的各种不确定性结合起来，并通过这种方法降低项目中的不确定性（不确定性基本上是不能去除的）。

5.2.4 目前预测数据的来源

政府及机构发布的文件，加之互联网的应用都为资源的获取提供了途径。表 5.3 简要概括了国际和英国的相关实例。

表 5.3 预测数据的一些来源：国际与英国

数据来源(国际)	内容简介
联合国环境规划署(UNEP)	UNEP《全球环境展望(GEO)项目 3》(2002 年，基于过去、现在及未来的视角)回顾了 1972～2002 年这一阶段的趋势，接着运用情景分析的方法探索了 2030 年可能会出现的情况，所运用的情景有：市场优先、政策优先、安全优先级和可持续优先。涉及的主题主要包括土地利用、人口、生物多样性、气候变化及水资源
经济合作与发展组织(OECD)	2008 年，OECD 在《关于 2030 年环境展望中》综述了经济与社会的发展会影响环境的变化。所考虑的环境影响包括气候变化、生物多样性的减少、水资源缺乏及污染的健康影响。这一研究可为减少有害的环境影响提供政策性建议
世界可持续发展企业委员会(WBCSD)	在《通往 2050 的路径：气候变化》(WBSCD，2005)中，委员会预测了社会-经济变化的大趋势并将其应用在了 2050 的项目中，作为中间年份的 2050 年是一个检查点，对美国、加拿大、中国、日本及欧盟等进行了区域性趋势的检查。预测趋势的范围主要包括：发电站、工业、制造业、移动产业、建筑业及消费者的选择等
政府间气候变化专门委员会(IPCC)	IPCC 通过将情境分析运用在气候变化、社会-经济及环境中，发现了到 2100 年为止社会-经济与温室气体排放之间的关系。IPCC 的数据分配中心拥有人口、人类发展、经济状况、土地利用、水、农业、能源及生物多样性的数据
欧洲环境署(EEA)	《欧洲展望声明及 2010 展望报告》(EEA，2010)包含了许多对于欧洲目前环境情形的评价，也包括对未来情形、全球化趋势的影响以及减少不利环境影响所需要采取的措施的评价。这份报告主要的关注点是 2020 年，但也包括直到 2100 年的对气候变化与洪水的情景分析。 在另一份报告中，EEA 着重考虑了环境与可持续之间的关系(EEA，2007)，综述了之前的一些预测报告及所用的技术方法，这些技术方法包括：头脑风暴、情景分析、新兴问题分析、预测及模型。同时也识别了一些与不确定性相关的主要趋势
数据来源(英国)	内容简介
食物和农村事物部门(DEFRA)	一个由 DEFRA 委任的基线调查项目(快速未来，2005)预测分析了可能对 DEFER 的工作区域产生影响的趋势及问题。其中优先进行 2015 年的长期预测及趋势识别，主要分类及次级分类如下。

续表

数据来源(英国)	内容简介
食物和农村事物部门(DEFRA)	•社会方面:人口统计、价值观、生活方式及文化。 •经济方面:产量、劳动力及贸易。 •政治方面:统治、政策、法律及规章制度。 •环境方面:生物圈、岩石圈、水圈及大气圈。 •科学方面:基础研究、技术及健康
能源和气候变化部门(DECC)	2050路径分析(DECC,2010)力图说明如何实现2050年温室气体减排80%这一目标。它考虑到了不同的经济部门,甚至涉及未来可能选用的能源及其温室气体的排放。它也通过计算方法(基于从"无措施"到"最大目标"一系列情景)展示了不同水平的能源使用所产生的不同影响
业务创新和技术部门(DBIS)	DBIS及其之前的一些组织已经实施了一些具有远见的项目,从2002年开始,已经对未来80年进行了预测。这些项目运用专家建议法给出了一些可能的结果,以帮助决策者进行决定。该报告中两个最近的例子如下所述。 •未来土地的利用,2010年2月。 •改善我们的生活:能源可持续性管理及环境建设,2008年9月
自然英格兰(NE)	自然环境的现状(主要是生物多样性及景观)强调了已存在的归趋及未来可能发生的变化(NE,2008)。《面向2060年的全球驱动力变化》(NE,2009)及英国《2060年的自然环境》(NE,2009)中通过四个场景展现了2060年世界可能会有的面貌。关于这些场景的话题非常多,包括:生长与繁荣、全球关系、居住、人口及人口统计、社会结构及融合、统治、可获得资源、应对气候变化、通信及运输、食物和农业、就业、创新的速度与方向、环境价值以及休闲和旅游业
环境署(EA)	EA网站上公布了许多环境方面的信息。它也包括了一些关于如何平行调查未来新兴问题的信息。EA认为此类平行调查与常规的预测是不同的,它基于较少的潜在变化与风险。比如,传统的气候变化预测在很大程度上是基于趋势预测,然后通过特定变量进行修正,而平行调查的方法则倾向于预测、检查较少变化的预测效应

5.3 评估

5.3.1 EIA程序中的评估

影响一旦被预测出来，就需评价它们的显著性，以此来告知决策者这些影响是否可以接受。显著性的标准包括影响的大小和可能性、影响的空间分布和时间范围、受影响环境的可恢复程度、受影响环境的价值、公众关注程度以及政策响应等。与预测类似，评估方法的选择与评价目标和可获得的资源有关。EIA程序的大多数阶段都涉及评估，但是所使用的方法却不尽相同。例如，根据替代方案的数目、信息集中水平或者相关团体的类型与数目来选择评估方法（如内部磋商和外部咨询）。

评估方法的种类很多，有简单的也有复杂的，有正规的也有非正规的，有定量的也有定性的，有集中的也有分散的（见 Maclaren 和 Whitney，1985；Voogd，1983）。目前许多EIA中影响的显著性评估都比较简单，而且更加关注实效性，常常借助经验和专家意见，而不是借助那些复杂深奥的分析。表5.4所示的是一个关键因子应用的实例，其地点是在EIA系统特别发达的澳大利亚西部（见第10章）。表5.4所列各因素也可以增加其可逆性范围。表 5.4 中的公共利益或公众认知也是一个重要的考虑因素，过去和现在对重大特殊问题和影响的认识可以提高其在评估中的地位。

表 5.4 环境重要性的决定因素

环西澳大利亚环境保护局运用专家判断法(通过环境影响评价中知识记忆经验的运用)针对一项提议是否会对环境产生重大影响做出决策。为了确定一个项目的发展是否会对环境产生重大影响,环境保护局需要考虑以下因素:

(a) 可能受影响环境的价值、敏感度和质量;
(b) 可能受影响的程度(强度、持续时间、等级和地理足迹);
(c) 影响(或变化)产生的结果;
(d) 环境受体应对变化的弹性;
(e) 其他项目的累积影响;
(f) 影响预测的置信度;
(g) 可以用来否决可被评估权利的行为、政策、指导方针、程序和标准;
(h) 公众认知;
(i) 现有的战略规划政策框架;
(j) 其他法定决策过程满足环保局的环境目标和环境影响评价的原则

资料来源：西澳大利亚环境保护局，2010。

最常用的评价方法就是将可能产生的影响与立法要求和标准（如空气质量标准、建筑法规）进行对比分析。表 5.5 列出了在英国用于评估项目交通噪声影响的标准。表 5.6 是欧盟委员会关于旅游开发通常使用的标准指南和环境优先权指南的实例介绍。当然，对于包括社会-经济方面在内的一些影响类型还没有明确的标准。社会-经济影响分析是评价“模糊”现象的典型例子，其重要与否的界限已经超出了通常的基于认知和事实的价值范畴。

表 5.5 英国项目交通噪声影响相关标准实例

- BS 7445 是描述和测量环境噪声的标准。包括三个部分:第一部分为数量与程序指南;第二部分为数据获取指南;第三部分为噪声限制应用指南。
- 噪声在指定频率下所测得的分贝(dB)值。只是一种测量声压的客观方法,使用校准的声压表进行测量。人们的听觉范围在 0～140dB 之间。

分贝噪声实例:

<40dB	安静的卧室
60dB	繁忙的办公室
72dB	7m 距离处以 60km/h 行驶的汽车
85dB	7m 距离处以 40km/h 行驶的重型货车(HGV)
90dB	持续暴露将会危害听觉
105dB	上空 250m 处飞行的喷气式飞机
120dB	极端痛苦

- 交通噪声即使分贝很低也是令人讨厌的。噪声来源于轮胎和地面的摩擦声、发动机、排气装置、刹车以及重型货车队。保养不好的公路和车辆以及拙劣的驾驶都会增加公路噪声。交通流量的不断增长和重型货车比例的不断上升也会增加噪声水平。一般,令人讨厌的是交通流量比例中 55dB 以上的噪声。人们对每 1dB 噪声的改变都很敏感(1dB 大约为 25%的交通流量的改变)。
- 交通噪声评价依据道路两旁 300m 内的影响进行评估。EIA 对车流量及相关地点(如人行道和体育用地)进行评估,并根据与线路的距离分为几个带:0～50m,50～100m,100～200m,200～300m。然后根据环境噪声的背景值(<50dB,50～60dB,60～70dB,>70dB)以及噪声的增加值(1～3dB,3～5dB,5～10dB,10～15dB,>15dB)进行分类。
- 地面噪声在距地面 1.7m、建筑物 1m、道路 10m 处测定,通常采用交通运输局的公路交通噪声计算(Department of Transport's Calculation of Road Traffic Noise,CRTN)方法,该方法通过测定 dB(A)$L_{A10,18h}$ 进行预测。在 6:00 与 24:00 之间所测得噪声等级数据中有 10%是超标的。地面的声压级大约要比建筑物外 10m 的声压级高 2dB。PPG13 采用 dB(A)$L_{Aeq,16h}$,它介于 7:00～23:00 之间。大多数交通噪声仪采用 dB(A)L_{A10},其近似转换关系如下:

$$L_{Aeq,16h}=L_{A10,18h}-2dB$$

续表

- DTP 推荐居住区的噪声最高上限为 72dB $L_{A10,18h}$（相当于 70dB $L_{Aeq,16h}$）。对于距公路 300m 以内的居民区，声压级增加 1dB(A)以上，若导致 $L_{Aeq,16h}$ 高于 67.5dB(A)，则需要对居民进行补偿。
- DTP 考虑对 30%轻微的、60%中等的和 90%严重的噪声做出修改。PPG13 认为有 5%是很重要的。
- 下面是居住区的 4 类噪声对应的情况：

	白天(16h)	夜晚(8h)	
A	<55 L_{Aeq}	<42 L_{Aeq}	申请不受限制
B	55～63 L_{Aeq}	42～57 L_{Aeq}	需要采取噪声控制措施
C	63～72 L_{Aeq}	57～66 L_{Aeq}	强烈反对
D	>72 L_{Aeq}	>66 L_{Aeq}	通常拒绝此类申请

对于夜间噪声(除非该噪声已经属于 D 类)日常单一事件(如重型货车的行驶)产生的噪声 L_{Aeq} 大于 82dB 的也应列入 C 类

资料来源：Bourdillon，1996。

表 5.6 欧盟指南在评价亚洲、加勒比海、太平洋等国旅游项目影响显著性中的应用实例

某种环境影响的显著性可以通过预测影响的大小，并通过与相关的环境标准或价值的对比进行评定。而对于旅游项目来说，影响的显著性也应该通过采用的环境优先权和社会认为的优先选项进行评价，不过这并不是一个量化的指标。需要特别注意环境优先选项以及由项目造成的直接影响。

环境标准

- 水质标准

—饮用水的供给标准[国家适用标准，见 1.3.2 部分，WHO(1982)饮用水水质指导方针 80/778/EEC 和 75/40/EEC]

—废水的排放标准(废水和渔业用水的国家适用标准，见 76/160/EEC 和 78/659/EEC)

- 国家和地方规划原则

—土地利用变更规则

—区域/地方土地利用规划(对受保护地区和沿海地区要进行特殊管理规划)

- 受国家法律保护的某些区域

—国家公园

—森林保护区

—自然保护区

—重要自然、历史或文化遗址

- 受国际公约保护的地区

—世界遗产公约

—拉姆塞尔(湿地)条约

- 对物种进行保护/维持，使其免于被出售或受到观光活动的危害

—国家立法

—国际协议

—濒危物种交易的 CITES 协议

环境优先权和优先选择

- 受项目影响的人的参与，以确定环境保护优先考虑的内容，包括：

—公众健康

—动、植物保护区(如文化/药用价值、视觉景观)

—为实施当地环境减缓措施而进行的技术培训

—饮用水供给的保护

—保护湿地/热带雨林的服务功能和产品，如猎取的野生动物、鱼类

—生产和工作中可持续性的收入问题(包括性别重要性，见 WID 指南)

- 环境保护的政府方针(包括适当地将国家环境研究/环境行动规划等目标结合)
- 旅游局和代表旅游承办商的商业协会的环境优先权

资料来源：CEC，1993。

社会-经济影响没有一个公认的标准。某种发展的预测性影响不一定可以被准确评估，

没有可以简单轻易地就适用于评价一个当地社会状况的标准。当地失业人数的减少可以看作是开发带来的正面影响，而当地犯罪频率的增加是其带来的负面影响，没有绝对的标准。对于经济影响的重要性的观点，如某一项目中当地工作者的比例与类型，通常都是政治的且专制的。然而，有时候在一个区域关于社会-经济的描述中，可能会识别出被称作"临界值或者步骤变化"的一些地方。例如，能够识别出可以对当地劳动力市场产生威胁并使其陷入困境，产生崩溃的情景的预测影响。也可能识别出区域预期收益大量流失的可能性，这很难被轻易接受。如果参与者可以定义一个可用于不同水平影响分析的标准将是非常有意义的，这至少可以为进行有见地的辩论提供基础。表 5.7 提供了一个核电站项目影响等级评价的示例。

表 5.7　一个较大能源项目当地影响等级评价方法示例：社会-经济影响方面

影响的类型	当地情况	不利影响	较小影响	中等影响	较大影响
人口统计的影响					
当地人口水平的变化	人口增长(2001～2009)	当地人口数量变化小于±0.25%	当地人口数量变化为±0.25%～1%	当地人口数量变化为±1%～2%	当地人口数量变化超过±2%
直接及非直接就业的影响					
当地经济中就业水平的变化	就业增长(ABI估计;2001～2007)	基于当地经济就业基线变化小于±0.25%	基于当地经济就业基线变化范围为±(0.25%～1%)	基于当地经济就业基线变化范围为±(1%～2%)	基于当地经济就业基线变化超过±2%
当地经济中失业水平的变化	最初的失业率(2010 年 6 月)	与初始失业水平相比变化小于±2%	与初始失业水平相比变化范围为±(2%～5%)	与初始失业水平相比变化范围为±(5%～10%)	与初始失业水平相比变化超过±10%
住房压力及其改善					
当地住宅存量的变化	住宅存量增长(2001～2008 年)	基于住宅存量基线变化小于±0.25%	基于住宅存量基线变化范围为±(0.25%～1%)	基于住宅存量基线变化范围为±(1%～2%)	基于住宅存量基线变化超过±2%

资料来源：作者，许多咨询研究案例。

社会-经济影响可以对评估特征进行归类，"谁盈利，谁亏损"（Glasson，2009；Vanclay，1999）。除了利用标准和法律要求外，影响的显著性（潜在的或明确的）评价都要根据影响的重要程度而定（有些评价要比其他的重要），其中也包括对判断的说明和应用。这种判断可以通过各种方法和模型使其更加合理，但是几乎所有内容都是主观的。Parkin（1992）认为这种判断是分析模型和直观模型的综合体。实际上，许多判断都是这个综合体的直观表现，但是这种判断并没有从分析中受益，它们是有缺陷的、矛盾的或具有偏见的。资源分配决策的社会影响非常广泛，以至于该决策很难从政策审查的不透明程序中"凸显"出来（Paikin，1992）。分析模型寻求引入一种合理的方法来进行评估。

环境影响评价的方法主要可以分为两类：一种是设定一个普遍而且实用的规范，使用单一的评估标准（货币）；另一种是以个人的效用尺度为基础，使用多种标准。费用-效益分析是用货币单位来表达影响的方法，可归于第一种；多目标分析、决策分析和目标实现这一类方法，可归于后一种。EIA的快速发展在一定程度上反映出费用-效益分析的局限性以及货币评估方法在环境影响评价中产生的问题。然而，经过了20多年的发展，环境费用-效益价值体系得到了更新（DoE，1911）。多标准/多属性方法包括计分和权重系统也不是没有问题。本书对这些方法进行了概述。在实践中，这两类方法会有很多交叉的方法，这在各类中都有所涉及。

5.3.2 费用-效益分析和货币评估技术

费用-效益分析（cost-benefit analysis，CBA）是项目和规划环评的一种方法，是运用货币价值来评估费用和效益（Lichfield等，1975）。局部分析方法就是其中的一种，如经济评估、成本最小化、成本效果法，这些方法仅仅考虑相关因子的一部分，或者全部规划和项目的一部分。经济评估通常被开发商局限在经济成本和投资产生的利润这一狭小的范围内。成本效果法就是选择一种方法使达到目标所花费的成本最小，如涉及一种成本最低的方式使海滩游泳用水满足CEC蓝旗（Blue Flag）标准。但是，当有很多个目标或者有为达到既定目标的多种行为时，成本效果法可能会有一些麻烦（Winpenny，1991）。

费用-效益分析的范围非常广泛，它要对项目进行长期（近期和更远期）和广泛的（考虑副作用）调查。它以福利经济为基础，试图涵盖所有相关的费用和效益，以此来评估项目的社会净收益。在20世纪60年代和70年代初期，这种方法在英国公共部门项目中得到广泛的应用，其中最显著的例子是伦敦第三机场项目（HMSO，1971）。费用-效益分析方法有以下几个步骤：项目定义，识别和计算项目的费用和效益，评估费用和效益，评估结果的折现和表达。某些步骤与EIA十分相似。基本的评估原则是用货币来测算，这是因为货币作为一般等价物，公众和决策者最容易理解它的价值，然后将费用和效益降低到相同资产或年度基础。成本和效益年流量常常折算成净现值（表5.8）。利率可以显示变化分析的灵活性。如果社会净收益减去成本是正的，那么可以支持该项目。不过，最终的结果往往都不太明确。结果表述应该区分为有形的和无形的损益，相应也应当允许决策者进行平衡选择。

表5.8 费用-效益分析：结果——有形资产和无形资产 单位：英镑

类别	选项1	选项2
有形资产		
年收益	B_1 B_2 B_3	b_1 b_2 b_3
年度总收益	$B_1+B_2+B_3$	$b_1+b_2+b_3$
年成本	C_1 C_2 C_3	c_1 c_2 c_3
年度总成本	$C_1+C_2+C_3$	$c_1+c_2+c_3$
“m”年每年X%[①]费用-效益的净现值贴现(NPDV)	D	E

续表

类别	选项 1	选项 2
无形资产		
无形资产可能包含的损益	I_1 I_2 I_3 I_4	i_1 i_2 i_3 i_4
无形资产合计(未贴现)	$I_1+I_2+I_3+I_4$	$i_1+i_2+i_3+i_4$

① 如 NPDV (Alt1)

$$D=\sum\left[\frac{B_1}{(1+X)^1}+\frac{B_1}{(1+X)^2}+\cdots+\frac{B_1}{(1+X)^n}+\frac{B_2}{(1+X)^1}+\cdots+\frac{B_2}{(1+X)^n}+\frac{B_3}{(1+X)^1}+\cdots+\frac{B_3}{(1+X)^n}\right]-$$
$$\sum\left[\frac{C_1}{(1+X)^1}+\frac{C_1}{(1+X)^2}+\cdots+\frac{C_1}{(1+X)^n}+\frac{C_2}{(1+X)^1}+\cdots+\frac{C_2}{(1+X)^n}+\frac{C_3}{(1+X)^1}+\cdots+\frac{C_3}{(1+X)^n}\right]$$

对于费用-效益分析方法，大家意见不一，有人支持（如 Dasgupta 和 Pearce，1978；Pearce，1989；Pearce 等，1989），也有人反对（如 Bowers，1990）。Hanley 和 Splash (2003) 提供了一个关于 CBA 与环境的有趣评论。CBA 确实还存在不少问题，包括识别、计算和无形资产的货币化等。很多环境影响就属于无形的，例如珍稀物种的消失、乡村景观的城市化和人类生活救助系统等。货币单位和非货币单位的不可兼容性给决策带来了困难 (Bateman，1991)。另一个问题是贴现率的选择，例如，是否应该在分析中使用较低的比率以防止未来费用和收益的快速贬值？比率的选择对于后代人使用资源的评价能产生深远的影响。对于使用单一的货币评价标准，还是潜在和基础的问题，因为它假设 1 英镑对任何人的价值都是一样的，不论他是流浪汉还是百万富翁，不论他是富人区的居民还是偏远乡村的穷人。此外，在评估整个社会福利的变化时，费用-效益分析也忽视了分配效率和费用与效益的聚集效应。

规划（收支）平衡表（planning balance sheet，PBS）是费用-效益分析法的一个变异方法，它尝试在识别、计算和评估受体之间的费用和效益分配方面超越费用-效益方法，同时也承认了无形影响货币化的难度很大。Lichfield（1975）等通过对比城镇规划的替代方案，使其不断发展。PBS 是一套基本的社会账目表，由参与各种交易的“生产者”和“消费者”构成。这种交易可能会产生负面影响，如飞机场的噪声（生产者）对当地居民（消费者）的影响；也可能会产生有益影响，如高速公路的修建（生产者）为驾驶者（消费者）节约了时间。对每一组生产者和消费者而言，每次交易中的费用和效益用货币等形式量化，并根据涉及的数量进行计算。将结果用表格形式表现出来，以便于决策者进行权衡，不过现在已经开始加入了一些关于期权的分配影响指南（图 5.7）。最近，Lichfield（1996）在考虑把 EIA 和 PBS 进一步整合成一种方法，他称之为社区影响评价（community impact evaluation，CIE）。

出于对 CBA 中部分“无形化”问题的响应，人们一直关注着货币评估技术的发展，以提高无形环境影响的经济衡量尺度（Barde 和 Pearce，1994；DoE，1991；Winpenny，1991）。广义上，这些技术可以分为直接和间接技术，它们参与环境优先权的度量，而非环境内在价值的计算。直接方法是直接衡量环境所具有的货币价值，例如，较好的大气质量或改良的自然景观。间接方法是通过“剂量-反应”关系的确定来衡量某种影响。表 5.9 中总结了已有的各种技术的直接方法和间接方法。这些技术对于开发行为或项目的经济总价值是

	计划A				计划B			
	效益		费用		效益		费用	
	资金	每年度	资金	每年度	资金	每年度	资金	每年度
生产者								
X	£$_a$	£$_b$	—	£$_d$	—	—	£$_b$	£$_c$
Y	i_1	i_2	—	—	i_3	i_4	—	—
Z	M_1	—	M_2	—	M_3	—	M_4	—
消费者								
X′	—	£$_e$	—	£$_f$	—	£$_g$	—	£$_h$
Y′	i_5	i_6	—	—	i_7	i_8	—	—
Z′	M_1	—	M_3	—	M_2	—	M_4	—

注：£为货币化了的效益和费用；M为评估的货币价值；i为无形化的。

图 5.7　规划（收支）平衡表的结构实例（PBS）

很有用的，它不仅包括使用价值（使用环境资源的优先人群，如河流用于捕鱼），也包括非使用价值（人们评估资源却不使用它，在今后可能会用到）。当然，这些技术也有其自身的不足。例如，人们在应用意愿调查评价法（CVM）时会有潜意识的偏好（一个典型的例子见 Willis 和 Powe，1998）。不过，通过各种环境特性确定价值，这类技术还能帮助我们加强理解，即环境的这些特性并不是“免费”的物品，也不应该被当作免费的物品来对待。

表 5.9　环境货币评价技术概要

直接家庭生产函数(HPF)

HPF 方法用来确定商品的支出，这些商品是环境特性的替代品或补充，并以此来评估该特性的变化。其类型包括以下几种。

- 防护支出法：为环境的改变找到各种替代情况而得到支出(例如将噪声屏蔽以用来评定宁静安逸的环境价值)。
- 旅行费用方法：去某特定地点(例如娱乐地点)旅行而产生的时间和金钱消费，这被认为是评估当地优良环境产生价值的一种方法(例如对该地点的利用而提升的价值)。

直接享乐评价法(HPM)

HPM 方法试图通过对现实市场的调查来确定环境属性的固有价值，这里的市场指具有交易环境属性的市场。这种方法有以下两种主要的类型。

- 住房土地价格：通过对有代表性的数据进行分析(例如不同地区住房的出售价格)，用来衡量“清洁的空气”和“静谧环境”之类的属性。
- 风险奖励：对某些高风险的行业支付额外的报酬，来审视高发病率和高死亡率职业(一般指人类)的价值变化。

模拟市场法

调查确定非交易类货物的个体价值。通过建立模拟市场来揭示人们对某种环境变化的价值评估。在调查的过程中可能会遇到以下两类问题。

- 条件成本法(CVM)：询问人们是否愿意为维持 X(例如一个好的风景、一个历史建筑)或阻止 Y 而支付费用(WTP)，或者是否愿意接受(WTA)损失 A 或忍受 B。
- 意愿排列法(CRM 或设定优先权)：要求人们为不同的环境要素排列优先权，然后参考市场中已有物质的真实价格(例如住宅价格)来估计该环境要素在市场交易的真实价格。

间接方法

间接方法试图通过判断“剂量”(例如大气环境改善)和效应(例如健康状况的改善)之间的关系来确定优先权，主要包括以下几种方法。

- 间接市场价格法：剂量-响应方法试图通过使用产出的市场价值来评估环境变化(例如渔场的油污染)所造成的影响(例如养鱼业损失的价值)。
- 重置成本法(replacement-cost approach)：利用受损资源复原或修复的成本测定环境修复(例如一座受污染、磨损和撞击的旧石桥)的效益。
- 促进生产法：在商品或服务市场存在的地方，环境影响可以通过由它带来的产量变化的价值来重现。这种方法广泛应用于发展中国家，这也是剂量-响应法的扩展

资料来源：DoE，1991；Winpenny，1991；Pearce 和 Markandya，1990；Barde 和 Pearce，1991，Nijkamp，2004。

5.3.3 计分和权重法与多标准方法

多标准/多属性方法试图克服费用-效益分析方法的不足，它们考虑了社会的多元论观点，由具有不同目标、对相关环境变化有不同估价的各“相关者”组成。大多数方法使用一些简单的计分和权重系统（有时也会误用），这种系统产生了许多争议。这里讨论一些成功案例中的关键因素，同时就分析人员可以使用的多标准/多属性方法范围做一个概述。

计分系统可以应用于定量或定性分析，其根据是所考虑影响的可获得的资料。Lee (1987) 给出了怎样用不同系统计算影响程度的实例 [表 5.10，在这个例子中，噪声用 L_{10}dB (A) 单位进行度量]。为了进行对比分析，这些系统试图使影响的计算标准化。当定量的数据无法得到时，我们就可以选择其他途径，例如使用字母（A、B、C 等）或描述性语言（不重要、重要、很重要）。

表 5.10　不同计分系统的比较关系

方法	替代方案				计分基础
	A(零作为)	B	C	D	
比值	65	62	71	75	绝对 L_{10}dB(A)的方法
间隔	0	−3	+6	+10	与 L_{10}dB(A)不同，以方案 A 为基础
顺次	B	A	C	D	根据 L_{10}dB(A)价值升序排列
二进制	0	0	1	1	0=小于 70L_{10}dB(A)，1=70L_{10}dB(A)及以上

注：基于 Lee (1987)。

权重系统试图确定不同类别影响间的相对重要性，这对一些类别的影响计算是有用的（例如，确定水污染影响的相对重要性及其对濒危花卉的影响）。权重系统可以为不同的影响分配权重（例如把总共 10 点分配给 3 种影响），但这由谁来实现呢？

多标准/多属性方法认可大多数人的观点和他们使用方法的权重。Delphi 方法也通过个体的权重来得到具有相同属性的组的权重。不过，在很多研究中，权重一般由技术小组给出。实际上，决策者们由于害怕破坏他们在谈判中的优势位置，而不愿意暴露他们的个人偏好。赋权的主观化不会影响权重的应用，但它的确强调了计分和权重系统分类的重要性，尤其在 EIA 中对权重来源的识别。只要可能，计分和权重就应该用于显示特定项目或替代方案的影响的平衡。例如，表 5.11 显示了一个方案中植物群落的影响与另一方案中噪声的影响之间平衡的主要问题。

表 5.11　权重、计分和平衡

影响	权重(w)	方案 A		方案 B	
		计分(a)	(aw)	计分(b)	(bw)
噪声	2	5	10	1	2
植物灭绝	5	1	5	4	20
空气污染	3	2	6	2	6
合计			21		28

一些对影响的计分和赋权方法在第 4 章中已经做了介绍。Leopold 矩阵除了能测定影响

的显著性（用1～10级别表示），还能测定影响的大小。这种矩阵方法经过一定的改进后，可以用来识别不同地理区域或不同团体间影响的分布（图5.8）。专家小组应用定量的环境评估（EES）方法和水资源评价方法（WRAM）确定了不同环境参数的权重。权重方法也可应用于叠图法中，以识别最具发展潜力的地区。这类方法的局限性在第4章中也有叙述。

环境要素组成	项目行为							
	建设阶段行为				运营阶段行为			
	A	B	C	D	a	b	c	d
第一组（例如本地大于45岁的人口）类别								
• 社会								
• 自然								
• 经济成分								
第二组（例如本地小于45岁的人口）类别								
• 社会								
• 自然								
• 经济成分								

图5.8　确定影响分布的简单矩阵

多目标/多属性类别中的其他方法包括决策分析法、目标核查表法（GAM）、多属性效用理论（MAUT）以及判断分析法。多目标决策分析（MCDA）技术已经成为解决自然资源管理问题的主要方法（Herath和Prato，2006；DCLG，2009），在EIA中的使用也越来越多，尤其是在大型项目中。MCDA是一种工具，尤其适用于单一规范方法不适合的案例。它允许决策者在解决无形的非货币化属性货币化过程中困难的同时，将环境价值、社会价值、经济价值与利益相关者的偏好整合在一起。习惯上，一个方法要明确目标，选择衡量目标的标准，明确提出替代方案，将标准尺度转化为单位尺度，加权其标准以反映其相对重要性，选择及应用数学方法将替代方案分级以及选择替代方案。评估可以使用各种定量、半定量和定性评价及调查方法（Figueira等，2005）。前者用数学方法来选择标准，这能够使计算结果像上面记录的一样。后者，定性的方法，主要是用主观判断及分级的方法，不使用数字，因此没有计算。总体上说，每种方法都有它们的长处。定量/半定量方法是系统的、可重复的，其输入、输出都是可以证实的。定性的方法在采集不同的信息上是有效的，尤其是隐性消息及见解。但是它们可重复性不好，它们通过提供文字叙述来解释结果。

尽管MCDA可能相当复杂，但它仍可以通过非常简洁的总结的方式呈现出来。这对决策者来说是非常有吸引力的，但是这种方式也会将问题过于简化，而且涉及平衡问题。常见的方法是使用交通信号灯的方式（绿色——正面，黄色——中立，红色——负面）来表示特定替代方案满足各种环境、社会及经济目标的程度。图5.9提供了一个大型项目替代点的定性MCDA的简易矩阵的图解示例［用了从深红到深绿的5种底纹（见二维码）］。在图形中设置了一些关键的元素及问题。据相关准则/标准，红色的底纹表示的是一种很不利的选择表现，而绿色表示的是一种有利的选择表现。在这个示例中，项目选择3是较好的选择。

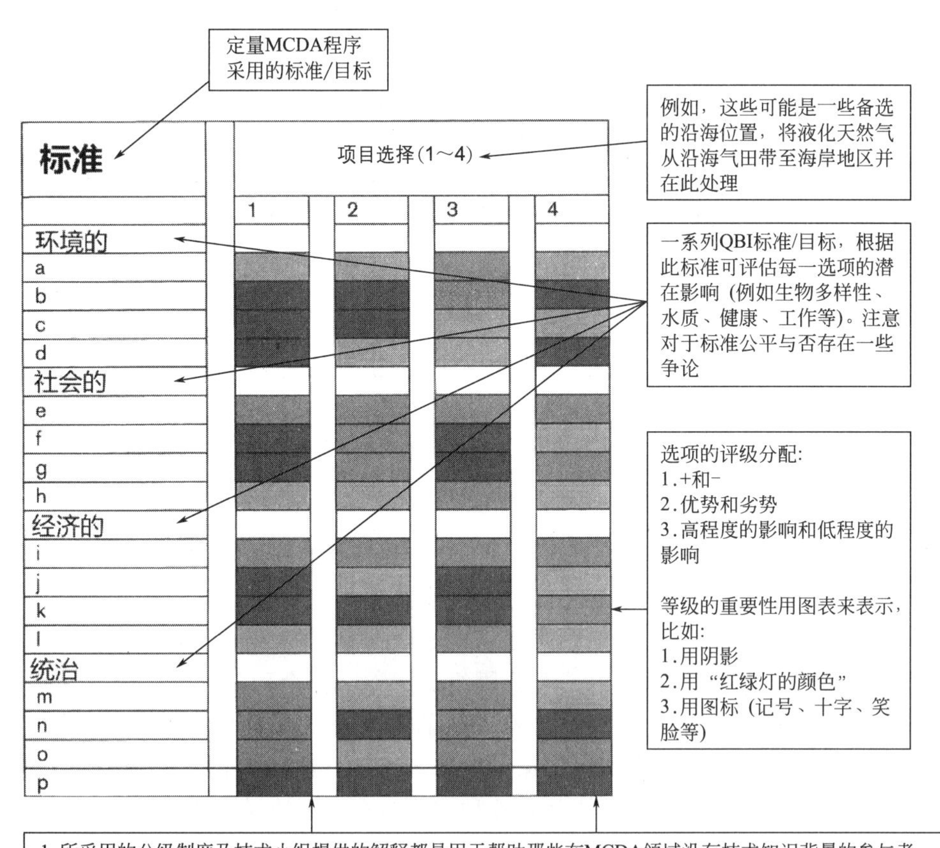

图 5.9　定量 MCDA 总结矩阵的图解示例

Hill（1968）把 GAM 发展为一种规划工具，用以克服 PBS 方法的缺陷。GAM 使项目/规划的目的和目标清晰明确，并通过测定达到既定目标的实现程度来完成对替代方案的评估。各种不同目标的存在导致了权重系统的形成。由于各利益团体存在政治地位上的不平等性，所以它们也应当被赋予权重。最终的结果是一个加权目标和加权效益/作用形成的矩阵（图 5.10）。目标和价值权重不仅用于评价经济效益，也用于社会利益规划的评价。同时，这种方法也为公众参与提供了机会。不幸的是，这种方法的复杂性限制了它的使用，并且使用的权重和目标常常反映的是分析人员的观点，而不是相关利益人群和政府机构的观点。

最后，简单介绍一下 Delphi 法，这种方法可以用来把不同参与者的观点整合到评价过程中。该方法是一种收集专家观点的方法，然后将专家的不同论点形成一致意见。在取得专家个人意见方面，这种方法具有一定的优势，它可以保证匿名，避免产生来自相同团体的意见失真的现象。与其他方法相比，这种方法也更加快捷，费用也更低。

在 EIA 中 Delphi 法有很多好的应用（Green 等，1989，1990；Richy 等，1985）。

目标描述	α			β		
相对重要性	2			3		
影响范围	相对重要性	成本	效益	相对重要性	成本	效益
a组	1	A	D	5	E	1
b组	3	H	J	4	M	2
c组	1	L	J	3	M	3
d组	2	—	J	2	V	4
e组	1	—	K	1	T	5
		Σ	Σ	Σ	Σ	

图 5.10　目标绩效矩阵（片段）

（资料来源：改编自 Hill，1968）

Green 等使用这种方法来评价 Bradford 地区著名盐厂的迁址和恢复的环境影响。这个方法包括拟建一个 Delphi 座谈小组，在盐厂的案例中，最初的小组由具有该项目背景的专家（例如规划师、旅游局官员）、议员、职工、学者、当地居民和商人等 40 人组成。这样做可以平衡专家与各利益团体的观点。Delphi 通常由三个阶段组成：①一份全面的问卷，要求小组的成员们识别重要的影响（积极的和消极的）；②第一轮的调查表，要求小组成员们对第一阶段识别的一系列影响按其重要性进行分级；③第二轮的调查表，要求小组成员们根据第一轮的结果对每种影响的重要性进行再评价。不过，这种方法也有一定的局限性。潜在使用者应该意识到在第一阶段拟建一个"平衡的"小组的困难性，并且要避免由于各阶段中小组成员的退出率变化和组织者的过分热情而导致的评估失真。至于另一种应用及其评判，见 Breidenhann 和 Butts（2006）以及 Landeta（2006）。

5.4 减缓措施

5.4.1 减缓措施的类型

减缓措施在欧盟 97/11 号指令中定义为"为避免、减少和尽可能减轻重大不利影响而设计的方法"（CEC，1997）。与此相似，美国环境质量委员会（US CEQ）在其实施《国家环境政策法》（NEPA）的条款中把减量化定义为：

不采用特殊的行动；限制拟议开发行为及其执行；补救、修复或建设影响的环境；在开发行为周期内体现并保留该行动；替代或提供可替代的资源或环境。（CEQ，1978）

表 5.12 列出了英国政府提供的减缓措施指南。该表并没有列出能够利用的减缓措施的类型，而是仅提供了几个叙述自然环境和社会-经济影响的实例。对于一些特殊影响类型的减缓措施，读者可以参考 Morris & Therivel（2009）和 Rodriguez-Bachiller 以及 Glasson（2004）提出的对特定影响类型的缓解措施的有用范围。对于与拟提议开发项目相类似的开发项目的 EISs 审查，也可以为开发提供一些有益的减缓措施。

表 5.12　英国程序指南中描述的减缓措施

重大不利影响一旦被识别出来，要采取一些措施来避免、减少或修正这些影响，例如：
(a) 方位规划。
(b) 技术措施，例如 （ⅰ）程序选择； （ⅱ）循环利用； （ⅲ）污染控制和处理； （ⅳ）密封装置（例如仓储码头）。
(c) 美学和生态学措施，例如 （ⅰ）基地； （ⅱ）图案、颜色等； （ⅲ）景观； （ⅳ）植树； （ⅴ）保护特殊栖息地或创建备用栖息地的措施； （ⅵ）记录古迹； （ⅶ）有历史意义的建设或场所的维护措施。
[评定] 减缓措施的可能效果

资料来源：DETR，2000。

影响的预测和评估可以反映负面影响的严重程度，在最严重的情况下，唯一有效的减缓措施就是全面放弃提议。当情况不太严重或比较正常时，减缓措施可以通过修改开发行为来消除各种影响。消除影响的方法实例如下：

- 将固体和液体废物回收再利用，或者把它们从对环境敏感的地方运送到其他地方以达到控制的目的。
- 使用指定的运输路线，仅选择白天作业，以避免公路建设和夜间作业对村落的干扰。
- 建立缓冲区并最小限度地使用有毒物质以避免对当地生态的影响。

一些负面影响很不容易消除，也不必完全消除。降低负面影响的方法实例如下：

- 对建筑物进行细致的设计，使用简洁的外观、本地的建材和柔和的色调以减少开发的视觉影响，使用景观来掩饰影响或者使其融入当地的环境。
- 在项目的建设期一般要雇用许多工人，利用建设地的旅馆和长途旅行客车，以减少对当地住房供给市场和公路的影响。
- 在干旱季节，利用淤泥地或阀来拦截泥沙、种植临时性覆盖物以及合理安排行动，以减少土壤侵蚀的沉积。

在项目生命周期内的一个或多个阶段中，某一特定环境要素可能会暂时性丧失或损坏。受影响的要素可能需要依据其受影响的程度来进行修正、复原和修复。例如：

- 建设期间用于物资存储的农业用地可能完全复原；用于沙粒筛选的土地可以修复为农业用地，但是这需要经历一个很长的时期，并且填埋使用的物质的性质还会产生混合影响。
- 由于公路项目建设而改道的河流可以不用改变其河流模式，或重新确定与其尽可能相似的河流模式。
- 当地社区为一个新的旅游设施提供路线，可以通过支线线路的建设来减轻交通对该地区的负面影响（这当然也会引入新的影响）。

总有一些负面影响是不能减轻的。在这种情况下，有必要对受到负面影响的人们进行补偿。例如：

- 对于公共娱乐空间或野生动植物栖息地的丧失，要在其他地方提供娱乐设施用地或创建自然保留地。

• 对于新建道路两旁住宅的私人空间、安静和安全环境的损失，应由开发商出资在道路两侧设置噪声屏障。

减缓措施应该在开发商和地方规划管理局（LPA）之间进行讨论，同时与“规划收益（planning gain）”连接起来。Fortlage（1990）曾谈及与这些讨论有关的一些潜在的复杂因素，并指出有必要区分减缓措施与规划受益：

在减缓措施提出前，开发商和当地规划机构必须就哪些是不利的或完全不利的影响达成一致，并确保补救工作的费用，否则整个过程就可能成为对双方都不利的讨价还价的游戏……

规划批准通常包括如下条件：要求开发商提供规划收益，并弥补开发导致的当地环境恶化。但是必须要清楚地区分这种收益是通过补偿负面影响的方式产生的，还是通过正常规划内容产生的。为了保障补偿措施的实施，当地规划机构可能会在规划条文中明确地提出补偿计划，因此开发商应该谨慎提出那些他们不打算实施的提案。

减缓措施必须以整体的、持续的方式进行规划，以保证它们的有效性，即保证相互间不冲突，也不仅仅是把问题从一种媒介转嫁给另一种媒介。

最近，在对英国 EIA 减缓措施实践的调查表明（DETR，1997），减缓措施有了相当明显的变化，如重视有形的措施，而忽视了对运行和管理的控制；缺少对建筑的影响和实行减缓措施后残留影响的关注。

表 5.13 提供了一个项目的减缓措施的宽泛分类，它是根据减缓的等级、减缓的层次和项目的阶段来划分的。减缓措施的等级水平与项目设计期间的决定有关，最下面两个级别反映的是，通过一些非物理的措施可以实现有效的减缓。减缓措施的层次关键在于预防而不是治理，至少从理论上讲，应首先尝试列表中较为靠前的选项，之后再尝试靠后的选项。与项目生命周期有关的项目各阶段已在第 1 章中进行了讨论，任何特殊的减缓措施都可以视为是三种方式的结合，例如，在建设阶段，根据自然法则的设计措施可以使项目对资源的影响最小化（DETR，1997）。

表 5.13 减缓措施的分类

减缓等级	减缓层次	项目的阶段
• 替代方案（战略、备选地点和程序） • 自然法则和设计措施 • 项目管理方措施 • 预留的减缓措施	• 从源头上避免 • 在源头上最小化 • 原位消除 • 受体消除 • 修复 • 补偿（in kind） • 其他补偿和改进措施	• 建设 • 试运营 • 运营 • 退役 • 修复、善后/安置

资料来源：DETR，1997。

5.4.2 EIA 过程中的减缓措施

正如表 5.13 中所示，与 EIA 的其他因子一样，减缓措施不只是局限在评价中。虽然逻辑上减缓措施是从评价和预测相关影响的重要性中得到的，但实际上，减缓措施与 EIA 程序的各个方面都有内在联系。对项目最初设计的修改可能是依据减缓措施的变化，也可能是依据早期 LPA 或当地社区的咨询结果。通过对替代方案、最初确定范围、本底数据研究和影响识别研究仔细思考，也可能分析得出进一步的减缓措施。虽然更深层次的研究可能会产生新的影响，但减缓措施可以减轻其他影响。因此，预测和评价工作可以集中在一定范围内的潜在影响上。

EIS 的每一部分通常都会讨论减缓措施（如大气质量、视觉质量、交通、就业）。这些讨论应当阐明减缓措施可以补偿相应负面影响的程度。表 5.14 对预期结果进行了明确的有效回顾总结，可能为规划提供一些有用的信息。该表认为，应该识别那些剩余的、未减缓的或仅仅部分减缓的影响，并根据其严重程度加以区分，例如可分为“次要影响”和“重大不可避免的影响”。

表 5.14　影响与减缓措施汇总

影响	减缓措施	减缓后的显著性水平
1.为化工厂提供用地，乡村将损失 400 英亩[①]优质农业用地	唯一完美的减缓措施就是放弃该项目	重大不可避免的影响
2.在高速公路盘山地段的额外货车和客车交通流量预计会在当前趋势的基础上增加 10%～20%	由开发商在高速公路上部投资建设货车慢速车道，将有助于缓解交通量，不过该影响只是局部的、短期的	重大不可避免的影响
3.项目可能会阻碍陆地物种从该项目东侧的山区向西侧湿地的迁移	应该沿着贯穿项目地点的现有河流，建立生物走廊，并进行维护。走廊的宽度最小应为 75 英尺[②]。应清理河床淤泥，并使河床通过偶发的淤泥得以加强。缓冲区应该种植土著的河岸植被，包括无花果树和柳树等	次要影响

① 1 英亩≈4046.86m^2。

② 1 英尺＝0.3048m。

缓解措施如果不加以实施，就没有什么价值可言。通过执行或管理计划可以证实有关缓解的承诺。这些计划可以采用全面的环境管理计划（EMP）或环境行动计划（EAP）的形式（见第 12 章）；它们还可能包括针对特定影响类型的更具体的辅助方案，例如以下有关实现社会经济利益的劳动力和采购计划部分。当项目获得批准并进入建设和运营阶段时，缓解和监测效果之间也有明显的联系。事实上，纳入一个明确的监测方案可能是最重要的缓解措施之一。第 7 章讨论的监测必须包括缓解措施和加强措施的有效性或其他情况。因此，在制定这些措施时必须要考虑到监测；缓解措施必须足够清晰以便对其有效性进行检查。个别缓解措施的使用也可以借鉴以往在其他相关和可比较的案例中开展的监测活动中有关有效性的经验。

5.4.3　提高潜在效益

英国指南（DETR 1997）也说明了在环评中纳入措施以创造环境效益的重要性。效益提升是环评中日益重要的一项内容，对大型项目来说尤其如此。这些改进可能包括生物物理行动——例如，在项目场地附近的废弃采石场建立一个自然保护区，该采石场已被开发商收购。然而，它们往往更多地是与社会经济问题相关的十个社会经济行动。一个项目可能给一个地区带来可观的利益，这些利益通常是社会经济方面的：在确定这些利益时，至少应该关注确保它们确实发生且效果不会被降低，并可能被加强。例如：

• 通过提供适当的技能培训项目、学徒制，以及“一站式”的当地招聘设施，可以鼓励和提高项目在当地的潜在就业效益。对于项目的施工阶段，这可能被整合到由开发人员和当地关键的利益相关者共同开发的施工人员管理计划中。该计划的实施具有明确的指标和目标，是项目主办区域将就业福利内化的重要手段，从而减少与重大项目相关的就业福利流失。

• 同样，采购管理计划可以帮助增加当地承包商从项目中获益的机会。它可以包括供应商活动，以提供有关当地与项目接触机会的信息，为开发商提供改进的当地供应信息，如当地供应商的在线数据库，并雇用一名供应链负责人，以增进与当地商业部门的交流互动。

• 在房地产领域，各种用地安排、建筑工地宿舍/校园可能对于当地有遗留用途。一个高质量的校园建筑选址可以用作教育、娱乐甚至酒店设施。空置的建筑工人住房可能为当地有需要的人提供有价值的可负担的住房。

• 类似的遗留地使用可能来自与建设相关的运输活动，也可能来自大型项目的运营。例如，为了减少项目施工现场的汽车穿梭，可能会与 LPAs 达成协议来建立一个公园和游乐设施，并在该地点连通公共汽车。在工作开始和结束时间之间，这些公共汽车可能被用于其他当地的需求——例如作为当地的校车。此外，如果设施地点位于城镇/城市附近，公园和游乐设施可以留在原地，在项目使用结束后供社区使用。

• 对于一些大型项目来说，在生活方面总有一些不太容易直接解决的间接干扰影响和变化。为了对这些影响提供补偿，CEGB 早在 1987 年发表了一份关于核电站 Sizewell B 的社会政策声明（CEGB 1987），其中包括提供补贴金用于各种慈善和社会中有益于当地社区和娱乐项目，这也是一系列“改善措施”中的一部分。这些措施受到当地社区的好评。在其他地方，更多是近年来，越来越多的人开始关注社区利益协议/社区影响协议的制定，为当地受影响的社区提供一系列缓解措施（Baxamusa 2008 年）。

就缓解措施来说，在评估过程的早期阶段，应在广泛咨询利益相关者的基础上，考虑加强环评的影响。应该明确规定改进措施，并在管理计划中确定以便后期的绩效监测。

5.5 总结

在 EIA 研究中影响预测和影响重要性评价常常组成了一个“黑箱”。在直觉上，它好像只是预先做好的专家意见，而不能为影响预测与评价阶段提供一个坚实的、经得起推敲的基础。其实，在这个阶段可供分析人员使用的方法很多，从简单到复杂的各种方法对分析人员来说都是有用的，这些方法可以加强分析的基础。减缓措施就是在这个阶段被引入的。不过，一些方法日趋完善，尤其是公共参与，减小了 EIA 过程中的一些风险。第 6 章将讨论公众参与的作用（很重要，但目前还很薄弱）、合理呈现的意义以及 EIS 评审和决策制定的方法。

问 题

以下问题旨在帮助读者关注本章的关键问题。

1. 影响的大小并不总与影响的重要性是同义的。从你的经验中举例说明这一点。

2. 将专家判断作为环境影响评价的关键预测方法进行案例评价。

3. 在环境影响评价中使用因果网络分析检查案例。

4. 如何在环评中处理影响预测的不确定性？考虑不同方法的优点。

5. 考虑图 5.9 所示的定性多准则决策分析（MCDA）的价值，对于不同的利益相关者群体，评估他们之间的权衡取舍。

5

Impact prediction, evaluation, mitigation and enhancement

5.1 Introduction

The focus of this chapter is the central steps of impact prediction, evaluation, mitigation and enhancement. This is the heart of the EIA process, although, as we have already noted, the process is not linear. Indeed the whole EIA exercise is about prediction. It is needed at the earliest stages, when a project, including its alternatives, is being planned and designed, and it continues through to mitigation and enhancement, monitoring and auditing. Yet, despite the centrality of prediction in EIA, there is a tendency for many studies to underemphasize it at the expense of more descriptive studies. Prediction is often not treated as an explicit stage in the process; clearly defined models are often missing from studies. Even when used, models are not detailed, and there is little discussion of limitations. Section 5.2 examines the dimensions of prediction (what to predict), the methods and models used in prediction (how to predict) and the limitations implicit in such exercises (living with uncertainty). It also includes a brief summary of some useful international and UK forecasting sources.

Evaluation follows from prediction and involves an assessment of the relative significance of the impacts. Methods range from the simple to the complex, from the intuitive to the analytical, from qualitative to quantitative, from formal to informal. CBA, monetary valuation techniques and multi-criteria/multi-attribute methods, with their scoring and weighting systems, provide a number of ways into the evaluation issue. The chapter concludes with a discussion of approaches to the mitigation of significant adverse effects. This may involve measures to avoid, reduce, remedy or compensate for the various impacts associated with projects. There is also a discussion of the increasingly considered aspect of impact enhancement—that is, where possible, developing the significant beneficial impacts of projects.

5.2 Prediction

5.2.1 Dimensions of prediction (what to predict)

The object of prediction is to identify the magnitude and other dimensions of identified change in the environment *with* a project or action, in comparison with the situation *without* that project or action. Predictions also provide the basis for the assessment of significance, which is discussed in Section 5.3.

One starting point to identify the dimensions of prediction in the UK is the *legislative requirements* (see Table 3.6, Parts 1 and 2). These basic specifications are amplified in guidance given in *Environmental assessment: a guide to the procedures* (DETR 2000) as outlined in Table 5.1. As already noted, this listing is limited on the assessment of socio-economic impacts. Table 1.3 provides a broader view of the scope of the environment, and of the environmental receptors that may be affected by a project.

Table 5.1 Assessment of effects, as outlined in UK guidance

Assessment of effects (including direct and indirect, secondary, cumulative, short-, medium-and long-term, permanent and temporary, and positive and negative effects of project)

Effects on human beings, buildings and man-made features

1 Change in population arising from the development, and consequential environmental effects.

2 Visual effects of the development on the surrounding area and landscape.

3 Levels and effects of emissions from the development during normal operation.

4 Levels and effects of noise from the development.

5 Effects of the development on local roads and transport.

6 Effects of the development on buildings, the architectural and historic heritage, archaeological features, and other human artefacts, e.g. through pollutants, visual intrusion, vibration.

Effects on flora, fauna and geology

7 Loss of, and damage to, habitats and plant and animal species.

8 Loss of, and damage to, geological, palaeotological and physiographic features.

9 Other ecological consequences.

Effects on land

10 Physical effects of the development, e.g. change in local topography, effect of earth-moving on stability, soil erosion, etc.

11 Effects of chemical emissions and deposits on soil of site and surrounding land.

12 Land-use/resource effects:

(a) quality and quantity of agricultural land to be taken;

(b) sterilization of mineral *resources*;

(c) other alternative uses of the site, including the 'do nothing' option;

Continued

(d) effect on surrounding land uses including agriculture;

(e) waste disposal.

Effects on water

13 Effects of development on drainage pattern in the area.

14 Changes to other hydrographic characteristics, e. g. groundwater level, watercourses, flow of underground water.

15 Effects on coastal or estuarine hydrology.

16 Effects of pollutants, waste, etc. on water quality.

Effects on air and climate

17 Level and concentration of chemical emissions and their environmental effects.

18 Particulate matter.

19 Offensive odours.

20 Any other climatic effects.

Other indirect and secondary effects associated with the project

21 Effects from traffic (road, rail, air, water) related to the development.

22 Effects arising from the extraction and consumption of material, water, energy or other, resources by the development.

23 Effects of other development associated with the project, e. g. new roads, sewers, housing power lines, pipelines, telecommunications, etc.

24 Effects of association of the development with other existing or proposed development.

25 Secondary effects resulting from the interaction of separate direct effects listed above.

Source: DETR 2000

Prediction involves the identification of potential change in indicators of such environment receptors. Scoping will have identified the broad categories of impact in relation to the project under consideration. If a particular environmental indicator (e. g. SO_2 levels in the air) revealed an increasing problem in an area, irrespective of the project or action (e. g. a power station), this should be predicted forwards as the baseline for this particular indicator. These indicators need to be disaggregated and specified to provide variables that are measurable and relevant. For example, an economic impact could be progressively specified as

direct employment→local employment→local skilled employment

In this way, a list of significant impact indicators of policy relevance can be developed.

An important distinction is often made between the prediction of the likely *magnitude* (i. e. size) and the *significance* (i. e. the importance for decision-making) of the impacts. Magnitude does not always equate with significance. For example, a large increase in one pollutant may still result in an outcome within generally accepted standards in a 'robust environment', whereas a small increase in another may take it above the applicable standards in a 'sensitive environment' (Figure 5. 1). This also highlights the distinction between *objective* and *subjective* approaches. The prediction of the magnitude of an impact should be an objective exercise, although it is not always easy. The determination of significance is often a more subjective exercise, as it normally involves value judgements.

As Table 1.4 showed, prediction should also identify *direct* and *indirect* impacts (simple cause-effect diagrams can be useful here), the *geographical extent* of impacts (e.g. local, regional, national), whether the impacts are *beneficial* or *adverse*, and the *duration* of the impacts. In addition to prediction over the life of a project (including, for example, its construction, operational and other stages), the analyst should also be alert to the *rate of change* of impacts. A slow build-up in an impact may be more acceptable than a rapid change; the development of tourism projects in formerly remote or undeveloped areas provides an example of the damaging impacts of rapid change. Projects may be characterized by non-linear processes, by delays between cause and effect, and the intermittent nature of some impacts should be anticipated. The *reversibility* or otherwise of impacts, their permanency, and their *cumulative* and synergistic impacts should also be predicted. Cumulative (or additive) impacts are the collective effects of impacts that may be individually minor but in combination, often over time, can be major. Such cumulative impacts are difficult to predict, and are often poorly covered or are missing altogether from EIA studies (see Chapter 12).

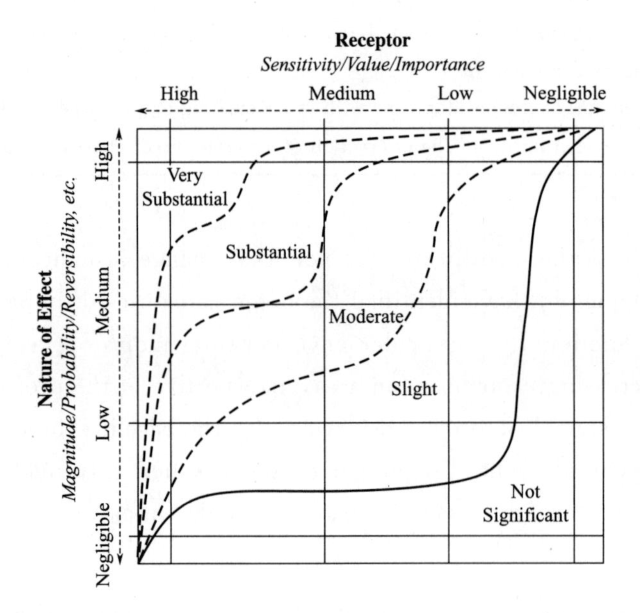

Figure 5.1 Example of generic significance matrix (e.g. 'impact magnitude' vs. 'environmental sensitivity')
Source: IEMA 2011

Another dimension is the unit of measurement, and the distinction between *quantitative* and *qualitative* impacts. Some indicators are more readily quantifiable than others (e.g. a change in the quality of drinking water, in comparison, for example, with changes in community stress associated with a project). Where possible, predictions should present impacts in explicit units, which can provide a basis for evaluation and tradeoff. Quantification can allow predicted impacts to be assessed against various local, national and international standards. Predictions should also include estimates of the *probability* that an impact will occur, which raises the important issue of uncertainty.

5.2.2 Methods and models for prediction (how to predict)

There are many possible methods to predict impacts; a study undertaken by Environmental Resources Ltd for the Dutch government in the early 1980s identified 150 different prediction methods used in just 140 EIA studies from The Netherlands and North America (VROM 1984). None provides a magic solution to the prediction problem.

> All predictions are based on conceptual models of how the universe functions; they range in complexity from those that are totally intuitive to those based on explicit assumptions concerning the nature of environmental processes ... the environment is never as well behaved as assumed in models, and the assessor is to be discouraged from accepting off-the-shelf formulae. (Munn 1979)

Predictive methods can be classified in many ways; they are not mutually exclusive. In terms of *scope*, all methods are *partial* in their coverage of impacts, but some seek to be more *holistic* than others. Partial methods may be classified according to type of project (e.g. retail impact assessment) and type of impact (e.g. wider economic impacts). Some may be *extrapolative*, others may be more *normative*. For extrapolative methods, predictions are made that are consistent with past and present data. Extrapolative methods include, for example, trend analysis (extrapolating present trends, modified to take account of changes caused by the project), scenarios (common-sense forecasts of future state based on a variety of assumptions), analogies (transferring experience from elsewhere to the study in hand) and intuitive forecasting (e.g. the use of the Delphi technique to achieve group consensus on the impacts of a project) (Green *et al*. 1989; Briedenhann and Butts 2006). Normative approaches work backwards from desired outcomes to assess whether a project, in its environmental context, is adequate to achieve them. For example, a desired socio-economic outcome from the construction stage of a major project may be 50 per cent local employment. The achievement of this outcome may necessitate modifications to the project and/or to associated employment policies (e.g. on training). Various scenarios may be tested to determine the one most likely to achieve the desired outcomes. Methods can also be classified according to their *form*, as the following types of model illustrate.

Mathematical and computer-based models

Mathematical models seek to represent the behaviour of aspects of the environment through the use of mathematical functions. They are usually based upon scientific laws, statistical analysis or some combination of the two, and are often computer based. The underpinning functions can range from simple direct input-output relationships to more complex dynamic mathematical models with a wide array of interrelationships. Mathematical models can be spatially aggregated (e.g. a model to predict the survival rate of a cohort population, or an economic multiplier for a particular area), or more locationally based, predicting net changes in detailed locations throughout a study area. Of the latter, retail impact models, which predict the distribution of retail expenditure using gravity model principles, provide a simple example; the comprehensive land-use locational models of Harris, Lowry, Cripps and others provide more holistic examples (Bracken, 2008). Mathematical models can also be divided into deterministic and stochastic models. Deterministic models, like the gravity model, depend on fixed relationships. In contrast, a stochastic model is probabilistic, and indicates 'the degree of probability of the occurrence of a certain event by specifying the statistical probability that a certain number of events will take place in a given area and/or time interval' (Loewenstein 1966).

There are many mathematical models available for particular impacts. Reference to various EISs (especially from the USA), and to the literature (e. g. Bregman and Mackenthun 1992; Hansen and Jorgensen 1991; Rau and Wooten 1980; Suter 1993; US Environmental Protection Agency 1993; Westman 1985) reveals the availability of a rich array. For instance, Kristensen *et al.* (1990) list 21 mathematical models for phosphorus retention in lakes alone. An example of a deterministic mathematical model, often used in socio-economic impact predictions, is the multiplier (Lewis 1988), an example of which is shown in Figure 5. 2. The injection of money into an economy—local, regional or national—will increase income in the economy by some multiple of the original injection. Modification of the basic model allows it to be used to predict income and employment impacts for various groups over the stages of the life of a project (Glasson *et al.* 1988). The more disaggregated (by industry type) input-output member of the multiplier family provides a particularly sophisticated method for predicting economic impacts, but with major data requirements, and limitations.

$$Y_r = \frac{1}{1-(1-s)(1-t-u)(1-m)} J$$

where

Y_r = change in level of income (Y) in region (r), in £

J = initial income injection (or multiplicand)

t = proportion of additional income paid in direction taxation and National Insurance contributions

s = proportion of income saved (and therefore not spent locally)

u = decline in transfer payments (e.g. unemployment benefits) which result from the rise in local income and employment

m = proportion of additional income spent on imported consumer goods

Figure 5. 2 A simple multiplier model for the prediction of local economic impacts

Statistical models use statistical techniques such as regression or principal components analysis to describe the relationship between data, to test hypotheses or to extrapolate data. For instance, they can be used in a pollution-monitoring study to describe the concentration of a pollutant as a function of the stream-flow rates and the distance downstream. They can compare conditions at a contaminated site and a control site to determine the significance of any differences in monitoring data. They can extrapolate a model to conditions outside the data range used to derive the model (e. g. from toxicity at high doses of a pollutant to toxicity at low doses) or from data that are available to data that are unavailable (e. g. from toxicity in rats to toxicity in humans).

Physical/architectural models and experimental methods

Physical, image or architectural models are illustrative or scale models that replicate some element of the project—environment interaction. For example, a scale model (or computer graphics) could be used to predict the impacts of a development on the landscape or built environment. Photo-montages can be used to show the views of the project site from the 'receptor' areas, with images of the project superimposed to give an impression of visual impact. The image could be a photograph of a model of the project, or a simple 'wire-line' profile of the project as it will appear to the viewer, showing just its skyline or a more sophisticated 3D impression. More sophisticated representations, where resources permit, can include 'fly-through' computer graphics, showing a proposed project in its proposed setting from a variety of perspectives.

Field and laboratory experimental methods use existing data inventories, often supplemented by special surveys, to predict impacts on receptors. Field tests are carried out in unconfined conditions, usually at approximately the same scale as the predicted impact; an example would be the testing of a pesticide in an outdoor pond. Laboratory tests, such as the testing of a pollutant on seedlings raised in a hydroponic solution, are usually cheaper to run but may not extrapolate well to conditions in natural systems.

Expert judgements and analogue models

All predictive methods in EIA make some use of expert judgement. Such judgement can make use of some of the other predictive methods, such as mathematical models and cause-effect networks or flow charts, discussed below. Expert judgement can also draw on analogue models-making predictions based on analogous situations. They include comparing the impacts of a proposed development with a similar existing development; comparing the environmental conditions at one site with those at similar sites elsewhere; comparing an unknown environmental impact (e. g. of wind turbines on radio reception) with a known environmental impact (e. g. of other forms of development on radio reception). Analogue models can be developed from site visits, literature searches or the monitoring of similar projects. The Internet now provides a wealth of information on potential comparative projects.

Other methods for prediction

The various impact identification methods discussed in Chapter 4 may also be of value in impact prediction. For example, overlays can also be used to predict spatial impacts.

Choice of prediction methods

The nature and choice of prediction methods do vary according to the impacts under consideration, and Rodriguez-Bachiller with Glasson (2004) have identified the following types:

• *Hard-modelled impacts*: areas of impact prediction where mathematical simulation models play a central role. These include, for example, air and noise impacts. Air pollution impact prediction has been dominated by approaches based on the so-called 'Gaussian dispersion model' which simulates the shape of the pollution plume from the development under concern (Elsom 2009).

• *Soft-modelled impacts*: areas of impact prediction where the use of mathematical simulation modelling is virtually non-existent. Examples here include terrestrial ecology and landscape. Terrestrial ecology depends very much on field sample survey for plant and animal species, where the expert's perception of what requires sampling plays an important role (Morris and Thurling 2001). Perception is also important in landscape assessment, but simple photomontages, and the use of GIS, can help in the prediction of impacts (Knight 2009; Wood 2000). Figure 5.3 provides an outline of key steps in landscape assessment.

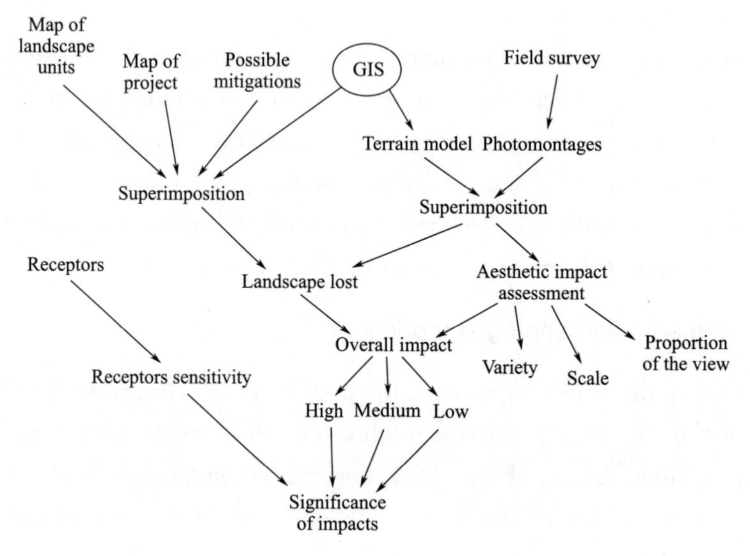

Figure 5.3 Key steps in landscape impact assessment

Source: Rodriguez-Bachiller with Glasson 2004

• *Mixed-modelled impacts*: areas of impact prediction where simulation modelling is complemented (and sometimes replaced) by more technically lower-level approaches. Traffic impacts make considerable use of modelling, but often with some sample survey input. Socio-economic impacts may use simple flow diagrams, and mathematical models (as in Figures 5.2 and 5.3) particularly for economic impacts, but they tend to build a great deal on survey methods and expert judgement. This is particularly so with regard to social impacts.

When choosing prediction methods, an assessor should be concerned about their appropriateness for the task involved, in the context of the resources available. Will the methods produce what is wanted (e.g. a range of impacts, for the appropriate geographical area, over various stages), from the resources available (including time, data, range of expertise)? In addition, the criteria of replicability (method is free from analyst bias), consistency (method can be applied to different projects to allow predictions to be compared) and adaptability should also be considered in the choice of methods. In many cases, more than one method may be appropriate. For instance, the range of methods available for predicting impacts on air quality is apparent from the 165 closely typed pages on the subject by Rau and Wooten (1980). Table 5.2 provides an overview of some of the methods of predicting the initial emissions of pollutants, which, with atmospheric interaction, may degrade air quality, which may then have adverse effects, for example on humans.

Table 5.2 Examples of methods used in predicting air quality impacts

Sources	• Original project design data on activity and emissions
↓	
POLLUTANT EMISSIONS	• Published emission data for similar projects
↓	• Emission factor models
↓	• Emission standards
Atmospheric interactions	• Gaussian dispersion models(interactive programmes)
↓	
DEGRADED AIR QUALITY	• Wind tunnel models
↓	• Water analogue simulation models
↓	• Expert opinion
↓	• Mathematical deposition models
EFFECTS ON RECEPTORS	• Laboratory or field experimental methods
e. g. humans	• Inventories/surveys
	• Dose-response factors

Sources: VROM 1984; Rau and Wooten 1980

In practice, there has been a tendency to use the less formal predictive methods, and especially expert opinion (VROM 1984). Even where more formal methods have been used, they have tended to be simple, for example the use of photo-montages for visual impacts, or of simple dilution and steady-state dispersion models for water quality. However, simple methods need not be inappropriate, especially for early stages in the EIA process, nor need they be applied uncritically or in a simplistic way. Lee (1987) provides the following illustration:

(a) a single expert may be asked for a brief, qualitative opinion; or

(b) the expert may also be asked to justify that opinion (i) by verbal or mathematical description of the relationships he has taken into account and/or (ii) by indicating the empirical evidence which supports that opinion; or

(c) as in (b), except that opinions are also sought from other experts; or

(d) as in (c), except that the experts are also required to reach a common opinion, with supporting reasons, qualifications, etc.; or

(e) as in (d), except that the experts are expected to reach a common opinion using an agreed process of consensus building (e.g. based on 'Delphi' techniques; Golden *et al*. 1979).

The development of more complex methods can be very time-consuming and expensive, especially since many of these models are limited to specific environmental components and physical processes, and may only be justified when a number of relatively similar projects are proposed. However, notwithstanding the emphasis on the simple informal methods, there is scope for mathematical simulation models in the prediction stage, especially where the assessment requires the handling of large numbers of simple calculations, some processes are time-dependent, and some assessment relationships can only be defined in terms of statistical probabilities.

Causal networks in EIA prediction

An important element in prediction, as noted in some of the previous methods, is the cause and effects relationship. But such relationships are often poorly expressed, if at all, in the EIA process. Yet there would seem to be a good case for the use of causal networks—diagrams that demonstrate causal relationships between their elements—as vehicles that can easily relate and transparently demonstrate cause and effects (Perdicoulis and Glasson 2006, 2009). The special identifiers of causal networks are a diagrammatic representation of relationships among elements and the attribution of causality to those relationships; the networks are abstract diagrams with nodes and links. Both the network logic and the causality logic of causal networks seem to tie in well with the EIA process. They do presuppose that (a) there are links between individual elements of the environment and projects (network logic) and (b) when one element is specifically affected this will have an effect on those elements that interact with it (causality logic) (CEC 1999). A key strength of causal networks is their capability to follow impacts to several levels through sequences of interactions—a fact that also gives them the alternative name of 'sequence diagrams' (Canter 1996). Two drawbacks are their difficulty in dealing with time and space, and the potential risk for increased complexity (CEC 1999). When they become too complex, they tend to be simplified in ad-hoc ways or ignored altogether (Goldvarg and Johnson-Laird 2001).

Two examples of causal networks are illustrated here. The digraph, or directed graph (Figure 5.4), is perhaps the simplest form of causal network. The elements are nodes and directional links (uni-directional arrows), with optional additional information marked directly on those elements. The + and − symbols are used in the sense of accompanying change (+) or reacting to change (−).

Cause and effect flow diagrams are directed graphs, but their elements are stated textually in various shapes—mostly rectangles. Causal relationships are marked by uni-directional arrows, usually carrying no quantitative information. In general, they are more elaborated graphically, but can be less rich in information than the simple digraphs. Cause and effect flow diagrams are mainly used in EIA for the identification and prediction of impacts related to development projects (CEC 1999; Glasson 2009). Figure 5. 5 provides a simple flow diagram for the prediction of the local socio-economic impacts of a power station development. Key determinants in the model are the details of the labour requirements for the project, the conditions in the local economy, and the policies of the relevant local authority and developer on topics such as training, local recruitment and travel allowances. The local recruitment ratio is a crucial factor in the determination of subsequent impacts.

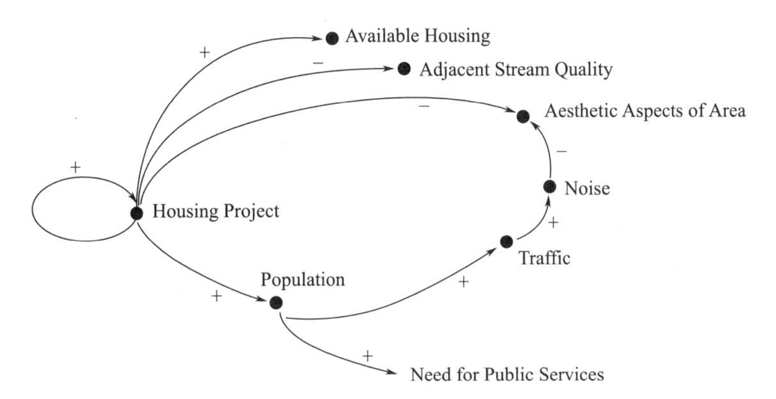

Figure 5. 4 Example of a digraph, or directed graph

Source: Adapted from Canter 1996

5. 2. 3 *Living with uncertainty*

Environmental impact statements often appear more certain in their predictions than they should. This may reflect a concern not to undermine credibility and/or unwillingness to attempt to allow for uncertainty. Yet all predictions have an element of uncertainty, and this should be acknowledged in the EIA process (Beattie 1995; De Jongh 1988). The amended EIA Directive (CEC 1997) and subsequent UK guidance (DETR 2000) include 'the probability of the impact' in the characteristics of the potential impact of a project that must be considered. There are many sources of uncertainty relevant to the EIA process as a whole. In their classic works on strategic choice, Friend and Jessop (1977) and Friend and Hickling (1987) identified three broad classes of uncertainty: uncertainties about the physical, social and economic working environment (UE), uncertainties about guiding values (UV) and uncertainties about related decisions (UR) (Figure 5. 6). All three classes of uncertainty may affect the accuracy of predictions, but the focus in an EIA study is usually on uncertainty about the environment. This may include the use of inaccurate and/or partial information on the project and on baseline-environmental conditions, unanticipated changes in the project during one or more of the stages of the life cycle, and oversimplification and errors in the application of methods and models. Socio-economic conditions may be particularly difficult to predict, as underlying societal values may change quite dramatically over the life, say 30—40years, of a project.

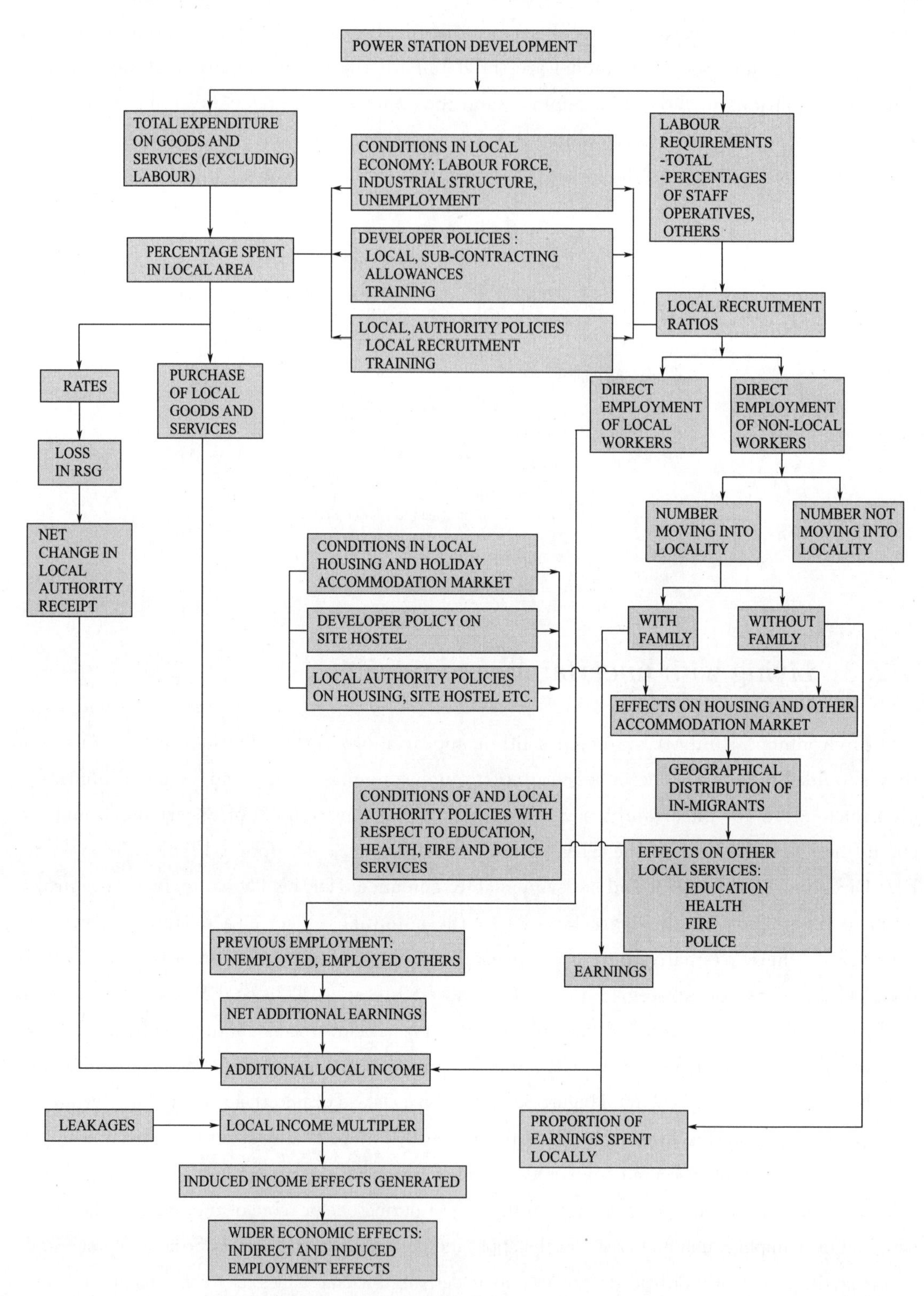

Figure 5.5 A cause-effect flow diagram for the local socio-economic impacts of a power station proposal

Source: Glasson *et al*. 1987

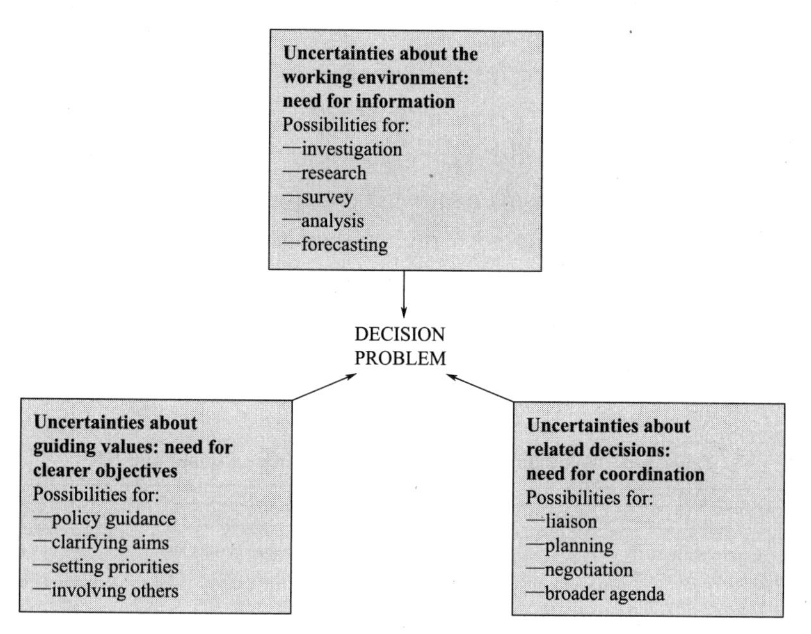

Figure 5. 6 The types of uncertainty in decision-making

Source: Friend and Hickling 1987

Uncertainty in EIA predictive exercises can be handled in several ways. The assumptions underpinning predictions should be clearly stated; issues of probability and confidence in predictions should be addressed, and *ranges* may be attached to predictions within which the analyst is *n* per cent confident that the actual outcome will lie. For example, scientific research may conclude that the 95 per cent confidence interval for the noise associated with a new industrial project is 65—70 dBA, which means that only 5 times out of 100 would the dBA be expected to be outside this range. Tomlinson (1989) draws attention to the twin issues of probability and confidence involved in predictions. These twin factors are generally expressed through the same word. For example, in the prediction 'a major oil spill would have major ecological consequences', a high degree of both probability and confidence exists. Situations may arise, however, where a low probability event based upon a low level of confidence is predicted. This is potentially more serious than a higher probability event with high confidence, since low levels of confidence may preclude expenditure on mitigating measures, ignoring issues of significance. Monitoring measures may be an appropriate response in such situations. It may also be useful to show impacts under 'peak' as well as 'average' conditions for a particular stage of a project; this may be very relevant in the construction stage of major projects.

Sensitivity analysis may be used to assess the consistency of relationships between variables. If the relationship between input A and output B is such that whatever the changes in A there is little change in B, then no further information may be needed. However, where the effect is much more variable, there may be a need for further information. Of course, the best check on the accuracy of predictions is to check on the outcomes of the implementation of a project after the decision. This is too late for the project under consideration, but could be useful for future projects. Conversely, the monitoring of outcomes of similar projects may provide useful information for the project in hand. Holling (1978), who believes that the 'core issue of EIA is how to cope with decision-making under uncertainty', recommends a policy of *adaptive EIA*, with periodic reviews of the EIA through a project's life cycle. Such adaptive assessment, using a 'predict, monitor and manage' approach is a valuable and sensible response to dealing with the inherent problem of uncertainty in prediction in impact assessment. Another procedural approach would be to require an *uncertainty report* as one step in the process; such a report would bring together the various sources of uncertainty associated with a project and the means by which they might be reduced (uncertainties are rarely eliminated).

Table 5.3 Some forecasting data sources: international and UK

Source(international)	Brief summary of content
United Nations Environment Programme(UNEP)	The UNEP *Global environment outlook* 3(GEO)project(2002:past,present and future perspectives)reviews trends over the period 1972 to 2002,and then uses scenarios to explore possible futures to 2030. The scenarios are:market first,policy first,security first,and sustainability first. Key topics covered include land use,population,biodiversity,climate change and water resources.
Organisation for Economic Co-operation and Development (OECD)	In *Environmental outlook to* 2030(OECD 2008),the organization reviews economic and social developments that will influence environmental changes. Key environmental impacts considered include climate change, biodiversity loss, water scarcity and the health impacts of pollution. The research can help in recommending policies to reduce detrimental environmental impacts.
World Business Council for Sustainable Development (WBCSD)	In *Pathways to* 2050:*energy and climate change*(WBCSD 2005),the Council identifies megatrends in socio-economic variables and uses them to project to 2050, with 2025 as an intermediate checkpoint. Regional trends are examined in the USA and Canada,China,Japan and the EU. Identified megatrends include:power generation,industry and manufacturing,mobility,buildings and consumer choices.
Intergovernmental Panel on Climate Change(IPCC)	The IPCC undertakes scenario analysis for climate change,socio-economics and the environment,assessing the interactions between socio-economic parameters and greenhouse gas emissions up to 2100. The IPCC Data Distribution Centre(DDC)holds data on population and human development,economic conditions,land use,water,agriculture,energy and biodiversity.
European Environment Agency(EEA)	The *European outlook state and outlook* 2010 report(EEA 2010)includes a set of assessments of the current state of Europe's environment,likely future state,effects of global megatrends, and actions needed to reduce detrimental environmental effects. The report primarily focuses on 2020,but scenarios to 2100 are included for climate change and flooding. Another report,*EEA research foresight for environment and sustainability*(EEA 2007),reviews previous forecasting reports and techniques,including:brainstorming, scenarios,emerging issues analysis,forecasts and modelling. It also identifies some key trends,and associated uncertainties.
Source(UK)	**Brief summary of content**
Department for Food and Rural Affairs(DEFRA)	A DEFRA-commissioned *Baseline scanning project*(Fast Futures 2005)identified trends and emerging issues that could affect the DEFRA work area. Longest term projections were made to 2051 onwards;trends were also prioritised. Key categories and sub categories were:

Continued

Source(UK)	Brief summary of content
Department for Food and Rural Affairs(DEFRA)	• Social:demographics,values,lifestyles and culture. • Economic:production,labour and trade. • Political:governance,policies,laws and regulations. • Environmental:biosphere,geosphere,atmosphere and hydrosphere. • Science:basic research,technology and health.
Department of Energy and Climate Change(DECC)	A 2050 *pathways analysis*(DECC 2010)seeks to illustrate how an 80% reduction in greenhouse gas emissions by 2050 can be achieved. It considers different economic sectors,possible future energy choices and subsequent emissions. It also includes a calculator tool to show impacts of different levels of energy use—with scenarios ranging from' little effort' to 'extremely ambitious'.
Department for Business Innovation and Skills(DBIS)	DBIS and its predecessors have been undertaking foresight projects, since 2002, 'forecasting' for periods of up to 80 years into the future. The projects use expert advice to outline a range of possible outcomes,to assist decision makers. Two recent examples of reports include: • *Land use futures*,February 2010. • *Powering our lives:sustainable energy management and the built environment*, November 2008.
Natural England(NE)	*The state of the natural environment*(NE 2008)highlights existing trends and likely future changes, primarily in terms of biodiversity and landscapes. *Global drivers of change to* 2060(NE 2009)and *England's natural environment in* 2060(NE 2009)include the development of four scenarios of how the world might look in 2060. Topics covered in the scenarios are wide ranging,including:growth and prosperity,global relations,settlements,population and demographics,social structure and cohesion,governance,resource availability,response to climate change,mobility and transport,food and farming,employment skills,pace and direction of innovation,environmental values,and leisure and tourism.
Environment Agency(EA)	The EA website includes an array of environmental information. It also includes information on horizon scanning the future for emerging issues. The EA sees such horizon scanning as different from routine forecasting in that it is predicated on less certain potential changes and risks. For example, whereas 'traditional' climate change forecasting is based largely upon extrapolation of trends, as modified by certain variable, horizon scanning involves methods of divining and examining less predictable effects.

5.2.4 Some current data forecasting sources

Published government and agency documents, plus of course the Internet, provide access to a wide range of sources. Some examples of key international and UK sources are briefly noted in Table 5.3.

5.3 Evaluation

5.3.1 *Evaluation in the EIA process*

Once impacts have been predicted, there is a need to assess their relative significance to inform decision-makers whether the impacts may be considered acceptable. Criteria for significance include the magnitude and likelihood of the impact and its spatial and temporal extent, the likely degree of the affected environment's recovery, the value of the affected environment, the level of public concern, and political repercussions. As with prediction, the choice of evaluation method should be related to the task in hand and to the resources available. Evaluation should feed into most stages of the EIA process, but the nature of the methods used may vary, for example, according to the number of alternatives under consideration, according to the level of aggregation of information and according to the number and type of stakeholder involved (e. g. 'in-house' and/or 'external' consultation).

Evaluation methods can be of various types, including simple or complex, formal or informal, quantitative or qualitative, aggregated or disaggregated (see Maclaren and Whitney 1985; Voogd 1983). Much, if not most, current evaluation of significance in EIA is simple and often pragmatic, drawing on experience and expert opinion rather than on complex and sophisticated analysis. Table 5. 4 provides an example of key factors used in Western Australia, where there is a particularly well-developed EIA system (see Chapter 10 also). To the factors in Table 5. 4 could also be added scope for reversibility. The factor of public interest or perception ((h) in Table 5. 4) is an important consideration, and past and current perceptions of the significance of particular issues and impacts can raise their profile in the evaluation.

Table 5. 4 Determinants of environmental significance

A decision by the(West Australian)EPA(Environmental Protection Authority)as to whether a proposal is likely to have a significant effect on the environment is made using professional judgement, which is gained through knowledge and experience in the application of EIA. In determining whether a proposal is likely to have a significant effect on the environment, the EPA may have regard to the following:
(a)*the values, sensitivity and quality of the environment which is likely to be impacted*;
(b)*the extent(intensity, duration, magnitude and geographic footprint)of the likely impacts*;
(c)*the consequence of the likely impacts(or change)*;
(d)*resilience of the environment to cope with change*;
(e)*the cumulative impact with other projects*;
(f)*level of confidence of the impacts predicted*;
(g)*objects of the Act, policies, guidelines, procedures and standards against which a proposal can be assessed*;
(h)*the public concern*;
(i)*presence of strategic planning policy framework*; or
(j)*the extent to which other statutory decision-making processes meet the EPA's objectives and principles for EIA*.

Source: West Australian Environmental Protection Authority 2010

The most formal evaluation method is the *comparison of likely impacts against legal requirements and standards* (e. g. air quality standards, building regulations). Table 5. 5 illustrates some of the standards that may be used to evaluate the traffic noise impacts of projects in Britain. Table 5. 6 provides an example of more general guidance on standards and on environmental priorities and preferences, from the European Commission, for tourism developments. Of course, for some type of impacts, including socio-economic, there are no clear-cut standards. Socio-economic impacts provide a good example of 'fuzziness' in assessment, where the line between being significant or not significant extends over a range of values that build on perceptions as much as facts.

Socio-economic impacts do not have recognized standards. There are no easily applicable 'state of local society' standards against which the predicted impacts of a development can be assessed. While a reduction in local unemployment may be regarded as positive, and an increase in local crime as negative, there are no absolute standards. Views on the significance of economic impacts, such as the proportion and types of local employment on a project, are often political and arbitrary. Nevertheless it is sometimes possible to identify what might be termed *threshold or step changes* in the socio-economic profile of an area. For example, it may be possible to identify predicted impacts that threaten to swamp the local labour market, and that may produce a 'boom-bust' scenario. It may also be possible to identify likely high levels of leakage of anticipated benefits out of a locality, which may be equally unacceptable. It is valuable if the practitioner can identify possible criteria used in the analysis for a range of levels of impacts, which at least provides the basis for informed debate. Table 5. 7 provides an example assessing impact magnitude from nuclear power station projects.

Table 5. 5 Examples of standards in relation to impacts of projects on traffic noise in Britain

- BS 7445 is the standard for description and measurement of environmental noise. It is in three parts: Part 1: Guide to quantities and procedures; Part 2: Guide to acquisition of data; and Part 3: Guide to application of noise limits.
- Noise is measured in decibels(dB) at a given frequency. This is an objective measure of sound pressure. Measurements are made using a calibrated sound meter. Human hearing is approximately in the range 0—140 dBA.

dB Example of noise

<40 quiet bedroom

60 busy office

72 car at 60 km/h at a distance of 7 m

85 Heavy goods vehicle(HGV) at 40 km/h at a distance of 7 m

90 hazardous to hearing from continuous exposure

105 jet flying overhead at 250 m

120 threshold of pain

- Traffic noise is perceived as a nuisance even at low dB levels. Noise comes from tyres on the road, engines, exhausts, brakes and HGV bodies. Poor maintenance of roads and vehicles and poor driving also increase road noise. Higher volumes of traffic and higher proportions of HGVs increase the noise levels. In general, annoyance is proportional to traffic flow for noise levels above 55 dB(A). People are sensitive to a change in noise levels of 1 dB(about 25% change in flow).
- Assessment of traffic noise is assessed in terms of impacts within 300 m of the road. The EIA will estimate the number of properties and relevant locations(e. g. footpaths and sports fields) in bands of distance from the route: 0—50 m, 50—100 m, 100—200 m, 200—300 m, and then classify each group according to the baseline ambient noise levels(in bands of <50, 50—60, 60—70, >70 dB(A)) and the increase in noise(1—3, 3—5, 5—10, 10—15 and >15 dB(A)).
- Façade noise levels are measured at 1. 7 m above ground, 1 m from façade or 10 m from kerb, and are usually predicted using the Department of Transport's Calculation of road traffic noise(CRTN), which measures dB(A)$L_{A10,18\,h}$.

Continued

This is the noise level exceeded 10% of the time between 6:00 and 24:00. Noise levels at the façade are approximately 2 dB higher than 10 m from the building. PPG 13 uses dB(A) $L_{Aeq,16\ h}$. This is between 7:00 and 23:00. Most traffic noise meters use dB(A) L_{A10}, and an approximate conversion is:

$L_{Aeq,16\ h} = L_{A10,18\ h} - 2dB$.

- The DTP recommends an absolute upper limit for noise of 72 dB(A) $L_{eq,18\ h}$ (= 70 dB(A) $L_{A10,18\ h}$) for residential properties. Compensation is payable to properties within 300 m of a road development for increases greater than 1 dB(A) which result in $L_{A10,18\ h}$ above 67.5.

- The DTP considers a change of 30% slight, 60% moderate and 90% substantial. PPG 13 considers 5% to be significant.

- There are four categories of noise in residential areas:

Day (16 h)	*Night* (8 h)	
A $<55\ L_{Aeq}$	$<42\ L_{Aeq}$ Not determining the application	
B 55—63 L_{Aeq}	<42—57 L_{Aeq}	Noise control measures are required
C 63—72 L_{Aeg}	57—66 L_{Aeq}	Strong presumption against developer
D $>72\ L_{Aeq}$	$>66\ L_{Aeg}$	Normally refuse the application

For night-time noise, unless the noise is already in category D, a single event occurring regularly (e.g. HGV movements) where $L_{Aeg} > 82dB$ puts the noise in category C.

Source: Bourdillon 1996

Table 5.6 Example of EC guidance on assessing significance of impacts for tourism projects for Asian, Caribbean and Pacific countries

The significance of certain environmental impacts can be assessed by contrasting the predicted magnitude of impact against a relevant environmental standard or value. For tourism projects in particular, impact significance should also be assessed by taking due regard of those environmental priorities and preferences held by society but for which there are no quantifiable objectives. Particular attention needs to be focused upon the environmental preferences and concerns of those likely to be directly affected by the project.

Environmental Standards

- Water quality standards
  - —potable water supplies (*apply country standards; see also Section* 1.3.2, *WHO* (1982) *Guidelines for Drinking Water Quality Directives* 80/778/*EEC and* 75/440/*EEC*)
  - —wastewater discharge (apply country standards for wastewaters and fisheries; see also 76/160/EEC and 78/659/EEC)
- National and local planning regulations
  - —legislation concerning change in land use
  - —regional/local land-use plans (particularly management plans for protected areas and coastal zones)
- National legislation to protect certain areas
  - —national parks
  - —forest reserves
  - —nature reserves
  - —natural, historical or cultural sites of importance
- International agreements to protect certain areas
  - —World Heritage Convention
  - —Ramsar Convention on wetlands
- Conservation/preservation of species likely to be sold to tourists or harmed by their activities
  - —national legislation
  - —international conventions
  - —CITES (Convention on International Trade in Endangered Species)

Environmental Priorities and Preferences

- Participation of affected people in project planning to determine priorities for environmental protection, including:
  - —public health

—revered areas, flora and fauna(e. g. cultural/medicinal value, visual landscape)
—skills training to undertake local environmental mitigation measures
—protection of potable water supply
—conservation of wetland/tropical forest services and products, e. g. hunted wildlife, fish stocks
—issues of sustainable income generation and employment(including significance of gender—*see WID manual*)
• Government policies for environmental protection (including, where appropriate, incorporation of objectives from country environmental studies/environmental action plans, etc.)
• Environmental priorities of tourism boards and trade associations representing tour operators

Source: CEC 1993

Socio-economic impacts can raise in particular the distributional dimension to evaluation, 'who wins and who loses' (Glasson 2009; Vanclay 1999). Beyond the use of standards and legal requirements, all assessments of significance either implicitly or explicitly apply weights to the various impacts (i. e. some are assessed as more important than others). This involves interpretation and the application of judgement. Such judgement can be rationalized in various ways and a range of methods are available, but all involve values and all are subjective. Parkin (1992) sees judgements as being on a continuum between an analytical mode and an intuitive mode. In practice, many are at the intuitive end of the continuum, but such judgements, made without the benefit of analysis, are likely to be flawed, inconsistent and biased. The 'social effects of resource allocation decisions are too extensive to allow the decision to "emerge" from some opaque procedure free of overt political scrutiny' (Parkin 1992). Analytical methods seek to introduce a rational approach to evaluation.

Table 5.7 Example of an approach to assessing the local impact magnitude of a major energy project: socio-economic impacts dimension

Type of impact	Local context	Negligible impact	Minor impact	Moderate impact	Major impact
Demographic impacts					
Change in local population level	Population growth (2001 to 2009):	Change in local population of less than ±0.25%	Change in local population of ±0.25—1%	Change in local population of ±1—2%	Change in local population of more than ±2%
Direct and indirect employment impacts					
Change in employment level in local economy	Employment growth (ABI estimates 2001 to 2007):	Change of less than ±0.25% on baseline employment levels in the local economy	Change of ±0.25—1% on baseline employment levels in the local economy	Change of ±1—2% on baseline employment levels in the local economy	Change of more than ±2% on baseline employment levels in local economy
Change in unemployment level in local economy	Claimant % unemployment rates (June 2010):	Change of less than ±2% in claimant unemployment level	Change of ±2—5% in claimant unemployment level	Change of ±5—10% in claimant unemployment level	Change of more than ±10% in claimant unemployment level
Accommodation pressures and development					
Change in stock of local housing	Housing stock growth (2001 to 2008)	Change of less than ±0.25% on baseline housings tock	Change of ±0.25—1% on baseline housing stock	Change of ±1—2% on baseline housing stock	Change of more than ±2% on baseline housing stock

Source: Authors, drawing on various consultancy studies

Two sets of methods are distinguished: those that assume a common utilitarian ethic with a single evaluation criterion (money), and those based on the measurement of personal utilities, including multiple criteria. The CBA approach, which seeks to express impacts in monetary units, falls into the former category. A variety of methods, including *multi-criteria analysis*, *decision analysis*, and *goals achievement*, fall into the latter category. The very growth of EIA is partly a response to the limitations of CBA and to the problems of the monetary valuation of environmental impacts. Yet, after several decades of limited concern, there is renewed interest in the monetizing of environmental costs and benefits (DoE 1991; HM Treasury 2003). The multi-criteria/multi-attribute methods involve scoring and weighting systems that are also not problem-free. The various approaches are now outlined. In practice, there are many hybrid variations between these two main categories, and these are referred to in both categories.

5. 3. 2 *Cost-benefit analysis and monetary valuation techniques*

Cost-benefit analysis (CBA) itself lies in a range of project and plan appraisal methods that seek to apply monetary values to costs and benefits (Lichfield *et al*. 1975). At one extreme are *partial* approaches, such as financial-appraisal, cost-minimization and cost-effectiveness methods, which consider only a subsection of the relevant population or only a subsection of the full range of consequences of a plan or project. *Financial appraisal* is limited to a narrow concern, usually of the developer, with the stream of financial costs and returns associated with an investment. *Cost effectiveness* involves selecting an option that achieves a goal at least cost (for example, devising a least-cost approach to produce coastal bathing waters that meet the CEC Blue Flag criteria). The cost-effectiveness approach is more problematic where there are a number of goals and where some actions achieve certain goals more fully than others (Winpenny 1991).

Cost-benefit analysis is more *comprehensive* in scope. It takes a long view of projects (farther as well as nearer future) and a wide view (in the sense of allowing for side effects). It is based in welfare economics and seeks to include all the relevant costs and benefits to evaluate the net social benefit of a project. It was used extensively in the UK in the 1960s and early 1970s for public sector projects, the most famous being the third London Airport (HMSO 1971). The methodology of CBA has several stages: project definition, the identification and enumeration of costs and benefits, the evaluation of costs and benefits, and the discounting and presentation of results. Several of the stages are similar to those in EIA. The basic evaluation principle is to measure in monetary terms where possible—as money is the common measure of value and monetary values are best understood by the community and decision-makers—and then reduce all costs and benefits to the same capital or annual basis. Future annual flows of costs and benefits are usually discounted to a net present value (Table 5. 8). A range of interest rates may be used to show the sensitivity of the analysis to changes. If the net social benefit minus cost is positive, then there may be a presumption in favour of a project. However, the final outcome may not always be that clear. The presentation of results should distinguish between tangible and intangible costs and benefits, as relevant, allowing the decision-maker to consider the trade-offs involved in the choice of an option.

Table 5.8 Cost-benefit analysis: presentation of results-tangibles and intangibles

Category	*Alternative* 1	*Alternative* 2
Tangibles		
Annual benefits	£$B1$	£$b1$
	£$B2$	£$b2$
	£$B3$	£$b3$
Total annual benefits	£$B1+B2+B3$	£$b1+b2+b3$
Annual costs	£$C1$	£$c1$
	£$C2$	£$c2$
	£$C3$	£$c3$
Total annual costs	£$C1+C2+C3$	£$c1+c2+c3$
Net discounted present value (NDPV) of benefits and costs over 'm' years at X%*	£D	£E
Intangibles		
Intangibles are likely to include costs and benefits	$I1$	$i1$
	$I2$	$i2$
	$I3$	$i3$
	$I4$	$i4$
Intangibles summation (undiscounted)	$I1+I2+I3+I4$	$i1+i2+i3+i4$

* e.g NPDV (Alt 1)

$$D=\sum\left[\frac{B1}{(1+X)^1}+\frac{B1}{(1+X)^2}+\cdots+\frac{B1}{(1+X)^n}+\frac{B2}{(1+X)^1}+\cdots+\frac{B2}{(1+X)^n}+\frac{B3}{(1+X)^1}+\cdots+\frac{B3}{(1+X)^n}\right]-$$

$$\sum\left[\frac{C1}{(1+X)^1}+\frac{C1}{(1+X)^2}+\cdots+\frac{C1}{(1+X)^n}+\frac{C2}{(1+X)^1}+\cdots+\frac{C2}{(1+X)^n}+\frac{C3}{(1+X)^1}+\cdots+\frac{C3}{(1+X)^n}\right]$$

Cost-benefit analysis has excited both advocates (e.g. Dasgupta and Pearce 1978; Pearce 1989; Pearce *et al*. 1989) and opponents (e.g. Bowers 1990). Hanley and Splash (2003) provide an interesting review of CBA and the environment. CBA does have many problems, including identifying, enumerating and monetizing intangibles. Many environmental impacts fall into the intangible category, for example the loss of a rare species, the urbanization of a rural landscape and the saving of a human life. The incompatibility of monetary and non-monetary units makes decision-making problematic (Bateman 1991). Another problem is the choice of discount rate: for example, should a very low rate be used to prevent the rapid erosion of future costs and benefits in the analysis? This choice of rate has profound implications for the evaluation of resources for future generations. There is also the underlying and fundamental problem of the use of the single evaluation criterion of money, and the assumption that £1 is worth the same to any person, whether a tramp or a millionaire, a resident of a rich commuter belt or of a poor and remote rural community. CBA also ignores distribution effects and aggregates costs and benefits to estimate the change in the welfare of society as a whole.

The *planning balance sheet* (PBS) is a variation on the theme of CBA, and it goes beyond CBA in its attempts to identify, enumerate and evaluate the distribution of costs and benefits between the affected parties. It also acknowledges the difficulty of attempts to monetize the more intangible impacts. It was developed by Lichfield *et al.* (1975) to compare alternative town plans. PBS is basically a set of social accounts structured into sets of 'producers' and 'consumers' engaged in various transactions. The transaction could, for example, be an adverse impact, such as noise from an airport (the producer) on the local community (the consumers), or a beneficial impact, such as the time savings resulting from a new motorway development (the producer) for users of the motorway (the consumers). For each producer and consumer group, costs and benefits are quantified per transaction, in monetary terms or otherwise, and weighted according to the numbers involved. The findings are presented in tabular form, leaving the decisionmaker to consider the trade-offs, but this time with some guidance on the distributional impacts of the options under consideration (Figure 5.7). Subsequently, Lichfield (1996) sought to integrate EIA and PBS further in an approach he called *community impact evaluation* (CIE).

	Plan A				Plan A			
	Benefits		Costs		Benefits		Costs	
	Capital	Annual	Capital	Annual	Capital	Annual	Capital	Annual
Producers								
X	£a	£b	—	£d	—	—	£b	£c
Y	i_1	i_2	—	—	i_3	i_4	—	—
Z	M_1	—	M_2	—	M_3	—	M_4	—
Consumers								
X′	—	£e	—	£f	—	£g	—	£h
Y′	i_5	i_6	—	—	i_7	i_8	—	—
Z′	M_1	—	M_3	—	M_2	—	M_4	—

£= benefits and costs that can be monetized
M= where only a ranking of monetary values can be estimated
i = intangibles

Figure 5.7 Example of structure of a planning balance sheet (PBS)

Partly in response to the 'intangibles' problem in CBA, there has also been considerable interest in the development of *monetary valuation techniques* to improve the economic measurement of the more intangible environmental impacts (Barde and Pearce 1991; DoE 1991; Winpenny 1991, Hanley and Splash 2003). The techniques can be broadly classified into direct and indirect, and they are concerned with the measurement of preferences about the environment rather than with the intrinsic values of the environment. The direct approaches seek to measure directly the monetary value of environmental gains—for example, better air quality or an improved scenic view. Indirect approaches measure preferences for a particular effect via the establishment of a 'dose-response' -type relationship. The various techniques found under the direct and indirect categories are summarized in Table 5.9. Such techniques can contribute to the assessment of the total economic value of an action or project, which should not only include user values (preferences people have for using an environmental asset, such as a river for fishing) but also non-user values (where people value an asset but do not use it, although some may wish to do so some day). Of course, such techniques have their problems, for example the potential bias in people's replies in the contingent valuation method (CVM) approach (for a fascinating example of this, see Willis and Powe 1998). However, simply through the act of seeking a value for various environmental features, such techniques help to reinforce the understanding that such features are not 'free' goods and should not be treated as such.

Table 5.9 Summary of environmental monetary valuation techniques

Direct household production function(HPF)

HPF methods seek to determine expenditure on commodities that are substitutes or complements for an environmental characteristic to value changes in that characteristic. Subtypes include the following:

1. Avertive expenditures: expenditure on various substitutes for environmental change(e. g. noise insulation as an estimate of the value of peace and quiet).
2. Travel cost method: expenditure, in terms of cost and time, incurred in travelling to a particular location(e. g. a recreation site) is taken as an estimate of the value placed on the environmental good at that location(e. g. benefit arising from use of the site).

Direct hedonic price methods(HPM)

HPM methods seek to estimate the implicit price for environmental attributes by examining the real markets in which those attributes are traded. Again, there are two main subtypes:

1. Hedonic house land prices: these prices are used to value characteristics such as 'clean air' and 'peace and quiet', through cross-sectional data analysis(e. g. on house price sales in different locations).
2. Wage risk premia: the extra payments associated with certain higher risk occupations are used to value changes in morbidity and mortality(and implicitly human life) associated with such occupations.

Direct experimental markets

Survey methods are used to elicit individual values for non-market goods. Experimental markets are created to discover how people would value certain environmental changes. Two kinds of questioning, of a sample of the population, may be used:

1. Contingent valuation method(CVM): people are asked what they are willing to pay(WTP) for keeping X(e. g. a good view, a historic building) or preventing Y, or what they are willing to accept(WTA) for losing A, or tolerating B.
2. Contingent ranking method(CRM or stated preference): people are asked to rank their preferences for various environmental goods, which may then be valued by linking the preferences to the real price of something traded in the market(e. g. house prices).

Indirect methods

Indirect methods seek to establish preferences through the estimation of relationships between a 'dose'(e. g. reduction in air pollution) and a response(e. g. health improvement). Approaches include the following:

1. Indirect market price approach: the dose-response approach seeks to measure the effect(e. g. value of loss of fish stock) resulting from an environmental change(e. g. oil pollution of a fish farm), by using the market value of the output involved. The replacement-cost approach uses the cost of replacing or restoring a damaged asset as a measure of the benefit of restoration(e. g. of an old stone bridge eroded by pollution and wear and tear).
2. Effect on production approach: where a market exists for the goods and services involved, the environmental impact can be represented by the value of the change in output that it causes. It is widely used in developing countries, and is a continuation of the dose-response approach.

Sources: Adapted from DoE 1991; Winpenny 1991; Pearce and Markandya 1990; Barde and Pearce 1991; Nijkamp 2004

5.3.3 Scoring and weighting and multi-criteria methods

Multi-criteria and multi-attribute methods seek to overcome some of the deficiencies of CBA; in particular they seek to allow for a pluralist view of society, composed of diverse 'stakeholders' with diverse goals and with differing values concerning environmental changes. Most of the methods use—and sometimes misuse—some kind of simple scoring and weighting system; such systems generate considerable debate. Here we discuss some key elements of good practice, and then offer a brief overview of the range of multi-criteria/multi-attribute methods available to the analyst.

Scoring may use quantitative or qualitative scales, according to the availability of information on the impact under consideration. Lee (1987) provides an example (Table 5.10) of how different levels of impact (in this example noise, whose measurement is in units of L_{10}dBA) can be scored in different systems. These systems seek to standardize the impact scores for purposes of comparison. Where quantitative data are not available, ranking of alternatives may use other approaches, for example using letters (A, B, C, etc.) or words (not significant, significant, and very significant).

Table 5.10 A comparison of different scoring systems

Method	Alternatives				Basis of score
	A(no action)	B	C	D	
Ratio	65	62	71	75	Absolute L_{10}dB(A) measure
Interval	0	−3	+6	+10	Difference in L_{10}dB(A) using alternative A as base
Ordinal	B	A	C	D	Ranking according to ascending value of L_{10}dB(A)
Binary	0	0	1	1	0=less than 70L_{10}dB(A) 1=70L_{10}dB(A) or more

Source: Based on Lee 1987

Weighting seeks to identify the relative importance of the various impact types for which scores of some sort may be available (for example, the relative importance of a water pollution impact, the impact on a rare flower). Different impacts may be allocated weights (normally numbers) out of a total budget (e.g. 10 points to be allocated between 3 impacts) — but by whom?

Multi-criteria/multi-attribute methods seek to recognize the plurality of views and weights in their methods; the Delphi approach also uses individuals' weights, from which group weights are then derived. In many studies, however, the weights are those produced by the technical team. Indeed the various stakeholders may be unwilling to reveal all their personal preferences, for fear of undermining their negotiating positions. This internalization of the weighting exercise does not destroy the use of weights, but it does emphasize the need for clarification of scoring and weighting systems and, in particular, for the identification of the origin of the weightings used in an EIA. Wherever possible, scoring and weighting should be used to reveal the trade-offs in impacts involved in particular projects or in alternatives. For example, Table 5.11 shows that the main issue is the trade-off between the impact on flora of one scheme and the impact on noise of the other scheme.

Table 5.11 Weighting, scoring and trade-offs

Impact	Weight(w)	Scheme A		Scheme B	
		Score(a)	(aw)	Score(b)	(bw)
Noise	2	5	10	1	2
Loss of flora	5	1	5	4	20
Air pollution	3	2	6	2	6
Total			21		28

Several approaches to the scoring and weighting of impacts have already been introduced in the outline of impact identification methods in Chapter 4. The matrix approach can also be usefully modified to identify the distribution of impacts among geographical areas and/or among various affected parties (Figure 5.8). Weightings can also be built into overlay maps to identify areas with the most development potential according to various combinations of weightings. Some of the limitations of such approaches have already been noted in Chapter 4.

Group environmental component	Project Action							
	Construction stage actions				Operational stage actions			
	A	B	C	D	a	b	c	d
Group 1 (e.g. indigenous population ≥ 45 years old) various								
• Social								
• Physical								
• Economic components								
Group 2 (e.g. indigenous population < 45 years old) various								
• Social								
• Physical								
• Economic components								

Figure 5.8 Simple matrix identification of distribution of impacts

Other methods in the multi-criteria/multi-attribute category include decision analysis, the goals achievement matrix (GAM), multi-attribute utility theory (MAUT) and judgement analysis. *Multi-criteria decision analysis* (MCDA) techniques have emerged as a major approach for resolving natural resource management problems (Herath and Prato 2006;

DCLG 2009), and are becoming increasingly used in EIA, especially for major projects. MCDA is a tool that is particularly applicable to cases where a single criterion approach is inappropriate. It allows decision makers to integrate the environmental, social and economic values and preferences of stakeholders, while overcoming the difficulties in monetizing the more intangible non-monetary attributes. Typically the approach defines objectives, chooses the criteria to measure the objectives, specifies alternatives, transforms the criterion scales into units, weights the criteria to reflect relative importance, selects and applies a mathematical algorithm for ranking alternatives, and chooses alternatives. The evaluation can use a variety of quantitative/semi-quantitative and qualitative assessment, and survey methods (Figueira *et al.* 2005). The former applies metrics to the selected criteria, which allows the calculation of outcomes as noted above. The latter, qualitative approach, uses subjective judgements and rating methods; there are no numbers used, and hence no calculations. In summary, both have their strengths. Quantitative/semi-quantitative tools are systematic, repeatable and inputs and outputs can be verified. Qualitative tools are effective at capturing diverse information, particularly intangible information and insights; while they are less repeatable, they provide narratives to explain results.

While MCDA can be quite complex, it can also be presented in very simple summary fashion, which can be very appealing to decision makers-but such presentations can also oversimplify the issues and trade-offs involved. A familiar approach is the use of a traffic lights colour approach (green-positive; amber-neutral; red-negative), indicating the extent to which the specified alternatives satisfy the various environmental, social and economic objectives. Figure 5.9 provides a schematic example of a summary matrix for a qualitative MCDA (with 5 shadings from deep red to deep green), for alternative locations for a major project. Some of the key elements, and issues, are set out in the figure. The red shadings indicate a more disadvantageous option performance against the relevant criterion/objective; the green shadings indicate more advantageous option performance. In this example, project option 3 is the preferred option by virtue of the spread of good overall performance against the QBL criteria.

The GAM was developed as a planning tool by Hill (1968) to overcome the perceived weaknesses of the PBS approach. GAM makes the goals and objectives of a project/plan explicit, and the evaluation of alternatives is accomplished by measuring the extent to which they achieve the stated goals. The existence of many diverse goals leads to a system of weights. Since all interested parties are not politically equal, the identified groups should also be weighted. The end result is a matrix of weighted objectives and weighted interests/agencies (Figure 5.10). The use of goals and value weights to evaluate plans in the interests of the community, and not just for economic efficiency, has much to commend it. The approach also provides an opportunity for public participation. Unfortunately, the complexity of the approach has limited its use, and the weights and goals used may often reflect the views of the analyst more than those of the interests and agencies involved.

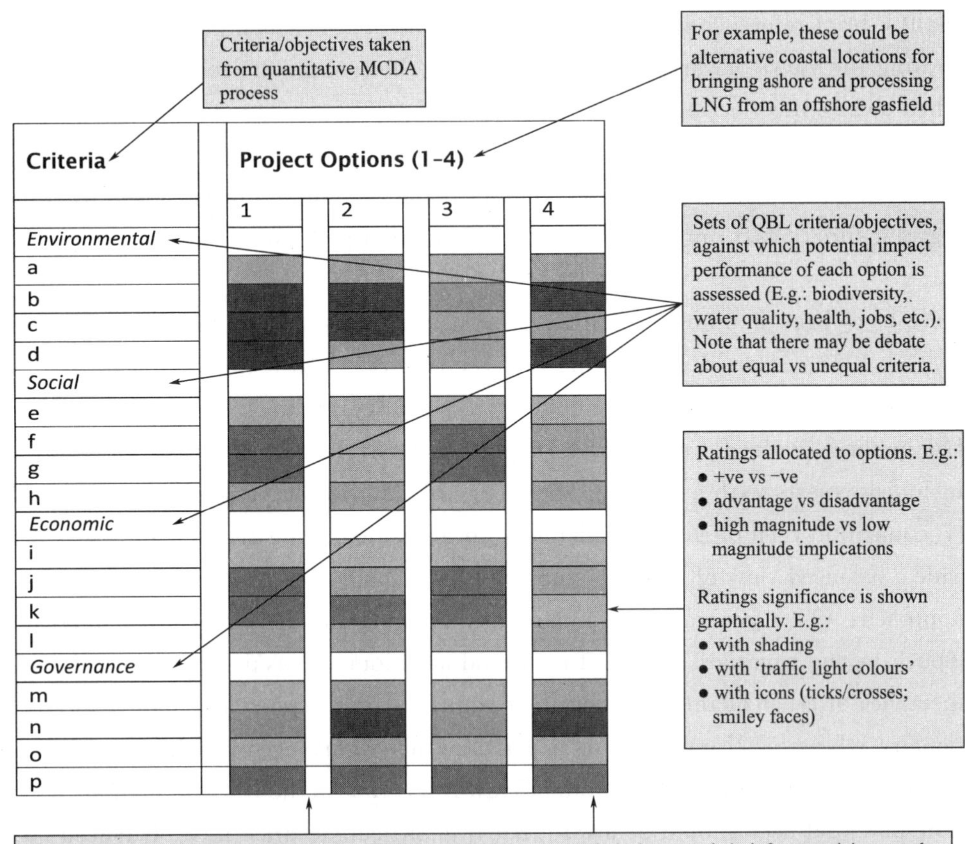

Figure 5.9 Schematic example of a summary matrix for a qualitative MCDA

Goal description:	α			β		
Relative weight:	2			3		
Incidence	Relative weight	Costs	Benefits	Relative weight	Costs	Benefits
Group a	1	A	D	5	E	1
Group b	3	H	J	4	M	2
Group c	1	L	J	3	M	3
Group d	2	—	J	2	V	4
Group e	1	—	K	1	T	5
		Σ	Σ	Σ	Σ	

Figure 5.10 Goals achievement matrix (GAM)

Source: Adapted from Hill 1968

Finally, brief reference is made to the *Delphi method*, which provides another way of incorporating the views of various stakeholders into the evaluation process. The method is an established means of collecting expert opinion and of gaining consensus among experts on various issues under consideration. It has the advantage of obtaining expert opinion from the individual, with guaranteed anonymity, avoiding the potential distortion caused by peer pressure in group situations. Compared with other evaluation methods it can also be quicker and cheaper.

There have been a number of interesting applications of the Delphi method in EIA (Green *et al.* 1989, 1990; Richey *et al.* 1985). Green *et al.* used the approach to assess the environmental impacts of the redevelopment and reorientation of Bradford's famous Salt Mill. The method involved drawing up a Delphi panel; in the Salt Mill case, the initial panel of 40 included experts with a working knowledge of the project (e. g. planners, tourism officers), councillors, employees, academics, local residents and traders. This was designed to provide a balanced view of interests and expertise. The Delphi exercise usually has a three-stage approach: (1) a general questionnaire asking panel members to identify important impacts (positive and negative); (2) a first-round questionnaire asking panel members to rate the importance of a list of impacts identified from the first stage; and (3) a second-round questionnaire asking panel members to re-evaluate the importance of each impact in the light of the panel's response to the first round. However, the method is not without its limitations. The potential user should be aware that it is difficult to draw up a 'balanced' panel in the first place, and to avoid distorting the assessment by the varying drop-out rates of panel members between stages of the exercise, and by an overzealous structuring of the exercise by the organizers. For other application and critique, see Breidenhann and Butts (2006), and Landeta (2006).

5.4 Mitigation and enhancement

5.4.1 *Types of mitigation measures*

Mitigation is defined in EC Directive 97/11 as 'measures envisaged in order to avoid, reduce and, if possible, remedy significant adverse effects' (CEC 1997). In similar vein, the US CEQ, in its regulations implementing the NEPA, defines mitigation as including:

> not taking certain actions; limiting the proposed action and its implementation; repairing, rehabilitating, or restoring the affected environment; presentation and maintenance actions during the life of the action; and replacing or providing substitute resources or environments. (CEQ 1978)

The guidance on mitigation measures provided by the UK government is set out in Table

5. 12. It is not possible to specify here all the types of mitigation measures that could be used. Instead, the following text provides a few examples, relating to biophysical and socio-economic impacts. The reader is also referred to Morris and Therivel (2009) and Rodriguez-Bachiller with Glasson (2004) for useful coverage of mitigation measures for particular impact types. A review of EISs for developments similar to the development under consideration may also suggest useful mitigation measures.

Table 5. 12 Mitigation measures, as outlined in *UK guide to procedures*

Where significant adverse effects are identified, [describe] the measures to be taken to avoid, reduce or remedy those effects, e. g:

(a) site planning;

(b) technical measures, e. g. :

  (i) process selection;

  (ii) recycling;

  (iii) pollution control and treatment;

  (iv) containment(e. g. bunding of storage vessels).

(c) aesthetic and ecological measures, e. g. :

  (i) mounding;

  (ii) design, colour, etc. ;

  (iii) landscaping;

  (iv) tree plantings;

  (v) measure to preserve particular habitats or create alternative habitats;

  (vi) recording of archaeological sites;

  (vii) measures to safeguard historic buildings or sites.

[Assess] the likely effectiveness of mitigating measures.

Source: DETR 2000

At one extreme, the prediction and evaluation of impacts may reveal an array of impacts with such significant adverse effects that the only effective mitigation measure may be to abandon the proposal altogether. A less draconian (and more normal) situation would be to modify aspects of the development action to avoid various impacts. Examples of methods to *avoid* impacts include:

- The control of solid and liquid wastes by recycling on site or by removing them from the site for environmentally sensitive treatment elsewhere.
- The use of a designated lorry route, and daytime working only, to avoid disturbance to village communities from construction lorry traffic and from night construction work.
- The establishment of buffer zones and the minimal use of toxic substances, to avoid impacts on local ecosystems.

Some adverse effects may be less easily avoided; there may also be less need to avoid them completely. Examples of methods to *reduce* adverse effects include:

- The sensitive design of structures, using simple profiles, local materials and muted colours, to reduce the visual impact of a development, and landscaping to hide it or blend it into the local environment.
- The use of construction-site hostels, and coaches for journeys to work to reduce the

impact on the local housing market, and on the roads, of a project employing many workers during its construction stage.

• The use of silting basins or traps, the planting of temporary cover crops and the scheduling of activities during the dry months, to reduce erosion and sedimentation.

During one or more stages of the life of a project, certain environmental components may be temporarily lost or damaged. It may be possible to *repair*, *rehabilitate* or *restore* the affected component to varying degrees. For example:

• Agricultural land used for the storage of materials during construction may be fully rehabilitated; land used for gravel extraction may be restored to agricultural use, but over a much longer period and with associated impacts according to the nature of the landfill material used.

• A river or stream diverted by a road project can be unconverted and re-established with similar flow patterns as far as is possible.

• A local community astride a route to a new tourism facility could be relieved of much of the adverse traffic effects by the construction of a bypass (which, of course, introduces a new flow of impacts).

There will invariably be some adverse effects that cannot be reduced. In such cases, it may be necessary to *compensate* people for adverse effects. For example:

• For the loss of public recreational space or a wildlife habitat, the provision of land with recreation facilities or the creation of a nature reserve elsewhere.

• For the loss of privacy, quietness and safety in houses next to a new road, the provision of sound insulation and/or the purchase by the developer of badly affected properties.

Mitigation measures can become linked with discussions between a developer and the local planning authority (LPA) on what is known in the UK as 'planning gain'. Fortlage (1990) talks of some of the potential complications associated with such discussions, and of the need to distinguish between mitigation measures and planning gain:

> Before any mitigating measures are put forward, the developer and the local planning authority must agree as to which effects are to be regarded as adverse, or sufficiently adverse to warrant the expense of remedial work, otherwise the whole exercise becomes a bargaining game which is likely to be unprofitable to both parties ...

Planning permission often includes conditions requiring the provision of planning gains by the developer to offset some deterioration of the area caused by the development, but it is essential to distinguish very clearly between those benefits offered by way of compensation for adverse environmental effects and those which are a formal part of planning consent. The local planning authority may decide to formulate the compensation proposals as a planning condition in order to ensure that they are carried out, so the developer should beware of putting forward proposals that he does not really intend to implement.

Mitigation measures must be planned in an integrated and coherent fashion to ensure that they are effective, that they do not conflict with each other and that they do not merely shift a problem from one medium to another. The results of a research project on the treatment of mitigation within EIA (DETR 1997) found that UK practice varied considerably. For example, there was too much emphasis on physical measures, rather than on operational or management controls, and a lack of attention to the impacts of construction and to residual impacts after mitigation.

Table 5.13 provides a wider classification of mitigation, adopted in the project, by levels of mitigation, mitigation hierarchy and project phase. The levels relate to broad decisions that are made during the design of a project, with the last two reflecting the fact that effective mitigation can be achieved through measures other than physical ones. The *mitigation hierarchy* focuses on the principle of prevention rather than cure where, in principle at least, the options higher in the list should be tried before those lower down the list. The project phases relate to the life cycle of the project first discussed in Chapter 1. Any particular mitigation measure can be classified in a combination of the three ways—for example, physical design measures can be used to minimize an impact at source, during the construction phase (DETR 1997).

5.4.2 *Mitigation in the EIA process*

Like many elements in the EIA process, and as noted in Table 5.13, mitigation is not limited to one point in the assessment. Although it may follow logically from the prediction and assessment of the relative significance of impacts, it is in fact inherent in all aspects of the process. An original project design may already have been modified, possibly in the light of mitigation changes made to earlier comparable projects or perhaps as a result of early consultation with the LPA or with the local community. The consideration of alternatives, initial scoping activities, baseline studies and impact identification studies may suggest further mitigation measures. Although more in-depth studies may identify new impacts, mitigation measures may alleviate others. The prediction and evaluation exercise can thus focus on a limited range of potential impacts.

Table 5.13 A wider classification of mitigation

Levels of mitigation	Mitigation hierarchy	Project phase
• Alternatives (strategic, alternative locations and processes) • Physical design measures • Project management measures • Deferred mitigation	• Avoidance at source • Minimize at source • Abatement on site • Abatement at receptor • Repair • Compensation in kind • Other compensation and enhancement	• Construction • Commissioning • Operation • Decommissioning • Restoration, afteruse/aftercare

Source: DETR 1997

Mitigation measures are normally discussed and documented in each topic section of the EIS (e.g. air quality, visual quality, transport, employment). Those discussions should clarify the extent to which the significance of each adverse impact has been offset by the mitigation measures proposed. A summary chart (Table 5.14) can provide a clear and very useful overview of the envisaged outcomes, and may be a useful basis for agreement on planning consents. Residual unmitigated or only partially mitigated impacts should be identified. These could be divided according to the degree of severity: for example, into 'less than significant impacts' and 'significant unavoidable impacts'.

Table 5.14 Example of a section of a summary table for impacts and mitigation measures

Impact	Mitigation measure(s)	Level of significance after mitigation
1. 400 acres of prime agricultural land would be lost from the county to accommodate the petrochemical plant.	The only full mitigating measure for this impact would be to abandon the project.	Significant unavoidable impact
2. Additional lorry and car traffic on the adjacent hilly section of the motorway will increase traffic volumes by 10—20 per cent above those predicted on the basis of current trends.	A lorry crawler lane on the motorway, funded by the developer, will help to spread the volume, but effects may be partial and short-lived.	Significant unavoidable impact
3. The project would block the movement of most terrestrial species from the hilly areas to the east of the site to the wetlands to the west of the site.	A wildlife corridor should be developed and maintained along the entire length of the existing stream which runs through the site. The width of the corridor should be a minimum of 75 ft. The stream bed should be cleaned of silt and enhanced through the construction of occasional pools. The buffer zone should be planted with native riparian vegetation, including sycamore and willow.	Less than significant impact

Mitigation measures are of little or no value unless they are implemented. Commitment to mitigation can be demonstrated through implementation or management plans. These may take the form of an all-encompassing Environmental Management Plan (EMP) or Environmental Action Plan (EAP) (see Chapter 12); they may also include more specific sub-plans for particular impact types-see for example the following section for workforce and procurement plans to deliver socio-economic benefits. There is also a clear link between mitigation and the monitoring of outcomes, when a project is approved and moves to the construction and operational stages. Indeed, the incorporation of a clear monitoring programme can be one of the most important mitigation measures. Monitoring, which is discussed in Chapter 7, must include the effectiveness or otherwise of mitigation and enhancement measures. The measures must therefore be devised with monitoring in mind; they must be clear enough to allow for the checking of effectiveness. The use of particular mitigation measures may also draw on previous experience of relative effectiveness, from previous monitoring activity in other relevant and comparable cases.

5.4.3 *Enhancement of potential benefits*

UK guidance (DETR 1997) also notes the importance of including measures in EIA to create environmental benefits. Benefit enhancement is becoming an increasingly important element in EIA, especially for major projects. Such enhancements can include biophysical actions—for example creating a nature reserve from an abandoned quarry that lies adjacent to a project site and which has been acquired by the developer. However they tend to be more often socio-economic actions related to socio-economic issues. A project may bring considerable benefit to an area, often socio-economically; where such benefits are identified, as a minimum there should be a concern to ensure that they do occur and do not become diluted, and that they may be enhanced. For example:

- The potential local employment benefits of a project can be encouraged and enhanced by the offer of appropriate skills training programmes, apprenticeships, plus a 'one-stop-shop' local recruitment facility. For the construction stage of a project, this might be brought together in a Construction Workforce Management Plan, developed between the developer and key local stakeholders. The implementation of such a plan, with clear indicators and targets, can provide an important means of internalising employment benefits to the project host area, reducing employment benefit leakage often associated with major projects.

- Similarly, a procurement management plan could help to enhance opportunities for local contractors to benefit from a project. It could include supplier events to provide information on local contact opportunities with the project, improved local supply information for the developer such as an online database of local suppliers, and the employment of a supply chain officer to improve interactions with the local business sector.

- In the housing domain, various tenure arrangements, construction site hostels/campuses might have *legacy use* for the local area. A high quality construction site campus might have legacy use as an educational, recreational or even hotel facility. Vacated construction worker housing might provide valuable affordable housing for local people in need.

- Similar legacy use might flow from transport activities associated with the construction, and possibly also the operation, of a major project. For example, to minimize car travel to a project construction site, there might be agreement with LPAs to build a park and ride facility, with connecting buses to the site. The buses might be used for other local needs between work start and end times-for example as local school buses. Further, if conveniently located near to a town/city, the park and ride facility could be left in place for the use of the community after the end of project use.

- For some large projects there are always likely to be some indirect disturbance effects and changes in lifestyle which are less easy to address directly. In an attempt to offer some compensation for such impacts with regard to Sizewell B nuclear power station, the CEGB as long ago as 1987 issued a Social Policy Statement (CEGB 1987) which included the provision for grants to be made available for various charitable, social and recreational projects of benefit to the local community, as part of a package of 'ameliorative measures'. Such measures were very well received by the local community. Elsewhere, and much more recently, there has been increasing focus on the development of Community Benefits Agreements/Community Impacts Agreements to bring together packages of measures for locally impacted communities (Baxamusa 2008).

As for mitigation, the consideration of enhancement of impacts in EIA should be built in at an early stage of the process, building on wide stakeholder consultation. Enhancement measures should be clearly specified, and identified in management plans for subsequent monitoring of performance.

5.5 Summary

Impact prediction and the evaluation of the significance of impacts often constitute a 'black box' in EIA studies. Intuition, often wrapped up as expert opinion, cannot provide a firm and defensible foundation for this important stage of the process. Various methods, ranging from simple to complex, are available to the analyst, and these can help to underpin analysis. Mitigation and enhancement measures come into play particularly at this stage. However, the sophistication of some methods does run the risk of cutting out key actors, and especially the public, from the EIA process. Chapter 6 discusses the important, but currently weak, role of public participation, the value of good presentation, and approaches to EIS review and decision-making.

SOME QUESTIONS

The following questions are intended to help the reader focus on the key issues of this chapter.

1 Magnitude of impact is not always synonymous with significance of impact. Provide examples from your experience to illustrate this point.

2 Assess the case for using expert judgement as a key prediction method in EIA.

3 Similarly, examine the case for using causal network analysis in EIA.

4 How can uncertainty in the prediction of impacts be handled in EIA? Consider the merits of different approaches.

5 Consider the value of the qualitative multi-criteria decision analysis (MCDA) exemplified in Figure 5.9, for various stakeholder groups, for assessing the trade-offs between different types of impacts.

6 Examine the application of the mitigation hierarchy to the impacts of a major project with which you are familiar. What constraints might there be in following the logical steps in that hierarchy in practice?

7 The enhancement of beneficial impacts has had a low profile in EIA until recently. Why do you think this has been so, and why is the situation now changing?

8 Consider what might be included in a Community Benefits Agreement for

(a) a major wind farm development in a remote rural location; and

(b) the redevelopment of a major football (soccer) stadium in a heavily populated urban area.

References

Barde, J. P. and Pearce, D. W. 1991. *Valuing the environment: six case studies*. London: Earthscan.

Bateman, I. 1991. Social discounting, monetary evaluation and practical sustainability. *Town and Country Planning* 60 (6), 174-6.

Baxamusa, M. 2008. Empowering communities through deliberation: the model of community. *Journal of Planning Education and Research* 27, 61-276.

Beattie, R. 1995. Everything you already know about EIA, but don' t often admit. *Environmental Impact Assessment Review* 15.

Bourdillon, N. 1996. *Limits and standards in EIA*. Oxford: Oxford Brookes University, Impacts Assessment Unit, School of Planning.

Bowers, J. 1990. *Economics of the environment: the conservationists' response to the Pearce Report*. British Association of Nature Conservationists.

Bracken, I. 2008, *Urban Planning Methods: Research and Policy Analysis*. London: Routledge.

Bregman, J. I. and Mackenthun, K. M. 1992. *Environmental impact statements*. Boca Raton, FL: Lewis.

Briedenhann, J. and Butts, S. 2006. Application of the Delphi technique to rural tourism project evaluation, *Current Issues in Tourism*, 9 (2), 171-90.

Canter, L. 1996. *Environmental impact assessment*. McGraw-Hill International Editions.

CEC 1993. *Environmental manual: sectoral environmental assessment sourcebook*. Brussels: CEC, DG VIII.

CEC 1997. Council Directive 97/11/EC of 3 March 1997 amending Directive 85/337 EEC on the assessment of certain public and private projects on the environment. *Official Journal*. L73/5, 3 March.

CEC 1999. *Guidelines for the assessment of indirect and cumulative impacts as well as impact interactions*. Luxembourg: Office for Official Publications of the CEC.

CEQ (Council on Environmental Quality) 1978. *National Environmental Policy Act*, Code of Federal Regulations, Title 40, Section 1508. 20.

Dasgupta, A. K. and Pearce, D. W. 1978. *Cost-benefit analysis: theory and practice*. London: Macmillan.

DCLG 2009. *Multi-criteria analysis: a manual*. London: DCLG.

De Jongh, P. 1988. Uncertainty in EIA. In *Environmental impact assessment: theory and practice*. P. Wathern (ed), 62-83. London: Unwin Hyman.

DETR (Department of the Environment, Transport and the Regions) 1997. *Mitigation measures in environmental statements*. London: DETR.

DETR 2000. *Environmental impact assessment: guide to the procedures*. Tonbridge, UK: Thomas Telford.

DoE 1991. *Policy appraisal and the environment*. London: HMSO.

Elsom, D. 2009. Air quality and climate. In *Methods of environmental impact assessment*, P. Morris and R. Therivel (eds), 3rd edn (Ch. 8). London: Routledge.

Figueira, J., Greco, S. and Ehrgott, M. 2005. *Multi-criteria decision analysis: state of the art surveys*. New York: Springer.

Friend, J. K. and Hickling, A. 1987. *Planning under pressure: the strategic choice approach*. Oxford: Pergamon.

Friend, J. K. and Jessop, W. N. 1977. *Local government and strategic choice: an operational research approach to the processes of public planning*, 2nd edn. Oxford: Pergamon.

Glasson, J. 2009. Socio-economic impacts 1: overview and economic impacts. In *Methods of Environmental Impact Assessment*, P. Morris and R. Therivel (eds), 3rd edn (Ch. 2). London: Routledge.

Glasson, J., Elson, M. J., Van der Wee, M. and Barrett, B. 1987. *Socio-economic impact assessment of the proposed Hinkley Point C power station*. Oxford: Oxford Polytechnic, Impacts Assessment Unit.

Glasson, J., Van der Wee, M. and Barrett, B. 1988. A local income and employment multiplier analysis of a proposed nuclear power station development at Hinkley Point in Somerset. *Urban Studies* 25, 248-61.

Golden, J., Duellette, R. P., Saari, S. and Cheremisinoff, P. N. 1979. *Environmental impact data book*. Ann Arbor, MI: Ann Arbor Science Publishers.

Goldvarg, E., and Johnson-Laird, P. N. 2001. Naive causality: A mental model theory of causal meaning and reasoning. *Cognitive Science*, 25, 565-610.

Green, H., Hunter, C. and Moore, B. 1989. Assessing the environmental impact of tourism development-the use of the Delphi technique. *International Journal of Environmental Studies* 35, 51-62.

Green, H., Hunter, C. and Moore, B. 1990. Assessing the environmental impact of tourism development. *Tourism Management*, June, 11-20.

Hanley, N. and Splash, C. 2003. Cost-benefit analysis and the environment. Cheltenham: Edward Elgar.

Hansen, P. E. and Jorgensen, S. E. (eds) 1991.

Introduction to environmental management. New York: Elsevier.

Herath, G and Prato, T. 2006 *Using multi-criteria decision analysis in natural resource management*. Aldershot: Ashgate.

Hill, M. 1968. A goals-achievement matrix for evaluating alternative plans. *Journal of the American Institute of Planners* 34, 19.

HMSO 1971. *Report of the Roskill Commission on the third London Airport*. London: HMSO.

HM Treasury 2003. 'Green Book' *Appraisal and evaluation in central government*. Available at: www. greenbook. treasury. gov. uk.

Holling, C. S. (ed) 1978. *Adaptive environmental assessment and management*. New York: Wiley.

IEMA (Institute of Environmental Management and Assessment) 2011. *The state of environmental impact assessment practice in the UK*. Lincoln: IEMA.

Knight, R. 2009. Landscape and visual. In *Methods of environmental impact assessment*, P. Morris and R. Therivel (eds), 3rd edn, London: Routledge.

Kristensen, P., Jensen, J. P. and Jeppesen, E. 1990. *Eutrophication models for lakes*. Research Report C9. Copenhagen: National Agency of Environmental Protection.

Landeta, J. 2006. Current validity of the Delphi method in social sciences, *Technological forecasting and social change*, 73, 467-82.

Lee, N. 1987. *Environmental impact assessment: a training guide*. Occasional Paper 18, Department of Town and Country Planning, University of Manchester.

Lewis, J. A. 1988. Economic impact analysis: a UK literature survey and bibliography. *Progress in Planning* 30 (3), 161-209.

Lichfield, N. 1996. *Community impact evaluation*. London: UCL Press.

Lichfield, N., Kettle, P. and Whitbread, M. 1975. *Evaluation in the planning process*. Oxford: Pergamon.

Loewenstein, L. K. 1966. On the nature of analytical models. *Urban Studies* 3.

Maclaren, V. W. and Whitney, J. B. (eds) 1985. *New directions in environmental impact assessment in Canada*. London: Methuen.

Morris, P. and Therivel, R. (eds) 2009. *Methods of environmental impact assessment*, 3rd edn, London: Routledge.

Morris, P. and Thurling, D. 2001. Phase 2-3 ecological sampling methods. In *Methods of environmental impact assessment*, P. Morris and R. Therivel (eds), 2nd edn, (Appendix G). London: Spon Press.

Munn, R. E. 1979. *Environmental impact assessment: principles and procedures*. New York: Wiley.

Nijkamp, P. 2004. *Environmental Economics and Evaluation* . Cheltenham: Edward Elgar.

Parkin, J. 1992. *Judging plans and projects*. Aldershot: Avebury.

Pearce, D. 1989. Keynote speech at the 10th International Seminar on Environmental Impact Assessment and Management, University of Aberdeen, 9-22 July.

Pearce, D. and Markandya, A. 1990. *Environmental policy benefits: monetary valuation*. Paris: OECD.

Pearce, D., Markandya, A. and Barbier, E. B. 1989. *Blueprint for a green economy*. London: Earthscan.

Perdicoulis, A and Glasson, J. 2006. Causal Networks in EIA, *Environmental Impact Assessment Review* 26, 553-69.

Perdicoulis, A. and Glasson, J. 2009. The causality premise of EIA in practice. *Impact Assessment and Project Appraisal*, 27 (3), 247-50.

Rau, J. G. and Wooten, D. C. 1980. *Environmental impact analysis handbook*. New York: McGraw-Hill.

Richey, J. S., Mar, B. W. and Homer, R. 1985. The Delphi technique in environmental assessment. *Journal of Environmental Management* 21 (1), 135-46.

Rodriguez-Bachiller, A. with JGlasson 2004. *Expert Systems and Geographical Information Systems*. London: Taylor and Francis.

Suter II, G. W. 1993. *Ecological risk assessment*. Chelsea, MI: Lewis.

Tomlinson, P. 1989. Environmental statements: guidance for review and audit. *The Planner* 75 (28), 12-15.

US Environmental Protection Agency 1993. *Sourcebook for the environmental assessment (EA) process*. Washington, DC: EPA.

Vanclay, F. 1999. Social impact assessment. In *Handbook of environmental impact assessment*, J. Petts (ed). Oxford: Blackwell Science (vol. 1, Ch. 14).

Voogd, J. H. 1983. *Multicriteria evaluation for urban and regional planning*. London: Pion.

VROM 1984. *Prediction in environmental impact assessment*. The Hague: The Netherlands Ministry of Public Housing, Physical Planning and Environmental Affairs.

West Australian Environmental Protection Authority 2010. *Environmental Impact Assessment: Administrative Procedures 2010*. Perth: EPA.

Westman, W. E. 1985. *Ecology, impact assessment and environmental planning*. New York: Wiley.

Willis, K. G. and Powe, N. A. 1998. Contingent valuation and real economic commitments: a private good experiment. *Journal of Environmental Planning and Management* 41 (5), 611-19.

Winpenny, J. T. 1991. *Values for the environment: a guide to economic appraisal*. Overseas Development Institute. London: HMSO.

Wood, G. 2000. Is what you see what you get? Post development auditing of methods used for predicting the zone of visual influence in EIA. *Environmental Impact Assessment Review* 20 (5), 537-56.

6

公众参与、EIA 呈现和评审

6.1 导言

开展环境影响评价（EIA）的主要目的之一就是为开发商、公众、法定咨询机构和决策者提供拟建项目可能产生的环境影响信息，以便做出更好的决策。在 EIA 程序中，咨询公众及法定咨询机构不仅可以提高 EIA 质量、提升其全面性和有效性，而且还可确保决策过程充分考虑了各类人群的意见。咨询和公众参与在 EIA 的大多数环节中都非常有用，如：

- 确定 EIA 范围时；
- 为选址提供专业知识时；
- 推荐替代方案时；
- 评估所有可能影响的相对重要性时；
- 提出减缓措施时；
- 确保 EIS 客观性、真实性和完整性时；
- 监控开发协议实施情况时。

同样，以何种方式呈现信息，利益相关方如何利用信息，决策者怎样整合信息，这些都是环评程序的主要内容。

英国传统的决策体系主要依靠行政裁量，且具有保密性，限制公众的介入（McCormick，1991）。但近年来已逐步允许更多公众参与决策制定，特别是扩大了公众对信息的获取途径。在环境保护领域，1990 年英国颁布的《环境保护法》（EPA）要求环境事务部和地方权力机构对潜在污染过程的信息建立公众登记制度；政府社区数据统计网站（neighbourhood statistics，data. gov. uk）以及一些其他政府数据库以公开形式向大众提供环境类数据；《奥胡斯公约》、2003 年修订的 EIA 与 SEA 指令以及 2008 年颁布的规划法中都对公众参与提出了相关要求，允许公众获得以前未汇总的或者被认为是机密的信息；欧盟 2003/4 号指令要求成员国免费提供环境信息，英国已经借助 2004 年通过的《环境信息条例》加以实施；2009 年对 2003 年修订版 EIA 指令实施情况的回顾总结（CEC2009）为公众参与提供了很好的反馈，尤其是对于一些新的成员国来说。总体上成员国认为 EIA 指令提高了公众参与权利，增加了决策透明度，确保公民基本权利对民主发展的巩固具有直接的贡献。

然而，尽管 EIA 程序中的广泛咨询和公众参与都呈现出积极的发展趋势，EIA 研究结论的交流也有所增加，英国在这方面仍旧相对欠发达。很少有开发商会在递交授权申请和环境影响报告书（EIS）之前真正尽力获取公众意见；很少有主管部门有时间和资源能在做出决策前充分评估公众意见；很少有 EIS 能体现出对公众参与的鼓励。

本章讨论了如何促进公众（6.2 节）与法定咨询机构（6.3 节）的咨询和参与，如何将咨询与参与结果用于完善拟建项目和加速授权；6.4 节讨论了 EIS 的有效呈现；6.5 节探讨了 EIS 评审及评价的准确性和全面性；最后以讨论决策制定以及制定决策后的法律质疑为结束。

6.2 公众咨询与参与

这部分主要讨论如何促进公众参与实现“最佳实践”。[1] 首先要考虑公众参与的优缺点，然后讨论有效公众参与的必要条件和审查方法，最后讨论英国开展公众参与的方法，包括规划法（2008）所带来的挑战。有关公众参与的更多信息，读者可参考审计署（Audit Commission 2000）和环境管理评价协会（IEMA 2002）的资料；亚洲开发银行（ADB）2006 年举办的 NGO 论坛也提供了公众参与编制 EIA 的优秀示例。

6.2.1 公众参与的优缺点

开发商通常不支持公众参与，认为公众参与可能会影响他们与地方规划管理局的友好关系，可能将项目置于引人注目的位置上，同时伴随着时间和金钱的耗费。公众参与可能不会主导项目的最终决策，因为不同的相关利益群体关心的重点和优先考虑的内容不同，且这种决策可能更多地反映的是最有影响力的利益群体观点，而不是普通公众的意见。大多数开发商仅在规划项目申诉和咨询的阶段才与公众接触，但这个时期的公众参与常常演变为有计划地阻止项目实施，因此很多开发商根本就看不到公众参与积极的一面。

历史上，公众参与曾一度具有极端主义、争执、延期、阻碍发展这类含义。在美国，NEPA 相关法律诉讼曾阻止了一些重点开发项目，其中包括怀俄明州石油和天然气的开发、加利福尼亚州滑雪胜地的开发及阿拉斯加伐木业的开发项目等（Turner，1998）。20 世纪 60

[1] 尽管此章节中将公众咨询与参与统称为“公众参与”，但事实上二者是分开的。咨询本质上是针对被动受众来征求其意见，但回应对最终决策并没有什么积极影响；相反，公众参与中公众扮演一个主动的角色，对项目修订及最终决策都有所影响。

年代末 70 年代初，在日本发生了一起为阻止修建 Narita 机场的武力暴乱，导致六人死亡，使得该项目的建设推迟了 5 年之久。英国历史上轰动世界的一次公众“参与”是抗议者们戴着防毒面具聚集在核电站，在即将修建第三条 Heathrow 高速公路的地方搭建了营地，最终被强行从纽伯里支路的隧道和树屋中驱逐了出去，这使得项目在正式动工前已花费 600 多万英镑用于维持治安。更有代表性的是，似乎所有的规划者都已经对这种激烈的公众集会和“取缔项目”运动习以为常了。公众参与已演变成故意阻挠开发的一种手段，但却是合法的，其取胜的有效手段就是使项目被迫延期。

另一方面，从开发商的角度看，公众参与对于传达项目信息、消除误解有一定的积极意义，为更好地理解并处理相关问题提供了可能，并且能在项目的早期规划阶段识别和处理争议。考虑和回应当地居民或某些特定利益群体的特殊贡献可以为开发商提供建议措施，以避免当地人反对和产生环境问题。与开发商自己提出的建议相比，这些建议可能更新颖、更可行、更容易为公众所接受；在规划过程前期或者规划开始前对项目进行修正，远比在规划后期进行调节更容易，调节价格更低；与那些不得不走向质询阶段的项目相比，无需质询的项目费用较低。前期的公众参与也能在一定程度上防止项目受挫和矛盾的升级，从而避免出现更加暴力的“参与”方式。图 6.1 中展示了三种完全不同的 EIA 公众参与的例子。如果拟建项目得到了当地居民的赞同，则项目的推进将更加顺畅，抗议更少，还能节约更多的资金，增加更多劳动力，对不良影响如噪声和交通方面的抱怨也会更少。

(a)

(b)

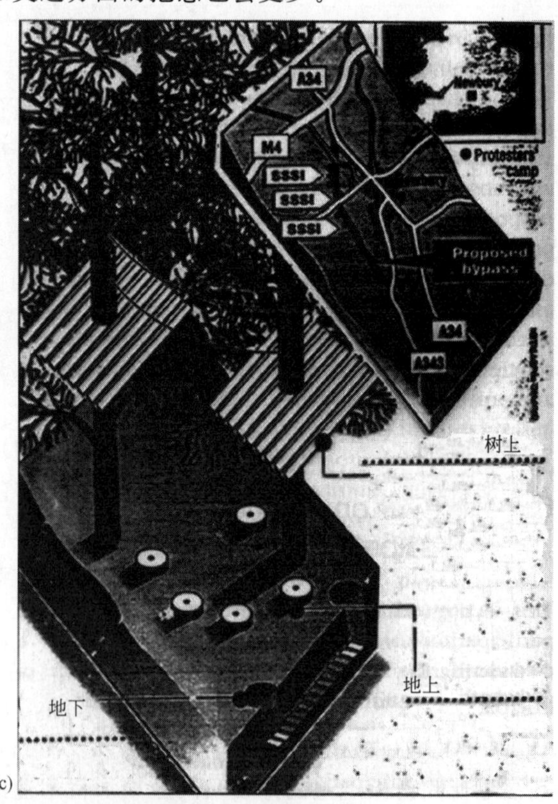

(c)

图 6.1　公众参与的不同形式

(a) 关于 Hinkley“C”核电站提案的公开会议；(b) 反对南安普顿秸秆发电厂提案的抗议；

(c) 纽伯利支路上的一些抗议方法

[资料来源：(a) EDF Energy，2009；(b) Daily Echo，2011；(c) Guardian，1996]

例如，欧洲（当时）最大的锌/铅矿的环保管理者发现：

通过恰当限定和广泛应用，EIA已成为一种有效的管理方式而不是威慑。它是一种确保所有与拟建项目相关的问题和受影响人群能够得到及时、有序考虑的机制，这也是EIA的真正益处所在。我们已经进入了由公众做决定的时代，因此，开发商应当确保公众相信他们的利益会得到关注，他们未来的生活能得到保障，从而愿意支持项目实施（Dallas，1984）。

类似的，赛车跑道建设项目的开发商也注意到：

EIS是说服当地议员、居民和相关利益群体时一个最重要的因素，在新提案的构想中，它要确保制订缓解赛车带来的环境影响的措施。同时大量的环境研究是对反对者进行强有力反击的基础，且为诸如居民委员会和环保局等独立实体提供了保证。如果案例中没有这些内容则该项目无疑要受到公众质询（Hancock，1992）。

从公众的角度看，参与EIA程序可以提高公众在决策制定过程中的话语权，从而完善管理功能，使决策制定更加民主。例如，O' Faircheallaigh（2010）引用的案例中，本土居民拒绝提供信息，除非他们能够充分了解拟建项目，而且有机会与开发商和政府部门共同商议其文化遗产的管理问题。反过来，这样可以改变那些传统上被忽略的支持群体的权利平衡。表6.1总结了EIA中公众参与的主要目的。

表6.1 EIA中公众参与的主要目的

一般目的	具体目的及行为
把获得的公众意见纳入决策	1. 向公众提供信息 2. 填补信息空缺 3. 信息的竞争性/检验开发商提供的信息 4. 解决问题，增强社会学习能力
分享公众参与下制定的决策	5. 反映民主原则/EIA过去常常获得那些受到影响的群体的赞同 6. 民主实践/公众参与具有教育意义 7. 多元化的体现
改变决策中决策权力的分配和决策结构	8. 包括处在社会边缘的群体（或巩固边缘化） 9. 转变决策制定轨迹，如在开发商和当地居民间达成一致

资料来源：O' Faircheallaigh，2010。

6.2.2 有效公众参与的要求和方法

联合国环境规划署（UNEP）列出五个与有效公众参与相关的要素：

- 识别与拟建项目利益相关或受到影响的群体/个人；
- 提供准确、易懂、中肯且及时的信息；
- 在决策者和受影响人群之间开展对话；
- 吸取公众对决策的意见；
- 提供有关措施实施和公众影响决策程度的反馈信息（Clark，1994）。

后文将依次讨论这些内容。

相关利益群体的识别看起来十分简单，但实际上困难重重。“公众”这个简单的词语实际上代表着各相关利益群体的复杂结合，且随着时间和项目的变化会发生变化。大致上，可以将公众分为两大类：第一类由志愿组织、准法定团体或有针对性的团体组成，他们关注的是环境的某一特定方面或整个环境状况；第二类由居住在拟建项目附近可能受到直接影响的人群组成。这两类公众具有的财力和资源明显不同：有组织的团体拥有大量的资金和丰富的专业资源可用于方案的处置，他们重点关注的是开发过程的某些方面，把公众参与作为获得政治支持或全国性宣传的一种方式；而当地居民一般缺少专业技术、教育或经济支撑，也不熟悉有效表达自己观点的相关程序，但他们却是受开发影响最直接的人群（Mollison，1992）。这两类公众的背景迥异，提出的观点自然也不同，"公众"的多样性因此而存在，每个群体都有自己的观点，都可能与其他群体或 EIA"专家"的意见相冲突。

目前，对于这两类公众是否都应当参与决策还存在争议，例如，"坐落在荷兰的 NGO 组织——国际绿色和平组织中拥有高度发言权的成员，是否有权利对地球另一边的某个项目决策表达观点，并试图影响决策"（Clark，1994）。实际中，可以通过法律法规限定具有参与资格的群体或组织，或根据标准判定受到项目直接影响的群体（如距离开发项目一定范围内的居民），并具有针对性地采取一些控制手段。例如，规划法（2008）中指出，拥有或经营拟建项目所占土地的受影响人群与一般的"相关利益群体"区别对待。

EIA 法中对公众做出如下区分：①"公众"是指了解开发许可相关要求、可以有效获取相关环境信息、具有可能决策的性质以及得到详细公众咨询（6.2 节）的人群；②"相关公众"，是有权知晓主管部门发布的环境声明、重点建议、报告以及其他可获得相关信息的人群，而且主管部门必须向他们提供早期参与环境决策制定的机会（6.3 节和 6.4 节）。虽然看起来欧盟成员国都是采用前者这种广义的定义，但实际上也有很多是按后者进行限定。不同成员国是否允许 NGO 参与 EIA 程序取决于这些团体成立了多久、覆盖区域有多大、环境保护是否是其目标之一以及是否合法（COWI 2009）。

拟建项目性质方面信息的缺失和误导也会妨碍公众参与，导致人们对项目的抱怨和批评。因此，公众参与的目标之一就是提供有关开发项目以及可能产生的影响的信息。在 EIS 编制之前，相关信息应该通过公开会议、展览或电话热线等方式提供给公众，这些信息应尽可能公正、真实，人们对回避性、片面性信息警惕性很高，如果不保证信息的公正、真实，会让他们有所担心。早期开展咨询能够影响决策，但咨询开展太早，很多信息可能不真实，不具备讨论的基础，因而要努力维持一种平衡状态。例如，在吸取了 EIS 前期咨询引起问题的教训后，英国的一个开发商决定开展非常具体的咨询活动，但只是在 EIS 发布之后（McNab，1997）。基础设施规划委员会（IPC）在网站上公布了它与相关群体间交流的所有内容，以避免出现任何涉及偏见和幕后处置的控告。

信息传递的方式可以影响公众参与。技术含量较高的信息只有少数公众能够理解，通过不同媒介（例如报纸、广播）传播的信息也只能被部分公众接收。确保公众参与的群体中包括少数族裔和低收入群体（一般不参与决策），可能是一个需要特别关注的问题，尤其是要按照布伦特兰委员会（Brundtland Commission）所强调的，要保证代际公平与参与。Ross（2000）给出一个发生在加拿大的典型案例，案例中由于语言障碍而使技术信息传达极为困难：

据说有一次在一个土著居民区里讨论有机氯化物的扩散时，其中有一位讲克里语的长者只能通过翻译人员翻译参与讨论。棘手的问题是如何将"有机氯化合物"这样一个词语译成克里语。翻译的同行中有一个会讲克里语的成员 Jim Bouncher，他将该词译为"有害的毒药"，解决了这个问题。

在美国，Williams 和 Hill（1996）指出许多传统的环境信息传递方式与少数族裔或低收入群体需求不一致的情况，如：

• 有些机构只关注于书面的研究，而不是积极地为这些群体服务；

• 有些机构通常不了解当下的权利体系，以致它们对社区领导（如低收入教堂的传教士）或协会领导不以为然；

• 有些机构在与目标团体不相关的地方举行会议，例如，在远离项目所在地的市中心召开会议；

• 有些机构举行会议的地点选在一些“阔气”的地方，而不是当地教堂、学校或社区中心，剥夺了公众应有的权利，使他们感觉自己被排除在外；

• 有些机构采用报纸告示、官方杂志出版物和大规模宣传邮件代替电话通知或学校散发宣传单；

• 有些机构准备了很厚的报告，想以此蒙混过关；

• 某些机构利用形式上的表达技巧掩盖不足，例如可升降讲台和幻灯片放映等。

以上几点表明在信息传递过程中应广泛利用各种方法，尤其是那些对很少参与进来的群体有用的技术方法：通过配有图片和地图的 EIS 总结、技术报告、非正式场合会议以及建立社区联系网络、派发宣传单、报纸新闻等与公众联系。

EIA 中公众参与的另一个目标就是在公众和决策者（包括项目的倡导者和授权机构）之间开展对话，确保决策者在决策过程中听取公众意见。它可以帮助人们对 EIS 中的假设、其平等真实性以及替代方案提出质疑（O’ Faircheallaigh，2010）；公众参与能够帮助识别当地居民所关注的问题，这些问题通常不同于开发商或外界专家的观点。因此，公众参与应该实现信息的双向流动，允许当地居民发表意见；公众在参与的同时也可以很好地识别开发商和不同群体需求间的矛盾，理想的结果是解决这些矛盾，并且能够就反映各群体共同目标的未来行动方针达成一致（Petts，1999，2003）。

有效的公众参与方法包括协商技术（如小组座谈会、德尔菲事务委员会、咨询委员会等）和适当的资源（可能通过资金介入）。Petts（2003）指出一些协商参与或对话交流方式的可能性及问题，并强调这类参与是为了使 EIA 程序更加完整而不是一种“扩展”。Balram 等（2003）提出一种有趣的德尔菲方法——基于 GIS 的空间协作德尔菲法（Collaborative Spatial Delphi），空间分析方法在 EIA 公众参与中可能具有巨大的应用潜力，当然互联网和社会网络体系持续快速的升级优化也具有巨大的潜力。由 7.3 节中有关中国香港案例研究的讨论可见，互联网可促进公众参与在 EIA 程序中多个阶段的应用。

Arnstein（1971）定义了“公民参与的八级阶梯”，从不参与（操纵、治疗）到象征性行为（通知、咨询、安抚）再到公民权利（合伙、委托、公民控制），梯级越高，决策中公众意见的体现越多。类似的，Westman（1985）把公众参与按公众权利递增的方式分为四个等级：信息反馈、咨询、联合规划和权利委托。表 6.2 列出了各层次公众影响力优缺点。

表 6.2 各层次公众影响力优缺点

途径	公众权利在决策中的力度	优点	缺点
信息反馈 幻灯片或影片演示、电视、信息工具包、报纸报道或广告、记者招待会、新闻发布、印刷材料、技术报告、网络、公告等	无	信息量大，快速呈现偏差	无信息反馈

续表

途径	公众权利在决策中的力度	优点	缺点
咨询 公开听证会、简报、监察专员或代表、调查、采访、反馈表等	低	允许双向信息传递；允许一定限制下的讨论	不支持持续沟通；有些费时
联合规划 咨询委员会、研讨会、非正式会议、角色扮演、座谈、互动投票、未来计划论坛等	中	允许持续的信息输入和反馈；提高了市民受教育的机会和参与的程度	非常费时；依赖于规划者提供什么样的信息
权利委托 市民评审理事会、市民规划委员会等	高	信息获取途径便捷；允许对决策的选择权和时限做出更有效的调控	需要长期的时间保证；很难在小范围内具有广泛代表性

资料来源：Westman，1985；公众参与国际协会，2001。

公众参与有很多不同的方式。表 6.3 列出了其中的一部分，同时也指出了它们在提供信息、满足特定利益、双向交流以及影响决策方面作用的大小。

表 6.3　公众参与方法及其有效性

方法	提供信息	满足特定利益	双向交流	影响决策
说明性会议、幻灯片或影片演示	√	1/2	1/2	—
小组展示	√	√	√	1/2
公众演示、展览、模型	√	—	—	—
新闻发布、合法公告	1/2	—	—	—
书面评论	—	1/2	1/2	1/2
投票	1/2	—	√	—
外地办事处	√	√	1/2	—
实地考察	√	√	—	—
咨询委员会、特别小组、社区代表	1/2	1/2	√	√
核心人员组成的团队	√	1/2	√	√
市民审查小组	1/2	1/2	√	√
公众质询	√	1/2	1/2	√/—
诉讼	1/2	—	1/2	√/—
示威游行、抗议、暴乱	—	—	1/2	√/—

资料来源：Westman，1985。

但是，不同的利益相关者对于一个特定的 EIA 程序如何有效影响决策制定有着不同的观点。Hartley 和 Wood（2005）就 EIA 中四个废弃物处置点采访了 22 个利益相关者——包括规划官员、一个开发商、地方行动组成员以及普通公众。虽然不同类型的利益相关者对于 EIA 公众参与程序很多方面的观点大致相似，但对于影响决策制定的观点则大相径庭：

案例研究中的所有规划官员都指出，在决策制定过程中要慎重考虑公众代表的意见，制定规划条件时常常要用到公众的建议……但是，行动组成员却坚信他们对于最终决策的影响是有限的，规划项目往往涉及政治，很多决定是在公众了解前已经确定的。

最后，对于任何决策、采取的措施以及公众意见的信息反馈如何影响决策是有效公众参与一个必不可少的部分。例如，在美国，有关 EIS 草案的评论和政府机构的回应也将一起纳入最终的 EIS 中。例如：

• 评论：我强烈反对在森林中使用除草剂，森林不应该受到毒物的污染。

• 回应：您关于反对使用除草剂的意见已纳入全部评论的分析内容中，但 EIS 的证据表明，如果恰当控制，选用低风险的除草剂是可行的，只要严格执行减缓措施，除草剂引起的风险可以降至最低。

如果没有这些信息的反馈，人们很可能会质疑他们的投入都用于何处、他们的参与是否有作用，这个可能影响他们对待后续项目的方式和对正在研究项目的看法。

6.2.3　英国的公众参与程序

欧洲共同体 85/337 号指令第 6 条和第 10 条（a）（后经 97/11 号指令和 2003/35/EC 修订）要求成员国做出以下保证。

• 应在环境决策制定过程前期将开发协议的相关要求、项目接受 EIA 的真实情况以及如何参与到 EIA 程序中告知公众。

• 向公众提供环境信息和其他与决策相关信息的复印件。

• 在开发项目被批准之前，公众关注的是给他们表达观点的机会，此类信息和咨询的具体安排由各成员国自己决定，可根据具体项目特征或场地位置展开：

① 确定相关公众；

② 指定信息咨询的地点；

③ 指定公众获得通知的途径，例如在一定范围内邮寄，在当地的报纸上发表，组织展览相关规划、图、表及模型；

④ 确定咨询公众的方式，例如书面陈述或公众调查；

⑤ 确定程序中各阶段合理的时间限制，以确保在合理的时间范围内做出决策。

• 利益较为相关的公众或权利受损的群众在法庭要求执行决策审核程序。

在英国，这一程序已通过各种 EIA 法规（具有较小的差别）演变成以下的基本要求：在提交开发申请及 EIS 前，必须在当地的报纸上发表公告，并粘贴于拟建项目的所在地，公示 7 天以上。公告中应对拟建开发项目的基本情况进行描述，声明向公众提供 EIS 副本及与项目申请相关的文件，给出获取 EIS 副本的地方和费用，并声明公告粘贴至少 21 天后再向主管部门提交书面申请，EIS 费用的核算一定要合理，包括印制和发放的费用。

《环境影响评价：程序指南》（DETR，2000），针对开发商的政府手册中提到：

在环境报告编制过程中，开发商应考虑是否向普通公众和有关的非法定实体咨询环境问题，这类群体也许能提供一些专业知识……在规划申请未获得批准前，开发商并没有公开提案的义务，此时向当地服务组织和普通民众进行咨询有助于关键环境问题的识别，并且会使开发商处于比较有利的位置，通过减轻负面影响，识别地区性环境问题并完善项目方案。同时，咨询过程可以帮助开发商提前发现问题，这一点可能在正式申请阶段非常重要，例如公众质询过程。

尽管英国大致上已经达到了 EIA 指令中关于公众参与的最低要求，但仍旧是敷衍行事，没有进行扩展（“镀金过程”）。欧洲委员会（EC，2010a）也曾对英国限制公民对环境决策的合法性提出质疑，并警告其将要付出一定的代价，最终可能还是要走向司法审查阶段。但英国这种情况也不足为奇：最近一个关于 EIA 指令（CEC，2009）应用性和有效性的评论指出，纵观各成员国公众参与的执行方式各不相同，但大多数也仅完成了指令的最低要求。

国家重大基础建设项目的情况与上述内容有所不同，规划法（2008）要求这类项目决策应由基础设施规划委员会制定，制定时要依据国家的政策，要有更大的初期投资和更为严格的公众参与要求。[1] 这类项目的开发商需要证明他们已开展了公众咨询，并已在向 IPC 提交申请前按照公众反馈采取了相应措施。之后 IPC 有 28 天的时间决定接受或拒绝该提案，咨询不够充分是拒绝的标准之一。如果某提案被接受，那么公众可以注册登录向 IPC 表达观点，参与公开会议及专题听证会。所有与 EIA 相关的文件均会上传到 IPC 网站。

6.3 向法定咨询机构和其他国家咨询

法定咨询机构对一个国家不同地区的环境状况更为熟悉，因此能对项目的适宜性和可能产生的影响做出更有价值的反馈（Wende，2002；Wood 和 Jones，1997），但是，这些机构可能有自己的偏好，可能导致他们对 EIS 的回应存在偏见。

指令 85/337（修订版）中第 6 条第一款指出：

成员国应采取必要的措施，以确保可能与项目有关的、具有特定环境职责的权力机构有机会就开发商提供的信息和开发协议中的其他要求表达意见。归根结底，成员国应指定被咨询的权力机构……收集信息……反馈给那些机构。咨询的具体安排各成员国可自行拟定。

在英国，不同的法定咨询机构被指派给不同的下放机构（英格兰、威尔士、苏格兰和北爱尔兰），对应不同的开发项目。例如，对于英国的规划项目，法定咨询机构应该是项目所在地的主要委员会、地方规划管理局、英国自然署、海洋管理局和相关环境机构。

EIA 指令同时还要求咨询其他欧盟成员国。当某成员国意识到另一个成员国的某项目可能产生重大的环境影响或可能受到严重影响时，那么承担项目的成员国必须将有关项目、可能的影响和将要制定的决策等信息递交给受影响的成员国。如果受影响成员国表示之后希望参与 EIA 程序，那么后期必须将其作为一个法定咨询机构。在英国，这种跨界咨询机构数量有限，截止到 2009 年共有 12 个，而爱尔兰共和国有 43 个（COWI，2009）。

理想情况下，应该在确定评价范围阶段就已经向这类机构进行了咨询。此外，法律要求要在决策制定前开展此类咨询活动。EIS 一旦完成，开发商或主管部门就可以直接向法定咨询机构提供副本。实际上，很多主管部门只向咨询机构递送 EIS 的部分章节（如只向考古学专家递送有关考古的章节）。但是，这样会限制咨询顾问对于整个项目背景和更大范围内影响的了解。因此，一般情况下应向咨询机构递送完整的 EIS 副本。

---

[1] 联合政府 2010 年对此进行调整，虽然本书编写时没有得到详细信息（Spring，2011）。IPC 与规划督察团将合并，IPC 将为相关制定决策的国务大臣提供建议。

6.4 EIA的呈现

尽管EIA法明确规定了EIS中必须包含的内容，但对于呈现形式并没有给出详细的标准。EIS从过去只有3页的装订报告，发展到电脑制作图表，装订成精装本，再到现在有专门装订设计的多卷文本。本节重点讨论EIS的内容、组织结构、表述的清楚度和呈现方式。

6.4.1 内容和组织结构

一份EIS应该保证其全面性，它的内容至少要满足相关EIA法规的要求。正如我们将要在第8章讨论的，过去的EIS并没有完全达到这些要求。然而，如今这种情况已得到改善，某种程度上是因为LPA越来越重视那些他们认为论证不充分的专题。如果已识别出其他重大影响，那么一份好的EIS应该比最低要求更进一步。多数EIS大致上可分为四个部分：非技术性总结；项目描述、替代方案和环境背景摘要；方法讨论，包括评价范围确定和公众参与；接下来是一系列针对专题的讨论章节，如大气、水、基准情景、可能的环境影响、减缓措施的提出以及遗留影响分析等。理想情况下，EIS还应包括规划条件、环境管理体系及监控方案的提议。《环境影响评价：程序指南》（DETR，2000）附录5给出了大部分或全部信息，见表3.7。表1.1提供了一个很好的EIS大纲示例。

EIS中应该解释清楚为什么有些影响未处置，通过界定“非显著环境影响”来解释某些影响被界定为非显著影响的原因。例如，如果某开发项目不会造成气候影响，那么就应该给出得出该结论的理由。EIS应突出关键性问题，这些问题应该在评价范围界定阶段就已经明确，但在EIA执行过程中也可能出现其他附加问题，那么EIS应说明这些问题的背景情况。同时，EIS中还应列出开发商名称、相关咨询者、相关LPA和咨询机构以及联系人信息，重点规划相关问题及法律法规也应当加以说明。EIS还应列出相关的参考资料，并在结尾给出参考书目。

非技术性总结是EIS中很重要的一部分，公众和决策制定者通常只阅读这一部分文件，因此应总结概括EIS主要结论，包括项目描述、替代方案和减缓措施的提出。第4章给出了EIS一些识别、归纳影响方法的示例，表6.4说明了如何概括项目影响。

表6.4 EIS实例摘要表——影响显著性的权衡

主题范围	影响描述	问题重要性的地理范围					影响	属性	显著性
		I	N	R	D	L			
人类	交通与噪声对当前情况的干扰				*		不利	St,R	较高
	现有居民点的合并			*			不利	Lt,IR	较高
动植物	当地自然保护区有价值草场的减少					*	不利	Lt,IR	较低
	新栖息地的建立					*	有利	Lt,R	较低
	SSSI的休闲压力递增		*				不利	Lt,R	较低

续表

主题范围	影响描述	问题重要性的地理范围					影响	属性	显著性
		I	N	R	D	L			
土壤和地质	120hm^2 耕地的减少(等级 3B)			*			不利	Lt,IR	较低
水体	地表水流失速度加快				*		不利	Lt,IR	较低
	地下水补充减少			*			不利	Lt,R	较低
字符含义	I 国际 N 国内 R 地区 D 区域 L 地方	St 短期 Lt 长期 R 可逆 IR 不可逆							

资料来源：DoE，1995。

注：此处仅列举了部分关键问题。

理想情况下，EIS 应该是一个统一的文件，附录作为第二卷。法院已确定一份 EIS 不应该变成单纯的“纸张追逐”(Berkeley vs. Sos & Fulham 足球俱乐部，2001)：

指令关注的环境声明重点是申请人在申请初期针对相关环境信息和非技术性总结完成的单一、易懂的编辑文档。

EIS 组织结构常见的问题来源于如何评价环境影响。开发商（或协调 EIA 的顾问）常常将 EIA 报告分包给专门从事该领域工作的专家顾问，如生态专家、景观顾问等。他们编写的报告篇幅、格式不一，对项目和未来可能的环境情况做出了一系列假设，提出了多种不同甚至可能冲突的减缓措施。解决此问题的一种方法是在 EIS 主题报告中总结影响预测，然后把详细的报告作为主题报告的附录；另一种方法是以多卷形式呈现 EIS，每一份报告都附上公司封面，每一卷单独讨论一个类型的影响。但这两种方法均存在问题：以附录形式呈现使得很多结论的精华大打折扣，而多卷的形式不便于阅读及携带。两种方法都不能以紧密的方式呈现结论，也没有强调出至关重要的影响，没有整理出清晰条理的减缓及监控措施。一份好的 EIS 应该将各承包单位的报告整合为一份清晰条理的文件，并且采用统一的假设和一定的减缓措施。

EIS 应在清楚表达必要信息的基础上尽可能简练。EIS 主题文本应该包括影响的所有相关分析，而附录应该只表述附加的数据和文件。在美国，一份 EIS 的篇幅不多于 150 页。

6.4.2 表述的清楚度

Weiss (1989) 认为一份难以理解的 EIS 是一种环境风险：

核心问题是文件的质量、在支持环境法规目标方面的有效性以及委托给科研人员进行环境管理工作的质量……难以理解的 EIS 不仅破坏了环境保护法律和环境本身，也会使真心付出的环保工程师变成“污染者”。

一份 EIS 应向各类读者传递信息，从决策者到环保专家到普通基层群众。尽管它可能不能满足所有读者的所有期望，但至少应向着希望大众广泛接受的方向努力。因此应至少做到编写认真、拼写正确、标点恰当、结构清晰、标题鲜明、逻辑合理。在正文前应列出包含页码的内容目录，以方便查找相关信息，基本观点可在文本内容之前或之后以表格形式呈现。

（1）EIS 报告应避免使用过多专业术语，涉及的专业术语均应在文章中加以解释或以脚注形式进行说明。以下案例均来源于真实的 EIS 案例。

• 不妥的表述：Brithdir 缝隙的隔水层特性有所下降，在 Brithdir 与下面的 Rhondda 河底间已经汇集了一定量的地下水……山谷侧面已出现大量渗流，蓄积状态取决于表层沉积物。

• 恰当的表述：目前可接受的评估某地区水禽（如涉禽和野禽）重要性的方法为“1%标准”法。即如果该地区水禽的常规保有量至少占到该物种在英国预估总数的 1%，那么就应该将该地区列为国家级重点保护区。这里的“常规”是指该野禽的拥有数量（通常为年鉴记录中的峰值）至少连续 5 年以上的平均值。

（2）EIS 应清晰说明影响预测所基于的假设，以下分别为不妥的表述和恰当的表述。

• 不妥的表述：当拟建项目延伸出一些潜在的［考古学］问题时，应建立一种方法，以确保考古实验顺利进行。

• 恰当的表述：对每一步操作，都需要对涉及的设备类型和数量做出假设，包括拆除设备——2 台风力破碎机、装载设备；挖掘设备——挖土机、轨道铲车……

（3）EIS 应当具有针对性。尽管判断某一影响的显著性和可能性并不难，也具有一定的保护意义，但 EIS 并不是对未来可能趋势的简单罗列。

• 不妥的表述：在［单轨铁路］的选址和设计过程中，如果选址和范围能与当地环境相协调，将有利于景观的保护。

• 恰当的表述：在道路的指定地区，有很多风轮机置于架空索上，尽管塔台看起来都很小、很模糊，但风轮的转动还是能引起路人的注意。拟建项目造成的景观上的变化对视觉产生了主要影响，其中风轮的密度是主要影响因素。

（4）在可能的情况下，影响的预测应尽量定量化，或者给出一定的范围和定性描述，对类似“一些”“很少”等词语应给予解释说明。

• 不妥的表述：要想使对某住宅区的影响降至最小，那么与其最近的另一住宅区应至少相距 200m。

• 恰当的表述：如果没有支路，城镇中心的交通量到 2008 年将增加 50%～75%；然而，如果有支路，整体的交通量可以下降到 1986 年水平的 65%～75%。

（5）另外，预测应当给出某种影响发生的概率以及预测实现的可信度。对于不确定的情形，EIS 应给出最坏的情况的情景分析。

• 恰当的表述：考虑到交通流量的产生，“最坏情况”是停车场使用率为 100%……，在更接近实际情况的分析中，可以假定停车场的使用率为 50%。

（6）最后，EIS 应当是客观、无偏见的。1991 年一份关于当地权力机构的评论中指出：“许多被调查者认为环境报告更关注的是对拟建项目的支持而不是其影响，因此缺乏客观公正性（Kenyan，1991）。”O’ Faircheallaigh（2010）中提出 EIS 缺乏客观性仍旧是一个问题：EISs 用于证明决策英明而不是评估决策合理性。开发商不可能自己得出项目有重大环境影响应该停止这样的结论。然而，想要在报告编制过程中解决所有重大环境问题也是不太可能的。

• 不妥的表述：拟建项目选址与环礁湖、湿地、沙滩相邻，形成了四块有科研价值的特殊区域（原文如此）。鸟类栖息地的丧失不是主要问题，因为周围还有类似可利用的栖息地。

6.4.3 呈现方式

虽然一份好的EIS重点在于其内容和表述的清晰度，但事实上其呈现形式对它能否被公众接受起着重大作用。EIA是一种间接上与公众相关的行为，EIS则可看作是开发商的宣传文件。好的呈现形式能更好地传达出其对环境的关注、影响分析方法的严谨以及对待公众的积极态度；反过来，呈现形式不佳，就会显得不够关注，或缺少经济支持。类似的，好的呈现形式可以更清楚地传达信息，而不好的呈现，就算EIS内容、组织结构都很好，也会因此受到负面影响。

EIS的呈现形式能在一定程度上体现出开发商的意图。比如纸张类型的选用，是否选用可回收纸张或光滑型纸张，选用较轻的还是较重的，图表做成彩色的还是黑白的，甚至章节之间是否有分隔符等都将对项目形象产生影响。绿色环保型公司会选用可回收纸张双面打印，而一些奢华的公司则会选择强力黏合剂装订的重型光滑纸张。通常情况下，强力黏合剂装订方式更适合于EIS，除非报告书很薄，否则多次翻阅螺旋装订很容易拉断或弯曲。类似的，用订书钉装订的文件也很容易被撕毁。多卷文件很难放置，最好提供一个盒子将它们放在一起。EIS还可通过互联网或制作成CD来提供给公众阅读（图6.2）。

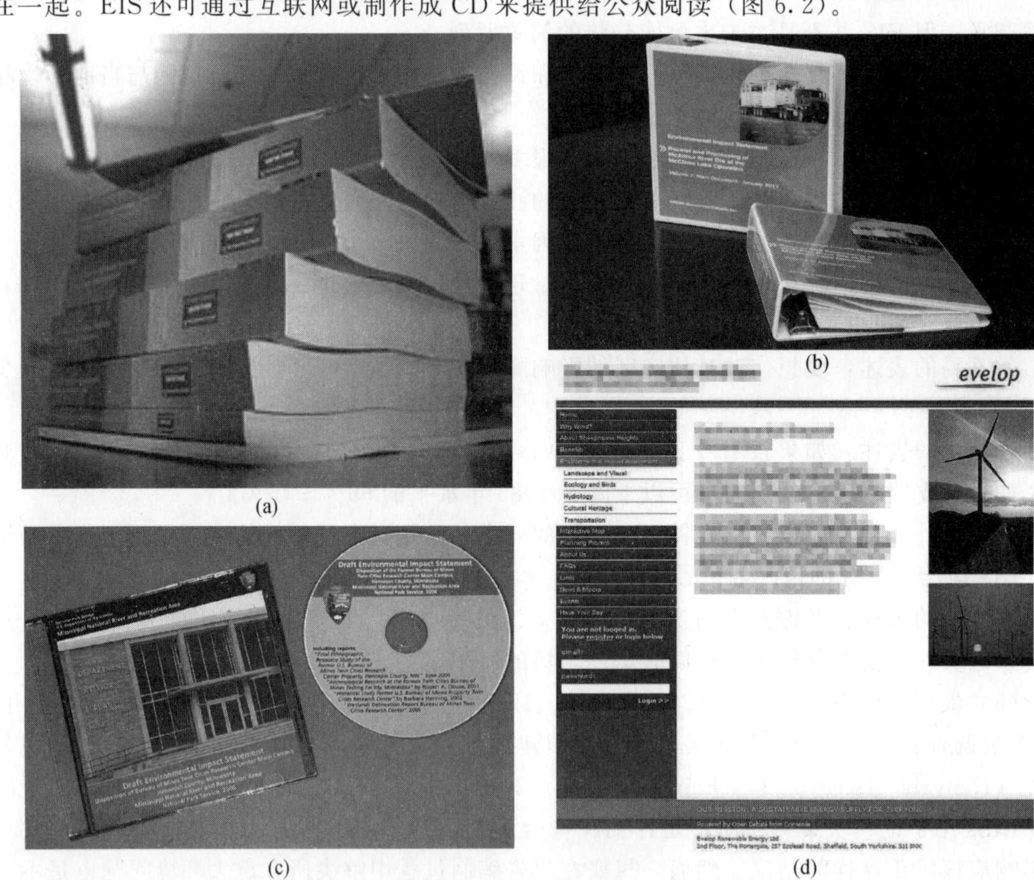

图6.2 EIS的不同呈现类型

［资料来源：(a) Metropolitan Council (Minneapolis/St Paul)；(b) AREVA Resources Canada Inc.；(c) Griff Wigley；(d) Evelop］

采用地图、图表、照片剪辑、图示及其他可视化交流形式可以在很大程度上提升 EIS 的层次。正如我们在第 4 章中提到的，用分布图、项目规划设计图及工程进度表对开发项目进行恰当描述是十分必要的。例如，地图可以简洁明了地显示可见的影响范围、指定区域的具体位置或农业用地类型等信息；在传达数字信息时，曲线图要比一般的表或图形更为有效。这些更为形象的表达形式突破了页面的限制，增加了读者的阅读兴趣。现在越来越多的开发商选择将报告书做成 CD 或上传到互联网上。

6.5 EIS 评审

有一些利益相关者提出希望检验 EIS 的准确性与全面性；开发商则期望能确保不容易受到法律挑战的影响；计划的反对者则想要静观开发商是否会受到法律挑战；法定咨询机构希望确保 EIS 涵盖了所有有关利益公平性话题的相关信息；地方规划机构则想要确保公布决策是否“符合目的”。

DCLG（2011）指南中指出主管部门自身必须对 EIS 足够满意，但未给出相关信息指示应该如何做到：

提交的 ES 首先要让 LPA 自身满意，提交的 ES 应包含法规列表 4 第 1 部分指定的信息以及第 1 部分中的相关信息，然后可对申请人编制 EIS 提出合理要求。为避免 EIA 在申请批准过程中耽搁，需要提前准备更多信息，必要的话，应尽早提出要求。汇集所有恰当评价重大环境影响所需要的相关信息非常重要，这是 EIA 程序的一部分。如果确定重大环境影响的可能性和范围需要测试或调查，那么测试、调查结果也应作为 ES 的一部分……是否因未能将调查结果纳入 ES 而违反了条例，取决于 LPA 已经掌握的信息水平，以及这是否足以使当局能够对可能产生的重大影响做出判断。

许多 EIS 甚至都没有达到法规的最低要求，很少能提供出决策制定所需的完整信息，这些将会在第 8 章展现。在一些国家，如荷兰、加拿大、马来西亚、印度尼西亚等，EIA 委员会或专门的评价小组已经建立了 EIS 评审制度以保证 EIA 程序的质量。但英国目前还没有强制要求决策前对 EIS 进行评审以确保其完整性和准确性。如果 EIS 的不充分、不完整使得规划的申请不能简单地判定无效：主管部门可能会要求进一步完善信息或冒着被上诉的风险拒绝批准。[1]

许多主管部门并不完全具备评价 EIS 充分性和完整性的专业知识和技术。有些机构，尤其是那些很少接到 EIS 的部门，在处理一些技术复杂的 EIS 时，会遇到很大的困难。因而，有时会请咨询机构帮助开展 EIS 评审。还有一些机构是环境管理与评价协会（IEMA）的成员，具有评审资格。其余的机构则很不情愿购买外围专业技术开展评审，尤其是地方支出有限的时候。国际影响评价协会（IAIA）所提倡的技术方法，虽然在实际应用中不常见，但它实现了公众在内的多个团体共同开展 EIS 评审而不仅仅是主管部门（Partidario，

[1] Weston（1997）指出 LPA 需要意识到他们有权要求提供更多信息，否则会被认为默认了当前所提供的信息。例如，在苏格兰采石场扩张被上诉的案例中，记者发现 LPA 在项目申请阶段并未要求提供更多信息，也没有对 EIS 提出反对意见，直到开发时被上诉。

1996)。

为了填补政府遗留的空白，一些组织机构建立了非强制性的评审标准，旨在确保 EIS 分析、体现所有相关信息，并能（在较小程度上）确保信息的准确性。这些评审标准还可以帮助读者快速熟悉拟建项目、评价影响的显著性、决定是否需要采取减缓措施。同时，间接地可以使评审者对 EIA 程序更加熟悉。此外，任何利益相关者均可参与评审过程。

Lee 和 Colley（1990）制定了一个分层级的评审标准。在层级的顶端是一个对报告整体的综合评分（从 A＝充分完整到 F＝很不充足），该评分基于给定的四个子标题的分值确定：一是对开发项目、当地环境和基准环境条件的描述；二是关键影响的识别和评价；三是影响的替代方案和减缓措施；四是结果的交流讨论。反过来，每个标题都是以更下一层的更具体主题或问题为基础。Lee 和 Colley 的评审标准已经直接或经（IEMA）修正后用于英国一些 EIS 的评审。附录 4 给出了该框架的具体内容。

欧盟委员会也发布了一套评审标准（CEC，2001）。该标准与 Lee 和 Colley 提出的标准类似，但采用了七个子标题而不是 Lee 和 Colley 的四个，包含了更多特殊问题，在项目背景、决策制定的重要性以及在 EIS 中存在的相关信息的基础上进行评判。附录 4 中给出的评审标准由牛津布鲁克斯大学的影响评价小组（IAU）提出，是 Lee 和 Colley 框架与欧盟标准的融合与延伸。Rodriguez-Bachiller 和 Glasson（2004）也提出一个 EIS 评审的专家系统方法。

要求每一份 EIS 都完全符合所有的评审标准是不太可能的，同样，一套标准也不可能适用于所有的项目。但是，它们可以作为检查清单来评判那些准备参与评审或正在评审的 EIS 是否是良好实践。表 6.5 列出了很多应用这些标准的可行方法。例（a）涉及最低要求、补充说明或其他关键信息；例（b）包括一个简单的分级，利用 Lee 和 Colley 的方法框架将每种标准分为 A～F 级（此处以一个标准为例）；例（c）采用欧盟标准的形式，对相关信息做出评价，判定决策制定是否完整、充分（不太完整但是不会妨碍决策制定的进程）或不充分。

表 6.5 EIS 评审标准举例

(a)最低法律条件检测			
标准	有/无(页码)	包括的信息	缺少的关键信息
描述提案,包括设计、规模及范围	√(p. 5)	规划选址,现有操作,途径	工作模式,交通工具,修复计划
指出项目开发的实体存在形式	×		建筑位置(位置、规模),修复
(b)各标准简单分级			
标准	有/无(页码)	注释	等级
解释开发目的及意义	√(p. 11)	引言部分简要概括,更多细节见第二部分	A
给出预期建设时间等阶段	√(p. 12)	尚未停运	B
(c)相关性完整性检测			
标准	是否相关(是/不是)	判断(C/A/I)	注释
考虑闲置的替代方案、备选流程等	Y	A	讨论替代场址,不是替代流程
如已识别意料之外的严重负面影响,则需要重新评估替代方案	N		充分了解沙地/砾石场地作业的影响

注：C＝完整；A＝充分；I＝不充分。

如果LPA认为EIS没有提供他们要求的、能使他们对拟建项目可能的环境影响有全面认识的所有信息，那么LPA必须要求开发商进一步完善信息，且任何添加的信息都必须再次公布并经过公众咨询过程。[1]

6.6 项目决策

6.6.1 EIA和项目授权

批准或驳回项目的决策者可分为几个等级：

位于顶端的是国家的相关部长……然后是主督察员，有时也称为书记员（苏格兰）；再往下依次为议员，地区民选议员，县、乡、单一的或大城市的自治区委员会；处在底层的是首席或高级规划官员，他们负责处理“委托的决策”……基本上项目越大，决策者在金字塔中所处的等级越高。(Weston，1997)

在英国，不同的管辖区应用不同的决策制定规则。例如，《城乡规划法案》1990（对比苏格兰的法律法规）要求规划决策的制定必须考虑到“与开发规划相关的准备工作，从材料到应用软件再到其他物质需求”。虽然还没有成形的物料规划，但通常包括可见的便利设施、噪声、交通、自然保护及其他EIA涵盖的内容。我们希望的是在国家规划政策框架的指导下进行调整，而不是介绍“支持可持续发展的推测”，换言之：

在符合自然环境、建筑、经济及社会标准，符合当地规划，并能上缴一定税费以补偿社区便利设施的损失和附加基础设施的成本的条件下，个人或企业有权修建房屋和其他地方建筑。(Parliament 2011)

针对国家重点基础设施项目申请，基础设施规划委员会（IPC）必须参考当地的影响报告、规定条款以及其他IPC认为与决策相关的重要事项做出决策；必须确定该申请与其他国家政策声明是否一致，尤其是拟建项目的不利影响是否大过有利影响（规划法2008）。但是，在所有的案例中，只要项目需要开展IEA，决策者就必须在做出决策时考虑“环境信息”，这些信息包含在EIS、法定咨询机构评审材料、公众陈述及其他材料中。

主管部门要求授权申请书必须与EIS一同上交。除非开发商同意延期，否则主管部门要在指定时间内完成批复（如地方机构规划申请的批复时间为16周）。不论在何种标准下，开发项目的决策制定都是一件很复杂的事情。项目决策对EIAs的要求使其也逐渐趋于复杂化，因为项目越大越复杂，其涉及的相关利益团体可能就越多：

利益冲突不仅仅发生在开发商和咨询机构之间，也可能发生在咨询机构与咨询机构之间，这样使得夹在中间的开发商很难满足各利益团体的需求。此时，“主管部门”需要建立新的平衡关系。(Weston，1997)

然而即使完成了EIA，决策制定过程前期基本上还是暗箱操作，后期往往会再做些努力试图使过程看起来严谨、透明。近期，曼彻斯特大学（Wood和Jones，1997）和牛津布鲁克斯大学（Weston等，1997）重点研究了环境信息是如何纳入英国决策制定过程的，这将

[1] 如果项目未被授权就开工建设，那么主管部门在决定是否拆除前要将环境信息纳入考虑范围。

在第8章中详细讨论。政府优秀实践导则中关于规划项目中环境信息的评估首先要从评估的定义说起（DoE，1994）：

在环境评价内容中，有很多不同的评价阶段和评价等级，主要涉及以下几方面内容：

① 检验ES及其他信息来源所提供的信息充分与否；

② 衡量单个环境影响的大小、重要性、显著性以及对周围特定区域的影响；

③ 制定整体的环境“权重”和其他需要考虑的指标，为规划决策的制定奠定基础。

该导则建议，在审查EIS和项目申请、公布项目建议和EIS、得出相关结果之后，规划机构应当进行决策制定的两个阶段：评估单个环境因素的影响和效应，权衡各方面信息做出决定。影响与效应的评估首先应核实EIS中所述是否属实，突出显示相关陈述并与开发商进行讨论；其次核实特殊影响的属性和特点，要么EIS已经进行了这方面的分析（如表6.4所示），要么由案例工作者制定这样的表格；最后，权衡该影响的显著性和重要性，将受影响区域的程度、范围、可能性、减缓措施的范围以及问题的重要性等纳入考虑范围之内。

对任何项目授权申请均有多种决策选项：主管部门可以（有条件或无条件地）批准或拒绝项目，也可以建议其商议后进一步提出减缓措施，并尝试与开发商就这些问题进行磋商。如果项目被拒绝，开发商可就该决议提出申诉；如果项目被批准，公众或组织也可以就此提出质询。相关的国务大臣也可以以不同理由“召回”申请，公众质询也许会有相应结果。

但决策制定不是临床试验。决定函常常被描述成“一封给失败者的信”，体现了决策制定者和决策接受者之间的个人关系。在一个加拿大的案例中，Ross（2000）解释了他和同事Mike Fanchuk是如何通过编写报告来解释他们是否批准关于纸浆厂的决策的：

Mike Fanchuk是一个农民，住在纸浆厂北部……在我和Mike的工作中，我们讨论了什么情况下我们才能对报告满意，从而愿意签署它……我认为Mike的方法是我遇到的最好的，他认为，如果未来几年之后，他能够开心地告诉他八岁的孙女他曾经参与了这个小组并批准了这个报告，那么他就愿意签署。从理论上讲，这种代际公平很好地阐述了可持续发展的原则……更重要的是，它说明以自己所做的工作而骄傲也是人类的基本精神需求之一。

初始决策还需要经过其他人的同意，在此阶段也有可能被推翻。在一个规划申请案例中，规划官员的意见也将递交给规划委员会，然后做出最终决策。在英国，IPC的决定将会提交给有关的国务大臣。其他程序将会应用到其他管辖区域。

6.6.2 EIA和公众质询

如果开发商质疑驳回的规划许可，或者最终决策未能按时做出，或国务大臣因为一些原因还有其他考虑，那么国务大臣就可能“召回”规划申请。如果开发商或地方规划机构提出要求，那就必须进行公众质询。质询过程中，各团体均可提供证据，并与其他团体进行交叉询问。这一过程对任一方来说都花费较高，且时间战线较长。例如，希思罗机场五号航站楼的公众质询持续了将近4年。Weston（1997）论证了为什么所有EIA涉及的团体都尽力避免公众质询：

当一个项目成为公众质询的焦点时，双方都被吸引，听证会将变成反对者和拥护者争论的主战场，项目反对者有专家指导并且专家指导费用很高；项目支持者们则受过专业培训，教唆证人服从或反对某意见。这些争论很少是理性的，或者说与系统性、可重复性、强调合作等特点的EIA关系并不大。当鉴定证人和中立的辩护律师得出全部调查结果并受到公众质询时，一切已经晚了。

尽管这样，还是已经有上百例的 EIA 项目已走到公众质询的阶段。

拟建项目的环境影响，特别是涉及交通、景观和便利设施等问题时，在任何质询中都必然会受到仔细审核。EIA 法规允许质询督察员和国务大臣提出要求：①如果他们认为合适，就可以在公众质询前提交 EIS；②如果他们认为 EIS 不能充分表达他们的观点，可以要求开发商提供更多信息。实际上，进行涉及 EIAs 的公众质询前，督察员通常会收到一份包含 EIS 的拟建项目案卷，通过核实这些文件来决定是否需要提供更多信息。如果督察员认为需要更多信息，那么就要召开前期质询会议。这类会议或许可以帮助开发商和主管部门在质询开始前就一些问题达成一致，同样也可以避免质询过程中出现不必要的延迟。在质询阶段，督察员通常会要求提供更多信息，如果不能在指定时间内呈交的话，质询会被延期举行。EIS 所包含的内容也属于提交材料的考虑范畴。但是，EIS 不充分不能作为阻止授权甚至拖延质询的正当理由。[1]

一份对 10 个已编制 EIS 的项目开展的公众质询的分析表明（Jones 和 Wood，1995），大多数督察员对 EIS 和 EIS 中的质询给出“中等”或“相当大的”重要性的权衡意见，环境信息对于是否批准申请的决策制定具有“合乎情理”的重要性。但是，接下来对 54 封督察员所出的决定函的研究显示，EIA 对质询过程的影响微乎其微：在 2/3 的案例中，督察员和国务大臣的决策依据是国家或地方的土地利用政策，其余案例中，传统的规划问题占主导地位。

主导督察员和国务大臣决策书的主题是传统规划所需要考虑的项目。例如生活便利设施、各种形式的风险、交通和供给，也包括一些动植物、噪声、景观等因素，这些因素会被分别讨论。（Weston 1997）

6.6.3 挑战决策：司法评审

英国的规划体系对于未被批准申请的上诉并没有官方规定。但如果申请被批准，第三方可以在法庭上按照司法审查程序或通过欧盟委员会对决策提出质疑。

在英国，要在 6 周或 3 个月内（取决于决策类型）及时、有针对性地提出关于决策的司法审查申请。司法审查程序首先要求第三方证明其具有起诉权，即其因一些特殊原因而与项目紧密相关，或其情况不同于其他的团体（如财政利益或健康利益）。起诉权的确定是司法审查的主要难点之一。[2] 起诉权一旦确立，第三方必须在法庭上证明主管部门未按照相关 EIA 程序执行，法庭的目的是审查主管部门做出决策的过程，不会评判案例的功过。

法庭仅会在以下情况下撤销主管部门的决策：主管部门的行为未经授权或超过权限，或未遵守自然公正原则，或应用了错误的法律条文，或决策太不合理。（Atkinson 和 Ainsworth，1992）

在这一过程中可能出现各种情况。主管部门可能未要求应进行 EIA 的项目开展 EIA；

---

[1] 例如，在苏格兰采石场扩张被上诉的案例中提到（苏格兰政府，P/PPA/SQ/336，6/1/1992）：ES 遭到了强烈的批评……它无法证明对环境影响进行了恰当的分析……尽管该 ES 还存在一些缺陷，但大体上符合 EIA 法规的相关要求。

[2] 例如，一个欧盟委员会法庭的案例中裁定，绿色和平组织对于使用地方基金在加那利群岛修建发电站所提出的质疑在个别问题上欠考虑［Greenpeace vs. Commission of the European Communities，Journal of Environmental Law，8 (1996)，139］。英国也出现过相似的评判。COWI（2009）从不同成员国的角度提供了更多信息。

可能有不具备正式授权权力的规划官员做出规划决策；规划当局可能在未被授权、未充分考虑所有相关环境信息的情况下做出决策。这样一来，决策是没有意义的。还有部分案例是一直围绕 EIS 中规划申请描述应详细到什么程度而徘徊不前，详见 8.6 节。虽然近期英国法庭上解释了 EIA 指令要求涉及范围较宽、应用广泛，但它仍无法与美国法庭在与 NEPA 关系中扮演的角色相媲美。

6.6.4 挑战决策：欧盟委员会

欧盟委员会是第三方对主管部门批准项目开发或对不要求 EIA 的项目提出质疑的另一个途径。此类案例需要证明英国没有履行其作为《罗马条约》成员国的义务，未完全执行欧盟的法规，在这里主要为指令 85/337。这种情况下，条约中 169 条允许申报不符合欧洲法庭的判决。此时，起诉权的确定就不再是问题，因为欧盟委员会可以通过其自身的主动权或任何人提出的申诉来启动程序。但要采用这种机制，委员会必须首先向成员国清楚陈述案例并征求他们的意见，然后提出一个合理的观点。若在规定时间内，成员国还是不能认同，那么该案例就按照程序接受欧洲法庭的审判。

根据《罗马条约》第 171 条，如果欧洲法庭的判决认为某成员国未按照规定履行义务，那么该成员国需要依照判决采取必要的措施。根据第 186 条，欧盟委员会可以采取临时措施要求成员国终止行动，直到对重要问题都做出了决策。但是，若想达到该目的，委员会必须证明紧急处置的必要性，并且一旦证明这些措施未采取，相关利益团体将遭受不可挽回的损失。关于该程序的更多信息，读者可参考 Atkinson 和 Ainsworth（1992）、Buxton（1992）以及 Salter（1992a，b，c）的研究成果。

在欧盟委员会处理的环境侵害案例中，10%与环境影响评价有关，2008～2009 年间，案件数量最多的国家是意大利、西班牙、爱尔兰、法国和英国（EC，2010b）。

6.7 小结

积极的公众参与、与各方咨询机构的全面沟通、清晰的呈现，是一个成功的 EIA 中几个非常重要的方面。尽管《奥胡斯公约》的地位有所转变，但至今这些仍未得到足够的重视。环境信息的呈现已经有所改善，法定咨询机构对 EIA 程序逐渐熟悉，但公众参与仍是比较薄弱的一方面，除非开发商和主管部门发现公众参与的利益大于成本。

评审可在一定程度上确保在决策制定阶段充分考虑 EIS，虽然关于决策制定过程中环境信息利用有一些非强制性的评审框架或政府建议，但事实上却很少开展正式的 EIA 评审。EIS 质量与规划决策价值的联系将在第 8 章中进一步讨论。

那些反对批准开发项目或反对主管部门未能较好执行 EIA 程序的申诉，可向英国法庭或欧盟委员会递交。历史上，英国法庭对于 EIA 法规中各项要求的解释相对狭隘，近年来有所改观。相反，欧盟委员会已经就英国政府对于指令 85/337 的执行情况提出质疑，而且很多具体的决策就来源于该指令的执行。

可以肯定的是，一个良好的 EIA 程序中接下来的一步是监控开发过程的实际影响以及实际影响和预测影响的对比。此内容将在下一章中讨论。

问 题

下列问题旨在帮助读者聚焦本章要点。

1. 公众参与与公众咨询常在运用中作为同义词出现，二者有什么区别？

2. 不同的利益相关者需要通过不同的技术方法了解一个项目的环境影响并为项目规划过程提供最佳的投入。假定你负责设计一个关于风电场（或你熟悉的其他类型的项目）EIS 的公众参与计划。以表 6.3 为基础，对于当地居民，你将选择哪三种方法？同理对于以景观为理由反对风力发电的国家层级的非政府组织和负责生物多样性保护的政府机构，你将分别采用哪三种方法？（注意：不同组别可以采用相同的方法）

3. EIA 指令的引言部分提到公众参与时描述如下："公众及私人项目开发申请应在预评估之后才能同意；而评估必须基于开发商提供的适当信息，当局与其他关注项目的人可进行补充。"那么如何改写可将公众参与在 EIA 中的地位上升到更高的层次？

4. 6.2.3 节中列举了一些英国开展公众参与在程序上的要求，它们是否与 EIA 指令引言部分所提出的公众参与的层次相吻合？（参考问题 3）

5. 根据你选择的国家和规划类型，确定咨询机构。

6. 用一页纸的篇幅，用非技术性方法解释 EIA 程序。

7. 解释 6.4.2 中不妥的表述中的不恰当之处，并试着将其改写为恰当的表述。

8. 图 6.2 中显示了一份 EIS 的不同呈现形式，针对以下群体，你认为哪种最有效：(1) 技术性咨询机构；(2) 能够使用网络及不能够使用网络的人群；(3) 阅读小号印刷字有困难的人群（或完全不能的）；(4) 难以支付所谓印刷出版 EIS 所需"合理成本"的人群。

9. 表 6.5 中各分级方法［(a) ～ (c)］分别用于什么情况下？

6

Participation, presentation and review

6.1 Introduction

One of the key aims of the EIA process is to provide information about a proposal's likely environmental impacts to the developer, public, statutory consultees and decision-makers, so that a better decision may be made. Consultation with the public and statutory consultees in the EIA process can help to improve the quality, comprehensiveness and effectiveness of the EIA, as well as ensuring that the various groups' views are adequately taken into consideration in the decision-making process. Consultation and participation can be useful at most stages of the EIA process:

- in determining the scope of an EIA;
- in providing specialist knowledge about the site;
- in suggesting alternatives;
- in evaluating the relative significance of the likely impacts;
- in proposing mitigation measures;
- in ensuring that the EIS is objective, truthful and complete; and
- in monitoring any conditions of the development agreement.

As such, how the information is presented, how the various interested parties use that information, and how the final decision incorporates the results of the EIA and the views of the various parties, are essential components in the EIA process.

Traditionally, the British system of decision-making has been characterized by administrative discretion and secrecy, with limited public input (McCormick 1991). However, there have been moves towards greater public participation in decision-making, and especially towards greater public access to information. In the environmental arena, the UK Environmental Protection Act of 1990 requires the Environment Agency and local authorities to establish public registers of information on potentially polluting processes; *Neighbourhood statistics*, *data.gov.uk* and other government data sources provide environmental data in a publicly available form; the public participation requirements of the Aarhus Convention, 2003 amendments to the EIA and SEA Directives, and Planning Act 2008 allow greater public access to information previously not compiled, or considered confidential; and Directive 2003/4/EC, which requires Member States to make provisions for freedom of access to information on the environment, has been implemented in the UK through the Environmental Information Regulations 2004. The 2009 review on the operation of the 2003 amended EIA Directive (CEC 2009) provided some positive feedback on public participation, especially from the new Member States, which, overall, see the EIA directive as contributing directly to the consolidation of democratic development by securing fundamental rights based on improvements in public participation rights and adding transparency in decision making.

However, despite the positive trends towards greater consultation and participation in the EIA process and the improved communication of EIA findings, both are still relatively underdeveloped in the UK. Few developers make a real effort to gain a sense of the public's views before presenting their applications for authorization and EISs. Few competent authorities have the time or resources to gauge public opinion adequately before making their decisions. Few EISs are presented in a manner that encourages public participation.

This chapter discusses how consultation and participation of both the public (Section 6.2) and statutory consultees (Section 6.3) can be fostered, and how the results can be used to improve a proposed project and speed up its authorization process. The effective presentation of the EIS is then discussed in Section 6.4. The review of EISs and assessment of their accuracy and comprehensiveness are considered in Section 6.5. The chapter concludes with a discussion about decision-making and post-decision legal challenges.

6.2 Public consultation and participation

This section discusses how 'best practice' public participation can be encouraged.[1] It begins by considering the advantages and disadvantages of public participation. It then discusses dimensions of effective public participation and reviews methods for such participation. Finally, it discusses the UK approach to public participation, including the changes brought about by the Planning Act 2008. The reader is also referred to the Audit Commission (2000) and IEMA (2002) for further information on public participation; and to the NGO Forum on

ADB (2006) for a good example of a guide to EIA participation written with the public in mind.

6.2.1 *Advantages and disadvantages of public participation*

Developers do not usually favour public participation. It may upset a good relationship with the LPA. It carries the risk of giving a project a high profile, with attendant costs in time and money. It may not lead to a conclusive decision on a project, as diverse interest groups have different concerns and priorities; the decision may also represent the views of the most vocal interest groups rather than of the general public. Most developers' contact with the public comes only at the stage of planning appeals and inquiries; by this time, participation has often evolved into a systematic attempt to stop their projects. Thus, many developers never see the positive side of public participation.

Historically, public participation has also had connotations of extremism, confrontation, delays and blocked development. In the USA, NEPA-related lawsuits have stopped major development projects, including oil and gas developments in Wyoming, a ski resort in California and a clearcut logging project in Alaska (Turner 1988). In Japan in the late 1960s and early 1970s, riots so violent that six people died delayed the construction of the Narita Airport near Tokyo by five years. In the UK, perhaps the most visible forms of public 'participation' have been protesters wearing gas masks at nuclear power station sites, being forcibly evicted from tunnels and tree houses on the Newbury bypass route (which cost more than £6 million for policing before construction even began), and setting up a protest camp at Simpson where a third Heathrow runway would have been built. More typically, all planners are familiar with acrimonious public meetings and 'ban the project' campaigns. Public participation may provide the legal means for intentionally obstructing development; the protracted delay of a project can be an effective method of defeating it.

On the other hand, from a developer's point of view, public participation can be used positively to convey information about a development, clear up misunderstandings, allow a better understanding of relevant issues and how they will be dealt with, and identify and deal with areas of controversy while a project is still in its early planning phases. The process of considering and responding to the unique contributions of local people or special interest groups may suggest measures the developer could take to avoid local opposition and environmental problems. These measures are likely to be more innovative, viable and publicly acceptable than those proposed solely by the developer. Project modifications made early in the planning process, before plans have been fully developed, are more easily and cheaply accommodated than those made later. Projects that do not have to go to inquiry are considerably cheaper than those that do. Early public participation also prevents an escalation of frustration and anger, so it helps to avoid the possibility of more forceful 'participation'; Figure 6.1 shows three quite different examples of participation in EIA. The implementation of a project generally proceeds more cheaply and smoothly if local residents agree with the proposal, with fewer protests, a more willing labour force, and fewer complaints about impacts such as noise and traffic.

Figure 6.1 Different forms of public 'participation':

(a) public meeting about the proposed Hinkley 'C' nuclear power station;

(b) protest against the proposed Southampton biomass plant;

(c) some Newbury bypass route protest methods

Sources: (a) EDF Energy 2009; (b) Daily Echo 2011; (c) Guardian 1996

For instance, the conservation manager of Europe's (then) largest zinc/lead mine noted that:

> properly defined and widely used, [EIA is] an advantage rather than a deterrent. It is a mechanism for ensuring the early and orderly consideration of all relevant issues and for the involvement of affected communities. It is in this last area that its true benefit lies. We have entered an era when the people decide. It is therefore in the interests of developers to ensure that they, the people, are equipped to do so with the confidence that their concern is recognized and their future life-style protected. (Dallas 1984)

Similarly, the developers of a motor-racing circuit noted:

> The [EIS] was the single most significant factor in convincing local members, residents and interested parties that measures designed to reduce existing environmental impacts of motor racing had been uppermost in the formulation of the new proposals. The extensive environmental studies which formed the basis of the statement proved to be a

robust defence against the claims from objectors and provided reassurance to independent bodies such as the Countryside Commission and the Department of the Environment. Had this not been the case, the project would undoubtedly have needed to be considered at a public inquiry. (Hancock 1992)

From the public's point of view, participation in an EIA process can increase people's say in decision-making, thus improving governance and making decision-making more democratic. For instance, O' Faircheallaigh (2010) cites cases where indigenous people would not release information unless they were fully informed about the proposed project and given the opportunity to negotiate with the development and government on management of their cultural heritage. This, in turn, can change the power balance in favour of groups that have traditionally been marginalized. Table 6.1 summarizes the main aims of public participation in EIA.

Table 6.1 Purposes for public participation in EIA

Broad purpose	Specific purposes and activities
Obtain public input into decisions taken elsewhere	1 Provide information to public
	2 Fill information gaps
	3 Information contestability/testing information provided by the developer
	4 Problem-solving and social learning
Share decision making with public	5 Reflect democratic principles/EIA used to obtain the consent of those affected
	6 Democracy in practice/participation as an educative function
	7 Pluralist representation
Alter distribution of power and structures of decision making	8 Involve marginalised groups(or, alternatively, entrench marginalization)
	9 Shift the locus of decision making, e.g. to agreements between developers and local people

Source: Adapted from O' Faircheallaigh 2010

6.2.2 Requirements and methods for effective participation

The United Nations Environment Programme lists five interrelated components of effective public participation:

- identification of the groups/individuals interested in or affected by the proposed development;
- provision of accurate, understandable, pertinent and timely information;
- dialogue between those responsible for the decisions and those affected by them;
- assimilation of what the public say in the decision; and
- feedback about actions taken and how the public influenced the decision. (Clark 1994)

These points will be discussed in turn.

Although the *identification of relevant interest groups* seems superficially simple, it can be fraught with difficulty. The simple term 'the public' actually refers to a complex amalgam of interest groups, which changes over time and from project to project. The public can be broadly classified into two main groups. The first consists of the voluntary groups, quasi-statutory bodies or issues-based pressure groups that are concerned with a specific aspect of the environment or with the environment as a whole. The second group consists of the people living near a proposed development who may be directly affected by it. These two groups can have very different interests and resources. The organized groups may have extensive financial and professional resources at their disposal, may concentrate on specific aspects of the development, and may see their participation as a way to gain political points or national publicity. People living locally may lack the technical, educational or financial resources, and familiarity with relevant procedures to put their points across effectively, yet they are the ones who will be the most directly affected by the development (Mollison 1992). The people in the two groups, in turn, come from a wide range of backgrounds and have a wide variety of opinions. A multiplicity of 'publics' thus exits, each of which has specific views, which may well conflict with those of other groups and those of EIA 'experts'.

It is debatable whether all these publics should be involved in all decisions; for instance, whether 'highly articulate members of the NGO, Greenpeace International, sitting in their office in Holland, also have a right to express their views on, and attempt to influence, a decision on a project that may be on the other side of the world' (Clark 1994). Participation may be rightly controlled by regulations specifying the groups and organizations that are eligible to participate or by criteria identifying those considered to be directly affected by a development (e. g. living within a certain distance of it). For instance, under the Planning Act 2008, 'affected persons' who own or manage the land proposed for development are treated differently from more generally 'interested parties'.

The EIA Directive distinguishes between (1) 'the public', who must be informed of the request for development consent, the availability of accompanying environmental information, the nature of the possible decision, and details of arrangements for public consultation (Article 6. 2); and (2) 'the public concerned', to whom the environmental statement, main advice and reports issued to the competent authority, and other relevant information must be made available, and who must be given an early and effective opportunity to participate in the environmental decision-making procedures (Articles 6. 3 and 6. 4). Although all Members States seem to use a broad definition for the former, many are more restrictive for the latter. Different Member States permit NGOs to participate in the EIA process depending on how long they have existed, their regional coverage, whether environmental protection is one of their objectives, or whether they are legal entities (COWI 2009).

Lack of information, or misinformation, about the nature of a proposed development prevents adequate public participation and causes resentment and criticism of the project. One objective of public participation is thus *to provide information* about the development and its likely impacts. Before an EIS is prepared, information may be provided at public meetings, exhibitions or telephone hotlines. This information should be as candid and truthful as possible: people will be on their guard against evasions or biased information, and will look for confirmation of their fears. A careful balance needs to be struck between consultation that is early enough to influence decisions and consultation that is so early that there is no real information on which to base any discussions. For instance, after several experiences of problematic pre-EIS consultation, one UK developer decided to conduct quite elaborate consultation exercises, but only after the EISs were published (McNab 1997). The Infrastructure Planning Commission publishes all of its communications from and with any parties on its website, to avoid any accusations of bias and behind-the-scenes dealings.

The way information is conveyed can influence public participation. Highly technical information can be understood by only a small proportion of the public. Information in different media (e. g. newspapers, radio) will reach different sectors of the public. Ensuring the participation of groups that generally do not take part in decision-making—notably minority and low-income groups—may be a special concern, especially in the light of the Brundtland Commission's emphasis on intragenerational equity and participation. Ross (2000) gives a compelling example of the difficulties of communicating technical information across a language barrier in Canada:

> At one of the hearings in an aboriginal community, there was a discussion of chlorinated organic emissions involving one of the elders, who was speaking in Cree through a translator. The translator needed to convey the discussions to the Elder. The difficult question was how to translate the phrase 'chlorinated organic compounds' into Cree. Fellow panel member Jim Boucher … who spoke Cree, listened to the translator, who had solved the problem by using the translation 'bad medicine'.

Williams and Hill (1996) identified a number of disparities between traditional ways of communicating environmental information and the needs of minority and low-income groups in the US. For instance agencies:

- focus on desk studies rather than working actively with these groups;
- often do not understand existing power structures, and so do not involve community leaders such as preachers for low-income churches, or union leaders;
- hold meetings where the target groups are not represented, for instance in city centres away from where the project will be located;
- hold meetings in large 'fancy' places which disenfranchised groups feel are 'off-limits', rather than in local churches, schools or community centres;
- use newspaper notices, publication in official journals and mass mailings instead of telephone trees or leaflets handed out in schools;
- prepare thick reports which confuse and overwhelm; and
- use formal presentation techniques such as raised platforms and slide projections.

These points suggest that a wide variety of methods for conveying information should be used, with an emphasis on techniques that would be useful for traditionally less participative groups: EIS summaries with pictures and maps as well as technical reports, meetings in less formal venues, and contact through established community networks, as well as through leaflets and newspaper notices.

Public participation in EIA also aims to *establish a dialogue* between the public and the decision-makers (both the project proponent and the authorizing body) and to ensure that decision-makers *assimilate the public's views* into their decisions. It can help to challenge the underlying assumptions behind an EIS, its balance and veracity, and the alternatives it considers (O' Faircheallaigh 2010). Public participation can help to identify issues that concern local residents. These issues are often not the same as those of concern to the developer or outside experts. Public participation exercises should thus achieve a two-way flow of information to allow residents to voice their views. The exercises may well identify conflicts between the needs of the developer and those of various sectors of the community; but this should ideally lead to solutions of these conflicts, and to agreement on future courses of action that reflect the joint objectives of all parties (Petts 1999, 2003).

Effective public participation methods could include deliberative techniques, such as focus groups, Delphi panels and consultative committees, plus appropriate resourcing, perhaps through intervenor funding. Petts (2003) highlights some of the possibilities and problems of deliberative participation, or communication through dialogue; she also stresses the need for such participation to be integral to the EIA process rather than an 'add-on'. Balram *et al.* (2003) provide an interesting development of the Delphi approach, Collaborative Spatial Delphi, using a GIS-based approach, and there may be considerable potential for using spatial technology in participation in EIA. There is then of course the great potential of the continuously and rapidly evolving Internet and social networking systems. As discussed in the Hong Kong of China case study in Section 7.3, the Internet can be used to facilitate participation at several stages in the EIA process.

Arnstein (1971) identified 'eight rungs on a ladder of citizen participation', ranging from nonparticipation (manipulation, therapy), through tokenism (informing, consultation, placation), to citizen power (partnership, delegated power, citizen control).

Assimilation of what the public say in the decision is likely to be higher the further up the ladder one goes. Similarly, Westman (1985) has identified four levels of increasing public power in participation methods: information-feedback approaches, consultation, joint planning and delegated authority. Table 6.2 lists advantages and disadvantages of these levels.

Table 6.2 Advantages and disadvantages of levels of increasing public influence

Approaches	Extent of public power in decision-making	Advantages	Disadvantages
Information feedback			
Film or Powerpoint presentation, television, information kit, newspaper account or advertisement, news conference, press release, print materials, technical report, website, notice, etc.	Nil	Informative, quick presentation subject to bias	No feedback

Continued

Approaches	Extent of public power in decision-making	Advantages	Disadvantages
Consultation			
Public hearing, briefing, ombudsperson orrepresentative, survey, interview, responsesheets, etc.	Low	Allows two-way information transfer; allows limited discussion	Does not permit ongoing communication; somewhat time-consuming
Joint planning			
Advisory committee, workshop, informal meeting, role playing, panels, interactive polling, future search conference, etc.	Moderate	Permits continuing input and feedback; increases education and involvement of citizens	Very time-consuming; dependent on what information is provided by planners
Delegated authority Citizens' review board, Citizens' planning commission etc	High	Permits better access to relevant information; permits greater control over options and timing of decision.	Long-term time commitment; difficult to include wide representation on small board.

Source: Adapted from: Westman 1985; International Association for Public Participation 2001

There are many different forms of public participation. A few are listed in Table 6.3, along with an indication of how well they provide information, cater for special interests, encourage dialogue and affect decision-making.

However, different stakeholders may have very different views on how effective a given EIA process is in influencing a planning decision. Hartley and Wood (2005) interviewed 22 stakeholders—planning officers, a developer, local action group members, and members of the public—involved in the EIAs for four waste disposal sites. Although the different types of stakeholders had broadly similar views on many aspects of the EIA public participation process (e.g. timing, information provision), they differed significantly in their views of its influence on decision-making:

> The planning officers in all the case studies indicated that public representations are carefully considered in the decision-making process and that suggestions from the public are often used when formulating planning conditions … However, members of action groups believed that their influence on the final decision was limited, that the planning process was too political and that decisions had largely been made before they were informed…

Finally, an essential part of effective public participation is *feedback about any decisions* and actions taken, and how the public's views affected those decisions. In the US, for instance, comments on a draft EIS are incorporated into the final EIS along with the agency's response to those comments. For example:

- *Comment*: I am strongly opposed to the use of herbicides in the forest. I believe in a poison-free forest!

Table 6.3 **Methods of public participation and their effectiveness**

	Provide information	Cater for special interests	Two-way communication	Impact decision-making
Explanatory meeting, slide/film presentation	√	1/2	1/2	—
Presentation to small groups	√	√	√	1/2
Public display, exhibit, models	√	—	—	—
Press release, legal notice	1/2	—	—	—
Written comment	—	1/2	1/2	1/2
Poll	1/2	—	√	—
Field office	√	√	1/2	—
Site visit	√	√	—	—
Advisory committee, task force, community representatives	1/2	1/2	√	√
Working groups of key actors	√	1/2	√	√
Citizen review board	1/2	1/2	√	√
Public inquiry	√	1/2	1/2	√/—
Litigation	1/2	—	1/2	√/—
Demonstration, protest, riot	—	—	1/2	√/—

Source: Adapted from Westman 1985

• *Response*: Your opposition to use of herbicides was included in the content analysis of all comments received. However, evidence in the EIS indicates that low-risk use of selected herbicides is assured when properly controlled-the evaluated herbicides pose minimal risk as long as mitigation measures are enforced.

Without such feedback, people are likely to question the use to which their input was put, and whether their participation had any effect at all; this could affect their approach to subsequent projects as well as their view of the one under consideration.

6.2.3 UK procedures

Articles 6 and 10 (a) of EC Directive 85/337 (as amended by Directives 97/11 and 2003/35/EC) requires Member States to ensure that:

• The public is notified early in the environmental decision-making procedures about the request for development consent, the fact that the project is subject to EIA, and information about how they can participate in the EIA process.

• Copies of the environmental information and other information relevant to the decision are made available to the public.

• The public concerned is given the opportunity to express an opinion before development consent is granted. The detailed arrangements for such information and consultation are determined by the Member States which may, depending on the particular characteristics of the projects or sites concerned:

—determine the public concerned;

—specify the places where the information can be consulted;

—specify the ways in which the public may be informed, for example, by bill posting within a certain radius, publication in local newspapers, organization of exhibitions with plans, drawings, tables, graphs and models;

—determine the manner in which the public is to be consulted, for example by written submissions or by public enquiry; and

—fix reasonable time limits for the various stages of the procedure in order to ensure that a decision is taken within a reasonable period.

- Members of the public who have a sufficient interest or whose rights are impaired have access to a decision review procedure before a court of law.

In the UK, this has been translated by the various EIA regulations (with minor differences) into the following general requirements. Notices must be published in a local newspaper and posted at a proposed site at least seven days before the submission of the development application and EIS. These notices must describe the proposed development, state that a copy of the EIS is available for public inspection with other documents relating to the development application, give an address where copies of the EIS may be obtained and the charge for the EIS, and state that written representations on the application may be made to the competent authority for at least 21 days after the notice is published. When a charge is made for an EIS, it must be reasonable, taking into account printing and distribution costs.

Environmental impact assessment: guide to the procedures (DETR 2000), the government manual to developers, notes:

> Developers should also consider whether to consult the general public, and non-statutory bodies concerned with environmental issues, during the preparation of the environmental statement. Bodies of this kind may have particular knowledge and expertise to offer... While developers are under no obligation to publicise their proposals before submitting a planning application, consultation with local amenity groups and with the general public can be useful in identifying key environmental issues, and may put the developer in a better position to modify the project in ways which would mitigate adverse effects and recognize local environmental concerns. It will also give the developer an early indication of the issues which are likely to be important at the formal application stage if, for instance, the proposal goes to public inquiry.

This suggests that, although the UK has broadly implemented the EIA Directive's minimal requirements for public participation, this has been done half-heartedly at best, with no extension ('gold plating') of these requirements. The European Commission (EC 2010a) has also formally warned the UK about the prohibitive expense for members of the public who wish to challenge the legality of decisions on the environment, and this may still result in a judicial review. However the UK is not unusual in this respect: a recent review of the application and effectiveness of the EIA Directive (CEC 2009) concluded that public participation practice varies widely across European Member States, with most Member States adhering only to the Directive's minimum requirements.

The situation for nationally significant infrastructure projects is different from that described above. The Planning Act 2008 requires that decisions about such projects are made by a new Infrastructure Planning Commission in accordance with new National Policy Statements, with more 'front loaded' and exigent public participation requirements.[2] Developers for such projects need to demonstrate that they have undertaken public consultation and acted on public feedback before they submit an application to the IPC. The IPC then has 28 days to accept or reject the proposal, and inadequate consultation is one of the criteria for rejecting proposals. If an application is accepted, the public can register to provide their views in writing to the IPC, and to participate in open-floor and special topic hearings. All of the EIA-related documents are put on the IPC website.

6.3 Consultation with statutory consultees and other countries

Statutory consultees have accumulated a wide range of knowledge about environmental conditions in various parts of the country, and they can give valuable feedback on the appropriateness of a project and its likely impacts (Wende 2002; Wood and Jones 1997). However, the consultees may have their own priorities, which may prejudice their response to the EIS.

Article 6 (1) of Directive 85/337 (as amended) states:

> Member States shall take the measures necessary to ensure that the authorities likely to be concerned by the project by reasons of their specific environmental responsibilities are given an opportunity to express their opinion on the information supplied by the developer and other requests for development consent. To this end, Member States shall designate the authorities to be consulted… The information gathered… shall be forwarded to those authorities. Detailed arrangements for consultation shall be laid down by Member States.

In the UK, different statutory consultees have been designated for different devolved administrations (England, Wales, Scotland and Northern Ireland) and different types of development. For planning projects in England, for instance, the statutory consultees are any principal council to the area in which the land is situated (if not the LPA), Natural England, and the MMO and the Environment Agency where relevant.

The EIA Directive also requires consultation of other European Member States. Where one Member State is aware that a project is likely to have significant environmental impacts in another Member State—or if requested by a Member State that is likely to be significantly affected—the first Member State must send to the affected Member State information on the project, its likely impacts and the decision that will be taken. If the affected Member State subsequently indicates that it wants to participate in the EIA process, then it subsequently must be treated essentially as though it was a statutory consultee. The UK has had a limited number of these transboundary consultations—12 by 2009, compared to the Republic of Ireland, which had the most at 43 (COWI 2009).

Ideally, consultees should already be consulted at the scoping stage. In addition, it is a legal requirement that the consultees should be consulted before a decision is made. Once the EIS is completed, copies can be sent to the consultees directly by the developer or by the competent authority. In practice, many competent authorities only send particular EIS chapters to the consultees (e. g. the chapter on archaeology to the archaeologist). However, this may limit the consultee's understanding of the project context and wider impacts; generally consultees should be sent a copy of the entire EIS.

6.4 EIA presentation

Although EIA legislation specifies the minimum contents required in an EIS, it does not give any standard for the presentation of this information. Past EISs have ranged from a three-page typed and stapled report to glossy brochures with computer graphics and multi-volume documents in purposedesigned binders. This section discusses the contents, organization, and clarity of communication and presentation of an EIS.

6.4.1 Contents and organization

An EIS should be *comprehensive*, and it must at least fulfil the requirements of the relevant EIA legislation. As we shall discuss in Chapter 8, past EISs have not all fulfilled these requirements; however, most EISs nowadays are adequate, in part because LPAs are increasingly likely to require information on topics they feel have not been adequately discussed in an EIS. A good EIS will also go further than the minimum requirements if other significant impacts are identified. Most EISs are broadly organized into four sections: a nontechnical summary; a description of the project, alternatives considered, and a brief outline of the environmental context; a discussion of methodology, including scoping and public participation; and then a series of chapters that discuss, for individual topics such as air and water, baseline conditions, likely environmental impacts, proposed mitigation measures and residual impacts. Ideally, an EIS should also include the proposals for planning conditions or environmental management systems, and monitoring. It could include much or all of the information given in Appendix 5 of *Environmental impact assessment: guide to the procedures* (DETR 2000): see Table 3.7. Table 1.1 provides an example of a good EIS outline.

An EIS should *explain why some impacts are not dealt with*. It should include a scoping section to explain why some impacts may be considered insignificant. If, for instance, the development is unlikely to affect the climate, a reason should be given explaining this conclusion. An EIS should *emphasize key points*. These should have been identified during the scoping exercise, but additional issues may arise during the course of the EIA. The EIS should set the context of the issues. The names of the developer, relevant consultants, relevant LPAs and consultees should be listed, along with a contact person for further information. The main relevant planning issues and legislation should be explained. The EIS should also indicate any references used, and give a bibliography at the end.

The *non-technical summary* is a particularly important component of the EIS, as this is often the only part of the document that the public and decision-makers will read. It should thus summarize the main findings of the EIS, including the project description, alternatives and proposed mitigation measures. Chapter 4 gave examples of techniques for identifying and summarizing impacts, and Table 6.4 shows how the impacts of a project can be summarized.

Table 6.4 Extract from an EIS summary table showing relative weights given to significance of impacts

Topic area	Description of impact	Geographical scale of importance of issue					Impact	Nature	Significance
		I	N	R	D	L			
Human beings	Disturbance to existing properties from traffic and noise				*		Adverse	St, R	Major
	Coalescence of existing settlements			*			Adverse	Lt, IR	Major
Flora and fauna	Loss of grassland of local nature conservation value					*	Adverse	Lt, IR	Minor
	Creation of new habitats					*	Beneficial	Lt, R	Minor
	Increased recreational pressure on SSSI		*				Adverse	Lt, R	Minor
Soil and geology Water	Loss of 120 hectares agricultural soil (grade 3B)			*			Adverse	Lt, IR	Minor
	Increased rates of surface water run-off				*		Adverse	Lt, IR	Minor
	Reduction in groundwater recharge			*			Adverse	Lt, R	Minor

Key: I International — St Short term
N National — Lt Long term
R Regional — R Reversible
D District — IR Irreversible
L Local

Source: DoE 1995

Note: only a selection of key issues are given here

An EIS should ideally be one *unified document*, with perhaps a second volume for appendices. The courts have stated that an EIS should not be a 'paper chase' (Berkeley vs. Sos and Fulham Football Club, 2001):

> the point about the environmental statement contemplated by the Directive is that it constitutes a single and accessible compilation, produced by the applicant at the very start of the application process, of the relevant environmental information and the summary in non-technical language.

A common problem with the organization of EISs stems from how environmental impacts are assessed. The developer (or the consultants co-ordinating the EIA) often subcontracts parts of the EIA to consultancies that specialize in those fields (e. g. ecological specialists, landscape consultants). These in turn prepare reports of varying lengths and styles, making a number of (possibly different) assumptions about the project and likely future environmental conditions, and proposing different and possibly conflicting mitigation measures. One way developers have attempted to circumvent this problem has been to summarize the impact predictions in a main text, and add the full reports as appendices to the main body of the EIS. Another has been to put a 'company cover' on each report and present the EIS as a multi-volume document, each volume discussing a single type of impact. Both of these methods are problematic: the appendix method in essence discounts the great majority of findings, and the multi-volume method is cumbersome to read and carry. Neither method attempts to present findings in a cohesive manner, emphasizes crucial impacts or proposes a coherent package of mitigation and monitoring measures. A good EIS would incorporate the information from the subcontractors' reports into one coherent document, which uses consistent assumptions and proposes consistent mitigation measures.

The EIS should be kept as *brief* as possible while still *presenting the necessary information*. The main text should include all the relevant discussion about impacts, and appendices should present only additional data and documentation. In the US, the length of an EIS is generally expected to be less than 150 pages.

6.4.2 *Clarity of communication*

Weiss (1989) suggests that an unreadable EIS is an environmental hazard:

> The issue is the quality of the document, its usefulness in support of the goals of environmental legislation, and, by implication, the quality of the environmental steward-ship entrusted to the scientific community ... An unreadable EIS not only hurts the environmental protection laws and, thus, the environment. It also turns the sincere environmental engineer into a kind of 'polluter'.

An EIS has to communicate information to many audiences, from the decision-maker to the environmental expert to the lay person. Although it cannot fulfil all the expectations of all its readers, it can go a long way towards being a useful document for a wide audience. It should at least be *well written*, with good spelling and punctuation. It should have a clear structure, with easily visible titles and a logical flow of information. A table of contents, with page numbers marked, should be included before the main text, allowing easy access to information. Principal points should be clearly indicated, perhaps in a table at the front or back.

An EIS should *avoid technical jargon*. Any jargon it does include should be explained in the text or in footnotes. All the following examples are from actual EISs:

• *Wrong*: It is believed that the aquiclude properties of the Brithdir seams have been reduced and there is a degree of groundwater communication between the Brithdir and the underlying Rhondda beds, although… numerous seepages do occur on the valley flanks with the retention regime dependent upon the nature of the superficial deposits.

• *Right*: The accepted method for evaluating the importance of a site for waterfowl (i. e. waders and wildfowl) is the '1 per cent criterion'. A site is considered to be of National Importance if it regularly holds at least 1 per cent of the estimated British population of a species of waterfowl. 'Regularly' in this context means counts (usually expressed as annual peak figures), averaged over the last five years.

The EIS should clearly *state any assumptions* on which impact predictions are based:

• *Wrong*: As the proposed development will extend below any potential [archaeological] remains, it should be possible to establish a method of working which could allow adequate archaeological examinations to take place.

• *Right*: For each operation an assumption has been made of the type and number of plant involved. These are: demolition: 2 pneumatic breakers, tracked loader; excavation: backacter excavator, tracked shovel…

The EIS should be *specific*. Although it is easier and more defensible to claim that an impact is significant or likely, the resulting EIS will be little more than a vague collection of possible future trends.

• *Wrong*: The landscape will be protected by the flexibility of the proposed [monorail] to be positioned and designed to merge in both location and scale into and with the existing environment.

• *Right*: From these [specified] sections of road, large numbers of proposed wind turbines would be visible on the skyline, where the towers would appear as either small or indistinct objects and the movement of rotors would attract the attention of road users. The change in the scenery caused by the proposals would constitute a major visual impact, mainly due to the density of visible wind turbine rotors.

Predicted impacts should be *quantified* if possible, perhaps with a range, and the use of non-quantified descriptions, such as severe or minimal, should be explained:

• *Wrong*: The effect on residential properties will be minimal with the nearest properties… at least 200 m from the closest area of filling.

• *Right*: Without the bypass, traffic in the town centre can be expected to increase by about 50-75 percent by the year 2008. With the bypass, however, the overall reduction to 65-75 percent of the 1986 level can be achieved.

Even better, predictions should give an *indication of the probability* that an impact will occur, and the degree of confidence with which the prediction can be made. In cases of uncertainty, the EIS should propose worst-case scenarios:

- *Right*: In terms of traffic generation, the 'worst case' scenario would be for 100 per cent usage of the car park... For a more realistic analysis, a redistribution of 50 per cent has been assumed.

Finally, an EIS should be *honest and unbiased*. A 1991 review of local authorities noted that '[a] number of respondents felt that the environmental statement concentrated too much on supporting the proposal rather than focusing on its impacts and was therefore not sufficiently objective' (Kenyan 1991). O' Faircheallaigh (2010) suggests that lack of objectivity is still a problem, with EISs being used to justify, not assess, decisions. Developers cannot be expected to conclude that their projects have such major environmental impacts that they should be stopped (otherwise one would hope that they would already have been stopped). However, it is unlikely that all major environmental issues will have been resolved by the time the statement is written.

- *Wrong*: The proposed site lies adjacent to lagoons, mud and sands which form four regional Special Sites of Scientific Interest [*sic*]. The loss of habitat for birds is unlikely to be significant, owing to the availability of similar habitats in the vicinity.

6.4.3 *Presentation*

Although it would be good to report that EISs are read only for their contents and clarity, in reality, presentation can have a great influence on how they are received. EIAs are, indirectly, public relations exercises, and an EIS can be seen as a publicity document for the developer. Good presentation can convey a concern for the environment, a rigorous approach to the impact analysis and a positive attitude to the public. Bad presentation, in turn, suggests a lack of care, and perhaps a lack of financial backing. Similarly, good presentation can help to convey information clearly, whereas bad presentation can negatively affect even a well-organized EIS.

The presentation of an EIS will say much about the developer. The type of paper used-recycled or not, glossy or not, heavy or light weight-will affect the image projected, as will the choice of coloured or black-and-white diagrams and the use of dividers between chapters. The ultra-green company will opt for double-sided printing on recycled paper, while the luxury developer will use glossy, heavyweight paper with a distinctive binder. Generally, a strong binder that stands up well under heavy handling is most suitable for EISs. Unless the document is very thin, a spiral binder is likely to snap or bend open with continued handling; similarly, stapled documents are likely to tear. Multi-volume documents are difficult to keep together unless a box is provided. EISs can also be made public on the Internet or through CDs (see Figure 6.2).

The use of maps, graphs, photo-montages, diagrams and other forms of visual com-

Figure 6.2　Different types of EIS presentation

Sources: (a) Metropolitan Council (Minneapolis/St Paul); (b) AREVA Resources Canada Inc.; (c) Griff Wigley; (d) Evelop

munication can greatly help the EIS presentation. As we noted in Chapter 4, a location map, a site layout of the project and a process diagram are essential to a proper description of the development. Maps showing, for example, the extent of visual impacts, the location of designated areas or classes of agricultural land are a succinct and clear way of presenting such information. Graphs are often much more effective than tables or figures in conveying numerical information. Forms of visual communication break up the page, and add interest to an EIS. Increasingly some developers are also producing the EIS as a CD, and putting it on the Internet.

6.5　Review of EISs

A range of stakeholders will want to check the accuracy and comprehensiveness of an

EIS. The developer will want to ensure that they are not vulnerable to legal challenge. Opponents of the scheme will want to see whether it is. Statutory consultees will want to ensure that it covers all the relevant information on their topics of interest in a fair manner. The local planning authority will want to confirm that it is 'fit for purpose' for informing its decision.

The DCLG (2011) guidance notes that the competent authority must satisfy itself of the adequacy of the EIS, but gives no information on how it should do this:

> LPAs should satisfy themselves that submitted ES contains the information specified in Part II of Schedule 4 to the Regulations and the relevant information set out in Part I of that Schedule that the applicant can reasonably be required to compile. To avoid delays in determining EIA applications, the need for any further information should be considered and, if necessary, requested as early as possible. It is important to ensure that all the information needed to enable the likely significant environmental effects to be properly assessed is gathered as part of the EIA process. If tests or surveys are needed to establish the likelihood or extent of significant environmental effects, the results of these should form part of the ES ... Whether there is a breach of the regulations through failing to include the outcomes of surveys in an ES depends on the level of information already at the LPA's disposal, and whether this is sufficient to enable the authority to make an informed judgment as to the likely significant effects.
>
> As will be shown in Chapter 8, many EISs do not meet even the minimum regulatory requirements, much less provide comprehensive information on which to base decisions. In some countries, for example the Netherlands, Canada, Malaysia and Indonesia, EIA Commissions or panels have been established to review EISs and act as a quality assurance process. However, in the UK there are no mandatory requirements regarding the predecision review of EISs to ensure that they are comprehensive and accurate. A planning application cannot be judged invalid simply because it is accompanied by an inadequate or incomplete EIS: a competent authority may only request further information, or refuse permission and risk an appeal. [3]

Many competent authorities do not have the full range of technical expertise needed to assess the adequacy and comprehensiveness of an EIS. Some authorities, especially those that receive few EISs, have consequently had difficulties in dealing with the technical complexities of EISs. In some cases, consultants are brought in to review the EISs. Other authorities have joined the Institute of Environmental Management and Assessment (IEMA), which reviews EISs. Others have been reluctant to buy outside expertise, especially at a time of restrictions on local spending. A technique advocated by the International Association for Impact Assessment (IAIA), although not seen often in practice, is to involve parties other than just the competent authority in EIS review, especially the public (Partidario 1996).

In an attempt to fill the void left by the national government, several organizations have devised non-mandatory review criteria that aim to ensure that EISs analyse and present all relevant information and (to a lesser extent) that this information is accurate. Such a review also allows the reader to become familiar with the proposed project, assess the significance of its effects, and determine whether mitigation of its impacts is needed. Indirectly, it also makes the reviewer more familiar with the EIA process. The review process can be used by any of the stakeholders in the process.

Lee and Colley (1990) developed a hierarchical review framework. At the top of the hierarchy is a comprehensive mark (A = well-performed and complete, through to F = very unsatisfactory) for the entire report. This mark is based on marks given to four broad sub-headings: description of the development, local environment and baseline conditions; identification and evaluation of key impacts; alternatives and mitigation of impacts; and communication of results. Each of these, in turn, is based on two further layers of increasingly specific topics or questions. Lee and Colley's criteria have been used either directly or in a modified form (e. g. by the IEMA) to review a range of EISs in the UK. Appendix 4 gives the Lee and Colley framework.

The European Commission has also published review criteria (CEC 2001). These are similar to Lee and Colley's, but use seven sub-headings instead of four, include a longer list of specific questions, and judge the information based on relevance to the project context and importance for decision-making as well as presence/absence in the EIS. The review criteria given in Appendix 4 are an amalgamation and extension of Lee and Colley's and the EC's criteria, developed by the Impacts Assessment Unit at Oxford Brookes University. Rodriguez-Bachiller with Glasson (2004) have also devised an expert system approach to EIS review.

Table 6.5 Examples of possible uses for EIS review criteria

(a) Test of minimum legal requirements

Criterion	Presence/absence (page number)	Information	Key information absent
Describes the proposed development, including its design, and size or scale	√ (p. 5)	Location (in plans), existing operations, access	Working method, vehicle movements, restoration plans
Indicates the physical presence of the development	×		Site buildings (location, size), restoration

(b) Simple grading for each criterion

Criterion	Presence/absence (page number)	Comments	Grade
Explains the purposes and objectives of the development	√ (p. 11)	Briefly in introduction, more details in Sec. 2	A
Gives the estimated duration of construction etc. phases	√ (p. 12)	Not decommissioning	B

Continued

(c) Test of relevance and then completeness

Criterion	Relevant? (Y/N)	Judgement(C/A/I)*	Comment
Considers the 'no action' alternative, alternative processes, etc.	Y	A	Alternative sites discussed, but not alternative processes
If unexpectedly severe adverse impacts are identified, alternatives are reappraised	N		Impacts of sand/gravel working well understood

C=complete; A=adequate; I=inadequate

It is unlikely that any EIS will fulfil all the criteria. Similarly, some criteria may not apply to all projects. However, they should act as a checklist of good practice for both those preparing and those reviewing EISs. Table 6.5 shows a number of possible ways of using these criteria. Example (a), which relates to minimum requirements, amplifies the presence or otherwise of key information. Example (b) includes a simple grading, which could be on the A—F scale used by Lee and Colley, for each criterion (only one of which is shown here). Example (c) takes the format of the EC criteria, which appraise the relevance of the information and then judge whether it is complete, adequate (not complete but need not prevent decision-making from proceeding) or inadequate for decision-making.

Where a local planning authority believes that an EIS does not provide the information they require to allow them to give full consideration of the proposed development's likely environmental effects, it must require the developer to provide further information. Any further information must be publicized and consulted on again.[4]

6.6 Decisions on projects

6.6.1 *EIA and project authorization*

Decisions to authorize or reject projects are made at several levels:

> At the top of the tree are the relevant Secretaries of State... ; below them are a host of Inspectors, sometimes called Reporters (Scotland); further down the list come Councillors, the elected members of district, county, unitary or metropolitan borough councils; and at the very bottom are chief or senior planning officers who deal with 'delegated decisions' ... [as] a rough guide, the larger the project the higher up the pyramid of decision makers the decision is made. (Weston 1997)

Different decision-making rules apply for different UK jurisdictions. For instance, the Town and Country Planning Act 1990 (and parallel legislation in Scotland) requires planning decisions to be made with 'regard to the provisions of the development plan, so far as material to the application, and to any other material considerations'. There is no definitive list of material planning considerations, but they commonly include visual amenity, noise, traffic generation, nature conservation and other topics covered by EIA. This is expected to change under the proposed National Planning Policy Framework that would, instead, introduce a 'presumption in favour of sustainable development', namely that:

> individuals and businesses have the right to build homes and other local buildings provided that they conform to national environmental, architectural, economic and social standards, conform with the local plan, and pay a tariff that compensates the community for loss of amenity and costs of additional infrastructure. (Parliament 2011)

When determining an application for a nationally significant infrastructure project, the Infrastructure Planning Commission must have regard to any local impact report, other prescribed matters, and other matters that the IPC thinks are important and relevant to the decision; and it must decide the application in accordance with any relevant national policy statements unless, *inter alia*, the adverse impacts of the proposed development would outweigh its benefits (Planning Act 2008). However, in all cases, where a project requires EIA, the decision-maker must take into account the 'environmental information' in the decision. This is the information contained in the EIS and other documents, and any comment made by the statutory consultees and representations from members of the public.

The decision on an application with an EIS must be made within a specified period (e. g. 16 weeks for a local authority planning application), unless the developer agrees to a longer period. By any standards, making decisions on development projects is a complex undertaking. Decisions for projects requiring EIAs tend to be even more complex, because by definition they deal with larger, more complex projects, and probably a greater range of interest groups:

> The competition of interests is not simply between the developer and the consultees. It can also be a conflict between consultees, with the developer stuck in the middle hardly able to satisfy all parties and the 'competent authority' left to establish a planning balance where no such balance can be struck. (Weston 1997)

Whereas the decision-making process for projects with EIA was initially accepted as being basically a black box, attempts have subsequently been made to make the process more rigorous and transparent. Research by the University of Manchester (Wood and Jones 1997) and Oxford Brookes University (Weston *et al*. 1997) have focused on how environmental information is used in UK decision-making; this is discussed further in Chapter 8. A government good practice guide on the evaluation of environmental information for planning projects (DoE 1994) begins with a definition of evaluation:

> ... in the context of environmental assessment, there are a number of different stages or levels of evaluation. These are concerned with:
>
> —checking the adequacy of the information supplied as part of the ES, or contributed

from other sources;

—examining the magnitude, importance and significance of individual environmental impacts and their effects on specific areas of concern…;

—preparing an overall 'weighing' of environmental and other material considerations in order to arrive at a basis for the planning decision.

The guide suggests that, after vetting the application and EIS, advertising the proposals and EIS, and relevant consultation, the planning authority should carry out two stages of decision-making: an evaluation of the individual environmental impacts and their effects, and weighing the information to reach a decision. The evaluation of impacts and effects first involves verifying any factual statements in the EIS, perhaps by highlighting any statements of concern and discussing these with the developer. The nature and character of particular impacts can then be examined; either the EIS will already have provided such an analysis (e.g. in the form of Table 6.4) or the case-work officer could prepare such a table. Finally, the significance and importance of the impacts can be weighed up, taking into consideration such issues as the extent of the area affected, the scale and probability of the effects, the scope for mitigation and the importance of the issue.

The range of decision options are as for any application for project authorization: the competent authority can grant permission for the project (with or without conditions) or refuse permission. It can also suggest further mitigation measures following consultations, and may seek to negotiate these with the developer. If the development is refused, the developer can appeal against the decision. If the development is permitted, individuals or organizations can challenge the permission. The relevant Secretary of State may also be able to 'call in' an application, for a variety of reasons. A public inquiry may result.

But decision-making is not a clinical exercise. Decision letters have been described as 'a letter to the loser' (Des Rosiers 2000), suggesting a type of personal relationship between decision-maker and decision-receiver. In a Canadian context, Ross (2000) explains how he and fellow panel member Mike Fanchuk wrote the report that explained their decision about whether to permit a pulp mill:

> Mike Fanchuk [is] a farmer from just north of the pulp mill site…During my work with Mike, we discussed when we would be satisfied with the report, and thus when we would be willing to sign it…I believe Mike's approach is the best I have ever encountered. He would only be willing to sign the report when he felt that, in future years, he would be pleased to tell his eight-year old granddaughter that he had served on the panel and authored the report. In academic terms, this intergenerational equity illustrates very well the principles of sustainable development…More importantly, however, it illustrates the basic human need to be proud of work one has done.

An initial decision may need to be ratified by others, and may be overturned at that stage. In the case of a planning application, the planning officer's recommendations will go to the planning committee, which makes the final decision. In the UK, the IPC's decision will be passed to the relevant Secretary of State. Other procedures will apply to other jurisdictions.

6.6.2 *EIA and public inquiries*

A Secretary of State may 'call in' a planning application if a developer challenges a refusal of planning permission, a planning decision is not reached within a given time limit, or if the Secretary of State wants to consider the application for other reasons. A public inquiry must then be held if the developer or local planning authority requests one. At the inquiry, various parties can provide evidence and may be able to cross-question other parties. Public inquiries are expensive to all parties, and can be very drawn out. For instance, the public inquiry for Heathrow Terminal 5 lasted nearly four years. Weston (1997) compellingly discusses why all parties involved in EIA try to avoid public inquiries:

> By the time a project becomes the subject of a public inquiry the sides are drawn and the hearing becomes a focus for adversarial debate between opposing, expensive, experts directed and spurred on by advocates schooled in the art of cajoling witnesses into submission and contradictions. Such debates are seldom rational or in any other way related to the systematic, iterative and cooperative characteristics of good practice EIA. By the time the inquiry comes around, and all the investment has been made in expert witnesses and smooth talking barristers, it is far too late for all that.

Nevertheless, hundreds of projects involving EIA have gone to inquiry.

The environmental impact of proposals, especially traffic, landscape and amenity issues, will certainly be examined in detail during any inquiry. The EIA regulations allow inquiry inspectors and the Secretary of State to require (a) the submission of an EIS before a public inquiry, if they regard this as appropriate, and (b) further information from the developer, if they consider the EIS is inadequate as it stands. In practice, before public inquiries involving EIAs the inspector generally receives a case file, including the EIS, which is examined to determine whether any further information is required. Pre-inquiry meetings may be held where the inspector may seek further information. These meetings may also assist the developer and competent authority to arrive at a list of agreed matters before the start of the inquiry; this can avoid unnecessary delays during it. At the inquiry, inspectors often ask for further information, and they may adjourn the inquiry if the information cannot be produced within the available time. The information contained in the EIS will be among the material considerations taken into account. However, an inadequate EIS is not a valid reason for preventing authorization, or even for delaying an inquiry. [5]

An analysis of ten public inquiries involving projects for which EISs had been prepared (Jones and Wood 1995) suggested that in their recommendations most inspectors give 'moderate' or 'considerable' weight to the EIS and consultations on the EIS, and that environmental information is of 'reasonable' importance to the decision whether to grant consent. However, a subsequent study of 54 decision letters from inspectors (Weston 1997) suggested that EIA has had little influence on the inquiry process: in about two-thirds of the cases, national or local land-use policies were the determining issues identified by the inspectors and the Secretary of State, and in the remaining cases other traditional planning matters predominated:

> The headings which dominate the decision letters of the Inspectors and Secretaries of State are the traditional planning material considerations such as amenity, various forms of risk, traffic and need, although some factors such as flora and fauna, noise and landscape do tend to be discussed separately. (Weston 1997)

6.6.3 *Challenging a decision: judicial review*

The UK planning system has no official provisions for an appeal against development consent. However, if permission is granted, a third party may challenge that decision through judicial review proceedings in the UK courts, or through the European Commission.

In the UK, an application for judicial review of a decision should be made promptly, typically within 6 weeks or 3 months of the decision (depending on the type of decision). Judicial review proceedings first require that the third party shows it has 'standing' to bring in the application, namely sufficient interest in the project by virtue of attributes specific to it or circumstances, which differentiate it from all other parties (e.g. a financial or health interest). Establishing standing is one of the main difficulties in applying for judicial review. [6] If standing is established, the third party must then convince the court that the competent authority did not act according to the relevant EIA procedures. The court does not make its own decision about the merits of the case, but only reviews the way in which the competent authority arrived at its decision:

> The court will only quash a decision of the [competent authority] where it acted without jurisdiction or exceeded its jurisdiction or failed to comply with the rules of natural justice in a case where those rules apply or where there is an error of law on the face of the record or the decision is so unreasonable that no [competent authority] could have made it. (Atkinson and Ainsworth 1992)

Various possible scenarios emerge. A competent authority may fail to require an EIA for project that should have had one; a planning officer without formal delegated authority may make a planning decision; a planning authority may make a planning decision without having access to, or having adequately considered, all relevant environmental information. In such a case, its decision would be void. Several court cases have also revolved around the level of detail needed in EISs of outline planning applications. Section 8.6 discusses these in more detail. Although recent UK court cases have interpreted the requirements of the EIA Directive as having a 'wide scope and broad purpose', it is very unlikely that the UK courts will ever play as active a role as those in the US did in relation to the NEPA.

6.6.4 *Challenging a decision: the European Commission*

Another avenue by which third parties can challenge a competent authority's decision to permit development, or not to require EIA, is the European Commission. Such cases need to show that the UK failed to fulfil its obligations as a Member State under the Treaty of Rome by not properly implementing EC legislation, in this case Directive 85/337. In such a case, Article 169 of the Treaty allows a declaration of non-compliance to be sought from the European Court of Justice. The issue of standing is not a problem here, since the European Commission can begin proceedings either on its own initiative or based on the written complaint of any person. To use this mechanism, the Commission must first state its case to the Member State and seek its observations. The Commission may then issue a 'reasoned opinion'. If the Member State fails to comply within the specified time, the case proceeds to the European Court of Justice.

Under Article 171 of the Treaty of Rome, if the European Court of Justice finds that a Member State has failed to fulfil an obligation under the Treaty, it may require the Member State to take the necessary measures to comply with the Court's judgement. Under Article 186, the EC may take interim measures to require a Member State to desist from certain actions until a decision is taken on the main action. However, to do so the Commission must show the need for urgent relief, and that irreparable damage to community interests would result if these measures were not taken. Readers are referred to Atkinson and Ainsworth (1992), Buxton (1992) and Salter (1992a, b, c) for further information on procedures.

Of the environmental infringement cases handled by the European Commission, about 10 per cent relate to environmental impact assessment. The largest number of environmental infringements in 2008 and 2009 were against Italy, Spain, Ireland, France and the UK (EC 2010b).

6.7 Summary

Active public participation, thorough consultation with relevant consultees and good presentation are important aspects of a successful EIA process. All have been undervalued to date, despite the transposition of the Aarhus Convention. The presentation of environmental information has improved, and statutory consultees are becoming increasingly familiar with the EIA process, but public participation is likely to remain a weak aspect of EIA until developers and competent authorities see the benefits exceeding the costs.

Formal reviews of EIAs are also rarely carried out, despite the availability of several non-mandatory review guidelines and government advice on the use of environmental information in decision-making. Such reviews can help to ensure that the EIS is fully taken into consideration in the decision-making stage. The links between the quality of an EIS and that of the planning decision are discussed in Chapter 8.

Several appeals against development consents or against competent authorities' failure to require EIA have been brought to the UK courts or the EC. The UK courts have historically taken a relatively narrow interpretation of the requirements of the EIA regulations, but this has recently been changing. The EC, by contrast, has challenged the UK government on its implementation of Directive 85/337 and on a number of specific decisions resulting from this implementation.

More positively, the next step in a good EIA procedure is the monitoring of the development's actual impacts and the comparison of actual and predicted impacts. This is discussed in the next chapter.

SOME QUESTIONS

The following questions are intended to help the reader focus on the key issues of this chapter.

1 Public participation and public consultation are often used synonymously. What is the difference between them?

2 Different stakeholders require different techniques to learn about a project's environmental impacts and provide optimum input to the project planning process. Assume that you are devising a public participation programme for a wind farm EIS (or a different kind of project with which you are familiar). Using Table 6.3 as a basis, which three techniques would you use for local residents? Which three would you use for a national-level non-governmental organization opposed to wind power on landscape grounds? Which three would you use for the government agency responsible for biodiversity? (NB: some of them might be the same from group to group).

3 The introductory section to the EIA Directive refers to public participation in the following way: 'Whereas development consent for public and private projects… should be granted only after prior assessment…; whereas this assessment must be conducted on the basis of the appropriate information supplied by the developer, which may be supplemented by the authorities and by the people who may be concerned by the project in question.' How might it be rephrased to promote a higher level of public participation in EIA?

4 Section 6.2.3 lists the UK procedural requirements for public participation. Do they match the level of participation set out in the introductory section to the EIA Directive (see question 3 above)?

5 For a country and type of plan of your choice, identify the statutory consultees.

6 In one page, explain the EIA process in a non-technical way.

7 Explain what is wrong with the 'wrong' quotes in Section 6.4.2. Try to rephrase

them in a 'right' way.

8 Figure 6.2 shows different ways of presenting an EIS. Which do you think would work best for: (i) technically minded consultees; (ii) people with access to the Internet, and those without access; (iii) people who have problems reading small print (or at all); (iv) people who might struggle to pay for the 'reasonable cost' of printing and distributing the EIS?

9 Under what circumstances might each of the grading approaches (a—c) in Table 6.5 be used?

Notes

1 Although this section refers to public consultation and participation together as 'public participation', the two are in fact separate. Consultation is in essence an exercise concerning a passive audience: views are solicited, but respondents have little active influence over any resulting decisions. In contrast, public participation involves an active role for the public, with some influence over any modifications to the project and over the ultimate decision.

2 The coalition government of 2010 will change this, although details are not yet available at the time of writing (spring 2011). The IPC and the Planning Inspectorate will be merged, and the IPC will make recommendations to the relevant secretary of state, who will make the final decision.

3 Weston (1997) notes that LPAs need to be aware that they have the power to ask for further information, and that failure to use it could later be seen as tacit acceptance of the information provided. For instance, when deciding on an appeal for a Scottish quarry extension, the Reporter noted that it was significant that the LPA had not requested further information when they were processing the application, and had not objected to the EIS until the development came to appeal.

4 Where the project has already been built without authorization, the competent authority considers the environmental information when determining whether the project will be demolished or not.

5 For instance, in the case of a Scottish appeal regarding a proposed quarry extension (Scottish Office, P/PPA/SQ/336, 6 January 1992), the Reporter noted that: 'The ES has been strongly criticised… [it] does not demonstrate that a proper analysis of environmental impacts has been made… Despite its shortcomings, the ES appears to me to comply broadly with the statutory requirements of the EA regulations.'

6 An EC court case, for instance, ruled that Greenpeace had insufficient individual concerns to contest a decision to use regional funds to help build power stations in the Canary Islands (Greenpeace vs. Commission of the European Communities, *Journal of Environmental Law*, 8 (1996), 139). Similar judgements have been made in the UK context. COWI (2009) provides further information on standing in different Member States.

References

ADB (NGO Forum on ADB) 2006. *The Advocacy Guide to ADB EIA Requirement, Philippines*. Available at: www. forum-adb. org/BACKUP/pdf/guidebooks/ EIA%20Guidebook. pdf.

Arnstein, S. R. 1971. A ladder of public participation in the USA. *Journal of the Royal Town Planning Institute*, April, 216-24.

Atkinson, N. and Ainsworth, R. 1992. Environmental assessment and the local authority: facing the European imperative. *Environmental Policy and Practice* 2 (2), 111-28.

Audit Commission 2000. *Listen Up! Effective Community Consultation*. Available at: www. audit-commission. gov. uk/SiteCollectionDocuments/AuditCommission Reports/NationalStudies/listenup. pdf.

Balram, S., Dragicevic, S. and Meredith, T. 2003. Achieving effectiveness in stakeholder participation using the GIS-based Collaborative Spatial Delphi methodology. *Journal of Environmental Assessment Policy and Management* 5 (3), 365-94.

Buxton, R. 1992. Scope for legal challenge. In *Environmental assessment and audit: a user's guide*, Ambit (ed), 43-4. Gloucester: Ambit.

CEC (European Commission of Communities) 2001. *Guidance on EIA: EIS review. DG* XI. Brussels: CEC.

CEC 2009. On the application and effectiveness of the EIA Directive (Directive 85/337/EEC, as amended by Directives 97/11/EC and 2003/35/EC). Available at: www. eur-lex. europa. eu/LexUriServ/LexUriServ. do? uri=COM: 2009: 0378: FIN: EN: PDF.

Clark, B. 1994. Improving public participation in environmental impact assessment. *Built Environment* 20 (4), 294-308.

COWI 2009. *Study concerning the report on the application and effectiveness of the EIA Directive*, Final report to European Commission DG ENV. Kongens Lyngby, Denmark: COWI.

Daily Echo 2011. Southampton biomass plant plans 'beyond belief' says New Forest District Council. 14 April. Available at: www. dailyecho. co. uk/news/ 8974234. Biomass _ plans beyond _ belief _.

Dallas, W. G. 1984. Experiences of environmental impact assessment procedures in Ireland. In *Planning and Ecology*, R. D. Roberts and T. M. Roberts (eds), 389-95. London: Chapman & Hall.

DCLG (Department for Communities and Local Government) 2009. *Publicity for planning applications: consultation*. London: DCLG.

DCLG (Department for Communities and Local Government) 2011. *Guidance on the Environmental Impact Assessment (EIA) Regulations 2011 for England*. London: DCLG.

Des Rosiers 2000. From telling to listening: a therapeutic analysis of the role of courts in miniority-majority conflicts. *Court Review*, Spring.

DETR 2000. *Environmental impact assessment: a guide to the procedures*. London: HMSO.

DoE (Department of the Environment) 1994. *Evaluation of environmental information for planning projects: a good practice guide*. London: HMSO.

DoE 1995. *Preparation of environmental statements for planning projects that require environmental assessment*. London: HMSO.

EC (European Commission) 2010a. Environment: Commission warns UK about unfair cost of challenging decisions. Available at: www. europa. eu/ rapid/pressReleasesAction. do? reference=IP/10/312&type=HTML.

EC 2010b. Statistics on environmental infringements. Available at: www. ec. europa. eu/environment/ legal/law/statistics. htm.

EDF Energy 2009. *Hinkley Point C: Consultation on initial proposals and options*. Available at: www. hinkley point. edfenergyconsultation. info/websitefiles/PPS _ SW _ XXXX _ EDF _ HINK _ POINT _ BDS _ 12. 09 _ 1 _ pps. pdf.

Hancock, T. 1992. Statement as an aid to consent. In *Environmental assessment and audit: a user's* guide, Ambit (ed), 34-35. Gloucester: Ambit.

Hartley, N. and Wood, C. 2005. Public participation in environmental impact assessment-implementing the Aarhus Convention. *Environmental Impact Assessment Review* 25, 319-40.

IEMA (Institute of Environmental Management and Assessment) 2002. *Perspectives: participation in environmental decision-making*. Lincoln: IEMA.

International Association for Public Participation 2001. IAP2's Public Participation Toolbox. Available at: www. iap2. affiniscape. com/associations/4748/ files/06Dec _ Toolbox. pdf.

Jones, C. E. and Wood, C. 1995. The impact of environmental assessment in public inquiry decisions. *Journal of Planning and Environment Law*, October, 890-904.

Kenyan, R. C. 1991. Environmental assessment: an overview on behalf of the R. I. C. S. *Journal of Planning and Environment Law*, 419-22.

Lee, N. and Colley, R. 1990. *Reviewing the quality of environmental statements*. Occasional Paper no. 24. Manchester: University of Manchester, EIA Centre.

McCormick, J. 1991. *British politics and the environment*. London: Earthscan.

McNab, A. 1997. Scoping and public participation. In *Planning and EIA in practice*, J. Weston (ed), 60-77. Harlow: Longman.

Mollison, K. 1992. A discussion of public consultation in the EIA process with reference to Holland and Ireland (written for MSc Diploma course in Environmental Assessment and Management). Oxford: Oxford Polytechnic.

O' Faircheallaigh, C. 2010. Public participation and environmental impact assessment: purposes, implications and lessons for public policy making. *Environmental Impact Assessment Review* 30, 19-27.

Parliament (2011) The Localism Bill. Available at: www. publications. parliament. uk/pa/cm201011/cmselect/cmenvaud/799/79903. htm.

Partidario, M. R. 1996. *16th Annual Meeting, International Association for Impact Assessment: Synthesis of Workshop Conclusions*. Estoril: IAIA.

Petts, J. 1999. Public participation and EIA. In *Handbook of environmental impact assessment*, J. Petts (ed), vol. 1. Oxford: Blackwell Science.

Petts, J. 2003. Barriers to deliberative participation in EIA: learning from waste policies, plans and projects. *Journal of Environmental Assessment Policy and Management* 5 (3), 269-94.

Rodriguez-Bachiller, A. with J. Glasson 2004. *Expert Systems and Geographical Information Systems for Impact Assessment*. London: Taylor and Francis.

Ross, W. A. 2000. Reflections of an environmental assessment panel member. *Impact Assessment and Project Appraisal* 18 (2), 91-8.

Salter, J. R. 1992a. Environmental assessment: the challenge from Brussels. *Journal of Planning and Environment Law*, January, 14-20.

Salter, J. R. 1992b. Environmental assessment-the need for transparency. *Journal of Planning and Environment Law*, March, 214-21.

Salter, J. R. 1992c. Environmental assessment-the question of implementation. *Journal of Planning and Environment Law*, April, 313-18.

Turner, T. 1988. The legal eagles. *Amicus Journal*, winter, 25-37.

Weiss, E. H. 1989. An unreadable EIS is an environmental hazard. *Environmental Professional* 11, 236-40.

Wende, W. 2002. Evaluation of the effectiveness and quality of environmental impact assessment in the Federal Republic of Germany. *Impact Assessment and Project Appraisal* 20 (2), 93-99.

Westman, W. E. 1985. *Ecology, impact assessment and environmental planning*. New York: Wiley.

Weston, J. (ed) 1997. *Planning and EIA in practice*. Harlow: Longman.

Weston, J., Glasson, J., Therivel, R., Weston, E., Frost, R. 1997. *Environmental information and planning projects*, Working Paper no. 170. Oxford: Oxford Brookes University, School of Planning.

Williams, G. and Hill, A. 1996. Are we failing at environment justice? How minority and low income populations are kept out of the public involvement process. *Proceedings of the 16th Annual Meeting of the International Association for Impact Assessment*. Estoril: IAIA.

Wood, C. M. and Jones, C. 1997. The effect of environmental assessment on UK local authority planning decisions. *Urban Studies* 34 (8), 1237-57.

7 决策后的监测与审计

7.1 导言

诸如道路、机场、发电站、废弃物处置站、矿产开发和度假村等重点工程项目，其整个生命周期都包括几个重要阶段（图 1.5）。该生命周期可能涵盖一个很长的时间段（例如，一个化石燃料发电站从设计、建造、运行直至“退役”，可能需要 50～60 年时间）。正如目前英国及其他许多国家的实践情况一样，EIA 更多地只与决策前的过程相关。更糟糕的是将 EIA 生搬硬套地用于一些具体项目而不是决策过程，这种 EIA 往往由开发商内部决定，没有任何公众参与。这种目光短浅的“前建后忘”的方法具有一定的风险（Culhane，1993）。但是，EIA 不应该停止于决策阶段。它不仅应该是帮助规划获得许可的一项辅助工具，更应该成为在项目整个生命周期中有效进行环境管理的一种方法。这就意味着在 EIA 程序中应该包含监测和审计。另一个持续性风险是过分强调决策前分析，它会使 EIA 脱离环境保护的关键目标。EIA 应该尽力使持续提升的潜力最大化。把资源用于基准研究和预测，实际上价值并不大，除非有检验预测结果的方法并能判断是否采取了恰当的减缓措施和改善措施（Ahammed 和 Nixon，2006）。实践中得出的经验和优秀的 EIA 程序都应该被记录下来；可参考 Morrison-Saunders 和 Arts（2004）给出的例子以及关于 EIA 跟踪评价的杂志专刊《影响评价与项目评价》（IAPAY 2005）。

7.2 节阐述了一些相关定义，如监测和审计以及它们之间的关系，并概括了它们在 EIA 中的重要作用；7.3 节概述了如何更好地将监测融入 EIA 过程中，特别引入了一些国际实践经验；接下来讨论了环境影响审计的一系列方法，包括对一些国家所开展的 EIS 审核尝试的总结回顾；最后一部分以英国修建 Sizewell B 压水反应堆（PWR）核电站为例，重点对其建设过程造成的对地方社会-经济影响开展具体的监测和审计研究，同时简述了时间较近的伦敦 2012 年奥林匹克项目的监测工作。

7.2 监测和审计在 EIA 程序中的重要性

在 EIA 的很多方面都存在一些重要的语义上的发展和相关术语的扩充。本章节早期的版本中，我们重点关注的是监测和审计，它们是决策制定之后或跟踪评价过程中的关键因素，这一点被加拿大、澳大利亚等国家所认同。第 7 章还会对此进行重点阐述，同时对其他关键因素进行讨论，包括管理和沟通（Arts 等，2001；Marshall 等，2005）。这些因素可以帮助我们更好地从实际中吸取经验，避免 EIA 流于形式，在项目决策制定后就几乎没有影响力了。

监测包括测量、记录与项目开发影响有关的自然、社会及经济方面的变化（如交通流量、空气质量、噪声、就业水平）。该项工作力求提供时间和空间上变量的特性、功能上的信息，尤其是关于突发事件及其影响程度的信息。监测能够优化项目管理。例如，它可以作为早期预警系统，在无法补救之前识别出对一个地区的不良影响。同时，它还有助于识别和修正预料之外的影响。此外，监测可以提供一个公认的数据库，可以用于调解相关利益团体纠纷。因此，对源头、路径和终点进行监测，可以弄清楚是哪一部分的责任，工业区灰尘就是一个例子。监测对于成功的环境影响审计来说是十分必要的，并且也是确定承担义务和实施减缓措施最有效的保证之一。

本章讨论的环境影响审计包括对比 EIS 预测的影响与项目实施后的实际影响，以此评价预测是否令人满意（Buckley，1991）。也有些文献中把这一步骤称为评估。审计包括影响预测（该预测好在哪里），也可以是减缓措施、改善措施和开发条件（减缓措施、改善措施是否有效，条件能否得到满足）。这种审计方法与环境管理审计形成对比，后者更关注环境管理中公共及私有的法定机构与程序以及与之相关的风险及负债情况。对此我们将在第 12 章进一步讨论。

管理是决策制定的重要组成部分，采取恰当的措施对监测和审计过程中出现的问题进行响应（Marshall 等，2005）。例如，一个重点项目建设阶段人员的招募，监测结果显示实际水平可能低于预计水平。那么进行地方培训管理时可能要付出双倍的努力。沟通是告知利益相关者 EIA 监测、审计及管理活动的结果。这种沟通可以来源于支持者，也可以来源于监管部门，并且希望二者之间是合作关系。理想情况下，社区利益相关者的角色不只是跟踪活动的被动接受者，更是直接参与跟踪活动及过程的合作人之一。

总之，这些因素对今后项目规划和 EIA 开展具有重要贡献（图 7.1）。为了从已有实践中吸收经验，我们有必要对反馈机制进行介绍；我们必须避免 EIA 中出现“白费力气做重复的工作”的现象（Sadler，1998）。对结果进行监测和审计以及由此产生的管理响应和相关沟通都有助于优化 EIA 的各个方面：从了解基础情况到制定有效的减缓及改善措施。此外，Greene 等（1985）注意到，通过让所有的参与者学习以往的实践案例，监测和审计应该减少 EIA 所消耗的时间和资源，同时也有助于全面提升支持者、监管部门和 EIA 程序的可信度。我们对判定 EIA 程序在实践应用中的有效性非常感兴趣，但是，到目前为止仍存在一些重要的问题，极大地限制了监测和审计工作的开展。监测和审计的这些问题及其解决方法成为当下讨论的重点。

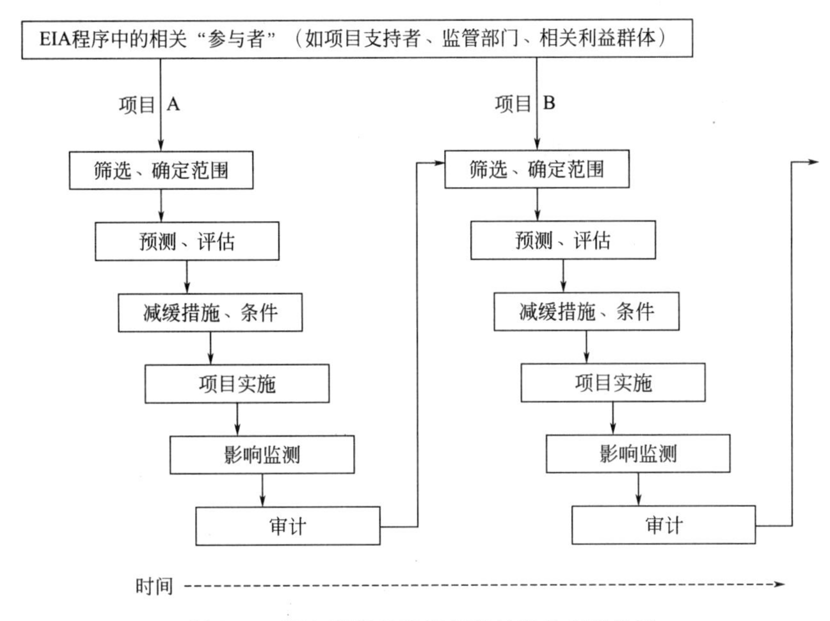

图 7.1　EIA 程序的监测与审计以及实践学习

（资料来源：Bisset 和 Tomlinson，1988；Sadler，1988）

7.3　监测的实际应用

7.3.1　关键因素

监测意味着在较长一段时间内系统地大量收集的相关信息。这类信息不仅应包括传统指标（如环境空气质量、噪声级、劳动力规模），还应包括一些具有因果关系的潜在因素（如地方权力机构和开发商的决定和政策）。这些因素决定着所产生的影响，若要改变影响，就必须改变这些因素。当然，对于影响的不同观点同样重要。个体和“现实的社会构建”团体（IOCGP，2003）通常持观望态度，仅凭感觉或情绪，而不是根据现实情况衡量。但这些意见在决定项目是否批复时却有非常重要的影响。忽略或者低估它们就失去了方法的防御性，还可能引出更多的反对意见。监测还应该分析影响的公正性。影响的分布在不同的群体和区域间可能发生变化；由于诸如年龄、种族、性别和收入等因素的影响，某些群体可能更容易受到伤害。因此，系统地识别意见可作为监测研究的一项重要内容。

收集的信息需要储存与分析，并与 EIA 程序中相关参与者进行交流。因此，主要的要求是监测过程重点关注“那些可能产生显著影响的环境参数以及那些评价方法或基础数据达不到期望状态的参数”（Lee 和 Wood，1980）。

监测是 EIA 中一个必不可少的部分；本底数据获取、项目介绍、影响预测以及减缓、改善措施都应当与监测同时开展。EIS 中应当包含一个监测程序，该程序要有清楚的目标、时间和空间控制、足够的持续时间（如涵盖项目实施的主要阶段）、实践方法、充足的资金、明确的责任以及公开且正规的报告。理想情况下，监测行为还应包括相关群体间的合作；例如，信息的收集可包括开发商、地方权力机构和当地社区。同时，监测程序还应该能够适应环境的动态特征（Holling，1978）。

7.3.2 强制监测和自愿监测

令人遗憾的是，在很多EIA程序中，监测不是一个强制性步骤，包括英国的现行程序。欧盟委员会EIA法规也没有对监测提出明确要求。欧洲共同体85/337号法令（CEC，1993）评审也承认忽略了这一点。委员会积极提倡将正式的监测程序纳入EIS，但成员国更多的是持谨慎的和反对的态度。结果，修订后的指令中不包括强制监测❶。然而，这并没有阻止部分成员国实施强制监测。例如，荷兰要求主管部门根据开发商提供的信息，监测项目的实施过程，并公开监测信息。如果实际影响超出了预期，主管部门必须采取相应措施降低或者减缓这些影响。然而，尽管有相关的法律规定，但实际执行中由于各种限制，很少开展EIA后期监测和评估工作。有关荷兰EIA后评估的全面理解可参阅Arts（1998）的内容。

在其他成员国，正如第2章和CEC（2009）中所提到的，缺少强制监测要求是一个持续的、严重的、长期存在的问题。在缺乏强制程序的情况下，很难说服开发商，使他们相信继续执行EIA符合他们的利益。尤其当项目为一次性项目，不需要为未来项目的实施积累经验时，情况更是如此。幸运的是，我们尚能参考少数国家的一些成功经验。如第10章中对加拿大监测程序的概述。在澳大利亚西部（第10章），对开发造成的环境影响通常都要进行监测和汇报。如果条件不满足，政府将采取适当的措施。有趣的是，在澳大利亚西部将EIS视为一个“环境评审和管理程序”（Morrision-Saunders，1996）。

7.3.3 加利福尼亚州案例

在加利福尼亚州，依据《加利福尼亚州环境质量法》（CEQA）制定的项目监测程序受到广泛关注（加利福尼亚资源局，1998）。自1989年1月起，加利福尼亚州及地方政府被要求对减缓措施和项目变化执行监测及报告程序，并强制作为识别显著环境影响的条件。其目的是提供一种确保在项目批准期限内及时执行减缓措施的机制。监测是指观察和勘测项目施工现场减缓措施的执行，报告是指将监测结果提交给有关部门和公众。若一个项目分为几个阶段，那么减缓措施和之后的报告及监测也应当分为几个阶段。如果监测结果表明减缓措施被忽视或未执行，相关部门将加强监管力度，如责令“停工”、罚款和赔偿。项目的监测程序通常包括以下内容：

- 环境影响报告（EIR）对识别的显著影响进行概述总结；
- 针对各项显著影响提出的减缓措施；
- 各项减缓措施的监测要求、责任机构；

---

❶ 欧盟指令早期的初稿包括对EIA项目进行事前-事后评价的要求。1980版初稿（CEC，1980）第11部分中指出主管部门应当每隔一段时间检查一次与规划许可有关的条款是否符合或恰当，是否符合其他环境保护条款，是否要求采取额外的措施保护环境免受项目的影响。

- 为各项减缓措施监测负责的人或机构；
- 监测的时间和频率；
- 确保监测过程遵守监测程序的责任机构；
- 报告编写要求。

图 7.2 提供了一个加利福尼亚州的伯里克西部针对废木料转换设备而制定的监测程序的部分内容。

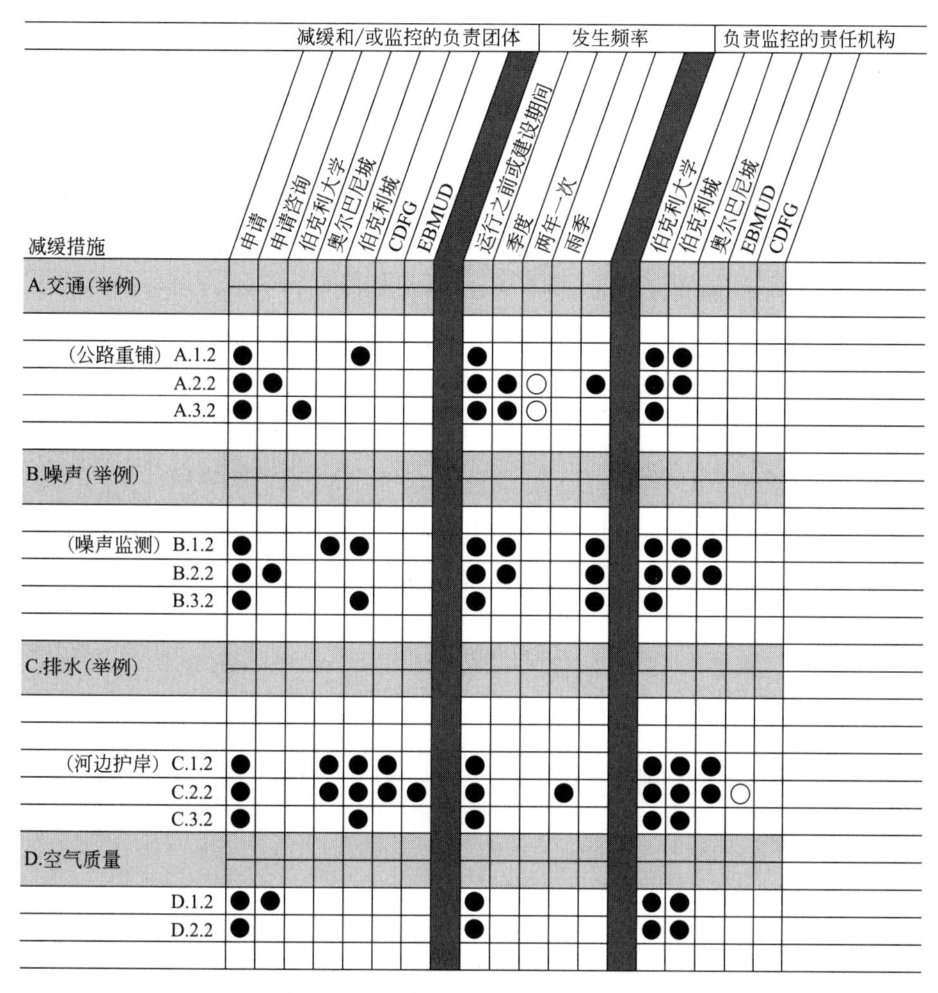

	减缓和/或监控的负责团体							发生频率					负责监控的责任机构				
减缓措施	申请	申请咨询	伯克利大学	奥尔巴尼城	伯克利城	CDFG	EBMUD	运行之前或建设期间	季度	两年一次	雨季		伯克利大学	伯克利城	奥尔巴尼城	EBMUD	CDFG
A.交通(举例)																	
(公路重铺) A.1.2	●				●			●					●	●			
A.2.2	●	●						●	●	○		●	●	●			
A.3.2	●		●					●	●	○			●				
B.噪声(举例)																	
(噪声监测) B.1.2	●			●	●			●	●			●	●	●	●		
B.2.2	●	●						●	●			●	●	●	●		
B.3.2	●				●			●				●	●				
C.排水(举例)																	
(河边护岸) C.1.2	●			●	●	●		●					●	●	●		
C.2.2	●			●	●	●	●	●			●		●	●	●	○	
C.3.2	●				●			●					●	●			
D.空气质量																	
D.1.2	●	●						●					●	●			
D.2.2	●							●					●	●			

图 7.2　加利福尼亚州监测程序示例

（资料来源：Baseline Environmental Consulting，1989）

7.3.4　中国香港案例

1990 年，中国香港为一些重点项目引进了一个系统的、综合的环境监测和审计体系。Chap Lap Kok 耗资 200 亿美元的新建机场是其重要的推动因素之一，该项目不仅包括机场的建设，还包括铁路、高速公路和立交桥的建设以及九龙填海工程项目。环境监测和审计(EM&A) 指南中将一个行动规划分为三个阶段：①触发等级，提供早期预警；②行动等级，在达到影响上限之前采取行动的等级；③目标等级，在该等级上，需执行事先计划的相应方案以避免或纠正问题。该方法事先通过取得支持者们对监测审计协议和事件行动规划的同意，从而将监测更好地融入项目决策过程中，但目前在执行方面仍存在一些问题（Au 和 Sanvicens，1996)。EM&A 要建立动态文件，在项目实施过程中定期审核、更新（如果必要的话）。

1998 年 4 月 EIA 法规开始生效，该法规对何时实施、怎样实施监测和审计等进行了详细规定（EPD1997，1998)。法规内容通常就是项目获得许可的条件。这为后续工作提供了法律基础，并且违反者要受到严厉的惩罚（高达 25 万美元罚款以及有期徒刑 6 个月)。近来，中国香港利用网络对大型项目的影响和许可条件的监测进行了改革。自 2000 年引入这些程序后，大型项目都必须建立监测网站。一些监测站还设置了联网摄像机来监测工程项目的重要部分。该网站设有对外开放途径，对项目比较关注的公众可以将他们对规划执行的意见反馈给政府和开发商（Hui 和 Ho，2002)。这将会是 EIA 系统的雏形吗？表 7.1 为香港直升机起飞降落场 EM&A 报告的部分内容。

表 7.1　香港直升机起飞降落场 EM& A 报告摘要：榕树湾直升机起飞降落场建设过程推荐采用的减缓措施实施计划（空气质量减缓措施）

EIA 相关内容	EM & A 相关内容	推荐采用的环境保护/减缓措施	推荐措施的目标以及重点关心的问题	措施的实施者	实施措施的位置/时限	措施应达到的要求及标准
S3.5.1	S2.5	《空气污染管制规例》中建议的所有适用的防尘措施都应实施	建设阶段的空气质量	承包商	所有施工地点，持续整个施工期	《环境影响评估程序技术备忘录》(EIAO-TM)，《空气污染管制（建造工程尘埃）规例》
	S2.4	典型防尘措施包括： • 在可行的情况下限制物料下落的高度，尽可能减小装载/卸载造成的扬尘	建设阶段的空气质量	承包商	所有施工地点，持续整个施工期	《环境影响评估程序技术备忘录》(EIAO-TM)，《空气污染管制（建造工程尘埃）规例》
		• 挖掘物和弃土堆积超过 $50m^3$，应将其圈起、覆盖或在干燥、多风时使其变湿润	建设阶段的空气质量	承包商	所有施工地点，持续整个施工期	《环境影响评估程序技术备忘录》(EIAO-TM)，《空气污染管制（建造工程尘埃）规例》
		• 采用有效的液体喷雾控制潜在的粉尘排放源，如未铺柏油的运输公路及正在施工的建设场所	建设阶段的空气质量	承包商	所有施工地点，持续整个施工期	《环境影响评估程序技术备忘录》(EIAO-TM)，《空气污染管制（建造工程尘埃）规例》

资料来源：CWE-ZHEC Joint Venture 2007。

7.3.5　英国的经验

虽然目前英国 EIA 法规中对开展监测尚无强制要求，但监测行为已比较普遍。最近，牛津布鲁克斯大学的一项研究（Frost，1997；Glasson，1994）试图提供一种通过“内容分析”和“实践分析”进行初步评价的方法。涉及监测意图的内容分析指分析将近 700 个具有代表性的案例的 EIS 和 EIS 摘要（来源于环境评价研究所的环境综述摘要)(EIA，1993)。

一些 EIS 中明确指出了监测部分；其他研究中，仅在相关减缓措施部分有所体现。还有一些案例中提到了一般的监测建议，例如，建议签约双方均要遵守合同约定。一般来说，约 30%的案例至少有一个涉及影响监测的建议。监测的类型最多有六种，即使在我们选择的几个案例中，建议也不可能详尽地说明所有影响监测类型。表 7.2 列出了 EIS 中的监测类型。水质监测引用频率高于空气质量监测；空气污染物和水体污染物排放的点源监测引用也较为频繁；涉及非生物物理影响（如社会-经济）的监测则非常有限。不同类型的项目其监测的类型也在变化，如燃气蒸汽联合循环（CCGT）发电站，通常建议进行大气污染物、空气质量以及建筑噪声的监测；而对于垃圾填埋项目来说，建议则更倾向于对渗滤液、填埋废气和水质的监测。

表 7.2 英国 EIS 影响监测的类型

类型	占被监测项目总数的百分比/%	类型	占被监测项目总数的百分比/%
水质	16	文物古迹	6
大气污染物排放	15	空气质量	5
水体污染物排放	13	结构调查	4
噪声	12	联络小组	3
一般项目	9	水位	3
其他	7	合计	100
生态	7		

资料来源：Glasson，1994。

由于 EIS 监测建议，实践分析使用了一个案例，包括 17 个具有代表性的项目，目前实践分析已经开始实施。LPA 的研究阐明了监测计划应包括：监测计划是否是在各项条款达成一致的情况下［如规划条件、S106 协议、综合污染物控制（IPC）条件、场所许可证］执行的；监测是否是自愿实施的等内容。结果显示，总体上 EIS 倾向于轻描淡写地陈述，实际执行监测的项目数量大约只占 30%。这可能是对决策过程中规划条件和协议的响应；也可能与其他相关的批准程序有关，例如 IPC。LPA 研究表明，不论何种案例中，EIS 中都应该包含监测，且要比 EIA 涉及的范围更为广泛，无可否认的是这一点常常受到限制。当然，结果并没有提供任何有关监测质量或预测准确性的信息。

7.4 审计的实践应用

审计已逐渐形成很多不同的种类。Tomlinson 和 Atkinson（1987a，b）用一系列术语对“标准”EIA 程序中审计的七个要点进行了标准化定义：

- 决策点审计（EIS 初稿）——在规划审批过程中由监管机构负责。
- 决策点审计（EIS 终稿）——同样是在规划审批过程中由监管机构负责。
- 执行审计——用于启动阶段；它可能包括政府和公众的详细审查，重点关注倡议者对减缓措施和其他强制条件的承诺。
- 绩效审计——用在整个操作阶段；也包括政府和公众的详细审计。

• 预测性的技术审计——将实际影响与预测影响的比较作为一种比较不同预测技术价值的手段。

• 项目影响审计——同样是比较实际影响与预测影响，从而为改善项目管理和将来的项目提供信息反馈。

• 程序审计——是指在 EIA 程序中政府和产业部门使用的程序外部评审（如公众）。

这些术语之间确实是相互交叉的。此处重点关注的是项目、绩效及执行的审计。但不论关注的焦点是什么，审计都将面临表 7.3 中列出的大量问题。

表 7.3　事后审计研究的相关问题

影响预测的性质
很多 EIS 几乎没有可测性预测；取而代之的是简单地识别一些潜在问题。
很多 EIS 预测是笼统的、不精确的、定性的。
可检测预测常与次要影响有关，主要影响仅为定性描述的条文。
项目修订
事后 EIS 项目修订使很多预测无效。
监测数据
监测数据和监测技术常常会证明审计目的不充分。
开发前本底值的监测即使完全进行，往往也是不充分的。
项目提倡者收集、提供了大部分监测数据，这样可能导致提供的信息存在一定的偏见。
广泛性
很多审计研究仅关心特定类型的影响（如生物物理影响而非社会-经济影响；运行阶段而非建设阶段），因此不能作为整个项目的 EIA 审计。
明确性
很少有审计研究能够清楚地说明判断预测准确性的标准；标准不明确阻碍了不同研究间的比较。
解释
大多数审计研究很少注意检查预测失误的根本原因：只有监测和审计工作在 EIA 程序中提供有效的反馈信息，才可能找出预测失误的根本原因。

资料来源：Chadwick 和 Glasson，1999。

这些问题可以在一定程度上从生态学的角度解释早期加拿大环境报告中存在不足的原因，因为准确的环境预测似乎成了个别例外而不是惯例（Beanlands 和 Duinker，1983）。还有一些案例，同样来自加拿大，其 EIA 均未能对重大影响做出准确预测。Berks（1988）指出在 James Bay mega-HEP（1971—1985）的案例中，EIA 未识别出汞在食物链中的连环影响，导致鱼体内的汞含量和当地居民发生汞中毒事件的增加。Dickman（1991）发现在加拿大最北端高浓度盐湖 Garrow 湖的案例中，EIA 未考虑到铅、锌矿渣量的增加对鱼类数量的影响。诸如此类的研究还有很多，加拿大在监测方面处于领先地位；这些研究有希望提高预测质量和水平。

英国早期开展了有限的审计活动，其结果并不令人满意。一项关于四个主要开发项目的研究——辛德兰岛 Sullom Voe 和奥克尼群岛 Flotta 石油港口、Cow Green 水库和 Redcar 炼钢厂——表明其中 88％的预测不可审计。而那些可被审计的项目中，预测准确率不到 50％（Bisset，1984）。Mill（1992）对英国近期五个主要开发项目（一条主干公路、两个风力农场、一个电站和一个露天煤矿）视觉影响的监测研究显示，EIS 中陈述的和实际产生的影响之间存在巨大的差异。在部分案例中，项目概述从根本上发生了变化，景观描述直接限制为项目开发场地的周围地区，且常常忽略美学方面的考虑。尽管如此，减缓措施还是执行得很好。

还有一些审计案例包括对 Toyota 工厂的研究（Ecotech 研究与咨询有限公司，1994）和各种风力发电厂的研究（Blandford，C. Associates，1994；ETSU，1994）。前者从广泛的角度对工厂的环境影响进行了研究；审计表明低估了使用和排放的影响，高估了住宅区的影响，对施工车辆的影响判断比较合理。另一个由 Blandford、C. Associates 共同完成的关于威尔士三个风力发电厂建设阶段的研究证实了生态影响较低的预测，但对视觉上的影响却比预测的要好，可见距离比预测的 15km 要远。但是也不排除后一研究结果与冬季审计有关，因为夏天有雾时可见距离可能会小一些。

澳大利亚的 Buckley（1991）开展了一项全国范围内环境影响预测审计准确性及准确度的研究，该研究是最为全面的研究之一。研究过程中，他发现从 1974 年到 1982 年的 1000 多份 EIS 报告中的大量用于检验预测的监测数据中，仅有 3%有效。一般而言，他发现可测性预测和监测数据仅对大而复杂的项目有效，因为这些项目常常是公众讨论的焦点，并且这些项目监测的主要目的是检验其是否达到标准，而不是检验影响预测的准确性，表 7.4 列出了 300 多项参与检验的重点及辅助预测。总的来说，Buckley 发现：定量预测、临界值预测及可测性预测的平均准确度的标准偏差为（44±5）%，且影响越严重，准确度越低，对地下水渗流的预测最不准确。当然，评价的准确性首先受到预测采用的精密度大小的影响。在这方面，用区间来表示影响预测的概率属性可能更为明智，且使监测是否合规更为直截了当，减少异议。Buckley 的全国性调查表明：影响预测的平均准确率不到 50%，这很难使大众满意。实际上，如果预测数据来自对相关运行企业的监测，那么预测结果可能比来源于大范围 EIS 数据得到的结果更加可信。另一方面，我们努力从实践中积累经验，而且近期很多 EIS 中预测越来越好，越来越准确。英国在最近一项研究中，Marshall（2001）对 41 份 EIS 中 1118 个减缓措施提议进行了分析。发现所有案例中有 38%（共 418 份）的案例，其倡议者不能将措施付诸实践。在这种情况下，减缓措施就没什么价值了，同时可能引起更多问题。

表 7.4　环境影响预测审计举例

构成要素/参数	开发类型	预测的影响	实际的影响	准确度/精确度
地表水质量、含盐量、pH 值	铝土矿	未检出河流盐度增加	未检出	正确
噪声	铝土矿	爆炸噪声<115dB(A)	只有 90%小于 115dB(A)	不正确；准确度 90%，较差
劳动力数量	炼铝厂	建设阶段 1500 人	达到 2500 人	不正确；准确度 60%，较差

资料来源：Buckley，1991。

直到最近，审计研究才开始强调预测技术审计的重要领域和特殊预测技术的价值。目前的研究趋向于关注识别与预测方法有关的错误，而不是设法解释那些错误。开发适用的审计方法是十分必要的，且项目实施越多，此类研究的范围就越广。在能见度、噪声和空气质量影响等领域研究的先驱 Wood，利用 GIS 进行审计并模拟了 EIA 误差，为今后的研究提供了一个很好的借鉴（Wood，1999a、b，2000）。

7.5 英国实例研究：Sizewell B PWR 建设项目对当地社会经济影响的监测与审计

7.5.1 案例研究背景

在英国，虽然影响的监测和审计并不是 EIA 程序中的强制步骤，但开发所造成的自然和社会-经济影响不容忽视。例如，很多公共机构对特殊污染物进行监测。LPA 监测一些与开发许可相关的条件。但是目前针对影响预测与减缓措施的监测与审计尚未形成系统的方法。该案例的研究报告试图引入一套更系统的监测方法和审计方法，尽管仍然不全面（以英国截至 2025 年的众多新能源项目提案为背景）。

在 20 世纪 70 年代和 80 年代早期，英国提出一项核电站修建计划。该计划在 1990 年修订后（现在更为活跃）包含一个委托事项，承诺修建一系列新的核电站（当时是压水反应堆电站）。其中第一个被批准的电站就是在 Anglia 东部修建的 Sizewell B 核电站。但这个项目在当时有很大争议，之后引发了英国历史上历时最长的公众质询。该项目于 1987 年动工，1995 年竣工。牛津布鲁克斯大学规划学院的 IAU 研究了发电站数量的影响，重点关注其社会-经济影响，并为 EIS 的编制做出很大贡献。针对相关的公用事业公司（英国中央电力局 CEGB），制定了相关提案：Sizewell B 的建设为项目实施阶段的详细监测提供了非常宝贵的机会，同时可以检验公众质询过程做出的各种预测以及项目批准时所需要的减缓条件。尽管预测还没有正式编入 EIS，但作为基于质询的一系列报告，它也是一个广泛的综合性研究成果（DoEn，1986）。CEGB（英国中央电力局）资助的一个监测研究开始于 1988 年，私有化之后公用事业的贷款方主要是核电/英国能源，近来主要是 EDF（法国电力公司），尽管英国核电方面的进一步发展尚存在不确定性，但仍有责任对其进行监测。目前，整个项目的建设阶段和运行阶段的监测报告已经完成（Glasson 等，1989～1997；关于项目的更多反思见 Glasson 2005）。

7.5.2 监测研究的运行特点

明确监测研究的目标十分重要，否则可能会因收集到很多不相关的信息而造成资源浪费。图 7.3 概括了监测研究的范围。对于 Sizewell B PWR（压水反应堆）1200MW 核电站的开发建设，研究重点是开发过程造成的社会-经济影响，也包括一些自然条件的影响。EIA 中的社会-经济元素包括“当环境已经受到开发活动或新政策的影响时，应该提前对人们和社区的日常生活质量所受的影响进行系统地评价”（Bowles，1981）。其中包括对就业、社会结构、消费和服务等方面的影响。虽然目前，尚没有将社会-经济的研究和影响评价研究紧密结合起来，它们在 EIS 报告中也仅占 1～2 章的篇幅，但它们仍旧很重要，至少考虑了对人类的影响，而人们可以据此来反驳或反对开发。

在研究中最优先考虑的是识别开发活动对当地就业的影响，重点反映就业影响引发其他影响时的关键作用，特别是在住宿和地方服务方面。此外，本研究可以为评价提供一个可更新、可改进的数据库信息；并针对 Sizewell B 项目在当地社区及审计影响预测方面的影响进

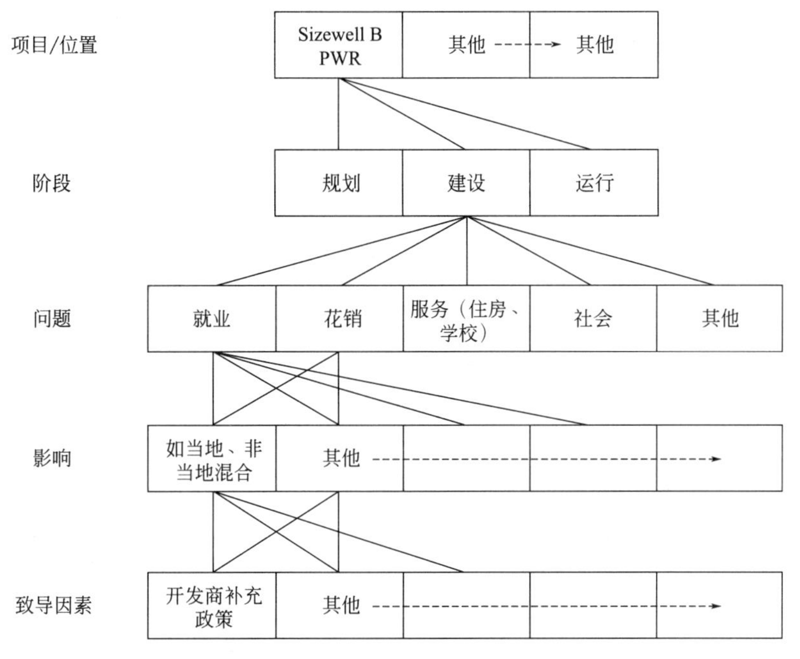

图 7.3 Sizewell B 监测研究的范围及数据库组织结构

（资料来源：Glasson 等，1989～1997）

行项目管理；同时还可监测和审计与批准电站建设相关的条件与承诺，包括对铁路和建筑运输专用线使用的承诺，以及使用当地劳动力、地方企业、当地服务设施及控制交通噪声的条件（DoEn，1986）。

监测研究包括各方面信息的收集、统计数据（如建设阶段本地和外地工人数量的混合比例、住房占有情况、工人的消费模式等）、决策、各方观点以及对影响的感知。研究的空间范围要扩展到建筑工人上下班往来的活动范围。研究的信息来源主要是开发商和现场主要承包人、相关地方当局和其他公共机构以及当地社区居民和建筑工人。当地高校地理系优秀学生也帮助收集了当地群众影响感知的数据，他们在建设地点邻近的莱斯顿镇进行了两年一次的问卷调查，同时每两年还会对 20%参与项目的劳动力进行一次调查，重点是了解社会-经济系统的特点与活动规律。最后 IAU 小组负责把数据收集起来。该研究得到了广泛的支持，其结果公开出版在年度监测报告和大幅报纸上，当地社区居民可免费获得此报告（Glasson 等，1989～1997）。

该研究突出强调监测和审计过程中方法上的困难。首先是如何从环境本底值的变化趋势中分离出与项目相关的环境影响。有很多变量的数据是可以获得的，它们能够反映当地的某些变化趋势，如失业率、交通量和犯罪率。但是当我们试图解释这些趋势时却遇到了困难。以下内容在何种程度影响趋势发展：①建设项目本身；②国家和区域因素；③其他与建设项目无关的地区性变化。虽然国家和地区因素很好分离，但建设项目及其他地方变化造成的影响却很难区分。这时可采用“控制”方法来区分与项目有关的影响。

第二个难题是对建设项目间接但必然影响的识别。间接影响可能很严重，但却不易被观察或测定，尤其是对就业的影响。例如，雇员离开当地去外地工作，就可能对就业产生间接影响。这些地方新增招聘雇员能替代之前的雇员吗？如果这样，他们是来自其他地区的雇员，还是当地未就业人员或者迁入工人呢？这类信息一般不可能获得。进一步的间接的就业影响还可能来源于当地商人在 Sizewell B 成为供应商或者承包商。他们需要负担额外的劳动力来满足附加的工作量。即使能从当地企业调查中获得某些有用的信息，也很难评估它们的影响程度（Glasson 和 Heaney，1993）。

7.5.3 一些研究结论

以下内容简要概括了一些研究结论（图 7.4）。

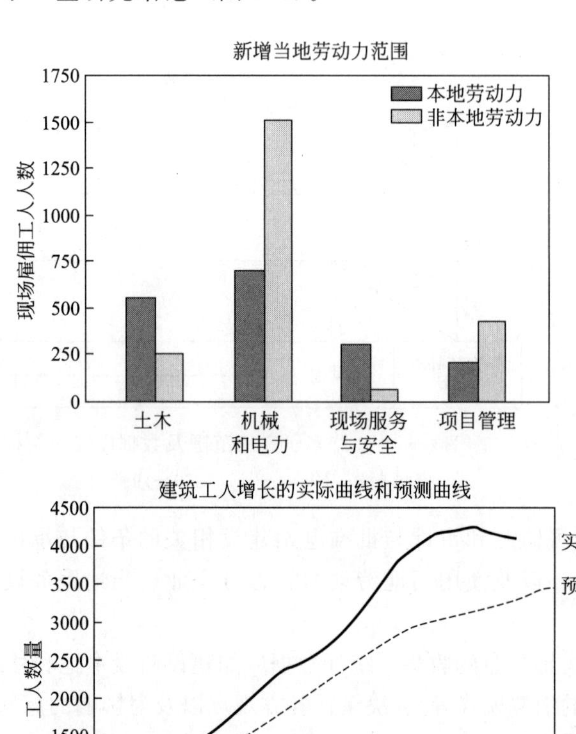

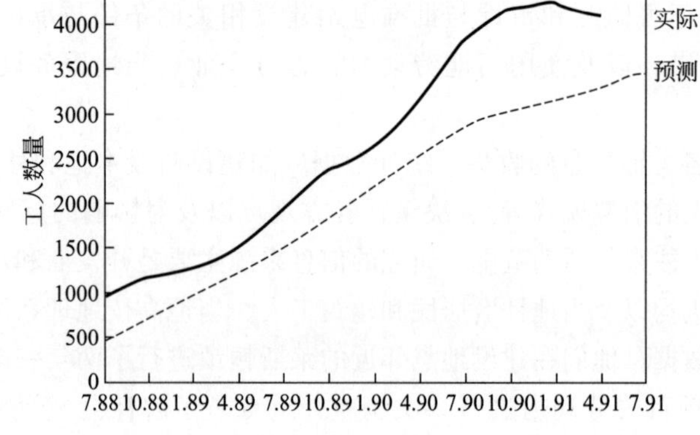

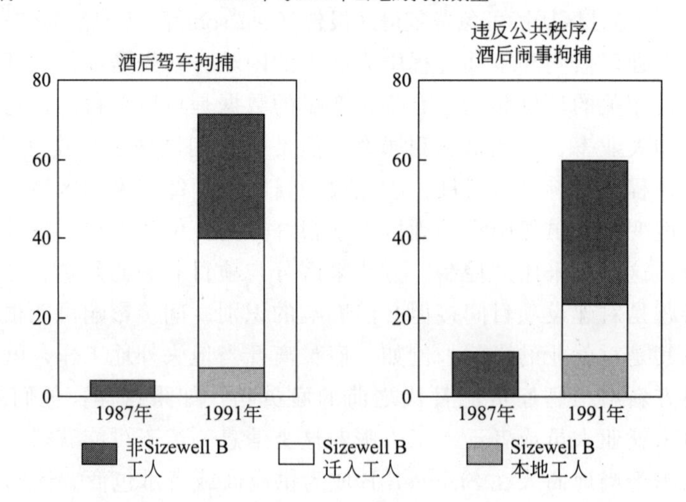

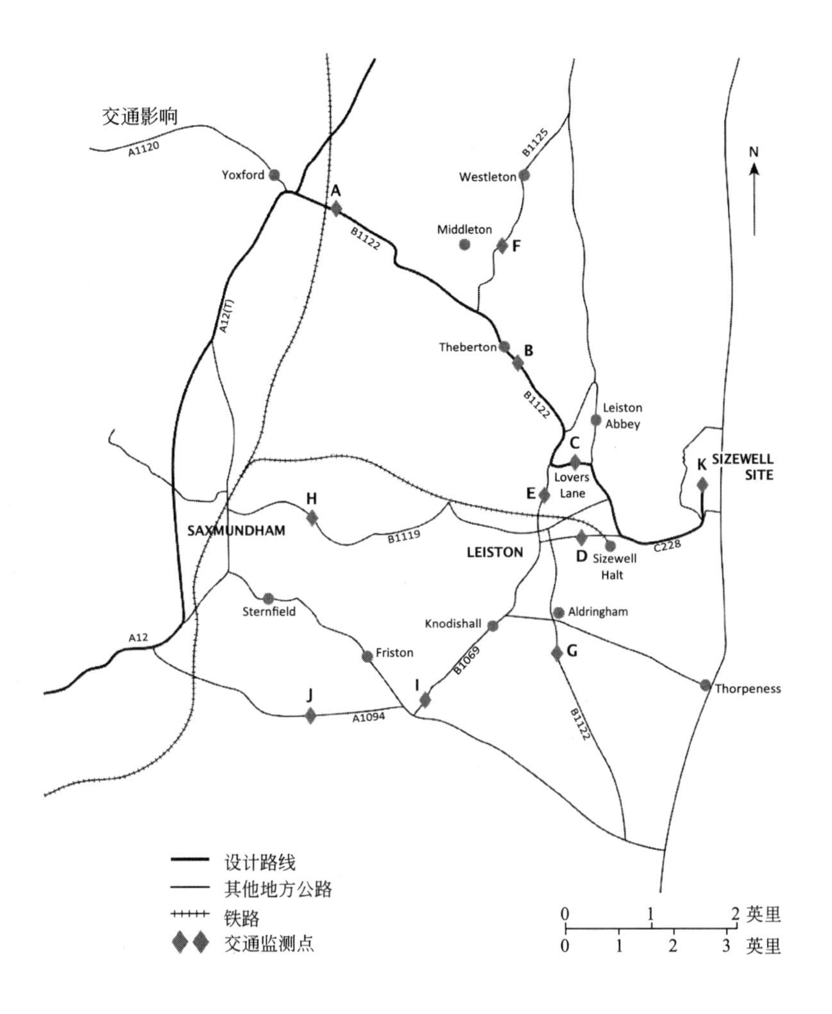

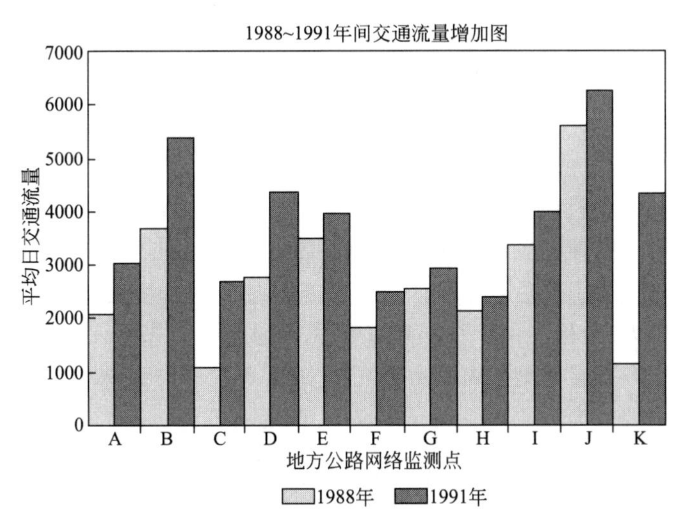

图 7.4 Sizewell B PWR 建设项目监测与审计研究结论概要

（资料来源：Glasson 等，1989～1997）

7.5.3.1 就业影响

一个很重要的预测和条件是建设项目的雇工中至少有50%是当地人（一天内能往返施工地点）。然而在农村地区可能确实存在这种情况，即当地人多数从事的是半技术或者非技术工作。随着该地区就业率的上升以及由土木工程向机械和电力工程的转变，这个50%的比例也将增大。1989年，附近的莱斯顿镇开办了一个培训中心，对80～120个当地失业人员进行了培训。

7.5.3.2 当地经济影响

一个大型项目可能对当地经济的影响具有经济乘数效应（multiplier effect）。截止到1991年年底，Sizewell B工人每周在萨福克和诺福克的花费约为50万英镑，核电公司已经与当地公司签订了价值超过4000万英镑的合约，同时制定了“友邻”政策，已经资助了一系列社区项目（包括在莱斯顿耗资190万英镑修建的游泳池）。

7.5.3.3 住房影响

大型项目往往会带来大量的移民劳动力，影响当地住房市场，在以旅游业为主的地区还会影响旅游住宿。Sizewell B采用的一个减缓措施是要求开发商提供一个较大的工地宿舍，能容纳600个工人（后来增至900个）。事实证明，该措施非常有效，到1991年已容纳了40%以上的移民劳动力，平均入住率超过85%，也降低了当地的住房需求。

7.5.3.4 交通和噪声影响

大型建设项目引起的交通问题会对当地城乡造成严重影响。为减缓这些影响，当局指定了一条通往Sizewell B的专用建设路线。通过对指定和非指定（控制）路线的交通流量进行监测，结果表明该措施已表现出一定的缓解效果。1988～1991年间指定路线上的4个监测点交通量大幅增加，而非指定路线上的7个监测点上交通量的增加幅度则要小很多。施工现场的噪声也是一个地方性问题，根据监测结果，建议对某些建设方法进行改进，尤其是铁路专用线的改进以及打桩方法的改变。

7.5.3.5 犯罪影响

通常在大型项目的建设阶段，当地的犯罪率会有所增加。从莱斯顿警局的记录可以看出，项目开工建设后，因某种犯罪行为被捕的人数将显著增加。大多数拘捕都会将未被雇佣的当地居民卷入其中，除了酒后驾车外，大多数案件以及犯罪率上升都归咎于Sizewell B的员工（尤其是迁入工人）。但是由于较早地发现了问题并及时采取了补救措施［包括为工人提供往返小型巴士服务、在工地宿舍开办酒吧、在工人就职培训课中强调酒后驾车的问题、从现场开除有严重不端行为或犯罪行为的工人（直接从Sizewell B岗位上开除）］，从项目开始以来，与工人相关的犯罪大幅减少，而且警方也认为项目的工人也相对很少制造麻烦，重大犯罪比预想的要少。

7.5.3.6 居民认知影响

根据对当地居民1989年和1991年认知调查可知，居民感知到的消极影响多于积极影响，其中主要的消极影响是交通量的增加以及工人的干扰。主要的积极影响是就业增加、额外贸易以及一些改善措施。项目建设在快速推进，监测结果表明，随着时间的推移，关于项目建设的抱怨大幅减少。

7.5.4 Sizewell B 和 Sizewell C 监测学习收获

表 7.5 列出了 Sizewell B 社会-经济预测的性质和可审计性。与之前关于事后审计的研究结果相反（Dipper 等，1998），Sizewell B 的大部分预测是以定量形式给出的。对影响的监测、对预测和减缓措施的审计表明（表 7.6），用于 Sizewell B 公众质询中的预测是合理准确的——尽管它低估了建筑工人数量的增加，高估了对地方经济的间接影响。对交通影响以及当地工人在建设项目雇佣劳动力所占比例的预测都比较接近实际数据。减缓措施也取得了一定的成效。总体上来说，大约有 60%的预测误差范围在 20%以内。造成预测与实际结果差距的原因包括不可避免的项目调整（特别是采用新技术的项目，在预测时很少甚至没有对照标准）以及长期的项目审批程序（从预测到建设完成差不多需要 10 年时间）。同时监测还揭示出一些其他地方性的问题，帮助参与者采取一些改善措施以更好地管理项目。不幸的是，在英国，这种系统化的监测方法仍然是自由决定采用与否，且很大程度上取决于开发商的意见。

表 7.5　Sizewell B 社会-经济预测的性质和可审计性

项目	预测编号	占总数的百分比/%
预测的性质		
定量的		
用绝对值表示	35	51
用百分数表示	21	30
定性的	11	16
将定性元素与定量元素相结合	2	3
总计：全部预测	69	100
预测的可审计性		
可审计：监测数据没有或几乎没有潜在错误	30	43
可审计：监测数据存在很大的潜在错误	28	41
不可审计	11	16
总计：全部预测	69	100

资料来源：Chadwick 和 Glasson，1999。

表 7.6　Sizewell B 预测审计的准确度

预测误差/%	预测编号	占总数的百分比/%
0：预测准确或在误差允许范围内	15	26
低于 10%	9	16
10%～20%	11	19

续表

预测误差/%	预测编号	占总数的百分比/%
20%～30%	5	9
30%～40%	5	9
40%～50%	2	3
超过 50%	8	14
预测不准确，但无法估计误差百分比	3	5
无法审计的预测	11	—
总计：全部预测	69	100

资料来源：Chadwick 和 Glasson，1999。

注：对于定量预测，计算误差时标准预测值为分母；对于非定量预测，其误差无法计算，因此根据研究小组的评价结果，将其分为“准确预测”和“不准确预测”。

监测所获得的信息对未来项目的规划和评价有重要的参考价值，尤其是当之后的项目与所监测的项目性质相同、区域相同时。Sizewell C 核电厂建设的审批和运营，与 Sizewell B 的情况完全相同。工作人员针对此项目编制了一个详细的 EIS（Nuclear Electric，1993），其对社会-经济影响的预测可直接采用 Sizewell B 的监测研究结果。然而这一提议最终因 20 世纪 90 年代早期英国放弃核电站建设计划而搁置。但 2007 年之后采取了一系列行动，涌现出越来越多超前的规划之后，英国核电站建设迈入一个新时代，其中包括新修建了 Sizewell C 核电站。在此背景下，事实证明 Sizewell B 的监测数据具有重大价值。

7.6 英国实例研究：伦敦 2012 奥运会项目对地方影响的监测

7.6.1 项目性质及其影响的生命周期

伦敦 2012 年奥运会项目是欧洲最大的项目之一，2011 年参与建设的劳动力人数峰值达到将近 12000 人。运动会场位于伦敦东部，距市中心约 5km，占地面积约 250 hm^2。奥运会、残奥会及遗留设施的设计不仅是为本项赛事提供专用场所，也是对原先下利亚谷区的重建，使其能够长期保留，相当于为伦敦建造了一个新的城区。该项目的生命周期以影响年表的方式在环境报告中突出体现（Symonds/EDAW，2004），图 7.5 对此进行了概括。同时报告还提供了该项目在生物物理及社会-经济影响方面的监测框架。

7.6.2 建设阶段的监测

该项目与伦敦奥运交付管理局（ODA）协调配合制订了一个详细的监测计划。以下为两个有关项目性质的详细监测示例。图 7.6 为整个施工现场每月的建筑噪声监测摘要，结果显示情况基本良好。表 7.7 提供了每月社会-经济监测摘要，重点关注施工工人的特征。

影响	奥运会开始前的建设阶段	奥运会期间	奥运后结束后遗留设施建设阶段	奥运后结束后遗留阶段
现有住房、工业、岗位和废弃物处置设施的过早亏损	//////	//////	//////	
考古学基准的潜在亏损				
拆除及环境变化对遗产保留造成的损害				
具有历史意义的地区区域特征的缺失				
城镇景观质量的提升		//////		//////
电力电缆地下铺设		//////	//////	//////
修复结果	//////	//////	//////	//////
建筑/结构中纳入的CCHP获得的能源效率及其他可持续/可再生能源特征		////// //////	////// //////	////// //////
创造奥运会工作岗位	//////	//////	//////	
“感觉良好”的因素，社会凝聚力及团体自豪感	//////	//////	//////	
鼓励积极参与体育运动/健康活动	//////	//////	//////	//////
地方交通设施的影响		//////		
大型建筑附近风力对队列及人群的影响（包括奥林匹克运动村）				
河流交叉口安全围栏造成的潜在洪涝风险				
新的公共区域已有污染造成的潜在影响				
额外的公用场地、开放场地及物资分配			//////	//////
创造额外的/替代的生境			//////	//////
可达性和渗透性的提高				//////
与相关技能或训练有关的遗留工作岗位的创造				//////
社区设施提高（学校、全托幼儿园、日托托儿所、医疗诊所等）				//////
建筑水平提高（如“全面共享”标准）				//////
对LB Newham生活垃圾处置设施的影响				//////

注：——显著有利；——显著不利；阴影线——有或没有明显的缓解；素色——没有明显的缓解。

图7.5 伦敦2012奥运会项目的生命周期影响

（资料来源：Symonds/EDAW，2004）

2011 年 1 月，两个主要工程（奥林匹克公园和运动员村）的劳动力数量接近 12000 人。其中，伦敦区当地的员工尤其是工地周围的居民占了很大的比例。监测从失业人员和黑色人种、亚洲人及少数族裔中招聘员工的过程，结果显示比计划目标要好，这与违反女性和残疾人招聘基准的不良表现形成了鲜明对比。同时，该项目实施了一系列员工发展措施以提高地方招聘的有利影响，如就业中介计划安排超过 1250 人（基本上是当地居民）在项目上获得工作，超额完成培训计划，共培训了 3250 人（截至 2011 年，计划目标是 2250 人），其中包括 400 个学徒（计划目标是 350 个）。

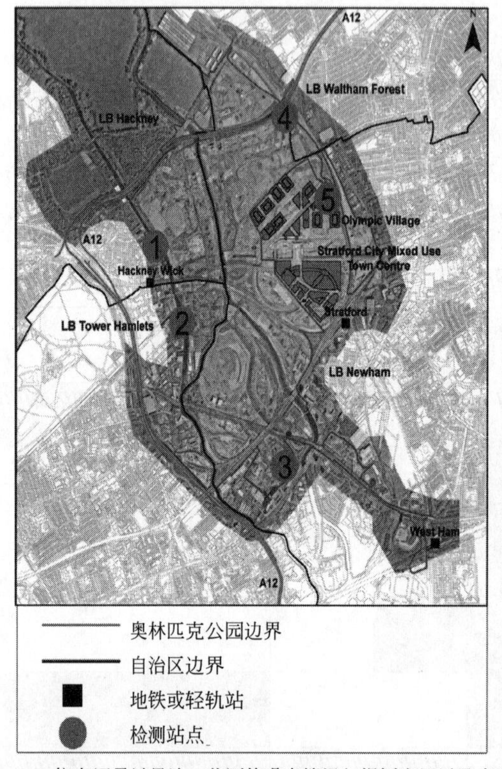

信息记录以月计，监测的噪声等级与规划当局以及东道主地区获得奥林匹克公园规划许可时确定的等级相比较，奥运村的规划许可单独获得，基于每小时的阈值。

①奥运村从属于不同的规划许可，其监测值以每天的小时水平计。

图 7.6　伦敦 2012 奥运会场址建设噪声监测（2010 年 8 月）

（资料来源：Olympics Delivery Authority，2010）

表 7.7　伦敦 2012 奥运会场馆建设劳动力/雇员监测（2010 年 12 月）

工地劳动力	奥林匹克公园		运动员村	
	6500	基准	5400	基准
当地居民/%	21	—	27	—
伦敦其他地区的居民/%	34	—	40	—
英国其他地方的居民/%	42	—	30	—
居住在英国以外的员工/或无信息可查的/%	3	—	3	—
之前失业的人员/%	12	7	10	7
女性/%	4	11	3	11

续表

工地劳动力	奥林匹克公园		运动员村	
	6500	基准	5400	基准
残障人士/%	1	3	0.5	3
黑色人种、亚洲人或少数族裔/%	19	15	13	15

资料来源：Employment and Skills Update，伦敦奥运交付管理局（2011 年 1 月）。

7.7 小结

在项目的整个生命周期中需要不断调节项目与环境的关系。环境影响评价的目的是确定项目实施条款及条件；但很少能持续跟进到这一阶段，甚至后续工作都很少进行。Arts（1998）指出，在对“EIA 进行事先事后评价”的全面检查之后，实际上它已经落后于 EIA 本身的实践。很少有国家会做一些继续跟踪的安排。对于那些已经进行了 EIA 评价的项目，经验也不是很有启发性——在一定程度上反映出过分夸大 EIS、跟踪技术落后、组织和资源限制、权力机构及项目参与者的支持有限等方面的缺陷。但许多项目延续时间较长，因而对其影响的监测也需要定期进行。Morrison-Saunders 等（2001）指出跟踪评价如何给不同的利益方带来积极的影响。

图 7.7 表明跟踪评价不仅会使项目支持者和参与者受益（例如对 Sizewell B 案例的研究），而且对于管理者而言——可以做出更好地决策，更好地完善 EIA 程序。这样监测不仅可以改进项目的管理，同时也有利于影响预测和减缓措施的审计。监测和审计可以提供重要

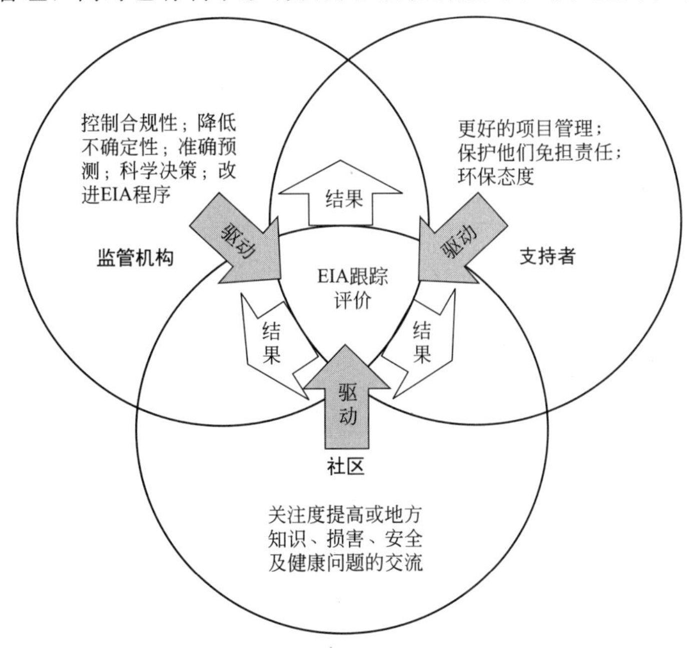

图 7.7 EIA 跟踪评价对于不同利益方的结果
（资料来源：Morrison-Saunders 等，2001）

的反馈信息，从而改进 EIA 程序，尽管在许多国家这依然是目前最薄弱的环节。自由裁定的措施肯定是不够的，监测和审计需要在强制性基础上，更全面地纳入 EIA 程序中。

问 题

以下问题旨在帮助读者更好地明确本章讨论的重点问题，逐步认识到监测与审计在 EIA 程序中的性质及重要性。

1. 如何理解 EIA 中“决策前”和“决策后”的区别？

2. 为什么对那些经过决策后已经开始动工的项目继续 EIA 程序很重要？

3. 如何理解 EIA 中“监测”和“审计”的区别？

4. 选择一个你熟悉的大型项目，考虑应收集什么类型的监测信息，包括传统指标、潜在因素、定量信息及定性信息。

5. 为什么很多国家 EIA 中强制监测和审计系统会受到诸多阻力？

6. 比较、对比美国加利福利亚州和中国香港监测系统的主要特点。

7. 回顾与事后审计研究有关的问题，如表 7.3 中所列，考虑该如何解决这些问题。

8. 以 Sizewell B 项目为例，指出监测中常常遇到的两个方法性问题的解决途径：(1) 从基线趋势中分离出与项目有关的影响；(2) 识别项目造成的间接影响。

9. 回顾可审计 Sizewell B 预测的相对准确性，如表 7.6 中所列。你认为导致这一结果的可能因素包括什么？

10. 图 7.5 显示了监测在一个项目的整个生命周期中的重要性，其中涉及的某些影响的监测可能比其他影响容易。快速浏览列表中的影响，尽可能快地识别伦敦 2012 年奥运会项目监测过程中可能用到的相关的影响指标。

11. 利用英国监测的两个实例研究，再加上你可能熟悉的项目，概述 EIA 监测/跟踪评价对图 7.7 中定义的相关利益方的潜在利益。

7

Monitoring and auditing: after the decision

7.1 Introduction

Major projects, such as roads, airports, power stations, waste processing plants, mineral developments and holiday villages, have a life cycle with a number of key stages (see Figure 1.5). The life cycle may cover a very long period (e.g. 50—60 years for the planning, construction, operation and decommissioning of a fossil-fuelled power station). EIA, as it is currently practised in the UK and in many other countries, relates primarily to the period *before* the decision. At its worst, it is a partial linear exercise related to one site, produced in-house by a developer, without any public participation. There has been a danger of a shortsighted 'build it and forget it' approach (Culhane 1993). However, EIA should not stop at the decision. It should be more than an auxiliary to the procedures to obtain a planning permission; rather it should be a means to obtain good environmental management *over the life* of the project. This means including monitoring and auditing fully into the EIA process. There is a continuing danger that emphasis on pre-decision analysis will keep EIA away from its key goal of environmental protection. EIA should seek to maximize the potential for continuous improvement. Resources spent on baseline studies and predictions may be rendered of little value unless there is some way of testing the predictions and determining whether mitigation and enhancement measures are appropriately applied (Ahammed and Nixon 2006). It is good to record that there is now more learning from experience and some good progress to note; see for example Morrison-Saunders and Arts (2004) and the special edition of the *Impact Assessment and Project Appraisal* journal (IAPAY 2005) on EIA Follow-Up.

Section 7.2 clarifies some relevant definitions, for example between monitoring and auditing, and outlines their important roles in EIA. An approach to the better integration of monitoring into the process, drawing in particular on international practice, is then outlined in Section 7.3. We then discuss approaches to environmental impact auditing, including a review of attempts to audit a range of EISs in a number of countries. The final section draws on detailed monitoring and auditing studies of the local socio-economic impacts of the construction of the Sizewell B pressurized water reactor (PWR) nuclear power station in the UK, and also briefly notes monitoring of the more recent London 2012 Olympics project.

7.2 The importance of monitoring and auditing in the EIA process

In many aspects of EIA there has been considerable semantic development and a widening of relevant terms. In earlier editions of this chapter we have focused primarily on *monitoring* and *auditing* as key elements in the after the decision process, or the follow-up process as it is known in some countries such as Canada and Australia. These elements are still crucial and the main focus of Chapter 7, but note should also be taken of a widening of those key elements to also include *management* and *communications* (Arts *et al.* 2001; Marshall *et al.* 2005). In total such elements can facilitate learning from experience, preventing EIA from becoming just a pro-forma exercise with little clout after the project decision has been taken.

Monitoring involves the measuring and recording of physical, social and economic variables associated with development impacts (e.g. traffic flows, air quality, noise, employment levels). The activity seeks to provide information on the characteristics and functioning of variables in time and space, and in particular on the occurrence and magnitude of impacts. Monitoring can improve project management. It can be used, for example, as an early warning system, to identify harmful trends in a locality before it is too late to take remedial action. It can help to identify and correct unanticipated impacts. Monitoring can also provide an accepted database, which can be useful in mediation between interested parties. Thus, monitoring of the origins, pathways and destinations of, for example, dust in an industrial area may clarify where the responsibilities lie. Monitoring is also essential for successful environmental impact auditing, and can be one of the most effective guarantees of commitment to undertakings and to mitigation and enhancement measures.

Environmental impact auditing, which is covered in this chapter, involves comparing the impacts predicted in an EIS with those that actually occur after implementation, in order to assess whether the impact prediction performs satisfactorily (Buckley 1991). In some of the literature this step is sometimes referred to as evaluation. The audit can be of both impact predictions (how good were the predictions?) and of mitigation and enhancement measures and conditions attached to the development (are the mitigation and enhancement measures effective; are the conditions being honoured?). This approach to auditing contrasts with *environmental management auditing*, which focuses on public and private corporate structures and programmes for environmental management and the associated risks and liabilities. We discuss this latter approach further in Chapter 12.

Management is an important element in terms of making decisions and taking appropriate action in response to issues raised from the monitoring and auditing activities (Marshall *et al*. 2005). For example, monitoring may show lower levels of local recruitment than predicted for the construction stage of a major project. A management response may be to redouble efforts on local training programmes. *Communication* is to inform stakeholders about the results of EIA monitoring, auditing and management activities. Such communication may emanate from both proponents and regulators; hopefully there may be a partnership approach between the two. Ideally the community stakeholder's role is more than that of passive recipient of follow-up activities, and rather more one of partner in the process, being involved directly in the follow-up activities.

In total, such activities can make important contributions to the better planning and EIA of future projects (Figure 7.1). There is a vital need to introduce feedback in order to learn from experience; we must avoid the constant 'reinventing of the wheel' in EIA (Sadler 1988). Monitoring and auditing of outcomes, and the resultant management responses and associated communication, can contribute to an improvement in all aspects of the EIA process, from understanding baseline conditions to the framing of effective mitigating and

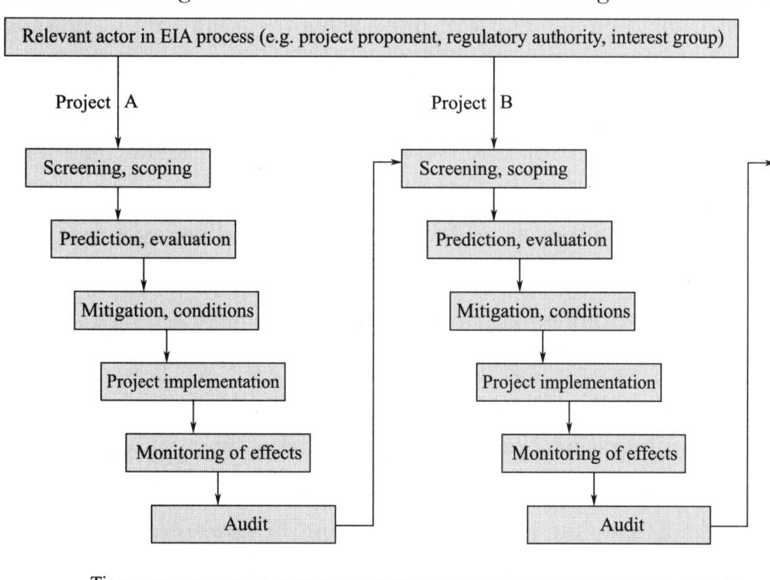

Figure 7.1 Monitoring, auditing and learning from experience in the EIA process
Source: Adapted from Bisset and Tomlinson 1988; Sadler 1988

enhancement measures. In addition, Greene *et al.* (1985) noted that monitoring and auditing should reduce time and resource commitments to EIA by allowing all participants to learn from past experience; they should also contribute to a general enhancing of the credibility of proponents, regulatory agencies and EIA processes. We are learning, and there is a considerable growth of interest in examining the effectiveness of the EIA process in practice. However, there are still a number of significant issues that have limited the use of monitoring and auditing to date. These issues and possible ways forward for monitoring and auditing in practice are now discussed.

7.3 Monitoring in practice

7.3.1 Key elements

Monitoring implies the systematic collection of a potentially large quantity of information over a long period of time. Such information should include not only the traditional *indicators* (e.g. ambient air quality, noise levels, the size of a workforce) but also *causal underlying factors* (e.g. the *decisions* and *policies* of the local authority and developer). The causal factors determine the impacts and may have to be changed if there is a wish to modify impacts. *Opinions* about impacts are also important. Individual and group 'social constructions of reality' (IOCGP 2003) are often sidelined as 'mere perceptions, or emotions', not to be weighted as heavily as facts. But such opinions can be very influential in determining the response to a project. To ignore or undervalue them may not be methodologically defensible and is likely to raise hostility. Monitoring should also analyse impact equity. The distribution of impacts will vary between groups and locations; some groups may be more vulnerable than others, as a result of factors such as age, race, gender and income. So a systematic attempt to identify opinions can be an important input into a monitoring study.

The information collected needs to be stored, analysed and communicated to relevant participants in the EIA process. A primary requirement, therefore, is to focus monitoring activity only on 'those environmental parameters expected to experience a significant impact, together with those parameters for which the assessment methodology or basic data were not so well established as desired' (Lee and Wood 1980).

Monitoring is an integral part of EIA; baseline data, project descriptions, impact predictions and mitigation and enhancement measures should be developed with monitoring implications in mind. An EIS should include a *monitoring programme* that has clear objectives, temporal and spatial controls, an adequate duration (e.g. covering the main stages of the project's implementation), practical methodologies, sufficient funding, clear responsibilities and open and regular reporting. Ideally, the monitoring activity should include a partnership between the parties involved; for example, the collection of information could involve the developer, local authority and local community. Monitoring programmes should also be adapted to the dynamic nature of the environment (Holling 1978).

7.3.2 Mandatory or discretionary?

Unfortunately, monitoring is not a mandatory step in many EIA procedures, including those current in the UK. In contrast to the more recent SEA regulations, European Commission EIA regulations do not specifically require monitoring. This omission was recognized in the review of Directive 85/337 (CEC 1993). The Commission is a strong advocate for the inclusion of a formal monitoring programme in an EIS, but EU Member States are normally more defensive and reactive. In consequence, the amended Directive does not include a mandatory monitoring requirement.[1] However, this has not deterred some Member States. For example, in The Netherlands the competent authority is required to monitor project implementation, based on information provided by the developer, and to make the monitoring information publicly available. If actual impacts exceed those predicted, the competent authority must take measures to reduce or mitigate these impacts. However, despite such legal provisions, practice has been limited and little post-EIA monitoring and evaluation has been carried out. See Arts (1998) for a comprehensive coverage of EIA follow-up in The Netherlands.

In other Member States, as noted in Chapter 2 and CEC (2009), the lack of a mandatory monitoring requirement is a continuing, serious and long-standing issue. In the absence of mandatory procedures, it is usually difficult to persuade developers that it is in their interest to have a continuing approach to EIA. This is particularly the case where the proponent has a one-off project, and has less interest in learning from experience for application to future projects. Fortunately, we can turn to some examples of good practice in a few other countries. A brief summary of monitoring procedures in Canada is included in Chapter 10. In Western Australia (also see Chapter 10), the environmental consequences of developments are commonly monitored and reported. If it is shown that conditions are not being met, the government may take appropriate action. Interestingly, there is provision in Western Australian procedures for an 'environmental review and management programme' (Morrison-Saunders 1996).

7.3.3 The case of California

The monitoring procedures used in California, for projects subject to the CEQA, are of particular interest (California Resources Agency 1988). Since January 1989, state and local agencies in California have been required to adopt a monitoring and/or reporting programme for mitigation measures and project changes that have been imposed as conditions to address significant environmental impacts. The aim is to provide a mechanism that will help to ensure that mitigation measures will be implemented in a timely manner in accordance with the terms of the project's approval. Monitoring refers to the observation and oversight of mitigation activities at a project site, whereas reporting refers to the communication of the monitoring results to the agency and public. If the implementation of a project is to be phased, the mitigation and subsequent reporting and monitoring may also have to be phased. If monitoring reveals that mitigation measures are ignored or are not completed, sanctions could be imposed; these can include, for example, 'stop work' orders, fines and restitution. The components of a monitoring programme would normally include the following:

- a summary of the significant impacts identified in the environmental impact report (EIR);
- the mitigation measures recommended for each significant impact;
- the monitoring requirements, and responsible agency (ies), for each mitigation measure;
- the person or agency responsible for the monitoring of the mitigation measure;
- the timing and/or frequency of the monitoring;
- the agency responsible for ensuring compliance with the monitoring programme; and
- the reporting requirements.

Figure 7.2 provides an extract from a monitoring programme for a woodwaste conversion facility at West Berkeley in California.

7.3.4 *The case of Hong Kong of China*

In Hong Kong of China, a systematic, comprehensive environmental monitoring and auditing system was introduced in 1990 for major projects. A major impetus for action was the construction of the new $20 billion airport at Chap Lap Kok, which included the construction of not only the airport island, but also a railway, highways and crossings and a major Kowloon reclamation project. The Environmental Monitoring and Audit (EM&A) manual includes three stages of an event action plan: (1) trigger level, to provide an early warn-ing; (2) action level, at which action is to be taken before an upper limit of impacts is reached; and (3) target level, beyond which a predetermined plan response is initiated to avoid or rectify any problems. The approach does build monitoring much more into project decision-making, requiring proponents to agree monitoring and audit protocols and event action plans in advance; however, enforcement has been problematic (Au and Sanvícens 1996). The EM&A is intended to be a dynamic document to be reviewed regularly and updated (as necessary) during the implementation of the project.

Since April 1998 there have been EIA regulations in force that stipulate in detail when and how environmental monitoring and auditing should be done (EPD 1997, 1998). The regulations normally result in permit conditions relating to project approval. This has provided a statutory basis for follow-up work, and offences carry stiff penalties (up to $250, 000 and six

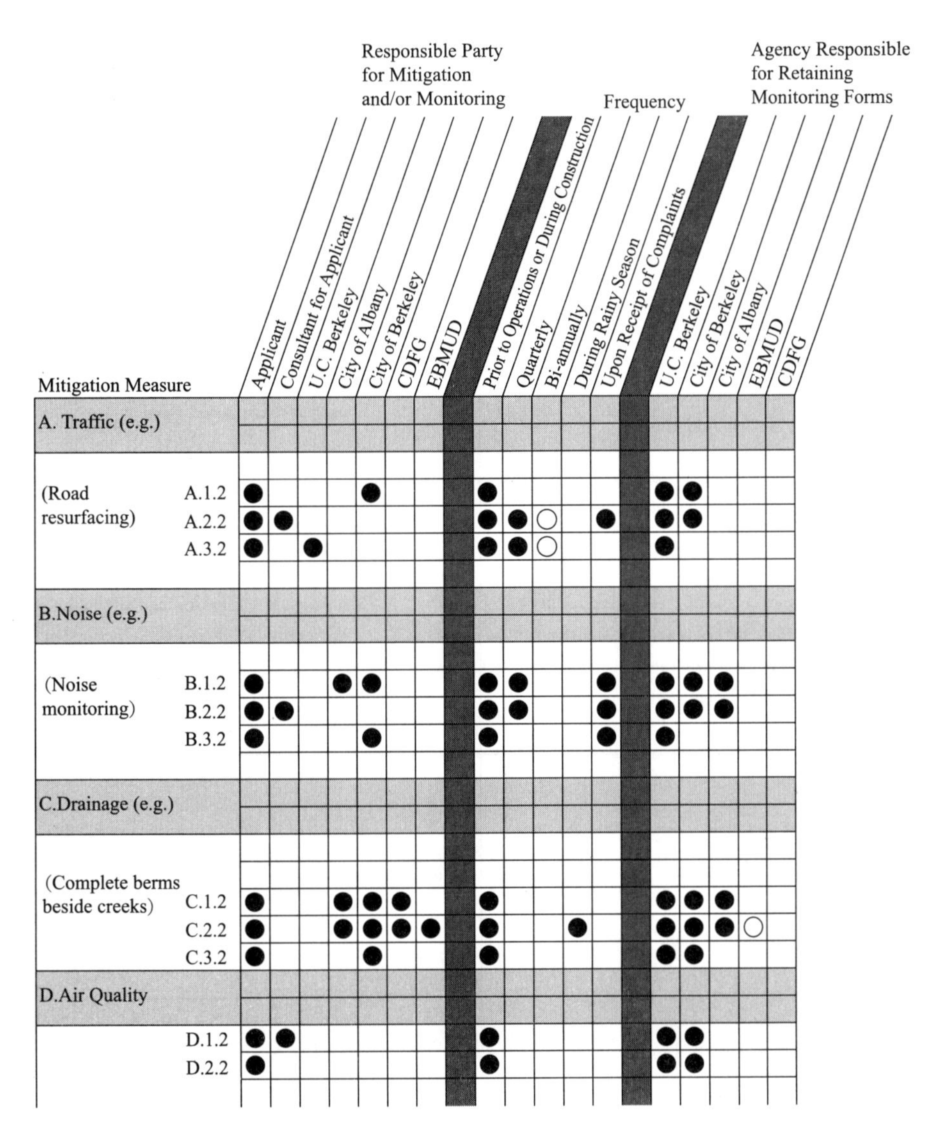

Figure 7. 2　Example of Californian monitoring programme

Source: Baseline Environmental Consulting 1989

months imprisonment). A recent and fascinating innovation in the Hong Kong of China system is the use of the Internet for monitoring the effects of large projects and of compliance with the permit conditions. Under procedures introduced since 2000 major projects must set up a monitoring website. Some sites include webcams focused on parts of the project. There is public access to the websites, and concerned members of the public can report their views on project performance back to both the government and the developer (Hui and Ho 2002). Is this the shape of things to come? Table 7. 1 provides an extract from an EM&A report for a Hong Kong of China helipad development.

Table 7.1 Extract from an EM&A report for a helipad project in Hong Kong of China: implementation schedule of recommended mitigation measures for construction of Yung Shue Wan Helipad (air quality mitigation measures)

EIA Ref.	EM&A Ref.	Recommended environmental protection/mitigation measures	Objectives of the recommended measures and main concerns to address	Who to implement the measure?	Location/timing of implementation measures	What requirements or standards for the measure to achieve?
S3.5.1	S2.5	All dust control measures as recommended in the Air Pollution Control Regulation, where applicable, should be implemented	Air quality during construction	Contractors	At all construction works sites through duration of construction works	EIAO-TM, Air Pollution Control (Construction Dust) Regulations
S3.5.1	S2.4	Typical dust control measures include: • restricting heights from which materials are dropped, as far as practicable, to minimize the fugitive dust arising from loading/unloading	Air quality during construction	Contractors	At all construction works sites through duration of construction works	EIAO-TM, Air Pollution Control (Construction Dust) Regulations
S3.5.1	S2.4	• all stockpiles of excavated materials or spoil of more than $50m^3$ should be enclosed, covered or dampened during dry or windy conditions	Air quality during construction	Contractors	At all construction works sites through duration of construction works	EIAO-TM, Air Pollution Control (Construction Dust) Regulations
S3.5.1	S2.4	• effective water sprays should be used to control potential dust emission sources such as unpaved haul roads and active construction areas	Air quality during construction	Contractors	At all construction works sites through duration of construction works	EIAO-TM, Air Pollution Control (Construction Dust) Regulations

Source: CWE-ZHEC Joint Venture 2007

7.3.5 *UK experience*

Although monitoring is still not a mandatory requirement under UK EIA regulations, there is monitoring activity. An early research study at Oxford Brookes University (see Frost 1997; Glasson 1994) sought to provide an estimate of the extent of such activity using a 'contents analysis' and a 'practice analysis'. The contents analysis of references to monitoring intentions used a representative sample of almost 700 EISs and summaries of EISs (taken from the Institute of Environmental Assessment's *Digest of environmental state-*

ments) (IEA 1993). For some EISs there was a clearly indicated monitoring section; for others, monitoring was covered in sections related to mitigation. In several cases there were generic monitoring proposals with, for example, a proposal to check that contractors are in compliance with contract specifications. Overall, approximately 30 per cent of the cases included at least one reference to impact monitoring. The maximum number of monitoring types was six, suggesting that impact monitoring was unlikely to be approaching comprehensiveness in even a select few cases. Table 7.2 shows the types of monitoring in EISs. Water quality monitoring was more frequently cited than air quality monitoring. Point of origin monitoring of air and aqueous emissions was also frequently cited. There was only very limited reference to the monitoring of non-biophysical (i. e. socio-economic) impacts. The type of monitoring varied between project types. For combined-cycle gas turbine (CCGT) power stations, proposals were often made for monitoring air emissions, air quality and construction noise; for landfill projects, the proposals were skewed towards the monitoring of leachate, landfill gas and water quality.

The practice analysis used a small representative sample of 17 projects, with EIS monitoring proposals, which had started. The LPAs were contacted to clarify monitoring arrangements including, for example, whether monitoring arrangements had been made operational under the terms of various consents (e. g. planning conditions, S106 agreements, integrated pollution control (IPC) conditions, site licence conditions) or whether monitoring was being carried out voluntarily. The findings revealed that overall EISs tended to understate, on average by about 30 per cent, the amount of monitoring actually undertaken. This may be a response to planning conditions and agreements resulting from the decision-making process; it may also relate to other relevant licensing procedures, such as IPC. Whatever the case, the findings do suggest that some monitoring proposals in EISs are carried out and are often more extensive than the, admittedly often limited, coverage in EIAs. The findings do not, of course, provide any information on the quality of the monitoring or about the accuracy of the predictions.

7.4 Auditing in practice

Auditing has developed a considerable variety of types. Tomlinson and Atkinson (1987a, b) attempted to standardize *definitions* with a set of terms for seven different points of audit in the 'standard' EIA process, as follows:

- Decision point audit (draft EIS): by regulatory authority in the planning approval process.

Table 7. 2 Types of impact monitoring in UK EISs

Type	*% of total monitoring proposals*	*Type*	*% of total monitoring proposals*
Water quality	16	Archaeological	6
Air emissions	15	Air quality	5
Aqueous emissions	13	Structural survey	4
Noise	12	Liaison group	3
General	9	Water levels	3
Others	7		100
Ecological	7		

Source: Glasson 1994

- Decision point audit (final EIS): also by regulatory authority in the planning approval process.
- Implementation audit: to cover start-up; it could include scrutiny by the government and the public and focus on the proponent's compliance with mitigation and other imposed conditions.
- Performance audit: to cover full operation; it could also include government and public scrutiny.
- Predictive techniques audit: to compare actual with predicted impacts as a means of comparing the value of different predictive techniques.
- Project impact audits: also to compare actual with predicted impacts and to provide feedback for improving project management and for future projects.
- Procedures audit: external review (e. g. by the public) of the procedures used by the government and industry during the EIA processes.

These terms can and do overlap. The focus here is on project, performance and implementation audits. Whatever the focus, auditing faces a number of major *problems* as outlined in Table 7. 3.

Table 7. 3 Problems associated with post-auditing studies

Nature of impact predictions

Many EISs contain few testable predictions; instead, they simply identify issues of potential concern.

Many EIS predictions are vague, imprecise and qualitative.

Testable predictions often relate to relatively minor impacts, with major impacts being referred to only in qualitative terms.

Project modifications

Post-EIS project modifications invalidate many predictions.

Monitoring data

Monitoring data and techniques often prove inadequate for auditing purposes.

Pre-development baseline monitoring is often insufficient, if undertaken at all.

Most monitoring data are collected and provided by the project proponent, which may give rise to fears of possible bias in the provision of information.

Comprehensiveness

Many auditing studies are concerned only with certain types of impacts (e. g. biophysical but not socio-economic; operational but not construction-stage impacts) and are therefore not full-project EIA audits.

Continued

Clarity

Few published auditing studies are explicit about the criteria used to establish prediction accuracy; this lack of clarity hampers comparisons between different studies.

Interpretation

Most auditing studies pay little attention to examining the underlying causes of predictive errors: this needs to be addressed if monitoring and auditing work is to provide an effective feedback in the EIA process.

Source: Chadwick and Glasson 1999

Such problems may partly explain the dismal record of the early set of Canadian EISs examined, from an ecological perspective, by Beanlands and Duinker (1983), for which accurate predictions appeared to be the exception rather than the rule. There are several examples, also from Canada, of situations where an EIA has failed to predict significant impacts. Berkes (1988) indicated how an EIA on the James Bay mega-HEP (1971—85) failed to pick up a sequence of interlinked impacts, which resulted in a significant increase in the mercury contamination of fish and in the mercury poisoning of native people. Dickman (1991) identified the failings of an EIA to pick up the impacts of increased lead and zinc mine tailings on the fish population in Garrow Lake, Canada's most northerly hypersaline lake. Such outcomes are not unique to Canada, which is a leader in monitoring; hopefully the incidence of such research is leading to improved and better predictions.

Findings from the early limited auditing activity in the UK were also not too encouraging. A study of four major developments—the Sullom Voe (Shetlands) and Flotta (Orkneys) oil terminals, the Cow Green reservoir and the Redcar steelworks—suggested that 88 per cent of the predictions were not auditable. Of those that were auditable, fewer than half were accurate (Bisset 1984). Mills's (1992) monitoring study of the visual impacts of five 1990s UK major project developments (a trunk road, two wind farms, a power station and an opencast coal mine) revealed that there were often significant differences between what was stated in an EIS and what actually happened. Project descriptions changed fundamentally in some cases, landscape descriptions were restricted to land immediately surrounding the site and aesthetic considerations were often omitted. However, mitigation measures were generally carried out well.

Other early examples of auditing included the Toyota plant study (Ecotech Research and Consulting Ltd 1994), and various wind farm studies (Blandford, C. Associates 1994; ETSU 1994). The Toyota study took a wide perspective on environmental impacts; auditing revealed some underestimation of the impacts of employment and emissions, some overestimation of housing impacts and a reasonable identification of the impacts of construction traffic. The study by Blandford, C. Associates of the construction stage of three wind farms in Wales confirmed the predictions of low ecological impacts, but suggested that the visual impacts were greater than predicted, with visibility distance greater than the predicted 15 km. However, the latter finding related to a winter audit; visibility may be less in the haze of summer.

One of the most comprehensive nationwide auditing studies of the precision and accuracy of environmental impact predictions was carried out by Buckley (1991) in Australia. At the time of his study, he found that adequate monitoring data to test predictions were available for only 3 per cent of the up to 1,000 EISs produced between 1974 and 1982. In general, he found that testable predictions and monitoring data were available only for large, complex projects, which had often been the subject of public controversy, and whose monitoring was aimed primarily at testing compliance with standards rather than with impact predictions. Some examples of over 300 major and subsidiary predictions tested are illustrated in Table 7.4. Overall, Buckley found the average accuracy of quantified, critical, testable predictions was 44 ± 5 per cent standard error. The more severe the impact, the lower the accuracy. Inaccuracy was highest for predictions of groundwater seepage. Accuracy assessments are of course influenced by the degree of precision applied to a prediction in the first place. In this respect, the use of ranges, reflecting the probabilistic nature of many impact predictions, may be a sensible way forward and would certainly make compliance monitoring more straightforward and less subject to dispute. Buckley's national survey, showing less than 50 per cent accuracy, provided no grounds for complacency. Indeed, as it was based on monitoring data provided by the operating corporations concerned, it may present a better result than would be generated from a wider trawl of EISs. On the other hand, we are learning from experience, and more recent EISs may contain better and more accurate predictions. Marshall (2001) reviewed a set of 1,118 mitigation proposals from 41 EISs in the UK. He found that in 38 per cent of the cases (418 in total), the mitigation proposals were expressed in such a way that the proponent could not be held to be committed to their implementation. In such cases mitigation is of little value, and there may be major compliance issues.

Table 7.4 Examples of auditing of environmental impact predictions

Component/parameter	Type of development	Predicted impact	Actual impact	Accuracy/precision
Surface water quality: salts, pH	Bauxite mine	No detectable increase in stream salinity	None detected	Correct
Noise	Bauxite mine	Blast noise <115dBA	Only 90 per cent < 115 dBA	Incorrect: 90 per cent accurate, worse
Workforce	Aluminium smelter	1,500 during construction	Up to 2,500	Incorrect: 60 per cent accurate, worse

Source: Buckley 1991

There has not, until recently, been much emphasis on auditing studies on the important area of predictive techniques audit, and on the value of particular predictive techniques. Where there have been studies, they have tended to focus on identifying errors associated with predictive methods rather than on explaining the errors. There is a need to develop appropriate audit methodologies, and as more projects are implemented there should be more scope for such studies. The pioneering study by Wood on visibility, noise and air quality impacts, using GIS to audit and model EIA errors, provides an example of a way forward for such work (Wood 1999a, b, 2000).

7.5 A UK case study: monitoring and auditing the local socio-economic impacts of the Sizewell B PWR construction project

7.5.1 *Background to the case study*

Although monitoring and auditing impacts are not mandatory in EIA procedures in the UK, the physical and socioeconomic effects of developments are not completely ignored. For example, a number of public agencies monitor particular pollutants. LPAs monitor some of the conditions attached to development permissions. However, there is no systematic approach to the monitoring and auditing of impact predictions and mitigation measures. This case study reports on one early and still very topical attempt (in the context of a raft of proposals for many new energy projects in the UK over the period to 2025) to introduce a more systematic, although still partial, approach to the subject.

In the 1970s and early 1980s, Britain had an active programme of nuclear power station construction. This included a commitment, revised in the 1990s (but now very much alive), to build a family of new nuclear plants (at the time they were PWR stations). The first such station to be approved was Sizewell B in East Anglia. The approval was controversial, and followed the longest public inquiry in UK history. Construction started in 1987, and the project was completed in 1995. The IAU in the School of Planning at Oxford Brookes University had studied the impacts of a number of power stations and made contributions to EISs, with a focus on the socio-economic impacts. A proposal was made to the relevant public utility, the Central Electricity Generating Board (CEGB), that the construction of Sizewell B provided an invaluable opportunity to monitor in detail the project construction stage, and to check on the predictions made at the public inquiry and on the mitigating conditions attached to the project's approval. Although the predictions were not formally packaged in an EIS, but rather as a series of reports based on the inquiry, the research was extensive and comprehensive (DOEn 1986). The CEGB supported a monitoring study, which began in 1988. To the credit of the utility, which became Nuclear Electric/British Energy, and latterly EDF (Électicité de France) following privatization, there was a continuing commitment to the monitoring study—despite the uncertainty about further nuclear power station developments in Britain. Monitoring reports for the whole construction period and on the project's operation were completed (Glasson *et al*. 1989—97; see also Glasson 2005 for further reflection on the project).

7.5.2 Operational characteristics of the monitoring study

It is important to clarify the *objectives of the monitoring study*, otherwise irrelevant information may be collected and resources wasted. Figure 7.3 outlines the scope of the study. The development under consideration was the construction stage of the Sizewell B PWR 1,200 MW nuclear power station. The focus was on the socio-economic impacts of the development, although with some limited consideration of physical impacts. The socio-economic element of EIA involves 'the systematic advanced appraisal of the impacts on the day to day quality of life of people and communities when the environment is affected by development or policy change' (Bowles 1981). This involves a consideration of the impacts on employment, social structure, expenditure, services, etc. Although socio-economic studies have often been the poor relation in impact assessment studies to date, meriting no more than a chapter or two in EISs, they are important, not least because they consider the impacts of developments on people, who can answer back and object to developments.

The highest priority in the study was to identify the impacts of the development on local employment; this emphasis reflected the pivotal role of employment impacts in the generation of other local impacts, particularly accommodation and local services. In addition to providing an updated and improved database to inform future assessments, assisting project management of the Sizewell B project in the local community and auditing impact predictions, the study also monitored and audited some of the conditions and undertakings associ-

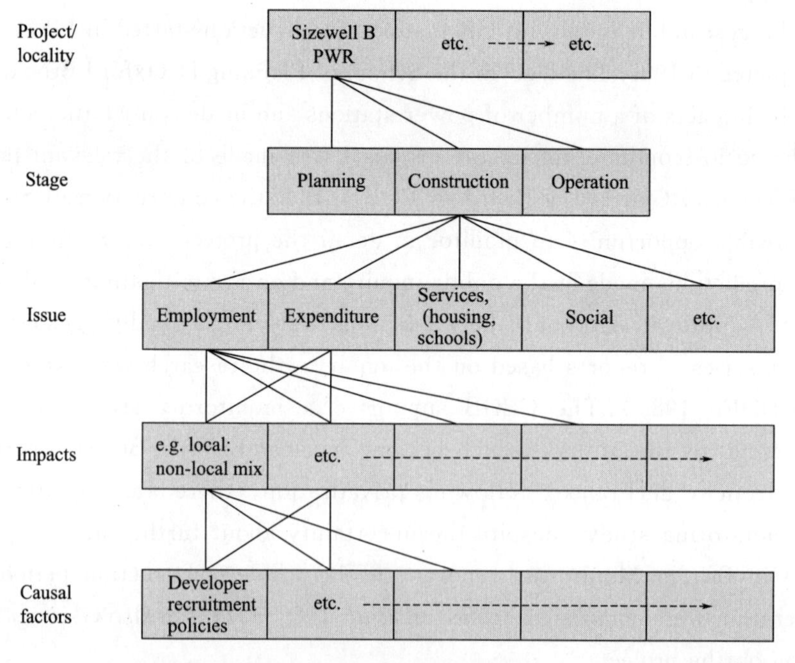

Figure 7.3 Scope of study and database organization: Sizewell B monitoring study

Source: Glasson *et al*. 1989—97

ated with permission to proceed with the construction of the power station. These included undertakings on the use of rail and the routeing of road construction traffic, as well as conditions on the use of local labour and local firms, local liaison arrangements and (traffic) noise (DOEn 1986).

The monitoring study included the collection of a range of information, including statistical data (e. g. the mixture of local and non-local construction-stage workers, the housing tenure status and expenditure patterns of workers), decisions, opinions and perceptions of impacts. The spatial scope of the study extended to the commuting zone for construction workers. The study included information from the developer and the main contractors on site, from the relevant local authorities and other public agencies, from the local community and from the construction workers. The local upper-school geography A-level students helped to collect data on the local perceptions of impacts via biennial questionnaire surveys in the town of Leiston, which is adjacent to the project site. A major survey of the socio-economic characteristics and activities of a 20 per cent sample of the project workforce was also carried out every two years. The IAU team operated as the catalyst to bring the data together. There was a high level of support for the study, and the results were made openly available in published annual monitoring reports and in summary broadsheets, which were available free to the local community (Glasson *et al*. 1989—97).

The study highlighted a number of *methodological difficulties with monitoring and auditing*. The first relates to the disaggregation of project-related impacts from baseline trends. Data are available that indicate local trends in a number of variables, such as unemployment levels, traffic volumes and crime levels. But problems are encountered when we attempt to explain these local trends. To what extent are they due to (a) the construction project itself, (b) national and regional factors, or (c) other local changes independent of the construction project? It is straightforward to isolate the role of national and regional factors, but the relative roles of the construction project and other local changes are very difficult to determine. 'Controls' are used where possible to isolate the project-related impacts.

A second problem related to the identification of the indirect, knock-on effects of a construction project. Indirect impacts-particularly on employment—may well be significant, but they are not easily observed or measured. For example, indirect employment effects may result from the replacement of employees leaving local employment to take up work on site. Are these local recruits replaced by their previous employers? If so, do these replacements come from other local employees, the local unemployed or in-migrant workers? It was not possible to obtain this sort of information. Further indirect employment impacts may stem from local businesses gaining work as suppliers or contractors at Sizewell B. They may need to take on additional labour to meet their extra workload. The extent to which this has occurred is again difficult to estimate, although surveys of local companies have provided some useful information on these issues (Glasson and Heaney 1993).

7.5.3 Some findings from the studies

A very brief summary of a number of the findings is outlined below and in Figure 7.4.

Employment impacts

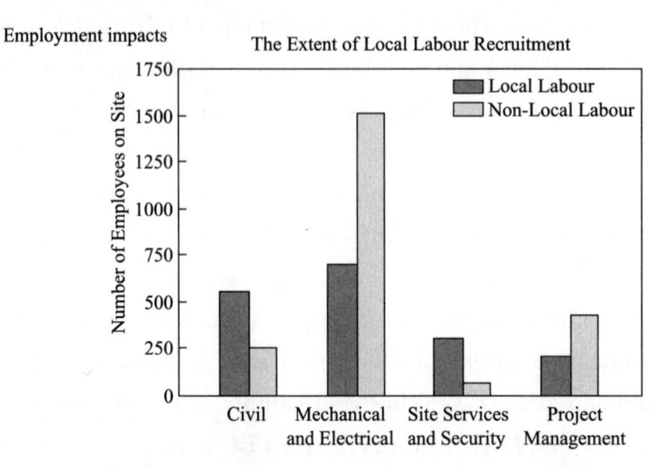

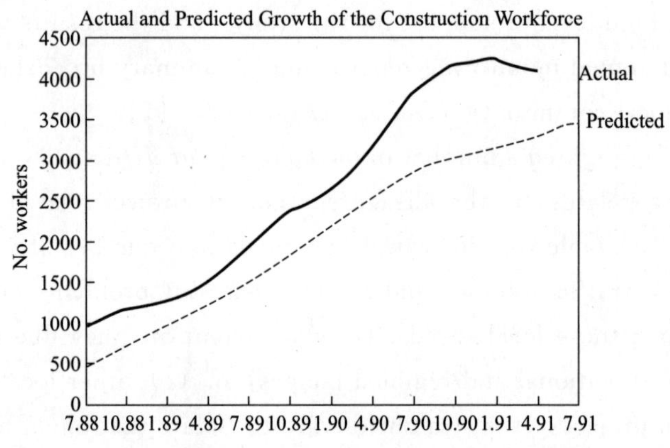

Social impacts

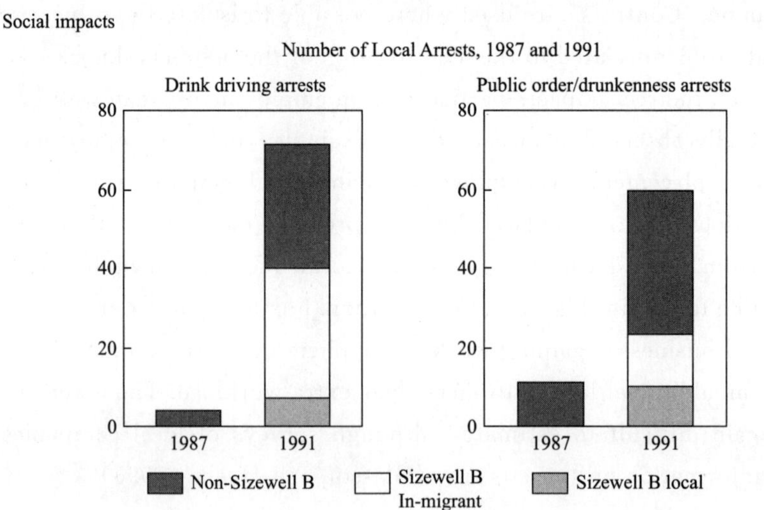

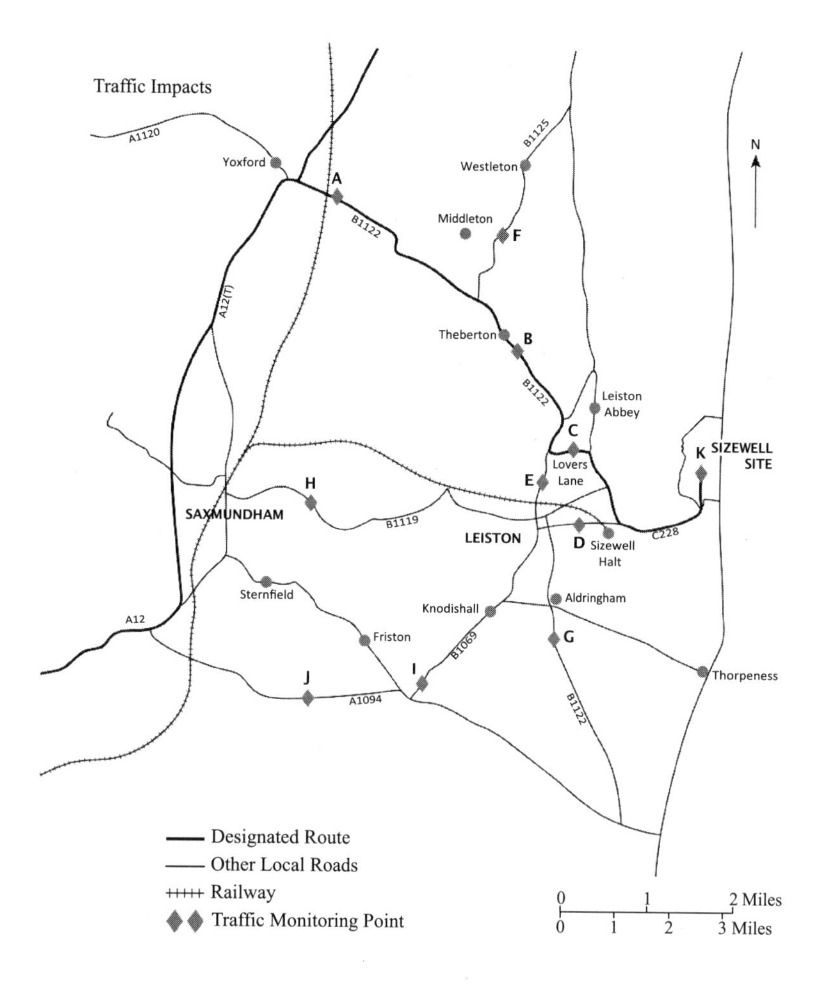

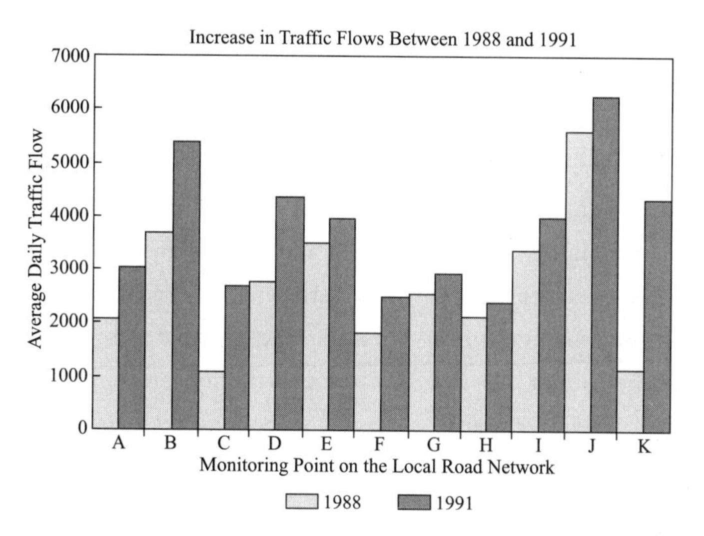

Figure 7.4　Brief summary of some findings from the Sizewell B PWR construction project monitoring and auditing study

Source: Glasson *et al*. 1989—97

Employment

An important prediction and condition was that at least 50 per cent of construction employment should go to local people (within daily commuting distance of the site). This was the case, although, predictably in a rural area, local people have the largely semi-skilled or unskilled jobs. As the employment on site increased, with a shift from civil engineering to mechanical and electrical engineering trades, the pressure on maintaining the 50 per cent proportion increased. In 1989, a training centre was opened in the nearest local town, Leiston, to supply between 80 and 120 trainees from the local unemployed.

Local economy

A major project has an economic multiplier effect on a local economy. By the end of 1991, Sizewell B workers were spending about £500,000 per week in Suffolk and Norfolk, Nuclear Electric had placed orders worth over £40 million with local companies and a 'good neighbour' policy was funding a range of community projects (including £1.9 million for a swimming pool in Leiston).

Housing

A major project, with a large in-migrant workforce, can also distort the local housing market, and tourism accommodation in tourist industry locations. One mitigating measure at Sizewell B was the requirement of the developer to provide a large site hostel. A 600-bed hostel (subsequently increased to 900) was provided. It was very well used, accommodating in 1991 over 40 per cent of the in-migrants to the development, at an average occupancy rate of over 85 per cent, and it helped to reduce demand for accommodation in the locality.

Traffic and noise

The traffic generated by a large construction project can badly affect local towns and villages. To mitigate such impacts, there was a designated construction traffic route to Sizewell B. The monitoring of traffic flows on designated and nondesignated (control) routes indicated that this mitigation measure was working. Between 1988 and 1991, the amount of traffic rose substantially at the four monitoring points on the designated route, but much less so at most of the seven points not on that route. Construction noise on site was a local issue. Monitoring led to modifications in some construction methods, notably improvements to the railway sidings and changes in the piling methods used.

Crime

An increase in local crime is normally associated with the construction stage of major projects. The Leiston police division did see a significant increase in the number of arrests in certain offence categories after the start of the project. However, local people not employed on the project were involved in most of the arrests, and in the increase in arrests, with the exception of drink-driving, for which Sizewell B employees (mainly in-migrants) accounted for most arrests and for most of the increase. However, the early diagnosis of the problems facilitated swift remedial action, including the introduction of a shuttle minibus service for workers, the provision of a large bar in the site hostel, the stressing of the problems of drink-driving at site-workers' induction courses, and the exclusion from the site (effectively the exclusion from Sizewell B jobs) of workers found guilty of serious misconduct or crime. After the early stages of the project, worker-related crime fell substantially, and the police considered the project workforce to be relatively trouble-free, with fewer serious offences than anticipated.

Residents' perceptions

Surveys of local residents in 1989 and 1991 revealed more negative than positive perceived impacts, increased traffic and disturbance by workers being seen as the main negative impacts. The main positive impacts of the project were seen to be the employment, additional trade and ameliorative measures associated with the project. The monitoring of complaints about the development revealed substantially fewer complaints over time, despite the rapid build-up of the project.

7.5.4 Learning from monitoring: Sizewell B and Sizewell C

Table 7.5 shows the nature and auditability of the Sizewell B socio-economic predictions. In contrast to the findings from previous post-auditing studies (see Dipper *et al.* 1998), a vast majority of the Sizewell B predictions were expressed in quantitative terms. The monitoring of impacts and the auditing of the predictions and mitigation measures revealed (Table 7.6) that many of the predictions used in the Sizewell B public inquiry were reasonably accurate—although there was an underestimate of the build-up of construction employment and an overestimate of the secondary effects on the local economy. Predictions of traffic impacts, and on the local proportion of the construction workforce, were very close to the actual outcomes. Mitigation measures also appeared to have some effect. Overall, approximately 60 per cent of the predictions had errors of less than 20 per cent. Explanations of variations from the predictions included the inevitable project modification (particularly associated with new-technology projects, with few or no comparators at the time of prediction), and the very lengthy project authorization process (with a gap of almost 10 years between the predictions and peak construction). Other local issues were revealed by the monitoring, allowing some modifications to manage the project better in the community (Glasson 2005). Unfortunately, such systematic monitoring is still discretionary in the UK and very much dependent on the goodwill of developers.

Table 7.5 Nature and auditability of the Sizewell B predictions

	No. of predictions	% of total
Nature of prediction		
Quantitative		
Expressed in absolute terms	35	51
Expressed in % terms	21	30
Qualitative	11	16
Incorporates quantitative and qualitative elements	2	3
Total: all predictions	69	100
Auditability of predictions		
Auditable: monitoring data subject to no or little potential error	30	43
Auditable: but monitoring data subject to greater potential error	28	41
Not auditable	11	16
Total: all predictions	69	100

Source: Chadwick and Glasson 1999

Table 7.6 Accuracy of auditable Sizewell B predictions

% error in prediction	No. of predictions	% of total
None: prediction correct or within predicted range	15	26
Less than 10%	9	16
10—20%	11	19
20—30%	5	9
30—40%	5	9
40—50%	2	3
Over 50%	8	14
Prediction incorrect, but % error cannot be calculated	3	5
Prediction cannot be audited	11	—
Total: all predictions	69	100

Source: Chadwick and Glasson 1999

Note: For quantified predictions, the predicted value was used as the denominator in the calculation of the % errors in the table. For nonquantified predictions, the % error could not be calculated and predictions were classified as either 'correct' or 'incorrect', based on assessment by the research team.

Information gained from monitoring can also provide vital intelligence for the planning and assessment of future projects. This is particularly so when the subsequent project is of the same type, and in the same location, as that which has been monitored. Nuclear Electric applied for consent to build and operate a replica of Sizewell B, to be known as Sizewell C. A full EIS was produced for the project (Nuclear Electric 1993). Its prediction of the socio-economic impacts drew directly on the findings from the Sizewell B monitoring study, but this proposed follow-on project fell victim to the abandonment in the early 1990s of the UK nuclear power station programme. However, since about 2007 there has been much activity, and increasingly advanced planning, for a new generation of UK nuclear power stations, including a new Sizewell C. In this context the monitoring data from the construction of Sizewell B is proving of considerable value.

7.6 A UK case study: monitoring the local impacts of the London 2012 Olympics project

7.6.1 *Nature of the project and its impacts life cycle*

The London Olympics 2012 project has been one of the largest projects in Europe, with a peak construction work force of almost 12,000 in 2011. The site of about 250 ha is located in the east of London, approximately 5 km from the centre of the city. The Olympic, Paralympic and Legacy Facilities have been designed to create not only an exceptional venue for the games, but also a lasting legacy to bring about the regeneration of the formerly rundown Lower Lea Valley-creating a new urban quarter for London. The life cycle of the project is reflected in the chronology of impacts highlighted in the environmental statement (Symonds/EDAW 2004), and summarized in Figure 7.5. This also provides a framework for the monitoring of the biophysical and socio-economic impacts of the project.

7.6.2 *Construction stage monitoring*

The project has a detailed monitoring programme, coordinated by the Olympics Delivery Authority (ODA). Two examples of the detailed nature of the monitoring are illustrated below. Figure 7.6 provides an extract from the monthly construction noise monitoring across the site, showing a generally good performance. Table 7.7 provides some extracts from the monthly socio-economics monitoring, with a focus on the characteristics of the construction workforce. For January 2011, it shows a workforce of almost 12,000 across the two main projects (park and athletes' village). A high proportion of the workforce is locally sourced from within the London boroughs, and much is very local to the site. Recruitment from the unemployed and from black, Asian and ethnic minority groups is also monitored as good against benchmark targets; this contrasts with poorer performance against benchmarks for female recruitment and for recruitment from those with disabilities. The project has used a range of construction workforce development activities to enhance beneficial local recruitment impacts, including: a job brokerage scheme that has placed over 1,250 people (primarily local borough residents) into employment on the project, and a training programme that has exceeded targets by training (up to 2011) 3,250 people (against a target of 2,250), including 400 apprentices (against a target of 350).

Impacts	Post-Olympic Construction Phase	Olympic Games Phase	Pre-Olympic Legacy Construction Phase	Post-Olympic Legacy Phase
Premature loss of existing housing, industry, jobs and waste management infrastructure	//////////////////////////////////	///////////	////////////////////////	
Potential loss of archaeological baseline				
Damage to built heritage from demolition and contextual changes				
Loss of district character of historic areas				
Improved quality of townscape and views		///////////		//
Undergrounding of power cables		///////////	////////////////////////	//
Consequences of remediation	//////////////////////////////////	///////////	////////////////////////	//
Energy efficiency gains from CCHP and other sustainable/ renewable energy features incorporated into buildings/ structures		/////////// ///////////	//////////////////////// ////////////////////////	// //
Creation of Olympic jobs	/////	///////////	/////	
'Feel good' factor, social cohesiveness and community pride	/////	///////////	/////	
Encouragement to participate in sporting/healthy activities	/////	///////////	////////////////////////	//
Impacts on local transport infrastructure		///////////		
Wind impacts on queues/crowd near large buildings (including Olympic Village)				
Potential flood risk due to Security Perimeter Fence at river crossings				
Potential impacts from existing contamination in newly public areas				
Additional parkland, open ground and allotments			////////////////////////	//
Additional /replacement habitat creation			////////////////////////	//
Improved accessibility/ permeability				//
Creation of Legacy jobs, with associated skills and training				//
Improved community facilities (schools, nurseries, creches, medical etc.)				//
Improved buildings (e.g. 'access for all" standards)				//
Impact on household waste management infrastructure of LB Newham				//

Notes: = Significant beneficial, = Significant adverse. Hatched = Significant with or without mitigation, Plain colour = Significant without mitigation

Figure 7.5 Life cycle of impacts for the London 2012 Olympics project

Source: Symonds/EDAW 2004

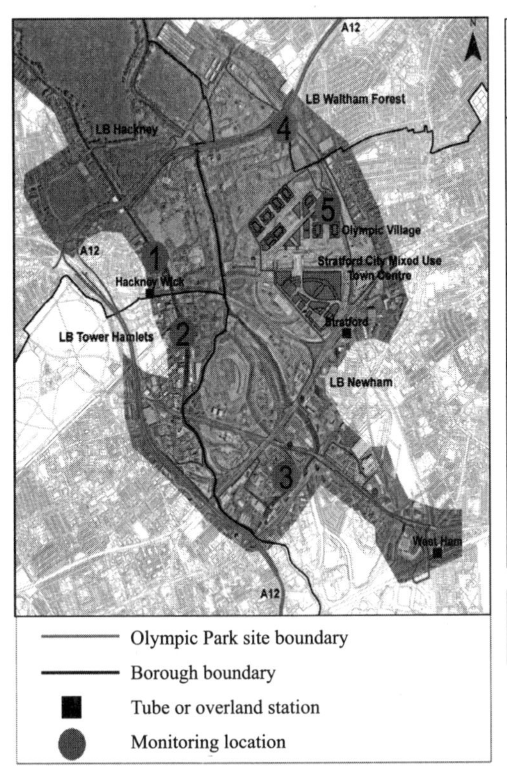

① Leabank Square, Hackney
No noise levels monitored from construction activities above agreed limits.

② Omega Flats, Tower Hamlets
No noise levels monitored from construction activities above agreed limits.

③ Marshgate Lane, Newham
No noise levels monitored from construction activities above agreed limits.

④ Onsite (150 metres from nearest home), Waltham Forest
One exceedance above agreed limits were noted which could be the result of construction activity.

⑤ Olympic Village onsite (120 metres from nearest home), Newham**
No noise levels monitored from construction activities above agreed limits.

Please refer to Noise FAQ for information on noise control measures implemented

"This information is recorded over the calendar month. Noise levels are monitored against levels agreed by the Planning Authority in conjunction with the Host Boroughs when planning permission was granted for the Olympic Park. Planning permission for the Olympic Village is based on hourly thresholds as this development has separate permission to the Olympic Park.

** Note: The Olympic Village is subject to a different planning permission which sets an hourly level during the day.

Figure 7.6 Construction noise monitoring across the London 2012 Olympics site (August 2010)

Source: Olympics Delivery Authority 2010

Table 7.7 Workforce/employment monitoring for the London 2012 Olympics site (December 2010)

Workforce on site	Olympic Park		Athletes' Village	
	6500	(benchmark)	5400	(benchmark)
% resident in host boroughs	21	—	27	—
% resident elsewhere in London	34	—	40	—
% resident elsewhere in UK	42	—	30	—
% residing outside UK/or no information	3	—	3	—
% previously unemployed	12	7	10	7
% women	4	11	3	11
% disabled	1	3	0.5	3
% black, Asian or ethnic minority	19	15	13	15

Source: Adapted from *Employment and Skills Update*, Olympics Delivery Authority (January 2011)

7.7 Summary

A mediation of the relationship between a project and its environment is needed throughout the life of a project. Environmental impact assessment is meant to establish the terms and conditions for project implementation; yet there is often little follow-through to this stage, and even less follow-up after it. Arts (1998) concluded, after a thorough examination of 'expost evaluation of EIA', that in practice it is lagging behind the practice of EIA itself. Few countries have made arrangements for some form of follow-up. In those that have, experience has not been too encouraging—reflecting deficiencies in often over-descriptive EISs, inadequate techniques for follow-up, organizational and resource limitations, and limited support from authorities and project proponents alike. Yet many projects have very long lives, and their impacts need to be monitored on a regular basis. Morrison-Saunders *et al.* (2001) show how this could bring positive outcomes for different stakeholders.

Figure 7.7 shows the benefits not only to the proponent and the community (as exemplified by the Sizewell B case study), but also to the regulator—in the form of a better decision and improvement of the EIA process. Such monitoring can improve project management and contribute to the auditing of both impact predictions and mitigating measures. Monitoring and auditing can provide essential feedback to improve the EIA process, yet this is still probably the weakest step of the process in many countries. Discretionary measures are not enough;

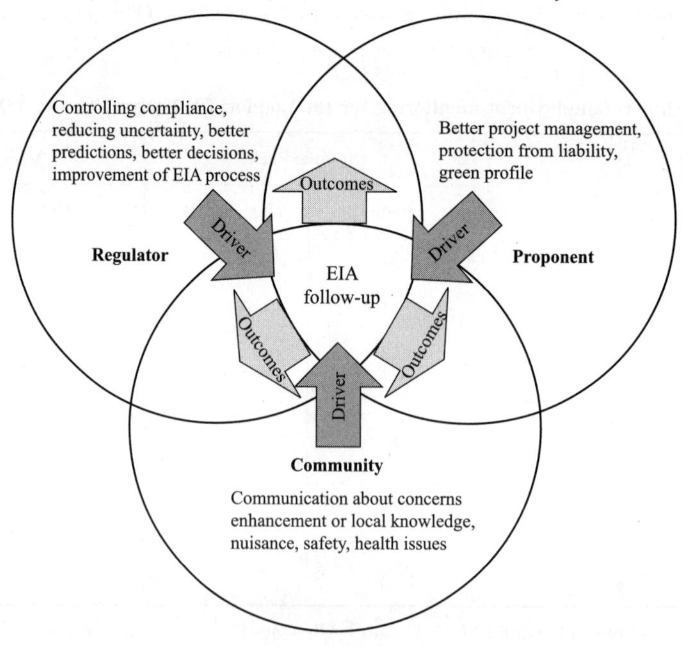

Figure 7.7 Outcome of EIA follow-up for different stakeholders

Source: Morrison-Saunders *et al.* 2001

monitoring and auditing need to be more fully integrated into EIA procedures on a mandatory basis.

SOME QUESTIONS

The following questions are intended to help the reader focus on the key issues of this chapter, and to start building some understanding of the importance and nature of monitoring and auditing in EIA.

1 What do you understand by the distinction between 'before the decision' and 'after the decision' in EIA?

2 Why is it important to continue the EIA process beyond the decision for those projects which proceed to implementation?

3 What do you understand by the distinction between monitoring and auditing in EIA?

4 Consider the sort of monitoring information which would be useful to collect for a major project with which you are familiar. Include indicators and underlying factors, and quantitative and qualitative information.

5 Why do you think that there has been such resistance in many countries to more mandatory monitoring and auditing systems for EIA?

6 Compare and contrast some of the key features of the monitoring systems in California and China Hong Kong.

7 Review the problems that can be associated with post-auditing studies, as set out in Table 7.3, and consider how some of these problems might be overcome.

8 As exemplified by the Sizewell B project, consider ways of overcoming the two highlighted methodological issues, often encountered in monitoring, of (i) disaggregating project-related impacts from baseline trends, and (ii) identifying the indirect impacts of projects.

9 Review the relative accuracy of auditable Sizewell B predictions, as displayed in Table 7.6. What factors might contribute to the findings set out there?

10 Figure 7.5 illustrates the importance of monitoring over the life cycle of a project. Some of the impacts noted in the figure are likely to be easier to monitor than others. Briefly run through the impact list and identify, as far as possible, relevant impact indicators that could be used in the monitoring process for the London 2012 Olympics project.

11 Drawing on the two UK monitoring case studies, plus any others with which you might be familiar, outline the potential benefits of EIA monitoring/follow-up to the relevant sets of stakeholders identified in Figure 7.7.

Notes

1 Early drafts of the EC Directive did include a requirement for an expost evaluation of EIA projects. Section 11 of the 1980 draft (CEC 1980) stated that the competent authority

should check at set intervals whether the provisions attached to the planning permission are observed or adequate, or other provisions for environmental protection are observed, and whether additional measures are required to protect the environment against the project's impacts.

References

Ahammed, A. K. M. Rafique, and Nixon, B. M. 2006, Environmental impact monitoring in the EIA process of South Australia, *Environmental Impact Assessment and Review*, 26, 426-47.

Arts, J. 1998. *EIA follow-up: on the role of expost evaluation in environmental impact assessment*. Groningen: Geo Press.

Arts, J., Caldwell, P and Morrison-Saunders, A. 2001. Environmental impact assessment follow-up: good practice and future directions-findings from a workshop at the IAIA 2000 Conference. *Impact Assessment and Project Appraisal*, 19 (3), 175-185.

Au, E. and Sanvícens, G. 1996. EIA follow-up monitoring and management. *EIA Process Strengthening*. Canberra: Environmental Protection Agency.

Baseline Environmental Consulting 1989. *Mitigation monitoring and reporting plan for a woodwaste conversion facility, West Berkeley, California*. Emeryville, CA: Baseline Environmental Consulting.

Beanlands, G. E. and Duinker, P. 1983. *An ecological framework for environmental impact assessment*. Halifax, Nova Scotia: Dalhousie University, Institute for Resource and Environmental Studies.

Berkes, F. 1988. The intrinsic difficulty of predicting impacts: lessons from the James Bay hydro project. *Environmental Impact Assessment Review* 8, 201-20.

Bisset, R. 1984. Post development audits to investigate the accuracy of environmental impact predictions. *Zeitschift für Umweltpolitik* 7, 463-84.

Bisset, R. and Tomlinson, P. 1988. Monitoring and auditing of impacts. In *Environmental impact assessment*, P. Wathern (ed). London: Unwin Hyman.

Blandford, C. Associates 1994. *Wind turbine power station construction monitoring study*. Gwynedd: Countryside for Wales.

Bowles, R. T. 1981. *Social impact assessment in small communities*. London: Butterworth.

Buckley, R. 1991. Auditing the precision and accuracy of environmental impact predictions in Australia. *Environmental Monitoring Assessment* 18, 1-23.

California Resources Agency 1988. *California eia monitor*. State of California.

CEC (Commission of the European Communities) 1980. Proposal for a council directive concerning the assessment of the environmental effects of certain public and private projects on the environment, 9 July 1980. *Official Journal of the EC*, L175, 40-49. Brussels: EC.

CEC 1993. *Report from the Commission of the implementation of Directive 85/337/EEC on the assessment of the effects of certain public and private projects on the environment*, vol. 13, *Annexes for all Member States*, COM (93), 28 final. Brussels: EC, Directorate-General XI.

CEC 2009. *Study concerning the report on the application and effectiveness of the EIA Directive: Final Report*. Brussels: DG Env.

Chadwick, A. and Glasson, J. 1999. Auditing the socio-economic impacts of a major construction project: the case of Sizewell B Nuclear Power Station. *Journal of Environmental Planning and Management* 42 (6), 811-36.

Culhane, P. J. 1993. Post-EIS environmental auditing: a first step to making rational environmental assessment a reality. *Environmental Professional* 5.

CWE-ZHEC Joint Venture, 2007, *Construction of Yung Shue Wan Helipad: EM&A Manual*, Hong Kong: Environmental Protection Department EIA Ordinance website (at 2011).

Dickman, M. 1991. Failure of environmental impact assessment to predict the impact of mine tailings on Canada's most northerly hypersaline lake. *Environmental Impact Assessment Review* 11, 171-80.

Dipper, B., Jones, C. and Wood, C. 1998. Monitoring and post-auditing in environmental impact assessment: a review. *Journal of Environmental Planning and Management* 41 (6), 731-47.

DOEn (Department of Energy) 1986. *Sizewell B public inquiry: report by Sir Frank Layfield*. London: HMSO.

Ecotech Research and Consulting 1994. *Toyota impact study summary*, unpublished.

EPD (Environmental Protection Department) 1997. *Environmental impact assessment ordinance and technical memorandum on environmental impact assessment*. Hong Kong: EPD Hong Kong Government.

EPD 1998. *Guidelines for development projects in Hong Kong-environmental monitoring and audit*. Hong Kong: EPDHK.

ETSU (Energy Technology Support Unit) 1994. *Cemmaes Wind Farm: sociological impact study*. Market Research Associates and Dulas Engineering, ETSU.

Frost, R. 1997. EIA monitoring and audit. In *Planning and eia in practice*, J. Weston (ed), 141-64. Harlow: Longman.

Glasson, J. 1994. Life after the decision: the importance of monitoring in EIA. *Built Environment* 20 (4), 309-20.

Glasson, J. 2005. Better monitoring for better impact management: the local socio-economic impacts of constructing Sizewell B nuclear power station. *Impact Assessment and Project Appraisal*, July .

Glasson, J. and Heaney, D. 1993. Socio-economic impacts: the poor relations in British environmental impact statements. *Journal of Environmental Planning and Management* 36 (3), 335-43.

Glasson, J., Chadwick, A. and Therivel, R. 1989-97. *Local socio-economic impacts of the Sizewell B PWR Construction Project*. Oxford: Oxford Polytechnic/Oxford Brookes University, Impacts Assessment Unit.

Greene, G., MacLaren, J. W. and Sadler, B. 1985. Workshop summary. In *Audit and evaluation in environmental assessment and management: Canadian and international experience*, 301-21. Banff: The Banff Centre.

Holling, C. S. (ed) 1978. *Adaptive environmental assessment and management*. Chichester: Wiley.

Hui, S. Y. M. and Ho, M. W. 2002. EIA follow-up: Internet-based reporting. In *Conference Proceedings of 22nd Annual Meeting IAIA June 15-21, The Hague*. IAIA.

IAPA 2005, Special issue on EIA Follow-up, *Impact Assessment and Project Appraisal*, 23 (3).

IEA 1993. *Digest of environmental statements*. London: Sweet & Maxwell.

IOCGP (Inter-Organisational Committee on Principles and Guidelines) 2003. Principles and guidelines for social impact assessment in the USA. *Impact Assessment and Project Appraisal*, 21 (3), 231-50.

Lee, N. and Wood, C. 1980. *Methods of environmental impact assessment for use in project appraisal and physical planning*. Occasional Paper no. 7. University of Manchester, Department of Town and Country Planning.

Marshall, R. 2001. Mitigation linkage: EIA follow-up through the application of EMPS in transmission construction projects. In *Conference Proceedings of 21st Annual Meeting IAIA*, 26 May-1 June, Cartagena, Colombia, IAIA.

Marshall, R., Arts, J. and Morrison-Saunders, A. 2005. International principles for best practice EIA follow-up. *Impact Assessment and Project Appraisal* 23 (3), 175-81.

Mills, J. 1992. *Monitoring the visual impacts of major projects*. MSc dissertation in Environmental Assessment and Management, School of Planning, Oxford Brookes University.

Morrison-Saunders, A. 1996. Auditing the effectiveness of EA with respect to ongoing environmental management performance. In *Conference Proceedings 16th Annual Meeting IAIA June 17-23 1996*,

M. Rosario Partidario (ed), vol 1, IAIA, Lisbon 317-22.

Morrison-Saunders, A. and Arts, J. (2004) (eds), *Assessing impact: handbook of EIA and SEA follow-up*. London: Earthscan.

Morrison-Saunders, A., Arts, J., Baker, J. and Caldwell, P. 2001. Roles and stakes in EIA follow-up. *Impact Assessment and Project Appraisal* 19 (4), 289-96.

Olympics Delivery Authority 2010. *Dust and Noise Monitoring*. London: Olympics Delivery Authority.

Olympics Delivery Authority 2011. *Employment and Skills Update: January 2011*. London: Olympics Delivery Authority.

Sadler, B. 1988. The evaluation of assessment: post-EIS research and process. In *Environmental Impact Assessment*, E. Wathern (ed). London: Unwin Hyman.

Symonds/EDAW 2004. *ES for Lower Lea Valley: Olympics and Legacy Planning Application*. Symonds/EDAW for London Development Agency.

Tomlinson, E. and Atkinson, S. F. 1987a. Environmental audits: proposed terminology. *Environmental Monitoring and Assessment* 8, 187-98.

Tomlinson, E. and Atkinson, S. F. 1987b. Environmental audits: a literature review. *Environmental Monitoring and Assessment* 8, 239.

Wood, G. 1999a. Assessing techniques of assessment: post-development auditing of noise predictive schemes in environmental impact assessment. *Impact Assessment and Project Appraisal* 17 (3), 217-26.

Wood, G. 1999b. Post-development auditing of EIA predictive techniques: a spatial analysis approach. *Journal of Environmental Planning and Management* 42 (5), 671-89.

Wood, G. 2000. Is what you see what you get? Postdevelopment auditing of methods used for predicting the zone of visual influence in EIA. *Environmental Impact Assessment Review* 20 (5), 537-56.

第三部分

实　　践

我们来确认一下：我们可以在石南荒野上养一些蜥蜴和蓝蝴蝶，作为回报，你可以在我们的绿带上建造3200座房子……

8 英国环境影响评价概述

8.1 导言

第 3 部分探讨 EIA 实践应该做什么和不应该做什么。本章回顾了自欧洲共同体 85/337 号指令生效以来的最初 20 年里英国的环境影响评价的实践进程。在第 9 章我们将参照一些具体案例进行深入研究。这些案例研究试图从本章及前章中所提出的 EIA 程序中提炼重点（例如，对替代方案的处理、公众参与以及更广阔的环境评价应该考虑其对社会和经济产生的影响）。这些案例大多是英国的项目环评，但也包括两个战略环境影响评价（SEA）的例子。第 10 章讨论了 EIA 在世界各国的实践情况，包括“最佳实践”体系、新兴的 EIA 体系以及国际基金组织（如世界银行）在 EIA 中的作用。

这些章节的设置是基于对 EIA 有效性研究的重要国际研究背景下提出来的，由 Sadler（1996，2012）撰写研究结果。在 2011 年的研究中，Sadler 提出测试 EIA 有效性的三个试验，制定了一个有效的“分检指数”（三个衡量方面）：①允许条件（什么是允许或者必须要做的；法律和制度框架以及研究方法的现实性）；②现实的状态（完成了什么，宏观和微观中最佳实践的案例——实现可能性的艺术）；③有效性和性能（所得的结果，对于决策以及环境效益的贡献）。Sadler 指出这些问题及调查它们所需要的技术必须包含在相关 EIA 系统运行的决策框架中。

第 8 章简要阐述了 Sadler 的前两个观点。8.2 节讨论了英国实施 EIA 的数量、类型及选址，以及 EISs 的收藏。8.3 节讨论了在 EIS 提交前 EIA 的各阶段和授权申请。8.4 节阐述了迄今为止，EIA 实践中最重要的研究方面，即 EIS 的质量。8.5 节介绍了 EIA 提交以后的各阶段，并讨论了 LPAs 和检查员如何将环境信息用到决策之中。8.6 节考虑了 EIA 面临的法律挑战，其中很多挑战促进了在第 3 章介绍的英国法律法规系统的改变。最后，在 8.7 节中讨论了根据不同预测得出的 EIA 的成本和效益。Sadler 的第三个观点在政府出版的《EIA 的实施和评审指南》（例如 DETR，1999a，2000；DCLG，2011a）得到了部分体现，这反映了英国政府在政策变更中，考虑了早期对 EIS 和 EIA 效果的研究结果。

本章所涉及的信息在 2011 年中期撰写时是正确的；但是随着更多 EIS 的实施，某些信息可能会出现明显的改变。

8.2 EIS 和项目的数量、类型

本节研究了已经编写了多少份 EIS，它们都针对哪些类型的项目。本节还简要介绍了 EIS 的收藏和储存，因为英国的 EIS 收藏和数据库是非常分散的（8.2.3 节）。某些问题使分析变得复杂了。第一，某些项目被归入了不止一个分类附录中，例如，矿物冶炼（清单 2.2）被填入固体废物栏（清单 2.11），而工业区/居民区开发（清单 2.10）也包含了娱乐方面的内容（清单 2.12）。第二，仅凭对项目的描述不足以确定 EIA 执行的规则，例如，根据电厂的规模不同，可将其归入到清单 1.2 或清单 2.3（a）中；公路也许会归入高速公路或是规划条例，这取决于它们是主干道还是地方公路。第三，很多 EIS 没有提及何时编写、由谁编写以及为谁编写。第四，1995 年以后，由于地方政府重组以及英格兰、苏格兰和威尔士的政府边界和属性发生诸多变化，使地区分析变得复杂。所有这些因素都会影响到分析本身。这一章起初是建立在 DCLG（2011b）和 IEMA（2011）的信息基础上的，同时还有 Wood 和 Bellanger（1998）、Wood（1996，2003）的信息支持。

8.2.1 EIS 的数量和分布

从 20 世纪 70 年代中期到 20 世纪 80 年代中期，英国每年大约会编写 20 份 EIS（Petts 和 Hills，1982）。在 85/337 号法令实施之后，这个数量显著上升，即使在最不景气的时候，如 20 世纪 90 年代早期，每年仍编写约 350 份 EIS。但是，从图 8.1 可以看出，在 20 世纪 90 年代中期 EIS 数量开始下降，这与在规划法案的影响下主要开发行为的减少有一定关系。但是，在 20 世纪 90 年代后期，EIS 数量又得以迅速回升，正如第 3 章所述的，自从实施了 1997 年修正案后，每年编写的 EIS 超过 600 份。这能反映出许多问题——修正案包括了更多的项目、更发达的英国经济以及开发商与 LPA 对涉及 EIA 法令的特定法庭判决的关注。到 2008 年底，在已经编制的近 9000 份 EIS 中，近 70%是在英格兰、威尔士和苏格兰制定的规划条例要求下编制的，其余的 EIS 是为北爱尔兰的项目编制的。更值得注意的是，如同在第 3 章讨论的那样，EIS 是为其他相关程序下的项目（例如高速公路、林业）编制的。然而，EIS 的数量在 2005 年以后开始下降，也说明了 EIS 活动可以被看作反映国家经济命运的一个有趣的晴雨表。

随着 EIS 数量的增多，EIA 的参与者也更加熟悉整个程序。在 20 世纪 90 年代中期，牛津布鲁克斯大学针对英国地方当局所做的调查指出：超过 80%的 LPA 至少接受过一份 EIS。按平均值计算，战略机构（国家和地区理事会和国家军事用地局）收到了 12 份 EIS，地方

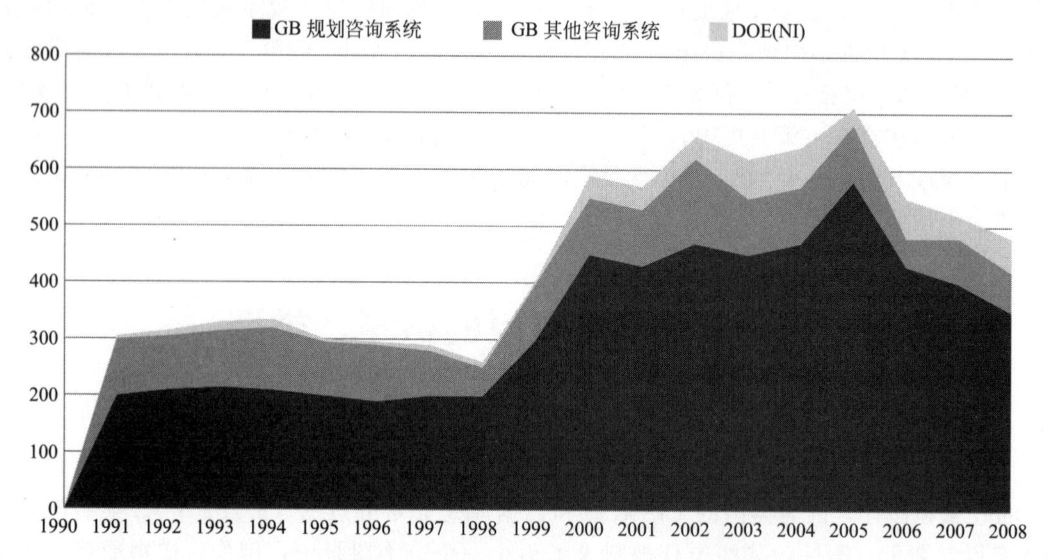

图 8.1 英国编写的 EIS（1991～2008）

［资料来源：DCLG（2011b）］

当局（自治区、特权城市、国家级特权城市以及发展公司）收到了 4 份 EIS。环境咨询调查（例如 Radcliff 和 Edward-Jones，1995；Weston，1995）发现大约有 1/3 的咨询机构都编写了 10 个或更多的 EIS。正如记录的那样，目前 EIS 的数量在 9000 份左右，而这一数目在 1995 年底的时候只有 2500 份，同时，与该程序相关的经验体会、LPA 和咨询活动也在相应增加。图 8.2 阐释了 EIS 在各个地区的分布。苏格兰只占英国 20%的人口，但是其 EIS 活动高于英国平均值，最近的活动包括苏格兰风力农场等。

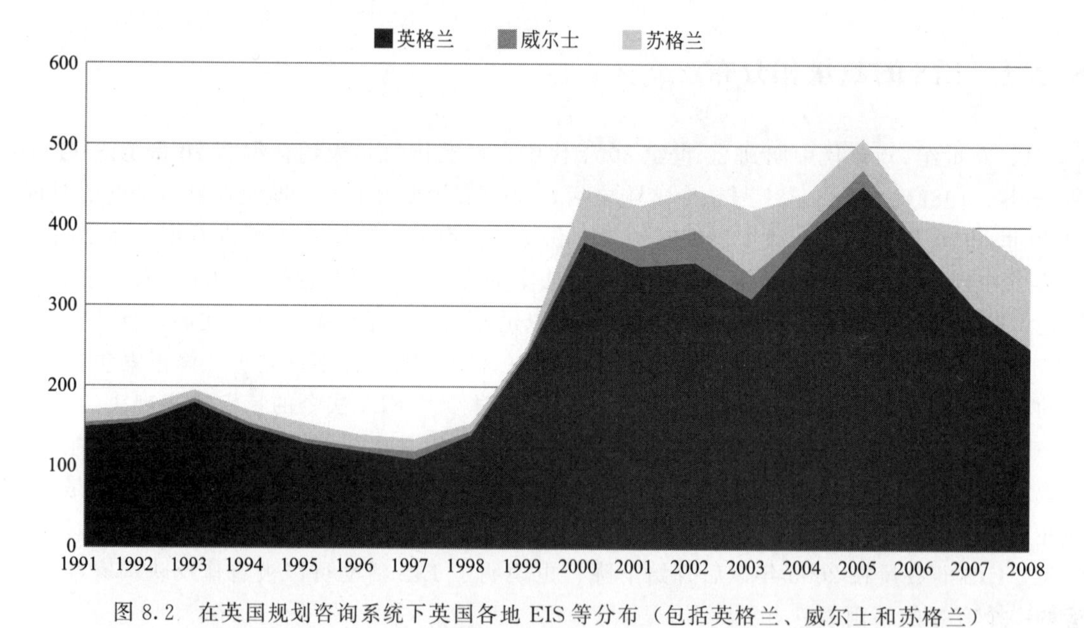

图 8.2 在英国规划咨询系统下英国各地 EIS 等分布（包括英格兰、威尔士和苏格兰）

［资料来源：DCLG（2011b）］

8.2.2 项目类型

图 8.3（a）表明了英国早期编制的 EIS 的几种项目类型，数量最大的项目类型集中在废物处理（大型垃圾填埋/堆积项目、污水处理项目以及焚烧项目）、城市/商业开发、公路、冶炼和能源项目上面（Wood 和 Bellanger，1998）。相反，2004～2010 年数据显示（仅英格兰），城市/商业开发是目前主要的项目。表 8.3（b）（也是仅英格兰）表明有些地区有所不同，例如在英国东北和东南的项目类型。

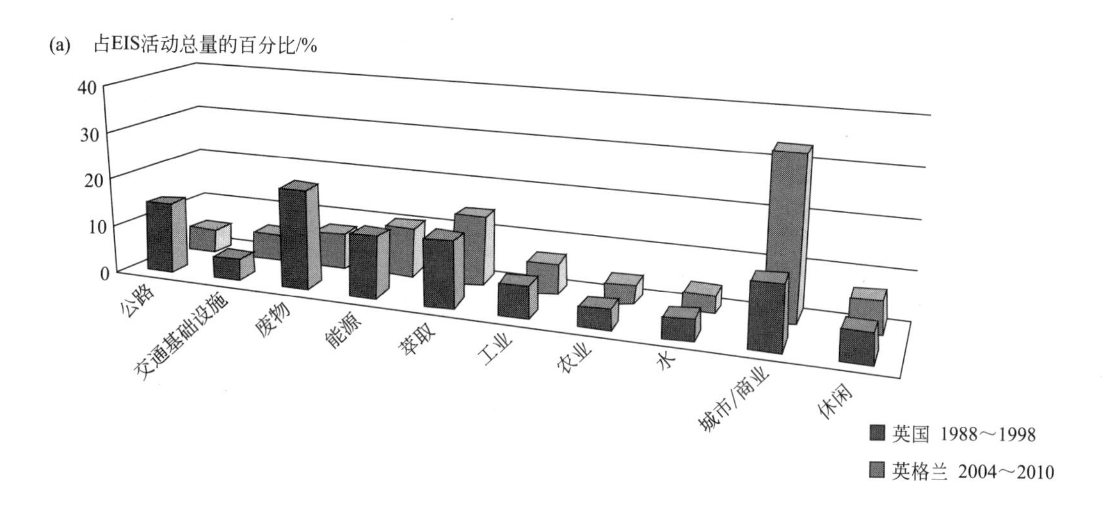

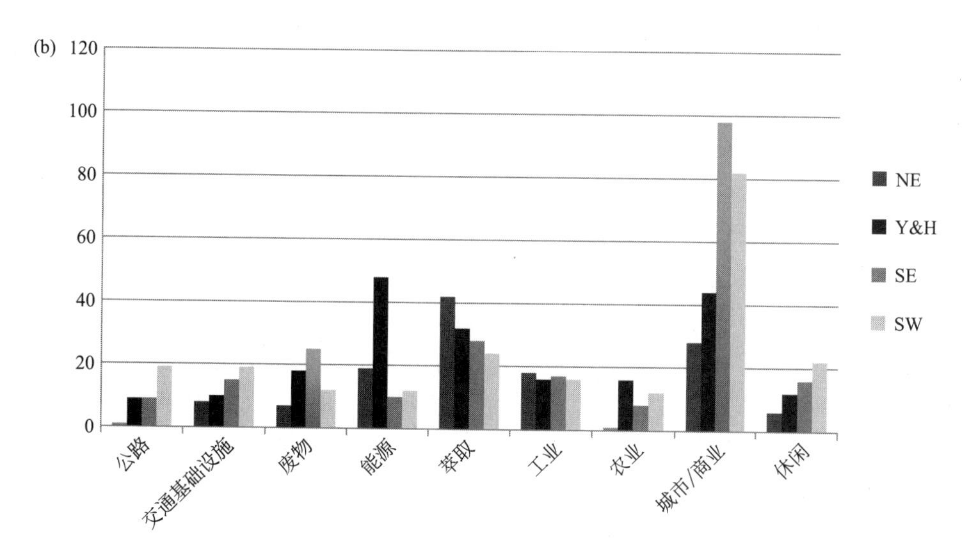

图 8.3 EIS 特定类型的趋势

（a）EIS 活动类型比较——英国（1988～1998）和英格兰（2004～2010）；

（b）项目类型——比较 EIS 产生的特定地区（2004～2010）

［来源：（a）Wood 和 Bellanger，1998；（a）（b）DCLG 图书馆数据库，2011］

表 8.1 显示了 EIS 在不同领域下的分布状况。这些基本可以分成三类项目，分别是关于

交通、农业、渔业以及能源的项目。高速公路是一个大类，虽然它的数量在21世纪拐点有所减少，但是从那时以后又有所回升。相反，林业、地面排水、渔业养殖在21世纪拐点是有所下降的。电力和管道也是一个重要的分类，尤其是最近天然气管道和离岸风力农场的兴起，这个类别大概包含600个EIS。如预期所示，在1999年法规之后，这里只有很少一部分关于核反应堆废弃的EIS，但是在未来，这里会有更多关于英国核反应堆在其完成寿命以后废弃的EIS。

表8.1 在其他规划咨询系统中产生环境影响评价报告（高速公路、森林、电力等）

年份	收到的环评报告							
	高速公路	海港工程	交通工程	森林	地面排水	渔业农场	核反应堆退役	电力管道工程（包括天然气管道和离岸风力农场）
1991	43	0		8	10	0		29
1992	38	2		9	11	0		23
1993	52	3	3	30	14	0		12
1994	61	5	3	18	9			23
1995	27	7	1	15	22	0		19
1996	19	8	3	12	29	0		17
1997	17	9	4	13	22	1		25
1998	6	1	10	13	20	2		12
1999	5	6	9	11	27	10	0	42
2000	3	12	8	7	17	13	0	48
2001	8	7	2	10	31	20	0	54
2002	21	4	7	8	37	32	2	42
2003	12	1	8	10	8	17	0	53
2004	17	2	6	3	3	11	2	64
2005	28	3	2	5	17	12	2	27
2006	24		1	1	0	0	1	25
2007	30		3	2	0	0	1	42
2008	11		3	2	0	0	2	42
总数	412	70	73	177	273	118	10	578

资料来源：DCLG，2011b。

在欧洲共同体85/337号指令实施后的最初几年内，有40%的EIS是针对公共部门而编写的，其余60%的EIS是针对私有部门而编写的（Wood，1991）。虽然私有部门的项目比例由于私有化而略有增加，但是由于政府在1990～2000年初投资新建公路编写了大量的EIS，导致该比例基本没变。还有一个非常有意思的部分，该部分的EIA机构既是项目的提案者，也是项目的主管部门（例如针对公路的高速公路管理局）。

8.2.3 EIS的来源

英国地方规划局（LPA）收到EIS后，需要把复本送到位于伦敦的社区暨地方政府部（DCLG）环境评价司图书馆。但是，这是一个漫长的过程。DCLG图书馆定期向公众预约开放，并允许公众在馆内影印。在威尔士，规划的EIS被转交到威尔士议会；在苏格兰，

EIS 全部被送到苏格兰议会；而在北爱尔兰则被送到北爱尔兰环境部。其他的政府机构，如高速公路管理局，也会录存它们管辖范围内的 EIS 及其清单。但是这些 EIS 除了有限地用于研究之外，并不对公众开放。

除了政府部门对 EIS 有所收集以外，环境评价与管理协会（IEMA）以及一些大学也对 EIS 有所录存（如牛津布鲁克斯特大学、曼彻斯特大学和东英吉利大学）。有些 EIS 可以在网络上找到，虽然这些报告的可见时间仅仅限于规划申请的时间。位于 Lincoln 的环境评价与管理协会（IEMA）已经收集了大量 EIS，但是这些 EIS 仅仅以预定的方式提供给所内职员使用，也主要供其企业会员所用。曼彻斯特大学的 EIA 中心保存了大量的 EIS 和 EIA 著作，并且向公众预约开放。牛津布鲁克斯大学收集的近 1000 份 EIS 也向公众预约开放，而且可以在馆内进行影印。其他组织，如英国皇家鸟类保护协会、陆地生态研究所、英格兰自然署和英格兰乡村保护组织以及诸多环境咨询机构也收集了一些 EIS，但是这些收集都以私人形式拥有，仅供组织内部使用，不对外开放。目前越来越多的 EIS 可以在当地政府或者开发商的网站上找到。

由于很难发现现存的 EIS，而分析 EIS 的成本也往往很高，这使得对 EIS 的获取和分析非常困难。很多组织（IEMA、曼彻斯特大学、牛津布鲁克斯大学）都要求在英国建立一个集合英国所有 EIS 于一处的中心资料库。2011 年春天开始启动的 IEMA 质量评分系统就是基于此的一个积极发展。为了取得质量加分，组织必须做出七个环境影响评价承诺（关于环评人员的能力、环评的内容以及环评的报告）。该项目的间接影响其中之一就是会使每年英国相当大的一部分 EIS 以及非技术总结（NTS）可以在线找到，这将会为研究和实践提供许多有价值的信息。

8.3 EIA 预提交程序

本节是讨论如何在英国进行 EIA 实践的三节内容中的第一节。重点讨论 EIS 的预提交阶段，也就是筛选、范围确定以及提交前的一些咨询。

8.3.1 筛选阶段

欧盟和英国的法律都要求对英国 EIA 的实施加强分析。根据原始法案，英国主管部门根据由 40 多项条例及附加导则组成的一个多重标准和阈值的框架，来自由决定附录 2 中哪些项目需要进行 EIA。整体上看，这个筛选程序运行良好（CEC，1993）。但是在英国，筛选过程中确实出现了一些特殊问题。因为筛选过程是一个相当自由的体系，所以 LPA 经常（大约占一半的时间）只在接收到规划申请之后才要求递交 EIS（DoE，1996）。出于同样的原因，不同的主管部门对筛选的需求也是千差万别。大多数不要求 EIA 的决定都是由下级职员给出的，他们从不考虑 EIA 的必要性，或者（错误地）认为在规划中被指定为特殊用途的土地，或者只是进行改建、扩建而非新建的场所就不需要进行环境影响评价。类似地，不同地方机构对于本质极其相似的开发项目也会做出不同决定（Gosling，1990）。

1999 年之后，由法令修正案建立起来的新筛选标准试图减少此类问题（3.4 节和 4.3

节）。一项由政府资助、牛津布鲁克斯大学IAU承办（IAU，2003）的研究，为1999年T& CP（EIA）法案所做出的LPA筛选决策的本质和特征提供了研究证据。该研究基于2002年在英格兰和威尔士的100多个LPA调查反馈，从筛选活动的频率在决定是否进行EIA时主要考虑的因素、不同筛选标准的重要性中获得了信息。对于一般的筛选活动和所有LPA来说，指示性阈值被认为是筛选决策时的主要考虑因素（约44%的LPA）。附录3的条件和相关导则（非常重要的主要开发项目，处于敏感地区的项目或者是会出现复杂/有害影响的项目），则需要一个专业的判断，排在第二位（占LPA的35%）。表8.2和表8.3概括了LPA最近的单项筛选政策。使用法规和阈值被看作是总体上最有效的方法，对于更有经验的LPA做专业判断也是同等重要的。

表8.2 在最近筛选程序中的有效方法

筛选程序中的有效方法	LPA/% $n=56$	<5 EIAs 56 ($n=26$)	>5 EIAs ($n=30$)
向自己的组织咨询	3.6	7.7	0.0
社区咨询	12.5	11.5	13.3
向部长直接要求筛选	0.0	0.0	0.0
在当地规划发展中按照筛选指南实施	1.8	3.8	0.0
在其他政策和规划中按照指南实施	7.1	11.5	3.3
用法规和阈值作为指南	35.7	38.5	33.3
咨询其他类似项目	1.8	0.0	3.3
使用专业人员进行判断	26.8	19.2	33.3
使用调查表寻找潜在影响	1.8	0.0	3.3
使用其他正式的科技	1.8	3.8	0.0
自有标准方法	0.0	0.0	0.0
潜在有争议的项目	0.0	0.0	0.0
其他	7.1	3.8	10.0

资料来源：IAU，2003。

表8.3 最近筛选过程中问题的重要性

问题($n=97$)	非常重要/%	重要/%
项目规模	47	40
接受距离	44	43
项目本质	42	32
交通/可进入性影响	33	32
生态影响	32	31
排放	31	31
景观影响	26	35
累积影响	20	26
经济影响	6	24
社会影响	5	22
争议/问题	9	16
事故风险	5	10
其他	3	1

资料来源：IAU，2003。

筛选决策中最重要的因素是项目的大小和规模，大约有87％的LPA指出这两个因素是“重要”或者是“非常重要”。接下来的两个重要因素是敏感的环境受体（76％）和项目的特性（74％）。另一个方面，只有15％的比例指出事故风险也是“重要”或者“非常重要”。对筛选决策产生主要限制作用的还有：资源的短缺（45％），时间的限制（44％），法规的不明晰（33％）和不确定性（如基础数据和项目特征不确定等）(32％)。总之，结果表明阈值在筛选决策中是非常重要的，但它们也需要以专业判断作为条件。换句话说，它们自己无法为筛选决策提供足够的证据（Weston，2000）。2011年指南也支持在EIA方面的这一观点，最基础的问题是：“这个项目是否可能有重大的环境影响?”对于大部分的拟建项目，都必须考虑其项目特点和拟建选址，这是为了确定其是否可能有重大的环境影响（DCLG，2011a）。

如第3章所示，筛选程序仍然被很多包括英国在内的欧盟成员国认为是一个问题较多的领域。在IEMA（2011）一份关于EIA从业者的研究报告中显示，英国的筛选程序是一个有效的、可以确保识别重大环境影响的工具；然而这里有一个重要的问题，即一些附录2中的项目可能有潜在的重大环境影响，却没有进行环境影响评价。随着时间的推移，将来可能会形成一些对法律的重大挑战，这些挑战将在8.6节进行讨论。这在某种程度上也解释了为什么与人口相似的欧盟其他国家相比，环境影响评价项目在英格兰和威尔士这种人口数量的地区比预想的低。

在过去的10年进程中EIS之所以发展很快，主要是以下几个方面结合在一起的影响：小规模规划发展的程序简化、阈值的使用、针对“切萨拉米香肠”式项目的管理创新（阈值以内的子项目）和改进申请筛选程序的指南。虽然DCLG（2011a）指南仍然援引筛选阈值，苏格兰政府引入了一套逐个审查筛选案例的实验方案（苏格兰政府，2007）。这里也有一些关于筛选程序自动化的进程。丹麦开发出来一个测试动物农场项目的电子模型，只需要在计算表格中填写所需要的数据，该程序就会计算并生成一副清晰的图画显示该拟建项目是否需要进行环境影响评价。在英国，Bachuller and Glasson（2004）在牛津布鲁克斯大学也开发了筛选专家系统和范围确定专家系统。

8.3.2 范围确定与预提交咨询

主管部门有充分的自由来确定EIA的范围。正如在第3章所讨论的：在英国法律中，引入的85/337号指令的附件Ⅲ中一部分要求强制执行，另一部分则可以自由选择。一项早期关于环境影响报告的研究（Jones等，1991）显示，虽然法律强制的部分都得以执行，但是自由选择的部分（如考虑替代方案、预测方法、间接次要的影响、范围确定等）通常执行得较少。之后的研究显示，尽管政府指导文件建议早期的范围确定应由开发商、从事环境影响评价工作的顾问、主管部门和其他咨询机构共同讨论完成，并认为这种讨论对有效的环境影响评价来说至关重要（Jones，1995；Sadler，1996），但是在实际操作中，提交前的顾问咨询只是偶尔为之。例如，一项针对环境顾问的调查（Weston，1995）表明，只有3％的项目在选址确定前就要求编写EIS，大约有28％的项目在详细设计前就要求编写EIS。有30％～70％的案例是在EIS提交之前开发商就咨询了LPA，虽然这个比例目前已经有所提高（DoE，1996；Lee等，1994；Leu等，1993；Radcliff和Edward-jones，1995；Weston，1995）。

如第2、3章所述，修正案和后来出台的英国法规已经勾勒出环境影响评价程序中范围确定的轮廓。之前提到的ODPM研究项目（IAU，2003；Wood，2003）也通过LPA、顾问以及法定顾问对范围确定的性质和特征进行了研究。有将近75%的LPA为如何进行范围确定出谋划策。所有这三类参与者在表达自己对范围确定的意见或做出相关报告时，都高度重视对项目选址的特点、减缓措施以及影响等级的初步评价。同样，组织内部咨询和各个等级的专业评价是影响识别的重要方法，使用法规和阈值对LPA和咨询组织来说也是非常重要的，就法定咨询而言，其重要性要稍逊一些。表8.4列出了最近的范围确定项目中（在研究调查中）每类参与者最关心的问题。LPA和咨询机构所强调的内容是非常相似的，其中对交通/运输、景观/视觉以及植物/动物等问题尤其关注。法定顾问也存在相似之处，但是他们更加关注其他环境问题（如其他排放物、废物的处理等）。令人吃惊的是，气候因素和事故风险（健康和安全）并未受到重视，这或许反映了那些更长期的、不可预测的问题的不确定性。

表8.4　最近项目范围确定观点/报告中主要关注的问题排序

项目	地方规划局($n=78$)	一般顾问($n=98$)	法定咨询机构($n=28$)
社会问题	10	20	11
文化遗产	19	29	18
经济	21	22	4
植物/动物	46	50	68
土壤	18	20	21
空气质量	35	21	14
噪声和振动	42	45	11
其他遗漏	26	31	39
气候因素	6	5	7
废物处理	13	19	32
水资源	37	31	43
岩土工程问题	18	21	29
景观/视觉	58	57	32
交通/运输	47	55	11
事故风险	5	9	7
以上各相间的关系	18	17	43
其他	3	9	4

资料来源：IAU，2003。

对于筛选和确定范围已经积累了相当多的经验。除了起初遇到的挫折，目前的筛选程序已经相对容易接受，而且通过1999年的修订已经更加精确。在所考虑的所有阶段中，范围确定通常被认为是环境影响评价中非常有价值且值得投入的一部分，同样地，经1999年再次修正后，其在英国环境影响评价程序中的地位也变得更加重要了。虽然如此，对于12个新的欧盟成员国中的10个国家，范围确定仍然不是法定必要程序，但是其仍然被当作最佳实践指南鼓励去进行。最佳实践指南在IPC受理的NSIPs类项目的筛选和项目范围确定在章节3.5.7中有所介绍。然而，人们对多个因素的结合仍然有一定顾虑，包括危险规避、规划不佳和商业现实，这些因素可能会引起过分的范围审视，就像谚语中说的“连厨房水槽都算上的一切”都被算作安全因素。表8.5显示了2010年IEMA研究的100份英国环境影响评价报告的环境主题的出现频率。

表 8.5　2010 年 100 份英国环境影响评价报告的环境主题的出现频率

环境主题(黑体显示主题表示其内容在第三条或者 EIA 指令附录Ⅳ中)	2010 年 100 份英国环境影响评价报告发生率
生态(**植物和动物**)	92%
噪声(和振动)	92%
水	90%
景观/城镇面貌/视觉分析	88%
交通	88%
文化/建筑**文化遗产(包括考古)**	82%
土壤和土地质量/地面状况	81%
空气	79%
社会-经济	64%
累积影响(**交流/内部关系**)	46%
废物	28%
气候变化	17%
EMP,残余影响和减缓措施总结	17%
人口/人类	13%
舒适性、可进入性、娱乐和通行权	12%
日光(太阳光)	11%
材料资产	10%
微观-**环境**/风	9%
电子干扰(广播和电视)	7%
可持续发展	7%
公共健康	6%
光	5%
航空	5%
地貌和海岸变迁	4%
能源	3%
阴影闪烁	3%

资料来源：IEMA，2011。

研究显示，90%以上的环境报告包括生态、噪声以及水影响（表 8.5）的章节，其他五个最严重的问题在另外 80%的报告均有涉及。总体来讲，这个研究显示，在 2010 年平均每个环境影响报告有 9.63 个关于环境的章节。然而，IEMA 的研究没有把范围扩展到评价是否每一个章节都是合适清晰的，而且很明显的是，在实践中范围的确定导致评价报告几乎很少集中在关键的环境问题上（IEMA1，2011）。

8.4　环境影响报告的质量

在 8.1 节中我们提到，高质量的 EIS 是保证 EIA 政策向实践有效转化的因素之一。在开始就提交高质量的 EIS，可以降低开发商与主管部门之间昂贵的互动费用（Ferrary，

1994)，也为公众参与提供了更好的基础（Sheate，1994），把责任适当地转移给开发商，这就全面增加了高效 EIA 的概率。保持环境信息的完整性十分重要，而且法定顾问的意见、公众的评论以及主管部门的专家意见都可以帮助克服低标准 EIS 带来的局限性（Brown，1993）。这个主题在后文 8.6 节法律裁决中会有所体现。

8.4.1 环境影响评价报告书质量的学术研究

在英国，环境影响报告书的质量受到有限的法规要求的影响，即使 EIS 不充分，规划申请也不会被驳回，EIA 程序中的重要环节（例如公众参与和监控）不能强制执行，再有就是，开发商可以为自己的项目做环境影响评价。本节首先在几项学术研究的基础之上讨论英国 EIS 的质量，接着简要探讨对 EIS 质量的其他理解，这是因为主管部门、法定的咨询机构以及开发商对 EIA 需求不同，看待 EIA 质量的观点也不同。本节最后将就可能会对 EIA 质量产生影响的各个因素进行讨论。

在环境评价指令颁布后的 10 年里面，有很多关于环境影响报告的研究，最典型的评估标准是使用 Lee 和 Colley 的标准（1990）(附录 4)。基于这些标准，EIS 被分为“满意”（分类等级为 A、B 或 C）和“不满意”（分类等级为 D 或 E）两类。表 8.6 总结了部分调查结果，结果表明开始时 EIS 质量并不令人满意，之后满意度逐渐上升。总体上讲，项目的描述、交流展示结果趋向于比确定主要影响、待选方案和减缓措施更加令人满意。

表 8.6 EIS 质量统计案例（满意率①）

项目	Lee 和 Colley 1990	Wood 和 Jones 1991	Lee 和 Brown 1992	Lee 等 1994	Jones 1995	Baker 和 Wood 1999	DoE 1996
案例规模	12	24	83	47	40	24	50
1988～1989	25	37	34	17			36
1989～1990			48				
1990～1991			60	47			
1991～1992					刚刚过半	58	
1992～1993							60
1993～1994							
1994～1995						66	
1995～1996							

① 满意度的成绩为 A、B、C 即为满意（1990 年或 1992 年）。

一些研究专注于特殊的项目类型：例如，Koubus 和 Lee（1993）以及 Pritchard 等（1995）评论了关于采掘工业项目的环境影响评价报告，Prenton-Jones 对养猪业和家禽业的开发的（Weston，1996）研究，Radcliff 和 Edward Jones（1995）对医疗废物焚烧项目的研究，Davision（1992）和 Zambellas（1995）对公路的研究，Edward Jones（1999）对林业的研究等。其他一些研究也分析了特定的 EIS 环境组成质量，例如，景观/视觉影响（如 Mills，1994）和社会-经济影响（如 Hall，1994）。研究结果都表明现在的 EIS 不够好，但是仍在改善中。

我们并没有发现最近有相似的、大规模的关于英国 EIS 质量的研究，但是形势表明目前 EIS 的质量在继续提高，尽管这种提高不是持续的：

为大型的、有争议的项目工作的咨询机构，最近几年不得不极大地提高他们工作的质量……这将对部分利益相关者以及代表当地居民和公众的团体进行审查报告有所帮助……虽然大型项目和有争议项目的环评报告质量有所上升，但是总体质量仍有不同。“凑合”一词多被用于描述现在的标准，包括环评报告质量以及当地政府审查的评价。(ENDS Directory, 2007)

8.4.2 谁的质量

但是这些发现必须结合背景来充分考虑“高质量到底为谁”？研究者发现：就算 EIS 的质量是一定的，但是相关的规划者或顾问对 EIS 的理解也各不相同。例如 DoE (1996)，Radcliff、Edward-Jones (1995) 和 Jones (1995) 发现，在规划者、顾问和研究者之间很少能就 EIS 的质量达成一致意见，唯一的一致性趋势是与规划者相比，顾问对 EIS 质量的要求更为严格。

在由 IAU (DoE, 1996) 主持的访谈中，规划官员认为编写 EIS 就是为了得到规划许可和减少潜在的影响。只有约 40%以上的人认为 EIS 质量已经得到改进，但是改进的程度通常很小。其他大部分的人认为很难对此做出评价，因为单个官员接触的 EIS 很少，而且他们所看到的 EIS 涉及不同类型的项目，其引起的问题也不尽相同。其中合理的范围确定和替代方案的讨论被认为是最主要的问题。EIS 被认为是“正在变好却也正在变大”。一些官员把 EIS 的质量与编写 EIS 顾问的名声联系起来，他们相信使用有经验而且名声好的顾问是得到高质量 EIS 的最好途径。

在 EIS 质量是否正在改进这一问题上，法定咨询机构也是意见不一。他们觉得 EIS 的客观性和清晰表述很重要，虽然正在改进，但是仍不理想。地方规划局、开发商、咨询机构和顾问通常认为综合、客观和清晰的标准正得到改进，但是整体仍然欠佳。开发商和顾问把 EIS 质量和获得规划许可的能力联系在一起。顾问们感到开发商越来越意识到进行环境保护的必要性，并开始在项目规划初期就引入了咨询，以保证项目能围绕环境保护的需要而设计。改进的原因之一也许就是压力团体对 EIA 越来越有经验，因此对 EIA 程序 (DoE, 1996) 的期望也越来越高。

8.4.3 EIS 质量的决定因素

以下因素会影响到 EIS 的质量，包括项目的规模和类型以及 EIA 程序中不同参与者的经验和特点。某些类型的项目 EIS 已经达到了较高质量。例如，附录 1 中的项目，它们通常具有较高的姿态，能够吸引更多的注意和资源，很可能有更好的 EIS (Lee 和 Brown, 1992; DoE, 1996)。在英国国家规划检察署基础设施组严格的预提交制度要求下，这个趋势在未来可能还会继续发展。

考虑到 EIA 参与者的经验和特点，由开发商内部编制的 EIS 往往比外来顾问编制的 EIS 的质量差：例如环境事务部 (1996) 的研究表明，内部编制的 EIS 质量的平均成绩通常为 D/E，由顾问编制的 EIS 的平均成绩为 C/D，而由两者共同参与编制的 EIS 的平均成绩为 B/C。Lee 和 Brown (1992) 通过对 83 份 EIS 进行分析得出以下结论：环境顾问编制的 EIS 中的 57%的质量是令人满意的，相比之下，内部编制的 EIS 的比例只有 17%。同样，由独立机构编制的EIS质量（C/D）比由地方政府为其自身工程编制的EIS质量（D/E）好一些（DoE，1996）。

开发商、顾问以及主管部门的经验也会影响EIS的质量。例如Lee和Brown（1992）的研究表明，由至少提交过一份EIS的开发商（没有顾问参与）编制的EIS中，有27%是令人满意的，而由无经验的开发商编制的EIS只有8%是令人满意的。Kobus和Lee（1993）引用的EIS中，满意率分别是43%和14%。Lee和Dancey（1993）的一项研究表明，有4次或更多经验者编制的EIS中，有68%是令人满意的，而对于无经验者，这个比例只有24%。环境事务部（1996）的研究表明，有5次或更少经验的顾问编制的EIS，其满意度只有50%，而有8次或更多经验的顾问编制的EIS，其满意度高达85%。无经验的地方当局编制的EIS，只有1/3令人满意，而有8次或更多经验的地方当局编制的EIS，有2/3是令人满意的（DoE，1996）。这表明当编写者有更多经验后EIS的质量也会有所提高。

其他决定EIS质量的因素包括EIA法规和导则的有效性，更多的导则［例如DETR，2000；DoT，1993；地方当局的导则，如肯特郡和艾塞克斯州（艾塞克斯州规划委员会，2007）］有助于得到高质量的EIS；项目的规划阶段提出的开发申请和EIA以及为详细规划申请编制的EIS，质量通常要比为申请大纲编制的EIS质量要高；在EIA程序中，与参与部门之间的相互作用相关的问题，包括EIA授权、EIA的资源分配和部门之间的交流等。在公众领域内，还有更多的EIS提供好的实践例证。

环境影响报告书的页数与EIS质量好坏也有着某些联系。例如，Lee和Brown（1992）指出，页数少于25页的EIS满意度只有10%，而页数多于100页的EIS满意度上升至78%。在环境事务部（1996）的研究中，页数少于20页的EIS，其质量等级平均只有E/F，当页数达到50页时，质量等级也上升至C。然而，当EIS超过150页时，其质量可能会产生一定的变化。尽管更多页数的报告包含了更多的信息，但是它们的篇幅也表明了组织和协调的低质量。IEMA（2001）发现有些环评报告长达350页，尤其是一些关系到国家重要基础设施项目的报告可能会更长。资料表明：

> 很多从业者对EIS的长度很沮丧，同时对提交给审批机构和包括公众在内的更宽泛的团体的文件价值也有所担心……25%的被调查者（一个在线调查）认为目前EIS的长度降低了评估结果对所有读者的价值，甚至包括那些在环境领域有专业知识的专家们，例如环境局。被调查者认为这种情况对非专业人士会更糟。多余2/5（42%）的受调查者认为，对于那些非法定咨询机构的利益相关者来说，冗长的环境报告一般会降低环境影响评价结果的价值。2/3（66.5%）的受调查者认为，对于当地社团，目前环境报告的长度降低了环境影响评价结果的价值。（IEMA，2011）

8.5 提交后的EIA程序

在主管部门收到EIS和项目授权申请后，必须对其进行评审，并向法定咨询机构以及公众进行咨询，然后对项目做出决定。本节对这些依次进行了介绍。

8.5.1 评审

规划官员很少看到需要 EIA 的项目和其他具有相似的复杂性和争议的项目之间的区别：申请一旦正式提出，开发程序就会被接管。主管部门在评审 EISs 时往往应用他们自己的知识和经验来判断出现的错误和局限性。评审 EIS 首先要通读全文，咨询主管部门中的其他官员，向外界咨询，并将 EIS 与相关法规进行比较（DoE，1996）。

尽管 Lee 和 Colley 的评审标准（1992）具有现实有效性，但是只有大约 1/3 的地方当局使用了其中的一种评审方法，而且通常只是用作指示标准来确定深入调查的内容，而不是作为正式方法来使用。有 10%～20%的 EISs 由外面的顾问或 IEMA 来进行评审。但即使是雇佣外面的顾问来评审 EIS，如果顾问之间可能存在竞争，就会对其所做评价的公正性产生影响。当项目决策的时间相当紧迫时，能否从评审顾问那里得到充分快捷的反馈就非常重要了（DoE，1996）。开发商正在使用一种创新的方法，这种方法需要顾问经过投标获得 EIA 资格，作为他们投标的一部分，“独立的”评审者或者“有批评色彩的友人”可以保证咨询工作的质量。

在英国 EIA 实行的早期，规划机关在 2/3 的案例中需要附加 EIA 信息（例如 Jones，1995；Lee 等，1994；Weston，1995），但是这通常是在未正式借助法规的情况下完成的。2/99 号指令再次强调了这一点：“……若开发商在 EIS 中没有提供足够的信息，就可做出拒绝申请的决定。”（法规 3，DETR，1996b）。如果规划权力机关认为申请者提交的信息不足时，可以继续要求申请者提供更多的 EIA 信息。即便如此，我们没有发现最近有更多类似的例子在发生。

8.5.2 提交之后的咨询和公众参与

主管部门通常会把 EIS 发给法定的或非法定的咨询机构，并让公众可以评论这些信息。在英国环境影响评价系统中公众参与只限于环评报告提交以后的几个星期，而且 EIS 公告一般也是地方性的公告，例如在当地报纸上、当地地界公告和社区宣传板。越来越多的 EISs 被发布到网络上，为了满足法律要求，地方政府或者图书馆也可以“合理有偿”地提供 EISs 的副本。

规划官员“非常信任咨询机构，使其至少对一部分 EISs 进行评审、修改和总结”（Kreuser 和 Hammersley，1999）。当地对项目有兴趣的团体在这个阶段，经常非常积极地对环评报告的质量和规划申请进行评审和建议。当地 EIS 包含的某一特殊环境的信息不充分时，主管部门通常会责令开发商和咨询机构进行直接接触，而不是他们亲自进行信息咨询（DoE，1996）。

正式的 EIS 可能只是规划申请中几个相似的规划文件中的一个。IEMA（2011）说：

> 在英国很多现行规划发展制度中，很多项目申请的环评报告和其他文件的信息是重复的。在英国最新的咨询制度下有个非常好的例子可以显示其不足，委员会专门制定一个精简的操作方式：申请国家重点基础设施项目。委员会接受了几个申请，其中有些包含环评报告。这些报告中解决了规划发展几方面的预测：洪水危险、历史环境、景观、废物和自然特征。所有这些问题都在环评报告中，并且在事件中解决了。

考虑到环境影响评价信息，规划申请程序可能也是众多减少法定咨询机构投入的时间和精力的制度中的一个。例如 Bird（1996）建议环境局（其在 1996 年成立，脱离了包括 HMIP 在内的一些机构）说：

在申请污染许可证的时候需要环境影响评价。因为在环境局作为 EIS 法定规划咨询机构的时候，如果收到另一份正好包含它们所关注内容的文件，（它们）就不会浪费时间去抱怨规划 EIS 所给定的不够详细的设计。Didcot B 案例的研究表明，即使 HMIP 认为 EIS 是满意的，它们之后也会要求对主要的设计进行调整。（如在利物浦海湾 Hamilton 项目石油气末期工程）HMTP 没有对 EIS 提出反对意见，之后却驳回了（综合污染控制）授权申请……

有关重新授权程序和 EIA 参与者讨论的问题将在第 9 章的案例研究中做进一步的论述。

8.5.3 决策制定

正如第 1 章所述，EIA 的主要目的之一就是辅助做出更好的决策，因此对环境影响评价绩效的评估也是非常重要的。所有的决策都必须要协调，这也是很重要的。包括就 EIA 程序中简单化和复杂化、全面和重点、紧迫性和对更多信息的需求、事实和价值、预测和评估、确定性和不确定性等的相互协调（Wood，2003；IEMA，2011）。此外还有更实质性的协调问题，尤其是就项目对社会-经济影响和对自然环境影响之间的协调——有时候可简化为“工作与环境”的抉择问题——在一个项目中获胜或者失败的群体之间的协调。专栏 8.1 说明了英国政府决定取消希思罗机场第三条跑道的例子。

在决策过程中，比起其他影响，一些影响可能更具有协调性。在社会经济和自然环境范畴内，Sippe（1994）对可协调影响和不可协调影响做了描述（表 8.7）。Sadler（1996）认为权衡选择是进行可持续发展决策的核心问题。

专栏 8.1　希思罗机场第三条规划跑道取消

希思罗机场原计划为增强交通吞吐能力而修建的第三条跑道，将会被新的联合政府取消。在保守党和自由民主党的公约中指出，取消第三条跑道是党派协商后的一个环境措施。该公约也承诺会拒绝 Gatwick 机场和 Stansted 机场跑道延长的申请。

但是商业代表马上警告新政府，表示限制伦敦交通吞吐量将会影响首都的竞争力，这将会是一个“错误的开始”。这个决定也会打击机场操作公司 BBA。首先，伦敦一些大企业的代表要求：假如不再扩建机场，政府必须拿出其他方案，并坚持认为高速铁路不能取代航空在国际运输中的地位。他说“考虑到已经使用了希思罗机场 99%的运营能力，这将会是影响伦敦竞争力的一个主要因素之一。”

商业集团对伦敦希思罗机场的扩建有长时间的争论。这个项目作为伦敦成为国际金融中心的一个关键支柱，曾经被前工党政府所批准。在近些年，频繁商务旅行的人也经常抱怨“希思罗的烦恼”。其中典型的问题包括日益复苏的恐怖警告、过分拥挤和过时的设施。

扩建的项目受到了议会、居民和环境团体的强烈反对。保守党和自由民主党在他们的执政纲领中都承诺放弃该项目。John Stewart 是反对希思罗机场扩建团体 Hacan Cleanskies 的主席，他表示“第三条跑道已经胎死腹中了”。在飞机跑道周围居住的居民十分高兴，他们不再会因此而搬家。项目的取消对伦敦整体也有好处，扩建将会对环境有害，除此之外，“经济幸福”并不需要这样的扩建。Norman Baker 是自由民主党的议员，他坦率地反对机场的扩建，他说对于生活在伦敦的人民来说，这项运动是公约达成的“第一个果实”。

资料来源：Financial Times，2010

表 8.7 判断的环境可接受性——权衡

项目因子	不可协商的影响	可以协商的影响[①]
生态(自然和生物系统因子)	基本生命支持系统退化； 降低保护地位； 生态完整性的不良影响； 损失生物多样性	没有恶化到超过承载能力； 生产系统没有恶化； 明智地使用自然资源
人类(人类作为个体活在社会组织中)	死亡； 公共健康和安全减少到不可接受的程度； 在人类居住区不合理地降低生命质量	社区的收益和成本以及出生地； 合理分配成本和收益； 合理的代际公平； 与定好的环境政策目标的兼容性

资料来源：Sippe（1994）。

① 根据纯环境利益。

在英国，在 EIA 程序中最为明显的决定往往受到主管部门规划决策影响最多。EIS 在其提交的前后可能会对规划官员的报告、规划委员会的决定以及项目的状况和修改等产生重大影响。但是 EIA 对决策的影响远比这些影响要大得多，例如，会影响到替代方案的选择、项目的设计和再设计以及减缓措施与监测的范围等（Glasson，1999）。事实上，一个有效的 EIA 系统会阻止一个不成熟的项目，同时也会阻止对环境造成巨大破坏项目的启动。

在第 3 章中我们确定了 EIA 程序中的不同参与者。这些参与者在决策中对 EIA 有着不同的看法。地方规划官员会非常关心决策中 EIA 的集权性（是否有所不同），中央政府可能关心全国范围内开发申请的一致性以及它们是否有助于可持续发展的有效进行；压力团体可能会关心这些标准和公平性（提供参与机会的公平性）以及项目周期和批准过程中的整体性(能够在多大程度上轻易避开 EIA)。很多研究者尝试去确定 EIA 和相关的咨询是否会影响决策，如是否批准，怎么批准一个项目。

对地方规划官（Kobus 和 Lee，1993；Lee 等，1994）的早期调查表明，在约一半的案例中，EISs 对于决策相当重要。对较大范围利益集团（DoE，1996）的采访发现，大约有 20%的被采访者感觉 EIS 对决策起“很大”作用，超过 50%的被采访者认为 EIS 对决策有“一些”作用，剩下的 20%～30%的人认为 EIS 作用很小或根本没有。Jones（1995）发现在规划官员、开发商以及公共利益集团中有大约 1/3 的人认为 EIS 会影响决策，而有大约一半的环境顾问和极少数咨询者认为 EIS 是有影响的。规划最终决策是由规划委员会的成员做出的。通过采访，得知他们对阅读 EIS 并不太感兴趣，而是根据官员的报告来总结主要问题（DoE，1996）。根据 Wood 和 Jones（1997）的研究，在 97%的案例中，规划委员会会采纳官员的建议。与 EIS 相关的咨询通常被认为与 EIS 本身一样重要（Jones，1995；Kobus 和 Lee，1993；Lee 等，1994；Wood 和 Jones，1997）。在另一方面，很多来自非法定组织的被访问者觉得他们被排除在决策程序之外，一个国家性的非法定野生动物组织抱怨说，如果自然管理委员会或国家委员会没有反对的话，他们自己的反对意见就会被忽视（DoE，1996）。

对早期 EIS 的研究（如 Kobus 和 Lee，1993；Lee 等，1994）表明，在项目授权的最终决策中，物质因素要比环境因素稍微重要一些。而后期的研究（Jones，1995）表明，环境是影响决策的主要因素，规划政策次之。Wood 和 Jones（1997）的研究表明，在37%的案例中，环境被看作是影响决策最为重要的因素。但是，只有极少数案例会在缺乏 EIS 的情况下得出不同的结论。通过对规划上诉案例的研究，Weston（2002）总结出：

与实行的最初几年相比，今天的 EIA 在英国已经逐渐地发展壮大。但是在 LPA 和规划检察员实际做出决策时，EIA 对其的影响仍然很小，因为决策是基于多个要素做出的，而这些影响要素在 EIA 引入之前就早已存在了……英格兰和威尔士的地方当局每年大约要处理 45 万份规划申请。对这些申请所做出的决策是在“物质因素”基础上完成的，包括：地方发展规划、国家规划政策导则和正式咨询程序的结果。在这些申请中，只有不到 0.1%受到了 EIA 的影响，而其他 99.9%的方案在实际做出决策时有无 EIA 并无不同。

综上所述，在英国，编写了 EIS 的项目申请和未编写 EISs 的项目申请在待遇上并无太大区别。尽管环境问题更为正式地被提出来，但在个别文件中，最终的决策程序并没有因 EIA 而改变太多。EIA 引起的主要程序上的差异是 EIS 咨询的必要性和更广泛的公众参与（在实际中并不经常使用）。主要的不同在于项目的调整以及在项目早期设计减免措施作为项目附加的或不同的条件，这也可能是主管部门对环境问题进行了全面的考虑。

8.6 法律挑战

有些项目应该提交 EIA 却没有提交，而有些项目 EIA 做的不够充分，最近 EIA 质量提升的主要驱动力来源于法律的威胁，包括 2011 年 EIA 法规和指南的颁布。法律挑战的重要性在于它们阐述了 EIA 法规体系的重要方面、制定 EIA 后续程序以及随着时间的推移，趋向于加强 EIA 的要求。一个供职于主要环境咨询机构的顾问解释说：

最近的案例法在我们在 EIA 领域有着意义深远的影响……对于反对者很容易根据立法去批评 EIA 过程，提出公众质询和流程运行不正确的地方。审查比以前变得更加频繁和强烈，至少在我们工作领域的这些类型的项目是这样的。Heritt 先生没有说某一个案例的法律质疑突然变得越来越多会加重法律挑战，而是强调 EIA 相关案例的累积影响让反对者提出反对 EIA 变得更加容易……，Heritt 先生说，连续不断地全面审视案例法作为 EIS 的法律依据是一件费时费力的事情，但是有经验的客户认为这是一个非常好的投资。为了取得规划许可的任何努力和/或者增加审批速度都是一件值得投资的事情。(ENDS Directory，2007)

这是一个审视英国部分法律案例的相对较长的小结，可以给读者在类似法律挑战中一个意见。这里包括几个关键的案例，这些案例引起了英国 EIA 指南的变革（Baker，Mellor）：

老的关键案例影响了过去EIA指南的变革和以后的法院的案例（Berkeley，Rochdele）以及一些其他在EIA法律中特殊领域的案例。这些问题分布在EIA程序的序列中，从筛选到应用减缓措施结束，它们包括引用较长的法律判决、法庭上用的逻辑分析和辩论，以及说明EIA法规如何使用和解读。

英国的法律系统是复杂的，有不同的层级和参与者。这一章节的目的，只限于与这个系统有关和被需要的主要参与者的信息。首先，原告（个人和组织启动法律诉讼，也被称为原告或者在某些案例中的上诉人）必须有“理由”来进行法律诉讼，也就是说他们必须受到他们决定诉讼的项目足够影响或者伤害。在EIA案例中的原告一般是规划许可给予某些项目的当地居民，或者相关主管部门要求某些项目进行EIA而受到影响的开发商。被告是原告起诉的个人或者组织。最后，基于法庭1947年一个复杂而又不相关的对法庭操作的决议，当一个公权机构的结论上诉到法庭裁决——所有法律挑战都在本章节中呈现——英国法庭使用“韦德内斯伯里不合理原则”。“韦德内斯伯里不合理原则”可以被解释为：只有在行政机关的决策非常不合理以致任何理性的机构都不会做出类似决策时，法院才能以行政机关未合理行使权力为由进行干预。

8.6.1 即使未能改变规划决策也需要环境评价

Berkeley VS. 环境部长和其他部门（2000）UKHL 36；（2000）3 ALL ER 897；（2000）3 WLR 420

“Berkeley 1”案例涉及1994年关于重新开发Fulham足球俱乐部的规划申请。当地规划机关没有要求提供EIS（也没有准备），但是项目咨询了很多的组织，同时也仔细地斟酌了拟建项目的优缺点。在1995年8月，部长参与到该项目中，他也没有要求提供EIS，并在1996年8月给出了规划许可。

Berkeley夫人在该项目附近居住，她质疑该项目的规划许可，并坚决主张该项目需要进行EIA。法官以足球俱乐部的建设不需要进行EIA而驳回了上诉。即使需要进行EIA，法官仍然认为不进行EIA是因为“质询的结果显示没有影响，所以就不可能有影响”。1998年法庭上诉判定该法官错误地决定了该项目不需要进行EIA，但是赞成他认为根据质询结果显示没有影响则不需要进行环评的判决。Berkeley夫人对此决定进行了再次上诉，然后“Berkeley 2”案例2000年在上议院被提出来。

部长律师这次接受了这个事实，即EIA不会改变规划决定的结果不可以作为法庭废弃该决定的充分理由，相反其坚持该判决完全是根据指令的要求去做的。

上议院总结该案例，根据附录2中第10款（b）的要求，重新开发足球俱乐部可以被认为是一个“城市发展项目”，该项目对泰晤士河生态影响意味着项目可能有严重的环境影响，这种情况下如果拟建项目对个人产生影响，个人有权要求部长在项目获得规划许可以前而非在法官判决之后考虑该项目是否需要进行EIA。

同时也规定了无论EIA是否能改变规划决定，其不仅仅只有通告规划的作用：

指令中规定的公民直接实施权不仅仅是一种在重要问题上可以获得通知的权利。它的实施必须建立在一个合适的基础上，同时要求包容性和指令中描述的针对公众的民主程序。然而有些误导和固执己见的观点可能被用于表达环境问题。

部长提出曾准备了与EIS相类似的报告的争论也同样被驳回：

[部长律师]声称类似的申请者的环境报告可根据《咨询程序法》找到其案例报告，通过读规划局的案例报告（交叉引用的好处），发现其向附属规划委员会提交了整理过的多个官员的报告，其整理的文件包括国家河流管理局和伦敦生态小组的文件，其文件可以作为咨询中的证据对公众可见。公众可以获取这些文件并且在咨询的过程中有权对文件发表意见……关于环境报告的观点在指令中有所考虑，其包括个体文件和可以获得的全面汇总文件，申请者在申请程序的开始就需要编写，其包含相关的环境信息以及非技术语言总结。

该规划许可被废止了。

8.6.2 未使用筛选程序的理由是规划条件使项目的影响不严重

R 申请 Lebus vs. 南剑桥地区 (2003)[1]

“Lebus 案例”涉及2000年2月关于一个拥有12000只自由放养母鸡农舍和其附属农业工人的住宅。2000年4月，原告律师给南剑桥地方政府写信声称该申请需要进行环评。规划官员非常迅速地给予回信，指出根据1999年EIA法规，虽然根据附录2中要求，该农舍建筑面积大于$500m^2$，但附录1中规定，该农舍少于6万只母鸡，所以不需要进行EIA。然而规划官员也指出：

拟建项目是否有严重的环境影响均应被考虑到，其考虑因素例如选址、影响、本质以及规模……当考虑是否需要环境评价时，以上因素以及下列问题需要考虑在内，例如空气污染、脏水、垃圾废弃物、生态、高速公路、可进入性和景观等。当局并不执着于索要环境影响报告，仅仅是因为其期望的信息已经足够，并已经考虑到上述所有要点足够详细，而不再需要正式地要求提供一份环境影响报告。

南剑桥地方政府在2002年1月份给予规划许可证。

Sullivan 法官总结出，当地政府没有要求准备任何文件也可以被明智地认为这是项目筛选意见的一种。然后他指出：

到目前为止，法律问题已经被解决了。解决的依据是规划条件将被推行，管理义务将按照106节执行。开发是否会像申请中描述的那样将有严重的环境问题，或者采取一定的减缓措施后开发项目是否还会有严重的环境影响，这些问题并没有考虑……法规的根本目的在于实行指令，当开发有潜在严重影响时，可以使用减缓措施来调查、阻止、减少或者在有条件的情况下抵消环境的有害影响。因此，可以让公众参与到评估任何减缓措施的有效性的程序中。虽然可能有严重的环境影响，然而个人在项目开始预审的时候自己去要求做一份筛选意见是不合适的，因为当项目实施了不同种类的规划条件后，其影响可能会被减小到不重要。在类似案例中，合适的方法是要求提供一份环境报告，并在环境报告中展示出严重的环境影响和可能减缓其严重性的措施。

---

[1] www. bailii. orgiew/cases/EWHC/Admin/2002/2009. html。

类似案例的考量曾经在南 Bellway 城市重建 vs. John Dillespie [2003] EWCA Civ 400 出现过。基于该项目会把一个有污染的土地重新利用，该地的去污可以通过规划措施来确保，当时的部长曾经给出一个煤气厂重新开发成为城区住宅的规划许可。评估中假定减缓措施已经制定但并不合适，原告成功地控诉部长没有考虑到项目有不可减缓的影响。

一个拟开发项目可能具有潜在严重影响，当做出与此项目相关的决定时，这些案例并不建议弥补措施被全部忽略，但是建议做这些决定以前一定要做出谨慎的判断。充分确定的、没有冲突的补救措施可能会得到很好的考虑，但是对于一些更复杂的项目，或者拟补救措施复杂或不容易被理解，这些补救措施就不合适了。另外，补救措施可能不会被用于阻止 EIA 指令或代替其需求。

8.6.3 考虑以前项目阶段的累积影响时的筛选程序

Roao Baker vs. Bath and North Somerest 议会 [(2009) EWHC 595][1]

“Baker 案例”涉及 2005 年 Hinton Organics（Wessex）有限公司申请扩建其堆制肥料生产基地 Charlton Fields 到 2.1hm^2。该地自从 2000 年就开始运营，Charlton Fields 把木质废料、纸板以及绿色废料合成肥料，申请扩展运营场地（不到 0.5hm^2）到大约 1km 以外一个叫 Lime Kiln 的地方。当堆制肥料的工作结束的时候，部分肥料废料将会转移到 Lime Kiln。规划机关在 2006 年给出规划许可并没有要求进行环境影响评价。这个决定被原告诉讼。

规划机关强调 Lime Kiln 的申请不需要进行 EIA，是根据英格兰和威尔士《城乡规划法案 1999（环境影响评价）》中规定：“在本附件中任何项目的改变或者扩建，都需要满足其设置在附件中的阈值。”相关附录中对于废弃物和阈值的解释是：“（ii）开发的面积超过 0.5hm^2……阈值和标准……在改变或者扩建中应用（开发不能进行改变或者扩建）。”

Collions 法官支持原告。根据两个欧盟法院案例——Liege 机场（C-2/07）和 Madrid 环形公路（C-142/07），他指出：

目前 Lime Kiln 的申请是让人担心的……这里有一个问题，在判决的时候是否应该考虑本项目和其他项目在一起有关联的情况下产生的累积影响，虽然说它的区域小于半公顷，达不到阈值的要求，这里依然有环境危害的可能……这里至少应该考虑到尽管未达到阈值的要求，但是事实上作为一个项目的开发，在决定是否需要进行 EIA 的时候是否要考虑到其累积影响……

我认为法规没有在此植入 [“（在开发的改变或扩建上没有相关规定）”] 准确的指令实施细则。这是因为在考虑改变或扩张的角度时，其筛选的目的考虑有限。所以我认为，根据指令的目的和解释，法庭的判决所展示的方法是与其相反的。

根据这项规定，英国政府现在采用了附录的阈值，一旦整个开发被修改，不仅仅是修改或者扩建，也增加了一个新的条款：现存的或附录 1 中已经批准的项目，如果进行任何改建或扩建都需要筛选是否需要进行环评（DCLG，2011a）。

[1] www.bailii.org/ew/cases/EWHC/Admin/2010/373.html。

与以前环境案例相似的 Rsv. Swale BC exp RSPB [1991] 1 PLR 6，16，考虑到建设一个将会充满 Lappel 河岸的集装箱储存地，但该地是鸟类迁徙的重要滩涂。Simon Brown 法官总结道：

拟建规划不应该被单独考虑，在现实中，这可以被合理地认为是一个整体的开发的一部分，可能会造成不可逆转的影响。对于我来讲，这是我对法规的合理理解，现在小规模的发展可能会助长大规模的发展，因此可能会对以后的环境影响造成侵害……开发商可能会战胜对逐步进行小规模开发的反对意见。

8.6.4 拆除的筛选程序

R（挽救英国文化遗产）vs. 社区与地方政府部部长（2011）EWCA Civ 334[1]

“挽救案例”涉及 Lancaster 市政府允许拆除历史建筑 Mitchll's Brewery 而不需要进行 EIA。两方均认为拆除将会对文化遗产造成严重影响。在 2010 年 5 月，高等法院规定拆除并不包含在指令规定的范围以内的“项目”，所以不需要进行 EIA。压力组织“挽救英国文化遗产”参与了此案并向法院上诉。在 2011 年 3 月的一个判决援引了最近的一个案例“委员会与爱尔兰 C-50/09”案件（Commission vs. Ireland C-50/09），上诉法院法官 Toulson 和 Sullivan 解释了 EIA 指令中项目的定义，如下：

—执行建设工程，或其他设施安装，或其他项方案；

—其他对自然环境和景观的干涉，包括涉及对矿产资源的开采……（1.2 条）。

（被告）乐意接受指令必须被有目的性地解释的提议。如果接受工程有能力造成严重环境影响，根据 1.2 条对“项目”的定义，假如有可能的话，可以解释成包括在内，而非剔除在外。使用这种办法的第一步就是分解定义 1.2 条，对于我来说执行拆除工程很自然地算在“执行建设工程，或其他设施安装，或其他项方案”以内……尽管这样，没有必要为了达到预期的结论而给予 1.2 条一个广泛而有目的性的建设，通俗来讲，拆除工作将会让一个地方变成一种新的状态，这保护了公众，维护了公共设施，这也可以被认为其作为“方案”的一种，就像 1.2 条所定义的目的。

我根本不会接受回应者提出的假设：“景观”必须被解释成乡村景观。在指令中规定的“景观”也可以是除了“周围自然环境”以外的东西。文中指令的目的地是确保在项目进行之前严重的环境影响被合理地评价，所以我根本没有理解为什么在 1.2 条中“景观”的定义必须是乡村景观……（被告）认为拆除可以不算在指令之内，即使其关于项目的定义在 1.2 条中有所提及，因为其没有在指令附件 1 和附件 2 的列表里面。（但是）列表中的项目包括了不需要进行工程建设的“项目”，例如“农业林业和水产养殖”的项目和一些“食品工业”的项目。（Commission vs. Ireland C-50/09 总结）附录提到了在部门分类中的项目，但没有准确地描述包括项目在内的工程性质。根据 1.2 条，如果拆除可以算作“方案”的一种，其也可以根据附录 2 第 10 条（b）被看作城市发展项目，即使项目仅仅包括该区域的拆除和重建。

---

[1] www. bailii. org/ew/cases/EWCA/Civ/2011/334. html。

8.6.5 不需要使用筛选程序时候的筛选报告

R vs. **环境事务部，交通和区域以及** Parceforce ex **单方面** Marson（1998）EWHC Admin 351;[1] **和** R oao Mellor，**社区与地方政府部部长**（2010）Env LR 2[2]

“Marson 案例”涉及 1999 年 Parceforce 公司申请规划许可在考文垂机场附近开发一个 $17hm^2$ 的分类和分解包裹的厂房。当地居民和地方政府向环境事务部部长要求进行环境评价。在 1998 年 2 月的部长信中指出项目在附录 2 中第 10 条（a）范围以内，并且开发可能不会产生严重的环境影响，所以不需要进行 EIA。申请者质疑这个决定，并建议必须给出拒绝进行 EIA 的原因。上诉法院法官 Nourse Pill 和 Mummery 完全不同意上诉，并援引特别指出他们没有法律义务去干涉，环境事务部部长的决定是经过深思熟虑的，申请者有机会通过其他方式去影响规划决策。

这个判决维持了多年，但是最近又因为“Mellor 案例”重新引起了质疑。Mellor 案例涉及 2004 年 Partnerships in Care（PIC）规划：申请在 HSM Forest Moor 建设一个中等封闭医院。在 2006 年 7 月，PIC 的顾问要求 Harrogate 自治区政府提供一个筛选意见。周围的居民，包括 Mellor 先生，写信给政府建议必须进行 EIA。当年 10 月政府决定需要进行 EIA。PIC 援引这个案例到时任部长，部长在 2006 年 12 月却决定不需要进行 EIA。

这件事情又被上诉到欧洲法庭，法律总顾问 Kokott 总结到，虽然 EIA 指令不需要给出消极筛选意见的原因，主管机构应该在做出该决定时使用强有力的法律方式使第三方能够满意。总之，她决定：①消极的筛选意见不需要包含任何特定原因；②但是，如果兴趣团体要求提供消极筛选意见的原因，规划权力机构有义务提供更多的信息和相关文件；③但是反过来讲，这些信息可以被简短汇报，不需要使用非常正式的方式。这些要求也被收录到最近英国新修改的 EIA 法规中（DCLG 2011a）。

8.6.6 虽然 EIA 在提纲咨询阶段没有被要求，但是在详细咨询中被要求

Roao Diane Baker vs. LB Bromley，C-290/03，（2007）Env LR 2

“Baker 案例”涉及一个两个阶段的规划申请，申请方是伦敦地区地产有限公司，征询在 Crystal Palace 公园开发一个综合娱乐中心的许可。伦敦 Bromley 自治区在其第一次申请的时候没有要求该公司提供 EIS，然后在 1998 年 3 月给予提纲规划许可。在正式项目开发以前，对其他事宜做出了给予规划的保留意见。开发商在 1999 年 1 月向政府提出申请，要求其对开发的保留事宜做出最终的批准，政府在 1999 年 5 月发表了同意开发的通知。

一个住在附近的居民 Baker 夫人，质疑第二个规划阶段的许可，因此该案例一直上诉到欧洲法院。广泛的质疑归结于以下问题：如果一个英国规划机关在提纲规划咨询阶段，以该项目不会有严重的环境影响为由同意不需要进行 EIA，然而在具体规划咨询阶段，又发现该项目可能会有严重的环境影响，规划权力机关是否“能够”和“必须”要求在具体规划咨询

---

[1] www. bailii. org/ew/cases/EWHC/Admin/1998/351. html。

[2] www. bailii. org/eu/cases/EUECJ/2009/C7508. html。

阶段要求提供EIA？注意这里“能够”和“必须”两词有很大的不同，因为Bromley自治区当局争辩即使他们很想去做，但是英国国内法律并不允许他们这样做。

欧洲法庭法律总顾问Léger支持Baker夫人并指出：

在85/337指令中1条2款定义“开发咨询”在指令中的目的是主管部门决定是否允许开发商进行项目。从85/337指令的方法和目标中可以明确看到，允许开发商进行项目施工的决定是适用于这项规定的（包括一个或者多个阶段）。考虑到那些问题，国家法院的目的是验证提纲规划许可和允许保留事宜的决定在主要程序中是否可以看作一个整体，是否可以作为85/337指令的“开发许可”……

当本国法律规定咨询程序允许有多个阶段时，一个涉及主要决定和其他决定的执行措施不能超过其主要决定设定的范围，项目在环境方面的影响一定会在做出主要决定的时间节点上进行识别和评价。只要这些影响没有在某程序决定实施之前被识别出，否则需要在该程序过程中进行评价。（在如此法律背景下）其法定主管部门，在某些情况下有义务去开展有些已经获得提纲开发许可的项目的环境影响评价……这些评价必须全面，需要涉及项目的所有没有被评价的方面，甚至需要一个新的评价。

（所以）85/337指令的第2条第1款和第4条第2款介绍，如果需要进行环境影响评价，需要获得多个阶段的规划许可的允许，因为在第二个阶段，根据项目的本质面积或者位置，可能会有非常明显的环境影响。

在2008年，这些要点也被收录到最近英国新修改的《城乡规划法》（EIA）中（DCLG 2010）。

8.6.7 在提纲规划申请中的和项目中存在不确定性的其他案例的EIA

“罗奇代尔（Rochdale）信封”

通过两个对Rochdale都市自治政府给予规划许可的案例工作人员建立了“Rochdale信封”的概念。

R. vs. Rochdale都市自治政府exparte Tew和其他（1999）[1] ——“Rochdale 1”——Wilson Bowden地产公司和英国伙伴公司1999年初做出开发一个213hm^{2}的商业园区的提纲规划申请。Rochdale MBC有一个由来已久的政策，即该地应该被开发成商业园区。规划申请包括用地指标的使用方案和建筑面积的数据以及一封信，该信提及：“总体规划准备的目的主要是指导性的……（它）目的是介绍总体开发的形式，综合描绘出拟建项目的入口、主要公路路线以及其他形式的土地使用……”Rochdale MBC给予了该项目的规划许可，但要求其满足一定的条件，包括以下几点。

- 条件1.3：“除非有地方规划局的书面同意，开发必须与提交申请时的环境报告中的减缓措施同时执行……”
- 条件1.7：“只有当开发方案（框架文件）被提交并被当地规划机关允许才可以进行开发，开发的方案描绘出总体设计以及拟建商业园区的分布，包括具体的开发阶段和开发阶段的时间表。框架文件应该规定开发类型的细节、开发的配置和在该区域的建筑景观。商业

[1] www.bailii.org/ew/cases/EWHC/Admin/1999/409.html（Ground 2）。

园区必须按照批准的框架文件进行建设，除非获得地方规划机关同意进行修改的书面允许。”

- 条件 1.11：“该许可不应该被认为是同意对指导性总体规划进行开发。”

原告质疑环境影响评价报告并不全面，因为它没有提供足够的关于“项目设计规模大小”的信息，而这些信息的作用是发现拟建项目的主要环境影响。他们同样质疑不仅仅提纲规划许可没有和基于环境影响报告的指导性总体规划相关联，也表达出不同的分支机构和不同的使用者将会作为框架文件中提到的开发的一部分的设想。他们还质疑关于该项目涉及EIA的规划许可必须全面考虑到项目潜在的环境影响，然而下一阶段信息并不容易充分获得。

政府争辩，像原告要求的这种详细规划设想是非常困难的，尤其是寻求大项目的规划许可是完全不现实的。通告 15/88 允许提纲规划许可需要 EIA。Sullivan 法官却支持原告，并在 1999 年 5 月做出判决：

条件 1.3……把减缓措施和环境报告（除非获得同意）相联系，但是这些措施是环境影响报告对开发指导性总体规划的一个反馈。我承认，关于类似的提纲规划申请程序，我不希望采用废弃总体规划的办法，只要总体规划是被关联在一起的，例如加强附加条件的实施申请开发许可。如果在描述潜在的严重影响以前，可以提供指导性建筑面积（公顷数），潜在环境影响评价可能会受限于一定的影响范围数据。需要采取限制条件来确保任何授予许可的开发可以在控制的范围以内。

在目前的案例中最基本的困难是……提纲规划许可没有以任何形式和这些文件相联系。条件 1.7 和条件 1.11 省略了总体规划，不恰当地用框架文件来代替它提交和获得许可。实施条件 1.11 的原因解释了总体规划仅仅作为指导性目的提交，在决定项目的分布上的细节是不充分的。如果是不全面的，这将很难去满足评价法规附录 3 中第 2 条（a）所要求的，提供全面描述开发的条款。

Rochdale Saga 迅速以“Rochdale 2”案进行上诉，R vs. Rochdale 都市自治政府 ex-parte Milne（2000）[1]。在 Rochdale 1 结束以后，开发商重新申请了商业园区提纲规划许可，提交了开发的方案、开发的框架和总体规划。详细的景观设计被保留（没有提交），同时保留了区域内大部分的交通安排设置点。这次规划申请附带了一份全面的环境影响报告。

Rochdale MBC 在 1999 年 12 月再一次给予了拟建项目的有条件规划许可，但是这次其条件更注重与环境影响报告的联系。例如，条件 1.7 指出：

“这个地区的开发应该严格按照规划布局安排，包括开发框架文件在内的文件”，因为“拟建商业园区的布局是环境影响评价的一个主题，在布局上任何性质的改变都可能对没有评价的程序有影响。”

条件 1.11 要求“除非在本申请中还有其他条件需要满足，开发应该根据在申请时提交的环境影响评价报告里面的减缓措施实行。”

原告再次质疑规划许可，声称：机关规划申请有所改变，但是其依然没有提供“拟开发的描述”。其保留事宜——设计和进入的安排，可以对环境产生严重的影响，例如使用的材料或者是否开发包括一些特别显著的“地标式”建筑。他建议，为了满足附录 3 中的要求，开发必须描述得尽可能详细，来确保没有任何会产生环境影响的内容可以被省略，因此提纲

---

[1] www.bailit.org/ew/cases/EWHC/Admin/2000/650.html。

规划申请基本上与对 EIA 的要求不一致。在这种情况下，Sullivan 法官站在了规划官员一边，然后撤销了司法审查的申请：

特别是这类项目，例如一个工业地产的开发项目……是非常自然的，不应该被开端局限，但是它的期望值会随着时间的推移而有所发展，这里不需要提供“项目描述”不能认识到未来发生的事件的原因。重要的是环境评价程序应该全面地考虑对于灵活因素环境需求的开端的含义……是权力机关有权利来授予开发许可……决定是否拟建项目的困难和不确定因素的活跃程度，对潜在环境影响是不可以接受的。

任何重要的开发项目都将会受到一定程度和细节的制约，并非所有的制约都在规划许可要求以内。项目对空气的污染物排放、对水的污染和废弃物的产生，都将会在环境相关法律法规的要求下有一定控制。在评价一个项目的潜在环境影响的时候，环境报告的作者和地方规划机关有资格根据控制污染条件来相信其权力机构的判断及其控制污染的力度……同样的办法应该被采用在当地规划机关有权去批准其保留事宜。

“Rochdale 信封”对提纲规划申请以及对以后所有拟建项目的细节有不确定性的案例都有重要的借鉴意义：

①“空泛”的提纲规划申请并有一定保留事宜未批复的案例，可能非常不满足 EIA 的要求。

② 开发商可以保持灵活性，但是对于 EIA，他们需要考虑项目潜在因素的范围以及他们必须足够详细地对可能存在的环境影响做出评价和拟减缓措施。

③ 灵活性不能被滥用，不能给开发商一个提供不充分项目描述的理由。规划权力机关必须考虑其项目的性质，其必须有“全面的知识”来对环境的潜在影响做出判断。

④ 提纲规划申请应该详细明确地定义项目随着发展可能变化的参数，并以条件和义务的形式表明。这些需要与环境报告中的环境信息相关联——这些参数的范围就是“Rochdale 信封”。

⑤ 没有在提纲许可下执行保留事宜是非法的（IPC，2011）。

8.6.8 Themes 司法审查的结果

直到最近，EIA 法规规定的对主管部门决议的司法审查被法庭限制在相对较小的阐述主管部门的义务方面。然而，一系列知名的案例——包括 Baker 案例、Mellor 案例、Barker 案例和 SAVE 案例，暗示法庭目前对于 EIA 指令的态度更加有前瞻性和宽泛性。法律判决可以被总结成以下几点：

① 在 Kraaijveld（Dutch Dykes）C-72/95 案例中，EIA 指令被法院解释称有“宽泛的范围和广阔的目的”。

② 一个项目不可能仅仅因为其没有在 EIA 指令条目清单或者其他执行法规以内就自动被排除在 EIA 要求以外。另外，因为项目可以被表述成多种方式，对于 EIA 筛选目的来讲，比起其标签，还有更好的方式考虑项目的范围和目的。

③ 当项目某些方面具有不确定性时，EIA 必须在未决策之前提供足够的信息去考虑这些不确定性。

④ 为 EIS 提供对等信息不等于实际进行 EIA，因为 EIA 中公众参与也是一个重要的因素。

⑤ 法庭声明 EIS 不是必须考虑备选方案，即使是规划权力机构认为应该考虑，或备选方案对拟建项目来说将会有比较小的环境影响。

⑥ 执行规划条件使项目不再有严重的环境影响也不等于进行了 EIA。EIA 筛选（不包括减缓措施）和环境评价（可以包括减缓措施）的不同性质是非常重要的。"一个目的性强的方法可能如下：一个头脑开明的顾问对主管部门或者公众质疑其潜在重要环境影响，希望系统提供 EIA 数据来判断拟议的缓解措施将如何有效？如果是这样，EIA 是必需的"（McCracken 2010）。

8.7 EIA 的成本和效益

人们早期对强迫接受 EIA 的抵制是由于人们在观念上普遍认为 EIA 会增加额外成本，而且会耽误规划程序。EIA 拥护者认为，EIA 的实际效益完全超过了它的成本。这一节主要讨论英国 EIA 对不同参与者的成本和效益。

8.7.1 EIA 的成本

对于那些想获得规划许可的开发商来说，EIA 会使成本略有增加。一份 EIS 的费用占项目费用的 0.01%～5%，英国的费用平均水平是 0.1%～1%（GHK，2010）。Weston（1995）对顾问的调查表明，顾问编写一份完整的 EIS 的平均费用是 34000 英镑，编写几个章节需要 4 万英镑，编写一节需要 14750 英镑，这份调查结果本身也揭示了成本的多变性。在 1997 年，（原）DETR 建议把 35000 英镑作为在新法规下编制 EIS 的最合适的平均费用（DETR，1997b）；在 2010 年，DCLG 建议平均的费用大概为 9000 英镑（10 万欧元）。目前还不确定这个不同是否是由于 EIA 实际费用的增长，或者别的原因。

令人担忧的是，由于顾问之间的竞争和降价、不负责任的顾问数量增多，开发商趋向于接受成本最低的 EIS 等现象都会对顾问的时间、专业和资质产生限制，从而可能会影响 EISs 质量。专家指出"除了特大项目外，几乎所有项目都会限制预算——环境影响评价只能在可用预算范围内扩大"（Radcliff 和 Edward-Jones，1995）。然而，Fuller（1992）指出，减少开支对开发商的长期经营可能是不利的：

> 低质量的报告书通常是系统延期的主要影响因素。因为必须要及时找到原始报告上那些没列出或只是简单提及的问题的额外信息……因此，将环境影响评价的费用降低到全面工作所需水平之下，这通常是一种虚假经济。

主管部门的 EIA 费用更加难以计算，而且到现在为止，对 EIA 费用的计算主要是考虑采访而不是系统的方法。英国对大型项目的规划决议往往需要很长时间，有些甚至需要数年。对于没有进行环境影响评价的项目，往往需要更多、更持续的时间做出决定（DoE，1996），而且这些趋于更大、更复杂、更具有政治敏锐性。

早期的研究（Lee 和 Brown，1992）表明，大约一半被采访过的官员都认为 EIS 并没有影响到决策的时间，其他人的看法均匀地分布在"可能加快"或"可能减缓"两种观点之间。在之后的采访中（DoE，1996），很多规划官员认为处理 EIS 和规划申请是一回事，都是“工作的一部分”。估计对EIS及相关的咨询进行评审的时间从5个小时到6～8个工作月不等。处理EIS的规划官员通常都是发展控制小组中较高级别的人，所以劳务费通常比普遍的规划申请高。在有些情况下，规划官员需要雇佣顾问来帮助他们评审和建议EISs，进而使得他们的成本增加。

最近几项研究关注了决定规划许可的时间（例如国家审计办公室，2008；DCLG，2008；Ball等，2008），其中没有任何研究表明环境影响评价是造成延期的原因。只有Killian Pretty在170页评价规划申请（DCLG，2008）的研究中提到了EIA，然后暗示其长度应该缩短。一些顾问认为EIA延长了决策过程时间，增加了开发商的成本，而且可以使LPA有机会提出不合理的要求，迫使开发商提供项目更多的信息，“而这些信息与规划制定并无严格的联系”（Weston，1995）。其他人则认为EIA不会使进程减缓：“有组织的方法会使效率变得更高，在某些情况下让一些问题及早解决，因此EIS可以加快系统进程”（DoE，1996）。

《规划法案2008》主要的目的就是提高规划系统的效率，在能源、交通、水、废水和废物处理开发等领域，建立一套新的国家政策报告系统，也在一些重点基础设施建设领域建立一套平行决策系统，帮助其加快实施。早期迹象表明，预申请阶段，包括EIS的准备，这些项目的审批时间正在延长；而且，至今没有确定政府在一年以内对项目做出决策的目标是否会在现实中实现（见3.5.7节）。

在对20项案例的研究中发现，法定咨询机构在EIA上所花费的时间从4小时到1天半不等，而非法定咨询机构在EIA上所花费的时间从1小时到2周不等。尽管一些咨询者，如规划师，争辩说“这就是我们在这儿的目的”，其他人则建议：出于时间和资源的限制，顾问们必须优先处理他们所参与的开发项目（DoE，1996）。在今天窘迫的环境下，这可能是一个十分严重的问题。

8.7.2 EIA的效益

EIA的效益多数都无法进行定量分析，所以把效益和费用直接进行比较是不可能的。衡量EIA是否有助于减少项目对环境影响的最简单的方式就是判断项目是否因EIA进行了修改。早期关于EIA效益的研究（如Kobus和Lee，1993；Tarling，1991；Jones，1995）表明，半数情况下，都需要根据EIA对项目进行修改，并且大部分修改都是很重要的。最近欧盟（2009）基于欧盟成员国的数据而非英国的数据显示环境影响评价使得更多项目得到改进。

除了修改项目之外，环境影响评价可能也有许多其他效益。对环境顾问（Weston，1995）的调查表明，他们中的3/4认为，环境影响评价使得环境保护问题有所改善，主要通过在项目的早期设计中加入减缓措施等以及开发商对环境问题的高度重视。但是，其他的顾问认为该体系只是“充满陈词滥调的欺骗”（Weston，1995）。开发商觉得“编写EIS耗费了他们太多的时间与金钱，而在实际的规划决策中，大量EIA工作并未产生切实的效益”（Prichard等，1995）。Jones等（1998）发现只有1/5开发商和咨询者认为EIA不能产生任何效益。据推测，这种观点目前为止可能没有大的改变。

主管部门认为不论是项目还是环境都从EIA中受益匪浅（Jones，1995；Lee等，1994）。他们认为EIA能够聚集智慧、突出重大问题、降低不确定性、以系统的方式考虑环境影响、无需通过规划官员自己收集信息从而节约了时间，能及早地发现问题并把它们委托给相应的负责人（DoE，1996；Jones，1995；Prichard 等，1995；Weston，2002）。一个规划官员指出："当该体系第一次出现时，我相当怀疑，因为我们需要将全部因素都考虑进去。但现在我是这个程序的忠实支持者，它使我在早期阶段就能注意每个方面的细节"（DoE，1996）。

顾问一致认为 EIA 创立了一种更有条理的方法来处理规划申请，EIS 为他们"提供信息而无需他们自己四处查询"。但是，当有些问题在 EIS 中没有涉及时，顾问就会陷入同没有 EIS 时一样的处境：他们提出反对意见并不是因为影响很坏，而是因为他们得不到任何有关该影响的信息或者得不到对为何在评估时不考虑某些特殊的影响的解释。顾问认为 EIA 会给他们提供有关位置的数据，而这些数据他们自己无法收集，而且 EIA 会使参与者不必进行频繁会面就可以达成共识（DoE，1996）。

8.8 总结

总之，所有参与者都认为，英国实施的 EIA 有助于改进项目和保护环境，尽管该体系还应该进一步壮大，这样 EIA 至少是部分地实现了其主要目的。虽然 EIA 会耗费时间和金钱，但是也会带来很多切实的利益，如对项目的修正和得到更有见地的决策。当问及 EIS 到底是利是弊时，"从规划官员和开发商/咨询者那里得到的绝大多数答复认为 EIS 确实是有效用的。只有少部分参与调查者认为 EIA 是一个障碍"（Jones，1995）。

EIA 的某些阶段（特别是早期的范围确定、向相关参与者的有效咨询以及编制一份清晰公正的 EIS）都被认为能够产生切实的好处和成本效益（DoE，1996；IEMA，2011）。第 9 章给出了英国的一些主要案例研究，这些研究试图为 EIA 中的一些问题找到例证，并就问题的某些方面做出答复。第 12 章将就英国及英国以外的 EIA 的未来走向进行讨论。

问　题

以下问题旨在帮助读者更好地明确本章讨论的重点问题。

1. 回顾图 8.1 和图 8.2 显示的英国 EIS 活动的性质，你如何解释活动的变化模式？

2. 如图 8.3 所示，如何解释 2000 年以前和 2000 年以后，英国 EIS 活动的主要部门的变化？

3. 表 8.2 中，是否有迹象表明，筛查方面从业者的环境影响评价经验存在显著差异？

4. 比较和对比表 8.4 和表 8.5 中的信息。确定并寻求解释与环境影响评估范围界定活动中各种环境组成部分的重要性有关的任何内容差异。

5. 鉴于 8.4.3 节有关 EIS 质量决定因素的信息，如果你是第一次管理环境影响评估流程，将如何尝试优化所得到的 EIS 质量？

6. 以 Heathrow 第三跑道为例，你认为机场可协商和不可谈判的影响是什么（表 8.7）？

7. 你是否同意在伯克利案（8.6.1 节）中法官所认为的，如果没有环境影响评估不会对调查结果产生影响，而且不可能这样做时，就不应要求进行环境影响评估？为什么或者为什么不呢？

8. 用你的话来说，“罗奇代尔的信封是什么?”（8.6.7 节）和为什么它是重要的环评实践？

9. 8.7 节建议环境影响评价的成本可以量化，但其效益不能量化。你如何确定环境影响评估的成本是否超过其收益？你认为这样做合适吗？

Part 3

Practice

LET'S MAKE SURE I'VE GOT THIS RIGHT. WE GET TO KEEP SOME LIZARDS AND BLUE BUTTERFLIES ON OUR HEATHLAND, AND IN RETURN YOU GET TO BUILD 3,200 NEW HOUSES ON OUR GREEN BELT....

8

An overview of UK practice to date

8.1 Introduction

Part 3 considers EIA practice: what is done rather than what should be done. This chapter provides an overview of the first 20 years or so of UK practice since EC Directive 85/337 became operational. We develop this further with reference to particular case studies in Chapter 9. The case studies seek to develop particular themes and aspects of the EIA process raised in this and in earlier chapters (for example, on the treatment of alternatives, of public participation and on widening environmental assessment to also consider social and economic issues). The case studies are largely UK-based, and project-focused, although two cases of SEA are also included. Chapter 10 discusses international practice in terms of 'best practice' systems, emerging EIA systems and the role of international funding agencies in EIA, such as the World Bank.

These chapters can be set in the context of international studies on EIA effectiveness, whose results have been written up by Sadler (1996, 2012). In the 2011 study, Sadler sets three effectiveness tests, setting out an effectiveness 'triage' (three clearance bars): (1) enabling conditions (what must or should be done; legal and institutional framework and methodological realities); (2) state of practice (what is done; macro and micro level cases of good practice-what is the art of the possible); and (3) effectiveness and performance (what is the outcome; contribution to decision-making and environmental benefits). Sadler notes that these questions and the attendant techniques for investigating them must be seen in the context of the decision-making framework in which the relevant EIA system operates.

Chapter 8 broadly addresses Sadler's first two points in sequence. Section 8.2 considers the number, type and location of projects for which EIAs have been carried out in the UK, as well as where the resulting EISs can be found. Section 8.3 discusses the stages of EIA before the submission of the EIS and application for authorization. Section 8.4 addresses what has, to date, been the most heavily studied aspect of EIA practice, the quality of EISs. Section 8.5 considers the postsubmission stages of EIA, and how environmental information is used in decision-making by LPAs and inspectors. Section 8.6 considers legal challenges to EIA, many of which have informed the recent changes to UK legislation and guidance discussed at Chapter 3. Finally, section 8.7 discusses the costs and benefits of EIA as seen from various perspectives. Sadler's third point is partially addressed by government-published good-practice guides on EIA preparation and review (e.g. DETR 1999a, 2000; DCLG 2011a), which over time have introduced some policy changes in response to research findings regarding EIS and EIA effectiveness.

The information in this chapter was correct at the time of writing in mid-2011; it will obviously change as more EISs are carried out.

8.2 Number and type of EISs and projects

This section considers how many, and for what types of projects, EISs have been produced. It concludes with a brief review of where collections of EISs are kept. UK EIS collections, and databases on EISs, are fragmented (see Section 8.2.3). Analysis is further complicated by several problems. First, some projects fall under more than one schedule classification; for example mineral extraction schemes (Schedule 2.2) that are later filled in with waste (Schedule 2.11), or industrial/residential developments (Schedule 2.10) that also have a leisure component (Schedule 2.12). Second, the mere description of a project is often not enough to identify the regulations under which its EIA was carried out. For instance, power stations may fall under Schedule 1.2 or 2.3 (a), depending on size. Roads may come under highways or planning regulations, depending on whether they are trunk roads or local highways. Third, many EISs do not mention when, by whom or for whom they were prepared. Fourth, locational analysis after 1995 is complicated by local government reorganization and many changes in the nature and boundaries of authorities in England, Scotland and Wales. All these factors affect the analysis. This chapter is based primarily on information from DCLG (2011b), IEMA (2011), supplemented by older information from Wood and Bellanger (1998) and Wood (1996, 2003).

8.2.1 Number and broad location of EISs

Between the mid-1970s and the mid-1980s, approximately 20 EISs were prepared annually in the UK (Petts and Hills 1982). After the implementation of Directive 85/337, this number rose dramatically and, despite the recession, about 350 EISs per year were produced in the early 1990s; but, as can be seen from Figure 8.1, this number began to drop in the mid-1990s partly as a result of a fall in major development activity under the planning regulations. However, the numbers quickly recovered in the late 1990s and, as noted in Chapter 3, there were over 600 per year for several years after the implementation of the 1997 amendments to the Directive. This probably reflected many factors-more projects, included in the amended Directive, a stronger UK economy and concern by developers and LPAs about certain court judgements involving the EIA Directive. By the end of 2008, over 9,000 EISs had been prepared, with approximately 70 per cent produced under the Planning Regulations for England, Wales and Scotland. The remainder are for projects in Northern Ireland and, more significantly, for projects under the other consent procedures (e.g. highways, forestry) discussed in Chapter 3. However the fall in numbers since 2005 illustrates that EIS activity can perhaps be seen as an interesting measure of the economic fortunes of a country.

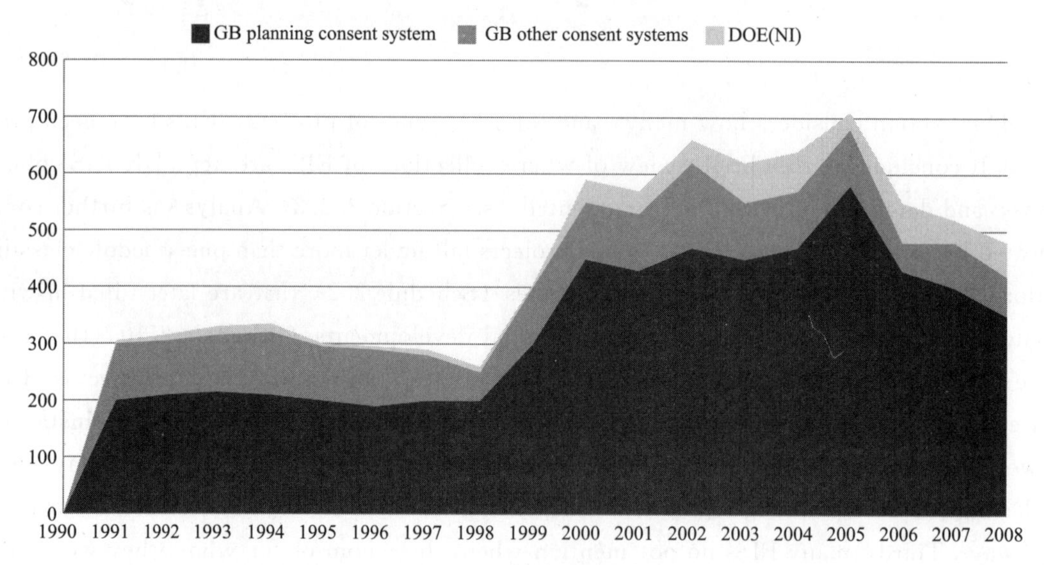

Figure 8.1 EISs prepared in the UK (1991—2008)

Source: DCLG (2011b)

In parallel with the increase in the number of EISs, the participants in EIA have become increasingly familiar with the process. Surveys of UK local authorities carried out by Oxford Brookes University in the mid-1990s showed that over 80 percent of LPAs even then had received at least one EIS. On average, strategic-level authorities (county and regional councils and national park authorities) had received 12 EISs and local-level authorities (district, borough, metropolitan boroughs and development corporations) had received four. Surveys of environmental consultants (e. g. Radcliff and Edward-Jones 1995; Weston 1995) found that about one-third of the consultancies surveyed had prepared 10 or more EISs. As noted, the total number of EISs is now over 9,000, compared with approximately 2,500 by the end of 1995, and LPA and consultancy activity and experience with the process has continued to grow accordingly. Figure 8. 2 shows the distribution of EISs by national authority. With only about 10 per cent of the UK population, the EIS activity in Scotland is often much higher than the UK average; recent activity includes many Scottish wind farm developments.

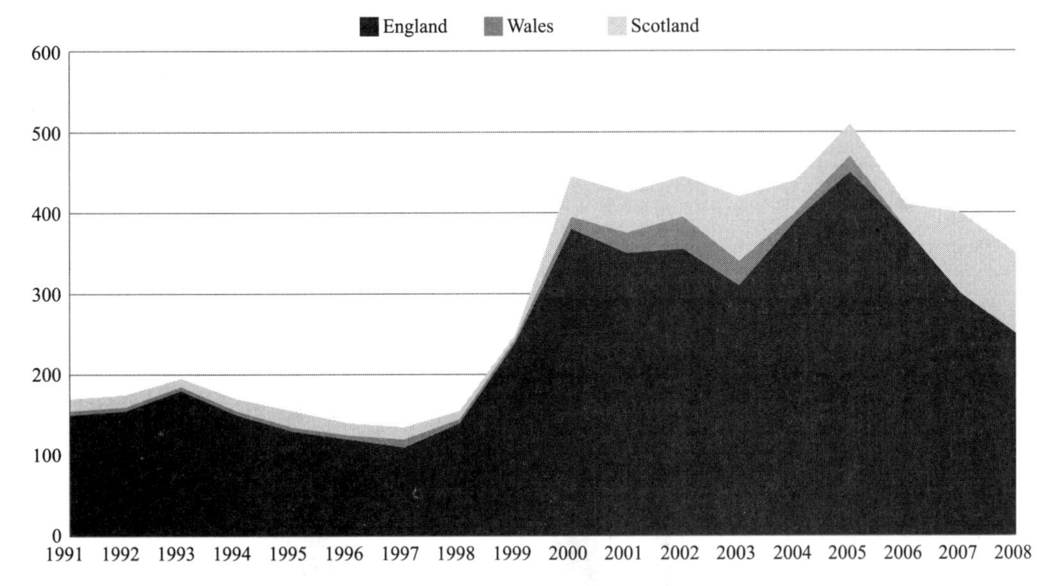

Figure 8. 2 Country distribution of EISs produced under the GB planning consent systems (including England, Wales and Scotland)

Source: DCLG (2011b)

8. 2. 2 Types of projects

Figure 8. 3 (a) shows the types of projects for which EISs were prepared in the early years of EIA in the UK. The largest numbers were for project types in waste (largely land-fill/raise projects, wastewater or sewage treatment schemes and incinerators), urban/retail developments, roads, extraction schemes and energy projects (Wood and Bellanger 1998). In contrast, data for 2004—2010 (for England only) highlights the predominance of urban/retail development projects. Figure 8. 3 (b) (also for England) illustrates some regional variations-for example between project types in the NE and SE of the country.

Table 8. 1 shows the distribution of EISs produced under other consent systems. There are basically three groups of projects relating to transport, agriculture and fisheries, and energy. The highway group is a large group, although numbers fell away around the turn of the century, but have increased again since. In contrast, forestry, land drainage and fish farming projects have recently fallen back from much higher numbers at the turn of the cen-

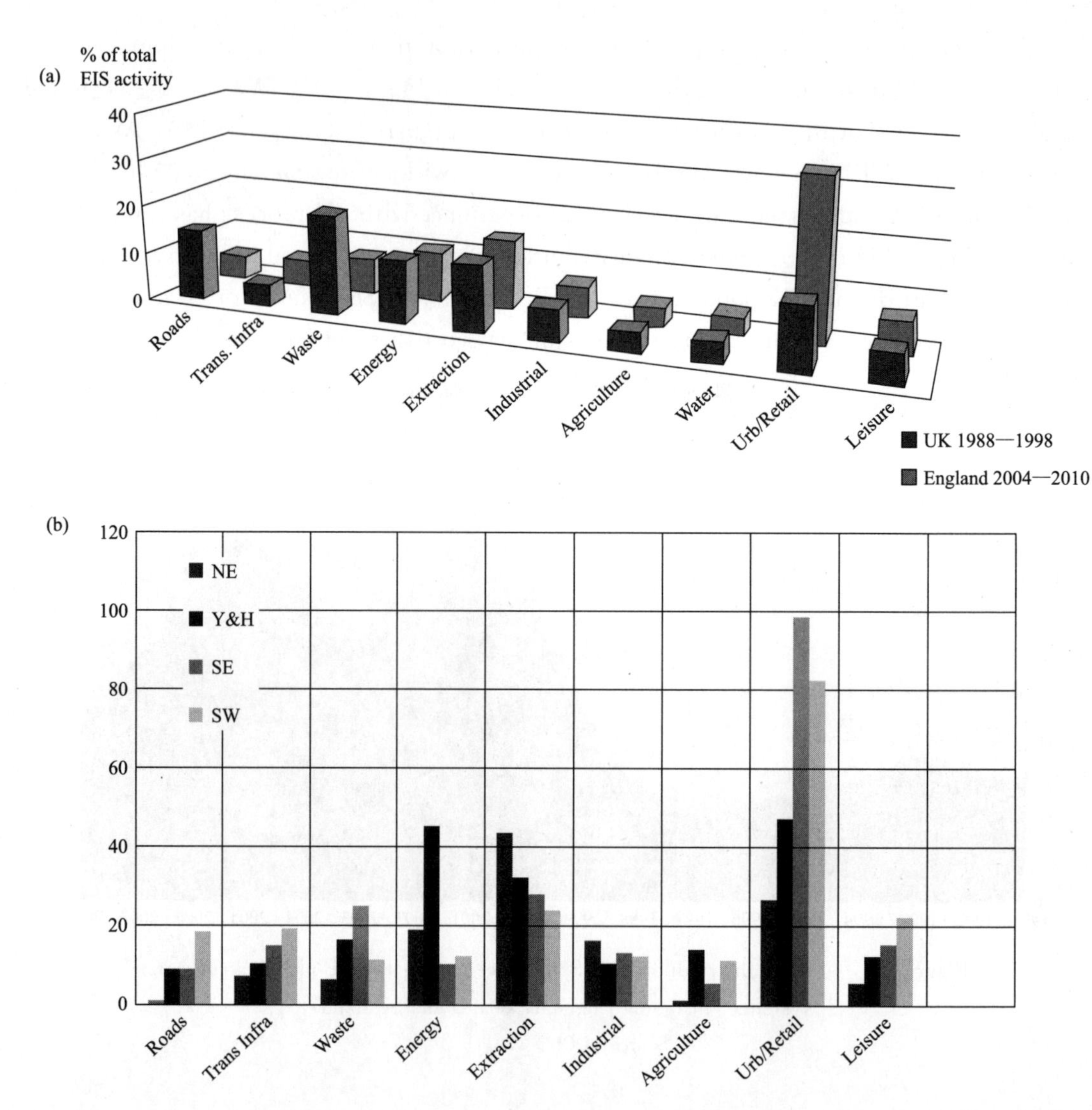

Figure 8.3 Trends in EISs for particular project types. (a) By project type % of total EIS activity-comparing UK (1988—1998) and England (2004—2010); (b) By project type-comparing numbers of EISs produced for particular English regions (2004—2010)

Source: (a) Wood and Bellanger 1998; (a and b) adapted from DCLG Library database 2011

tury. Electricity and pipeline works is a major category, boosted by the more recent addition of gas pipelines and offshore wind farms to the category; there have been about 600 EISs in total in this category. In contrast, and as expected, there have been only a handful of EISs for the decommissioning of nuclear reactors, following the introduction of legislation in 1999; but more will follow as the UK's ageing reactors reach the end of their operational life.

In the first few years following the implementation of Directive 85/337, 40 per cent of EISs were produced for the public sector and 60 per cent for the private sector (Wood 1991). The percentage of private sector projects has since increased considerably owing to privatization, but much of this was offset in the 1990s and early 2000s by the heavy government investment in-and consequently EISs for—new roads. A particularly interesting subset is that of those EIAs for which one agency acts as both the project proponent and the competent authority (e. g. the Highway Agency for roads).

Table 8.1 EISs produced under other consent systems (highways, forestry, electricity, etc.)

Year	EISs received							
	Highways	Harbour works	Transport and works	Forestry	Land drainage	Fish farming	Nuclear reactor decommissioning	Electricity and pipelines works (including gas pipelines and offshore wind farms)
1991	43	0		8	10	0		29
1992	38	2		9	11	0		23
1993	52	3	3	30	14	0		12
1994	61	5	3	18	9			23
1995	27	7	1	15	22	0		19
1996	19	8	3	12	29	0		17
1997	17	9	4	13	22	1		25
1998	6	1	10	13	20	2		12
1999	5	6	9	11	27	10	0	21
2000	3	12	8	7	17	13	0	42
2001	8	7	2	10	31	20	0	48
2002	21	4	7	8	37	32	2	54
2003	12	1	8	10	8	17	0	42
2004	17	2	6	3	3	11	2	53
2005	28	3	2	5	17	12	2	64
2006	24		1	1	0	0	1	27
2007	30		3	2	0	0	1	25
2008	11		3	2	0	0	2	42
Total	412	70	73	177	273	118	10	578

Source: DCLG 2011b

8.2.3 *Sources of EISs*

Copies of EISs received by English LPAs are forwarded to the Environmental Assessment Division of the Department of Communities and Local Government (DCLG) library in London once the application has been dealt with. However, this process can be a long one. The DCLG library is open to the public by appointment; photocopies can be made on the premises. In Wales, planning EISs are forwarded to the Welsh Assembly. In Scotland, all EISs are sent to the Scottish Assembly, while in Northern Ireland they are sent to the Northern Ireland DoE. Other government agencies, such as the Highways Agency, also hold collections and lists of the EISs that fall under their jurisdiction. These collections are, however, generally not publicly available, although limited access for research purposes may be allowed.

In addition to government collections, EIS collections can also be found in several universities (e. g. at Oxford Brookes University, Manchester University and University of East Anglia), and at the Institute of Environmental Assessment and Management (IEMA). Some EISs can also be found on the Internet, although this access may be fleeting and only as long as there is an active planning application. The Institute of Environmental Management and Assessment (IEMA), based in Lincoln, has a substantial collection of EISs, which are available by pre-arrangement with institute staff, but primarily for corporate bodies. The EIA Centre at the University of Manchester keeps a database of EISs and EIA-related literature: its collection of EISs is, like its database, open to the public, by appointment. Oxford Brookes University's collection of approximately 1,000 EISs is open to the public, by appointment, and photocopies can be made on the premises. Other organizations, such as the Royal Society for the Protection of Birds, the Institute of Terrestrial Ecology, Natural England and the Campaign to Protect Rural England, as well as many environmental consultancies, also have limited collections of EISs, but these are generally kept by individuals within the organization for in-house use only, and are not available to the public. EISs are also increasingly being made available on the websites of local authorities and/or developers.

The difficulty of finding out which EISs exist, and their often prohibitive cost, can make the acquisition and analysis of EISs arduous. Various organizations (e. g. IEMA, the University of Manchester and Oxford Brookes University), have called for one central repository for all EISs in the UK. One positive new development was the launch in Spring 2011 of the IEMA's Quality Mark system. To acquire the Quality Mark, organizations will have to sign up to seven EIA commitments (relating to EIA team capabilities, EIA content, EIA presentation etc.). One of the indirect benefits of the scheme is that it will gather a substantial proportion of the UK's annual EIS output and make their non-technical summaries (NTSs) available to search on line, which will provide a valuable resource for research and practice.

8.3 The pre-submission EIA process

This is the first of three sections that discuss how EIAs are carried out in practice in the UK. It focuses on some of the pre-EIS submission stages of EIA, namely screening, scoping and pre-submission consultation.

8.3.1 *Screening*

Underpinning any analysis of the implementation of EIA in the UK are the requirements of the EC and UK government legislation. Under the original legislation, competent authorities in the UK were given wide discretion to determine which Schedule 2 projects require EIA within a framework of varying criteria and thresholds established by the 40-plus regulations and additional guidance. Generally, this screening process worked quite well (CEC 1993). However, some specific problems did arise regarding screening in the UK. For example, because of the largely discretionary system for screening, LPAs often—about half of the time—required an EIS to be submitted only after they had received a planning application (DoE 1996). For the same reason, screening requirements varied considerably between competent authorities. In the early days of EIA, the decision not to require an EIA had often been taken by junior members of staff who had never considered the need for an EIA, or who thought (incorrectly) that no EIA was required if the land was designated for the type of use specified in the development plan, or if the site was being extended or redeveloped rather than newly developed. Similarly, different government regional offices gave different decisions on appeals for what were essentially very similar developments (Gosling 1990).

The screening criteria established by the amendments to the Directive sought to reduce these problems post-1999 (see Sections 3. 4 and 4. 3). A government-sponsored study, undertaken by the Oxford Brookes University's IAU (IAU 2003), provided some research evidence on the nature and characteristics of LPA screening decision-making under the T&CP (EIA) Regulations 1999. The research, based on survey responses from over 100 LPAs in England and Wales in 2002, sought information on frequency of screening activity, on main considerations in the LPA decision whether or not an EIA should be undertaken and on the importance of different screening criteria. For screening activity in general, and for LPAs in total, the indicative thresholds were identified as the main consideration in screening decision-making (44 per cent of LPAs). The criteria of Schedule 3 and associated guidance (project is a major development of more than local importance, is in a sensitive location or will have complex/hazardous environmental effects), which require a greater degree of professional judgement, were noted second most frequently (35 per cent of LPAs). Tables 8. 2 and 8. 3 relate to views on the LPAs' (then) most recent single screening decision. Using regulations and thresholds is seen as the most effective approach overall, but professional judgement is an equally important approach among the more experienced LPAs.

Table 8. 2 Most effective approach in most recent screening decision

Screening approach considered most effective	LPA(%) n=56	< 5 EIAs 56 (n=26)	>5 EIAs (n=30)
Consultation with own organization	3. 6	7. 7	0. 0
Community consultation	12. 5	11. 5	13. 3
Asked for screening direction from Secretary of State	0. 0	0. 0	0. 0
Followed screening guidance in local development plan	1. 8	3. 8	0. 0
Followed guidance in other plans/policies	7. 1	11. 5	3. 3
Used regulations and thresholds as guide	35. 7	38. 5	33. 3
Consulted examples of other similar projects	1. 8	0. 0	3. 3

Continued

Screening approach considered most effective	LPA(%) n=56	< 5 EIAs 56 (n=26)	>5 EIAs (n=30)
Used professional judgement/experience	26.8	19.2	33.3
Used checklist to identify possible impacts	1.8	0.0	3.3
Used other formal technique	1.8	3.8	0.0
Own standard approach	0.0	0.0	0.0
Likely controversy of project	0.0	0.0	0.0
Other	7.1	3.8	10.0

Source: IAU 2003

Table 8.3 Importance of issues in most recent screening decision

Issue(n = 97)	Very important(%)	Important(%)
Size/scale of project	47	40
Proximity to receptor	44	43
Nature of project	42	32
Traffic/access impacts	33	32
Ecological impacts	32	31
Emissions	31	31
Landscape impacts	26	35
Cumulative impacts	20	26
Economic impacts	6	24
Social impacts	5	22
Controversy/concern	9	16
Risk of accidents	5	10
Other	3	1

Source: IAU 2003

The most important factors in the screening decision are the size and scale of the project with 87 per cent of LPAs indicating that these are 'important' or 'very important'. Proximity to sensitive environmental receptors (87 per cent) and the nature of the project (74 per cent) are the next most important factors. At the other extreme, only 15 per cent indicated that risk of accidents was important or very important. The main constraints on screening decision-making were identified as lack of resources (45 per cent), time-frame constraints (44 per cent), lack of clarity of the regulations (33 per cent) and uncertainty over baseline data, project characteristics, etc. (32 per cent). Overall, the findings show that while thresholds are clearly important in the screening decision, they are often conditioned by professional judgement. In other words, in themselves, they do not provide sufficient justification for a screening decision (Weston 2000). This is supported by the 2011 guidance on EIA, which notes that the basic question to be asked is: 'Would this particular development be likely to have significant effects on the environment?' For the majority of development proposals, it will be necessary to consider the characteristics of the development in combination with its proposed location, in order to determine whether there are likely to be significant environmental effects. (DCLG, 2011a)

As noted in Chapter 3, screening is still considered a problematic area in several EU Member States, including the UK. A survey of UK practitioners reported in a recent IEMA study (2011) found that while most practitioners agreed that the UK's screening process was an effective tool to ensure that only projects likely to have significant environmental effects are subject to an EIA, there was still serious concern that EIA had not been required for some Schedule 2 projects with what they considered to be likely significant environmental effects. Over time, this has led to a number of important legal challenges, some of which are discussed in Section 8. 6. This might partly explain the lower level than might be expected of EIA activity for a country of England and Wales population size, and in comparison with other EU Member States with similarly large populations.

Progress has been made over the last decade, through a combination of some or all of the following: simplified procedures for small-scale development applications, adoption of thresholds, regulatory initiatives against the 'salami slicing' of projects (into sub-projects that then fall below threshold levels), and improved guidance on the application of screening procedures. The Scottish Government has introduced an example methodology for undertaking case by case screening (Scottish government 2007), although DCLG (2011a) guidance still refers to screening thresholds. There has also been progress on the automation of screening procedures. For example, in Denmark, an electronic model has been developed for intensive animal farming projects in which the developer simply, by inserting the required data in a calculation sheet, may get a clear picture of whether the proposed project will result in an EIA procedure or not. For the UK, see Rodriguez-Bachiller and Glasson (2004), for the Screen Expert System, and also for a Scope Expert System, developed at Oxford Brookes University.

8. 3. 2 *Scoping and pre-submission consultation*

Competent authorities also have much discretion to determine the scope of EIAs. As discussed in Chapter 3, the original Directive 85/337's Annex Ⅲ was interpreted in UK legislation as being in part mandatory and in part discretionary. A survey of early EIS output (Jones *et al*. 1991) showed that although the mandatory requirements of the legislation were generally carried out, the discretionary elements (e. g. the consideration of alternatives, forecasting methods, secondary and indirect impacts, and scoping) were, understandably, carried out less often. Subsequent studies showed that although early scoping discussions between the developer, the consultants carrying out the EIA work, the competent authority and relevant consultees were advised in government guidance and were considered increasingly vital for effective EIA (Jones 1995; Sadler 1996), in practice, presubmission consultation was carried out only sporadically. For instance, a survey of environmental consultants (Weston 1995) showed that only 3 per cent had been asked to prepare their EISs before site identification, and 28 per cent before detailed design. LPAs were consulted by the developer before EIS submission in between 30 and 70 per cent of cases, although this subsequently increased (DoE 1996; Lee *et al.* 1994; Leu *et al.* 1993; Radcliff and Edward-Jones 1995; Weston 1995).

As noted in Chapters 2 and 3, the amended EC 1997 Directive and subsequent UK regulations have raised the profile of scoping in the EIA process. The ODPM study noted earlier (IAU 2003; Wood 2003) also carried out research on the nature and characteristics of scoping activities by LPAs, consultants and statutory consultees. Nearly 75 per cent of the LPAs had been involved in producing scoping opinions. All three sets of stakeholders ranked very high the preliminary assessment of characteristics of the site, consideration of mitigation and consideration of impact magnitude in formulating the scoping opinion/report. Similarly, all ranked professional judgement and consultation within their own organization as key approaches to impact identification; use of legal regulations and thresholds were also very important for LPAs and consultancies, but much less so for statutory consultees. Table 8.4 shows the issues of most concern in the most recent scoping project (at the time of the survey) for each group of participants in the process. There is considerable similarity in emphasis between the LPAs and consultancies, with traffic/transport, landscape/visual and flora/fauna issues ranking particularly high. For statutory consultees, there are some similarities, but other environmental issues (e.g. other emissions, waste disposal) come more into play. Climatic factors and risk of accidents (health and safety) ranked surprisingly low, reflecting perhaps uncertainty about what were then seen as more long-term and less predictable issues.

Table 8.4 Ranking of issues of major concern in most recent project scoping opinion/report (% ranking of major concern)

	LPAs (n=78)	Consultancies (n=98)	Statutory consultees (n=28)
Social issues	10	20	11
Culture/heritage	19	29	18
Economic	21	22	4
Flora/fauna	46	50	68
Soil	18	20	21
Air quality	35	21	14
Noise and vibration	42	45	11
Other emissions	26	31	39
Climatic factors	6	5	7
Waste disposal	13	19	32
Water resources	37	31	43
Geo-technical issues	18	21	29
Landscape/visual	58	57	32
Traffic/transportation	47	55	11
Risk of accidents	5	9	7
Inter-relationships of above	18	17	43
Others	3	9	4

Source: IAU 2003

Considerable experience has been gained with screening and scoping. After initial hiccups, the screening process now seems to be relatively well accepted and has been refined after the 1999 amendments. Scoping is generally considered to be a very valuable and cost-effective part of EIA by all those concerned, and again following the 1999 amendments, has increased in significance in the UK EIA process although, in comparison with 10 of the 12 new EU Member States, for example, it is still not mandatory in the UK—but it is clearly encouraged by good practice guidance. The good practice guidance on screening and scoping by the IPC for NSIPs has already been set out in Section 3. 5. 7. However, there is still a real concern that a combination of factors, including risk aversion, poor planning and commercial reality, may be leading to over-broad and insufficiently focused scoping activity—with 'everything plus the kitchen sink' included to be on the safe side. Table 8. 5 shows the frequency of inclusion of environmental topic chapters in a sample of 100 UK environmental statements, submitted in the UK in 2010, which were reviewed by the IEMA (2011).

Table 8. 5 The frequency of inclusion of environmental topic chapters in a sample of 100 UK environmental statements from 2010

Environmental topic(bold text denotes that the topic is included in either Article 3 or Annex Ⅳ of the EIA Directive)	Occurrence rate in 100 UK Environmental Statements from 2010
Ecology(**flora and fauna**)	92%
Noise(and vibration)	92%
Water	90%
Landscape/townscape/visual analysis	88%
Transport	88%
Cultural/built **heritage**(**inc. archaeology**)	82%
Soil and land quality/ground conditions	81%
Air	79%
Socio-economic	64%
Cumulative effects(**interactions**/**inter-relationships**)	46%
Waste	28%
Climate change	17%
EMP, summary -residual effects and mitigation	17%
Population/**human beings**	13%
Amenity, access, recreation, rights of way	12%
Daylight/sunlight	11%
Material assets	10%
Micro-**climate**/wind	9%
Electronic interference(radio and TV)	7%
Sustainability	7%
Public health	6%
Lighting	5%
Aviation	5%
Geomorphology and coastal processes	4%
Energy	3%
Shadow flicker	3%

Source: IEMA 2011

The review found that over 90 per cent included chapters assessing ecology, noise and water effects (Table 8. 5), with a further five environmental issues being found to have their own chapter in nearly 80 per cent of all the ES reviewed. Overall this research found that the average UK environmental statement in 2010 had 9.63 environmental topic chapters. While IEMA's analysis did not extend to assessing whether the inclusion of each chapter was appropriate it is clear that current practice in scoping rarely leads to an assessment focused on a handful of key environmental issues (IEMA 2011).

8.4 EIS quality

As we mentioned in Section 8.1, the preparation of high-quality EISs is one component of an effective translation of EIA policy into practice. Submission of EISs of the highest standard from the outset reduces the need for costly interaction between developer and competent authority (Ferrary 1994), provides a better basis for public participation (Sheate 1994), places the onus appropriately on the developer and increases the chance of effective EIA overall. That said, the entirety of environmental information is also important, and the advice of statutory consultees, the comments of the general public and the expertise of the competent authority can help to overcome the limitations of a poor EIS (Braun 1993). This theme is considered further in Section 8.6 on judicial review.

8.4.1 *Academic studies of EIS quality*

Environmental impact statement quality in the UK is affected by the limited legal requirements for EIA and by the fact that planning applications cannot be rejected if the EIS is inadequate, that some crucial steps of the EIA process (e.g. public participation and monitoring) are not mandatory, and that developers undertake EIAs for their own projects. This section first considers the quality of EISs produced in the UK, based on several academic studies. It continues with a brief discussion of other perceptions of EIS quality, since competent authorities, statutory consultees and developers require different things from EIA and may thus have different views of EIA quality. It concludes with a discussion of factors that may influence EIS quality.

A range of academic studies of EIS quality were carried out in the first 10 years of the EIA Directive's implementation, typically using the criteria of Lee and Colley (1990) (see Appendix 4). Based on these criteria, EISs were divided into 'satisfactory' (i.e. marks of A, B or C) and 'unsatisfactory' (D or below). Table 8.6, which summarizes some of the findings, shows that EIS quality increased over time, but only after dismal beginnings. Generally, the description of the project, and communication and presentation of results, tended to be done better than the identification of key impacts, alternatives and mitigation.

Some studies focused on specific project types: for instance Kobus and Lee (1993) and Pritchard *et al.* (1995) reviewed EISs for extractive industry projects, Prenton-Jones for pig and poultry developments (Weston 1996), Radcliff and Edward-Jones (1995) for clinical waste incinerators, Davison (1992) and Zambellas (1995) for roads, and Gray and Edward-Jones (1999) for forestry projects. Other studies analysed the quality of specific EIS environmental components, for instance, landscape/visual (e. g. Mills 1994) and socio-economic impacts (e. g. Hall 1994). These studies also broadly suggested that EIS quality was not very good, but improving.

Table 8.6 Examples of aggregated EIS quality (percentage satisfactory*)

	Lee and Colley 1990	Wood and Jones 1991	Lee and Brown 1992	Lee *et al.* 1994	Jones 1995	Barker and Wood 1999	DoE 1996
Sample size	12	24	83	47	40	24	50
1988—89	25	37	34	17			36
1989—90			48				
1990—91			60	47			
1991—92					'just over half'	58	
1992—93							60
1993—94							
1994—95						66	
1995—96							

* Satisfactory means marks of A, B or C based on the Lee and Colley criteria (1990 or 1992)

We are not aware of similarly formal, large scale recent studies of UK EIS quality, but indications are that EIS quality has continued to improve since then, albeit not consistently:

> Consultants working on large or controversial projects have had to improve the quality of their work substantially in recent years … That has been one benefit of increased scrutiny of ESs on the part of statutory stakeholders and groups representing local residents and the public … but the consensus seems to be that although the quality of ESs prepared for large and/or controversial projects has gone up, overall quality remains highly variable. 'Patchy' is the word most frequently used to describe current standards, both in terms of the quality of ESs themselves and the scrutiny they are given by local authorities. (ENDS Directory 2007)

8.4.2 Quality for whom?

These findings must, however, be considered in the wider context of 'quality for whom?' Academics may find that an EIS is of a certain quality, but the relevant planners or consultees may perceive it quite differently. For instance, the DoE (1996) study, Radcliff and Edward-Jones (1995), and Jones (1995) found little agreement about EIS quality between planners, consultees and the researchers; the only consistent trend was that consultees were more critical of EIS quality than planners were.

In interviews conducted by the Impacts Assessment Unit (DoE 1996), planning officers thought EISs were intended to gain planning permission and minimize the implication of impacts. Just over 40 per cent felt that EIS quality had improved, although this improvement was usually only marginal. Most of the others felt that this was difficult to assess when individual officers see so few EISs and when those they do see tend to be for different types of project, which raises different issues. A lack of adequate scoping and discussion of alternatives was felt to be the major problem. EISs were seen to be getting 'better but also bigger'. Some officers linked EIS quality with the reputation of the consultants producing them, and believed that the use of experienced and reputable consultants is the best way to achieve good quality EISs.

Statutory consultees differed about whether EIS quality was improving. They generally felt that an EIS's objectivity and clear presentation were important and were improving, yet still wanting. LPAs, developers, consultants and consultees generally thought the key EIS criteria of comprehensiveness, objectivity and clear information were improving but still generally not good. Developers and consultants linked EIS quality with ability to achieve planning permission. Consultants felt that developers were increasingly recognizing the need for environmental protection and starting to bring in consultants early in project planning, so that a project could be designed around that need. One reason for this improvement may be that pressure groups were becoming more experienced with EIA, and thus had higher expectations of the process (DoE 1996).

8.4.3 *Determinants of EIS quality*

Several factors affect EIS quality, including the type and size of a project, and the nature and experience of various participants in the EIA process. Certain types of *project* have been associated with higher quality EISs. For instance, Schedule 1 projects, which generally have a high profile and attract substantial attention and resources, are likely to have better EISs (Lee and Brown 1992; DoE 1996). This trend is likely to continue under the rigorous pre-application requirements of the UK's National Infrastructure Unit of the Planning Inspectorate.

Regarding the *nature and experience of the participants* in the EIA process, EISs produced in-house by developers were generally of poorer quality than those produced by outside consultants: the DoE (1996) study, for instance, showed that EISs prepared in-house had an average mark of D/E, while those prepared by consultants averaged C/D, and those prepared by both B/C. Lee and Brown's (1992) analysis of 83 EISs concluded that 57 per cent of those prepared by environmental consultants were satisfactory, compared with only 17 per cent of those prepared in-house. Similarly, EISs prepared by independent applicants have tended to be better (C/D) than those prepared by local authorities for their own projects (D/E) (DoE 1996).

The experience of the developer, consultant and competent authority also affects EIS quality. For instance, Lee and Brown (1992) showed that of EISs prepared by developers (without consultants) who had already submitted at least one EIS, 27 per cent were satisfactory, compared with 8 per cent of those prepared by developers with no prior experience; Kobus and Lee (1993) cited 43 and 14 per cent respectively. A study by Lee and Dancey (1993) showed that of EISs prepared by authors with prior experience of four or more, 68 per cent were satisfactory compared with 24 per cent of those with no prior experience. The DoE (1996) study showed that of the EISs prepared by consultants with experience of five or fewer, about 50 per cent were satisfactory, compared with about 85 per cent of those prepared by consultants with experience of eight or more. EISs prepared for local authorities with no prior EIS experience were just over one-third satisfactory, compared with two-thirds for local authorities with experience of eight or more (DoE 1996). This suggests that EIS quality should improve over time simply by dint of practitioners gaining more experience.

Other determinants of EISs' quality include the availability of EIA guidance and legislation, with more guidance (e. g. DETR 2000; DoT 1993; local authority guides such as that of the Essex Planning Officers' Association 2007) leading to better EISs; the stage in project planning at which the development application and EIA are submitted, EISs for detailed planning applications generally being better than those for outline applications; and issues related to the interaction between the parties involved in the EIA process, including commitment to EIA, the resources allocated to the EIA and communication between the parties.

Environmental impact statement length also shows some correlation with EIS quality. For instance, Lee and Brown (1992) showed that the percentage of satisfactory EISs rose from 10 for EISs less than 25 pages long to 78 for those more than 100 pages long. In the DoE (1996) study, quality was shown to rise from an average of E/F for EISs of less than 20 pages to C for those of over 50 pages. However, as EISs became much longer than 150 pages, quality became more variable: although the very large EISs may contain more information, their length seems to be a symptom of poor organization and coordination. IEMA (2011) found the main text of many EISs to be more than 350 pages long, with those for nationally significant infrastructure projects typically being longer still. It noted that:

> many practitioners are frustrated with the length of ESs and have concerns about the value these documents give to consenting authorities, let alone wider groups such as the public… 25% of respondents [to an online survey] consider that the current length of ESs reduces the value of the assessment's findings to all audiences, even those with specialist environmental knowledge, such as the Environment Agency. Respondents indicate that this situation is worse in relation to less technical audiences. More than two fifths of respondents (42%) suggested that long ESs regularly reduce the value of an EIA's findings to stakeholders in non-statutory bodies and nearly two thirds (66.5%) believing the current length of the documents reduces the value of EIA to local communities. (IEMA, 2011)

8.5 The post-submission EIA process

After a competent authority receives an EIS and application for project authorization, it must review it, consult with statutory consultees and the public and come to a decision about the project. This section covers these points in turn.

8.5.1 *Review*

Planning officers generally see little difference between projects subject to EIA and other projects of similar complexity and controversy: once an application is lodged, the development process takes over. Competent authorities usually review EISs using their own knowledge and experience to pinpoint limitations and errors: the review is carried out primarily by reading through the EIS, consulting with other officers in the competent authority, consulting externally and comparing the EIS with the relevant regulations (DoE 1996).

Despite the ready availability of the Lee and Colley (1992) review criteria, only about one-third of local authorities use any form of review methods at all, and then usually as indicative criteria, to identify areas for further investigation, rather than in a formal way. About 10—20 per cent of EISs are sent for review by external consultants or by IEMA; but even when outside consultants are hired to appraise an EIS, it is doubtful whether the appraisal will be wholly unbiased if the consultants might otherwise be in competition with each other. There are also problems involved in getting feedback from the reviewing consultants quickly enough, given the tight timetable for making a project determination (DoE 1996). An innovative approach being used by some developers requires consultants who are bidding to carry out an EIA to include as part of their bid an 'independent' peer reviewer or 'critical friend' who will check and guarantee the quality of the consultants' work.

In the early days of EIA implementation in the UK, planning authorities required additional EIA information in about two-thirds of cases (e.g. Jones 1995; Lee *et al.* 1994; Weston 1995). This was usually done informally, without invoking the regulations. Circular 2/99 supports this by noting that '... if a developer fails to provide enough information to complete the ES, the application can be determined only by a refusal' (Regulation 3, DETR 1999b). We are not aware of similar more recent studies, although it is clear that planning authorities continue to require additional EIA information where they feel that the EISs do not provide adequate information.

8.5.2 *Post-submission consultation and public participation*

Competent authorities are required to send EISs to statutory and non-statutory consultees, and make them available to the public for comment. Public participation in the UK EIA system is typically limited to a few weeks following EIS submission, and notice of an EIS is normally in the form of an advertisement in a local newspaper, a site notice, and notification of neighbours. Increasingly, EISs are placed on the Internet, as well as being made available in hard copy form in council offices or libraries, and being available for purchase at a 'reasonable cost' as required by legislation.

Planning officers 'place great reliance on the consultees to review, verify and summarize at least parts of ESs' (Kreuser and Hammersley 1999). Local interest groups are often particularly active at this stage, reviewing and commenting on EIS quality as well as the planning application. Where the EIS contains insufficient information about a specific environmental component, competent authorities may put the developer and consultee in direct contact with each other rather than formally require further information themselves (DoE 1996).

The formal EIS may be only one of several similar documents presented as part of a planning application. IEMA (2011) note that

> In many of the UK's current development consent regimes there is considerable duplication of information between the Environmental Statement and other documents submitted alongside the application. A good example of this inefficiency can be seen in one of the UK's newest consent regimes, which was specifically designed to operate in a streamlined manner: applications for [Nationally Significant Infrastructure Projects]. Of the first few applications accepted by the Commission, a number have included an ES. However, included within these submissions were several additional assessment reports, addressing the predicted effects of the developments on: flood risk, sustainability, health, transport, carbon emissions, historic environment, landscape, waste and natural features. All of these issues were included within the ESs themselves and are regularly addressed in EIA practice.

The planning application process, including the consideration of EIA information, may also be only one of several consent regimes, diluting the amount of time and energy that the statutory consultees will put into it. For instance, Bird (1996) noted that the Environment Agency (which was formed in 1996 out of several agencies including HMIP):

> requires impact assessments to be supplied with pollution permit applications. Therefore in their role as statutory planning EIS consultees, [they] are unlikely to waste time complaining about the poorly detailed designs given in a planning EIS, if they will be receiving another type of EIA document which precisely covers their area of concern. The Didcot B case study showed that even though HMIP considered the EIS to be satisfactory, they later demanded major design changes. [In the case of the Hamilton Oil gas terminal project in Liverpool Bay] HMIP raised no objectives to the EIS, but then rejected the [Integrated Pollution Control] authorization…

This problem of duplicate authorization procedures and the issues relating to discussion between EIA participants will be discussed further in case studies in Chapter 9.

8.5.3 *Decision-making*

As we noted in Chapter 1, one of the main purposes of EIA is to help to make better decisions, and it is therefore important to assess the performance of EIA in relation to this purpose. It is also important to remember that all decisions involve trade-offs. These include trade-offs in the EIA process between simplification and complexity, comprehensiveness and focus, urgency and the need for better information, facts and values, and certainty and uncertainty (Wood 2003; IEMA 2011). There are also trade-offs of a more substantive nature, in particular between the socio-economic and biophysical impacts of projects—sometimes reduced to the 'jobs vs. the environment' dilemma—and between groups who would win and lose from a project. Box 8.1 illustrates these trade-offs in relation to the UK government's decision to cancel a third runway at Heathrow Airport.

PLANNED THIRD RUNWAY AT HEATHROW IS SCRAPPED

Plans to build a third runway to increase capacity at London's overstretched Heathrow Airport will be cancelled by the new Coalition government. The text of the pact between the Conservatives and Liberal Democrats specifically identifies 'the cancellation of the third runway at Heathrow' as one of the parties' agreed environmental measures. It also pledges to refuse extra runways at Gatwick and Stansted.

But business representatives immediately warned the new government that restricting London's transport capacity would hit the capital's competitiveness and was a 'bad way to start'. The decision will also be a blow to the airport's operator, BAA. London First, which represents big businesses in the capital, said the government must come up with a 'plan B' if it was ruling out airport expansion, and insisted that high speed rail was no substitute for adequate international air links. It said: 'Given that Heathrow is currently at 99 per cent capacity, this is one of the principal factors affecting the competitiveness of London.'

Business groups had long argued that the expansion of Heathrow, which was approved by the former Labour government, was critical to supporting the growth of London as an international financial centre. Frequent business travellers have regularly complained about 'Heathrow hassle' in recent years, typified by long queues in the wake of terror alerts, overcrowding and outdated facilities.

The expansion had been bitterly opposed by councils, residents and environmental groups. Both the Tories and Liberal Democrats had pledged to scrap the scheme in their manifestos. John Stewart, chairman of anti-Heathrow expansion group Hacan CleanSkies, said: 'The third runway is dead in the water. Residents under the flight path are delighted and people who stood to lose their homes are relieved. It is also good for London as a whole as it would have been bad for the environment and it was not needed for the city's economic well-being.' Norman Baker, the Lib Dem MP who is an outspoken opponent of Heathrow expansion, said the move was 'the first fruit of the agreement' for people in London.

Source: Financial Times 2010

Some impacts may be more tradeable in decision making than others. Sippe (1994) provides an illustration for both socio-economic and biophysical categories, of negotiable and nonnegotiable impacts (Table 8.7). Sadler (1996) identified such trade-offs as the core of decision-making for sustainable development.

Table 8.7 Judging environmental acceptability—trade-offs

	Non-negotiable impacts	Negotiable impacts*
Ecological (physical and biological systems components)	• Degrades essential life support systems • Degrades the conservation estate • Adversely effects ecological integrity • Loss of biodiversity	• No degradation beyond carrying capacity • No degradation of productive systems • Wise use of natural resources
Human (humans as individuals or in social groupings)	• Loss of human life • Reduces public health and safety unacceptably • Unreasonably degrades quality of life where people live	• Community benefits and costs and where they are borne • Reasonable apportionment of costs and benefits • Reasonable apportionment of intergenerational equity • Compatibility with defined environmental policy goals

Source: Sippe (1994)

* In terms of net environmental benefits

In the UK, the most obvious decision that is influenced by the EIA process is the competent authority's planning approval decision. The EIS may inform a planning officer's report, a planning committee's decision, and modifications and conditions to the project before and after submission. But the impact of EIA on decision-making may be much wider than this, influencing, for example, the alternatives under consideration, project design and redesign, and the range of mitigation measures and monitoring procedures (Glasson 1999). Indeed, the very presence of an effective EIA system may lead to the withdrawal of unsound projects and the deterrence of the initiation of environmentally damaging projects.

In Chapter 3 the various participants in the EIA process were identified. These participants will have varying perspectives on EIA in decision-making. A local planning officer may be concerned with the *centrality* of EIA in decision-making (does it make a difference?), central government might be concerned about *consistency* in application to development proposals across the country, and whether they help to deliver *sustainable development* in an *efficient* manner; pressure groups may also be concerned with these criteria, but also with *fairness* (in providing opportunities for participation) and *integration* in the project cycle and approval process (to what extent is EIA easily bypassed?). A number of studies have attempted to determine whether EIA and associated consultations have influenced decisions about whether and how to authorize a project.

Early surveys of local planning officers (Kobus and Lee 1993; Lee *et al*. 1994) suggested that EISs were important in the decision in about half of the cases. Interviews with a wider range of interest groups (DoE 1996) found that about 20 per cent of respondents felt that the EIS had 'much' influence on the decision, more than 50 per cent felt that it had 'some' influence, and the remaining 20—30 per cent felt that it had little or no influence. Jones (1995) found that about one-third of planning officers, developers and public interest groups felt that the EIS influenced the decision, compared with almost half of environmental consultants and only a very small proportion of consultees. For planning decisions, it is the members of the planning committees who make the final decision. Interviews suggest that they are not generally interested in reading the EIS, but instead rely on the officer's report to summarize the main issues (DoE 1996). According to Wood and Jones (1997), planning committees followed officers' recommendations in 97 per cent of the cases they studied. The consultations related to the EIS are generally seen to be at least as important as the EIS itself (Jones 1995; Kobus and Lee 1993; Lee *et al*. 1994; Wood and Jones 1997). On the other hand, many non-statutory bodies feel excluded from the decision-making process, and one national non-statutory wildlife body complained that if a statutory consultee did not object, then their own objections went largely ignored (DoE 1996).

While studies of early EISs (e. g. Kobus and Lee 1993; Lee *et al*. 1994) suggested that material considerations were slightly more important than environmental considerations in the final decision on a project's authorization, a later study (Jones 1995) suggested that the environment was the principal factor influencing the decision, with planning policies given slightly less weight. Wood and Jones (1997) reported that the environment was seen to be the overriding factor influencing the decisions in 37 per cent of the cases they studied. However, only in very few cases would the final decision have been different in the absence of an EIS. Weston's (2002) study of planning appeal cases concludes:

> Procedurally EIA is much stronger today in the UK than it was in the early years of its implementation—yet the influence that EIA has on the actual decisions made by LPAs and planning inspectors remains weak, as those decisions are based on a complex web of factors that had evolved long before EIA was introduced ... Local authorities in England and Wales deal with around 450,000 applications for planning consent per year. The decisions on those applications are made on the basis of 'material considerations' including the local development plan, national planning policy guidance and the results of a formal consultation process. EIA cases make up less than 0.1% of those applications and for the most part the actual decision-making process for those cases will be little different to the other 99.9%.

Overall, then, in the UK project applications with EISs are not treated much differently from those without EISs. Although environmental issues are addressed more formally, in a discrete document, the final decision-making process is not changed much by EIA. The main procedural difference brought about by EIA is the need to consult people about the EIS, and the broader scope for public participation (not often used in practice) that it brings. The main substantive differences come in the form of modifications to projects and mitigation measures designed in early on, possibly additional or different conditions on the project, and generally a more comprehensive consideration of environmental issues by the competent authority.

8.6 Legal challenges

A key driver behind recent improvements to EIA quality, including the 2011 EIA regulations and guidance, has been the threat of legal challenge, either on the basis that the project should have been subject to EIA but wasn't, or that the EIA process was inadequate. Legal challenges are important in that they interpret aspects of EIA legislation, set precedents for subsequent EIAs and, over time, have tended to strengthen EIA requirements. A consultant from a major environmental consultancy noted that:

> 'Recent case law has had a profound impact on our EIA work… It's easy for objectors to criticise the EIA process on the basis of case law, to stand up at public inquiries and say the process has not been followed correctly. Scrutiny comes far more frequently and intensely than it used to, at least on the type of projects we're working on.' Mr Hewitt isn't pointing the finger at a single legal dispute that suddenly opened the floodgates to further legal challenges, rather he's arguing that the cumulative effect of EIA-related case law has been to make the raising of objections easier… Continuously referring back to case law in a bid to construct ESs that comprehensively take account of all relevant legal judgments is a laborious process, but one that experienced clients have come to view as a good investment, says Mr Hewitt. Anything that improves the chance of being granted planning approval and/or increasing the speed with which approval is granted is worth paying for. (ENDS Directory 2007)

This rather lengthy section reviews a selection of UK judicial review cases, to give the reader an idea of the kinds of issues that arise during such a legal challenge. It includes some of the key recent court cases that are triggering changes in UK EIA guidance (Baker, Mellor); key older cases that have influenced past guidance and subsequent court cases (Berkeley, Rochdale); and a few other cases that illustrate specific themes in EIA law. They are presented in the rough sequence within the EIA process in which the issue occurs, starting with the screening process and concluding with mitigation. They include quite lengthy quotes from the legal judgements, to give a flavour of the logic and arguments used by the courts, and to show how EIA legislation is used and interpreted.

The UK courts system is complex, with multiple layers and players. For the purposes of this section, however, only limited information about the system and main players is needed. First, the claimant (the person or organization initiating the legal challenge, also known as the plaintiff or in some cases the appellant) must have 'standing' to bring a legal challenge, in that they must be sufficiently affected or harmed by the decision that they are challenging. Claimants in EIA cases are typically local residents affected by a planning decision to allow a project, or developers affected by a competent authority's decision to require EIA. The defendant is the person or organization against which the claimant is making the legal challenge. Finally, where a public authority's decision is being judicially reviewed—as is the case for all of the legal challenges presented in this section—the UK courts use the test of 'Wednesbury unreasonableness', which is based on a completely unrelated court decision of 1947 involving operation of a cinema. A decision is 'Wednesbury unreasonable' if it is so unreasonable that no reasonable authority could have decided that way.

8.6.1 *Environmental assessment is required even if it would not change a planning decision*

Berkeley vs. Secretary of State for the Environment and others (2000) UKHL 36; (2000) 3 All ER 897; (2000) 3 WLR 420

The 'Berkeley 1' case involved a 1994 planning application for redevelopment of the Fulham Football Club. The local planning authority did not ask for an EIS (and none was prepared), but it did consult with a wide range of organizations, and carefully weighed up the advantages and disadvantages of the proposed scheme. In August 1995, the Secretary of State called in the application. He also did not require an EIS. He granted permission in August 1996.

Lady Berkeley, who lived near the site, challenged this planning permission, arguing that an EIA should have been carried out. Her appeal was dismissed, with the judge agreeing with the football club that upon the true construction of the regulations, no EIA was required. The judge also noted that, even if an EIA had been required, in his opinion the absence of the EIA 'had no effect on the outcome of the inquiry and could not possibly have done so'. A Court of Appeal decision of 1998 found that the judge was wrong to determine that no EIA was required, but upheld his decision on the basis of the EIA's lack of effect on the outcome of the inquiry. Lady Berkeley challenged the Court of Appeal decision, and the 'Berkeley 2' case came before the House of Lords in 2000.

The counsel for the Secretary of State this time accepted that the fact that an EIA would have made no difference to the outcome of a planning decision is not a sufficient reason for the courts not to quash that decision. Instead he argued that there had been substantial compliance with the requirements of the Directive.

The House of Lords concluded first that redevelopment of the football club could arguably be considered an 'urban development project' within paragraph 10 (b) of Schedule 2; that its effect on ecology of the River Thames meant that it was likely to have significant environmental effects; and that in these circumstances individuals affected by the proposed development had a directly enforceable right to have the need for an EIA considered by the Secretary of State prior to him granting planning permission and not afterwards by a judge.

It also ruled that the EIA process goes beyond simply informing a planning decision, and so matters even if the EIA would not have changed the decision:

> The directly enforceable right of the citizen which is accorded by the Directive is not merely a right to a fully informed decision on the substantive issue. It must have been adopted on an appropriate basis and that requires the inclusive and democratic procedure prescribed by the Directive in which the public, however misguided or wrongheaded its views may be, is given an opportunity to express its opinion on the environmental issues.

It also dismissed the Secretary of State's argument that an equivalent of an EIS had been prepared:

> [The Secretary of State's counsel] says that the equivalent of the applicant's environmental statement can be found in its statement of case under the Inquiry Procedure Rules, read (by virtue of cross-referencing) with the planning authority's statement of case, which in turn incorporated the comprehensive officers' report to the planning sub-committee, which in turn incorporated the background papers such as the letters from the National Rivers Authority and the London Ecology Unit and was supplemented by the proofs of evidence made available at the inquiry. Members of the public had access to all these documents and the right to express their opinions upon them at the inquiry… I do not accept that this paper chase can be treated as the equivalent of an environmental statement… The point about the environmental statement contemplated by the Directive is that it constitutes a single and accessible compilation, produced by the applicant at the very start of the application process, of the relevant environmental information and the summary in non-technical language.

The planning permission was quashed.

8.6.2 Screening out on grounds that planning conditions can make the project's impacts insignificant

R on the application of Lebus vs. South Cambridgeshire DC (2003)[1]

The 'Lebus case' concerned a February 2000 planning application for an egg production unit for 12,000 free range chickens in a 1,180 m^2 building, and an associated dwelling for an agricultural worker. In April 2000, the claimants' solicitor wrote to South Cambridgeshire

District Council claiming that the application required an EIA. The planning officer wrote back promptly, noting that the proposal did not require EIA under Schedule 1 of the 1999 EIA regulations since it housed less than 60,000 hens, but that it was a Schedule 2 proposal as the building would be more than 500 m^2. However the planning officer also noted that:

> consideration has to be given to each proposal on whether it would have significant effects on the environment by virtue of factors such as its location, impact, nature and size... When considering the need for an Environmental Assessment the above factors as well as issues such as: Airborne pollution; Dirty water and litter disposal; Ecology; Highways and access; and Landscape were taken into account. The Council did not wish to be drawn into requesting an Environmental Statement purely to get information it should rightfully expect anyway. It was considered all the above points could be covered in sufficient detail without formally requesting an Environmental Statement.

South Cambridgeshire granted planning permission in January 2002.

Justice Sullivan concluded that the local authority had not prepared any document that could be sensibly described as a screening opinion for the project. He then also noted that:

> in so far as the statutory question was addressed, it was addressed upon the basis that planning conditions would be imposed and management obligations would be enforceable under section 106. The question was not asked whether the development as described in the application would have significant environmental effects, but rather whether the development as described in the application subject to certain mitigation measures would have significant environmental effects... the underlying purpose of the Regulations in implementing the Directive is that the potentially significant impacts of a development are described together with a description of the measures envisaged to prevent, reduce and, where possible, offset any significant adverse effects on the environment. Thus the public is engaged in the process of assessing the efficacy of any mitigation measures. It is not appropriate for a person charged with making a screening opinion to start from the premise that although there may be significant impacts, these can be reduced to insignificance as a result of the implementation of conditions of various kinds. The appropriate course in such a case is to require an environmental statement setting out the significant impacts and the measures which it is said will reduce their significance.

Similar considerations applied in Bellway Urban Renewal Southern vs. John Gillespie [2003] EWCA Civ 400, where the Secretary of State gave permission for the redevelopment of a former gasworks into residential units, on the basis that the scheme would bring a contaminated site back into beneficial use, and that decontamination of the site could be secured through a planning condition. The claimant successfully argued that the Secretary of State had not considered the project's unmitigated effects, but rather had inappropriately assessed the effects only assuming that mitigation had been put in place.

These cases do not suggest that remediation measures must be totally ignored when decisions are made about the likely significant effects of a proposed development, but they do suggest that care and judgement has to be exercised. Well-established and uncontroversial remedial measures may well be taken into account, but for more complex projects, or where the proposed remediation measures are complex or less well understood, this may be less appropriate. Furthermore, the offer of remediation measures should not be used to frustrate the purpose of the EIA Directive or substitute for its requirements.

8.6.3 *Screening where there are cumulative impacts with previous phases*

Roao Baker vs. Bath and North East Somerset Council ([2009] EWHC 595)[2]

The 'Baker case' involved a 2005 application by Hinton Organics (Wessex) Limited for an extension to a 2.1 ha composting site called Charlton Fields, which it had been operating since 2000. The application was to compost wood waste and cardboard, as well as green waste, at Charlton Fields, and to expand operations to another smaller (less than 0.5 ha) site about 1 kilometre away from Charlton Fields, named Lime Kiln. Partly composted waste from Charlton Fields would be transported to Lime Kiln, where the composting process would finish. The planning authority granted planning permission in 2006 without requiring EIA. This was challenged by the plaintiff.

The planning authority argued that no EIA was required for the Lime Kiln application because of the wording of the Town and Country Planning Environmental Impact Assessment (England and Wales) Regulations 1999 that required EIA for 'Any change to or extension of development listed in this Schedule where such a change or extension meets the thresholds, if any, or description of development set out in this Schedule.' The relevant schedule for the disposal of waste included as thresholds: '(ii) the area of the development exceeds 0.5 of a hectare ... the thresholds and criteria ... applied to the change or extension (and not to the development as changed or extended).'

Mr Justice Collins ruled in favour of the plaintiff. Referring to two European court cases—Liege Airport (C-2/07) and Madrid ring road (C-142/07) -he noted that

> As far as the Lime Kiln application is concerned ... there is an issue as to whether that should be regarded as cumulative in the sense that it goes along with the other project, or whether in itself, albeit smaller than the threshold because it covers an area less than half an hectare, there is a likelihood of environmental damage ... there should have been, at the very least, consideration as to whether, notwithstanding the threshold had not been crossed, it was, indeed, an EIA development, whether or not it was to be regarded as cumulative ...
>
> I have come to the conclusion that the regulations do not in the passage in parenthesis ['(and not to the development as changed or extended)'] properly implement the Directive. This is because they seek to limit consideration for the purposes of screen-

ing to consideration of the change or extension on its own. That is, in my view, contrary to the purpose of and the language of the Directive and the approach that should be adopted as set out by the court.

As a result of this ruling, the UK government is now applying the thresholds in Schedule 2 to the whole of the development once modified, and not just to the change or extension; and has added a new provision that will require any change or extension to an existing or approved Schedule 1 project to be screened for the need for EIA (DCLG 2011a).

A similar early EIA case, R. vs. Swale BC exp RSPB [1991] 1 PLR 6, 16, concerned the construction of a storage area for cargo that would require the infill of Lappel Bank, a mudflat important for its wading birds. Justice Simon Brown concluded that:

> The proposals should not then be considered in isolation if, in reality, it is properly to be regarded as an integral part of an inevitably more substantial development. This approach appears to me appropriate to the language of the Regulations, the existence of the smaller development of itself promoting the larger development and thereby likely to carry in its wake the environmental effects of the latter… developers could otherwise defeat the object of the Regulations by piecemeal development proposals.

8.6.4 *Screening for demolition*

R (SAVE Britain's Heritage) vs. Secretary of State for Communities and Local Government, [2011] EWCA Civ 334[3]

The 'SAVE case' involved a decision by Lancaster City Council to permit the demolition of the historic Mitchell's Brewery building without requiring an EIA. Both parties agreed that the demolition would have a significant impact on cultural heritage. In May 2010, the High Court ruled that demolition was not a 'project' within the scope of the EIA Directive and so did not require EIA. The pressure group SAVE Britain's Heritage took the case to the Court of Appeals. In a judgement of March 2011 that reflected the recent case of Commission vs. Ireland C-50/09, Lord Justices Toulson and Sullivan took as a basis the EIA Directive's definition of 'project', namely:

> —the execution of construction works or of other installations or schemes;
>
> —other interventions in the natural surroundings and landscape including those involving the extraction of mineral resources;
>
> … (Article 1.2).
>
> [The defendant] readily accepted the proposition that the Directive must be interpreted in a purposive manner. If it is accepted that works are capable of having significant effects on the environment, the definition of 'project' in Article 1.2 should, if possible, be construed so as to include, rather than exclude, such works. Applying this approach to the first limb of the definition in Article 1.2, it seems to me that the

execution of demolition works falls naturally within 'the execution of … other … schemes' … [But] it is unnecessary to give Article 1. 2 a broad and purposive construction in order to reach the conclusion that, in ordinary language, demolition works which leave a site on completion in a condition which protects the public and preserves public amenity are capable of being a 'scheme' for the purposes of Article 1. 2 …

I would not accept the premise underlying the Respondent's submission: that 'landscape' in the second limb must be a reference to a rural landscape. For the purposes of the Directive 'landscape' is something other than 'natural surroundings'. In the context of a Directive the purpose of which is to ensure that significant environmental effects are properly assessed before projects proceed, I do not see why 'landscape' in Article 1. 2 should be confined to rural landscapes …

[The defendant] submitted that demolition could not fall within the Directive, even if it fell within the definition of project in Article 1. 2, because it was not included in the lists of projects in Annexes I and II to the Directive. [But] the lists of projects include 'projects' that do not necessarily involve construction works; see eg. the list of 'Agriculture, silviculture and aquaculture' projects, and some of the 'Food Industry' projects. [The Commission vs. Ireland C-50/09 case concluded] that the Annexes refer to sectoral categories of projects, and do not describe the precise nature of the works which may comprise such a project. If demolition is capable of being a 'scheme' for the purposes of Article 1. 2, it is also capable of being an 'urban development project' within paragraph 10 (b) of Annex Ⅱ, even though the project comprises only demolition and restoration of the site.

8. 6. 5 Need for a screening statement where a project is 'screened out'

R vs. Secretary of State for the Environment, Transport and Regions and Parcelforce exparte Marson [1998] EWHC Admin 351;[4] and Roao Mellor and Secretary of State for Communities and Local Government [2010] Env LR 2[5]

The 'Marson case' involved a 1996 planning application by Parcelforce to develop a 17 hectare site near Coventry Airport for sorting and handling of parcels. A group of local residents and parish councils sought a direction from the Secretary of State that environmental assessment was required. In a letter of February 1998, the Secretary of State noted that the development fell within 10 (a) of Schedule 2, but would not be likely to have significant environmental effects, and that no EIA would be required. The applicants challenged this decision, suggesting that reasons must be given for declining to require an EIA. Lord Justices Nourse, Pill and Mummery comprehensively disagreed, citing *inter alia* the lack of a stated legal duty to do so, the discretion afforded to the Secretary of State, the applicant's opportunity to influence the planning process through other means.

This finding held for many years, but has recently been called into question by the 'Mellor case'. The Mellor Case involved a 2004 planning application by Partnerships in Care (PIC) for a medium secure hospital at HSM Forest Moor. In July 2006, PIC's consultants asked Harrogate Borough Council for a screening opinion. Nearby residents, including Mr Mellor, wrote to the council arguing that an EIA was needed, and in October the council determined that an EIA was required. PIC referred the matter to the Secretary of State, who decided in December 2006 that EIA was not required.

The matter was referred to the European Court of Justice. There, Advocate General Kokott concluded that, although the EIA Directive does not require a negative screening opinion to include reasons, third parties should be able to satisfy themselves that the competent authority has come to its determination in a legal robust manner. In sum, she determined that (1) a negative screening opinion does not need to contain reasons; but (2) the planning authority has a duty to provide further information and relevant documents on the negative screening decision if this is requested by an interested party; but in turn that (3) further information can be brief and does not have to be formal. These requirements are are also included in recent changes to UK EIA regulations (DCLG 2011a).

8.6.6 *EIA may be required at the detailed consent stage even if it was not required at the outline consent stage*

Roao Diane Barker vs. LB Bromley, C-290/03, [2007] Env LR 2

The 'Barker case' involves a two-stage planning application by London & Regional Properties Ltd for first outline and then detailed consent to develop a leisure complex in Crystal Palace Park. The London Borough of Bromley did not require an EIS for the first application, and granted outline planning consent in March 1998, reserving certain matters for subsequent approval before development started. The developer applied to the council for final determination of these reserved matters in January 1999, and the council issued a notice of approval in May 1999.

A nearby resident, Ms Barker, challenged the second permission and ended up bringing her challenge to the European Court of Justice. Broadly the challenge can be boiled down to the question: if a UK planning authority has decided that EIA is not required at the outline planning permission stage because they believe that the project would not have significant environmental effects, and if at the detailed planning permission stage it emerges that the project could have significant environmental effects, can and must they the planning authority require EIA at the detailed stage? Note that there is a distinction between 'can' and 'must', since LB Bromley argued that UK national legislation did not allow them to do this even if they had wanted to do it.

Advocate General Léger ruled in favour of Ms Barker:

> Article 1 (2) of Directive 85/337 defines 'development consent' for the purposes

of the directive as the decision of the competent authority or authorities which entitles the developer to proceed with the project. It is apparent from the scheme and the objectives of Directive 85/337 that this provision refers to the decision (involving one or more stages) which allows the developer to commence the works for carrying out his project. Having regard to those points, it is therefore the task of the national court to verify whether the outline planning permission and decision approving reserved matters which are at issue in the main proceedings constitute, as a whole, a 'development consent' for the purposes of Directive 85/337 ...

[W] here national law provides for a consent procedure comprising more than one stage, one involving a principal decision and the other involving an implementing decision which cannot extend beyond the parameters set by the principal decision, the effects which a project may have on the environment must be identified and assessed at the time of the procedure relating to the principal decision. It is only if those effects are not identifiable until the time of the procedure relating to the implementing decision that the assessment should be carried out in the course of that procedure. [In such a legal context] it follows that the competent authority is, in some circumstances, obliged to carry out an environmental impact assessment in respect of a project even after the grant of outline planning permission, when the reserved matters are subsequently approved ... This assessment must be of a comprehensive nature, so as to relate to all the aspects of the project which have not yet been assessed or which require a fresh assessment.

[So] Articles 2 (1) and 4 (2) of Directive 85/337 are to be interpreted as requiring an environmental impact assessment to be carried out if, in the case of grant of consent comprising more than one stage, it becomes apparent, in the course of the second stage, that the project is likely to have significant effects on the environment by virtue *inter alia* of its nature, size or location.

The UK government changed the Town and County Planning (EIA) (England) Regulations in 2008 to reflect these points (DCLG 2010).

8.6.7 EIA for outline planning applications and in other cases where there is uncertainty about the project

The 'Rochdale envelope'

Two legal challenges to planning permissions by Rochdale Metropolitan Borough Council have established the concept of the 'Rochdale envelope'.

R vs. Rochdale Metropolitan Borough Council exparte Tew and others (1999) 6—'Rochdale 1' —concerned an outline planning application made in early 1999 by Wilson Bowden Properties Limited and English Partnership for a 213 hectare business park. Rochdale MBC had a long-standing policy that the area should be developed as a business park. The planning application was accompanied by an indicative land use schedule and floor space figures, and the accompanying letter noted that 'The master plan has been prepared for illustrative purposes... [It] aims to demonstrate the general form of development, showing the integration of the access proposals, principal highways alignment, and possible patterns of land use...' Rochdale MBC gave planning permission for the project subject to a range of conditions, including

- Condition 1.3: 'The development shall be carried out in accordance with the mitigation measures set out in the Environmental Statement submitted with the application, unless otherwise agreed in writing by the Local Planning Authority...'
- Condition 1.7: 'No development shall be commenced until a scheme (the Framework Document) has been submitted to and approved by the Local Planning Authority showing the overall design and layout of the proposed Business Park, including details of the phasing of development and the timescale of that phasing. The Framework Document shall show details of the type and disposition of development and the provision of structural landscaping within and on the perimeters of the site. The Business Park shall be constructed in accordance with the approved Framework Document unless the Local Planning Authority consents in writing to a variation or variations.'
- Condition 1.11: 'This permission shall not be construed as giving any approval to the illustrative Masterplan accompanying the application.'

The plaintiffs argued that the EIS was inadequate because it did not provide adequate information as to the 'design and size or scale of the project', and that this information was necessary to identify the proposed development's main environmental effects. They also argued that, not only was the outline planning permission not tied to the illustrative Masterplan on which the EIS was based, but that it expressly envisaged that a different layout and composition of users would emerge as part of the development of the Framework Document. They argued that a planning decision for an EIA project must be taken in the full knowledge of the project's likely environmental effects, and that it is not sufficient that full knowledge will be obtainable at a later stage.

The council, in turn, argued that the kind of detailed requirements envisaged by the plaintiffs would make it difficult, and possibly completely impractical, to seek planning permission for large projects; and that Circular 15/88 expressly allowed outline planning permission to be sought for projects requiring EIA. Justice Sullivan, however, agreed with the plaintiffs in a judgement of May 1999:

> Condition 1.3... ties the mitigation measures to the Environmental Statement (unless otherwise agreed), but those measures were a response to the environmental impacts of development in accordance with the illustrative Masterplan. Recognising, as I do, the utility of the outline application procedure for projects such as this, I would not wish to rule out the adoption of a Masterplan approach, provided the Masterplan was tied, for example by the imposition of condition, to the description of the development permitted. If illustrative floor space or hectarage figures are given, it may be appropriate for an Environmental Assessment to assess the impact of a range of possible figures

before describing the likely significant effects. Conditions may then be imposed to ensure that any permitted development keeps within those ranges.

The fundamental difficulty in the present case is that … the outline planning permission was not tied in any way to either of those documents. Conditions 1.7 and 1.11 dispensed with the Masterplan and replaced it with the Framework Document to be submitted and approved in due course. The reason given for the imposition of condition 1.11 explains that the Masterplan was submitted for illustrative purposes only and that it gave insufficient detail on which to determine the layout of the site. If it was inadequate for that purpose, it is difficult to see how it could have been an adequate description for the purposes of paragraph 2 (a) of Schedule 3 to the Assessment Regulations.

The Rochdale saga promptly continued as 'Rochdale 2', R vs. Rochdale Metropolitan Borough Council exparte Milne (2000).[7] After the results of Rochdale 1, the developers re-applied for outline planning permission for the business park, including a schedule of development, development framework and masterplan. Details of landscape and design were reserved (not provided), as were transport arrangements for most of the plots on the site. The planning application was accompanied by a comprehensive EIS.

Rochdale MBC again gave planning permission for the proposal in December 1999, but this time the planning conditions were much more closely tied to the EIS. For instance Condition 1.7 stated that:

'The development on this site shall be carried out in substantial accordance with the layout included within the Development Framework document', because 'The layout of the proposed Business Park is the subject of an Environmental Impact Assessment and any material alteration to the layout may have an impact which has not been assessed by that process.'

Condition 1.11 required that 'The development shall be carried out in accordance with the mitigation measures set out in the Environmental Statement submitted with the application unless provided for in any other condition attached to this permission.'

The plaintiff again challenged the planning permission, arguing that, despite the changes made to the planning application, it still did not provide 'a description of the development proposed'. The reserved matters—design and access arrangements—could significantly affect the environmental impacts of the project, for instance by the materials used or whether the development included a particularly striking 'landmark' building. He suggested that, to comply with the requirements of Schedule 3, the development must be described in enough detail to ensure that nothing is omitted that may be capable of having a significant environmental effect; and that outline planning applications are thus fundamentally inconsistent with the requirement for EIA. In this case, Justice Sullivan sided with the local authority and dismissed the application for judicial review:

If a particular kind of project, such as an industrial estate development project … is, by its very nature, not fixed at the outset, but is expected to evolve over a number of years depending on market demand, there is no reason why 'a description of the project' for the purposes of the directive should not recognize that reality. What is important is that the environmental assessment process should then take full account at the outset of the implications for the environment of this need for an element of flexibility… It is for the authority responsible for granting the development consent … to decide whether the difficulties and uncertainties are such that the proposed degree of flexibility is not acceptable in terms of its potential effect on the environment.

Any major development project will be subject to a number of detailed controls, not all of them included within the planning permission. Emissions to air, discharges into water, disposal of the waste produced by the project, will all be subject to controls under legislation dealing with environmental protection. In assessing the likely significant environmental effects of a project the authors of the environmental statement and the local planning authority are entitled to rely on the operation of those controls with a reasonable degree of competence on the part of the responsible authority… The same approach should be adopted to the local planning authority's power to approve reserved matters.

The Rochdale judgements have significant implications for outline planning applications and for cases where there is uncertainty about details of the proposed project:

- Applications for 'bare' outline permissions with all matters reserved for later approval are very unlikely to comply with EIA requirements.
- Developers can have flexibility, but for EIA purposes they need to consider the range of possible parameters within which the project might evolve, and these must be detailed enough to allow a proper assessment of the likely environmental effects and proposed mitigation.
- The flexibility should not be abused, and does not give developers an excuse to provide inadequate descriptions of their projects. The planning authority must satisfy itself that, given the nature of the project, it has 'full knowledge' of its likely environmental effects.
- The outline planning application should specify clearly defined parameters within which the project may evolve, in the form of conditions and obligations. These should be 'tied' to the environmental information of the ES—the range of these parameters is the 'Rochdale envelope'.
- Implementation of reserved matters consents granted for matters that are not in the outline consent will be unlawful (IPC 2011).

8.6.8 Themes in judicial review findings

Until recently, judicial reviews of competent authority decisions have been limited by the courts' relatively narrow interpretation of the duties of competent authorities under the EIA regulations. However, a series of high-profile recent cases—including Baker, Mellor,

Barker and SAVE—suggests that the courts are now taking a more proactive and wider view of the EIA Directive's requirements. Several conclusions can be drawn from these legal judgements:

- The EIA Directive is interpreted by the courts as having a 'wide scope and broad purpose', in line with the Kraaijveld (Dutch Dykes) Case C-72/95.
- A project cannot be automatically excluded from EIA requirements simply because it is not listed in the EIA Directive or implementing regulations. Furthermore, because projects can be described in different ways, for EIA screening purposes it is probably better to consider the project's scope and purpose rather than its label.
- Where aspects of a project are uncertain, the EIA must provide enough information to allow a decision to be made taking into account these uncertainties.
- Provision of information equivalent to that in an EIS is not equivalent to EIA, since the public participation requirements of EIA are also important.
- The courts have clarified that EISs do not have to consider alternatives, even if the planning authority thinks that they should have been considered or if the alternatives would have a less severe impact than the proposed project.
- The imposition of planning conditions to the point where the project would no longer have significant environmental impacts is also not equivalent to EIA. The distinction between EIA screening (not including mitigation) and impact assessment (which can include mitigation) is important. 'A purposive approach might be as follows. Would an open minded adviser to the competent authority or member of the public concerned about the potential [significant environmental effect] want the systematic assembly of the EIA data to judge how effective the proposed [mitigation measure] would be? If so EIA is required' (McCracken 2010).

8.7 Costs and benefits of EIA

Much of the early resistance to the imposition of EIA was based on the idea that it would cause additional expense and delay in the planning process. EIA proponents refuted this by claiming that the benefits of EIA would well outweigh its costs. This chapter concludes with a discussion of the costs and benefits of EIA to various parties in the UK.

8.7.1 *Costs of EIA*

Environmental impact assessment has slightly increased the cost to *developers* of obtaining planning permission. An EIS generally costs between 0.01 and 5 per cent of project costs, with 0.1 to 1 per cent being a rough average for the UK (GHK 2010). Weston's

(1995) survey of consultants showed that consultancies received on average £34,000 for preparing a whole EIS, £40,000 for several EIS sections, and £14,750 for one section: this itself highlights the variability of the costs involved. In 1997, the (former) DETR suggested £35,000 as a median figure for the cost of undertaking an EIA (DETR 1997b), and in 2010 DCLG suggested an average cost of almost £90,000 (€100, 000): it is unclear whether this difference is due to a strong rise in the actual costs of EIA, or to some other factor.

There has been some concern that competition and cost-cutting by consultancies, an increase in 'cowboy' consultancies and the tendency for developers to accept the lowest bid for preparing an EIS may affect the quality of the resulting EIAs by limiting the consultants' time, expertise or equipment. Consultants note that 'on all but the largest developments there is always a limited budget—an EA expands to fill the available budget, and then some' (Radcliff and Edward-Jones 1995). However, Fuller (1992) argued that cost-cutting may not be helpful to a developer in the long run:

> A poor-quality statement is often a major contributory factor to delays in the system, as additional information has to be sought on issues not addressed, or only poorly addressed, in the original… Therefore reducing the cost of an environmental assessment below the level required for a thorough job is often a false economy.

The cost of EIA to *competent authorities* is much more difficult to measure and has until now been based on interviews rather than on a more systematic methodology. UK planning decisions for the kinds of large projects that require EIA have always taken a long time, often years. They consistently take considerably more time than do decisions for projects that do not require EIA (DoE 1996), but then they tend to be larger, more complex and more politically sensitive.

An early study (Lee and Brown 1992) found that about half the planning officers interviewed felt that the EIS had not influenced how long it took to reach a decision; the rest were about evenly split between those who felt that the EIA had speeded up or slowed down the process. In later interviews (DoE 1996), many planning officers felt that dealing with the EIS and the planning application were one and the same, and 'just part of the job'. Estimates for reviewing the EIS and associated consultation ranged from 5 hours to 6—8 months of staff time! Planning officers handling EIS cases tend to be development control team leaders and above, so staff costs would generally be higher than for standard planning applications. In some cases, planning officers also hire consultants to help them review and comment on EISs, adding to their costs.

The time taken to decide planning applications has recently been the focus of several studies (e. g. National Audit Office 2008; DCLG 2008; Ball *et al*. 2008), none of which identified EIA as a factor leading to delays. Only Killian Pretty's 170 page review of planning applications (DCLG 2008) mentioned EIA at all, and then only very briefly to imply that their length should be shortened. Some consultants feel that EIA slows down the decision-making process and is a means through which LPAs can make unreasonable demands on developers to provide detailed information on issues 'which are not strictly relevant to the planning decision' (Weston 1995). However others feel that EIA does not necessarily slow things down: 'The more organised approach makes it more efficient and in some cases it allows issues to be picked up earlier. The EIS can thus speed up the system' (DoE 1996).

The Planning Act 2008, the primary aim of which was to speed up the planning system, set up a new system of National Policy Statements to provide policy guidance on energy, transport, water, waste water and waste developments. It also established a parallel decision-making stream for major infrastructure projects, to help speed up their delivery. Early indications are that the preapplication stage—including preparation of EISs—for these projects is lengthening; it is not yet clear whether the government's aim of then deciding on the projects within a year of application will be met in practice (see also Section 3.5.7).

In 20 case studies, the time spent by *consultees* on EIA ranged from four hours to one-and-a-half days for statutory consultees, and from one hour to two weeks for non-statutory consultees. Although some consultees, like planning officers, argued that 'this is what we are here for', others suggested that they needed to prioritize what developments they got involved in because of time and resource constraints (DoE 1996). This may well be even more of a problem in today's more straightened circumstances.

8.7.2 *Benefits of EIA*

The benefits of EIA are mostly unquantifiable, so a direct comparison with the costs of EIA is not possible. Perhaps the clearest way to gauge whether EIA helps to reduce a project's environmental impacts is to determine whether a project was modified as a result of EIA. Early studies on EIA effectiveness (e.g. Kobus and Lee 1993; Tarling 1991; Jones 1995) showed that modifications to the project as a result of the EIA process were required in almost half the cases, with most modifications regarded as significant. More recently, the European Commission (2009) reported that EIAs led to improvements for most projects, although this was based on information from Member States other than the UK.

Environmental impact assessment can have other benefits in addition to project modification. A survey of *environmental consultants* (Weston 1995) showed that about three-quarters of them felt that EIA had brought about at least some improvements in environmental protection, primarily through the incorporation of mitigation measures early in project design and the higher regard given to environmental issues. However, other consultants felt that the system is 'often a sham with EISs full of platitudes' (Weston 1995), and some *developers* felt that 'the preparation of the ES had cost them too much time and money, and that the large amounts of work involved in EA often yielded few tangible benefits in terms of the actual planning decision reached' (Pritchard *et al.* 1995). Jones *et al.* (1998) found that only one-fifth of developers and consultants felt that there had been no benefits associated with EIA. Presumably this view has not significantly changed in the interim.

Competent authorities generally feel that projects and the environment benefit greatly from EIA (Jones 1995; Lee *et al*. 1994). EIA is seen as a way to focus the mind, highlight important issues, reduce uncertainty, consider environmental impacts in a systematic manner, save time by removing the need for planning officers to collect the information themselves, and identify problems early and direct them to the right people (DoE 1996; Jones 1995; Pritchard *et al*. 1995; Weston 2002). One planning officer noted: 'when the system first appeared I was rather sceptical because I believed we had always taken all these matters into account. Now I am a big fan of the process. It enables me to focus on the detail of individual aspects at an early stage' (DoE 1996).

Consultees broadly agree that EIA creates a more structured approach to handling planning applications, and that an EIS gives them 'something to work from rather than having to dig around for information ourselves'. However, when issues are not covered in the EIS, consultees are left in the same position as with non-EIA applications: some of their objections are not because the impacts are bad but because they have not been given any information on the impacts or any explanation of why a particular impact has been left out of the assessment. Consultees feel that an EIA can give them data on sites that they would not otherwise be able to afford to collect themselves, and that it can help parties involved in an otherwise too often confrontational planning system to reach common ground (DoE 1996).

8.8 Summary

In summary, all the parties involved agree that EIA as practised in the UK helps to improve projects and protect the environment, although the system could be much stronger: EIA is thus at least partly achieving its main aims. There are time and money costs involved, but there are also tangible benefits in the form of project modifications and more informed decision-making. When asked whether EIA was a net benefit or cost, 'the overwhelming response from both planning officers and developers/consultants was that it had been a benefit. Only a small percentage of both respondents felt that EIA had been a drawback' (Jones 1995).

Some stages in EIA-particularly early scoping, good consultation of all the relevant parties, and the preparation of a clear and unbiased EIS-are consistently cited as leading to clear benefits and cost-effectiveness (e. g. DoE 1996; IEMA 2011). Chapter 9 provides a set of primarily UK case studies that seek to exemplify some of the issues of and responses to particular aspects of the EIA process. Suggestions for future directions in EIA in the UK and beyond are discussed in Chapter 12.

SOME QUESTIONS

The following questions are intended to help the reader focus on the key issues of this chapter.

1 Reviewing the nature of UK EIS activity displayed in Figures 8. 1 and 8. 2, how might you explain the changing patterns of activity?

2 What might explain the changes in the predominant sectors of UK EIS activity since 2000, compared with before 2000, as set out in Figure 8. 3?

3 From Table 8. 2, are there any indications of notable differences in approaches to screening in relation to amount of practitioner EIA experience?

4 Compare and contrast the information in Tables 8. 4 and 8. 5. Identify and seek to explain any differences in content in relation to the importance of various environmental components in EIA scoping activity.

5 Given the information from Section 8. 4. 3 about determinants of EIS quality, if you were managing an EIA process for the first time, how would you try to optimize the quality of the resulting EIS?

6 Using the third runway at Heathrow (Box 8. 1) as an example, what do you think are negotiable and non-negotiable impacts of an airport (Table 8. 7)?

7 Do you agree with the judge in the Berkeley case (Section 8. 6. 1) who felt that no EIA should be required where the absence of an EIA would have 'no effect on the outcome of the inquiry and could not possibly have done so'? Why or why not?

8 In your words, what is the 'Rochdale envelope' (Section 8. 6. 7) and why is it important for EIA practice?

9 Section 8. 7 suggests that the costs of EIA can be quantified, but its benefits cannot. How could you determine whether the costs of an EIA outweigh its benefits? Do you think that they do?

Notes

1 www. bailii. org/ew/cases/EWHC/Admin/2002/ 2009. html.
2 www. bailii. org/ew/cases/EWHC/Admin/2010/ 373. html.
3 www. bailii. org/ew/cases/EWCA/Civ/2011/334. html.
4 www. bailii. org/ew/cases/EWHC/Admin/1998/ 351. html.
5 www. bailii. org/eu/cases/EUECJ/2009/C7508. html.
6 www. bailii. org/ew/cases/EWHC/Admin/1999/ 409. html (Ground 2).
7 www. bailii. org/ew/cases/EWHC/Admin/2000/ 650. html.

References

Ball, M. , Allmendinger, P. and Hughes, C. 2008. Housing supply and planning delay in the South of England. research funded by Economic and Social Research Council, grant RES-000-22-2115, Reading: University of Reading.

Barker, A. and Wood, C. 1999. An evaluation of EIA system performance in eight EU countries. *Environmental Impact Assessment Review* 19, 387-404.

Bird, A. 1996. *Auditing environmental impact statements using information held in public registers of environmental information*. Working Paper 165. Oxford: Oxford Brookes University, School of Planning.

CEC (Commission of the European Communities) 1993. *Report from the Commission of the Implementation of Directive 85/337/EEC on the assessment of the effects of certain public and private projects on the environment*. COM (93), 28, final. Brussels: CEC.

Davison, J. B. R. 1992. *An evaluation of the quality of Department of Transport environmental statements*. MSc dissertation, Oxford Brookes University.

CLG (Communities and Local Government) 2008. The Killian Pretty review: planning applications-a faster and more responsive system: final report. Available at: www. communities. gov. uk/publications/ planningandbuilding/killianprettyfinal.

DCLG 2010. The Town and Country Planning (EIA) Regulations: Consultation on draft regulations. Available at: www. communities. gov. uk/documents/ planningandbuilding/pdf/1682192. pdf.

DCLG 2011a. Guidance on the Environmental Impact Assessment (EIA) Regulations 2011 for England. London: DCLG.

DCLG 2011b. Personal communication with Environmental Assessment Division on throughput of UK EISs.

DETR (Department of Environment, Transport and the Regions) 1997b. *Consultation paper: implementation of the EC Directive (97/11/EC) - determining the need for environmental assessment*. London: DETR.

DETR 1999a. *Town and Country Planning (EIA) Regulations*. London: HMSO.

DETR 1999b. *Town and Country Planning (EIA) Regulations*. Circular 2/99 London: HMSO.

DETR 2000. *Environmental impact assessment: a guide to the procedures*. London: DETR.

DETR (Department of Environmental, Transport and the Regions) 2000. Environmental impact assessment: A guide to the procedures. Available at: www. communities. gov. uk/documents/planningandbuilding/pdf/157989. pdf.

DoE 1996. *Changes in the quality of environmental impact statements*. London: HMSO.

DoT (Department of Transport) 1993. Design manual for roads and bridges, vol. 11, *Environmental assessment*. London: HMSO.

ENDS 2007. *Directory of Environmental Consultants 2006/2007*. London: Environmental Data Services.

Essex Planning Officers' Association 2007. *The Essex guide to Environmental Impact Assessment*.

Chelmsford: Essex County Council. Available at: www. essex. gov. uk/Environment%20Planning/ Planning/Minerals-Waste-Planning-Team/Planning-Applications/Application-Forms-Guidance-Documents/Documents/eia _ spring _ 2007. pdf.

European Commission 2009. Study concerning the report on the application and effectiveness of the EIA Directive. Brussels: EC. Available at: ec. europa. eu/environment/eia/pdf/eia _ study _ june _ 09. pdf.

Ferrari, C. 1994. Environmental assessment: our client's perspective. Environmental Assessment: RTPI Conference, 20 April. Andover.

Financial Times 2010. Planned third runway at Heathrow is scrapped. *Fnancial Times*. Available at: www. ft. com/cms/s/0/9b278458-5ddb-11df-8153-00144 feab49a. html# axzz1Ki8xefJV.

Fuller, K. 1992. Working with assessment. In *Environmental assessment and audit: a user's guide*, 14-15. Gloucester: Ambit.

Glasson, J. 1999. Environment impact assessment-impact on decisions. In *Handbook of environmental impact assessment*, J. Petts (ed), vol. 1, Oxford: Blackwell Science.

GHK 2010. Collection of information and data to support the impact assessment study of the review of the EIA Directive. London: GHK.

Gosling, J. 1990. *The Town and Country (assessment of environmental effects) regulations 1988: the first year of application*. Proposal for a working paper, Department of Land Management and Development, University of Reading.

Gray, I. M. and Edward-Jones, G. 1999. A review of the quality of environmental assessments in the Scottish forest sector, *Forestry* 72 (1), 1-10.

Hall, E. 1994. *The environment versus people? A study of the treatment of social effects in environmental impact assessment* (MSc dissertation, Oxford Brookes University).

IAU (Impacts Assessment Unit) 2003. *Screening decision making under the Town and Country Planning (EIA) (England and Wales) regulations* 1999. IAU: Oxford Brookes University.

IEMA (Institute of Environmental Management and Assessment) 2011. *The state of environmental impact assessment practice in the UK*. Lincoln: IEMA.

IPC (Infrastructure Planning Commission) 2011. Advice note 9: using the 'Rochdale Envelope'. Available at: www. infrastructure. independent. gov. uk/wp- content/uploads/2011/02/Advice-note-9. -Rochdale-envelope-web. pdf.

Jones, C. E. 1995. The effect of environmental assessment on planning decisions, *Report*, special edition (October), 5-7.

Jones, C. E., Lee, N. and Wood, C. 1991. *UK environmental statements 1988-1990: an analysis*, Occasional Paper no. 29. EIA Centre, University of Manchester.

Jones, C. E., Lee, N. and Wood, C. 1991. *UK environmental statements* 1988-1990: *an analysis*. Occasional Paper 29. Manchester: EIA Centre, University of Manchester.

Jones, C., Wood, C. and Dipper, B. 1998. Environmental assessment in the UK planning process. *Town Planning Review* 69, 315-19.

Kobus, D. and Lee, N. 1993. The role of environmental assessment in the planning and authorisation of extractive industry projects. *Project Appraisal* 8 (3), 147-56.

Kreuser, P. and Hammersley, R. 1999. Assessing the assessments: British planning authorities and the review of environmental statements. *Journal of Environmental Assessment Policy and Management* 1, 369-88.

Lee, N. and Brown, D. 1992. Quality control in environmental assessment. *Project Appraisal* 7 (1), 41-5.

Lee, N. and Colley, R. 1990 (updated 1992). *Reviewing the quality of environmental statements*, Occasional Paper no. 24. University of Manchester.

Lee, N. and Dancey, R. 1993. The quality of environmental impact statements in Ireland and the United Kingdom: a comparative analysis. *Project Appraisal* 8 (1), 31-6.

Lee, N., Walsh, F. and Reeder, G. 1994. Assessing the performance of the EA process. *Project Appraisal* 9 (3), 161-72.

Leu, W. S., Williams, W. P. and Bark, A. W. 1993. An evaluation of the implementation of environmental assessment by UK local authorities. *Project Appraisal* 10 (2), 91-102.

McCracken, R. QC 2010. EIA, SEA and AA, present position: where are we now? *Journal of Planning Law* 12, 1515-32.

Mills, J. 1994. The adequacy of visual impact assessments in environmental impact statements. In *Issues in environmental impact assessment*, Working Paper no. 144. School of Planning, Oxford Brookes University, 4-16.

National Audit Office 2008. Planning for homes: speeding up planning applications for major housing development in England. Available at: www. nao. org. uk/publications/0809/planning_for_homes_speeding. aspx.

Petts, J. and Hills, P. 1982. *Environmental assessment in the UK*. Nottingham: Institute of Planning Studies, University of Nottingham.

Pritchard, G., Wood, C. and Jones, C. E. 1995. The effect of environmental assessment on extractive industry planning decisions. *Mineral Planning* 65 (December), 14-16.

Radcliff, A. and Edward-Jones, G. 1995. The quality of the environmental assessment process: a case study on clinical waste incinerators in the UK. *Project Appraisal* 10 (1), 31-8.

Rodriguez-Bachiller, A. with J. Glasson 2004. *Expert systems and geographical information systems*. London: Taylor and Francis.

Sadler, B. 1996. *Environmental assessment in a changing world: evaluating practice to improve performance*. Final report of the international study on the effectiveness of environmental assessment, Canadian Environmental Assessment Agency.

Sadler, B. 2012. Latest EA effectiveness study (not available at time this book went to press).

Scottish Government 2007. Environmental impact assessment directive: questions and answers. Available at: www. scotland. gov. uk/Publications/ 2007/11/26103828/1.

Sheate, W. 1994. *Making an impact: a guide to EIA law and policy*. London: Cameron May.

Sippe, R. 1994. *Policy and environmental assessment in Western Australia: objectives, options, operations and outcomes*. Paper for International Workshop, Directorate General for Environmental Protection, Ministry of Housing, Spatial Planning and the Environment, The Hague, The Netherlands.

Tarling, J. P. 1991. *A comparison of environmental assessment procedures and experience in the UK and the Netherlands* (MSc dissertation, University of Stirling).

Weston, J. 1995. Consultants in the EIA process. *Environmental Policy and Practice* 5 (3), 131-4.

Weston, J. 1996. Quality of statement is down on the farm. *Planning* 1182, 6-7.

Weston, J. 2000. EIA, decision-making theory and screening and scoping in UK practice. *Journal of Environmental Planning and Management* 43 (2), 185-203.

Weston, J. 2002. From Poole to Fulham: a changing culture in UK environmental impact decision making? *Journal of Environmental Planning and Management* 45 (3), 425-43.

Wood, C. 1991. *Environmental impact assessment in the United Kingdom*. Paper presented at the ACSP-AESOP Joint International Planning Congress, Oxford Polytechnic, Oxford, July.

Wood, C. 1996. Progress on ESA since 1985-a UK overview. In *The proceedings of the IBC Conference on Advances in Environmental Impact Assessment*, 9 July. London: IBC UK Conferences.

Wood, C. 2003. *Environmental impact assessment: a comparative review*, 2nd edn. Harlow: Prentice Hall.

Wood, C. and Jones, C. 1991. *Monitoring environmental assessment and planning*, DoE Planning and Research Programme. London: HMSO.

Wood, C. and Jones, C. 1997. The effect of environmental assessment on local planning authorities. *Urban Studies* 34 (8), 1237-57.

Wood, G. and Bellanger, C. 1998. *Directory of environmental impact statements July 1988-April 1998*. Oxford: IAU, Oxford Brookes University.

Zambellas, L. 1995. *Changes in the quality of environmental statements for roads*. MSc dissertation, Oxford Brookes University.

9 环境影响评价案例研究

9.1 导言

本章在第8章的分析基础上，研究了一些环境影响评价实践案例。选择的案例主要是项目层次的环境影响评价，其中也包括了战略环境评价的案例。本章还研究了环境影响评价和其他类型的评价之间的关系，包括《欧盟栖息地指令》规定进行的“适宜性评价”和健康影响评价。本章选择的案例主要集中在英国，覆盖各行各业和不同的发展模式，包括能源（海岸风能、燃气发电站和高架输电线路）、交通运输（道路和机场）、废弃物（生活垃圾焚烧厂和污水处理厂）和基础设施（港口和防洪工程）。

选择这些研究案例是为了说明和环境影响评价实践相关的特定主题或问题，而且其中一些案例还与环境影响评价过程的特定时期相关联。这些案例是：

① 环境影响评价中的项目定义和分解项目带来的影响［威尔顿（Wilton）电站，9.2节］；

② 环境影响评价、欧洲保护栖息地和适宜性评价的关系（N21连接公路，9.3节）；

③ 环境影响评价中的公众参与方法（Portsmouth焚化炉，9.4节）；

④ 累积影响评价（Humber河口项目，9.5节）；

⑤ 健康影响评估（Stansted机场第二跑道，9.6节）；

⑥ 环境影响评价的减缓措施（Cairngorm山索道，9.7节）；

⑦ 国家层面的战略环评（英国海岸风能开发，9.8节）；

⑧ 地方层面的战略环评（Tyne和Wear交通计划，9.9节）。

本章所选的案例并不是最好的环境影响评价实践案例——实际上其中的两个案例因为对环境影响的评估不充分而遭到欧盟委员会的正式投诉。但这些案例确实包括了一些解决特定问题的创新或新颖的方法，如累积影响评价（Humber河口计划）、公众参与和风险交流的处理方法（Portsmouth焚化炉）。案例研究也注意到环境影响评价中遇到的一些实际困难、实践过程中的局限性以及未来的机会。这些研究加强了第8章对英国环境影响评价实践的批评，同时预先判定了第11章和第12章出现的环评新方向。

本章案例大都建立在作者或作者的牛津布鲁克斯大学同事的研究基础上（9.2节中电站的案例除外，该案例出自1995年出版的由William Sheate编写的《伦敦大学帝国学院环境影响评价读本》）。

9.2 威尔顿电站案例研究：环境影响评价中的项目定义

9.2.1 简介

案例研究最初是由Sheate于1995年编写的，它阐明了环境影响评价中项目定义的一些问题，特别指出案例总体规划中各个部分的批准程序是分开的。案例着重论述了全面评价拟议的英国电站发展影响的环境影响评价程序的失败之处。尤其是，环境影响评价程序没有在电站批准建设之前识别开发所需的大规模输电线路的环境影响。

本案例强调了由于电力生产和传输的批准程序分开而导致的英国能源部门项目环境影响评价中存在的基本问题。这个问题产生于1989年《电力法》宣布英国供电工业私有化之后。《环境影响评价指令》要求对开发项目所产生的直接、间接以及二次影响进行评价，本案例说明了同一项目的各个组成部分的批准程序分开如何导致与上述指令要求发生冲突。虽然本案例发生在英国和欧盟环境影响评价探索阶段（20世纪90年代初期），但所提出的问题大部分都未得到解决，且对英国以及其他地区仍具有现实意义。

9.2.2 Wilton电站项目

1991年初，报纸开始报道英国东北部Teesside的新电站和国家电网系统之间连接的高压输电线网项目的环境后果。令人惊讶的是，这些影响在电站项目批准时却没有被识别出来。对电站的环境报告（ES）的深入调查表明了这些问题在当时没有被识别出来，进而在电站批准程序中也没有反映出来。后来，英格兰农村保护委员会（CPRE）因担心该项目引发公众非议，1991年4月向在布鲁塞尔的欧盟委员会发起了对国家能源部部长的诉讼（Sheate，1995）。

项目发展到这个地步，我们很难说它是一个好的环境影响评价案例，但这一切到底是如何发生的呢？此次诉讼是针对坐落于Middlesbrough附近的Wilton的一座新建大型燃气电站的环境影响评价，该项目由Teesside能源公司规划开发。CPRE认为该项目在没有彻底分析其环境影响的情况下就得以批准。由于该项目的装机容量为1875MW，属于附件Ⅰ中的项目，因此需要强制执行环境影响评价。然而，除电站本身外，总体“项目”却被分割为若干个，包括：

① 新建天然气管道；

② 燃气接收和处理设施；

③ 处理厂与热电厂之间的热电联产（CHP）的燃料管线；

④ 新建400kV高架空电缆和系统升级（75～85km长，从电站所在地至北约克郡附近的Shipton）。

新建电站的电缆铺设工程的影响受到了特别关注，虽然项目的其他部分也有潜在的环境影响。电站所在地的克里夫兰乡村委员会（CC）认为，在电站获得批准之前应该对项目的所有组成部分的环境影响进行全面评价。郡规划官员这样评论：

委员会希望电站的批准决定能够推迟，直到所有环境影响得到全面考虑。但国家能源部部长（当时此类项目的审批部门的领导）并不准备这样做。这导致方案的不同部分（包括管道、气体净化设备、电站主体以及输电线路）在不同阶段经历不同的审批程序，因而缺乏整体性。

尽管有这些担忧，国家能源部部长还是于1990年11月批准了该电站。这个批准决定以该电站的环境报告信息为基础，而没有发挥公众质询的优势。重要的是，电站的环境报告中并没有对完整项目的各个部分（如管线、气体处理设施以及输电线路）的环境影响进行描述或评价，这些部分在分开的批准和环境影响评价程序下被视为单独的项目。虽然对这些项目的其他部分分别进行了环境影响评价，以便之后考虑其环境影响，但英格兰乡村保护委员会在向欧盟委员会的诉讼中提及：按照《环境影响评价指令》，该电站的环境影响评价应包括其相关开发的主要环境影响。他们还认为如不这样执行则会使得环境影响评价方法变得零碎，且与《环境影响评价指令》中的要求相抵触，即在项目被批准之前，要先对项目的所有直接的、间接的和二次影响进行全面的评价。

Sheate（1995）对英格兰乡村保护委员会所做的评论进行了总结：

令人担忧的是，国家能源部部长并没有在相关开发影响方面要求得到更多的信息，但其有权按照英国所贯彻的法令做出决策。尽管在提交给国家能源部部长的附件中并未体现出有关输电线缆、燃气管线、燃气处理设施以及热电联产管线的主要环境影响，开发商却成功取得了电站的批准。鉴于国家能源部部长并没有得到相关信息（当然他也没有要求提供这样的信息），如果能把所有相关信息提供给他，其决定可能会有所不同。由于这些信息没有包括在环境报告中，公众和受影响群体也未能警惕其产生的影响，否则很可能会进行公众质询，问题也将不可避免地被公开。

在英国，新电缆铺设的项目由国家电网公司（NGC）负责，而不是由电站开发商负责。国家电网公司有义务将新建电厂与国家电网连接起来——如果可能涉及重大的环境影响——就必须对新的高架线和现有线路的升级进行环境影响评价。因此，确实需要对电缆铺设项目和相关的其他建设内容进行环境影响评价。然而，本项目的环境影响评价程序是在电站批准建设之后进行的，因此它既不能决定项目是否可以建设，也不能决定电厂的选址（是建在远处还是距离现有电网更近的地方）从而使高架线的负面视觉影响减小。

本质上，Wilton案例考虑了以环境影响目标来定义“项目”的方法。Wilton电站的环境报告提到的“整体项目”应该包括电站本身及其相关建设，例如电缆铺设等。然而，由于不同的建设部分执行分开的审批程序，该项目就被拆分成若干独立的“分项目”，每个“分项目”的环境影响也是分开评价的，而且是在不同的时期执行不同的审批程序。英格兰乡村保护委员会认为，根据指令，脱离主体建设而去评估其他相关建设的环境影响是不合适的，这其中也包括了对后者的选址评价。

在答复英格兰乡村保护委员会的诉讼时，欧盟委员会原则上也认同对此类案例进行整体评价的必要性，而且也认为以这样的方式拆分项目是违背《环境影响评价指令》的：

我能证实委员会的观点，作为一般原则，不管是因为铺设电缆确保电站发挥其功能，还是作为连接，电站与电缆铺设需要同步建设。根据85/337/EEC号指令的第三章和第五章，如果电缆铺设会对环境有重大影响时，必须对电站和电缆所带来的环境影响进行评价。（楷体部分为1993年11月11日，欧盟委员会给英格兰乡村保护委员会的一封信）

由于升级计划将增加国家电网公司（NGC）从苏格兰至英格兰的电力输出，英国政府认为在Wilton案例中铺设线缆的首要作用并不是为了服务于新建电厂。但是，在后来的公众质询中，国家电网公司（NGC）提供的证据似乎与这一观点相矛盾。尽管升级计划将为国家电网公司（NGC）带来巨大的经济效益，但是通过提议的主要理由以及实际提议的方案，可以很明显地判断出项目成立的主要理由就是Wilton电站需求。政府同时也认为，指令允许电站建设和电缆铺设实行分开的环境影响评价程序，因为前者属于附件Ⅰ中的项目，而后者属于指令的附件Ⅱ中的项目（仅在可能有重大影响时要求做EIA）。然而，指令明确要进行直接影响和间接影响的评价，对该电站建设来说，这样的做法并不能保证电缆铺设项目在其环境影响评价范围内。委员会的答复很显然得到了此指令解释的支持。

尽管澄清了指令的目的，欧盟委员会决定在这个案例中不对英国政府采取侵权诉讼。而早些时候，欧盟委员会在关于海峡隧道连接线与Kings Cross终点站项目的环境影响评价中就采取了侵权诉讼，欧盟委员会强调由于在选址或路线选择时的相互影响，所以这两个工程是不可分割的：

将伦敦海峡隧道工程分成铁轨连接和终点站两部分，实际上是绕开85/337/EEC号指令，只有作为整体来进行评估，才能将选址对环境的影响降到最小。无论是终点站的选择会对铁轨连接造成影响还是铁轨连接会对终点站选址造成影响，二者都是密不可分的。完成终点站的评估后再去评估铁轨连接线，这种方式不会让终点站的评估被认可，因为这种做法没有考虑终点站的选址对铁轨连接路线的选择所造成的影响，同时与指令第三章的内容相悖。(1991年10月17日环境委员会委员致英国政府的信件)

同样的情况看起来同样适用于Wilton例子。电站和输电线缆也是不可分割的，因为如果没有新电站修建，就不需要更多的输电线，并且电站选址对输电线线路的选择也是至关重要的。

9.2.3 Lackenby-Shipton电线的公众质询

如上所述，新建电站的电缆铺设是由一独立开发商（即NGC）负责，并在之后进行了独立于电站主体的审批程序和环境影响评价。在该案例中，公众质询同时考虑了5个路线选择的替代方案。不过公众质询开始于1992年5月，即在电站批准后18个月左右，而实际上，在此阶段，电站的建设已经开始了。

拟建电站线路开始于邻近Wilton电站的Lackenby，通过南部线路或北部线路延伸到Picton，再从Picton经由西部、东部或中心线路向南延伸至York西北部的Shipton（图9.1）。根据目前选择的路线，新电站线路以及系统升级总长度为75～85km。NGC更希望选择较短的路线，即经由南部线路从电站到Picton以及经由西部线路从Picton到Shipton。所有的拟建路线都经由或邻近风景名胜区或自然保护区，包括North York Moors国家公园和

Howardian 山，一个自然风景区（AONB）。在该提议的公众质询中，主要反对者包括拟议路线途经的当地权力机构（北约克郡委员会、克里夫兰乡村委员会以及其他权力机构）、国家土地所有者组织、国家农户联盟、英国乡村保护委员会以及居民（包括农民和当地居民）。公众质询中提出的主要问题包括电缆塔及高架输电线的视觉影响、电磁辐射引起的潜在健康影响以及需求、替代方案对农田耕作的影响。

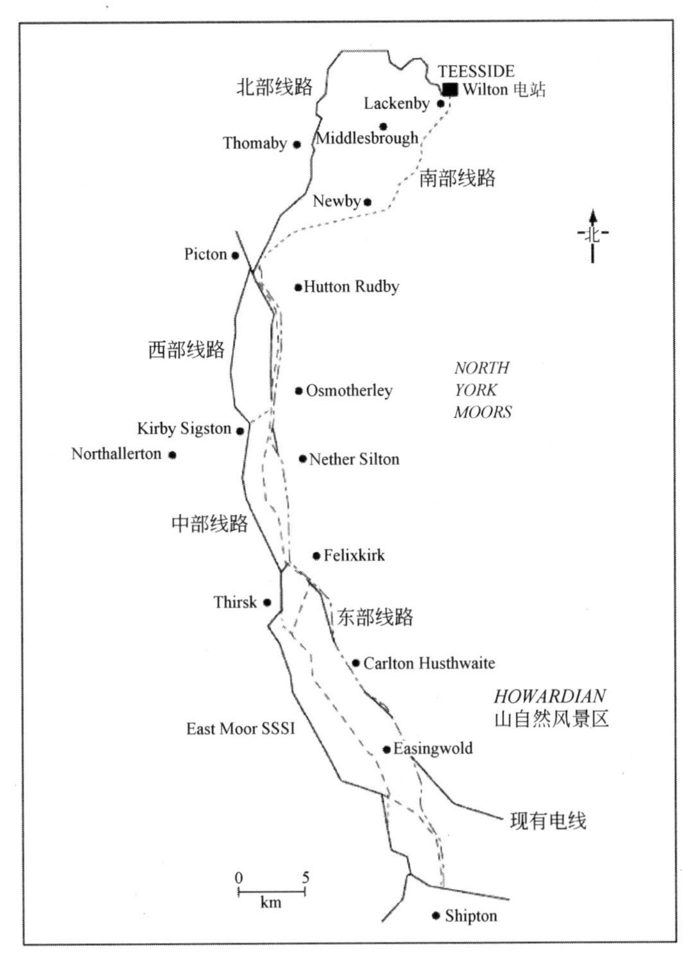

图 9.1　考虑了替代路线的北约克郡电站线路质询

在质询中，英国乡村保护委员会评论认为，这些做法所造成的视觉影响是不可接受的，并且在更早的阶段就应该被预料到。该委员会还强调质询监察员“不应该因为发电站的建设获得批准并且正在建设，就不得不同意电站线缆的施工”（Sheate，1995）。在这个电力生产和电力传输的审批分开进行的案例中，监察员也受邀对此类案例中现有的环境影响评价程序的不足之处进行评论。

该质询于 1992 年 12 月结束，经过长时间拖延后，监察员报告最终于 1994 年 5 月公布。该报告认为，在进行各种调整并将环境影响减至最小的前提下［例如，将 East Moor 周围的一个科学研究地点（SSSI）考虑在内］，批准国家电网公司对优先路线的选择，即南部路线从 Lackenby 到 Picton，西部路线从 Picton 到 Shipton。不过在下面的案例中，监察员也会同意英国乡村保护委员会关于环境影响评价程序的观点：

在我们看来，选择电站地址时没有考虑到所有相关因素，包括向用电地区输电等，这将很有可能导致高压电线延伸至目前尚未受影响的地区，而且会使已经受到影响地区的电线增加。毫无疑问，从塔架的规模和形状上看，这些路线将不可避免地严重扰乱和毁坏几乎所有的景观，以至受影响的地区不再被人们喜爱。

在我们看来，需要更加慎重考虑所有程序的执行，以保证在批准未来电站的建设时能考虑到其所带来的传输需求，并考虑到在拟议电厂和用电地区之间因国家电网的必要扩张和增加而带来的环境影响。(监察员总结，1993 年 9 月 23 日)

对 Wilton 电站解决电力传输问题的环境影响评价的失败导致“Teesside 电力有限公司(开发商) 既不用负责电站选址和电站发展所涉及的所有问题，也不用承担全部的经济和环境费用”(Sheate，1995)。因为成本的限制，剩下的向各个发电项目提供输电服务的成本会由国家电网公司补偿。由于将电线的环境影响最小化所带来的任何额外成本 (例如，采用的较长线路通过较少敏感区或将所有或部分线路布置到地下等) 都将由国家电网公司而不是由电站开发商承担，这就意味着国家电网公司要在压力下做最经济的选择。

9.2.4 环境影响评价的教训

Sheate (1995) 评论认为，Wilton 电站案例能充分证明当时供电行业审批程序与环境影响评价指令的内容和精神是相违背的。他认为，这种情况可以通过修改电力和管线项目环境影响评价标准进行补救。建议所做修改如下：

环境报告书应包括与发电站相关的可能对环境产生重大影响的输电线项目和其他基础设施所产生的影响的评价以及其所牵涉到的所有其他问题。(1993 年 9 月 22 日，CPRE 致 DTI 的信件)

修改后将会产生这样的效果：输电线及相关配套公共设施是否会对环境产生重大的影响，将成为国家能源部决定是否批准发电站建设的重要条件。

这样修改的结果将确保电站建设的支持者不得不考虑所牵涉的输电问题，这些问题也会成为环境影响评价以及后续公众质询的部分内容。该修改结果也减少定义项目和方案产生的困难。(Sheate，1995)

这个例子也指出了《环境影响评价指令》中广泛存在的问题，即“项目”这一术语的定义很模糊。正如我们所见，供电行业项目仅仅是一个例子，其他基础设施项目也存在相同的问题，特别是公路、铁路和其他交通运输规划。一个完整的项目能否被分成几个小的部分，从而能通过环境影响评价并得到批准，这一问题在 EC 中产生了很多争议。问题在于，和大多数欧盟国家一样，英国已经将环境影响评价作为已有的批准程序的一部分，如果将一个项目拆开，环境影响评价也会随之分开。这种做法是违背指令的目的的，指令中表明“对环境的影响应在所有技术、计划和决策过程的最早阶段考虑”(指令 85/337/EEC 导言)。正如本案例研究所述，如果环境影响评价仅仅应用于单个项目而不是项目整体，那么指令的目的将无法完成。

在该案例研究中所提出的问题仍然与目前的环境影响评价实践有关。本案例中对项目模糊的理解问题，在之后经过修改并指挥欧盟行动的《环境影响评价指令》中并没有得到解决，对于电站建设和它的输电线路建设依然执行分开的审批程序。欧盟《环境影响评价指令》的实施对解决项目的“香肠切片”问题能有多大帮助，目前仍不清楚。

9.3 N21连接公路，爱尔兰共和国：环境影响评价和欧洲栖息地保护

9.3.1 项目介绍

本案例最早是由 Weston 和 Smith（1999）提出的针对爱尔兰共和国凯瑞县公路的改进方案。规划公路线路穿越了欧洲的部分保护栖息地——著名的残余冲积森林 Ballyseedy Wood。尽管当时该方案并未被要求进行环境影响评价（主要由于当时生态现状并没有引起足够的重视），但不久后却经过了相关的程序［被称为“适宜性评价”（也称为生态调控评价，HRA）］，该程序隶属于《欧盟栖息地指令》。该指令规定项目若对欧盟认定的优先级栖息地有一定的环境影响，则必须进行相应的环境影响评价。该评价包括一系列的测试，只有通过这些测试，项目才能得到审批部门的批准并继续进行。本案例阐述了这些测试的性质和说明，并体现了欧盟在优先级较高的栖息地保护方面的重视程度（Weston 和 Smith，1999）。

9.3.2 提议

自 20 世纪 60 年代后期起，Kerry 乡村委员会就反对对爱尔兰西部 Kerry 郡和 Tralee 郡的 N21 主干道进行改进，直到 20 世纪 90 年代中叶，欧盟有望为这些改进筹集资金，规划的改进工作才有了实质性进展。1994 年该路线被纳入《爱尔兰政府实施交通运输计划（OTP）》。同年，OTP 作为爱尔兰欧盟支持框架的一部分，被欧盟委员会选为合作筹资对象。在该支持框架下，欧盟同意为 N21 连接公路的改进计划提供 85%的资金支持。

拟建项目包括对 Castleisland 和 Tralee 之间现有的 N21 高速公路中 12.5km 长的路段进行整修，其中包括在 Balleycarty 和 Tralee 之间新建一段 2.4km 长的双向车道（图 9.2）。规划中，双向车道经 Ballyseedy 森林通向现有高速公路南部。但在实施过程中发现，Ballyseedy 森林是隶属《欧盟栖息地指令》指定的其中一个优先栖息地。在 1994 年 7 月欧洲合作筹资公布后，Kerry 乡村委员会作为当地高速公路权威机构，开始对拟定规划进行设计。

9.3.3 规划及环境影响评价程序

N21 公路改进规划属于附件Ⅱ中的项目，这种项目仅当其可能有重大环境影响时才被要求进行环境影响评价。与大多数欧盟成员国一样，爱尔兰沿袭了这一系列限值来决定附件Ⅱ中项目是否应该进行环境影响评价。在这种拟定（乡村双向车道）公路规划案例中，规划路段长度超过 8km 或认为有重大环境影响时就要求对规划进行环境影响评价。在此类案例中，由当地高速公路权威机构进行环境影响评价并将评价结果提交给 DoE。经过一段时间的咨询和公众参与（若有的情况下）之后，环境事务部部长才对申请做出相应决策。如果附件Ⅱ中的规定低于限值（长度小于 8km）且被认定为不可能产生重大环境影响的项目，则

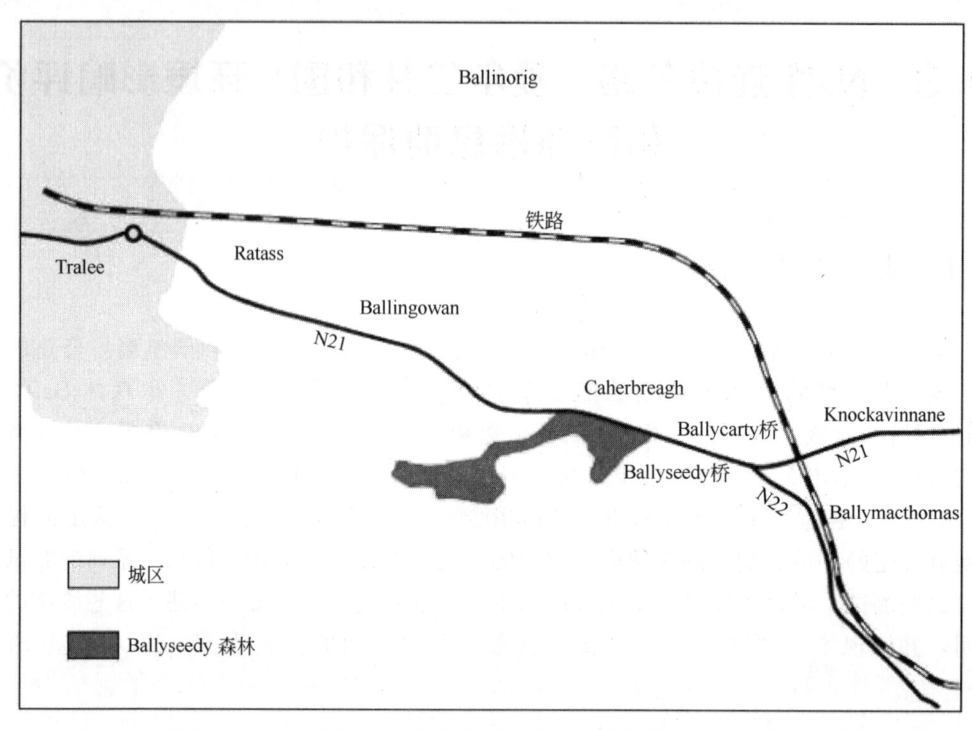

图 9.2　目前的 N21/N22 路网图

可不对其进行环境影响评价，只依照一般法规来处理。该规划会允许有一段时间的公众咨询环节，而关于是否批准规划的最后决定则取决于当地的相关权力机构。该公路规划是否需要进行环境影响评价是由项目开发商和作为当地高速公路管理部门的乡村委员会（CC）共同来裁定的。

在将环境顾问的报告应用到规划中之后，Kerry 乡村委员会决定不对该项目进行环境影响评价。这个决定是以规划中双向车道的长度为基础的（2.4km，远低于 8km 限值）。然而，在公布计划书之后，委员会收到诸多反对意见，这些意见的焦点主要集中在新建双向车道对 Ballyseedy 森林的环境影响上。在考虑这些反对意见以及随后提交给委员会报告后，委员会当选的成员决定继续执行该计划。然而，这并不是授权过程的终止，因为项目必须进行的强制收买指令（CPOs）仍然需要授权并进行公众质询（CPO 调查）。在爱尔兰的体系下，允许公众及相关利益集团在 CPO 质询中就环境问题提出证据。但是，质询以及后续决定（包括任何线路位置的替代方案）必须只能以土地的获得、通行权利和入口问题为基础。N21 规划的 CPO 质询于 1996 年 3 月举行。

在委员会决定继续执行该项规划之后，一个反对该计划的地方组织受委托为 Ballyseedy 森林做了相应的生态评价。评价显示该森林由残留积土森林组成，尽管目前缺少保护，但该森林仍符合 1992 年《欧盟栖息地指令》对优先栖息地的定义。这些结论得到了爱尔兰相关国家权力机构和 Kerry 乡村委员会的认可，之后按照指令该场所被建议定义为特殊保护区。Ballyseedy 森林重要生态现状的发现，导致了许多关于该项目合作筹资问题的正式诉讼被提交到欧盟委员会，这些诉讼认为，如果该项目对重要栖息地有潜在危害，委员会就应该重新考虑对该项目合作筹资的决定。这些诉讼导致欧盟授权开展了一项独立研究，为是否需要重新考虑规划的合作筹资提供建议。

9.3.4 欧盟栖息地指令

《欧盟栖息地指令》需要所有成员国制定特殊保护区来繁衍重要的野生物种。与依照鸟类指令制定的特殊保护区一起，这些地点将形成欧洲栖息地保护网，称为“自然 2000”。《欧盟栖息地指令》用于保护欧洲范围内此类地点组成的网络完整性，并为“自然 2000”和优先级栖息地提供保护。

当一个项目可能会对受保护区域产生重大影响时，指令规定必须“结合保护目标对该地区所涉及的问题进行适宜性评价”。根据指令，规划的适宜性评价结果只有满足以下两种情况之一时可能被批准：(a) 推断项目对该地点的完整性不产生负面影响；(b) 当预见到可能产生负面影响时，有相应的替代方案可减轻该负面影响且不与公众利益相冲突。指令的整体目的是“用替代方案来避免对有限栖息地造成的负面影响”(Weston 和 Smith，1999)。项目对一个有限栖息地的完整性有负面影响，但其替代方案若能通过相关测试，该项目就仍能继续执行。但在这种情况下，开发商必须提供补偿方案来补偿有限栖息地的损失。Huggett (2003) 对与一系列英国港口发展计划相关的此类措施的发展进行了讨论（见 12 章）。

指令中的测试是不完整的，它需要一定程度的解释。如“缺少替代方案”测试的书面解释是：应该选择任何可使受保护栖息地损失少于现有项目的替代方案，而不考虑成本或对其他利益的影响。《欧洲判例法》(*Case Law*) 为指令测试的适当解释提供了一些思路，Weston 和 Smith (1999) 对其进行了回顾评价，他们推断鉴于指令为优先级栖息地提供了重要的保护，“缺乏替代方案”及“公众利益因素”测试都应加以严格的解释。目前已将这些测试的探索应用于 N21 连接公路计划中。

9.3.5 拟建 N21 连接公路的适宜性评价

根据《欧盟栖息地指令》第六条，针对 N21 项目的适宜性评价涉及一系列的连续测试，包括：

① 项目对优先栖息地完整性的影响；

② 有无替代方案；

③ 不考虑公众利益规划而继续执行的迫切原因。

正如 Weston 和 Smith (1999) 报告的那样，参照欧盟授权的独立研究结果依次对这些测试进行了检测。

9.3.5.1 对优先栖息地完整性的影响

Ballyseedy 森林总占地面积达 41hm^2，而作为评价主体的优先栖息地仅占很小部分。森林的小块地区与《欧盟栖息地指令》中残留积土森林的定义相符。森林的北部由桤木和白蜡树组成，受周期性水患的影响。这类森林湿地在爱尔兰森林中很少见，在该区存活下来的可能性很小；优先级栖息地占地不到 0.5hm^2。该区域位于森林北部边缘地区，包括最具有生态价值的区域都将因规划中的双车道而失去。

规划直接占用的土地仅会损失掉优先级栖息地总面积的 3%。然而，这并不意味着不会对栖息地完整性产生重大影响。独立研究推断认为：

以多位环境顾问实施的评价及提交给CPO咨询的证据为基础，仅将栖息地损失评价认为是对多个栖息地完整性没有任何显著性影响是不客观的。树林被砍伐会导致河流改道（规划的另一要素）以及使得栖息地和水文模式发生改变。CPO调查的证据表明，在所占土地区域外的森林地区也将受到这些改变的影响。作为欧盟政策的一部分，需要应用预防的原则来评价项目对优先级栖息地的影响。应用这一原则，必定会推断出栖息地的完整性会因拟建公路而减少。(Weston 和 Smith，1999)

因此我们可以看出，该项目对优先栖息地的影响是负面的。但是，如果项目同时满足余下的两条测试，那它仍然可以继续进行。首先要考虑的是缺少替代方案的问题。

9.3.5.2 缺少替代方案

乡村委员会对拟议规划首选路线以及可能替代路线的识别部分是以限制性测绘工作为基础的，并识别出该地区的各种环境限制因素（如考古学遗址）。在选择首选路线时也突出消费因素，并进行了成本效益分析（CBA）。委员会准备了一份设计报告，该报告陈述了项目需要及考虑的替代方案。这表明委员会已经识别并调查了许多替代方案。Weston 和 Smith (1999) 为拟定规划找到了6个主要替代方案，一般都选择更靠北的路线。但某些替代方案会对其他方面造成影响，例如毁坏居民财产、农业割裂和重新安置 Ballyseedy 森林纪念碑——当地的战争纪念物。尽管存在着这些影响，显然许多可行的替代路线方案能有效地满足拟定规划目标。

乡村委员会测试了各种路线方案为“人类安全、容量及经济可行性”的问题提供最好的解决办法的能力。然而，即使在委员会意识到 Ballyseedy 森林的重要性之后，似乎也没有尝试系统的测试替代方案能否满足避免优先级栖息地损失的需要。独立研究的结论认为“这些替代方案没有像首选路线那样进行同等级的严格检测，而且它们也没有经过清楚的定义和质量认证。最终这些替代方案都被否定了”(Weston 和 Smith，1999)。

问题在于路线选择时乡村委员会没有考虑到拟建公路是为未来发展扩张的区域而服务的。在该例子中，这些区域是指 Tralee 北部边缘地区，该地区有新建的区域技术学院及配套工业区。这就意味着在位置上更靠北的双向车道能够避免对 Ballyseedy 森林的影响，而且与拟定规划相比，该定位可以容纳下计划开发而产生的交通增长量。拟定规划中服务于计划发展地区的交通线将通过 Tralee 的中心地区。这样会加剧城镇已有的交通问题，而且 Tralee 交通堵塞的增长可能会丧失因改进 N21 公路获得的时间效益。独立研究评论认为：

N21 和 N22 公路将计划用于未来发展的 Tralee 地区和公共基础设施连接起来。令人惊讶的是委员会并没有对这两条公路替代路线的位置进行充分的调查研究。由私人个体提出的满足双向车道标准的更北路线并没有依据实施运输计划的战略目标或依据其他利益（如避开 Ballyseedy 森林和维持 Ballyseedy 森林纪念碑附近地区的特殊性等）而做出充分的评价。对其他可能路线似乎也没有进行充分的考虑，例如没有考虑铁路南部或北部路线（通向现有 N21 公路的北边）。由于跨越铁路线、农业割裂及对财务的影响等因素，委员会的设计报告否决了这些线路，但是似乎并没有进行充分的调查和评价来彻底否决这些选项。

总之，没有迹象表明委员会的替代方案得到了与优先方案同等程度的测试。没有按照详细成本、调查、时间成本的节省、生态影响或实际满足建设筹资战略目标的能力等方面来考虑替代方案。所有的替代方案都缺乏针对清楚客观的决定标准所进行的严格测试，因此在此案例中就不能证明替代方案会毁坏优先级栖息地而不能达到 OTP 的目标。(Weston 和

Smith，1999）

第二项测试由于缺少替代方案，因此也失败了。按照《欧盟栖息地指令》，既然许多可行的替代方案在这个案例中是有效的，该项目就不能继续进行，因此就不必进行第三项测试（即为支持该计划不考虑公众利益的必要的理由）。但是，演示这项测试如何在实践中应用是大有裨益的。

9.3.5.3 公众利益问题

第三项测试涉及优先栖息地损失与必要的公众利益问题间的平衡。如果公众利益问题导致“比栖息地损失更重大的影响”（Weston 和 Smith，1999），它的重要性将超过栖息地损失。例如，如果预测到原规划仅对栖息地产生微量影响，而替代方案会给公众利益或经济带来很大损失，这时就可以说公众利益问题的重要性超过栖息地损失。相反，如果对栖息地的影响很大或难以确定而对公众利益问题影响小时，那么可优先考虑栖息地利益。

《欧洲判例法》为这类公众利益问题提供了指引，这些公众利益可被认为是“势在必行”的原因。案例中包括了经济社会结合的公众利益、人类健康、公共安全以及其他环境考虑。然而：

当公众利益大于栖息地损失时，他们必须同优先栖息地保护具有相似程度的重要性——就整个欧洲团体是有利的——而且是可论证的。（Weston 和 Smith，1999）

在 N21 规划案例中，乡村委员会认为许多公众利益问题是相关的，这些问题加起来的重要性就超过 Ballyseedy 森林有限栖息地的损失。该案例中出现的公众利益问题包括：

① 更广泛的 OTP（规划为其中一个部分）战略目标；

② 替代方案的成本；

③ 家庭住宅损失；

④ 公路安全问题；

⑤ 对 Ballyseedy 森林纪念碑遗址的影响；

⑥ 农业割裂；

⑦ 对考古学的影响。

对规划的独立研究总结：没有一个可以看作是势在必行的（即同栖息地的损失具有同等重要性）和高于一切的（即其受损害的程度足以不去考虑对栖息地的保护），因此第三项测试也失败了。其原因如下：

① 替代方案的存在。事实上很多路线替代方案都是有效的，这使得针对案例所产生的不可避免的公众利益问题的论证变得困难。例如，如果采纳某一替代方案，就需要重新安置 Ballyseedy 森林纪念碑，这就会引起当地人的极大关注。但是，6 个主要替代方案中仅有 1 个需要重新安置纪念碑，因此可通过采纳其他替代方案来避免。

② 同拟定的规划相比，替代方案不一定会带来更有害的影响。例如，除了一个路线方案之外，没有证据表明其他的任何替代方案会比拟定规划带来更多的家庭住宅损失。

③ 由于缺乏资料，许多公众利益问题是不可论证的。例如与拟定规划相比，关于替代路线的公路安全隐患的资料数据不足。在 CPO 调查会上来自乡村委员会的证据表明，委员会检查的所有替代方案是同等安全的。而且既然没有计算出所有替代路线方案的成本，就不可能证明拟定规划在成本上一定是最合理的。事实还表明“其中一个替代路线在建设上可能更便宜，因为它可减少对现有公路使用者的影响，减少建设对现有廊道性质的影响并且减少

减排成本”（Weston 和 Smith，1999）。另一个问题就是替代路线对农业割裂的影响。但是同样地，“没有证据表明该地区有过任何详细检查或以任何方式进行定量评价，或与拟定规划做过对比评价”（Weston 和 Smith，1999）。

④ 与大部分提出的公众利益问题相比，栖息地损失的利益更大。该案例所产生的大部分公众利益问题的重要性都不能与有限栖息地的重要性相等同，因此不能认为是“高于一切的”利益，如家庭住宅损失和农业割裂。尽管在当地这些是很重要的问题，但根据优先级栖息地在《欧盟栖息地指令》中的地位，这些问题不能被看作与保护优先栖息地的需求具有同等重要性。同样，相关的考古学影响，若要成为“高于一切的”利益，受影响的考古学特征必须在全欧洲范围内具有比优先栖息地更高的等级，但没有证据表明任何替代路线会产生此类影响。

一个看起来更重要的公共利益问题就是需要重置 Ballyseedy 森林纪念碑，而该问题仅出现在其中一个替代路线中。

委员会及当地社区团体普遍认为重新安置纪念碑是不可接受的，因为其被认为是爱尔兰最重要的现代纪念碑之一，然而该纪念碑没有受到国家或当地法律的保护。相反，按照《欧盟栖息地指令》，Ballyseedy 森林却在欧洲水平上受到法律保护。在此基础上，根据指令建立的保护优先栖息地推测，虽然纪念碑的重新安置清楚地成了非常重要的公众利益问题，但是纪念碑的重新安置不能视为“高于一切的”利益。（Weston 和 Smith，1999）

在“适宜性评价”程序中要求的三个测试都失败后，拟定 N21 连接公路的欧盟基金也被撤回。

9.3.6 结论

一旦确定项目对优先栖息地会产生消极影响，依据《欧盟栖息地指令》所进行的适宜性评价将是一个严格的过程，特别是很少有项目确实缺少可行替代方案，尤其是在对可能替代方案的调查已被广泛定义的前提下。而且，若要在重要性上超过优先栖息地的损失，公众利益的重要性必须在欧洲水平上等同于或高于对优先级栖息地的保护。这就意味着如果问题的重要性仅停留在地方或者国家水平，那并不足以超过栖息地的损失。综上所述，很难将缺少替代方案和必要的原因测试联系起来。虽然“替代方案存在可行性，但并没有高于一切公众利益来判定最优解决方案”（Weston 和 Smith，1999）。EC（2000，2001，2007）以及第 12 章提供了对适宜性评价程序的进一步指导。

9.4 Portsmouth 焚化炉——环境影响评价中公众参与的新方法

9.4.1 项目介绍

这个案例研究涉及了环境影响评价中公众参与的创新性方法，该方法是在对英国 Hampshire 郡 Portsmouth 市拟建的市政废物焚化炉的环境影响评价中提出的。在该案例中，开发商所采用的方法使得公众成员在规划申请和评价报告提交之前就有机会参与到关于拟建项目及其环境影响的有组织的讨论中。该方法扩展了公众参与，超出了欧盟《环境影响评价指令》的要求，在近几年里已被应用到英国废物部门的多起案例中，该方法不仅应用于项目水平上（本案例），也可应用于如地方废物管理政策和规划发展类战略水平上（Petts，1995，2003）。

这些方法的广泛使用反映出，在颇有争议的废物处置设施规划中，比较传统的公众参与形式已不能满足要求。但是，在为公众及其他利益相关方提供机会参与到环境影响评价和更广泛的发展程序中，该方法的有效性仍存在问题。这里描述的例子主要以 Chris Snary 的博士研究为基础，是 Chris Snary 与牛津布鲁克斯大学的 IAU 合作进行的（之前已被收录为文件 Snary，2002），其他材料来源于 Petts（1995，2003）。

9.4.2　公众参与和环境影响评价

该案例研究的大背景是附近居民对拟建的废物处理场几乎都持反对意见。此类公众反对意见通常都表现为"别在我家后院（NIMBY）"（即别影响我的生活环境）或是对潜在影响无依据的、不理智的恐惧，尤其是与排放或健康风险有关的影响，而这些意见常常被简单地忽略掉。这与 EIA 从业者对影响和风险实施的科学技术性评价是相对照的。但是，Snary（2002）指出最近的研究表明公众的反对基于更多考虑，包括"对废物管理方案的恰当性的关注、废物工厂的信赖度以及决策程序的公正性"。

许多评论员提倡在废物管理规划过程的各阶段都要与公众进行交流，以便更好地理解公众反对意见的内涵（ETSU，1996；IWM，1995；Petts，1999）。此类交流可以采用各种形式，包括开发商单方面向公众发布信息，或不同级别的咨询和公众参与（公众与开发商或批准政府之间可进行双向交流，公众的观点可被合理地加入决策过程中）。所有的这些交流方式都可被视作焚化炉和其他废物设施规划和环境影响评价程序的重要组成部分，正如 Snary（2002）所解释的：

健康风险需要关于污染物排放预测的综合性信息与咨询程序，使得公众的观点能够通过此程序来影响决策。评估废物处理场的能力和调节者的风险管理能力都需要公众参与，这样就可以公开表述公众的担心并就能力条件进行讨论。对基本政策问题和废物规划程序的合理性的讨论也需要公众参与，这样在废物焚烧规划的制定阶段就可达成一致意见。

很多评论家特别指出，对改进的公众参与方法的探求也是同对科学及专家社会信任度的提高相关联的（如 House of Lords，2000；Petts，2003；Weston，2003）。

9.4.3　拟定规划的背景

Hampshire 郡是英国南海岸的一个郡，拥有约 160 万人口。在 20 世纪 80 年代末期，该郡面临生活垃圾量日益增加的问题，与之相对应的是垃圾焚化炉的老化（不能达到最新的排放标准），而且更难找到新的、环境可接受的填埋场地。为了应对这种状况，郡委员会在 1989 年提出了废物管理计划，该计划提倡废物的综合管理，强调循环利用和废物最小化，减少对填埋的依赖。当时政府实施的财政制度（非石化燃料合约）会提供资金，以激励支持利用废物获取能源的计划而非废物填埋。乡村委员会认识到与建立几个小的焚烧厂相比，一个郡内建一个大的焚烧厂会取得很大的规模效益。由此，经过招标之后，郡委员会于1991年底提交了一份在该郡南部的Portsmouth建立一个大型废物再生能源焚化炉的申请(Petts，1995)。该厂的处理能力是40万吨/年，相当于该郡生活垃圾量的2/3（Snary，2002)。拟建的厂址在该郡的一个废弃的焚化炉处，这个焚化炉因不能满足最新的排放标准而于1991年关闭。

该计划遭到了来自当地居民和相关地方当局（Portsmouth委员会）的众多反对。反对的焦点集中在很多环境问题上，包括由工厂排放产生的健康风险、视觉、噪声和交通影响以及工厂的选址距居民区太近等。也出现了人们对政策的担忧，特别是作为首选的废物处理措施，集中焚烧会对在全郡范围内促进废物回收和废物最小化产生不利影响。Portsmouth委员会（当时负责批准此类项目的权威机构）考虑到计划太大，而且也没有为该地区形成更综合的废物管理体系，因此决定不支持该申请（Snary，2002)。拟定规划不能得到批准使得郡委员会改变了方法，正如Petts（1995）所解释的：

直到1992年夏天，郡委员会都没能使工厂得到批准，而又面临着为废弃物处理问题寻找解决办法的紧急任务。传统方法已经失败了。虽然支持废物再生能源的郡废物管理计划已经进行了公众咨询，并得到了很多支持意见，但现在该计划被认为是个很被动的过程，郡委员会似乎没有认识到社区最关心和需要优先考虑的因素。用集中处理来进行废物管理的方法并未获得强烈支持，对将废物中产生的能源“卖”给公众的方法也几乎无人重视。建议者对他们的能力过度乐观，他们认为自己有使用标准的、以信息为基础的方法使得项目通过的能力。

面对这些问题，郡委员会着手于制定更完整的而且可被公众接受的生活垃圾管理政策(Snary，2002)。委员会的新方法涉及在1993年发起的更积极的公众参与计划，以检查各种各样的生活垃圾处理方案。其目的是试图“为更换新设备提供广泛的公众支持”（Petts，1995)。作为程序的一部分，以公民专门小组形式为基础，在郡的三个地区建立了社团咨询讨论会。其成员由不同利益相关方和背景的人混合组成，也包括对废弃物问题有着极少知识的人，在该程序（持续了6个月）的最后，讨论会向郡委员会陈述了他们的结论。达成的广泛一致意见是：

① 应该在废弃物减量和循环利用上做更大的努力；

②《废物再生能源规划》将作为废物综合管理政策的一部分，但要对其环境影响及工厂的监测给予相当多的关注；

③ 垃圾填埋是最不受欢迎的选择（Petts，1995)。

Hampshire郡的公众参与导致了修改后的废物政策的出台（公布于1994年)，其计划建立三个小的由废物再生能源的焚化炉（每个处理能力为10万～16.5万吨)，来代替最初拟定的一个大焚化炉。新厂分别位于Portsmouth（地址与早期申请相同)、Basingstoke附近的Chineha和Southampton附近的Marchwood。1998年开始对这些拟定发展计划进行环境影响评价（Pens，2003)。这使得这些工厂第一次成为案例研究的焦点。

9.4.4 联络小组程序

对Hampshire的三个拟定焚化炉所进行的环境影响评价都用到了叫作“联络小组”的一种公众参与方法。该方法是公众质询的扩展，即在准备ES的过程中通过一个联络小组就一系列关键的当地利益问题进行公众质询。这些联络小组是由开发商（Hampshire郡废弃物服务部门，HWS）建立的，是郡委员会在合约中对公司所提要求的一部分（Petts，2003）。在ES提交给法定机构之前，该方法有能力确保公众拥有对ES所处理的问题进行再评估的能力，并且有能力改变项目中的建议和减缓措施。

Portsmouth联络小组的参考条款表明：①做出合理的决定；②协助开发商（HWS）以确保当地社区成员的观点能得到理解和响应（Snary，2002）。在该案例中，为扩展公众参与所做的安排已经超出了英国《环境影响评价法案》（在第6章讨论过）的法定要求，也是此类方法第一次在英国废物焚化炉的环境影响评价程序中使用（Snary，2002）。

在Portsmouth焚化炉案例中，联络小组包括10名公众成员——他们是由HWS选出来的，代表一系列当地利益团体，包括一名来自当地学校的代表、一名来自FoE地方部门的代表和一名来自Portsmouth环境讨论会的代表以及7名来自距离项目场所最近的6个邻近地区讨论会的代表。小组成员被鼓励去与他们邻近地区的当地居民进行网络联系。很明显的是作为小组成员并不意味着支持项目计划，而且事实上大部分参与者都持反对意见。

在1998年8月计划实施方案和环境评价提交之前的6个星期内，联络小组每星期会碰一次面。在会上联络小组考虑的问题包括：

① 由废物再生能源的焚化炉及环境影响评价；

② 工厂的设计；

③ 噪声和交通评价；

④ 视觉和生态问题；

⑤ 替代地点和噪声问题；

⑥ 空气质量问题和健康风险评估。

在会议上由HWS及其顾问就这些问题提供信息。在讨论中，参与者可以通过提问和建议来发表他们的观点。当天HWS就这些问题就会予以回答，并在下次会议上给出书面答案，也会召开闭幕会来讨论ES的结论。HWS指派一位独立的主席“保证所有的参与者有同等机会为会议出谋划策，并使问题得到公平的处理”（Snary，2002）。

9.4.5 程序评估

用于该案例中的公众参与方法效果如何？该程序的各类参与者又持有什么观点？在对参与者采访的基础上，Snary（2002）对Portsmouth案例中联络小组程序的成功应用做出了评价；Pens（2003）利用对三个联络小组的观察资料也对这个程序进行了评估。以下是总结出的关键结论，尤其着重于该程序在实践中的局限性。

① 环境影响评价程序中联络小组程序开始得太迟。在ES提交前的6个星期才召开联络小组会议，这个阶段环境影响评价的主体部分已经完成。这样就会使联络小组左右影响的定义、评价方式等的时机受到限制。特别是由排放物产生的健康风险仅在小组最后一次会议中才讨论。如果在环境影响评价范围确定阶段就开始，该程序将更加有效。然而，在实际确定范围时，仅限于向当地规划部门和法定咨询机构咨询，没有公众参与（Snary，2002）。

② 未给该程序提供充足的时间。许多参与者评论，会议试图在很短时间内讨论很多内容，这些内容通常都非常复杂。这就再次表明程序应该尽早开始，以使得所涉及的一系列问

题都能得到充分的解决。

③ 对环境影响评价顾问做出批判。一些参与者对参与准备环境报告的顾问进行了批判。批判意见包括：认为评价大多是以案头研究为基础，顾问对当地的详细资料缺乏了解；顾问不能总是对问题做出充分的回答；环境影响评价工作本应该由独立顾问承担并加以阐述。

④ 参与者应该对提议更加了解。几乎所有的参与者都表示由于参加了会议，他们了解了更多的与提议相关的问题。这并不让人惊讶，但开发商的项目主管也指出同传统的公众参与方法相比，该程序能更好地为重要的公众成员提供信息（Snary，2002）。可是，参与者对提供的与发展产生的健康风险相关的复杂信息表示怀疑。一位参加者评论说“我不是一名科学家，我觉得那很难理解。我感觉他们好像想用数据和技术条款来蒙蔽我。与我交谈的一些去看过环评报告的居民也有完全一样的感觉，他们不能真正理解这个评价”。这些批评与联络小组程序紧张的时间安排有一定关系，尽管没有哪位专家总是需要对环境影响评价所包含的技术信息提供置信度。Snary 建议可以通过采用独立的顾问或独立的三方来总结提出的信息并使信息有效化，以此增加此类信任。

⑤ 仅对发展建议产生有限的影响。联络小组会议似乎仅对规划建议产生了有限的影响。开发的项目主管表示程序改变了该项目的建筑（特别是建筑的颜色），并使交通评价有所改进。可是，除了这些相对很小的改变，许多参与者怀疑小组的观点对计划还能产生什么其他的影响。在规划的计划、设计及环境影响评价工作最后阶段才召开联络小组会议，在这种情况下，这些发现并不让人惊讶。

⑥ 对开发商和顾问的可信度较低。所有参与者都认为开发商和顾问的可信度很低或一般。其原因是感觉开发商不公正，因为他的目的就是使计划获得批准，并认为在健康风险和考察环境影响评价顾问的能力等方面，小组成员被告知的信息并不完整。

⑦ 程序没能解决提议所关注的基本问题。在联络小组程序的后期，除一个人之外几乎所有的参与者仍然对计划的风险性有着强烈的关注。因此：

尽管联络小组能够更好地将废物产生的风险告知给当地的重要利益相关者，但不能使小组中多数人相信风险是可接受的，也不能相信废物再生能源的焚化技术是恰当的废物管理方法（Snary，2002）。虽然如此，Portsmouth 焚化炉计划仍作为郡级的废物处理计划出现，并通过扩展的、创新的公众参与活动发展起来。Snary 将其归结于早期战略水平的、不够充分的咨询活动，导致对于废物再生能量的焚化计划在该郡废弃物政策中所应扮演的合适角色的问题未能达成一致意见，而之前多数的联络小组成员也并没有意识到这一点。对于在政策咨询会上陈述的观点如何影响该郡废物政策的发展也不甚明了。

通过对 Hampshire 郡的联络小组程序所做的评估，Petts（2003）取得非常相似的结论：

该程序确实将环境评价展示给小而有代表性的公众团体，并接受他们细致的询问。但可论证的是，该程序开始得太晚，在有限的调整阶段，联络小组成员不能对考虑和评估的问题进行构造和定义。通过作者本人对程序的观察和评估，显然可以针对评估方法提出一些问题。例如，Portsmouth 联络小组以当地公路循环知识为基础，识别出运输评价中存在的不足。由于这样一个公众质量保证机制，确实有再评估的事件发生。可是，这是有限的。参与者很珍惜给予他们的对评价进行评论的机会，但是他们怀疑其实结果早已决定好了。

9.4.6 结论

该案例研究阐述了扩展的公众咨询方法在环境影响评价中的使用，这种做法超出了欧盟《环境影响评价指令》的最低法律要求。同时也强调这些方法在实际应用中存在困难。就这个案例而言，主要的缺点是联络小组会议在整个环境影响评价程序中开始得太晚。在环境影响评价划定范围阶段介入公众参与可以帮助联络小组避免遇到类似的问题。

需要补充的是，2000年举行公众调查会议后，Portsmouth焚化炉计划——于该地建焚化炉的最初申请提交10年后——最终在2001年10月份获得批准。

9.5 Humber河口项目——累积影响的评价

9.5.1 项目介绍

在20世纪90年代后期，在英国Humberside郡Humber河口有众多项目同时施工，本案例研究了对这些项目造成的累积影响进行评价的例子。累积影响评价（cumulative effects assessment，CEA）由各项目的开发商合作进行，这在环境影响评价中相当少见。但确实存在一些例子，如在风能开发的案例中出现了提议在同一地区建许多风力发电场的情况。一般而言，累积影响评价被认为是项目环境影响评价的薄弱环节之一（Cooper和Sheate，2002；11.3节）。

累积影响评价研究存在着很多困难，案例研究将分析如何克服这些困难及克服到什么程度。根据各参与方的不同视角，我们讨论了从累积影响评价程序中获得的效益。此案例研究是以Jake Piper的研究为基础，作为其博士研究的一部分，该研究是Jake Piper与牛津布鲁克斯大学的IAU合作进行的，之前已被收录为Piper（2000）文件。Humber河口案例研究以及英国其他一些关于累积影响评价的案例研究，也在Piper（2001a，b；2002）中做了分析。

9.5.2 Humberside河口案例累积效益分析

该案例涉及一组邻近的项目，它们是由不同的开发商几乎在同一时间提出的。这些项目都需要进行环境影响评价。由于项目类型不同，项目最后被批准通过需要经过很多机构的审批。开发商们一致同意通过合作对他们的组合项目准备一份累积影响评价，并提交给审批机构。

在1996～1997年，沿Humber河口的北岸提出了5个独立开发项目，每个项目相距不到5km。这些项目包括：

① 为Hull市提供服务的新污水处理厂；

② 一个1200MW功率的燃气发电站；

③ 一个滚装海洋渡船停泊港；

④ 一个渡轮码头的修复工程；

⑤ 一个防洪工程。

这5个拟定项目牵涉到4个开发商和5个批准机构。项目附近的环境是很敏感的，有一个欧洲自然保护地——依照《欧盟鸟类指令》和《欧盟栖息地指令》为保护区的鸟类指定的特殊保护区（SPA）——距离开发区很近。由于存在着这样一个保护地，而且5个项目又几乎是同时提出的，这就促使了在该案例中对累积影响评价的研究。事实上，依照《欧盟栖息地指令》条例，累积影响评价是为了保证特殊保护区内拟定规划的影响能满足“适宜性评价”的要求（与9.3描述的过程很相似）。希望累积影响评价有助于避免在安全批准项目中的过分延误。Piper（2000）解释道：

采纳的政策假定通过提供一个统一的评价来回复5个参与主管部门的需要……会减少这些机构之间的讨论和相互影响以避免过分的延误……但是，这项政策也意味着与任何一个和项目有关的不可解决的问题都可能同时阻碍所有的批准申请。

为了指导累积影响评价过程，成立了一个指导小组，最初由开发商和两个相关的当地机构组成。后来包括环境局及英国自然（资源）局（EN）在内的其他主要的法定顾问也加入进来。但是非政府环保组织和公众并没有直接参与。

一个代表4个开发商的环境咨询机构编制了累积影响评价。报告草案在编制过程中向法定咨询机构和开发商咨询，其咨询是以回顾和评论的方式进行的。若要明确地提出对特殊保护区产生的潜在影响，与EN（EN是负责自然保护区的法定主体）的紧密联系是该程序的重要部分，确保向地方当局及其他批准机构提交文件是很重要的，而且该文件要满足法定咨询机构的要求。

指导小组参与了确定累积影响评价的范围，但在这个研究阶段是没有安排公众参与的。在划定范围的过程中识别出可能产生潜在累积影响的问题，这些问题包括：在建设阶段，对特殊保护区鸟类的影响和对交通的影响；在运行阶段，对河口流体动力学、河水质量和水生生态的影响。开发商为研究提供了有效的数据，包括来自已有环境影响评价工作的信息，并进行了一些建模工作。提供的信息包括每个项目工程建设活动的大概时间、这些活动所需的人力及配套的交通活动。本底数据包括在一年间不同时期里特殊保护区中鸟的种类范围以及它们最容易受到的干扰（Piper，2000a）。小组通过制作一系列图表来帮助预测累积影响，这些图表将每个项目中识别的影响级别和时间综合到一起，包括：

① 主要建设活动的综合时间表；

② 鸟类受干扰的可能性（一年中每个月的敏感期）；

③ 可能影响鸟类的建筑工程列表，每个月的敏感性；

④ 开发对水生动植物的潜在影响；

⑤ 预期的交通模式（建设工作中每个月每天的车辆）。

为了达到预期目的，小组决定使用开发商预测的最佳情景，而不是使用最坏情况进行情景分析（Piper，2000a）。

累积影响预测的结果：小组提出了许多附加的缓解措施（除了那些在规划被分开评价时提出的应该考虑的措施）。例如，安排某些产生噪声的建设活动时间，如在敏感期（如鸟类栖息时）之外打桩，引入交替工作时段以减少高峰时的交通量。同时建议邻近地区的项目应该以某种方式综合设计以使得它们对环境的影响最小化。例如，修改渡船码头结构的设计以补充水处理工程出水口，并增强河口水体的混合力度及针对鸟类和水生环境的累积影响的持续监测等建议。并提出由参与规划的开发商的一个下级小组共同负责项目的筹资（Piper，2000）。

9.5.3 累积影响评价程序的费用和效益分析

通过对包括开发商、当地相关权力机构、其他批准机构以及法定咨询机构在内的参与者的采访，Piper（2000）对Humber河口累积影响评价研究有关的费用和效益进行了评估。将各不同参与者的观点总结如下，首先是项目开发商的观点。

9.5.3.1 项目开发商的观点

① 对区域和潜在影响更加了解。4个开发商中的3个认为累积影响评价程序增加了他们对河口和拟议开发活动的潜在影响的了解。例如，电站开发商谈到了他对泥滩、鸟类和潜在的交通影响都有了更好的认识，而码头开发商强调了对流体动力学、河口的形态结构以及规划与特殊保护区之间的关系有了更好的了解。

② 其他效益。包括当地关系的发展，如与其他开发商、LPAs和法定咨询机构密切的工作关系；为减缓不利影响和监测建立了牢固的基础；提供了为河口持续监测工作分担费用的机会；对其中一个开发商来说，实际上累积影响评价程序加快了计划的批准。

③ 累积影响评价程序的经济成本。对所有开发商而言，进行累积影响评价的经济成本是相对较低的，虽然事实上大部分成本是由一个开发商（供水公司）承担的，该公司承担的累积影响评价成本约为其规划环境影响评价总成本的5%。对其他开发商来说，成本还是很低的。

④ 拟建项目和附加减缓措施的改变。如上所述，累积影响评价程序引起了最初的拟建项目和减缓措施的变化，如果项目被分开评价，这些变化就不会产生。例如，在电站建设期间改变打桩操作使噪声影响最小化，修改渡船码头建设以补偿河口其他鸟类栖息地的损失，改变某些建设活动时间和交叉工作时间以减少高峰期的交通流量。所有的开发者都表明由累积影响评价促生的额外减缓措施对整个发展计划增加了相对很少的成本。这也反映出了开发商之间为减缓措施分担成本的能力。若非如此，减缓措施效用可能会减弱或其成本会增加（Piper，2000）。

⑤ 由累积影响评价程序导致的延迟。关于累积影响评价程序对拟定规划的批准是节省时间还是浪费时间众说纷纭。这部分反映出了累积影响评价程序开始时每个开发商在批准程序中所处的阶段。延迟的时间范围从供水公司的1～2个月到码头开发商的6个月（这个延误归咎于一个法定咨询机构，虽然很早就被邀请了，但参与得很晚）；电站开发商认为其时间表没有受到影响。某些延迟可能归因于一些规划项目的大量初始咨询和评估工作全部完成之后才开始累积影响评价程序。这导致了一些工作的重复。

⑥ 其他问题。一位开发商提出了由累积影响评价程序引起的改变与若分开考虑每个规划可能会引起的变化二者之间的区分问题。还应该进一步关注在累积影响评价开始之后，对该地区出现的新项目的适当处理方法。这些项目是应该并入到这个累积影响评价程序中（意味着为该程序制订一个开放的时间表）还是应该为下一组规划开始一个新的累积影响评价。

9.5.3.2 当地规划部门和审批机构的观点

拟开发地区的两个负责当地规划的部门支持了累积影响评价研究，并从程序中识别了许多有利之处：

发现该研究有助于评价在一个相对较小的地理空间提出的一些主要项目的综合影响。该研究对影响的技术评价很有帮助，并对两个委员会了解可能的影响有很大的价值。

在向开发商阐明其自身可能的影响时可能有相同的价值，使得他们充分意识到建议的可能结果。（地方当局代表的评论，摘自 Piper，2000）

重点是地方规划管理部门缺乏对环境评价进行详细评价的专门技术和资源，因此要依靠 ES 的编写者和顾问一起来识别可能关注的区域。在这方面，支持累积影响评价程序的一个主要因素就是每个规划建设者的顾问都会帮助“监视其他的项目”，从而“产生更平衡的结果”（Piper，2000）。两个部门都对累积影响评价研究中缺乏公众参与做出了评价，其中一个认为部分原因是时间紧张，极少甚至没有实施公众咨询，这代表了程序的主要不足之处。

其他的批准机构包括了 3 个政府部门［贸易和工业部（DTI），环境、交通和区域事务部（DETR）和农业、渔业和食品部（MAFF）］。贸易和工业部（DTI）评论说该研究促进了决策，还表明“若没有累积影响评价，电站项目将被否决”（引自 Piper，2000）。累积影响评价方法将被推荐在其他类似的复合多项目中使用。

9.5.3.3 法定咨询机构的观点

英国自然（资源）局（EN）作为负责自然保护区的法定主体，是该案例的首席咨询机构，在很早阶段就开始参与累积影响评价程序。若在《欧盟栖息地指令》下确保对特殊保护区（SPA）的保护，累积影响评价必须满足 EN 的要求。这些要求在各种规划批准的附属规划条件中进行了表述：

这些条件包括建设工程（通过措施减少对鸟类的干扰和所有相关人员的行程，并服从于与累积影响评价相关的其他建设项目的工程计划）的减缓措施和对建设活动的监测。概要的监测规划将贯穿于整个建设过程及后续的 5 年中以观察水鸟的活动和数量变化。只要符合了这些规定条件，英国自然（资源）局认为无论是个体的还是联合的项目都不能对特殊保护区的保护目标造成不利影响。（Piper，2000）

英国自然（资源）局的评论认为许多因素（其中一些因素对本案例来说是特有的）有助于累积影响评价完成。这些因素包括规划所在地理空间相对较小；虽然一些特殊项目环境影响评价工作已经完成，但在程序开始时所有规划都处于早期阶段；开发商之间没有为首先获得规划批准而引起的直接竞争；其中一个开发商（供水公司）主动地承担研究的启动工作（Piper，2000）。按照《欧盟栖息地指令》由批准机构负责承担“适宜性评价”，在这个前提下，后者被视为是特别重要的。正如我们所看见的，在该例子中有不少于 5 个不同的批准部门。因此：

倘若累积影响评价战略并非由供水公司及其顾问进行，在精确地划分责任方面就会存在疑问。由于这些原因，英国自然（资源）局表示虽然在这个特殊例子中累积影响评价是“一个优秀的解决方法”，但它并不是直接方法，且没有广泛的适用性，它取决于每个案例所处的具体环境。（Piper，2000）

9.5.4 结论

累积影响评价不论在项目环评中（见 12.4 节）还是在战略环评中（见 9.8 节和 9.9 节）都被广泛认为是薄弱环节之一。本案例阐述了评价累积影响的新方法，在该例子中累积影响与许多邻近的拟议开发的影响叠加。由于诸多因素的作用，评价才能够得以实施。这些包括其中一个开发商积极主动地承担累积影响评价的启动工作、参与的开发商彼此之间有直接竞争。这些条件并非在所有的案例中都适用。无论如何，此类累积影响评价研究为开发商、批准机构和其他重要参与者带来了诸多益处。而且（至少在本例中被证实）似乎只增加了相对很少的额外成本。

9.6 斯坦斯特德（Stansted）机场第二条跑道：健康影响评价

9.6.1 导言

本案例研究是关于英格兰东南部一个重要机场的扩建方案的健康影响评价。健康影响评价（health impact assessment，HIA）是一个经常与环境影响评价同时进行的程序，这二者之间有很密切的关系且部分内容相重叠。健康影响评价的目的是识别和评价拟建项目、规划或方案的潜在的健康影响（有利的和不利的），并提出最大化有利健康影响和减少或消除不利健康影响的建议（见第 12 章）。在本案例研究完成时，健康影响评价仍不是英国规划程序中的强制性要求。但与航空运输相关的政府政策认为，该机场扩建计划应该进行环境影响评价和健康影响评价。

本案例研究的对象是位于埃塞克斯郡的斯坦斯特德机场第二跑道。这个案例之所以有趣，是因为它提供了一个对重大建设项目应用健康影响评价的例证，并揭示了健康影响评价和环境影响评价的密切联系。斯坦斯特德健康影响评价是典型的“综合健康影响评价”，其特点是进行了详细的科学文献总结，在此基础上得到了利益相关者的广泛参与并使用了丰富的评估方法。健康影响评价的弱项包括如何划定某个健康影响的范围，未能充分考虑整个机场规划增长的累积影响，也未提出替代方案，使得健康影响评价对该拟议项目的总体影响较为有限。

9.6.2 项目背景

斯坦斯特德机场位于伦敦埃塞克斯郡，在伦敦市中心东北方向 35km 处。它是英国仅次于希思罗机场和盖特威克机场的第三大繁忙机场，2010 年运输旅客量 19 万人次。第二跑道项目是在 2000～2010 年中期被提出来的，这期间机场的运送旅客量刚刚在上一个 10 年中经历了非常高速的增长，单条跑道容量即将达到极限。政府在 2003 年发布的白皮书《航空运输的未来》（DfT，2003）中提出了进一步扩建的建议。白皮书为接下来 30 年英国航空运输能力发展提出了国家战略政策框架。对斯坦斯特德机场和广大东南地区，政府的结论是：

① 迫切需要一条额外跑道增加东南方向的容量；

② 首要任务是使现有跑道达到最佳利用状态，包括斯坦斯特德机场和卢顿机场的剩余容量；

③ 截至2030年时，东南方向应该建起两条新的跑道；

④ 第一条新跑道应在斯坦斯特德机场且应尽快交付使用。

在该白皮书结论的支持下，政府邀请机场运营商提出增加机场容量的计划。白皮书里也指出了不论提出何种发展规划，都需要在进行具体计划之前进行全面的环境影响评价，并希望运营商进行"适当的健康影响评价"（DfT，2003）。

作为白皮书对斯坦斯特德机场扩建的支持的回应，机场运营商英国机场集团提出了未来进一步发展的两个阶段，并分别称这两个阶段为第1代（G1）和第2代（G2）提案。G1和G2提案分别独立地提出规划申请，并分别走各自的环境影响评价/健康影响评价程序（ERM，2006，2008）。G1提案试图解除现有规划条件对年发送旅客量和航班起降数量的限制（从每年25万人次到35万人次，241000班次到264000班次）。主要通过最大限度地利用现有跑道的容量和进行有限的物理扩建（如增加航站楼）来实现。2006年，英国机场集团提交了G1规划申请，但未获得地方规划部门的通过。英国机场集团后来表示拒绝接受这一结论，并在2007年举行的一次公众咨询中重新提出了该提案。这次咨询之后，G1规划申请最终于2008年10月获得了政府的批准。

G2发展规划包括修建第二条跑道、改善附近的相关铁路和公路运力等。提案还包括扩建乘客和飞机管理设施，包括一个新航站楼、相关酒店、餐饮和停车场等。和G1发展规划不同，G2将大幅拓展机场边界，并规划了一个周期为4年的建设项目，计划在2015年时将第二条跑道投入运行。人们预计该发展规划将在2030年时为机场额外增加发送旅客量68万人次/年。G2规划申请于2008年3月提交，并准备在中央政府举行的一个公众咨询中讨论。然而，在调查开始之前，新选举出的联合政府改变了国家机场政策。新政府于2010年5月宣布将拒绝在斯坦斯特德机场和盖特威克机场新增跑道的申请。该提案立刻失去了政府的支持，并被英国机场集团撤回。

9.6.3 健康影响评价的目标和评价范围

斯坦斯特德机场第二跑道建设方案按规定应进行环境影响评价。虽然环境影响评价和健康影响评价之间有许多重要的联系，但健康影响评价被作为一个独立的程序开展。英国机场集团和埃塞克斯战略卫生局就健康影响评价的总体目标达成了如下共识：

① 识别G2项目可能产生的区域潜在健康影响（有利的和不利的）；

② 评估重大区域健康影响发生的可能性和范围；

③ 根据得到的证据提出相应建议，使有利影响最大化、不利影响最小化，并酌情提出监测区域健康影响的要求。

ERM咨询公司主持开展了健康影响评价工作，地方和地区公共健康组织、地方规划部门和申请方（英国机场集团）共同组成了一个健康主题小组，为健康影响评价工作提供支持和建议。根据完成评估所需要的时间和协商开展的程度，健康影响评价又可以被分为"快速健康影响评价"和"综合健康影响评价"。斯坦斯特德机场进行的是综合健康影响评价，体现在该项目的健康影响的预测方法以及广泛利益相关者参与的范围和复杂程度上。该评估还采用了一个广义的健康定义，包括身体、心理和社会福祉等内容。健康影响评价的范围包括由以下项目的变化对当地居民健康产生影响：

① 空气质量；

② 噪声（空中和地面噪声）；

③ 交通；

④ 就业和收入；

⑤ 社会资本；

⑥ 非自愿搬迁；

⑦ 视觉效果和光污染；

⑧ 保健和社区设施。

扩建项目的某些部分和机场外的一些开发项目被排除在了健康影响评价范围之外。包括如下几个方面：

① 与 M11 走廊建设有关的扩建部分（因为该部分在区域规划战略中解决）；

② 围绕机场周围人口增长的卫生服务基础设施规划（因为这部分由当地基层医疗信托基金通过其战略和运营计划解决）；

③ 任何相关应急规划的实施（因为该问题被假设由现有的应急规划程序解决）；

④ 气候变化对人类健康的影响（因为其影响范围比区域影响要大得多，因此被排除在健康影响评价的范围之外）。

还有在开展健康影响评价时尚未获得批准的斯坦斯特德机场 G1 扩建提案也被排除在了评价范围之外（9.6.7 节将进一步讨论）。

评价的地理范围仅限于邻近机场的社区。实际操作时，这些区域覆盖了埃塞克斯郡和赫特福德郡临近机场且最可能受到健康影响的四个地方辖区政府。该区域内更小的内部区域被要求考虑某些特定的健康影响（例如基于污染物浓度和飞机噪声等声线的预测结果）。

评价主要包括以下几个关键阶段：

① 描述项目概况；

② 描述社区概况；

③ 利益相关者的参与；

④ 评价方法开发；

⑤ 健康影响评价；

⑥ 减缓和增强影响及其监控。

每一关键阶段将会在本案例研究小节的剩余部分被讨论。

9.6.4 描述项目概况

评价的第一阶段是找出项目可能对健康产生影响的主要航线或航迹。这些“健康航线”的识别包括核查拟建项目的特征并识别其对健康因子的潜在影响。识别出来的主要健康航线均与以下几个项目特征和活动有关。

① 结构：暴露于环境（例如增加的噪声、粉尘和交通量）所产生的影响。

② 土地征用：对社区功能和社区网络（“社会资本”）的影响，对医疗保健和交通服务

的可及性的影响。

③ 航空、铁路和公路运量的增加：增加了噪声、空气污染、交通可及性及潜在伤害的暴露所产生的影响。

④ 改变当地道路：分割社区、进出社区及社会资本变动所产生的影响。

⑤ 增加就业机会：提高社会经济福祉、减少失业和社会不公所产生的影响。

也有人指出开发活动可能会影响其他并不与物理变化相关但会反映无形/感知影响的方面。这些影响包括对社区网络、社会身份、社区访问和可及性以及类似影响。大部分识别出来的方面都在健康影响评价中进行了详细的评估，但也有一小部分在这个阶段被剔除出了评价范围。这些方面被判定为不会对健康因子产生足够影响以至于导致任何健康后果。例如扬尘、恶臭、燃料倾倒、破坏公共设施及增加的交通产生的视觉效果等。

9.6.5 描述社区概况

社区概况指的是要对可能受项目影响的社区进行描述。描述主要是基于四个地方辖区政府提供的信息。但也通过对比哪些社区更可能受开发活动的直接影响，识别了一些更详细的研究区域。这些区域主要通过以下四个主要标准来定义。

① 土地征用：识别扩建机场的征地边界。

② 飞机噪声：根据2030年的航班量预测结果绘制了54dB噪声等声线。该等声线被认为是社区可能会受到影响的最低噪声阈值水平。

③ 视觉影响：扩建机场的可见的范围被定义为视觉影响区域。

④ 次要社会-经济影响：对与扩建工程和相关服务、设施的施工工人有关的次要影响，识别可能受这些次要影响影响的区域。

由此得到的研究区域并不是一个简单的以机场为圆心定义的圆，而是一个反映主要健康影响的可能性分布情况的复杂区域。该区域包括了机场西南和东北方向的26个行政区，大致反映了飞机航线的方向和机场扩建噪声的轮廓。

社区概况基本上是基于现有的二手数据，这些资料提供了一系列指标，包括人口、教育、就业、收入、住房等信息。在此基础上识别出那些发展较不平等或健康水平较差的区域。识别出这些区域是非常重要的，因为这些区域的社区更容易受到健康影响。总体而言，现有机场的周边社区被认为是相对富裕的，且研究区域内并未发现严重贫困的区域。大部分健康指标的表现都显著优于全国平均水平。

9.6.6 利益相关者参与

利益相关者参与方案被作为健康影响评价的一部分。这要求考虑对那些可能会受拟议扩建项目影响的或对其有兴趣的群体的潜在健康影响。进一步说，参与程序需要识别以下这些问题：

① 利益相关者对项目及项目对健康、福祉的潜在影响的关注情况以及如何减小这些潜在影响；

② 利益相关者对项目收益以及如何提高项目收益的看法；

③ 受利益相关者关注程度较高的问题以及他们希望健康影响评价中应当着重关注的

问题。

公众参与的地理范围与健康影响评价的社区概况相似但不完全相同。参与区域被划分为内部和外部两部分。内部参与区域的边界主要根据以下三个因子划定：①机场扩建后的飞机噪声等声线预测结果；②空气质量可能发生变化的区域；③会被外界连接扩展所影响到的区域（交通和施工影响）。外部参与区域覆盖范围较广，包括那些预期机场扩建将需要劳动力而来此定居的人群。

参与的机制包括利益相关者访谈、讨论会及对当地居民进行问卷调查。访谈和讨论会主要在关键利益相关者中开展，即那些在健康、住房、教育、商业、交通领域战略性活跃的组织机构的负责人。参与机制不包括开放式社区讨论，因为该方式在斯坦斯特德机场G1期的参与程序中对公众的吸引力不高（虽然反对机场扩建的团体认为关注度不高的原因是缺乏前期宣传），因此，该项目采取了问卷调查的形式。内部和外部参与区域最后总共调查了9300个样本，问卷内容由健康影响评价小组、英国机场集团、健康主题小组和专家评审共同审议并定夺。

利益相关者参与得到的一些主要关注问题包括：

① 空气质量（特别是对弱势群体的潜在健康影响，包括儿童和老人）；

② 飞机噪声；

③ 道路交通/拥挤；

④ 对该地区的“社会资本”的影响（社区人口流动速度加快；缺乏认同感）；

⑤ 社会-经济问题（就业机会；施工工人导致的外来移民；对现有住房供应的影响）；

⑥ 医疗保健和社区设施（现有容量的充足性；对应急服务机构的影响；对医疗保健招聘的影响）。

这些问题和环境影响评价中涉及的问题有部分重叠，特别是在就业和社会-经济问题方面。

9.6.7 健康影响评价

健康影响评价报告为项目概况中识别出来的每一条健康航迹准备了一系列详细的“方法陈述”。这些方法陈述主要是在相关科学证据（如特定空气污染物对健康的影响）和利益相关者参与的基础上，将采取何种方法确定健康影响评价中健康后果的规模。在大部分情况下，这些方法都是基于环境影响评价研究中输入的数据，例如特定空气污染物的浓度预测结果、扩建工程前后的飞机噪声轮廓线等。这种联系意味着环境影响评价预测方法或假设中的缺陷也会被反映到健康影响评价工作中去（参见《停止斯坦斯特德机场扩建工程》2006中对这些问题的进一步讨论）。大部分健康影响的预测结果都是定量的，通过一定数量的人群的健康状态可能的变化结果来衡量。然而，也有一些影响只进行了定性讨论，某些健康影响的科学证据存在不确定性，无法进行量化估计。方法陈述中对这些不确定性、数据的局限性和实际操作中遇到的困难进行了清晰描述。

健康影响评价中考虑了以下健康影响：

① 空气质量：可吸入颗粒物和二氧化氮浓度增加对健康的影响（如寿命缩短；呼吸系统疾病和其他医疗负担；需要对哮喘进行全科医生会诊等）。

② 飞机噪声：烦躁；睡眠障碍；对儿童造成负面影响等。

③ 地面噪声：烦躁；睡眠障碍。

④ 交通：公路和铁路事故导致的伤亡。

⑤ 就业和收入：增加就业和收入的正面影响，包括对心理健康（如抑郁症）、健康自评（幸福量表）、慢性自限性疾病和死亡率等的影响。

⑥ 社会资本：改变区域的公民参与、社会网络和支持、社会参与、互惠和信任、满意度等（如建筑工人迁入和土地征用带来的改变）以及随之而来的对健康/福祉的影响。

⑦ 非自愿搬迁：由于土地征用而被迫迁出现住地所带来的紧张、焦虑、福祉下降。

⑧ 视觉效果和光污染：福祉下降。

⑨ 医疗保健和社区设施：对现有设施（如由于建筑工人迁入、事故、疾病传播等）的影响。

虽然健康影响评价中采取的方法和环境影响评价大相径庭，更关注识别影响的规模而不强调评价影响的重大程度，但健康影响评价考虑的内容仍和环境影响评价有某种程度上的重叠。例如，对于飞机噪声，环境影响评价试图识别噪声暴露的变化是否重大，而健康影响评价则把重点放在量化一定数量的人群的健康状态的变化结果上。如果有这些数据的话，这些定量评估结果将会与当地人群相关健康结果的广泛性进行比较，却并不直接评价健康结果的重大程度。健康影响评价报告中指出这是由于缺乏判断这些影响是否重大的评价标准。

在评估第二跑道（G2）提案的影响时，健康影响评价主要按以下几个情景考虑项目造成的健康结果。

① 未进行 G2 扩建（2015 年和 2030 年）前的基准情景：假设年发送旅客量 3500 万人次。

② G2 扩建后的 2015 年：这是第二跑道投入运营的日期。

③ G2 扩建后的 2030 年：预计机场年发送旅客能力已达 6800 万人次。

项目组对第二跑道施工期的健康影响也进行了评价。需要注意的是评估中所说的基准情景并不代表提出第二跑道扩建申请时的实际情况，它是一个预计的基准情景，即上一个扩建现有跑道容量的 G1 规划申请会得到批准的预测结果。这使得基准情景中的年发送旅客量为 3500 万人次，而不是 2008 年进行评估时的实际年发送旅客量 2200 万人次（如果 G1 扩建计划不被通过，基准情景的年发送旅客量将不超过 2500 万人次，因此该 2500 万人次也被作为基准情景的一个替代值）。因此，第二跑道健康影响评价假设的年发送旅客量增量总共为 3300 万人次（从 3500 万人次至 6800 万人次），与基准情景相比年发送旅客量增长 94%。

另一种替代方法是将 G1 和 G2 扩建提案视为一个更大的机场扩建计划的两部分。在这种情况下，年发送旅客量的增幅将显著提高，从基准情景时的 2200 万人次（或 2500 万人次）到 2030 年时的 6800 万人次（增幅为 172%～209%）。其产生的健康影响也将比单独考虑 G2 扩建提案时更大。因此，在健康影响评价中考虑 G1 和 G2 扩建提案的综合影响将很有必要。由于实际操作中并未尝试将 G1 和 G2 作为一个整体考虑其综合健康影响，因此在对第二跑道扩建提案进行健康影响评价时只能忽略 G1 扩建提案的健康影响。同样地，G1 扩建提案的健康影响评价也未考虑第二跑道的健康影响。对规划申请进行环境影响评价时也采取了这种分别应用评价程序的方式。在 G1 提案进行公众咨询时，这一方式得到了“规划巡视”的支持，他说“考虑到《环境影响评价法案》的目的，我接受英国机场集团的观点，即 G1 提案并不是一个更长远的开发活动不可分割的一部分”，他还说“当前的环境报告中不考虑 G1 和 G2 提案的联合影响并不会妨碍《环境影响评价法案》和指令实现其目的”（规划巡视，2008)。这一观点似乎适用于G1提案，但在之后的G2中却有点站不住脚，因为很明显其现在已经无法代表“一个更长远的开发活动”的一部分。

未能对G1和G2提案的健康影响进行综合评价意味着累积健康影响的识别很有可能是有缺陷的。这个问题和9.2节威尔顿电站案例研究中提到的问题非常相似，这两个案例的项目都是分别获批和采用评价程序的，在试图充分解决增加和累积影响时都遭到了失败。

9.6.8 减缓和增强影响及其监控

健康影响评价报告向英国机场集团和机场运营商提供了一份关于减缓和增强健康影响的建议清单。根据评价成果和利益相关者提出的建议，报告识别了一系列可行的减缓和增强措施。这份清单之后被交给健康主题小组和专家小组进行审查，最后得到一份推荐措施的最终清单。

健康影响评价得到的推荐减缓措施的关注点比较狭窄，且不包括拟议项目会导致的实质性改变，因为这些改变在健康影响评价中被作为“给定”的因素。正如健康影响评价所指出的，“除非改变开发活动自身的目的，否则不可能对一些和G2项目的特征有关的影响做出调整……这意味着有些影响将会比另一些更容易管理”(ERM，2008)。健康影响评价提出的一些减缓措施中包括建议和当地社区就健康影响的规模问题加强沟通；建议交给独立第三方而不是英国机场集团来控制噪声水平可能会更好。健康影响评价还推荐了一些通用的降低排放和从声源治理噪声的措施，包括提高技术标准、逐步禁止高噪声和高污染的飞机等。有人争辩到这些措施即使没有扩建活动也会被实施，因此应该被作为没有开发活动时未来基准情景的一部分，而不应当作为扩建项目的具体减缓措施的一部分。

健康影响评价并不考虑项目在早期规划时是否将健康影响明确纳入拟议项目的提案中，也不考虑项目在进行环境影响评价时环境报告中是否推荐了减缓措施。健康影响评价报告只提到“在选择跑道位置和评价运营模式时，全面考虑了许多方面的环境影响（以及健康影响）”(ERM，2008)。因此，在评估这些选择时如何进一步考虑健康影响、如何在项目最终提案中考虑这些健康影响的减缓措施，是十分有必要的。

这些约束在一定程度上反映出开展健康影响评价工作的时机是在项目设计和环境影响评价研究之后。这种安排是非常必要的，因为健康影响评价的评价方法严重依赖于环境影响评价的预测输入数据。然而，这也引发了人们对健康影响评价是否能影响到项目最终提案的疑问。这一评价时点也说明健康影响评价实际上不能对任何替代方案进行评价，只能对规划申请中已经被细化的最终提案的健康影响进行评价，也不能比较不同的设计或运营模式的健康影响。因此无法确定已选方案和其他可行方案比起来是否对当地社区能产生更多的有利影响(或更少的不利影响)。

9.6.9 总结

本案例研究是一个对重大建设项目进行健康影响评价的例子。健康影响评价越来越被作为与环境影响评价平行的一个程序开展，一些关键的数据输入却又依赖于环境影响评价。斯坦斯特德机场健康影响评价案例是一个典型的综合健康影响评价程序，开展了广泛的利益相关者参与工作，评价方法也是在对相关科学证据的详细总结的基础上确定的。虽然该项目采用了一个非常宽泛的健康和福祉的定义，评价重点却被严格限制在临近机场的当地社区内。这就排除了一些重要问题如气候变化对健康的影响。由其他一些机构或组织负责且会在其规划和战略中被解决的一些问题也被排除在评价范围之外。这样评价就只集中在第二跑道提案上，忽略掉斯坦斯特德机场扩建计划的其他方面，因此未能充分考虑机场的未来增长所增加和累积的健康影响。评价也未能比较不同的项目选择所产生的健康影响（并非没有任何扩建计划时的基准情景），这导致健康影响评价对项目提案的总体影响水平比较局限。

9.7 Cairngorm 山索道——环境影响评价的减缓措施

9.7.1 项目介绍

本案例是一个关于旅游业的项目，作为世界上最大的产业，旅游业发展迅速，而且其本身就包含着自毁因素，所以在此研究是比较合适的。在过去约 30 年内，人们日益意识到旅游可以毁掉旅游，关注的点从最初巨大的经济影响扩大到现在的一系列社会和环境影响（见 Glasson 等，1995；Flupter 和 Green，1995；Mathieson 和 Wall，2004）。旅游对山区的影响尤其敏感，包括步行、清雪及相关设施。该案例研究了一个特别有争议的项目——位于苏格兰高原的 Cairngorm 山索道，在对其影响和对影响的管理进行了冗长而拖沓的讨论之后，该索道于 2001 年开放。这个简要的案例研究主要关注影响的管理，将其作为环境影响评价中减缓措施的例证方法。

9.7.2 项目背景

Cairngorm 山滑雪区是苏格兰五大滑雪区之一。20 世纪 60～70 年代，它在与邻近的 Aviemere 社区合并之后得以迅速发展。它建造了滑雪升降机，可以在冬天将滑雪者送至高坡，在其他季节用来承载徒步行者。可是，该行业容易受到气候/天气的变化趋势和基础设施质量的影响。1993 年 Cairngorm 山滑雪升降机公司公布了 Cairngorm 山滑雪区发展规划，该规划设计了更新的设施，以减轻不利天气条件的影响，改善旅游者体验并提高了经济实力，同时确保充分考虑到所有相关环境问题。Cairngorm 山索道是该规划中的关键部分。

Cairngorm 山索道（图 9.3）是英国最高最快的山道。大约 2km 长，在 8min 内将乘客从现有的升降机站/Coire Cas 的停车场（610m）送至 Ptarmigan 顶站（1100m）。它包括在两终点间行驶的单线索轨上的两个车厢（或车厢列）。两车厢由绳索牵引，对向开出。一个车厢沿索轨下行时，另一个上行，它们在双索轨道处交错而过，该双索轨道的距离很短，位于整个轨道的中部。索轨支在升高的架子上，距地面最低 1m、最高 6m，最后 250m 经过一个以挖填方式修建的隧道。规划包括改造现有的机站底站及修建新机站来替代现有的机站顶站，新顶站包括一个可供应大约 250 人的饮食设施和一个新的解说中心，其中包括了各种展示和户外观赏台。作为发展规划的一部分，以前的机站和高塔已被移走。预计索道每年可乘载游客约 30 万人，其中非滑雪月份的人数占 2/3。这表明与 20 世纪 90 年代相比，乘升降机到达顶站的人数增加了 3～4 倍，是 20 世纪 70 年代人数的两倍。

(a)

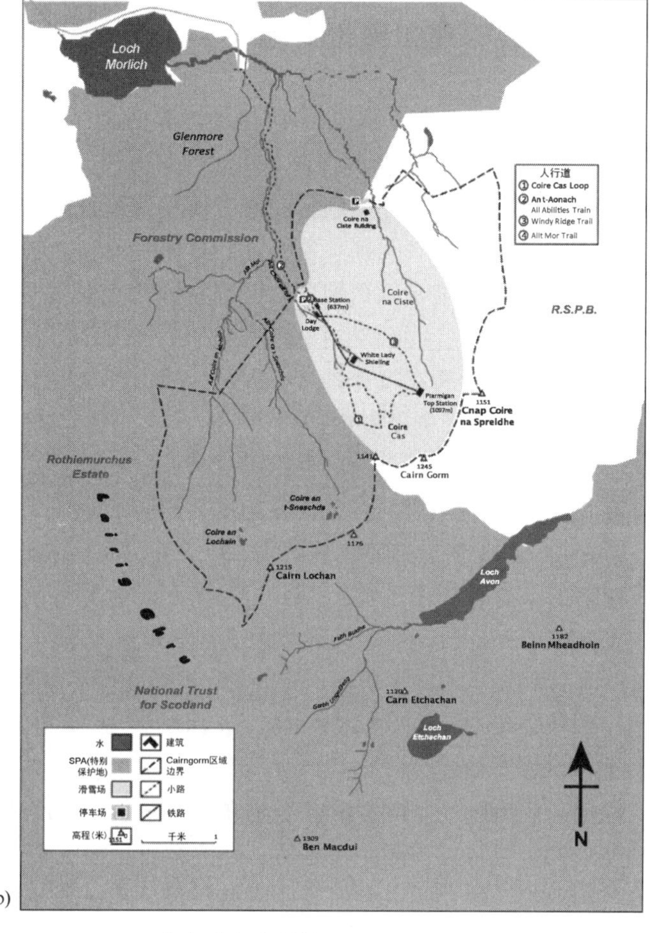

(b)

图 9.3 CairnGorm 山索道和大环境

(资料来源：HIE，2005)

9.7.3 环境影响评价和规划过程

1994 年 Cairngorm 山滑雪升降机公司提交了索道的最初规划申请和环境报告（土地利用咨询，1994）。1995 年初，又提交了修改建议和补充环境报告。该规划颇具争议性，且受到了多方反对。关注点主要集中在改善之后游客到达环境敏感的高原顶端所产生的潜在影响，该高原顶端被认为是欧洲候选的特殊保护区和 SPA。作为计划批准的条件，开发商必须满足苏格兰自然遗产保护协会（SNH)(负责苏格兰自然保护区法定主体）的相关规定，即提出配套的游客管理计划（VMP）和其他减缓措施以避免对高原顶端的不利影响。

根据第 50 节（现为第 75 节），通过与苏格兰自然遗产保护协会合作，在与开发商/权力机构（为游客和环境控制和管理成立的组织）达成一致后，高原委员会（批准机构）在 1996 年批准了规划申请，1999 年批准了站点建筑的修改设计，建设工作最后在 1999 年 8 月开始。在苏格兰自然遗产保护协会批准了拟定的游客管理计划之后，索道于 2001 年 12 月正式开放。

9.7.4 游客管理和减缓措施

附在计划批准书中的第50节协议是与苏格兰自然遗产保护协会（SNH）合作的规划机构、开发商/实施者和土地所有者之间的约束协议。内容包括（高原委员会，2003）：

① 在更大范围内对环境现状和游客使用情况进行本底数据调查；

② 一份实施计划，明确项目实施的具体时间和方法；

③ 年度监测制度，以识别变化并确定导致变化的原因（第一份监测报告已经准备好且正在讨论当中）；

④ 执行者要对所有行为进行必要的年度评估，以保证对位于高原顶端的欧洲指定的保护区的影响在可接受的限度内；

⑤ 违约事件的责任追溯问题；

⑥ 公众停止使用后的场址恢复问题。

Cairngorm山索道游客管理计划（Cairngorm Funicular Railway VMP）就是在该协议背景下制订的。游客管理计划（VMP）是为了保护邻近的《欧盟栖息地指令》和《鸟类指令》中指定区域的完整性，使这些区域免受因索道开发时非滑雪游客直接接触导致的潜在影响。游客管理计划经历了许多阶段，还在2000年进行了短期的公众咨询。该计划还提出了许多问题，包括创新性的或强制性的（依你的看法而定）“封闭系统”和相关的监控安排（SNH，2000）。

根据封闭系统，非滑雪游客不允许从Ptarmigan顶站进入Cairngorm山高原，这是游客管理计划的关键特点。取而代之的是，游客通过参观一系列室内解说展，户外观赏台，当然还有使用购物、饮食及卫生间等设施得到满足。在游客管理计划公众咨询中，该系统颇有争议，受到相当多的批评。一些人认为这样侵犯了游客的漫步自由，还有人认为这是为了确保商店和餐厅获利的一种颇具讽刺性的设计方案。其他人则认为倘若改进了从Ptarmigan顶站到Cairngorm山顶的小路，那么这个系统就是多余的。正如一个被咨询者说的“多年来我一直赞成石路。人们使用恢复地周围的道路和土地。现在通向顶端的道路也已完成，而大部分人也就不会使用它了”（SNH，2000）。另一个问题是如何让高原上不使用索道的步行者进入顶站的设施，同时防止非步行者利用该设施出去。计划对封闭系统提出了许多替代方法，包括进行巡查和限定入出时间等，但系统现已就位，并作为25年协议的一部分。提供给索道使用者的使用指南包括如下内容：

保护高山环境：脆弱的景观地和Cairngorm山栖息地的大片地区受欧洲法律的保护。Cairngorm山的限制条例是要确保娱乐活动在环境承受范围之内。因此，任何时候都不能使用索道进入非滑雪区的高山、高原。非滑雪季节，游客只能在Ptarmigan建筑和观赏台的范围内活动，并利用索道回到底站。欢迎爬山者从停车场开始步行并使用Ptarmigan站的各类设施，但回程时不能使用索道，游客需按要求在步行者入口处进行出入登记。

正如本书前几章所述，监测可以支持有效的减缓措施。就该项目而言，监测包括本底数据调查的所有内容，包括游客水平和行为、栖息地、鸟类、土壤和地貌等。使用可接受改变的限制（Limits Acceptable Change，LAC）方法，识别和监测可接受变化的指示器和变化程度，当达到限制程度时将激起管理响应（见Glasson等，1995）。为了对监测活动的独立性做出响应，SNH 和高原委员会联合指派独立报告官员向他们陈述每年的监测报告。

9.7.5 结论

Cairngorm 山索道目前已运行 10 年了。游客人数比 ES 中预料的要少，但比 20 世纪 90 年代以前的水平还是有了大幅增加。所有的条件都得以满足，第 50（75）节的协议得到了保证，并且在国家机构和执行机构之间建立了友好合作伙伴关系。通过多方努力建成了进入 Ptarmigan 顶站的路口，改进了路径系统并拆除了旧的 White Lady 滑雪升降机系统。最近，包括山地铁路乘客在内的 10 人以下的小旅游团，在 5 月至 9 月的滑雪季节之处，每天都可以在向导的引导下，在山路上步行 90 分钟到达山顶。

9.8 英国海岸风能开发——战略环境评价

9.8.1 项目介绍

该案例研究将战略环境评价（SEA）用于实践，他是英国政府关于海岸风能未来开发计划的战略环境评价。该项目是在 2002～2003 年间实施的，即在欧盟《战略环境评价指令》（2001/42/EC）生效之前。该指令的要求和战略环境评价将在本书的第 11 章做详细介绍。

之所以进行这个特殊的战略环境评价，是基于英国政府对可再生能源生产的宏伟目标以及英国政府《京都议定书》中大量减少 CO_2 排放量的承诺。在开展这个战略环境评价时（2002～2003 年），英国政府承诺到 2010 年有 10%的电力能源靠可再生能源提供，而到 2020 年该值上升为 20%。海岸风能被认为是实现这些目标的主要支撑因素（DTI，2003a）。英国政府也希望看到这个行业的迅速发展，它同时也遵循了战略环境评价的程序，以便于决定沿海哪些区域应该被开发，哪些区域不能开发，并引导开发商的投资计划。当时，海岸风能是一个发展迅速的新兴行业，因此其对环境的影响存在许多不确定因素，如潜在的累积影响。这表明战略环境评价工作很有难度。

9.8.2 英国海岸风能的开发

英国海岸风能开发项目涉及两个独立的申请和批准体系。申请体系由王冠地产（Crown Estate）来执行，他是英国海岸地产的所有者。申请需要一个竞争的过程，要求开发商对所有可能的风力发电厂所提交投标书。接下来由开发商自主决定可能的场址，然后由中标的开发商选址。对厂址详细的技术研究、咨询和环境影响评价工作都应在提交项目批准之前由开发商完成。在通过对影响最大的 LPAs、法定咨询机构以及公众进行咨询之后，由 DTI 和 DEFRA 授予必要的计划批准权。一旦获得批准，开发商可获得该场所 40～50 年的租赁期，然后就可以建设风力发电厂了。

英国的 Crown Estate 最早于 2001 年 4 月邀请开发商商谈场地租赁（第一轮项目许可），共提议了 18 个风力发电场开发项目，每个都不低于 30 个叶轮机。大部分项目在 2002～

2003 年获得了批准，并在 2003 年前就开始建设。在经过第一轮项目许可之后，英国政府发表了《未来海岸》的文件，表明进入了第二轮的许可计划（DTI 2002），这比前一回设想了更大的发展，表明将来的发展集中在三个“战略领域”——Thames 河口、Greater Wash 和利物浦海湾（图 9.4）。选择这些地区是因为它们有巨大的发展潜力，可以获得潜在的风力资源、海岸地区的探测技术、接近已有的电网连接，而且开发商对它们也很感兴趣（DTI，2003b）；然而环境因素似乎没有影响到战略领域的选择。

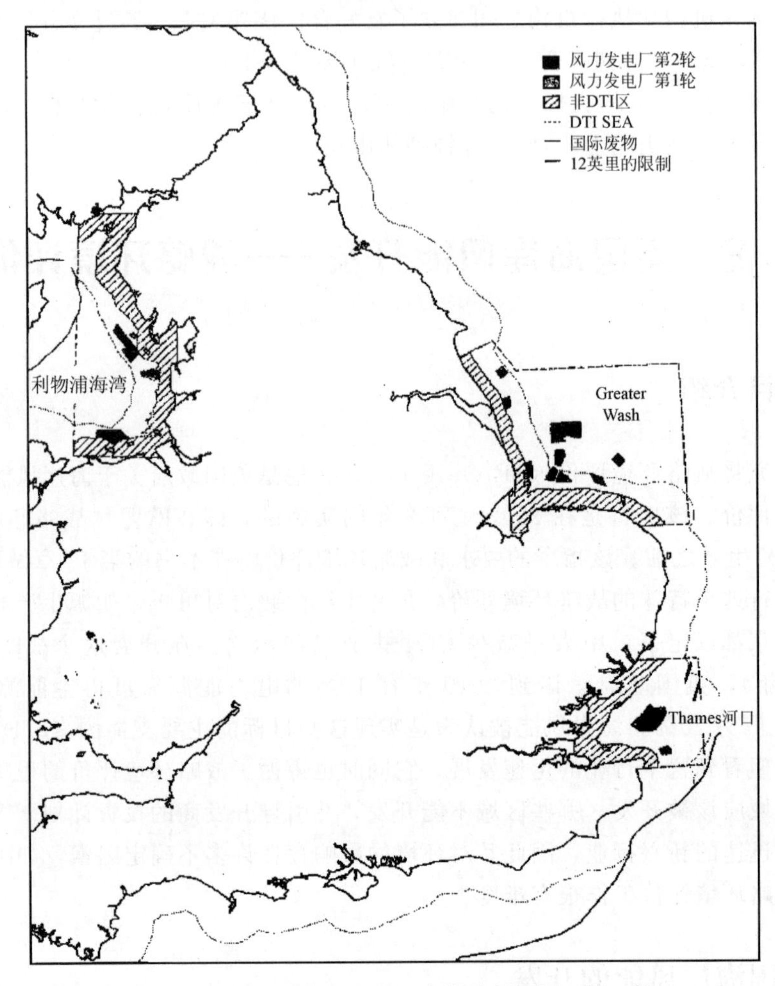

图 9.4　海岸风力发电厂的三个战略领域图（第二轮）

从 2002 年 11 月起，王冠地产针对《未来海岸》文件进行了长达 3 个月的咨询，同时对政府的第二轮许可计划开展了战略环境评价，也就是本案例研究的焦点，并于 2003 年 5 月提交战略环境评价环评报告（进行了 28 天的咨询）。尽管开展了这个战略环境评价，英国政府仍强烈要求维持海岸风能行业发展的速度。开发商向王冠地产提交对第二轮场所租赁意愿的截止日期是 2003 年 3 月底——也就是在战略环境评价报告书完成之前或收到回应报告的咨询之前。

对第二轮开发的中标项目最终于 2003 年 12 月公布。其中包括 15 个项目，总容量为 5.4～7.2GW，与第一轮批准的 1.2GW 相比，表明风电开发项目在英国取得了巨大进步。一些项目选址很快就引起了争议，主要是考虑英国皇家鸟类保护协会 RSPR（2003）中提出的对重要鸟类栖息地的潜在影响。

9.8.3 战略环境评价

该案例是对英国政府第二轮海岸风能开发计划草案的战略环境评价。该战略环境评价是由DTI依照欧盟《战略环境评价指令》的要求（尽管当时它还没生效）自愿进行的。规划期限从2003年至2020年，而且将到2010年与到2020年的开发分开进行评估（DTI，2003a）。规划考虑了两个可能的开发情景（“可能的”和“最可信的”），并且就其可能的影响进行了评价。“可能的”开发情景设想到2010年容量将开发至4.0GW，而“最可信的”的开发情景设想到2010年容量将达到7.5GW。根据这两种情景，预测至2020年该容量将分别增加至10.2GW和17.5GW。同时零开发方案也在考虑之中。

由一个指导小组来指导战略环境评价，其成员包括研究沿海/海事环境问题的专家、风能开发专家和战略环境评价的专业人士。指导小组成员包括相关政府部门的代表（DTI，DEFRA，ODPM）；Crown Estate的代表；英国风能协会（BWEA），即英国风能行业的实体代表；政府和非政府环境组织的代表，如RSPB、联合国自然保护委员会（JNCC）、威尔士乡村委员会、EN；能代表英国环境影响评价业的环境管理和评价协会（IEMA）的代表。

在战略环境评价的范围确定和设计阶段进行了咨询。2002年底召开范围确定的专题讨论会并形成了确定范围的报告。根据对于来自咨询会的响应，战略环境评价范围进行一些改变，例如许多顾问都认为一系列社会经济影响是很重要的（DTI，2003a）。环评报告编写于战略环境评价程序后期，提供了如下信息：

① 技术、环境和社会-经济限制的性质和程度，这些因素可能会影响到风力发电厂的建设或者使得风力发电厂受到一些负面影响。

② 在三个战略区域（Thames河口、Greater Wash和利物浦海湾）中选择最低限制水平的厂址。

③ 在最低限制水平下/在这些地区风力发电厂开发的不同实际规模下产生的环境和社会-经济影响的重要程度。

④ 提出对三个战略领域内开发风力发电厂所产生影响的管理建议。

综上所述，我们可以得出如下结论：

到2010年的“可能的”开发情景对于每一个战略区域来说都是可以实现的，而且不会引起重大影响风险，这些区域包括易受视觉影响的敏感区域、敏感海鸟的集结地、指定的和可能被指定的保护区以及MoD实践地区和主要航海交通区。

然而，2020年“可能的”开发情景只能解决对以下因素影响的不确定性：自然过程、鸟类、软骨皇（鲨鱼、鳐鱼等种类）和鲸类等。

到2020年，所有战略区域（尤其是Greater Wash和Thames河口）的“最可信的”开发情景可能受到制约，特别是累积影响以及与航海交通（商业和娱乐航海）的冲突；大规模开发将影响重要捕鱼场地的渔业开发，尤其是当它与其他海岸活动有关的割断地区重叠时。(DTI，2003a)

为了使环境影响减至最小，建议采用以下广泛的战略方法：

① 开发少量的容量在1GW（1000MW）以上的大型风厂，更远离海岸的地点一般更适合许多小规模的开发，尽管后者更适合在近海岸开发。

② 在所有战略领域内，避免在近海岸的高度视觉敏感地带进行大型开发。

③ 当开发可能靠近海岸时，优先选择低限制地区和优先考虑小规模开发。

④ 在监测研究结果得出之前，避免浅水区的开发，那里是鸟类（如常见的黑海番鸭和红颈潜鸟）和其他种类动物（包括海中哺乳动物）的聚集地（特别在利物浦海湾和 Greater Wash）。

⑤ 描述大范围影响的不确定性，特别是战略水平上的累积影响（DTI，2003a）。

环境报告经过了一个短期的公众咨询阶段（28 天）。政府认为这个报告以及收到的评论将成为政府第二轮许可决策的重要影响因素（DTI，2003a）。

9.8.4 战略环境评价方法

不得不承认，在战略环境评价中本底数据、影响的量化分析和表达的详细程度，都比项目环评要差。对于这个特殊的战略环境评价来说更是如此，它“更关注对限制、敏感和风险的评价，而不是对特殊影响特征的详细分析”（DTI，2003a）。用于战略环境评价的方法包括基于 GIS 的空间分析（严格的测绘工作）以及对所选发展情景（包括同时应用）可能的影响风险评价。下面将对每个方法做简要的描述，并举例说明。

9.8.4.1 空间分析

空间分析是综合利用各种技术、社会-经济和环境特征来绘制电子覆盖图，以识别三个战略区域内高或低限制区的海域。反映在地图上的主要特征包括：

（1）风力发电厂开发的技术限制

① 关于集料开采、废弃物处置和军事操作的已有和计划的许可地区；

② 石油和燃气构造（管线）和安全地带；

③ 文化遗址（船只失事和其他海床障碍物）；

④ 电缆；

⑤ 现有海运/航海航道；

⑥ 第一轮许可的拟建风力发电厂选址。

（2）社会-经济限制

① 海运；

② 渔业；

③ 贝类养殖。

（3）环境限制

① 海洋栖息地保护区（指定的和可能被指定的）；

② 海景敏感区；

③ 鱼类产卵区；

④ 鱼类生长区。

由于受到本底数据的限制，在 SEA 相对紧迫的时间内，并不能把所有的相关限制都绘制成图。特别是有许多重要的环境限制是不能绘制成图的，如区分特定鸟类、鱼类的种类和迁移路线。这些遗漏使来自强制测绘工作的结论是否失效受到了公开质疑（见下面对回复该问题的咨询的总结）。但这些无法绘制的因素可以在稍后的影响风险分析中予以考虑。

在绘制限制因素时采用评分体系，对所绘制的每个限制因素从0分至3分予以赋值（分数越高表明制约程度越高）。评分体系有助于在三个战略区域内选址，其中每个战略区都有许多限制因素（总得分高）以及总限制因素最少的（总体得分最低），不是所有的相关限制都可被绘制成图（见表9.1）。对每一个战略领域，从空间分析中得到的结论如下（DTI，2003a）：

表 9.1 绘制限制因素评分

渔业影响得分： • 0分 没有 • 1分 低(每年低于500h) • 2分 中(每年500～5000h) • 3分 高(每年超过5000h)
指定栖息地保护区得分： • 0分 不存在指定栖息地 • 1分 不适用的 • 2分 存在国家重点栖息地(包括还未指定的) • 3分 存在国际重点栖息地(包括还未指定的)
海景得分： • 0分 不敏感 • 1分 低敏感 • 2分 中敏感 • 3分 高敏感

① 利物浦海湾。由于鸟类、海洋栖息地保护区、海景、渔场和海洋交通的存在，大量限制因素和敏感因素都出现在该战略区的南部。在该区北部的海景限制因素也是很重要的。

② Greater Wash。与其他战略区相比 Greater Wash 的低限制区面积最大，为风力发电厂的开发提供了最大的潜在容量。在沿海地区，特别是该区的南部，限制因素和敏感因素最多，特别涉及了可视影响、沿海渔场、海洋哺乳动物、鸟类和海岸栖息地保护区等。

③ Thames 河口。其东边界有低限制区，还有与其他战略区比环境限制相对较少的地区。然而，许多河口和湿地是重要的鸟类栖息地。商业活动（如集料开采）和娱乐航海是其他的重要限制因素。

9.8.4.2 影响风险评价

对每一个战略区域，分别对两种发展情景（“可能的”和“最可信的”）可能影响进行风险评价。这个分析过程将先前空间分析中所绘制的部分因素与那些无法绘制成图的特殊个体（如特殊鸟类种类）结合起来。利用风险分析的方法可以尽可能地将影响定量化，或者定性描述，还可以评估它们的重要性。这基于以下分析：①影响发生的可能性；②（对受体影响）预期结果。利用空间分析法，评分体系用于评估影响的重要性，如表9.2和表9.3所示（DTI，2003a）。

表 9.2 影响结果得分

得分	说 明
5	严重（如影响将导致对关键的自然和/或生态过程不可逆转的或长期不利的改变；罕见的和濒危栖息地或物种的直接消失和/或对它们持续存在和生存的直接损害）

续表

得分	说　明
3	中等［如影响导致对自然和生态过程的中等期限（5～20年）的不利改变；一些栖息地（5%～20%）的直接损害，该栖息地对保护该地区物种的持续存在和生存和/或一些重要保护物种的死亡率至关紧要］
1	次要（如影响导致对自然和生态过程的短期的不利改变；对物种的暂时干扰；在极少或没有干涉的情况下，可在两年内自然恢复）
0	无（如影响被自然环境所吸收，没有产生可辨别的影响；不需要恢复或干涉）
+	积极（如该活动对环境的改变产生了净收益）

表 9.3　影响可能得分

得分	说　明
5	肯定（影响会发生）
3	可能（影响可能会在风力发电厂生命周期的某处发生）
1	不可能（影响不可能发生，但可能会在风力发电厂生命周期的某处发生）

每一个影响的重要性得分是结果和可能得分的乘积，从1分（较小后果和不可能）到25分（严重后果和肯定）。

9.8.5　在咨询反馈中提出的问题

环境报告在战略环境评价程序的最后阶段编写，它需要经过一个短期的咨询。对咨询反馈意见的分析能够显示出利益相关方所关注的关键问题，包括对战略环境评价程序的质量和有效性的关注以及它对下一阶段风能开发决策的影响。许多问题的产生是因为战略环境评价项目时间紧迫以及由此导致的实际困难。咨询所提出的主要问题将在下面做简短概述［在DTI（2003b）中有详细描述］。许多问题是相互联系的，如背景资料的缺乏是由于战略评价有限的咨询或有限的时间造成的。

① 三个战略区域的预选，缺乏国家水平的战略环境评价。在第二轮开发中重点选择的三个战略区域具有巨大的发展潜力，它们可以获得潜在的风能、拥有海岸地区的海洋测深技术、接近已有的电网连接，而且是开发商最初的利益所在（DTI，2003b）。然而，这个选择似乎没有清楚地考虑环境限制因素，这也是许多反馈者关注这个问题的原因。

② 战略环境评价紧迫的时间以及战略环境评价对未来发展决策影响的不确定性。战略环境评价的时间非常紧迫，以至于相关方不能有效地参与和咨询，也没有时间收集更多的本底数据。开发商为第二轮开发地点的投标往往在战略环境评价过程的环境报告完成之前进行，这也是引起关注的原因。

③ 考虑到海岸风能行业的快速发展，首先要正确考虑潜在的影响。一些反馈者认为海岸风能行业的发展太不成熟；在允许该行业大规模发展之前，应加大力度了解第一轮小规模发展带来的影响。

④ 需要更清楚的选址导则。对哪些区域适合未来海岸风能的发展以及哪些区域不适合（包括排除或禁止区的确定）要有清楚建议，这些在战略环境评价程序中并没有得到充分体现。

⑤ 关注战略环境评价的范围确定和方法体系。许多反馈者支持所有的战略环境评价方

法，包括对影响重要性的风险评价，但是也有人不赞同对特殊受体或者特殊地理区域进行详细赋值。

⑥ 缺乏有效的本底数据。在咨询反馈中反复出现的问题就是战略环境评价研究所用本底数据的局限性。包括对某些重要环境因素（不绘制成图的）的数据的丢失、在战略环境评价中采用的数据不是最精确或最准确的。一些反馈者认为这些数据的局限性足以严重到使战略环境评价中对高低限制区的识别无效。数据的缺失和影响的不确定性使反馈者强烈要求采取一些预防方法，而在《欧盟栖息地指令》下对划定海岸保护区的一再延误也迫切需要采取此类方法。其他的不确定因素包括小型风力发电厂的影响（第一轮发展许可）是否一定可以外推出未来在更远离海岸的地点建立的更大的风力发电厂的影响。

⑦ 与特定相关方的有限咨询。根据反馈意见，战略环境评价只对某些特定相关方进行了有限的咨询，如捕鱼者以及划船爱好者。缺乏咨询在一定程度上与战略环境评价有限的时间有关，这就是为什么战略环境评价中缺少特定问题的相关信息。

⑧ 没有充分考虑到累积和间接影响。没有充分注意战略环境评价有关沿海发展的影响，如传输连接。这个问题反映了 9.2 节第一个案例研究所关注的问题。在环境报告中应更加关注累积影响。环境报告中没有给出每一个战略区域的承载力，因此很难评价在两种发展情景下的累积影响大小。

⑨ 与项目环评的重叠之处。对于战略环境评价需要的详细程度以及哪些问题可以进行项目层次的环境评价存在着不同看法。例如，某一地区的鸟类分布数据可在更战略性的水平上合理地收集。

⑩ 对以后战略环境评价研究的责任。一些反馈者要求明晰谁将负责对战略环境评价提出的问题做进一步研究，包括收集附加资料以弥补数据空缺和持续监测。资料的共享也是很重要的。

9.8.6 结论

这个战略环境评价案例是在欧盟《战略环境评价指令》生效之前自愿执行的。它为如何在新兴的、发展迅速的行业里实施战略环境评价提供了范例。英国政府希望大规模开发海岸风能以满足其减少 CO_2 排放量的义务，使得这个战略环境评价的时间非常紧迫。然而，由于基础资料有限，而且对相关方的咨询和反馈时间都非常有限，这都成为该战略环境评价的明显缺陷。

9.9 战略环境评价案例：泰恩-威尔郡交通计划

9.9.1 介绍

这个案例研究涉及泰恩-威尔郡当地的运输计划，该郡位于英格兰东北部的一个大城市。该案例研究的有趣之处在于它是一个集成评估形式的例子，将各种其他类型的评估的结果运用到该战略环境评价过程，包括健康影响评估（HIA）、平等影响评估（EqlA）和栖息地管理评估（HRA）。对各种替代方案的评估是战略环境评价非常重要的特性，这也是为什么选择本案例的原因。

本地运输计划是《交通法案》(2000) 的强制性要求。该计划需要当地交通部门每五年准备一个当地运输计划备查。根据《英格兰运输法案》，本地运输计划需要从权力机构政策处罚，来促进和鼓励安全的、集成的、高效的和经济的交通设施和服务。本地运输计划涵盖所有形式的运输（客运、货运、行人、公共交通和私人交通），包括战略政策和一个相应的实现或交付更详细计划的建议书。

下面以泰恩-威尔郡市区的第三次本地运输计划（LTP3）为例介绍。该计划覆盖的时间是2011～2021年，由一个三年的实施计划支撑（泰恩-威尔郡综合运输管理局2011）。该计划是在2010～2011年提出的，同时进行的是战略环境评价过程（泰恩-威尔郡联合运输工作组2011 a, b）。本地运输计划的战略环境评价是《英国战略环境评价条例》所要求执行的。英国交通部（DfT）也提出健康影响评估和平等影响评估也是该项目需要开展的（DfT 2009a，b）。对于泰恩-威尔郡的本地运输计划，它尝试将这些评价形式都合并到战略环评中去。

9.9.2 本地运输计划背景

泰恩-威尔郡是英格兰东北部的城市，包括一个城市核心和众多农村地区。该地区包括泰恩河畔的主要城市中心，盖茨黑德和桑德兰，有110万人口。泰恩-威尔郡经历了历史性的经济危机，目前主要问题包括失业率居高不下、收入在平均收入水平以下、贫穷以及相关的社会和健康问题。本地区有相对较低水平的地区汽车保有量，和高于平均水平的公共交通工具使用率。城市有两个主干道路服务于本区域，A1和A19。公共交通包括泰恩-威尔郡地铁（轻轨）系统、密集的公交网络、南北渡轮和前往盖茨黑德地铁中心（零售和休闲综合体）和桑德兰的地方铁路服务。

本地运输计划的目的是帮助解决该地区面临的关键挑战。挑战包括经济发展和再生的需求，满足气候变化的减排目标的需求，建设安全、可持续社区的需求，保护自然环境的需求。为了应对这些挑战，计划采取了一个基于三种广泛使用方法的战略框架：①管理旅游需求（包括鼓励向更可持续的旅游模式的转变）；②管理和进一步整合现有的交通网络（尤其是继续鼓励现在的旅游模式和公共交通）；③有针对性寻找新的投资方案（包括投资电动汽车、公交走廊的改进、新公园和骑行道方案）。

9.9.3 战略环境评价过程概览

本地运输计划的战略环境评价是由来自泰恩-威尔郡综合运输管理局的咨询师阿特金斯编制的，当地交通部门负责泰恩-威尔郡各个区域。在这个过程中关键阶段包括：

① 范围、本底研究和战略环境评价方法：这涉及确定和咨询战略环境评价；底线和情境研究；战略环境评价方法，包括确定适当的战略环境评价目标并进行评估。

② 评估替代方案：开发的战略选择和它们对战略环境评价的评价目标的影响。

③ 计划草案的影响的评估：评估还处于草案阶段的政策的影响；对缓解或增强影响的建议；并以环境报告的形式公布结果。

④ 针对草案和环境报告举行座谈会：紧跟着就是对草案进行最终的校正。

⑤ 监控方案执行产生的影响。

战略环境评价的一个有趣的特性就是它试图整合其他类型的评价到SEA过程中，包括健康影响评估（HIA）、平等影响评估（EqIA）和栖息地管理评估（HRA）。相关政府指出健康影响评估（HIA）、平等影响评估（EqIA）是本地运输计划的必需组成部分，但这些评估不一定必须包含于战略环境影响评价。一个计划可能对特殊保护区域产生重大影响，那么栖息地管理评估（HRA）是一个额外的法定要求的评价。

对人类健康的影响在战略环境评价中被规定为一个战略环境评价需要考虑的环境主题。政府在本地运输计划中对健康影响评估的要求反映出的一种理解就是，本地运输计划政策和提案可能影响社区和个人的健康相关的因素。包括例如交通的普及和费用承受能力的变化、锻炼水平、空气和噪声污染、人身安全（安全感觉）和社区隔离。对于泰恩-威尔郡的本地运输计划，独立的健康影响评估没有进行，健康方面的评价放在了战略环境评价中。这种集成的健康评估发生在战略环境评价过程的所有阶段，例如：

① 在战略环境影响评价范围确定阶段，与当地的NHS和其他相关公共卫生机关共同商议确定评价范围；

② 在审查政策环境时制定计划和识别关键的可持续性问题的时候，把与健康有关的计划和程序囊括进来；

③ 有关卫生指标纳入基础数据采集；

④ 特定的健康战略环境评价目标（改善健康和幸福、减少健康中的不平等）。

平等影响评估不同的社会群体中政策、战略和规划的影响。其目的是为了确保该计划不会歧视特定团体，和在可能的情况下促进更大的平等。泰恩-威尔郡本地运输计划的平等影响评估是和战略环境影响评价同时进行的，不过其结果是单独报告的（泰恩-威尔郡联合运输工作组，2011c）。然而，平等影响评估也完全融入战略环境影响评价的不同阶段，这点与健康影响评估相似。特定的关于平等的战略环境影响评价目标也包含在战略环境影响评价框架中（“为所有公民提供平等的机会，实现一个更为公平的社会”战略环境影响评价框架在9.9.5节中进一步讨论）。

泰恩-威尔郡本地运输计划也接受栖息地管理评估审查。它与平等影响评估一样，和战略环境影响评价是平行的过程，但拥有独立的报告。对Natura 2000站点的影响也被包含并作为战略环境影响评价目标（来尽可能保护和加强欧洲站点）。

9.9.4 范围和基础研究

战略环境影响评价中评价的内容是基于战略环境影响评价指令中规定的因素。其中包括对生物多样性的影响、人口、人类健康、动植物、土壤、水、空气、噪声、气候因素、重大资产、文化遗产和景观的影响。作为范围界定阶段的一部分，范围报告在这阶段编写，并会征求法定机构和其他机构的意见（包括代表公共卫生、人权平等、多样性利益等的机构）。范围报告包括与其他运用于战略环境影响评价的主体的符合性，包括规划、政策、本地运输报告方案、初始基础信息、环境、心理、社会和健康问题的总结。此外，还将举办确定范围的研讨会，从而收集更多的信息，并讨论关键问题，提出战略环境影响评价主体（9.9.5节中进一步讨论这些目标）。

战略环境影响评价的本底研究包括政策背景综述、收集环境状态和关键环境问题的详细数据。建立本地运输计划政策背景需要其他相关计划和规划的情况，包括那些有关健康与平等的问题。这需要识别对环境、健康、和平等有关的问题的关键主体和政策目标。详细的本底信息是建立在环境的当前状态和在没有实现计划时环境可能的演变的基础上的。本底数据收集包括环境和社会指标，后者包括健康、不平等、连通性、可行性方面的数据。

本底数据差距和限制在战略环境影响评价报告没有明确标识，因此不清楚潜在的指标是否由于缺乏本底数据已经排除了。由正规咨询机构发布的环境报告初稿中的一些具有争议的数据也会在讨论会上被删除，因为本底数据的完整性和一致性得不到保障（包括生物多样性、农业土地质量和洪水风险区域）。战略环境影响评价咨询机构承认这些问题，但表示问题不在于他们，而是因为预算限制。对于很多本底因素，也不可避免地依赖一些过时的信息来源（例如使用2001年的人口普查数据，已经过去10年）。

本底信息和政策背景综述是用来识别本区域最关键的环境问题。这些涵盖了广泛的经济和社会可持续发展问题（例如历史性的子区域的经济危机和相关社会问题，包括贫困、贫困儿童、低收入和低水平的汽车拥有量）以及环境问题（如空气质量差、噪声、水质、气候变化、生物多样性的威胁、遗产和景观性格变化）。一般健康问题和不平等问题也被确定为关键问题。其他问题是直接关系到运输（如交通拥堵、通行能力和道路安全）。

9.9.5 战略环境影响评价方法和目的

评估方法的关键部分是一个“战略环境影响评价框架”的发展。这个框架由一系列的战略环境影响评价目标和相关指标组成，用于评价计划草案和替代选项的影响。战略环境影响评价的目标是通过一个反复过程发展而来的，依据对规划和方案的分析、相关本底数据，从可持续发展的角度分析本地运输计划是否可以解决问题。目标的列表包含特定的健康、平等和栖息地问题。这样集成的方式是为了确保健康影响评估、平等影响评估和栖息地管理评估能满足战略环境影响评价框架要求。最终确认了16个战略环境影响评价目标，包括空气质量、生物多样性、欧洲站点、气候变化、洪水风险、资源使用和浪费、水质、土地利用、历史和文化遗产、景观和城镇的风景质量、通行能力和社区隔离、噪声和光污染、健康和福祉、平等、道路安全、犯罪。对于每一个评价目标，战略环境影响评价框架提供了一组相关的指标（对目标进行客观的评估）和评估的问题列表（指导本地运输计划去评估评价目标的效果）。以空气质量作为例子，评价的问题列表及相关指标见表9.4。

表9.4 战略环评结构摘录

空气质量战略环评目的的问题	针对战略环评目的的建议评价指标
—减少交通拥堵，促进整个区域可持续交通模式，特别关注空气质量低的地区（例如空气质量管理领域）?	—国家空气质量的主要污染物的目标水平
—促进步行和骑自行车并为这些出行方式改善基础设施?	—在空气质量管理区域住宅的数量
—鼓励绿色出行计划和学校出行计划?	—在空气质量管理区/未来的Tyne and Wear空气质量管理区的欧洲引擎汽车的数量
—增加对最现代汽车的使用，包括公共汽车和私家车吗?	—有效利用宣传和营销活动中具有环保意识的人群比例
—认识到环保意识的重要性以及宣传营销活动促进该地区改善空气质量的问题?	—企业和学校的出行计划的数量

续表

空气质量战略环评目的的问题	针对战略环评目的的建议评价指标
—减少大城市城际干线公路网的拥挤情况？ —根据污染者付费原则（如拥堵付费、道路收费）采取财政激励手段？ —促进公共交通的使用？	—通过当地权威机构的注资和运作，减少氮氧化物和主要 PM_{10} 排放（国家指标 194）

在可能的情况下，战略环境影响评价的指标框架类似于之前的本底研究。通过分析这些指标的本底情况，提供了一种可以总结环境质量现状和预测的方式（在计划没有实施的情况下）。使用一个简单的三分规范规模来分析的战略环境影响评价目标，描述当前条件（好、中度或差）和预期未来趋势（改善、稳定或下降）。总的结论是：如果没有本计划的实现，那么战略环境影响评价的指标未来会出现负面的趋势。这些目标包括空气质量、交通造成的二氧化碳排放、噪声、光污染、交通基础设施对气候变化和洪水风险的影响、犯罪率。

9.9.6 对各种方案的识别和评估

战略环境影响评价指令要求考虑：指标和计划的地理范围的合理替代方案；选择替代方案的原因。战略环境影响评价编制团队工作人员对于泰恩-威尔郡的本地运输计划的战略环境影响评价，提出了三个战略替代方案。三个不同的方案分别能命名为“最小扰动”“现实主义”“理想主义”（或有足够的资金）。这些替代方案是根据对当地交通目标和挑战的认识而提出来的。三种方案的主要区别如下。

9.9.6.1 “最小扰动”方案

① 公共交通、公路管理、骑自行车、步行或陆运和海运没有额外干预；
② 高速公路通行能力计划已经确认进行或已确认资金；
③ 提高公共汽车和地铁的票价；
④ 铁路和地铁翻新；
⑤ 工作场所和学校旅行计划保持在当前水平。

9.9.6.2 “现实主义”方案

① 强调高速公路通行能力计划（主要改进平交道口）；
② 三个新的公园和自行车计划；
③ 提供 11 个地点的公交优先车道；
④ 集成“智能”公共交通票务服务；
⑤ 电动汽车充电点；
⑥ 强调步行和骑自行车优先性。

9.9.6.3 “理想主义”方案（在“现实主义”方案的基础上增加以下内容）

① 更加重视高速公路通行能力计划；

② 再提供11个地点的公交优先车道；

③ 再提供9个地点的公交优先车；

④ 更加重视铁路和地铁，包括新开的铁路；

⑤ 没有额外强调步行和骑自行车优先性。

对于选择这些特定的备选方案的原因，战略环境影响评价的环境报告只提供了有限的信息。例如，报告没有说明其他战略选择是不是在计划的早期准备阶段计划被考虑，然后被否决。"理想主义"方案面临的现状也被质疑，因为公共交通计划的资金已经被一再缩减。

为了比较这些替代方案的影响，一个对于它们在每一个指标上的执行效果的定性评估在战略环境影响评价框架中被提出。这个评估是基于一系列对于框架中定义为"提示问题"的。约束映射也同时使用：在已知一个特定地点的情况下，用地图显示相关的环境约束和选择的方案的关系（如公园和骑行方案位置）。评估备选方案对战略环境影响评价目标的影响是一个定性过程，使用一个简单的七个等级来评价（大、中度或轻微的有益效果；中性或没有效果；轻微、中度或大的负面影响）。中度或大的效应都被认为是重大的。结论中对每个战略环境影响评价目标可能的影响，都在环境报告中有评论或说明。然而，没有明确的标准来指导这个过程，这样的标准在其他的当地交通战略环境影响评价中被证明是有用的，而且能使这个案例的总结部分变得更加清楚（这样的标准运用的例子可以详见TRL 2004）。

对于战略规划的战略环境影响评价中重要性评价通常是非常困难的，因为缺乏具体细节计划、位置和时间尺度。在这种情况下，也很难衡量不同的战略所带来的短期和长期的影响。对于很多战略环境影响评价目标来说，都希望能将短期不利影响（如新基础设施的建设）和潜在的长期好处（如方案执行带来的交通压力减小）结合考虑。因为缺乏明确的标准来评估各种情况下遇到的困难，所以很难去平衡不同的影响。战略环境影响评价似乎也采用一种假设，那就是假设各种措施是行之有效的，比如将旅行行为改成方案所想的那样。准备采取的措施实际上没有任何明确的传递性和风险。

对各种战略的评估总结来看，"现实主义"的做法总体来讲在不利和有利影响之间更加平等（表9.5）。这个结论在环境报告中并没有明确提出，即便是相比起"最小扰动"方案它有更多的有利影响，相比"理想主义"方案它有更少的不利影响。总体来说，对于战略环境影响评价目标的预测的影响如下。

表9.5 Tyne and Wear LPT战略替代方案的评价总结

战略环评目的	替代战略1最小化方案	替代战略2现实主义方案	替代战略3乐观主义方案
环境目标			
1	/	0	++
2	/	/	−
3	/	/	−
4	/	0	/
5	−	−	/
6	/	−	−
7	0	−	−−
8	0	−	−−
9	/	−	−−
10	−	++	++

续表

战略环评目的	替代战略 1 最小化方案	替代战略 2 现实主义方案	替代战略 3 乐观主义方案
社会目标，包括健康和不平等问题			
11	0	+	++
12	0	0	/
13	—	++	+
14	/	+	++
15	0	+	+
16	0	+	+

效用度量(SE)

+++ 巨大好处　　++ 中等好处　　+ 微小好处

0 无影响

—— 巨大弊处　　— 中等弊处　　/ 微小弊处

序号	战略环评目的
1	确保当地空气质量良好
2	保护和提高生物多样性、多功能的绿色基础设施网络
3	保护和提升欧洲栖息地(HRA 特定目标)
4	通过减少碳排放缓解气候变化
5	为确保对气候变化和洪水风险影响的适应性
6	促进审慎使用自然资源、废物减量化和资源化
7	保护和提高地下水、河流和海洋水质
8	确保土地的有效利用，保护耕地资源
9	维护和提升历史文化遗产的质量和独特性
10	保护和加强自然景观和城市景观的特点
11	提高服务设施和娱乐设施的可用性，避免社区分散
12	减少噪声、振动和光污染
13	提升人类健康水平，减少不平等(HIA 特定目标)
14	为所有公民提供更加公平的机会，期望实现更为公平的社会(EqIA 特定目标)
15	提高道路安全
16	减少犯罪和对犯罪的恐惧并促进社区安全

资料来源：Tyne 和 Wear 联合运输工作组，2011a。

①“最小扰动”方案：没有重大有利影响，对三个战略环境影响评价目标有重大影响(均为中等不利)。

②“现实主义”方案：对 2 项战略环境影响评价目标有重大有利影响(均为中等有利)；对 5 项战略环境影响评价目标有重大不利影响(均为中等不利)。

③“理想主义”方案：对 4 项战略环境影响评价目标有重大有利影响(均为中等有利)；对 6 项战略环境影响评价目标有重大不利影响(3 项为中等不利，3 项为非常不利)。

没有明确的标准和方法来认定这些影响的权重是战略环境影响评价的一个致命伤。例如，是不是每一个战略环境影响评价目标都有相同的重要性并不明确。

9.9.7 减缓措施和影响

评估了替代方案之后，本地运输计划的优先选择就是团队进一步发展 SEA。作为结果的草稿计划主要是建立在“现实主义”方案下的，拥有 45 条独立的政策。为了简化评估，战略环境影响评价将这些政策分组编辑。在每一个组别中的政策都有相似的效果。共有 6 组，分别是：a. 提高安全；b. 维护和管理基础设施；c. 促进可持续交通模式；d. 停车；e. 货运；f. 主要计划。每一个政策的效果都会用战略环境影响评价目标来评估，采用的也是相似的七个等级来评价。

三个环境影响评价目标有重大不利影响：确保气候变化影响和洪水风险的恢复能力；确保有效地利用土地资源中富饶的土壤；减少噪声、振动和光污染。这些不利影响是由拟定方案中的停车、货运和主要计划造成的。根据评估的结果，环境报告中详述了改善可持续性的建议，包括：

① 政策中关于主要计划的部门应要求任何建设项目都要实施施工环境管理计划（CEMP）和废弃物管理计划。

② 在任何政策的计划草案中没有考虑适应气候变化。政策中关于主要计划和基础设施的维护必须考虑对气候变化的适应性。

③ 越野卡车停车和货运中心的选址必须考虑其生物多样性和地质多样性，从而减小对土地的影响。

④ 停车的规定中需要考虑对犯罪的防范措施，特别是出城停车场，合适的照明和安全人员可以适当缓解。

⑤ 停车的规定中需要考虑噪声和光污染问题，城外的停车场也可能会有上述问题。

草案中的大多数修订建议都被纳入最终的计划中。而这些最终的修订建议都使得战略环境影响评价目标得到了不同程度的改善，相比草案中的计划产生更少的不利影响。重大的不利影响仅仅在于主要计划对两个战略环境影响评价目标（气候变化和洪水风险的恢复能力；土地利用）的影响。最后的环境报告提出了更多的为了将这些影响最小化的措施，这些措施强调了主要计划在洪水危险区域实施的不可行性和如何减少计划内耕地流失问题。一些通用的缓解措施也在报告中被推荐（如施工项目环境管理中的最佳实践措施）。

9.9.8 小结

这个案例研究讲述了本地运输计划的战略环境影响评价。政府对战略环境影响评价的额外要求是需要同时进行健康评价和平等评价。本地运输计划同时要求筛查潜在的对欧洲保护栖息地的影响。泰恩-威尔郡本地运输计划将这些不同类型的评估集成在战略环境影响评价中，是通过集成健康、平等和栖息地问题与战略环境影响评价的范围和本底研究过程，同时在战略环境影响评价框架中囊括具体的目标。这种方式是创新的，本案例还是揭示了一些战略环境影响评价过程中的问题，特别是对替代方案的评估和对重要性的评估。使用更明确的标准来评估各预测影响的重要性，同时和代替方案进行比对有助于提高评估的准确性。

9.10 总结

本章研究了大量的环境影响评价的实践案例。尽管也讨论了战略环境影响评价的案例，但大部分案例仍是针对项目环评的。同时本章还探索了如何将环境影响评价与其他类型的评价相联系，如健康影响评价和栖息地指令下的适宜性评价。虽然这些案例研究并不代表环境影响评价的最佳案例，但是它们的确给出了解决环境影响评价特殊问题的一些新颖和创新的方法，例如扩大公众参与的方法和累积效应的处理方法。案例也关注了一些环评中遇到的实际困难和一些实践过程中的局限，第 11 章和 12 章会进一步讨论一些案例研究中的新方法。

问　题

以下问题旨在帮助读者关注本章环境影响评价和战略环境评价案例中提出的关键问题。

1. 在一些案例研究中，增量和累积相应的评估是一个薄弱的环节（9.2 节和 9.6 节）。为什么这些评估中的问题通常都会出现在项目层面的环境影响评价中？你可以提出什么解决办法吗？

2. 在 9.5 节中的 Humber 河口案例研究中，累积效应最有效的解决办法和哪些因素有关？

3. 根据影响欧洲优先栖息地的项目，哪些是在“适宜性评价”过程中所必须实施的测试？为了解释这些测试有哪些相关的指导？

4. 在 Portsmouth 焚烧炉的案例研究中，公众参与的主要优点和缺点是什么？公众参与过程的有效性如何提升？

5. 在 Stansted 机场的案例研究中（9.6 节），健康影响评价和环境影响评价的主要异同是什么？这两者又以什么方式联系在一起？

6. 9.8 节中，英国海岸风能战略环境影响评价的主要不足是什么？这些利益相关者的考虑如何能更有效地解决？

7. 如何评价在泰恩-威尔郡本地运输计划的战略环境影响评价中使用的评估和比较备选方案的方法？战略环境影响评价的各个部分应该如何优化？

9

Case studies of EIA in practice

9.1 Introduction

This chapter builds on the analysis in Chapter 8 by examining a number of case studies of EIA in practice. The selected case studies mainly involve EIA at the project level, although examples of SEA are also included. Links between EIA and other types of assessment are also examined, including a case study of the 'appropriate assessment' process required under the European Union Habitats Directive and an example of health impact assessment. The selected case studies are largely UK-based and cover a wide range of project and development types, including energy (offshore wind energy, gas-fired power stations and overhead electricity transmission lines), transport (road projects and airports), waste (municipal waste incinerator and wastewater treatment works) and other infrastructure projects (port development and flood defence works).

The case studies have been selected to illustrate particular themes or issues relevant to EIA practice, and some are linked to specific stages of the EIA process. These are:

- project definition in EIA and the effect of divided consent procedures on EIA (Wilton power station, Section 9.2);
- EIA, European protected habitats and appro-priate assessment (N21 link road, Section 9.3);
- approaches to public participation in EIA (Portsmouth incinerator, Section 9.4);
- assessment of cumulative impacts (Humber Estuary schemes, Section 9.5);

- health impact assessment (Stansted airport second runway, Section 9.6);
- mitigation in EIA (Cairngorm mountain railway, Section 9.7);
- SEA at the national level (UK offshore wind energy development, Section 9.8); and
- SEA at the local level (Tyne and Wear local transport plan, Section 9.9).

It is not claimed that the selected case studies represent examples of best EIA practice—indeed two of the cases were the subject of formal complaints to the European Commission regarding the inadequate assessment of environmental impacts. However, the examples do include some innovative or novel approaches towards particular issues, such as towards the assessment of cumulative effects (Humber Estuary) and the treatment of public participation and risk communication (Portsmouth incinerator). The case studies also draw attention to some of the practical difficulties encountered in EIA, the limitations of the process in practice, plus opportunities for the future. This reinforces some of the criticisms of UK EIA practice made in Chapter 8, and pre-empts new directions identified in Chapters 11 and 12.

The selected case studies are largely based on original research either by the authors or by colleagues in the IAU at Oxford Brookes University (the exception is the power station case study in Section 9.2, which was researched by William Sheate, Reader in Environmental Assessment at Imperial College, University of London, and published in 1995).

9.2 Wilton power station case study: project definition in EIA

9.2.1 *Introduction*

This case study, originally documented by Sheate (1995), illustrates the problems of project definition in EIA, particularly in cases in which consent procedures for different elements of an overall scheme are divided. The case highlights the failure of the EIA process to fully assess the impacts of a proposed UK power station development. In particular, the EIA process failed to identify prior to the power station consent decision the environmental implications of the extensive electricity transmission lines required to service the new development.

The case study highlights a basic problem within EIA for UK energy sector projects caused by the splitting of consent procedures for electricity generation and transmission. This situation arose after the privatization of the UK electricity supply industry by the 1989 Electricity Act. The case illustrates how the division of consent procedures for individual components of the same overall project can result in conflicts with the EIA Directive's requirement to assess the direct, indirect and secondary effects of development projects. Although the case study relates to early EIA practice in the UK and EU (in the early 1990s), the issues raised remain largely unresolved and are still relevant to current practice in the UK and elsewhere.

9.2.2 The Wilton power station project

Early in 1991, newspaper reports began to identify the environmental consequences of proposed high-voltage electricity transmission lines necessary to connect a new power station on Teesside, northeast England, to the National Grid system. To many, it was astonishing that these impacts had not been identified at the time the power station itself was proposed. Close inspection of the environmental statement (ES) produced for the power station revealed that such issues had barely been identified at the time and therefore did not feature in the consent process for the power station. Following considerable public uproar over the proposed power lines, in April 1991 the Council for the Protection of Rural England (CPRE) lodged a formal complaint with the European Commission (EC) in Brussels against the UK Secretary of State for Energy (Sheate 1995).

This state of affairs could hardly be regarded as an example of good EIA practice, but how did it come about? The complaint concerned the EIA for a large new gas-fired power station at Wilton, near Middlesbrough on Teesside, proposed by Teesside Power Limited. CPRE argued that consent had been granted for the power station without the full environmental impacts of the proposal having been considered. Because of its size (1875 MW), the power station was an Annex I project and EIA was mandatory. However, the overall 'project' consisted of a number of linked components, in addition to the power station itself, including:

- a new natural gas pipeline;
- a gas reception and processing facility;
- a combined heat and power (CHP) fuel pipeline from the processing facility to the CHP facility; and
- new 400 kV overhead transmission lines and system upgrades (75—85 km in length, running from the power station site to Shipton, near York).

It was the implications of the transmission connections required to service the new power station that were of particular concern, although the other project components also had the potential for environmental impacts. Cleveland County Council (CC), in whose area the power station was located, expressed the view that a full assessment of the implications of all project components should be undertaken before the consent decision on the power station was taken. The County Planning Officer commented:

> My council wanted the power station [consent decision] deferred until all the implications could be fully considered. But the Secretary of State [for Energy—the consenting authority for schemes of this type at the time] wasn't prepared to do this. The result is that different features of the scheme, which includes pipelines and a gas cleaning plant as well as the main station and its transmission lines, come up at different stages with different approval procedures. An overall view hasn't been possible.

Despite these concerns, consent for the power station was granted by the Secretary of State for Energy in November 1990. The decision was based on the information contained in the ES for the power station, and without the benefit of a public inquiry. However, crucially, the ES did not include a description or assessment of the effects of the other elements of the overall 'project', including the pipelines, gas processing facility and transmission lines, which were seen to be the responsibility of other companies under separate consent and EIA procedures. Although separate EIA procedures were in place for these other project components, so that their environmental impacts would subsequently be considered, CPRE in its complaint to the EC argued that, under the EIA Directive, the EIA for the power station should have included the main environmental effects of its associated developments. The failure to do so resulted in a piecemeal approach to EIA which, it was argued, contravened the requirement in the EIA Directive that all direct, indirect and secondary effects of a project should be assessed prior to consent being granted.

Sheate (1995) summarizes the argument made by CPRE:

> Concern was expressed that the Secretary of State did not see fit to require further information on these aspects [the impacts of associated developments], as he is entitled to do under the UK's own implementing legislation. [The developer] had successfully received consent for the power station even though the major impacts on the environment of the electricity transmission lines, the gas pipeline, the gas processing facility and the CHP pipeline did not feature in the accompanying documentation provided to the Secretary of State for Energy. Since the relevant information was not available to the Secretary of State—nor did he request such information—it was argued that his decision might not have been the same had all the relevant information been available to him. Since the information was not contained in the ES, neither the public nor interest groups had been alerted to these consequential impacts, which might otherwise have caused a public inquiry to be held where the issue would inevitably have been aired.

At the time of the case study, responsibility for new transmission lines in England rested with the National Grid Company (NGC), not the developer of the power station. NGC had an obligation to connect a new electricity generator into the national grid, and—if significant environmental impacts were likely—it was required to undertake its own EIA for new overhead lines or major upgrades of existing lines. EIA would therefore take place for new transmission lines (and for other types of associated development). However, this EIA process was undertaken after the power station had been given consent, and it was therefore unable to influence the decision over whether the power station should have been built in the first place, either in that location or somewhere closer to the existing transmission network, hence minimizing the adverse visual impacts of new overhead lines.

Essentially, the Wilton case revolved around the way in which 'projects' are defined for the purposes of EIA. The ES for Wilton power station referred to the 'overall project' as including both the power station and its associated developments, such as transmission lines. However, because of the fact that different elements of this overall project were subject to separate consent procedures, the project was divided into separate 'sub-projects', with the environmental impacts of each being assessed separately and at different time periods depending on the timescale of the various consent procedures. CPRE argued that, under the Directive, it was not appropriate to assess the impacts of associated developments in isolation from (and after) the main development, including the implications of the latter's location.

In their response to CPRE's complaint, the EC agreed in principle that, in such cases, combined assessment was necessary and that splitting of a project in this way was contrary to the EIA Directive:

> I can confirm that it remains the Commission's view that, as a general principle, when it is proposed to construct a power plant together with any power lines either (a) which will need to be constructed in order to enable the proposed plant to function, or (b) which it is proposed to construct in connection with the proposals to construct the power plant, *combined assessment of the effects of the construction of both the plant and the power lines in question will be necessary* under Articles 3 and 5 of Directive 85/337/EEC when any such power lines are likely to have a significant impact on the environment. (Letter from EC to CPRE, 11 November 1993; emphasis added)

The UK government had argued that the proposed transmission lines in the Wilton case were not required primarily to service the new power station, since the proposed upgrading would allow NGC to increase exports of electricity from Scotland to England. However, evidence presented by NGC to the subsequent public inquiry into the power line proposals appeared to contradict this view. Although the upgrading would have some wider benefits for NGC, it was clear that the primary justification for the proposals, and indeed for the specific routes proposed, was the needs of Wilton power station. The government also argued that the Directive allowed for separate EIA procedures for power stations and transmission lines, since the former tend to be Annex I projects while the latter fall within Annex II of the Directive (for which EIA is required only if significant effects are likely). However, the Directive requires an assessment of *direct and indirect* effects, which cannot be ensured for a power station scheme unless the transmission implications are included within the EIA. The Commission's response clearly supported this interpretation of the Directive.

Despite this clarification of the purpose and intention of the Directive, the EC decided against taking infringement action against the UK government in this case. Earlier, action had been taken by the EC in connection with EIA for the Channel Tunnel rail link and Kings Cross terminal 'project'. In that case, the EC had argued that these two projects were indivisible, because of the effect of each on the choice of site or route of the other:

> The effect of dividing the London-Channel Tunnel project into the rail link on the one hand, and the terminal on the other, leads to the circumvention of Directive 85/337/EEC, since the siting of the rail link in London is no longer capable of being

assessed and—for instance by the choice of another site for the terminal—its effects minimized during the consideration of the rail link route. Terminal and link are, because of the impact of the choice of the terminal site on the link, or the link on the site, indissociable. The intention to assess the link once the assessment of the impact of the terminal is [completed] does not therefore make acceptable the assessment of the terminal…, which failed, contrary to Article 3 of the Directive, to take into account the effects of its siting on the choice of [route for] the rail link. (Letter from the Environment Commissioner to the UK government, 17 October 1991)

The same argument seems to apply in the Wilton example. The power station and transmission lines were also indivisible, since the power lines would not have been required were it not for the new power station, and the location of the power station was critical to any subsequent decisions on the route of the power lines.

9.2.3 *The Lackenby-Shipton power lines public inquiry*

As noted above, the installation of power lines to service the new power station was the responsibility of a separate developer (the NGC) and was subject to separate—later—consent and EIA procedures. In the event, five alternative routing proposals were considered concurrently at a public inquiry, which started in May 1992—some 18 months after consent had been given for the power station; indeed, by this stage, construction of the power station had already begun.

The proposed power line routes started in Lackenby, adjacent to the Wilton power station site, and then proceeded south to Picton, via alternative southern and northern routes. From Picton, alternative western, eastern and central routes ran south to Shipton, northwest of York (Figure 9.1). The total length of new power lines and system upgrades required was between 75 and 85 km, depending on the route options selected. The NGC itself expressed a preference for the shorter southern route from the power station to Picton, and for the western route option from Picton to Shipton. All of the proposed routes passed through or adjacent to (and visible from) important protected landscapes, including the North York Moors National Park and the Howardian Hills, an Area of Outstanding Natural Beauty (AONB). Key objectors to the proposals at the public inquiry included the local authorities through which the proposed routes ran (North Yorkshire CC, Cleveland CC and others), the Country Landowners Association, the National Farmers Union and CPRE, as well as many individuals, including farmers and local residents. The principal issues considered at the inquiry included the visual impact of the pylons and overhead lines, potential health risks from electromagnetic radiation, issues of need and alternatives and effects on farming operations.

CPRE argued at the inquiry that the visual impacts of the proposals were unacceptable and should have been foreseen at a much earlier stage. It urged that the inquiry inspectors 'should not feel obliged to grant consent for the power lines simply because consent for the power

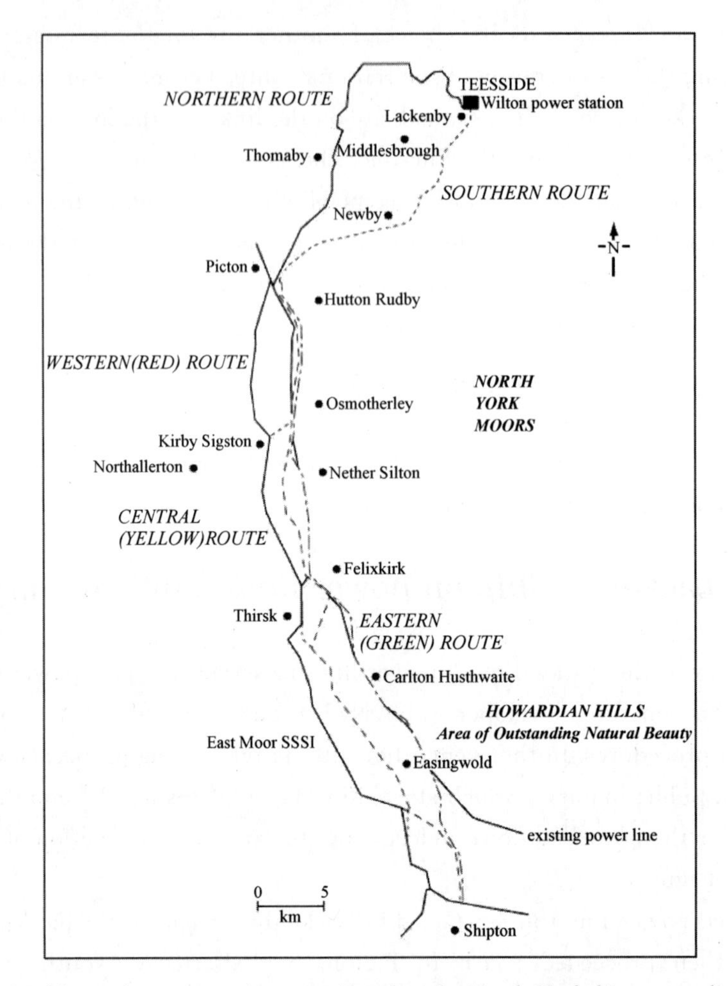

Figure 9.1 Alternative route options considered at the North Yorkshire power lines inquiry

station had already been granted and it was already being built' (Sheate 1995). It also invited the inspectors to comment on the inadequacy of the existing EIA procedures in such cases, in which consent for electricity generation is divided from consent for electricity transmission.

The inquiry ended in December 1992 and, after a long delay, the inspectors' report was published in May 1994. It recommended approval of NGC's preferred route options—the southern route from Lackenby to Picton and the western route from Picton to Shipton, subject to various detailed modifications to minimize the environmental impacts (e. g. around East Moor, a Site of Special Scientific Interest, SSSI). However, the inspectors agreed with CPRE's views on the EIA procedures in such cases:

> It seems to us that to site power stations without taking into account all relevant factors, including transmission to the areas of consumption, is likely to lead to the extension of high voltage power lines through areas currently not affected and the reinforcement of lines in areas already affected. It is not disputed that in the view of the scale and form of the towers these lines are inevitably highly intrusive and damaging to almost any landscape and as a result are unwelcome.

> It appears to us that there is a strong case for consideration to be given to the introduction of procedures to ensure that consents for future power stations take account of the resulting transmission requirements, and the environmental impacts of any necessary extension or reinforcement of the National Grid, between the proposed generating plant and areas of consumption. (Inspectors' conclusions, 23 September 1993)

The failure of the EIA for Wilton power station to address the implications of transmission connections resulted in a situation in which 'Teesside Power Limited [the developer] neither had to demonstrate the full implications of the siting and development of the power station, nor to bear the full economic and environmental costs' (Sheate 1995). This was because there were limits on the costs that could be recouped by NGC for the provision of transmission connections to individual generating projects. This meant that NGC was under commercial pressure to develop the cheapest options, since any additional costs incurred to minimize the environmental impact of power lines—such as taking a longer route through less sensitive areas or placing all or part of the route underground—would be borne by NGC rather than the power station developer.

9.2.4 *Lessons for EIA*

Sheate (1995) argues that the Wilton power station case provides powerful evidence that, at the time, the procedures for consent approval in the electricity supply industry ran counter to the letter and spirit of the EIA Directive. According to Sheate, the situation could have been remedied by an amendment to the Electricity and Pipeline Works EIA Regulations. The suggested amendment read as follows:

> An environmental statement shall include information regarding the overall implications for, and impact of, power transmission lines and other infrastructure associated with the generating station where these are likely to have significant effects on the environment. (CPRE, letter to DTI, 22 February 1993)

The effect would be that, in cases where power lines or other associated infrastructure were likely to have a significant effect on the environment, these impacts should be material considerations in whether consent for the power station should be given and the Secretary of State for Energy should be aware of these before giving consent.

> The consequence of such an amendment would be to ensure that power station proponents were forced to consider the transmission implications of their proposals, and that they would form part of the EIA and of any subsequent public inquiry. It would begin to reduce the difficulties that arise over the definition of projects and programmes. (Sheate 1995)

The case also highlights a wider problem within the EIA Directive concerning its ambiguous definition of the term 'project'. As we have seen, this is a particular issue for projects in the electricity supply industry, but it also applies to other infrastructure projects such as road, rail and other transport schemes. It has resulted in a number of complaints to the EC about whether a larger project can be split into a number of smaller schemes for the purposes of consenting and (therefore) EIA. The problem is that EIA in the UK—as in most EU Member States—has been implemented as part of existing consent procedures, and if these are divided for a project, then so is the requirement for EIA. This so-called 'salami-slicing' of projects runs counter to the purposes of the Directive, which states 'effects on the environment [should be taken] into account at the earliest possible stage in all the technical, planning and decision-making processes' (Preamble to Directive 85/337/EEC). As the case study illustrates, this purpose cannot be achieved if EIA is applied only to individual project components rather than to the project as a whole.

The issues raised by this case study remain relevant to current EIA practice. The issue of ambiguous project definition has not been resolved in the subsequent amendments to the EU EIA Directive, and consent procedures for electricity generation and transmission projects in the UK remain divided.

9.3 N21 link road, Republic of Ireland: EIA and European protected habitats

9.3.1 *Introduction*

This case study, researched by Weston and Smith (1999), concerns a proposed road improvement scheme in County Kerry, Republic of Ireland. The proposed route of the road passed through part of a European protected habitat, a residual alluvial forest known as Ballyseedy Wood. Although the proposal was not subject to EIA (largely because the ecological status of the site was not known at the time), it was later subjected to a related procedure known as 'appropriate assessment' (or Habitats Regulation Assessment, HRA), which operates under the EU Habitats Directive. The Habitats Directive requires that projects likely to have a detrimental impact on a European priority habitat must be subject to an assessment of that impact. This assessment involves a series of sequential tests that must be passed for the project to be allowed to proceed. The case study examines the nature and interpretation of these tests, and demonstrates the high level of protection afforded to designated habitats in the EU (Weston and Smith 1999).

9.3.2 *The proposals*

Improvements to the N21 main road into Tralee, County Kerry, in the west of Ireland, had been an objective of the local authority, Kerry County Council, since the late 1960s. However, it was not until the prospect of European funding for these improvements emerged during the mid-1990s that substantial progress was made in advancing the scheme. The route was included in the Irish government's Operational Transport Programme for Ireland (OTP) in 1994. In the same year, the OTP was adopted for cofunding by the EC as part of the EU's Community Support Framework for Ireland. Under this Framework, the EU agreed to provide 85 percent of the funding for the proposed improvements to the N21 link.

The proposed project comprised improvements to 12.5 km of the existing N21 highway between Castleisland and Tralee, including a short (2.4 km) new section of dual carriageway between Ballycarty and Tralee (Figure 9.2). The dual carriageway section of the scheme ran to the south of the existing highway and through Ballyseedy Wood, which was later discovered to be a priority habitat under the EU Habitats Directive. Following the announcement of European co-funding in July 1994, Kerry CC, as the local highways authority, began design work on the proposed scheme.

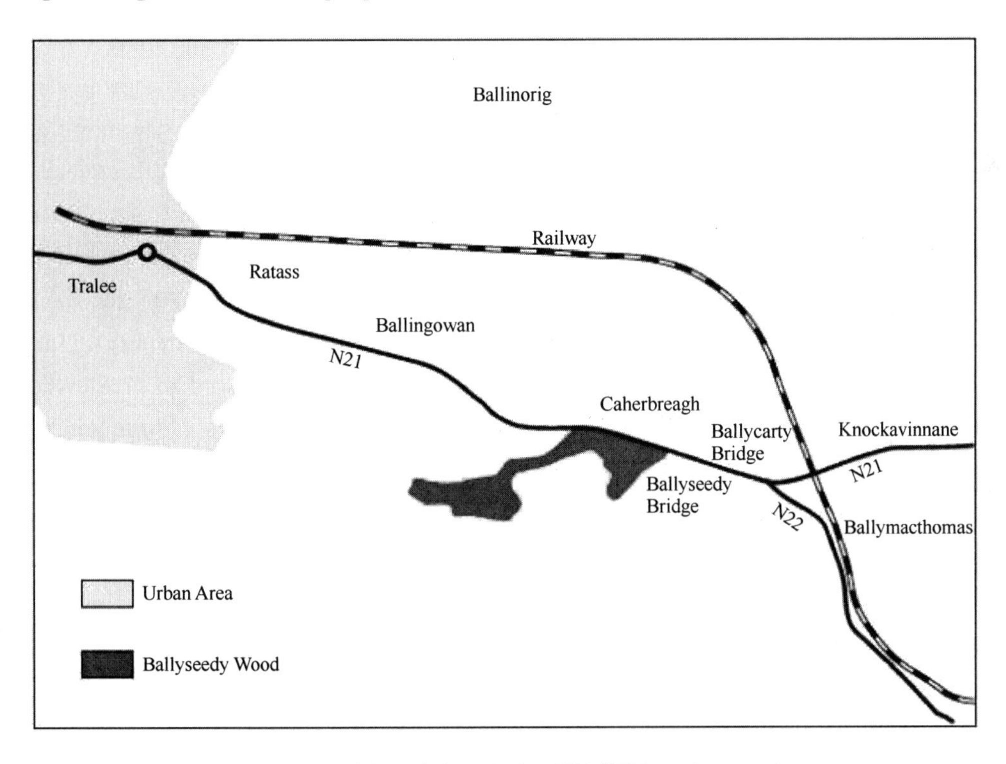

Figure 9.2 Map of the existing N21/N22 road network

9.3.3 The planning and EIA process

The N21 road improvement scheme was an Annex II project, for which EIA is required only if there are likely to be significant environmental effects. Like most EU Member States, at the time Ireland employed a series of size thresholds to help determine whether Annex II projects should be subject to EIA. In the case of road schemes of the type proposed (rural dual carriageways), EIA was required for schemes in excess of 8 km in length or if there were considered to be significant environmental effects. In such cases, EIA was carried out by the local highways authority and submitted to the DoE. After a period of consultation and a public inquiry (if one was held), the Minister for the Environment made a decision on the application. For Annex II schemes falling below the size threshold (less than 8 km in length) and not considered likely to cause significant environmental impacts, EIA was not required and the proposal was dealt with under normal planning legislation. This allowed for a period of public consultation, with the final decision as to whether to approve the scheme resting with the relevant local authority. For road schemes, it was the developer of the project, in this case the County Council as local highways authority, who determined whether or not EIA was required.

After commissioning a report from environmental consultants into the proposed scheme, Kerry CC decided that an EIA was not required in this case. This decision was based on the length of the dual carriageway section of the scheme (at 2.4 km, well below the 8 km threshold for such schemes) and on the belief that there were unlikely to be any significant environmental effects. However, following the publication of the proposals, the Council received a number of objections, mainly regarding the impact of the new dual carriageway on Ballyseedy Wood. After considering these objections and the subsequent report to the Council on the scheme prepared by the authority's officers, the Council's elected members decided to proceed with the proposals. However, this was not the end of the authorization process, since the compulsory purchase orders (CPOs) necessary for the scheme to proceed still had to be served and considered at a public inquiry (the CPO inquiry). Under the Irish system, members of the public and interested parties were allowed to give evidence at the CPO inquiry on environmental issues. However, the inquiry and the subsequent decision (including any alterations to the alignment of the route) was based solely on land acquisition, rights of way and access issues. The CPO inquiry for the N21 scheme was held in March 1996.

Following the Council's decision to go ahead with the scheme, a local organization objecting to the proposals commissioned an ecological assessment of Ballyseedy Wood. This assessment concluded that the wood comprised an area of residual alluvial forest that, although currently lacking protected status, complied with the description of a priority habitat as set out in the EU Habitats Directive of 1992. These conclusions were accepted both by the relevant Irish national authorities and Kerry CC, and the site was subsequently proposed as a Special Area of Conservation under the terms of the Directive. The revelation of the important ecological status of Ballyseedy Wood resulted in a number of formal complaints being submitted to the EC concerning its co-funding of the proposed scheme. It was argued that the Commission should reconsider its decision to co-finance the project, given its potentially damaging impacts upon a habitat of recognized European-wide importance. As a result of these complaints, the EC commissioned an independent study to provide advice on whether there was a need to re-consider the co-funding of the scheme.

9.3.4 *The EU Habitats Directive*

The EU Habitats Directive (92/43/EEC) requires all Member States to designate sites hosting important habitat types and species as Special Areas of Conservation. Together with Special Protection Areas (SPAs) designated under the Birds Directive (79/409/EEC), it is intended that these sites will form a network of European protected habitats known as 'Natura 2000'. The Habitats Directive is designed to protect the integrity of this European-wide network of sites, and includes provisions for the safeguarding of Natura 2000 sites and priority habitats.

In cases in which a project is likely to have a significant impact on a protected site, the Directive states that there must be an 'appropriate assessment of the implications for the site in view of [its] conservation objectives'. Under the terms of the Directive, consent can only be granted for such a project if, as a result of this appropriate assessment, either (a) it is concluded that the integrity of the site will not be adversely affected, or (b) where an adverse effect is anticipated, there is shown to be an absence of alternative solutions and imperative reasons of overriding public interest that the project should go ahead. The overall intention of the Directive is 'to prevent the loss of existing priority habitat sites whenever possible by requiring alternative solutions to be adopted' (Weston and Smith 1999). Projects that have a negative impact on the integrity of a priority habitat, but which are able to satisfy both the absence of alternatives and overriding-reasons tests, can go ahead. However, in such cases, the developer must provide compensatory measures to replace the loss of priority habitat. Huggett (2003) discusses the development of such measures in relation to a range of port-development proposals in the UK (see also Chapter 12).

The tests set out in the Directive are not absolute and require a degree of interpretation. For example, a literal interpretation of the 'absence of alternative solutions' test could be taken to imply that any alternative that is less damaging to the protected habitat than the proposed scheme should be selected, regardless of cost or impacts on other interests. European case law provides some indication as to the appropriate interpretation of the Directive's tests, and this is reviewed by Weston and Smith (1999). They conclude that both the 'absence of alternatives' and 'reasons of public interest' tests should be interpreted stringently, in view of the intention of the Directive to provide a significant level of protection to priority habitats. The application of these tests to the N21 link road proposal is now explored.

9.3.5 *Appropriate assessment of the N21 link road proposals*

Under Article 6 of the Habitats Directive, the appropriate assessment of the N21 project involved a series of sequential tests, concerning:

- the impact of the scheme on the integrity of the priority habitat;
- the presence or absence of alternative solutions; and
- the existence of imperative reasons of overriding public interest that the scheme

should go ahead.

Each of these tests is examined, drawing on the results of the independent study commissioned by the EC, as reported in Weston and Smith (1999).

Impact on the integrity of the priority habitat

Ballyseedy Wood covers 41 ha, although the priority habitat that was the subject of assessment represented only a very small part of this overall area. A small area in the northern corner of the wood accorded with the definition in the Habitats Directive of a residual alluvial forest; it consists of alder and ash and is subject to regular flooding. Wet woodland of this type is the least common type of Irish forest, and the surviving examples tend to be small in area; the priority habitat area covered less than half a hectare. It was this northern edge of the wood that was to be lost to the proposed dual carriageway, including the areas of greatest ecological interest.

The direct land take of the proposed scheme involved the loss of only 3 per cent of the total area of the priority habitat. However, this does not necessarily mean that the effect on the integrity of the habitat would be insignificant. The independent study concluded that:

> On the basis of the assessments carried out by a number of environmental consultants and the evidence presented to the CPO inquiry, the loss of habitat could not be objectively assessed as being of no significance to the integrity of the habitat as a whole. There will be change caused to the habitat as a result of the removal of trees, the change in the hydrological regime and the re-routing of the river [another element of the scheme]. Evidence to the CPO inquiry suggested that areas of the wood, outside of the land take, would also be affected by this change. As part of EU policy the precautionary principle also needs to be applied to the assessment of the impact of a project on a priority habitat. In applying that principle it must be concluded that there is a risk that the integrity of the [habitat] will be significantly diminished by the proposed road. (Weston and Smith 1999)

It was therefore concluded that the impact on the priority habitat would be negative. Nevertheless, the project could still go ahead if it satisfied both of the remaining two tests; the first concerned the absence of alternative solutions.

The absence of alternative solutions

The County Council's identification of the preferred route alignment for the proposed scheme, and possible alternative routes, was based partly on a constraint mapping exercise in which areas with various environmental constraints (such as archaeological remains) were identified. Cost factors also featured in the choice of the preferred route, and a form of cost-benefit analysis (CBA) was carried out. The Council prepared a Design Report, which set out the need for the scheme and the alternatives considered. This reveals that the Council had identified and investigated a number of alternatives. Weston and Smith (1999) identify a total of six main alternatives to the proposed scheme, including a do-minimum option. Most of the alternatives considered completely avoided Ballyseedy Wood, generally by following more northerly route alignments. However, other adverse impacts arose from some of these alternative schemes, such as the demolition of residential properties, farm severance and the relocation of Ballyseedy Monument, a local war memorial. Notwithstanding these impacts, a number of viable alternative route options were clearly available to meet the objectives of the proposed scheme.

The CC tested the various route options against their ability to provide the best solution 'in terms of human safety, capacity and economic viability'. However, there appears to have been no systematic attempt to test the alternatives against the need to avoid the loss of priority habitat, even after the Council became aware of the importance of Ballyseedy Wood. The conclusion of the independent study was that '[the] alternatives were not examined to the same rigour or degree as the preferred route and appear to have been rejected without clearly defined and quantified justification' (Weston and Smith 1999).

An issue that appears not to have been considered by the CC in its route selection was the need to serve those areas where future development growth was planned. In this case this was the northern edge of Tralee, which was the location of a new Regional Technical College and of allocated industrial areas. This suggests that a more northerly route alignment for the dual carriageway—which would have avoided the impacts on Ballyseedy Wood—may have been better placed than the proposed scheme to accommodate the growth in traffic generated by these planned developments. The proposed scheme would have involved traffic serving these planned growth areas passing through the town centre of Tralee. This would have added to existing traffic problems in the town, and may have resulted in the time benefits derived from the improved N21 being lost because of increased congestion in Tralee. The independent study comments:

> It is surprising therefore that an alternative alignment for both the N22 and N21, which links the infrastructure to the areas of Tralee where future development is planned, has not been more fully investigated. A northern route proposed by private individuals, which could be of dual carriageway standard, was not adequately assessed in terms of the strategic objectives of the Operational [Transport] Programme or in terms of its benefits such as avoiding Ballyseedy Wood and maintaining the existing distinctive quality of the area around the Ballyseedy Monument. There are other possible alignments that appear not to have been fully considered, such as routes south or north of the railway line [which runs to the north of the existing N21]. Although the Council's Design Report rejects such routes because of the problems of crossing the railway line, farm severance and the impact on property, there appear to have been insufficient investigations and assessment on which to base such an outright rejection of such options.
>
> Overall, there is little to suggest that the Council's alternatives have been tested to the same degree as the preferred option. The alternatives considered were not subjected to detailed costings, surveys, timesaving considerations, their ecological impacts or indeed their ability to meet the strategic objectives of Structural Funding. In the absence of the rigorous testing of

all alternatives against clear objectively determined criteria it cannot, in this case, be concluded that the objectives of the [OTP] cannot be achieved with an alternative solution to that which would damage the priority habitat. (Weston and Smith 1999)

The second test, an absence of alternative solutions, was therefore failed. Under the terms of the Habitats Directive, the project could not therefore proceed, since a number of viable alternative solutions were shown to have been available in this case. It was not therefore necessary to carry out the third test, the existence of imperative reasons of overriding public interest in favour of the scheme. However, it is useful to do so, since this demonstrates how this test is applied in practice in the appropriate assessment process.

Public interest issues

The third test involves the balancing of the loss of priority habitat against other imperative public interest issues. Public interest issues would outweigh the loss of habitat if they resulted in 'far greater adverse impacts than does the loss of habitat' (Weston and Smith 1999). So, for example, if only a minor impact on the habitat was anticipated and the alternative options would result in extreme economic or other public interest disbenefits, then the public interest issues could be said to outweigh the loss of habitat. Conversely, if the impact on the habitat was great or uncertain, and the impact on the public interest issues was small, then the interests of the habitat would take precedence.

European case law provides some guidance on the type of public interest issues that can be considered to be 'imperative' reasons. Examples include the public interest of economic and social cohesion, human health, public safety and other environmental concerns. However,

> for such public interest reasons to outweigh the loss of habitat they must be of a similar scale in importance [as the protection of the priority habitat] —that is of interest to the [European] Community as a whole-and be demonstrable. (Weston and Smith 1999)

In the case of the N21 scheme, it was the view of the CC that a number of public interest issues were relevant and that, when combined, the sum total of these issues outweighed the loss of the priority habitat at Ballyseedy Wood. The public interest issues arising in the case included:

- the strategic objectives of the wider OTP (of which the scheme was a component);
- the cost of alternative solutions;
- loss of family homes;
- road safety issues;
- heritage impacts on the Ballyseedy Monument;
- farm severance; and
- impacts on archaeology.

The independent study into the scheme concluded that none of these issues could be regarded as both imperative (that is, of equal importance as the loss of habitat) and overriding (that is, sufficiently damaging to override the protection of the habitat), and therefore this third test was also failed. The reasons for this conclusion included:

- *The existence of alternative solutions*. The fact that a range of alternative route options were available made it difficult to argue that the public interest issues arising in the case were unavoidable. For example, there was considerable local concern about the need to relocate the Ballyseedy Monument, a local war memorial, should one of the alternative routes be adopted. However, the need to relocate the Monument arose only with one of the six main alternatives considered and could therefore have been avoided by the adoption of one of the other alternative solutions.
- *The alternatives would not necessarily result in greater adverse impacts than the proposed scheme*. For example, there was no evidence that, apart from one of the route options, any alternative solution would result in the loss of more family homes than the proposed scheme.
- *Some of the public interest issues were not demonstrable, due to a lack of data*. For example, no quantified data was produced on the road safety implications of alternative routes, compared with the proposed scheme. Evidence from the CC at the CPO inquiry suggested that all alternatives examined by the Council were equally safe. Also, it was not possible to argue that the proposed scheme was necessarily the most cost-effective, since the costs of all the alternative route options had not been worked out in detail. Indeed, it was suggested 'that an alternative route may be cheaper to construct because of the decreased disruption to existing road users, the impact of construction on properties in the existing corridor and the reduction of some mitigation costs' (Weston and Smith 1999). Another issue raised was the impact of alternative routes on farm severance. Again, however, 'there is little hard evidence to show that this is an area that has either been examined in any great detail, been quantified in any way or has been comparatively assessed against the [proposed] scheme' (Weston and Smith 1999).
- *The loss of habitat was a superior interest compared to most of the public interest issues raised*. Most of the public interest issues arising in the case were not equivalent in importance to the loss of priority habitat, and could not therefore be regarded as 'overriding' interests. Examples include the loss of family homes and farm severance. Although important issues at a local scale, these cannot be seen as equal in importance to the need to protect the priority habitat, given the status of the latter in the EU Habitats Directive. Similarly, in relation to archaeological impacts, in order to be of 'overriding' public interest, the archaeological feature affected would need to rank higher than the priority habitat on a European scale. There was no evidence that such impacts would arise with any of the alternative routes.

One public interest issue that appeared to be of greater importance was the need to relocate the Ballyseedy Monument, which arose with one of the alternative route options.

> The Council and the local community generally consider the relocation of the Monument to be unacceptable as it is considered one of the most important modern monuments in Ireland. The Monument, however, has no national or local statutory protection, whereas [Ballyseedy Wood] has statutory protection [at European level] through

the [Habitats] Directive… On that basis, the relocation of the Monument, while clearly a very important public interest issue, cannot be seen as an 'overriding' public interest in terms of the presumption established by the Directive to protect the priority habitat. (Weston and Smith 1999)

Having failed all three of the tests required under the 'appropriate assessment' process, EU funding for the proposed N21 link road scheme was withdrawn.

9.3.6 Summary

The case study demonstrates that the process of appropriate assessment under the Habitats Directive, once a negative impact on a priority habitat has been established, is an exacting one. In particular, few projects are likely to have a genuine absence of viable alternatives, especially if the search for possible alternatives is widely defined. Also, to outweigh the loss of priority habitat, public interest reasons must be of equal or greater weight than the protection of priority habitats at European level. This means that issues of only local or even national importance would not be sufficient. Finally, as illustrated above, the absence of alternatives and imperative reasons tests are inextricably linked. 'While there remains the possibility of alternative solutions there are unlikely to be "imperative reasons of overriding public interest" to justify the preferred solution' (Weston and Smith 1999). Further guidance on the appropriate assessment process is provided in EC (2000, 2001, 2007) and Chapter 12.

9.4 Portsmouth incinerator: public participation in EIA

9.4.1 *Introduction*

This case study involves an innovative approach to public participation within the EIA process for a proposed municipal waste incinerator in Portsmouth, Hampshire, UK. The approach adopted by the developer in this case provided an opportunity for members of the public to take part in structured discussions about the project proposals and their environmental impacts before the submission of the planning application and environmental statement. This approach to extended public participation, beyond that required in the EU EIA Directive, has been used in a number of cases in the UK waste sector in recent years, not only at project level (as in this case) but also at more strategic levels in the development of local waste management strategies and plans (Petts 1995, 2003).

The increasing use of these methods reflects the perceived inadequacy of more traditional forms of public participation in the highly contentious arena of waste facility planning. However, questions remain about the effectiveness of such methods in providing genuine opportunities for the public and other interested stakeholders to participate in the EIA and wider development processes. The case described here is based largely on research carried out by Chris Snary as part of his PhD studies with the IAU at Oxford Brookes University (previously documented as Snary 2002), with additional material from Petts (1995, 2003).

9.4.2 *Public participation and EIA*

The wider context to the case study is the almost universal opposition towards proposed waste management facilities among those who live near proposed sites. Such public opposition is often dismissed simply as a NIMBY reaction or as being based on unjustified and irrational fears about potential impacts, particularly in relation to emissions and associated health risks. This is contrasted with the scientifically based technical assessments of impact and risk carried out by EIA practitioners. However, Snary (2002) points out that recent studies indicate that public opposition to such facilities is often based on a much wider range of considerations, including 'concern about the appropriateness of the waste management option, the trustworthiness of the waste industry and the perceived fairness of the decision-making process'.

Reflecting this improved understanding of the nature of public opposition, a number of commentators have called for better communication with the public at all stages of the waste management facility planning process (ETSU 1996; IWM 1995; Petts 1999). Such communication can take a variety of forms, ranging from a one-way flow of information from developer to public, through different levels of consultation and participation (in which there is a two-way exchange of views between the public and the developer and/or consenting authority, and the public's views are a legitimate input into the decision-making process). All of these types of communication are seen to be important components in the planning and EIA process for incinerators and other waste facilities, as Snary (2002) explains:

> Concerns about health risks require comprehensive information on the [predicted] emissions and a consultation process through which the public's views can affect the decision-making process. Concerns about the ability of the waste industry and regulators to manage risk competently require participation in a process through which their concerns can be openly addressed and conditions of competency discussed. Debate concerning fundamental policy issues and the legitimacy of the waste planning process [also] requires a public participation process through which a consensus may be built [at the plan-making stage of the waste incinerator planning process].

The search for improved methods of public participation is also linked to the growing social distrust of science and experts noted by a number of commentators (see, for example, House of Lords 2000; Petts 2003; Weston 2003).

9.4.3 *Background to the proposed scheme*

Hampshire is a county on the south coast of England, with a population of around 1.6

million. By the end of the 1980s, the county was faced with the problem of increasing volumes of household waste, set against a background of an ageing stock of incinerator plants (which failed to meet the latest emission standards) and growing difficulties in finding new and environmentally acceptable landfill sites. In response, the County Council's Waste Management Plan (1989) advocated an integrated approach to waste management, supporting recycling and waste minimization initiatives and emphasizing the need for a reduced reliance on landfill. Government financial regimes in operation at the time (the Non-Fossil Fuel Obligation) also provided cost incentives for the development of energy-from-waste schemes rather than landfill. It was also recognized that significant economies of scale could be obtained by developing a single large plant in the county rather than several smaller ones. As a result, following a tendering process, an application was submitted at the end of 1991 for a large energy-from-waste incinerator in Portsmouth, in the south of the county on a site selected by the CC (Petts 1995). The capacity of the plant was 400,000 tonnes per annum, which represented two-thirds of the household waste arising in the county (Snary 2002). The proposed location was on the site of one of the county's redundant incinerators, which had been closed in 1991 after failing to meet the latest emission standards.

The proposal met with much local opposition, from both local residents and ultimately the relevant local authority, Portsmouth City Council. Objections focused on a number of environmental issues, including the health risks posed by emissions from the plant; visual, noise and traffic impacts; and the close proximity of the site to residential areas. Policy concerns were also raised, in particular that, by concentrating on incineration as the preferred waste option, the promotion of recycling and waste minimization in the county would be adversely affected. In the event, the CC (the consenting authority for this type of project at the time) decided that it could not support the application, on the grounds that the proposal was too large and did not form part of a more integrated waste management strategy for the area (Snary 2002). The failure to gain approval for the proposed scheme resulted in a change of approach from the County Council, as Petts (1995) explains:

> By the summer of 1992 the County Council had failed to gain approval for the plant and was facing an urgent task to find a solution to the waste disposal problem. The traditional approach had failed. While the [county's waste management] plan which had supported the need for [energy-from-waste] had been subject to public consultation with relatively little adverse comment, this was now regarded as too passive a process and it seemed that the real concerns and priorities of the community had not been recognized by the County [Council]. There had not been strong support of the need for an integrated approach to waste management and there had been little recognition of the need to 'sell' [energy-from-waste] to the public. The proponents had been overly optimistic about their ability to push the project through with the standard, information-based approach to public consultation.

Faced with these problems, the CC embarked on the development of a more integrated and publicly acceptable household waste management strategy (Snary 2002). The Council's new approach involved an extensive proactive public involvement programme, launched in 1993, to examine the various options for dealing with household waste in the county. The aim was to attempt to establish 'a broad base of public support for a strategy which could be translated into new facilities' (Petts 1995). As part of this process, Community Advisory Forums were established in the three constituent parts of the county, based on the model of citizens' panels. Membership included a mix of people with different interests and backgrounds, including those with little prior knowledge of waste issues. At the end of the process (which lasted for six months), the forums presented their conclusions to the CC. The broad consensus reached was that:

- greater efforts should be made in waste reduction and recycling;
- energy-from-waste schemes would be needed as part of an integrated waste management strategy, but there was considerable concern about their environmental effects and the monitoring of plant; and
- landfill was the least preferred option (Petts 1995).

The public participation exercise in Hampshire resulted in the inclusion in the county's revised waste strategy (1994) of plans to build three smaller energy-from-waste incinerators (each with a capacity of 100,000—165,000 tonnes), rather than the single large incinerator originally proposed. The new plants were to be located in Portsmouth (on the same site as the earlier application), Chineham, near Basingstoke, and Marchwood, near Southampton. EIA work for these proposed developments began in 1998 (Petts 2003). It is the first of these plants that is the focus of this case study.

9.4.4 *The contact group process*

The EIA process for each of the three proposed incinerators in Hampshire involved a method of public participation known as the 'contact group' process. This involved an extended process of public questioning during the preparation of the ES for each site through a contact group involving a range of key local interests. These contact groups were established by the developer, Hampshire Waste Services (HWS), and were part of the contractual requirements placed on the company by the CC (Petts 2003). This approach had the potential to enable the public's views to result in reassessment of issues dealt with in the ES, and to changes in the project proposals and mitigation measures, prior to the submission of the ES to the competent authority.

The terms of reference for the Portsmouth contact group stated that it was designed (a) to allow key members of the public to develop informed decisions about waste issues and the proposal; and (b) to assist the developer (HWS) in ensuring that it understood and responded to the views of the members of the local community (Snary 2002). The arrangements for extended public participation in this case go beyond the legal requirements under the UK EIA Regulations (discussed in Chapter 6), and were the first time that such methods had been used in the UK EIA process for a waste incinerator (Snary 2002).

In the Portsmouth incinerator case, 10 members of the public were included in the contact group-they were selected by HWS to represent a range of local interests, and included a representative from the local school, the local branch of Friends of the Earth (FOE), and the Portsmouth Environmental Forum, plus seven representatives from the six neighbourhood forums in closest proximity to the project site. Group members were encouraged to network with the local residents in their neighbourhood. It was made clear to participants that membership did not imply support for the proposals, and indeed almost all of the participants were opposed to the development.

The contact group met once a week over a sixweek period immediately prior to the submission of the planning application and ES in August 1998. Issues covered by the contact group at these meetings included:

- waste-to-energy incinerators and EIA;
- design of the plant;
- noise and traffic assessments;
- visual and ecological issues;
- alternative sites and noise issues; and
- air quality issues and health risk assessment.

Information was provided on these issues by HWS and by its consultants at the meetings. During discussions, the participants were able to make their views known by raising questions, concerns and suggestions. Answers to questions were provided on the day and in written form at the next meeting. There was also a closing meeting to discuss the conclusions of the ES. An independent chairperson was appointed by HWS 'to ensure that all participants had an equal opportunity to contribute to the meetings and that issues were fairly addressed' (Snary 2002).

9.4.5 Evaluation of the process

How effective were the methods of public participation employed in this case, and what were the views of the various participants in the process? Snary (2002) has assessed the success of the contact group process in the Portsmouth case, based on interviews with those involved; the process has also been evaluated by Petts (2003), drawing on observation of all three contact groups. Key findings are summarized below, focusing in particular on the limitations of the process in practice.

- *The contact group process took place too late in the EIA process.* The contact group meetings started only six weeks prior to the submission of the ES, and by this stage the majority of the EIA work had been completed. The opportunity for the group to influence the way in which impacts were defined, assessed and evaluated was therefore very limited. This was particularly true of the health risks posed by emissions, which were discussed only at the last meeting of the group. The process would have been more effective if it had started during the scoping stage of the EIA. However, the scoping exercise was restricted to consultation with the local planning authority and statutory consultees, with no public involvement (Snary 2002).

• *Insufficient time was allowed for the process.* A number of participants commented that the meetings were attempting to cover too much information—often of a complex nature—in too short a time. Again, this suggests that the process should have been started earlier to allow the wide range of issues involved to be dealt with adequately.

• *Criticisms were made of the EIA consultants.* Participant criticisms included the view that assessments were based too much on desk studies and that the consultants lacked detailed knowledge of the locality; that the consultants did not always provide adequate answers to questions; and that the EIA work should have been undertaken and presented by independent consultants (i. e. not employed by the developer).

• *Participants were generally better informed about the proposals.* Almost all participants stated that they felt better informed about the issues relating to the proposal as a result of attending the meetings. This is hardly surprising, but the developer's project manager also argued that the process had 'informed key members of the public better than the traditional methods of public involvement could ever have done' (Snary 2002). However, doubts were expressed about the complex nature of the information provided about the health risks posed by the development. One participant commented: 'I am not a scientist and I found it very difficult to understand. I felt as though they were trying to blind me with figures and technical terms. The residents that I have spoken to who went to have a look at the environmental statement felt exactly the same; they didn't really understand the assessment.' These criticisms are partly related to the tight timescale for the contact group process, although nonexperts will always need to have a degree of trust in those providing technical information in EIA. Snary suggests that such trust could have been increased by the use of independent consultants or an independent third party to summarize and validate the information presented.

• *Limited impact on the development proposals.* The project manager for the development stated that the process had resulted in changes to the architecture of the scheme (in particular the colour of the buildings) and improvements to the traffic assessment. However, apart from these relatively minor changes, many participants were sceptical about how else the views of the group had affected the proposals. These findings are not surprising, given the fact that the meetings took place at such a late stage in the planning, design and EIA work for the scheme.

• *Low levels of trust in the developer and consultants.* Reasons included a feeling that the developer was bound to be biased because its aim was to gain planning permission, a view that group members were only being provided with part of the information about health risks and concerns over the competency of the EIA consultants.

• *The process failed to resolve fundamental concerns about the proposal.* All but one of the participants still had relatively strong risk-related concerns about the proposal at the end of the contact group process. Therefore, although the contact group was able to better inform key local stakeholders about the risks posed by emissions, it was unable to convince the majority of the group that the risks were acceptable and that waste-to-energy incineration was an appropriate waste management solution (Snary 2002). This is despite the fact that the Portsmouth incinerator proposal emerged as part of a county-wide waste strategy that was developed through an extensive and innovative public involvement exercise. Snary attributes this to inadequacies in the earlier strategic-level consultation exercise, which had failed to reach a consensus on the appropriate role of waste-to-energy incineration in the county's waste strategy and which most of the contact group members had been unaware of prior to joining the group. It was also unclear how the views expressed in the strategic consultation had influenced the county's developing waste strategy.

In her evaluation of the Hampshire contact group process, Petts (2003) reaches broadly similar conclusions:

> While the process did open up the environmental assessment to detailed questioning by a small but representative group of the public, it arguably started too late in the limited regulatory process to allow the Contact Group members to frame and define the problems to be considered and assessed. During the author's own observation and evaluation of the process, it was evident that questions about the assessment methods were able to be raised (for example, the Portsmouth Contact Group identified deficiencies in the transport assessment based upon knowledge of cycling on the local roads). Some re-assessment did take place as a result of such a public quality assurance mechanism. However, this was limited. Participants valued the opportunity provided to them to review the assessment but were suspicious that outcomes had already been decided.

9.4.6 Summary

This case study has illustrated the use of extended methods of public consultation in EIA, which go beyond the minimum legal requirements in the EU EIA Directive. These methods are not without their practical difficulties, and these have been highlighted. The main weakness in this case appears to have been that the contact group meetings started too late in the overall EIA process. Public involvement at the scoping stage of the EIA may have helped to avoid some of the problems encountered. As a postscript, after a public inquiry was held in 2000, planning permission for the Portsmouth incinerator was finally granted in October 2001—some 10 years after the initial application for an incinerator on the site had been submitted.

9.5 Humber Estuary development: cumulative effects assessment

9.5.1 *Introduction*

This case study provides an example of an attempt to assess the cumulative impacts of a number of adjacent concurrent projects in the Humber Estuary, Humberside, UK, undertaken in the late 1990s. This type of cumulative effects assessment (CEA), which was undertaken collaboratively by the developers involved in the various projects, is relatively uncommon in EIA. However, a number of other examples do exist, for example, in wind energy development cases in which several wind farms have been proposed in the same area. More generally, the assessment of cumulative impacts is widely regarded as one of the weak elements in project-level EIA (see, for example, Cooper and Sheate 2002; also Chapter 12).

Cumulative effects assessment studies of the type described here present a number of difficulties, and the case study examines how and to what extent these were overcome. The benefits derived from the CEA process are also discussed, from the viewpoint of the various stakeholders involved. This case study is based on research carried out by Jake Piper as part of her PhD studies with the IAU, Oxford Brookes University, and has previously been documented as Piper (2000). The Humber Estuary case study, together with a number of other examples of cumulative effects assessment in the UK, is also examined in Piper (2001a, b, 2002).

9.5.2 The Humber Estuary CEA

This case study involved a cluster of adjacent projects, proposed at around the same time by different developers. Each of the proposed projects required EIA, and because of the variety of project types, more than one consenting authority was involved in approving the projects. However, the developers concerned agreed to collaborate in the preparation of a single CEA of their combined projects, which was presented to each of the consenting authorities simultaneously.

In 1996—97, five separate developments were proposed along the north bank of the Humber Estuary, within a distance of 5 km of each other. The projects included:

- a new wastewater treatment works serving the city of Hull;
- a 1200 MW gas-fired power station;
- a roll-on/roll-off sea ferry berth;
- reclamation works for a ferry terminal; and
- flood defence works.

The five proposed projects involved four separate developers and five consenting authorities. The environment in the vicinity of the projects was a sensitive one, with a European site for nature conservation—an SPA designated for its bird interest under the EU Birds Directive and EU Habitats Directive—located within a short distance of the developments. It was the presence of this site, and the almost concurrent timing of the projects, that prompted the CEA study in this case. Indeed, the CEA was designed to satisfy the requirements for an 'appropriate assessment' of the effects of the proposed schemes on the SPA, under the terms of the Habitats Directive (similar to the process described in Section 9.3). It was also hoped that the CEA would help to avoid lengthy delays in securing approval for the projects, as Piper (2000) explains:

> The strategy adopted assumed that, by providing a common assessment to answer the needs of each of five competent authorities involved…, the amount of interplay and discussion required between these authorities would be reduced, avoiding lengthy delays… The strategy means, however, that any insoluble problems associated with any one project could tie up all consent applications simultaneously.

In order to guide the CEA process, a steering group was established consisting initially of the developers and the two local authorities concerned. Other key statutory consultees, including the Environment Agency and English Nature, joined the steering group later, but non-governmental environmental organizations and the public were not directly involved.

A single environmental consultancy prepared the CEA, acting equally on behalf of all four developers. Draft reports were prepared in consultation with the statutory consultees and developers, with opportunities for review and comment. Close liaison with EN (the statutory body responsible for nature conservation) was an important element in the process, given the need to specifically address the potential impacts on the SPA. It was important to ensure that the document presented to the local authorities and other consenting authorities also fulfilled the requirements of this statutory consultee.

The steering group was involved in determining the scope of the CEA, but no public participation was arranged for this stage of the study. The scoping exercise identified those issues where there was potential for cumulative effects to occur. These included, during the construction phase, effects on bird species on the SPA site and on traffic, and during the operational phase, effects on estuary hydrodynamics, water quality and aquatic ecology. Data was made available for the study by the developers, including information from existing EIA work already undertaken; some additional modelling work was also carried out. The information provided included the probable timing of activities within the construction programmes for each project, the manpower requirements for these activities and associated traffic movements. Existing baseline data available included the range of bird species present at different times of year in the SPA, and their vulnerability to disturbance (Piper 2000). Prediction of cumulative impacts was assisted by the production of a series of tables and matrices, which brought together the levels and timing of impacts identified for each project. These included:

- a combined timetable of major construction works;
- bird disturbance potential (sensitivity in each month of the year);
- timetable of construction work potentially affecting birds, and monthly sensitivity;
- potential aquatic impacts of the developments; and
- predicted traffic patterns (vehicles per day, for each month of the construction works).

In arriving at predictions, it was decided to use the developers' best estimates, rather than a worst-case scenario approach (Piper 2001a).

As a result of the cumulative impacts predicted, a number of additional mitigation measures were proposed (in addition to those measures that would have been considered had the schemes been assessed separately). Examples included the scheduling of certain noise-generating construction activities such as piling outside sensitive periods (e. g. bird roosting), and the introduction of staggered working hours to reduce peak traffic volumes. It was also proposed that the design of adjacent projects should be integrated in such a way as to minimize environmental impacts. An example was revisions to the design of the ferry berth structure to complement the design of the outfall from the water treatment works, and so enhance mixing of water in the estuary. Finally, recommendations were made for continued monitoring of the cumulative effects on birds and the aquatic environment. Responsibility for funding this work was shared among a sub-group of the developers involved in the proposed schemes (Piper 2000).

9.5.3 Costs and benefits of the CEA process

Piper (2000) has assessed the costs and benefits associated with the Humber Estuary CEA study, drawing on a series of interviews with those involved in the process, including the developers, the relevant local authorities, other consenting authorities and statutory consultees. The views of these different stakeholders are summarized below, beginning with the developers of the proposed schemes.

Views of developers

- *Greater understanding of the area and potential development impacts*. Three of the four developers felt that the CEA process had increased their understanding of the estuary and the potential impacts of the proposed developments. For example, the power station developer referred to better understanding of the impacts to the mudflats and birds and potential traffic impacts, while the dock developer emphasized greater understanding of the hydrodynamics and morphology of the estuary and the relationship between the schemes and the SPA.
- *Other benefits*. These included the development of local relationships, including closer working relationships with the other developers, LPAs and statutory consultees; the establishment of a consistent basis for mitigation and monitoring; the opportunity to share the costs of ongoing monitoring work in the estuary; and—for one of the developers—the fact that the CEA process had facilitated the rapid achievement of planning approval.
- *Financial costs of the CEA process*. The financial cost of undertaking the CEA was relatively low for all of the developers, although the majority of the cost was in fact borne by a single developer (the water utility company). The cost of the CEA to this company represented around 5 per cent of the total cost of the EIA work for its proposed scheme. Costs were much lower for the other developers.

• *Changes to the project proposals and additional mitigation.* The CEA process resulted in some changes to the original project proposals and additional mitigation measures, which would not have occurred if the projects had been assessed separately. Examples included changes to piling operations during the power station construction to minimize noise impacts, modifications to the ferry berth construction to compensate for loss of bird habitats elsewhere in the estuary, changes to the timing of certain construction activities and staggering of working hours to minimize peak traffic flows. All the developers indicated that the additional mitigation prompted by the CEA had added relatively little to the costs of the overall development. This may reflect the ability to share the costs of mitigation measures between the developments. Without this opportunity, mitigation might have been less effective or more costly (Piper 2000).

• *Delays caused by the CEA process.* Views differed about whether the CEA process had resulted in a saving or loss of time in obtaining consent for the proposed schemes. In part, this reflected the stage in the planning approval process reached by each developer at the start of the CEA process. Delays ranged from one to two months for the water utility company to six months for the dock developer (this last delay was attributed to the late involvement of a statutory consultee, despite an earlier invitation to join the study); the power station developer felt that its timetable had not been affected. Some delay may have been caused by the fact that the CEA process began after the bulk of the initial consultation and assessment work on some of the schemes had been completed. This resulted in some duplication of effort.

• *Other issues.* One developer noted the problem of distinguishing between those changes that resulted from the CEA process and those that would have occurred anyway through the proper consideration of each scheme in isolation. A further issue concerned the appropriate treatment of new projects that may come forward in the area after the initiation of the CEA. Should such projects be incorporated into the CEA process, implying an open-ended timescale for the process, or should a new CEA be started for the next group of schemes?

Views of local planning authorities and consenting authorities

The two local planning authorities responsible for the area in which the developments were located were supportive of the CEA study and identified a number of benefits from the process:

> The study was found to be helpful in assessing the overall impact of several major projects proposed for a relatively small geographic area. The study was very helpful in its technical assessment of impacts. The study was definitely of great value for both [councils] in understanding likely impacts. [It] was probably of equal value in demonstrating the likely impacts to the developers themselves, making them fully aware of the potential consequences of their proposals. (Comments from local authority representatives, quoted in Piper 2000.)

The point was made that local planning authorities lack the technical expertise and resources to carry out detailed review of environmental assessments, and therefore rely on the integrity of ES authors and consultants to identify areas of potential concern. In this respect, 'a major factor in favour of the CEA [process] is that the advisers of each scheme proponent help "to monitor the others", thus "producing a more balanced product" ' (Piper 2000). Both authorities commented on the lack of public participation in the CEA study. One noted that, partly due to the tight timescales involved, there had been little or no public consultation, and that this represented the main weakness in the process.

Other consenting authorities included three government departments (DTI, DETR and MAFF). The DTI commented that the study had facilitated decision-making, stating that 'without the CEA, the power station project would have been refused' (quoted in Piper 2000). The CEA approach would be recommended in similar cases of multiple projects elsewhere.

Views of statutory consultees

English Nature, as the statutory body responsible for nature conservation, was the principal consultee in this case and was involved in the CEA process from an early stage. It was necessary for the CEA to satisfy the requirements of EN, given its responsibilities under the Habitats Directive to ensure the protection of the SPA. These requirements were expressed in a number of planning conditions attached to the consents for the various schemes:

> The conditions covered the mitigation of construction works (via measures to reduce disturbance of birds, a code of practice for personnel and compliance with a programme of works designed to take account of other CEA-related construction projects) and the monitoring of construction. A monitoring scheme was outlined which will last throughout construction and for 5 years subsequently and will observe the movements and ranges of population of waterfowl. Provided these stipulated conditions are met, English Nature was of the opinion that the various projects would not, individually or [in combination], adversely affect the conservation objectives of the Special Protection Area. (Piper 2000)

English Nature commented that a number of factors—some of which were unique to this case—had assisted the completion of the CEA. These included the relatively small geographical area covered by the schemes; the fact that all schemes were at an early stage of development at the start of the process, although some project-specific EIA work had already been completed; the absence of direct competition between the developers to be the first to obtain planning consent; and the willingness of one of the developers (the water utility) to take the initiative in getting the study underway (Piper 2000). The latter was seen as particularly important, given that responsibility for undertaking the 'appropriate assessment' under the Habitats Directive properly rested with the consenting authority. As we have seen, in this case there were no fewer than five different consenting authorities. It was suggested that:

it would have been problematic to sort out exactly where responsibility lay, had the CEA strategy not been devised by the water utility and its advisers. For these reasons English Nature indicated that, whilst CEA was 'an excellent solution' [in this particular case], it is not a method of immediate and general applicability but depends upon the circumstances encountered in each case. (Piper 2000)

9.5.4 *Summary*

The consideration of cumulative effects is widely regarded as one of the weak areas in EIA, both at project level (see Section 12.4) and in some SEA studies (see Sections 9.8 and 9.9). This case study has demonstrated a novel approach to the assessment of cumulative effects, in this case associated with the impacts of a number of adjacent proposed developments. The assessment process was made possible by a number of factors, including the willingness of one of the developers to take the initiative in starting the CEA study and the fact that the developers involved were not directly in competition with each other. These circumstances may not apply in all such cases. Nevertheless, CEA studies of the type described have a number of benefits, for developers, consenting authorities and other key stakeholders, and—at least based on the evidence in this case—appear to involve relatively little additional cost.

9.6 Stansted airport second runway: health impact assessment

9.6.1 *Introduction*

This case study provides an example of health impact assessment for expansion proposals at a major airport in the southeast of England. Health impact assessment (HIA) is frequently undertaken as a parallel process alongside EIA and there are close links and partial overlaps between the two processes. The purpose of HIA is to identify and assess the potential health effects (adverse and beneficial) of a proposed project, plan or programme, and to provide recommendations that maximize beneficial health effects and reduce or remove adverse health impacts or inequalities (see also Chapter 12). At the time of the case study, HIA was not a regulatory requirement in the UK planning process. However, relevant government policy on air transport indicated that both EIA and HIA would be expected for major airport expansion proposals.

The case study concerns proposals for a second runway at Stansted airport in Essex. It provides an interesting example of the application of HIA to a major project proposal and reveals the close linkages between HIA and EIA. The Stansted HIA was an example of 'comprehensive HIA', characterized by extensive stakeholder engagement and assessment methodologies based on a detailed review of the scientific literature. Weaker aspects of the assessment included the scoping out of certain health effects and a failure to adequately consider the cumulative effects of the airport's overall planned growth. There was also no consideration of alternatives in the assessment and as a result the overall level of influence of the HIA on the project proposals appears to have been limited.

9.6.2 Background to the proposals

Stansted airport is located in Essex, around 35 miles north east of central London. It is the third busiest airport in the UK, after Heathrow and Gatwick, with almost 19 million passengers in 2010. The second runway proposals at Stansted date from the mid-2000s. At this time, passenger numbers at the airport had experienced very rapid growth during the previous decade and the capacity of the existing single runway was expected soon to be reached. Further impetus for the proposals was provided by a government White Paper, 'The future of air transport', published in 2003 (DfT 2003). The White Paper set out a national, strategic policy framework for the development of UK airport capacity for the next 30 years. In relation to Stansted Airport and the wider southeast region, the government concluded that:

- there was an urgent need for additional runway capacity in the southeast;
- the first priority was to make best use of the existing runways, including the remaining capacity at Stansted and Luton;
- provision should be made for two new runways in the southeast by 2030; and
- the first new runway should be at Stansted, to be delivered as soon as possible.

The government invited airport operators to bring forward plans for increased airport capacity in the light of the White Paper's conclusions. The White Paper also stated that, in all cases where development was envisaged, full EIA would be required when specific proposals were brought forward. Operators would also be expected to undertake 'appropriate health impact assessment' (DfT 2003).

In response to the White Paper's support for expansion at Stansted, the airport's operator, BAA, brought forward proposals for two future phases of development. The first of these phases was known as Generation 1 (or G1), followed by Generation 2 (or G2). The G1 and G2 proposals were submitted as separate planning applications and were subject to separate EIA/HIA processes (ERM 2006, 2008). The G1 proposals sought to lift existing planning conditions limiting the annual number of passengers and flight movements at the airport (from 25 to 35 million passengers and from 241,000 to 264,000 air traffic movements per annum). This was to be achieved by making maximum use of the capacity of the existing runway, along with limited physical development (e.g. expanded terminal buildings). The G1 planning application was submitted in 2006. Permission was refused by the local planning authority, but BAA appealed against this refusal and the proposals were considered at a public inquiry held in 2007. Following the inquiry, planning permission was granted by the government in October 2008.

The G2 development involved proposals for a second runway and associated rail and road improvements in the immediate vicinity. Expanded passenger and aircraft handling infrastructure was also proposed, including a new terminal building, hotels, catering and car parks. Unlike the G1 development, the proposals also involved a substantial extension to the perimeter of the airport. A four year construction project was envisaged, with the second runway becoming operational in 2015. The additional capacity provided was expected to result in an increase in passenger numbers to 68 million by 2030. The G2 planning application was submitted in March 2008. The application was 'called in' for determination by central government following a public inquiry. However, before the start of the inquiry, the election of a new coalition government signalled a change in national airport policy. The new government announced in May 2010 that it would refuse permission for additional runways at Stansted and Gatwick. Consequently, on the basis that there was no longer government support for the proposals, BAA withdrew its application.

9.6.3 *Aims and scope of the HIA*

The Stansted second runway development proposals were subject to EIA. Health impact assessment was undertaken as a separate process, although there were important links between the EIA and HIA work. The overall objectives of the HIA were agreed between BAA and Essex Strategic Health Authority as to:

- identify the potential local health effects (positive and negative) from the G2 project;
- assess the likelihood and scale of the key local health effects; and
- make evidence based recommendations, which maximize positive effects and minimize negative effects and, as appropriate, recommend local requirements for monitoring local health effects.

The HIA was undertaken by consultants ERM, with support and advice from a Health Topic Group comprising representatives from local and regional public health organizations, the local planning authority and the applicant (BAA). HIA is normally categorized as either 'rapid HIA' or 'comprehensive HIA', depending on the time taken to complete the assessment and the extent of consultation undertaken. The Stansted HIA was an example of comprehensive HIA, reflected in the range and complexity of the methods used to predict the health consequences of the project and the extensive stakeholder engagement undertaken. The assessment also adopted a broad definition of health, encompassing physical, mental and social well-being. The scope of the HIA included the health effects on local residents arising from project-induced changes in the following:

- air quality;
- noise (air and ground noise);

- transport;
- employment and income;
- social capital;
- involuntary relocation;
- visual effects and light pollution; and
- health care and community facilities.

Certain aspects of the project and associated developments outside the airport were excluded from the HIA scope. These included the following issues:

- expansion associated with development in the M11 corridor (as this was addressed in the regional planning strategy);
- health service infrastructure planning for population expansion around the airport (as this was being considered separately by the local Primary Care Trust through its strategic and operational plans);
- any implications for emergency plans (as it was assumed that these would be addressed through existing emergency planning processes); and
- the effects of climate change on human health (as this was regarded as a wider than local impact and was therefore deemed to be outside the scope of the HIA).

The health effects of the Stansted G1 development proposals, which had yet to be approved at the time of the assessment, were also excluded (this is discussed further in Section 9.6.7).

The geographical scope of the assessment was confined to those communities adjacent to the airport. In practice, this area covered four local authority districts in Essex and Hertfordshire which were closest to the airport and considered most likely to experience health effects. A smaller inner zone within this area was also defined for the consideration of certain health effects (e.g. based on predicted pollutant concentrations and aircraft noise contours).

The assessment consisted of the following key stages:

- project profile;
- community profile;
- sakeholder engagement;
- development of the assessment methodology;
- assessment of health impacts; and
- mitigation/enhancement and monitoring.

Each of these key stages is discussed in more detail in the remainder of the case study.

9.6.4 Project profile

This first stage of the assessment was designed to identify the main routes or pathways through which the project might have implications for health. Identification of these 'health pathways' involved an examination of the characteristics of the development proposals and identification of their potential influence on health determinants. The key health pathways identified were associated with the following project features or activities:

• Construction: implications for exposure to environmental influences (e. g. increased noise, dust and traffic movements).

• Land take: implications for the functioning of and networking within communities ('social capital'), and access to health care and transport services.

• Increased aircraft, rail and road traffic movements: implications for increased exposure to noise and air pollutants, accessibility and the potential for injury.

• Changes to local roads: implications for community severance, access and social capital.

• Increased employment opportunities: implications for improved socio-economic well-being, reductions in unemployment and reduced inequalities.

It was also acknowledged that the development may influence additional pathways not associated with physical changes but reflecting intangible and/or perceived effects. These could include effects on social networks, community identity, access and accessibility and well-being. Most of the identified pathways were taken forward for detailed assessment in the HIA, although some minor health pathways were scoped out at this stage. These pathways were judged to have insufficient influence on health determinants to have health outcomes of consequence. Examples included the generation of dust, odour, fuel dumping, disruption to utilities and the visual effect of additional vehicles.

9.6.5 *Community profile*

A community profile was drawn up to provide a description of the communities that might be affected by the project. The profile was based primarily on information for the four immediate local authority districts. However, a more detailed study area was also defined, comprising those parishes most likely to be directly affected by the development. This area was defined using four main criteria:

• Land take: defined by the proposed boundary of the expanded airport.

• Aircraft noise: defined as the 54 dBA noise contour for the proposed development, based on air traffic predictions for 2030. This contour represented the lowest threshold noise level at which community annoyance was considered likely to be experienced.

• Visual impacts: defined as the zone of visual influence, from which the expanded airport would be visible.

• Secondary socio-economic effects: defined as those areas most likely to be affected by secondary effects associated with the construction workforce for the development and the services and facilities needed.

The resulting study area was not simply a circular zone defined by a radius from the airport, but a more complex area reflecting the likely distribution of key health effects. This area comprised 26 parishes in a zone extending to the south-west and north-east of the airport. This distribution largely reflected the orientation of flight paths and the resulting noise contours with the expanded airport.

The community profile was based on existing secondary data sources and provided information on a range of indicators, including population, education, employment and income, housing, crime and health. Areas with relatively high levels of deprivation or poor existing health were identified. The identification of such areas was important as these communities are more likely to be susceptible to health effects. Overall, the communities surrounding the existing airport were found to be relatively affluent and there were no severely deprived neighbourhoods within the study area. Performance on most health indicators was also significantly better than the national average.

9.6.6 *Stakeholder engagement*

A stakeholder engagement programme was undertaken as part of the HIA. This sought the views on potential health effects of those interested in or affected by the proposed development. More specifically, the engagement process sought to identify:

- stakeholder concerns regarding the project and its potential effects on health and well-being, and how to minimize such effects;
- stakeholder perception of the benefits that could arise from the project and how such benefits could be enhanced; and
- the priority issues and concerns of the stakeholder and what recommendations they would like to see noted within the HIA.

The geographic scope of engagement was similar but not identical to that used for the HIA community profile. Inner and outer engagement zones were defined. The boundaries of the inner zone were based on three main factors: (a) predicted aircraft noise contours for the expanded airport; (b) areas that could experience air quality changes; and (c) areas affected by the surface access development (traffic and construction effects). The outer engagement zone covered a wider area that also included those settlements that were expected to supply the bulk of the workforce for the expanded airport.

Engagement mechanisms included stakeholder interviews, workshops and a questionnaire survey of local residents. The interviews and workshops were carried out with key stakeholders representing organizations with strategic responsibilities in relation to health, housing, education, business, transport and other relevant areas. Open community workshops were not undertaken, as these had attracted a low level of interest from the public during the engagement process for the earlier Stansted G1 HIA (although groups opposing the development argued that the low response was due to a lack of advance publicity). Members of the public were engaged instead through a questionnaire survey. This was distributed to a sample of around 9,300 households in the inner and outer engagement zones. The content of the questionnaire was subject to a process of review by the HIA team, BAA, the Health Topic Group and expert reviewers.

Key issues of concern to emerge from the stakeholder engagement included:

• air quality (particularly the potential health effects on vulnerable groups, including children and the elderly);

• air traffic noise;

• road traffic/congestion;

• effects on the 'social capital' of the area (more rapid pace of neighbourhood change; loss of identity);

• socio-economic issues (employment opportunities; inward migration of construction workers; effects on existing housing provision); and

• healthcare and community facilities (adequacy of existing capacity; effects on emergency services; effects on healthcare recruitment).

There was some overlap in the issues raised with those addressed in the EIA, particularly with respect to employment and wider socio-economic issues.

9.6.7 Assessment of health impacts

The HIA report included a series of detailed 'methodology statements' for each of the health pathways identified in the project profile. These methodology statements describe the methods used in the HIA to determine the likely scale of health outcomes, based on a review of the relevant scientific evidence (e.g. on the health effects of specific air pollutants) and the stakeholder engagement. In most cases, the methods were dependent on input data from the EIA studies, such as predictions of the concentrations of specific air pollutants or of air noise contours with and without the proposed development. This interdependence means that any weaknesses in the EIA's predictive methodologies or assumptions will also have been reflected in the HIA work (see Stop Stansted Expansion 2006 for a more detailed discussion of these issues). Most of the predicted health effects were quantified, generally by estimating the number of people likely to experience specific health outcomes. However, some impacts were discussed only qualitatively, in cases where the scientific evidence on health effects was less certain and did not allow quantified estimates to be made. Uncertainties, data limitations and other practical difficulties encountered were clearly outlined in the method statements.

The health effects considered in the assessment included the following:

• Air quality: health effects associated with increases in particulate matter and NO_2 concentrations (e.g. years of life lost; respiratory and other hospital admissions; GP consultations for asthma).

• Air noise: annoyance; sleep disturbance; cognitive effects on schoolchildren.

• Ground noise: annoyance; sleep disturbance.

• Transport: injuries and fatalities from road and rail accidents.

• Employment and income: positive effects of additional employment and income, including effects on mental health (e.g. depression), self-rated 'good health' (well-being), long-term limiting illness and mortality.

• Social capital: changes in civic participation, social networks and support, social participation, reciprocity and trust, and satisfaction with the area (e. g. associated with inward migration of construction workers and land take), and consequent health/well-being effects.

- Involuntary relocation: stress, anxiety and reduced well-being for those moving from existing residential properties due to land take.
- Visual effects and light pollution: reduction in well-being.
- Health care and community facilities: effects on existing facilities (e. g. due to inward migration of construction workforce; accidents; transmission of infectious diseases).

There was some degree of overlap with the impacts considered in the EIA, although the approach adopted in the HIA was rather different, with a greater emphasis on identifying the magnitude of effects and less focus on the evaluation of significance. For example, in relation to aircraft noise, whereas the EIA sought to identify the significance of changes in noise exposure, the HIA placed much more emphasis on quantifying the numbers of people likely to experience specific health outcomes. These quantified estimates were compared with the prevalence of the relevant health outcomes in the local population, where such data was available. However, the significance of the estimated health outcomes was not directly assessed. The HIA report stated that this was due to a lack of recognized significance assessment criteria for these effects.

In assessing the impacts of the second runway (G2) proposals, the HIA considered the health outcomes arising in the following scenarios:

- Base case with no G2 development (in 2015 and 2030): annual passenger numbers were assumed to be 35 million in this scenario, in both 2015 and 2030.
- With G2 development, in 2015: this was the date at which the second runway was expected to become operational.
- With G2 development, in 2030: at this date the enlarged airport was expected to have reached its capacity of 68 million passengers per annum.

Health effects arising during the construction of the second runway were also assessed. It is important to note that the base case for the assessment did not represent the situation at the time of the second runway application. Instead it was a projected baseline, which assumed that the earlier G1 planning application, to expand the capacity of the existing runway, would be approved. This resulted in a base case of 35 million passengers per annum rather than the actual baseline at the time of assessment in 2008, which was only 22 million (without approval of the G1 development, annual passenger numbers would have been limited to no more than 25 million in both 2015 and 2030, and therefore a figure of 25 million might also have been used as an alternative base case). The overall increase in passenger numbers assumed in the second runway HIA was therefore 33 million (from 35 to 68 million); this represented a 94 per cent increase in annual passenger numbers compared with the base case.

An alternative approach would have been to consider the G1 and G2 development proposals as two parts of a larger planned expansion of the airport. In this case, the resulting increase in passenger numbers would be significantly larger, from a baseline of 22 (or 25) million to 68 million by 2030 (or an increase of 172—209 per cent). The resulting health effects would also be larger than for the G2 development alone. A consideration of the combined effects of the G1 and G2 proposals in the HIA would therefore have been useful. There was in fact no attempt to assess the combined health impacts of the G1 and G2 proposals as a whole. The health effects of the G1 development were therefore ignored in the HIA for the second runway development. Similarly, the health effects of the second runway had been ignored in the earlier HIA for the G1 development proposals. This use of separate assessment processes also applied to the EIA's for both planning applications. This approach was supported by the Planning Inspector at the public inquiry into the G1 proposals, when he concluded that 'for the purposes of the EIA Regulations, I accept BAA's view that the G1 proposals are not an integral part of an inevitably more substantial development' and that 'the lack of consideration of the combined impacts of the G1 and G2 proposals in the current ES does not frustrate the aims of the EIA Regulations and Directive' (The Planning Inspectorate 2008). Even if this view is accepted with respect to the G1 development, it seems less tenable for the later G2 proposals as these clearly did now represent part of a 'more substantial development'.

The failure to include an assessment of the effects of the combined G1 and G2 expansion proposals means that identification of cumulative health effects was likely to have been deficient. The issues raised here are similar to those highlighted in the earlier case study on Wilton power station in Section 9. 2; in both cases the use of divided project consent and assessment procedures resulted in a failure to adequately address incremental and cumulative impacts.

9. 6. 8 *Mitigation and enhancement of impacts*

The HIA report provided a list of recommendations to BAA, the airport's operators, for the mitigation and enhancement of health effects. Feasible options for mitigation and enhancement measures were identified from the results of the assessment phase and the suggestions made by stakeholders. These options were then subjected to a review by the Health Topic Group and an expert panel, before arriving at the final list of recommended measures.

The recommended mitigation measures in the HIA were relatively narrow in focus and did not include substantive changes to the actual project proposals. These were taken as a 'given' in the HIA. As the HIA notes, 'some of the effects are associated with features of the G2 project that cannot be adjusted without changing the purpose of the development itself… Inevitably, this means that some effects are more amenable to management than are others' (ERM 2008). A number of the proposed mitigation measures involved suggestions for improved communication about the scale of health effects with local communities; it was also recommended that monitoring of noise levels might be better undertaken by an independent third party rather than BAA. A number of generic mitigation measures to reduce emissions and noise at source were also recommended, including, for example, introduction of increasingly stringent technical standards, improved operational practices and the progressive withdrawal of the noisiest and dirtiest aircraft. It could be argued that such measures would have been implemented even in the absence of the proposed development, and should therefore be regarded as part of the 'no-development' future baseline rather than as project-specific mitigation.

The HIA did not identify whether, and in what ways, the earlier planning and EIA work on the project had incorporated an explicit consideration of health impacts into the development of the project proposals and of the mitigation measures recommended in the environmental statement. The HIA report simply states that 'many of the effects on the environment (and by extension, health) have been considered very thoroughly at the planning stage where options for runway location and mode of operation were evaluated' (ERM 2008). Further information on how health effects were taken into account in the evaluation of these options, and in what ways mitigation of health effects had been incorporated into the final project proposals, would have been useful.

These limitations partly reflected the timing of the HIA work, which was undertaken following the project design and EIA studies. This was necessary since the HIA's assessment methods were heavily dependent on data inputs from the EIA predictions, for example in relation to the predicted changes in air quality and noise levels. However, this raises questions about the overall ability of the HIA to influence the final project proposals. The timing of the assessment also explains the failure of the HIA to include any evaluation of alternative options; only the health effects of the final proposals, as detailed in the planning application, were assessed in the HIA. There was no consideration of the comparative health effects of alternative designs or modes of operation. It is therefore difficult to assess whether the chosen options delivered more favourable health effects (or smaller adverse effects) for the local communities than other feasible alternatives.

9.6.9 Summary

This case study has provided an example of the application of health impact assessment to a major project proposal. HIA is increasingly undertaken as a parallel process alongside EIA and is often dependent on the EIA for critical data inputs. The Stansted HIA was an example of a comprehensive HIA process. Extensive stakeholder engagement was undertaken and the assessment methodology was based on a detailed review of the relevant scientific evidence. Although a wide definition of health and well-being was adopted, the focus of the assessment was restricted to effects on the local communities adjacent to the airport. This resulted in the exclusion of important issues such as the effects of climate change on health. Other issues were also scoped out of the assessment on the basis that they were the responsibility of other organizations and would be addressed in their evolving plans and strategies. By focusing only on the second runway proposals and ignoring other aspects of the Stansted expansion plans, the assessment failed to adequately consider the incremental and cumulative health effects of the airport's planned future growth. There was also no consideration of the comparative health effects of alternative project options (other than the no-development baseline), and as a result the overall level of influence of the HIA on the project proposals appears to have been limited.

9.7 Cairngorm mountain railway: mitigation in EIA

9.7.1 *Introduction*

It is appropriate that one of our case studies includes a tourism project, for tourism is the world's largest industry, it is growing apace and it contains within itself the seeds of its own destruction. That tourism can destroy tourism has become increasingly recognized over the last 30 years or so, with a focus of concern widening from initially largely economic impacts to a now wider array that includes social and biophysical impacts (see Glasson *et al*. 1995; Hunter and Green 1995; Mathieson and Wall 2004). Mountain areas can be particularly sensitive to tourism impacts, including from walking, skiing and associated facilities. This case study takes a particularly controversial project, the Cairngorm mountain railway, in the Highlands of Scotland, which was opened in 2001 after a long and protracted debate about its impacts and their management. This brief case study focuses on the latter aspect as an example of approaches to mitigation and monitoring in EIA.

9.7.2 *The project*

The Cairngorm Ski Area is one of five ski areas in Scotland. It developed rapidly in the 1960s and 1970s in combination with the adjacent settlement of Aviemore. Chairlift facilities were built to take skiers to the higher slopes in winter, and also to carry walkers in other times of the year. However, the industry has been vulnerable to climate/weather trends and to the quality of the infrastructure. In 1993 the Cairngorm Chairlift Company published a Cairngorm Ski Area Development Plan designed to upgrade facilities, to give better access to reliable snow-holding in the area, to reduce vulnerability to adverse weather conditions, to improve the quality of visitor experience and to improve economic viability, while ensuring that all relevant environmental considerations were taken fully into account. The Cairngorm Funicular Railway was a key element in the plan.

The Cairngorm Funicular is the UK's highest and fastest mountain railway. It is approximately 2 km in length and takes visitors in eight minutes from the existing chairlift station/car park base at Coire Cas (610 m) to the Ptarmigan top station (1100 m). It comprises two carriages (or trains of carriages) running on a single-line railway track between two terminal points (see Figure 9.3). The carriages, which start at opposite ends of the track, are connected by a hauling rope. As one carriage descends the track, the other travels upwards and they pass each other at a short length of double track midway. The track is carried on an elevated structure, a minimum of 1 m and a maximum of 6 m above ground level. The final 250 m runs in a 'cut and cover' tunnel. The development has also included a major remodelling of the existing chairlift base station, and replacement of the existing top station with a new development, which includes catering facilities for about 250 people, and a new interpretative centre, including various displays and an outdoor viewing terrace. The previous chairlift and towers have been removed as part of the development. It was anticipated that the railway would carry approximately 300,000 visitors a year, with two-thirds in the non-skiing months. This would represent a three-to fourfold increase over 1990s numbers reaching the top station by the chairlift, and a doubling of numbers from the early 1970s.

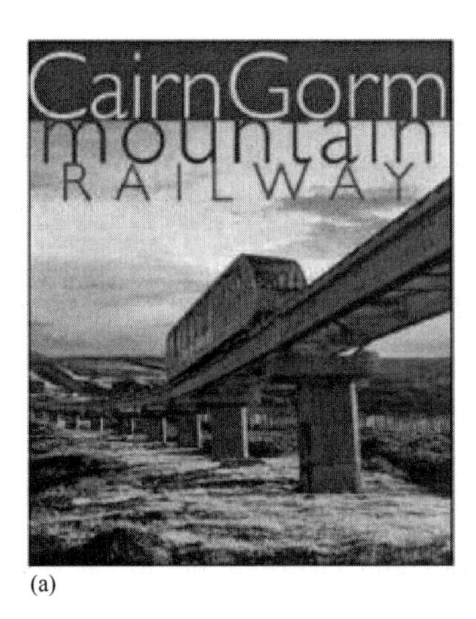

(a)

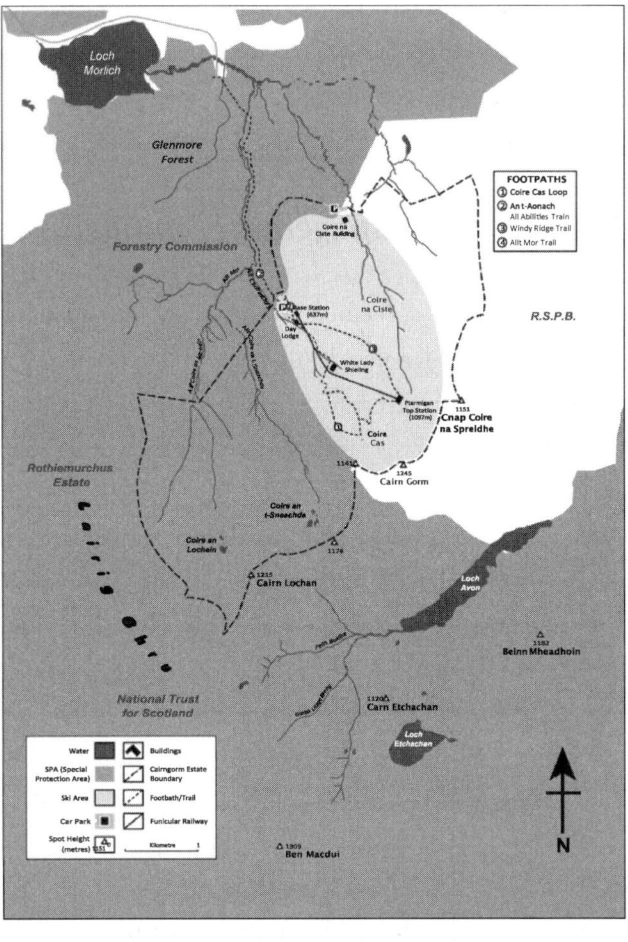

(b)

Figure 9.3 (a) Cairngorm mountain railway; (b) the wider environment

Source: HIE 2005

9.7.3 The EIA and planning process

The original planning application for the Funicular Railway was submitted in 1994, with an ES (Land Use Consultants 1994). Revised proposals and a supplementary ES were submitted in early 1995. The scheme was very controversial, with much opposition. Particular concerns focused on the potential impact of improved visitor access to the sensitive environment of the summit plateau, which is recognized as a European candidate Special Area of Conservation and an SPA. As a condition of the planning approval, it was necessary for the developer to satisfy Scottish Natural Heritage (SNH, the statutory body responsible for nature conservation in Scotland) that a visitor management plan (VMP) and other mitigation measures would be put in place that would avoid adverse impacts on the summit plateau.

The planning application was approved by the Highland Council (the consenting authority) in 1996. This was subject to a Section 50 (now Section 75) planning agreement to create, in partnership with the SNH and in agreement with the developer/authority, a regime for visitor and environmental monitoring and management. Amended designs for the station buildings were approved in 1999, and construction work finally began in August 1999. The railway opened in December 2001, following the approval of the proposed VMP by SNH.

9.7.4 Visitor management, mitigation and monitoring measures

The Section 50 agreement attached to the planning approval is a legally binding agreement between the planning authority, in partnership with SNH, and the developer/operator and landowner. The agreement provides for:

- a baseline survey of current environmental conditions and visitor usage in the wider locality;
- an implementation plan providing details of the timing and means of implementation of the development with particular reference to reinstatement following construction;
- an annual monitoring regime to identify changes and establish causes and consequences;
- an annual assessment by the operator of any actions necessary to ensure acceptable impacts to the European designated conservation sites on the summit plateau;
- fall-back responsibilities in the event of default; and
- eventual site restoration if public use of the development ceases (Highland Council 2003).

The Cairngorm Funicular Railway VMP was produced in the context of this agreement. The objective of the VMP is to protect the integrity of the adjacent areas that have been designated or proposed under the European Habitats and Birds Directives from the potential impacts of non-skiing visitors as a direct consequence of the funicular development. The VMP went through several stages and was subject to a short period of public consultation in 2000. Many issues were raised, including the innovative or repressive (according to your perspective) 'closed system', and the associated monitoring arrangements (SNH 2000).

The closed system, whereby non-skiing visitors are not allowed access to the Cairngorm plateau from the Ptarmigan top station, is a key feature of the VMP. Instead visitors must be content with a range of inside interpretative displays and access to an outside viewing terrace-plus, of course, shopping, catering and toilet facilities! This system proved very contentious, and received considerable criticism, in the public consultation on the VMP. Some saw it as violating the freedom to roam; for others, it was a cynical device for extracting economic benefit in shops and catering outlets. Others considered it unnecessary, given the recent improved pathway from the Ptarmigan top station to the summit of Cairngorm, as noted by one respondent: 'For years I have been advocating stone paths. People use the paths and the ground round about recovers. Now that the path up to the summit is pretty well complete most people will be barred from using it!' (SNH 2000). Another issue has been how to allow ingress to the facilities of the top station from non-railway-using walkers on the plateau, while preventing egress from non-walkers. Alternatives were suggested at the time to the closed system including ranger-led walks and time-limited access, but the system was put in place and is part of a 25-year agreement. The guide leaflet for the funicular users includes the following:

> Protecting the Mountain Environment: large areas of the fragile landscape and habitats of the Cairngorms are protected under European Law. Cairngorm Mountain Limited is committed to ensuring that recreational activities are environmentally sustainable. For this reason the Railway cannot be used to access the high mountain plateau beyond the ski area at any time. Outwith the ski season, visitors are required to remain within the Ptarmigan building and viewing terrace, returning to the base station using the railway. Mountain walkers are welcome to walk from the car park and use the facilities at the Ptarmigan, but may not use the railway for their return journey and are asked to sign in and out of the building at the walkers' entrance.

Monitoring can support effective mitigation measures. For this project, monitoring covers all topics subject to baseline surveys—including visitor levels and behaviour, habitats, birds, soils and geomorphology. It uses the limits of acceptable change (LAC) method, whereby indicators and levels of acceptable change are identified, monitored and, when levels are reached, management responses can be triggered (see Glasson *et al.* 1995). In response to a concern about the independence of the monitoring activity, the annual monitoring reports are presented to the SNH and the Highlands Council by an independent reporting officer jointly appointed by them.

9.7.5 *Conclusions*

The Cairngorm Funicular has been operational for over 10 years. Visitor numbers have been less than the predictions in the ES, but still represent a substantial increase on previous levels in the 1990s. Conditions have been complied with, the Section 50 (75) agreement has been secured, and a good working partnership has been established between public authorities and the operator; there is access for all abilities to the Ptarmigan top station, an improved footpath system and the old White Lady chairlift system has been removed. Recently provision has been made for small groups of up to 10 visitors, including mountain railway passengers, to enjoy a 90 minute guided walk on a mountain trail path to the summit, outside the skiing season, every day between May and September.

9.8 SEA of UK offshore wind energy development

9.8.1 *Introduction*

This case study provides an example of the application of SEA to plans and programmes at a national level. It concerns the SEA of the UK government's plans for the future development of offshore wind energy. The SEA was carried out during 2002—03, prior to the implementation of the EU SEA Directive (2001/42/EC). Further information on the requirements of this Directive, and on SEA more generally, can be found in Chapter 11.

The context for this particular example of SEA was ambitious government targets for renewable energy generation, linked to the achievement of the UK's commitments in the Kyoto Protocol to significantly reduce CO_2 emissions. At the time of the SEA (2002—03), the UK government was committed to supplying 10 per cent of electricity needs from renewable sources by 2010, rising to 20 per cent by 2020. Offshore wind energy was seen as a major contributor towards these targets (DTI 2003a), and the UK government wished to see rapid development of the industry. But it was also committed to an SEA process, which was intended to influence decisions on which areas of the sea should be offered to developers (and which should be excluded), as well as to guide decisions on bids for development licences submitted by individual developers. At the time, offshore wind energy was a new industry undergoing rapid development, and there were therefore many uncertainties about environmental impacts, including potential cumulative effects. This presented difficulties for the SEA work.

9.8.2 Development of offshore wind energy in the UK

The development of offshore wind energy in the UK involves separate licensing and consent systems. The licensing system is operated by the Crown Estate, in its role as landowner of the UK sea bed. Licensing takes place under a competitive tendering process in which developers submit bids for potential wind farm sites. It is left to the developers themselves to identify potential sites, from within broad areas defined by the Department of Energy and Climate Change (DECC, previously DTI). The developers submitting successful bids are then offered an option on their proposed site. Detailed technical studies, consultation and EIA work on the site is then undertaken by the developer, prior to the submission of a consent application. The necessary planning consents are granted by DECC and DEFRA, following consultation with the LPAs most closely affected, statutory consultees and the public. Once the necessary consents have been obtained, developers are granted a lease of 40—50 years on the site and can then begin construction of the wind farm.

In the UK, the Crown Estate's first invitation to developers for site leases for offshore wind development (Round 1 of licensing) took place in 2001. This resulted in 18 planned developments, each of up to 30 turbines. Most of these schemes obtained planning consent in 2002—03 and were installed from 2003 onwards. After this first round of licensing, the government published 'Future offshore', a document setting out its plans for the second licensing round (DTI 2002). This envisaged much larger developments than in the previous round, and stated that future development was to be focused in three 'strategic areas' —the Thames Estuary, the Greater Wash and Liverpool Bay (Figure 9.4). These areas were selected as having the greatest development potential, based on the potential wind resource available, the bathymetry of the offshore area, proximity to existing grid connections and initial expressions of interest from developers (DTI 2003b); however, environmental constraints appeared to have had less influence on the choice of strategic areas.

A three-month consultation period on the 'Future offshore' document started at the end of 2002. The SEA of the government's plans for Round 2 licensing, which is the focus of this case study, started at the same time, with the resulting SEA Environmental Report submitted in in May 2003 (for a 28-day consultation period). Despite this SEA process, the government was keen to maintain the pace of development in the offshore wind energy industry, and the deadline for developers to submit expressions of interest for Round 2 site leases to the Crown Estate was the end of March 2003 (i.e. prior to the completion of the SEA Environmental Report or the receipt of consultation responses on this report).

The successful bids for Round 2 developments were finally announced in December 2003. These included 15 projects with a total capacity of between 5.4 and 7.2 GW—this compares with the 1.2 GW consented under Round 1, and so represented a step change in the development of the industry in the UK. Some of the selected sites soon proved controversial, with concerns about the potential impacts on important bird habitats raised by the RSPB (2003).

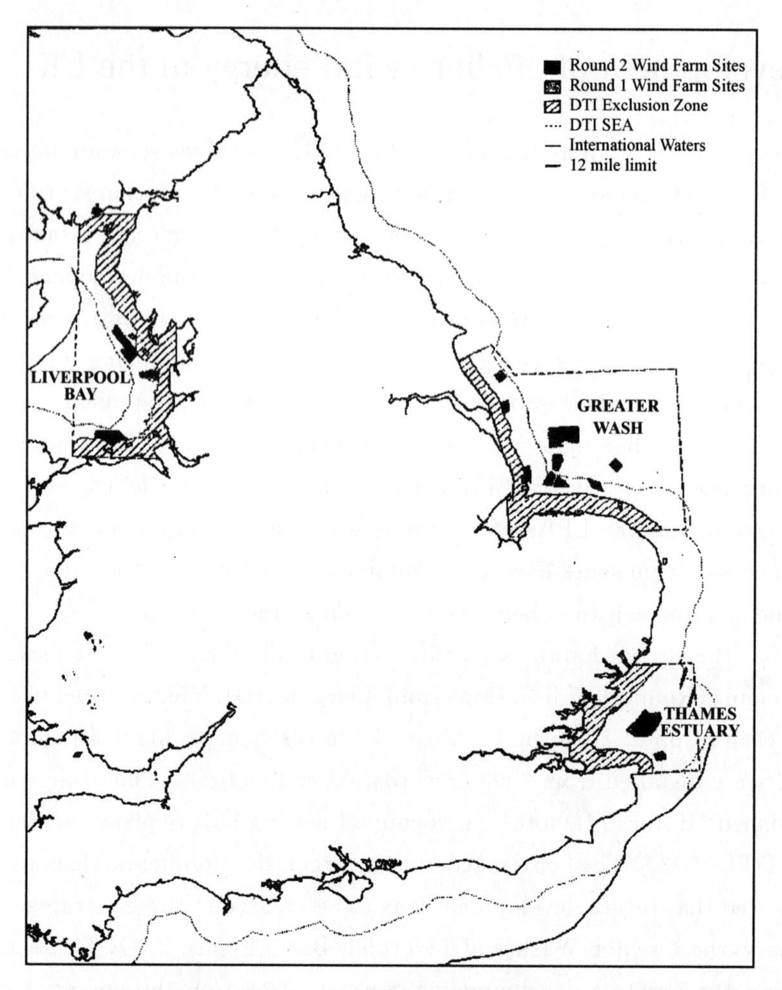

Figure 9.4 Map of the three strategic areas for offshore wind farm development (Round 2)

9.8.3 The SEA approach

The SEA in this case was of the UK government's draft programme for the second licensing round of offshore wind energy development. The SEA was commissioned by the DTI voluntarily, in accordance with the requirements of the EU SEA Directive (although this had not yet been implemented at the time). The timescale under assessment was from 2003 until 2020, with separate assessments undertaken of development up to 2010 and 2020 (DTI 2003a). Two potential development scenarios ('likely' and 'maximum credible') were considered and their likely impacts assessed. The 'likely' scenario envisaged the development of 4.0 GW of capacity by 2010, while the 'maximum credible' scenario envisaged 7.5 GW. By 2020, these figures were expected to increase to 10.2 and 17.5 GW respectively. A no-development option was also considered.

A steering group was used to guide the SEA process, with membership drawn from specialists in coastal/marine environmental issues, wind energy development and SEA. Steering group members included representatives from relevant government departments (DTI, DEFRA, ODPM); the Crown Estate; the British Wind Energy Association (BWEA), the body representing the UK wind energy industry; government and non-governmental environmental organizations, such as the RSPB, Joint Nature Conservancy Council (JNCC), Countryside Council for Wales, and EN; and the IEMA, the body representing the UK EIA 'industry'.

Consultation was undertaken on the scope and design of the SEA. A scoping workshop was held towards the end of 2002 and a scoping report was produced. Some changes to the scope of the SEA were introduced as a result of the consultation responses received, for example by including a wider range of socio-economic impacts that had been identified as important by a number of consultees (DTI 2003a). The environmental report produced at the end of the SEA process provided information on:

- The nature and extent of the technical, environmental and socio-economic constraints that may preclude or be affected by wind farm development.
- The identification of locations within the three strategic areas (the Thames Estuary, the Greater Wash and Liverpool Bay) with the lowest levels of constraint.
- The significance of the environmental and socio-economic impacts arising from different realistic scales of wind farm development in those areas with the lowest levels of constraint.
- Recommendations for managing the impacts of wind farm development in the three strategic areas.

Overall, it was concluded that:

> The likely development scenario, to 2010, is achievable for each Strategic Area without coming into significant conflict with the main significant impact risks, namely areas of high sensitivity to visual impact, concentrations of sensitive seabirds, designated and potentially designated conservation sites, MoD Practice and Exercise Areas and main marine traffic areas.
>
> [However], the 2020 likely development scenario would only be achievable subject to resolving the uncertainties concerning impacts on: physical processes, birds, elasmobranchs (shark, skate and ray species) and cetaceans.
>
> The maximum credible scenario for all Strategic Areas, particularly the Greater Wash and Thames Estuary, for 2020, may be compromised by constraints, particularly cumulative impacts and conflict with marine traffic (commercial and recreational navigation); and large scale development could exclude fisheries from significant areas of fishing grounds, particularly if it were to coincide with severance areas associated with other offshore activities. (DTI 2003a)

In order to minimize environmental impacts, the following broad strategic approach was recommended:

- The development of fewer large wind farms, of around 1 GW (1,000 MW) or more capacity, located further offshore is generally preferable to several small-scale developments, though the latter would be preferable for development closer to the coast.

- In all strategic areas, avoid the majority of development within the zone of high visual sensitivity close to the coast.
- Where development might occur close to the coast, preferentially select low constraint areas and consider small-scale development.
- Pending the outcome of monitoring studies, avoid development in shallow water where birds such as common scoter and red-throated diver, and other species (including marine mammals) are known to congregate (particularly in Liverpool Bay and the Greater Wash).
- Address the uncertainties of large-scale impacts, particularly cumulative effects, at a strategic level (DTI 2003a).

The environmental report was subject to a short period of public consultation (28 days). The government argued that the report and the comments received would be 'a significant input to government decision-making on the nature of the second licensing round' (DTI 2003a).

9.8.4 SEA methods

It must be accepted that, in an SEA, the level of detail that can be analysed and presented, in respect of both baseline data and quantification of impacts, is less than in a project-level EIA. This was true of this particular SEA, which 'focuses more on assessing constraints, sensitivities and risks instead of detailed analysis of the characteristics of specific impacts' (DTI 2003a). The methods used in the SEA included a GIS-based spatial analysis (constraint mapping exercise), followed by a risk-based analysis of the likely impacts of the selected development scenarios (including the cumulative implications). Each of these methods is described briefly below, with selected examples included to illustrate the approach used.

Spatial analysis

The spatial analysis made use of electronic overlay mapping of a variety of technical, socio-economic and environmental features to identify areas of the sea with high or low constraints within each of the three strategic areas. Examples of the main features mapped are listed below:

Technical constraints to wind farm development:

- existing and planned licensed areas for aggregate extraction, waste disposal and military operations;
- oil and gas structures (pipelines) and safety zones;
- cultural heritage sites (wrecks and other sea bed obstructions);
- cables;
- existing shipping/navigation lanes; and

- proposed wind farm sites from the first round of licensing.

Socio-economic constraints:

- shipping;
- fishing effort; and
- shell-fishery areas.

Environmental constraints:

- marine habitats of conservation interest (designated and potentially designated);
- seascape sensitivity;
- fish spawning areas; and
- fish nursery areas.

Because of baseline data limitations, not all relevant constraints could be mapped within the relatively tight timescale of the SEA. In particular, it was not possible to map a number of important environmental constraints, such as the distribution of certain bird and fish species and migration routes. Whether these omissions invalidate the conclusions drawn from the constraint mapping exercise is open to question (see below for a summary of consultation responses on this issue). However, those factors that could not be mapped were considered in the later risk-based analysis of impacts.

A scoring system was used in the mapping of constraints, in which each area was awarded a score between 0 and 3 for each mapped constraint (with higher scores indicating greater constraints). The scoring system allowed the identification of locations within each of the three strategic areas that had several constraints (a high total score) and those with fewer overall constraints (a lower overall score), subject to the qualification that not all relevant constraints could be mapped (see Table 9. 1). Broad conclusions from the spatial analysis are summarized below, for each strategic area (DTI 2003a):

Table 9. 1 Scores for mapping of constraints

Scores for fishing effort	
0	None
1	Low(less than 500 hours per annum)
2	Medium(500—5,000 hours per annum)
3	High(over 5,000 hours per annum)
Scores for designated habitats of conservation interest	
0	Designated habitats are absent
1	Not applicable
2	Nationally important habitats are present(including those not yet designated)
3	Internationally important habitats are present(including those not yet designated)
Scores for seascape	
0	No sensitivity
1	Low sensitivity
2	Medium sensitivity
3	High sensitivity

- *Liverpool Bay*. Overall, the greater amount of constraint and sensitivities occured in the southern part of this strategic area, due to the presence of bird interests, marine habitats of conservation interest, seascape, fisheries and marine traffic. Seascape constraints in the north of the area were significant.

• *Greater Wash*. The Greater Wash had the largest area of low constraint in comparison with the other strategic areas and offered the greatest potential capacity for wind farm development. Inshore areas, particularly in the southern part of the area, had the greatest amount of constraint and sensitivity, particularly with respect to visual impacts, inshore fisheries, marine mammals, birds and offshore habitats of conservation interest.

• *Thames Estuary*. This region included areas of low constraint on its eastern boundary, and had fewer environmental constraints than the other strategic areas. However, several estuaries and marshes were important bird habitats. Commercial activities (e. g. aggregate extraction) and recreational navigation were other important constraints.

Risk-based analysis of impacts

For each strategic area, the likely impacts of the two development scenarios ('likely' and 'maximum credible') were assessed. This analysis incorporated factors that were mapped as part of the earlier spatial analysis, plus specific receptors that could not be mapped, such as particular bird species. Impacts were quantified wherever possible, or otherwise described qualitatively, and their significance evaluated using a risk-based approach. This was based on an assessment of (a) the likelihood of the impact occurring, and (b) the expected consequences (impact on the receptor). As with the spatial analysis, a scoring system was used in the evaluation of impact significance, as shown in Tables 9. 2 and 9. 3 (DTI 2003a).

Table 9. 2 Scores for impact consequence

5	Serious(e. g. impacts resulting in irreversible or long-term adverse change to key physical and/or ecological processes; direct loss of rare and endangered habitat or species and/or their continued persistence and viability)
3	Moderate(e. g. impacts resulting in medium-term(5—20 years) adverse change to physical and ecological processes; direct loss of some habitat(5—20 per cent); crucial for protected species' continued persistence and viability in the area and/or some mortality of species of conservation significance)
1	Minor(e. g. impacts resulting in short-term adverse change to physical and ecological processes; temporary disturbance of species; natural restoration within two years requiring minimal or no intervention)
0	None(e. g. impact absorbed by natural environment with no discernible effects; no restoration or intervention required)
+	Positive(e. g. activity has net beneficial effect resulting in environmental improvement)

Table 9. 3 Scores for impact likelihood

5	Certain(the impact will occur)
3	Likely(impact is likely to occur at some point during the wind farm life cycle)
1	Unlikely(impact is unlikely to occur, but may occur at some point during wind farm life cycle)

The impact significance scores for each impact were calculated as the product of the consequence and likelihood scores, ranging from 1 (minor consequence and unlikely) to 25 (serious consequence and certain).

9.8.5 Issues raised in consultation responses

The environmental report produced at the end of the SEA process was subject to a short period of consultation. An analysis of the consultation responses reveals a number of key issues that were raised by interested stakeholders. These include a range of concerns about the quality and effectiveness of the SEA process, and its influence on decisions for the next phase of wind energy developments. Many of these concerns arose from the tight timescale for the SEA work and the resulting practical difficulties encountered. The main points raised in consultation are highlighted briefly below (a fuller discussion can be found in DTI, 2003b). Many of the issues raised are interlinked; for example, weaknesses in baseline data may be due to limited consultation or a tight timescale in which to complete the SEA.

- *Pre-selection of the three strategic areas, and lack of a national-level SEA*. The three strategic areas in which Round 2 development was to be focused were selected as having the greatest development potential, based on the potential wind resource available, the bathymetry of the offshore area, proximity to existing grid connections and initial expressions of interest from developers (DTI 2003b). However, the selection did not appear to have taken explicit account of environmental constraints, and this was a cause of concern to a number of respondents.
- *The tight timescale for the sea and uncertainty over the influence of the SEA process on decision-making for future developments*. The timescale for the SEA was considered too tight to allow effective stakeholder engagement and consultation, or to allow additional baseline data to be collected. The fact that developer bids for Round 2 sites were invited before the completion of the SEA Environmental Report was also a source of concern.
- *Concern over the rapid development of offshore wind energy, prior to the proper consideration of potential impacts*. Some respondents argued that the development of the offshore wind energy industry was too rapid and premature; greater efforts should be made to understand the impacts of the smaller Round 1 developments before allowing large-scale expansion of the industry.
- *The need for clearer locational guidance*. There was felt to be a need for clearer recommendations on suitable and unsuitable locations for future offshore wind energy development (including the definition of exclusion zones or 'no-go' areas), and it was considered that these had not emerged sufficiently from the SEA process.
- *Concerns about the scope and methodology of the SEA*. Most respondents were supportive of the overall methodology of the SEA, including the risk-based approach to the assessment of impact significance, but there was some disagreement over the detailed scores awarded to specific receptors or geographical areas.
- *Weaknesses in the available baseline data*. A recurrent theme in the consultation re-

sponses was limitations in the baseline data available to the SEA study. This included missing data for certain important environmental constraints (which could not be mapped) and areas in which the data used in the SEA was not the most accurate or appropriate. Some respondents thought that these data limitations were sufficiently serious as to invalidate the identification of areas of high and low constraints in the SEA. Data gaps and uncertainties about impacts also led respondents to urge a precautionary approach; the need for such an approach was also strengthened by ongoing delays in the designation of offshore areas of conservation interest under the EU Habitats Directive. Other uncertainties included doubts about whether the impacts of smaller wind farms (from the first round of licensing) could necessarily be extrapolated to larger wind farms further offshore.

- *Limited consultation with certain stakeholders*. According to some respondents, the SEA had involved only limited consultation with certain stakeholders (e. g. fisheries and recreational boating interests). This lack of consultation, again partly linked to the tight timescale for the exercise, helped to explain some of the data weaknesses on certain issues in the SEA.
- *Insufficient attention to cumulative and indirect effects*. It was considered that insufficient attention was given to the impact of related onshore development in the SEA, such as transmission connections. This concern echoes the issues highlighted in the first case study in Section 9. 2. More attention also needed to be devoted to cumulative impacts in the environmental report. There was no indication in the report of the carrying capacity of each of the strategic areas, and it was therefore difficult to assess the significance of the cumulative impacts arising under the two development scenarios.
- *Overlaps with project-level EIA*. There was some disagreement over the level of detail needed in the SEA, and which issues could be left to project-level EIA for individual sites. For example, bird distribution data was considered to be one area in which survey data could reasonably be collected at a more strategic level.
- *Responsibility for future SEA studies*. Some respondents requested clarification about who would be responsible for progressing further studies arising from the SEA, including additional data collection to fill existing data gaps and ongoing monitoring. Arrangements for the sharing of such data were felt to be important.

9. 8. 6 Summary

This case study of SEA was undertaken voluntarily, prior to the implementation of the EU SEA Directive. It provides an example of how SEA can be applied, within the context of a new, rapidly developing industry. The UK government's commitment to large-scale development of offshore wind energy to meet international obligations to reduce CO_2 emissions dictated a tight timescale for this SEA. However, the resulting limitations in baseline data, and restricted timescale for stakeholder consultation and feedback, were identified as particular weaknesses in this case.

9.9 SEA of Tyne and Wear local transport plan

9.9.1 *Introduction*

This case study concerns the SEA of the Local Transport Plan for Tyne and Wear, a metropolitan subregion in the northeast of England. The case study is interesting in that it provides an example of an integrated form of assessment, in which an attempt was made to incorporate the results of various other types of assessment into the SEA process. These include health impact assessment (HIA), equality impact assessment (EqIA) and habitats regulation assessment (HRA). The assessment of alternative options is an important feature of SEA and this is therefore also a particular focus of the case study.

Local transport plans (LTPs) were introduced as a statutory requirement in England by the Transport Act 2000. This required local transport authorities to prepare a Local Transport Plan every five years and to keep it under review. Under the terms of the Act, LTPs are required to set out the authority's policies 'for the promotion and encouragement of safe, integrated, efficient and economic transport facilities and services to, from and within their area'. LTPs cover all forms of transport (passenger, freight and pedestrian; public and private) and include strategic policies and an associated implementation or delivery plan outlining more detailed proposals.

The case study concerns the third local transport plan (LTP3) for the Tyne and Wear metropolitan area. The LTP comprises a strategy covering 2011—21, supported by a delivery plan for the first three years of this period (Tyne and Wear Integrated Transport Authority 2011). The plan was prepared during 2010—11, with a parallel SEA process undertaken (Tyne and Wear Joint Transport Working Group 2011a, b). SEA of LTPs is a statutory requirement under the UK's SEA Regulations. Guidance from the UK Department for Transport (DfT) also indicates that there is a requirement for HIA and EqIA for LTPs (DfT 2009a, b). For the Tyne and Wear LTP, an attempt was made to integrate these assessments within the SEA.

9.9.2 *Background to the plan*

Tyne and Wear is a city-region in the northeast of England, encompassing an urban core plus a more rural hinterland. The area includes the major urban centres of Newcastle-upon-Tyne, Gateshead and Sunderland, and has a population of 1.1 million. Tyne and Wear has suffered from historic economic weaknesses and this is currently reflected in high levels of unemployment, below average income levels, deprivation, and related social and health problems. There are relatively low levels of car ownership in the area and higher than average levels of public transport use. Two main trunk roads, the A1 and A19, serve the region. Public transport includes the Tyne and Wear Metro (light rail) system, an extensive bus network, the North Shields to South Shields cross-Tyne ferry and local rail services to the Gateshead MetroCentre (a major retail and leisure complex) and Sunderland.

The LTP was intended to help address key challenges facing the area. These included the need for economic development and regeneration, the need to meet climate change targets for emissions reductions, the need for safe and sustainable communities, and protection and enhancement of the natural environment. In response to these challenges, the plan adopted a strategic framework based on three broad intervention types: (1) managing the demand for travel (including encouragement of modal shift towards more sustainable travel modes); (2) management and further integration of existing networks (with a particular focus on the encouragement of active travel modes and public transport); and (3) targeted new investment in key schemes (including, for example, investment in electric vehicles, bus corridor improvements and new park and ride schemes).

9.9.3 *Overview of the SEA process*

The SEA of the local transport plan was undertaken by consultants Atkins for the Tyne and Wear Integrated Transport Authority, the local transport authority responsible for the Tyne and Wear subregion. Key stages in the process included:

- Scoping, baseline studies and SEA methods: this involved determining and consulting on the scope of the SEA; baseline and contextual studies; and development of the SEA methodology, including the identification of a set of appropriate SEA objectives to be used in the assessment.
- Assessment of alternatives: the development of strategic alternatives and their appraisal against the SEA objectives.
- Assessment of the effects of the draft plan: assessment of the effects of the policies in the draft plan; recommendations for mitigation or enhancement of impacts; and publication of the results of the assessment in an Environmental Report.
- Consultation on the draft plan and Environmental Report: followed by revisions to the draft plan as appropriate and publication of the final plan.
- Monitoring the effects of plan implementation.

An interesting feature of the SEA was its attempt to integrate other types of assessment within the SEA process. These included HIA, EqIA and HRA. Relevant government guidance states that HIA and EqIA are required for LTPs, although these assessments do not necessarily need to be formally incorporated within the SEA (DfT 2009a). HRA is an additional statutory requirement in cases in which a plan contains proposals that are likely to have a significant effect on a Special Protection Area or Special Area of Conservation (collectively known as Natura 2000 sites).

Impacts on human health are identified in the SEA Directive as one of the environmental topics to be considered in SEA. The requirement for a specific HIA in the government guidance on LTPs reflects the understanding that LTP policies and proposals may impact on factors influencing the health of communities and individuals. This could include for example changes in the accessibility and affordability of transport, levels of physical activity, air and noise pollution, personal safety (or perception of safety) and community severance. For the Tyne and Wear LTP, a separate HIA process was not undertaken; health considerations were instead fully integrated within the SEA process. This integration of health considerations took place at all stages in the SEA process and included, for example:

- involvement of the local NHS Primary Care Trust and other relevant public health organizations in the consultation on the scope and methods of the SEA;
- inclusion of health-related plans and programmes in the review of the policy context for the plan and in the identification of key sustainability issues;
- inclusion of relevant health indicators in the baseline data collection; and
- inclusion of a specific SEA objective on health ('to improve health and well-being and reduce inequalities in health').

EqIA involves an assessment of the impact of policies, strategies or plans on different social groups. Its purpose is to ensure that the plan does not discriminate against particular groups and where possible promotes greater equality. The EqIA of the Tyne and Wear LTP was undertaken as a parallel exercise to the SEA and its results were reported separately (Tyne and Wear Joint Transport Working Group 2011c). However, equalities issues were also fully integrated into the various stages of the SEA in the same ways as for health considerations. A specific SEA objective on equality was also included in the SEA Framework ('to promote greater equality of opportunity for all citizens, with the desired outcome of achieving a fairer society'; the SEA Framework is discussed further in Section 9. 9. 5).

The Tyne and Wear LTP was also subject to HRA screening. As with the EqIA, this was undertaken as a parallel process to the SEA and reported separately, although the findings were incorporated in the SEA environmental report. The impact on Natura 2000 sites was also included as a specific SEA objective ('to protect and where possible enhance the European sites').

9. 9. 4 *Scoping and baseline studies*

The topics assessed in the SEA were based on the list of factors identified in the SEA Directive. These included impacts on biodiversity, population, human health, fauna and flora, soil, water, air, noise, climatic factors, material assets, cultural heritage and landscape. As part of the scoping stage, a Scoping Report was produced and comments invited from statutory and other consultees (including those representing public health, equality and diversity interests). The Scoping Report included contextual information on other plans, policies and programmes relevant to LTP3, initial baseline information, a summary of key environmental, social and health issues emerging from the initial work and a preliminary framework of objectives to be used in the SEA assessment process. A scoping workshop was also held in order to gather additional information and discuss the key issues and proposed SEA objectives (these objectives are discussed further in Section 9.9.5).

Baseline studies for the SEA included a review of the policy context, collection of a detailed evidence base on the state of the environment and identification of key environmental issues. Establishing the policy context to the LTP involved a review of other relevant plans and programmes, including those related to health and equality issues. This allowed identification of key themes and policy objectives, for environmental, health and equality-related issues. Detailed baseline information was collected on the current state of the environment and its likely evolution without the implementation of the plan. Baseline data collected included both environmental and social indicators, the latter including data on health, inequality, connectivity and accessibility.

Baseline data gaps and limitations were not explicitly identified in the SEA Environmental Report. It was therefore not clear whether potential indicators had been excluded due to a lack of baseline data. Consultation responses to the draft Environmental Report from one of the statutory consultees also included some criticism of the completeness and consistency of baseline data (including data on biodiversity, agricultural land quality and flood risk zones). The SEA consultants acknowledged these deficiencies, but stated that they could not be addressed due to 'budgetary constraints'. For a number of baseline indicators, there was also an unavoidable reliance on somewhat dated sources of information (e.g. 2001 Census data, which was almost 10 years out of date at the time of the assessment).

The baseline information and review of the policy context was used to identify a list of key environmental issues for the plan area. These covered a wide range of economic and social sustainability issues (e.g. the historic economic weakness of the sub-region and related social issues, including deprivation, child poverty, low incomes and low levels of car ownership), as well as environmental issues (e.g. poor air quality, noise, water quality, climate change, biodiversity threats, heritage and landscape character change). Problems of general health and health inequalities were also identified as key issues. Other issues were directly related to transport (e.g. congestion, accessibility and road safety).

9.9.5 SEA method and objectives

A key part of the assessment methodology was the development of an 'SEA framework'. This framework comprised a series of SEA objectives and associated indicators that were used in the assessment of the effects of the draft plan and of alternative options. The SEA objectives were developed through an iterative process, based on the review of relevant plans and programmes, the evolving baseline information, the key sustainability issues identified and consideration of which of these issues could potentially be addressed by the LTP. The list of objectives incorporated specific health, equalities and habitats issues. This was designed to ensure the integration of the HIA, EqIA and HRA screening processes within the SEA, while also meeting the requirements of the SEA Directive. A final list of 16 SEA objectives was drawn up, covering air quality, biodiversity, European sites, climate change, flood risk, resource use and waste, water quality, use of land, historic and cultural heritage, landscape and townscape quality, accessibility and community severance, noise and light pollution, health and well-being, equality, road safety and crime. For each of these objectives, the SEA framework provided a set of associated indicators (to measure performance against the objective) and a list of assessment 'prompt questions' (to guide the assessment of the effect of LTP policies or proposals on this objective). An example of these prompt questions and indicators, for the SEA objective on local air quality, is provided in Table 9.4.

Table 9.4 Extract from SEA framework for Tyne and Wear LTP3 SEA

Prompt questions for air quality SEA objective: will LTP proposals	Suggested indicators to measure performance against SEA objective
—Reduce traffic levels and promote more sustainable transport patterns across the area, particularly focusing on areas with low air quality (e.g. Air Quality Management Areas)? —Promote walking and cycling and improve infrastructure for these forms of travel? —Encourage Green Travel Plans and school travel plans? —Promote operation of the most modern vehicles, including buses and private cars? —Recognize the importance of awareness and marketing campaigns promoting the issue of improving air quality in the region? —Reduce congestion on the inter-urban trunk road network in large urban areas? —Instigate financial incentives and measures on the basis of the polluter' pays principle (e.g. congestion charge, road pricing)? —Promote the use of public transport?	—Levels of main pollutants for national air quality targets —Number of residential properties within AQMAs —Number of Euro engine buses operating in AQMAs/future AQMAs in Tyne and Wear —Effective use of awareness and marketing campaigns percentage of the population reached by awareness campaigns —Number of business and School Travel Plans —Reduction in NO_x and primary PM_{10} emissions through local authority's estate and operations (National Indicator 194)

Where possible, the indicators in the SEA framework were similar to those used in the earlier baseline studies. Analysis of the baseline situation for these indicators provided a means of summarizing current environmental conditions and predicted future trends (without implementation of the plan). This analysis was carried out for each of the SEA objectives using a simple three-point normative scale to describe both current conditions (good, moderate or poor) and expected future trends (improving, stable or declining). The overall conclusion was that, without the implementation of the plan, performance against a number of the SEA objectives was predicted to decline. These objectives included those relating to air quality, transport related CO_2 emissions, noise and light pollution, the resistance of transport infrastructure to climate change and flood risk, and crime levels.

9.9.6 *Identification and assessment of alternatives*

The SEA Directive requires a consideration of 'reasonable alternatives taking into account the objectives and the geographical scope of the plan or programme' and 'an outline of the reasons for selecting the alternatives dealt with'. For the Tyne and Wear LTP SEA, three broad strategic alternatives were considered. These comprised a 'do-minimum' scenario, a so-called 'realistic' scenario and an 'optimistic' (or aspirational funding) scenario. These alternative scenarios were developed by the LTP team in response to the identified local transport objectives and challenges. The main differences between the alternative options are summarized below.

Do-minimum scenario:

- no additional interventions in public transport, highway management, cycling and walking or freight and ports;
- highway capacity schemes that are already underway or have confirmed funding to go ahead;
- bus and metro fares to increase;
- rail and metro refurbishment; and
- workplace and school travel planning to remain at current levels.

Realistic scenario:

- emphasis on highway capacity schemes (mainly junction improvements);
- development of three new park and ride schemes;
- bus priority lanes to be provided at eleven locations;
- integrated 'smart' ticketing on public transport;
- electric vehicle charging points; and
- emphasis on walking and cycling initiatives.

Optimistic scenario (as for realistic scenario, plus the following):

- much greater emphasis on highway capacity schemes;
- development of a further eleven park and ride schemes;
- provision of an additional nine bus priority lanes;
- greater emphasis on rail and metro, including re-opening a number of rail lines; and
- no additional emphasis on walking and cycling initiatives.

The SEA Environmental Report provided only limited information on the reasons for the selection of these particular alternatives. For example, it was not clear whether other strategic options had been considered at an earlier stage in the plan preparation process and subsequently rejected. The realism of the optimistic scenario (based on aspirational funding levels) at a time of significantly reduced public funding for transport schemes could also be questioned.

In order to compare the impacts of these alternative strategies, a qualitative assessment was made of their performance against each of the objectives in the SEA framework. This assessment was based on the series of assessment 'prompt questions' defined in the framework. Use was also made of constraint mapping, with a range of maps produced showing relevant environmental constraints and selected schemes overlaid where specific locations were known (e. g. park and ride scheme locations). The assessment of the effects of the alternatives on the SEA objectives was a qualitative exercise, employing a simple seven point scale (large, moderate or slight beneficial effect; neutral or no effect; slight, moderate or large adverse effect). Moderate or large effects were deemed to be significant. The conclusions on the likely scale of effects on each SEA objective were supported by a detailed commentary or explanation in the environmental report. However, there was no use of explicit significance criteria to guide this process; such criteria have proved useful in other SEAs of local transport plans and would have helped to more clearly justify the conclusions on significance in this case (see TRL 2004 for an example of the application of such criteria).

Evaluation of significance is of course often difficult in SEA of strategic plans due to the lack of detail about specific schemes, locations and timescales. In this case, there was also a difficulty in balancing the short-term and longer-term effects of the different strategies. For a number of SEA objectives, there was expected to be a mix of shortterm adverse effects (e. g. associated with the construction of new infrastructure schemes) and potential longer-term benefits (e. g. due to reduced traffic and congestion levels once the schemes were operational). The absence of explicit criteria with which to assess significance presented some difficulties in these circumstances, as it was difficult to see how these different effects had been balanced. The SEA also appeared to adopt an implicit assumption that the proposed measures or interventions would prove to be effective, for example in changing travel behaviour in the desired direction. There was no explicit consideration of the deliverability or risk associated with proposed measures.

The assessment of the alternative strategies concluded that, overall, the realistic scenario provided the best balance between adverse and beneficial effects (Table 9. 5). This conclusion was not clearly explained in the environmental report, although it appeared to reflect the absence of beneficial effects with the do-minimum scenario and the larger number of more severe (i. e. large rather than moderate) adverse effects with the optimistic scenario. Overall, predicted effects against the SEA objectives were as follows:

Table 9. 5 Assessment summary of strategic alternatives for Tyne and Wear LTP

SEA Objective	Strategic Alternative 1 The 'Do Minimum' Scenario	Strategic Alternative 2 The Realistic Scenario	Strategic Alternative 3 The Optimistic Scenario
Environmental Objectives			
1	-	0	++
2	-	-	—
3	-	-	—
4	-	0	+

Continued

SEA Objective	Strategic Alternative 1 The 'Do Minimum' Scenario	Strategic Alternative 2 The Realistic Scenario	Strategic Alternative 3 The Optimistic Scenario
5	—	—	-
6	-	—	—
7	0	—	——
8	0	—	——
9	-	—	——
10	—	++	++
Social Objectives, Including Health and Inequality Issues			
11	0	+	++
12	0	0	-
13	—	++	+
14	-	+	++
15	0	+	+
16	0	+	+

Scale of Effect (*SE*)

+++Large beneficial ++Moderate beneficial +Slight beneficial

0 Neutral or no effects

——Large adverse —Moderate adverse -Slight adverse

Those effects which are either moderate or major are deemed to be significant

No.	SEA Objective
1	Ensure good local air quality for all
2	To protect and where possible enhance biodiversity, geodiversity and the multi-functional green infrastructure network
3	Protect and where possible enhance the European sites (*HRA specific objective*)
4	To mitigate against climate change by decarbonizing transport
5	To ensure resilience to the effects of climate change and flood risk
6	Promote prudent use of natural resources, waste minimization and movement up the waste hierarchy
7	Protect and enhance the quality of the area's ground, river and sea waters
8	To ensure efficient use of land and maintain the resource of productive soil
9	Maintain and enhance the quality and distinctiveness of the area's historic and cultural heritage
10	Protect and enhance the character and quality of landscape and townscape
11	To improve accessibility to services, facilities and amenities for all and avoid community severance
12	Reduce noise, vibration and light pollution
13	Improve health and well-being and reduce inequalities in health (*HIA specific objective*)
14	To promote greater equality of opportunity for all citizens, with the desired outcome of achieving a fairer society (*EqIA specific objective*)
15	Improve road safety
16	Reduce crime and fear of crime and promote community safety

Source: Tyne and Wear Joint Transport Working Group 2011a

• Do-minimum scenario: no significant beneficial effects; significant adverse effects against 3 SEA objectives (all moderate adverse).

• Realistic scenario: significant beneficial effects against two SEA objectives (both moderate beneficial); significant adverse effects against five SEA objectives (all moderate adverse).

• Optimistic scenario: significant beneficial effects against four SEA objectives (all moderate beneficial); significant adverse effects against six SEA objectives (three moderate adverse and three large adverse).

The absence of explicit methods or criteria with which to sum or weigh these effects was a weakness in the SEA. For example, it was unclear whether all of the SEA objectives were regarded as of equal importance or weight in the overall assessment.

9.9.7 *Effects of the plan and mitigation*

Following the appraisal of strategic alternatives, the preferred option for the LTP was developed further by the plan team. The resulting strategy in the draft plan was based largely on the realistic scenario and included 45 separate policies. In order to simplify the assessment, the SEA grouped related policies into a smaller number of 'policy components'. Policies within these components were expected to have broadly similar effects. Six main policy components were identified in the draft plan, including those concerned with (1) improving safety; (2) maintaining and managing infrastructure; (3) promoting sustainable transport modes; (4) parking; (5) freight; and (6) major schemes. The effects of these policy components were assessed against each of the SEA objectives, in a similar way as for the assessment of strategic alternatives using the same seven-point qualitative scale.

Significant adverse effects were identified against three of the SEA objectives (to ensure resilience to the effects of climate change and flood risk; to ensure efficient use of land and maintain the resource of productive soil; and to reduce noise, vibration and light pollution). These adverse effects were associated with the draft plan's policies on parking, freight and major schemes. Based on the results of the assessment, recommendations to improve the overall sustainability performance of the plan were made in the environmental report. These included:

• Policies relating to major schemes should include a requirement for a Construction Environmental Management Plan (CEMP) and a Site Waste Management Plan for any scheme that requires construction.

• There was no consideration of climate change adaptation within any of the policies in the draft plan. Policies relating to major schemes and maintaining infrastructure should be updated to include reference to climate change adaptation.

• Policies on identifying suitable sites for off-road lorry parking provision and freight consolidation centres should include reference that the site would not be located in a sensitive location for biodiversity/geodiversity and would minimize land take and loss of productive land.

• Policies on parking should include a reference to the fear of crime that may occur in car parks, particularly out of town car parks; suitable lighting and security presence may be appropriate forms of mitigation.

• Policies on parking should also make reference to the consideration of noise and light pollution that may be introduced at out of town car parks.

Most of these suggested amendments to the draft plan were incorporated into the policies in the final plan. As a result of these amendments, the performance of the final plan against the SEA objectives was improved, with fewer significant adverse effects identified than for the draft plan. Significant adverse effects were identified only for the major schemes policy component of the final plan, against two of the SEA objectives (resilience to climate change and flood risk and the efficient use of productive land). Further mitigation measures were recommended in the final environmental report in order to minimize these effects; these emphasized the need to avoid the location of major schemes in flood risk areas and to minimize the loss of agricultural land in the design of such schemes. A number of generic mitigation measures were also recommended (e. g. environmental management best practice measures to be adopted during construction projects).

9.9.8 Summary

This case study has examined the application of SEA to local transport plans. UK government guidance indicates that LTPs should be subject to HIA and EqIA, in addition to the requirement for SEA. LTPs also require screening for potential impacts on European protected habitats (habitats regulation assessment). In the case of the Tyne and Wear local transport plan, an attempt was made to integrate these different types of assessment within the SEA. This was achieved by incorporating health, equalities and habitats issues in the scoping and baseline stages of the SEA, and by the inclusion of specific objectives on these issues in the SEA assessment framework. Despite this innovative approach, the case study also reveals some weaknesses in the SEA process, particularly in relation to the identification and assessment of alternative options and in the evaluation of significance. The use of more explicit criteria with which to assess the significance of predicted effects and to compare alternatives would have helped to improve the clarity of the assessment.

9.10 Summary

This chapter has examined a number of case studies of EIA in practice. Most of the cases involve EIA at individual project level, although examples of SEA have also been discussed. Links with other related types of assessment, such as health impact assessment and appropriate assessment under the Habitats Directive, have also been explored. While it is not claimed that the selected case studies represent examples of best EIA practice, they do include examples of some novel and innovative approaches towards particular issues in EIA, such as extended methods of public participation and the treatment of cumulative effects. But the case studies have also drawn attention to some of the practical difficulties encountered in EIA, and to some of the limitations of the process in practice. Chapters 11 and 12 provide further discussion of a number of the new approaches considered in the case studies.

SOME QUESTIONS

The following questions are intended to help the reader focus on the key issues raised by the EIA and SEA case studies in this chapter.

1 The assessment of incremental and cumulative effects is revealed to be a weak area in a number of the case studies (Section 9. 2 and 9. 6). Why is assessment of these impacts often problematic in project-level EIA, and what solutions might you suggest to remedy the problem?

2 What factors accounted for the more successful treatment of cumulative effects in the Humber Estuary case study in Section 9. 5?

3 For projects affecting European priority habitats, summarize the tests that must be carried out as part of the 'appropriate assessment' process (as discussed in Section 9. 3). What guidance is available on the interpretation of these tests?

4 What were the main strengths and weaknesses of the public participation approaches adopted in the Portsmouth incinerator case study (Section 9. 4)? How could the effectiveness of the public participation process have been improved?

5 Summarize the key similarities and differences between the HIA and EIA processes, as revealed by the Stansted airport case study (Section 9. 6). In what ways are the two processes linked?

6 Summarize the main weaknesses in the SEA for UK offshore wind energy, as revealed in the stakeholder consultation (Section 9. 7). How might these stakeholder concerns have been more effectively addressed?

7 Comment on the approach used to assess and compare alternatives in the SEA of the Tyne and Wear Local Transport Plan (Section 9. 8). How could this element of the SEA have been improved?

References

Cooper, L. M. and Sheate, W. R. 2002. Cumulative effects assessment-a review of UK environmental impact statements. *Environmental Impact Assessment Review* 22 (4), 415-39.

DfT (Department for Transport) 2003. *The future of air transport-White Paper*. December 2003.

DfT 2009a. *Guidance on Local Transport Plans*. July 2009.

DfT 2009b. *Transport analysis guidance 2. 11D: strategic environmental assessment for transport plans and programmes. Draft guidance*, April 2009.

DTI (Department for Trade and Industry) 2002. *Future offshore-a strategic framework for the offshore wind industry*.

DTI 2003a. *Offshore wind energy generation-phase 1 proposals and environmental report*. Report prepared by BMT Cordah for the DTI. April 2003.

DTI 2003b. *Responses to draft programme for future development of offshore windfarms and the accompanying environmental report-summary of comments, and DTI response*. June 2003.

EC (European Commission) 2000. *Managing Natura 2000 sites: the provisions of Article 6 of the 'Habitats' Directive 92/43/EEC*.

EC 2001. *Assessment of plans and projects significantly affecting Natura 2000 sites: methodological guidance on the provisions of Article 6. 3 and 6. 4 of the Habitats Directive 92/43/EEC*.

EC 2007. *Guidance document on Article 6 (4) of the 'Habitats Directive' 92/43/EEC: clarification of the concepts of alternative solutions, imperative reasons of overriding public interest, compensatory measures, overall coherence: opinion of the Commission*.

ERM (Environmental Resources Mangement) 2006. *Health impact assessment of Stansted Generation 1: final report*. June 2006.

ERM 2008. *The Stansted Generation 2 project: a health impact assessment*. April 2008.

ETSU (Energy Technology Support Unit) 1996. *Energy from waste: a guide for local authorities and private sector developers of municipal solid waste combustion and related projects*. Harwell: ETSU.

Glasson, J., Godfrey, K. and Goodey, B. 1995. *Towards visitor impact management*. Aldershot: Avebury.

HIE (Highlands and Islands Enterprise) 2005. *Cairngorms estate management plan: 2005-2009*. Inverness: HIE.

House of Lords Select Committee on Science and Technology 2000. *Science and society*. Third report. London: HMSO.

Huggett, D. 2003. *Developing compensatory measures relating to port developments in European wildlife sites in the UK*.

Hunter, C. and Green, H. 1995. *Tourism and the environment: a sustainable relationship?* London: Routledge.

IWM (Institute of Waste Management) 1995. *Communicating with the public: no time to waste*. Northampton: IWM.

Land Use Consultants 1994. *Cairngorm Funicular Project—Environmental Statement (and Appendices)*. Glasgow: LUC.

Mathieson, A. and Wall, G. 2004. *Tourism: economic, physical and social impacts*. London: Longmans.

Petts, J. 1995. Waste management strategy development: a case study of community involvement and consensus-building in Hampshire. *Journal of Environmental Planning and Management* 38 (4), 519-36.

Petts, J. 1999. Public participation and environmental impact assessment. In *Handbook of environmental impact assessment*, J. Petts (ed), vol. 1, Process, methods and potential. Oxford: Blackwell Science.

Petts, J. 2003. Barriers to deliberative participation in EIA: learning from waste policies, plans and projects. *Journal of Environmental Assessment Policy and Management* 5 (3), 269-93.

Piper, J. M. 2000. Cumulative effects assessment on the Middle Humber: barriers overcome, benefits derived. *Journal of Environmental Planning and Management* 43 (3), 369-87.

Piper, J. M. 2001a. Assessing the cumulative effects of project clusters: a comparison of process and methods in four UK cases. *Journal of Environmental Planning and Management* 44 (3), 357-75.

Piper, J. M. 2001b. Barriers to implementation of cumulative effects assessment. *Journal of Environmental Assessment Policy and Management* 3 (4), 465-81.

Piper, J. M. 2002. CEA and sustainable development-evidence from UK case studies. *Environmental Im-*

pact Assessment Review 22 (1), 17-36.

The Planning Inspectorate 2008. *Stansted G1 Inquiry: Inspector's Report*. January 2008.

Report to the Department for Transport and Somerset County Council, July 2004.

RSPB (Royal Society for the Protection of Birds) 2003. *Successful offshore wind farm bids raise serious concerns for birds*. RSPB press release, 18 December.

Sheate, W. R. 1995. Electricity generation and transmission: a case study of problematic EIA implementation in the UK. *Environmental Policy and Practice* 5 (1), 17-25.

Snary, C. 2002. Risk communication and the waste-to-energy incinerator environmental impact assessment process: a UK case study of public involvement. *Journal of Environmental Planning and Management* 45 (2), 267-83.

SNH (Scottish Natural Heritage) 2000. *Cairngorm Funicular Railway-Visitor Management Plan*. Paper presented to the SNH Board. SNH 100/5/7.

Stop Stansted Expansion 2006. *SSE response to BAA health impact assessment*. August 2006.

TRL (Transport Research Laboratory) 2004. *Strategic environmental assessment guidance for transport plans and programmes: Somerset County Council local transport plan SEA pilot: Alternatives and significance*. July.

Tyne and Wear Integrated Transport Authority 2011. *LTP3: The third Local Transport Plan for Tyne and Wear: strategy 2011-2021*. March 2011.

Tyne and Wear Joint Transport Working Group 2011a. *Local Transport Plan 3: strategic environmental assessment-environmental report*. April 2011.

Tyne and Wear Joint Transport Working Group 2011b. *Local Transport Plan 3: strategic environmental assessment-SEA statement*. April 2011.

Tyne and Wear Joint Transport Working Group 2011c. *LTP3 Equality impact assessment: final report*. April 2011.

Weston, J. 2003. Is there a future for EIA? Response to Benson. *Impact Assessment and Project Appraisal* 21 (4), 278-80.

Weston, J. and Smith, R. 1999. The EU Habitats Directive: making the Article 6 assessments. The case of Ballyseedy Wood. *European Planning Studies* 7 (4), 483-99.

10 国际实践对比

10.1 导言

世界上许多国家已制定了环境影响评价法规，并且开展了项目环评。不过，这些法规有着很大的不同，如一些具体实践中的细节，这是因为各国有着不同的政治、经济和社会背景。

环境影响评价制度也在世界范围内迅速发展。例如，1999 年，在编写本书的第二版时，许多非洲国家和转型中的国家只在近期颁布了环境影响评价法规。而到目前为止，这些国家已经积累了相当多的环境影响评价经验，而且它们正在开发更详细的评价导则和法规。自本书 2005 年版之后，书中所提到的 7 个国家中的 3 个国家（秘鲁、波兰和加拿大）经历了环境影响评价体系的巨大变革，10.9 节中讨论的国际组织有一半更新了他们的环境影响评价导则。

本章旨在以实例说明当前的环境影响评价体系，并与此前谈到的英国和欧盟体系进行对比分析，首先回顾了世界各国环境影响评价的实施情况，分析了影响环境影响评价发展的因素；然后讨论了 6 大洲中 7 个国家的环境影响评价体系，重点是反映每个体系的特定方面：

① 贝宁拥有非洲最先进的环境影响评价体系，具有较高的透明度和大量的公众参与，综合考虑了环境问题、国家计划和有力的行政与制度手段。

② 秘鲁的环境影响评价体系是许多南美洲国家的典型，表现在对特定行业的定位、公众参与透明度的相对缺乏以及介入项目规划的时期晚等。

③ 波兰与许多其他转型中的国家相类似，其环境影响评价体系发生了巨大变化。在 2000 年，波兰制定了全新的条例，使其环境影响评价体系达到了欧盟的要求。

④ 中国的环境影响评价体系之所以被讨论是因为，在未来，中国的任何一项环境政策都很可能带来全球性的影响。中国的环境政策受到需要协调经济发展计划的限制。

⑤ 加拿大因其不断发展的环境政策而闻名。其联邦环境影响评价体系在调解和公众参与方面非常突出。该体系目前正发生重大变化。

⑥ 与加拿大相类似，澳大利亚的环境影响评价体系分别在联邦政府和州政府被执行。西澳大利亚州是一个非常有意思的优秀的州级环境影响评价体系的例子，该体系体现出许多创新性特征。

本章也讨论了国际机构在其投资和支持的项目与计划中发展和推广好的环境影响评价实践的作用。

10.2 世界各国环境影响评价的基本情况

据作者所知，2011 年中期世界各国环境影响评价制度情况如表 10.1 和图 10.2 所示。专栏 10.1 列出了世界各国环境影响评价信息来源的初步清单，其中 140 多个国家有某种形式的环境影响评价法规。

表 10.1 世界各国现有的环境影响评价制度（及开始实施年份）

国家/地区	实施	国家/地区	实施	国家/地区	实施
西欧		匈牙利	2005	科特迪瓦	1996
奥地利	1993,2000	哈萨克斯坦	1997	刚果	2002(部分)
比利时	1985～1992	科索沃	2009	埃及	1995
丹麦	1989,1999	吉尔吉斯斯坦	1997	埃塞俄比亚	2002
芬兰	1994,2004	拉脱维亚	2004	加蓬	1979
法国	1976	立陶宛	1996～2005	冈比亚	2005?
德国	1990,2005～2006	马其顿	2005	加纳	1999
希腊	1986,2002	摩尔多瓦	1996	几内亚	1987
冰岛	2000,2005	黑山共和国	2005	伊朗	1994～2005
爱尔兰	1989～2000	波兰	1990～2008	伊拉克	1997(部分)
意大利	1986～1996	罗马尼亚	1995	以色列	1982,2003
卢森堡	1994	俄罗斯	2000(部分)	约旦	1995
荷兰	1987,2002	塞尔维亚	2004	肯尼亚	2002～2003
挪威	1990	斯洛伐克	1994	科威特	1990,1995～1996
葡萄牙	1987	斯洛文尼亚	1996	黎巴嫩	发展中
西班牙	1986,2008	塔吉克斯坦	2006	莱索托	2003
瑞典	1987～1991	土耳其	1983,1997	利比里亚	2003
瑞士	1985	土库曼斯坦	2001	马达加斯加	1997,2004
英国	1988	乌克兰	1995	马拉维	1997(导则)
中欧及东欧		乌兹别克斯坦	2000	马里	2008
阿尔巴尼亚	2004	**非洲及中东**		毛里塔尼亚	2004(部分)
亚美尼亚	1995	阿尔及利亚	1983,1990	毛里求斯	2002
阿塞拜疆	1996	安哥拉	2004	摩洛哥	2003
白俄罗斯	1992	巴林岛	?	莫桑比克	1998
波黑	2003	贝宁	2001	纳米比亚	2011
保加利亚	2002,2009	博茨瓦纳	2005	尼日尔	1998
克罗地亚	2006	布基纳法索	1997	尼日利亚	1992
捷克共和国	1991,2001	喀麦隆	2005	阿曼	1982,2001
爱沙尼亚	1993,2005	科摩罗	1994	巴勒斯坦	2000
格鲁吉亚	2002				

续表

国家/地区	实施	国家/地区	实施	国家/地区	实施
非洲及中东		智利	1997	柬埔寨	1999(部分)
卢旺达	2005	哥伦比亚	1997～2005	中国	2002
奥塔尔	2002	哥斯达黎加	1998(部分)	印度	1994
塞内加尔	2001	古巴	1999	印度尼西亚	1999
塞舌尔	1994	多米尼加	2002	日本	1997
塞拉利昂	2008	厄瓜多尔	1999	韩国	1977～2000
南非	1997	萨尔瓦多	1998	老挝	2000
斯威士兰	2002	危地马拉	2003	马来西亚	1987
叙利亚	2002,2008	圭亚那	1996	蒙古国	1998
坦桑尼亚	2004～2005	洪都拉斯	1993	尼泊尔	1997
多哥	1988	墨西哥	1988	巴基斯坦	1997,2000
突尼斯	1988,1991	尼加拉瓜	1994	巴布亚新几内亚	1978(部分)
乌干达	1998	巴拿马	2000	菲律宾	2003
阿拉伯联合酋长国	1993,2009	巴拉圭	1994	斯里兰卡	1988
也门	1995	秘鲁	2009	泰国	1992
赞比亚	1997	乌拉圭	1994	越南	1994
津巴布韦	1997(部分)	美国	1969	**澳洲以及其他地方**	
美洲		委内瑞拉	1976	南极洲	1991
阿根廷	1994～1996	**亚洲**		澳大利亚	1999
伯利兹城	1995	阿富汗	2007	新西兰	1991
玻利维亚	1995	孟加拉国	1995		
巴西	1986,1997	不丹	2000		
加拿大	1992				

注：这份清单代表了我们所知的全球环境影响评价法规现状，虽然我们无法确定其准确性。在本书上一个版本之后很多环境影响评价法规都已更新或调整，一般框架的法规被具体的环境影响评价法规所补充。这份清单包括了2005年后的上述大部分变化。

资料来源：Badr，2009；CISDL，2009；非洲经济委员会，2005；世界银行，2006；很多其他网站和期刊文章来源。

专栏10.1　世界范围内环境影响评价制度的一些参考资料

虽然其中一些参考资料是针对某个国家的特例，但是它们讨论了与区域相关更广泛的问题。

➢ 总体上：Lee和George (2000)，Wood (2003)；国际影响评价协会的成员和分支机构，环境影响评价维基百科出版物，由经合发组织、联合国环境规划署和世界银行出版；以及发表在例如《环境影响评估评价》(*Environmental Impact Assessment Review*)、《影响评价与项目评估》(*Impact Assessment and Project Appraisal*)、《环境评估政策与管理杂志》(*the Journal of Environmental Assessment Policy and Management*) 期刊上的文章。

➢ 西欧：欧盟 (2009)，GHK (2010)，Wood (2003)。

➢ 中东欧：Kovalev et al. (2009)，Rzeszot (1999)，Unalan和Cowell (2009)，世界银行 (2002)。

➢ 澳大利亚、新西兰：Elliott和Thomas (2009)，Wood (2003)。

➢ 亚洲：Briffett (1999)，Vidyaratne (2006)，世界银行 (2006)。

➢ 北美：Wood (2003)。

➢ 南美：Brito和Verocai (1999)，Chico (1995)，国际可持续发展法律中心 (2006)，Kirchhoff (2006)。

➢ 非洲和中东：Almagi et al. (2007)，Appiah-Opoku (2005)，Bekhechi and Mercier (2002)，非洲经济委员会 (2005)，El-Fadl和El-Fadel (2004)，Kakonge (1999)，Marara et al. (2011)，南非环境评价研究所 (2003)。

然而，环境影响评价实践仍然没有突破国界的约束。图 10.1 总结了一个代表性国家的环境影响评价发展过程：它以最初有限的环境影响评价开始，其实施是为临时响应公众的关注，或者是应资助者的要求而执行的，或者是有环境影响评价要求国家的企业在没有环境影响评价要求的国家里对其采取的活动进行环境影响评价。随着时间的推移，该国制定了环境影响评价导则或法规。与大多数欧盟国家情况相同，这将促使该国的环境影响评价数量快速增加。然而，这些法规可能仅被应用于少量的项目或者往往被忽视，从而导致环境影响评价数量很少的增加。又过了一段时间，这些法规可能得到了调整或补充，每年实施的环境影响评价数量趋于稳定甚至会有所减少，环境影响评价体系也变得更加“成熟”。目前，各国的环境影响评价状况可大致描绘在一个连续图谱上。图 10.2 笼统地显示了各大洲的情况，个别国家可能会与此不同。

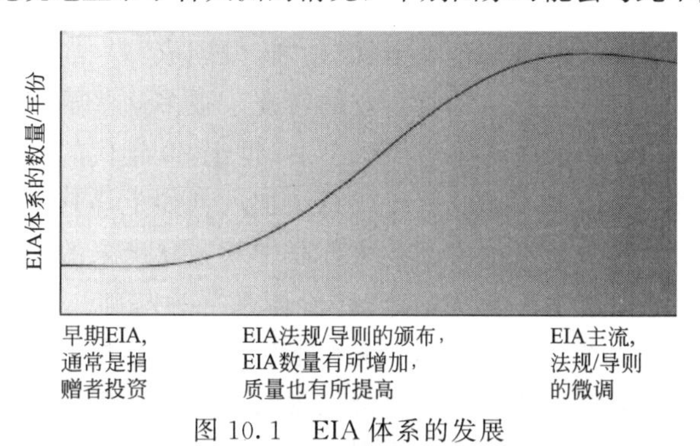

图 10.1　EIA 体系的发展

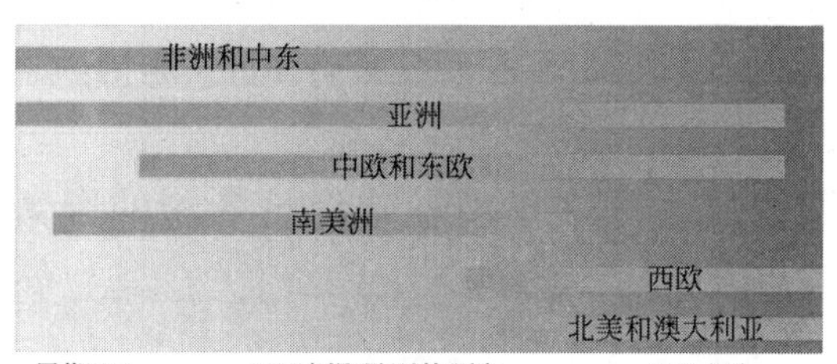

图 10.2　当前世界各国 EIA 体系状况

近来，伴随着许多国家环境影响评价具体法规的制定，非洲的形势变化很快。这都是一系列努力的结果，包括 1995 年非洲部长级环境会议，该会议委托非洲环境部长促成环境影响评价的应用正式化；一些旨在改善环境影响评价能力和非洲国家间合作的组织的建立（如非洲环境影响评价能力发展和联系组织、非洲环境评价合作组织以及国际影响评价协会的几个非洲分会）。另一方面，非洲的环境影响评价仍受困于缺乏训练有素的专业人员、缺乏经费、对环境影响评价可能阻碍经济发展的担忧以及政治意愿的缺失（Appiah-Opoku，2005；Kakonge，1999）。即使环境影响评价法规存在，也不意味着它能够被很好地付诸实践、有公众参与或者被强制执行（Okaru 和 Barannik，1996）。综上，虽然一些非洲国家如贝宁、塞舌尔和突尼斯等有较好的环境影响评价制度和实践，其他国家如布隆迪、几内亚比绍等仍没有任何的 EIA 的迹象（远远地落在图 10.2 上的最左边）。

南美和中美的几乎所有国家都有不同形式的环境保护法律体系，至少包括环境影响评价的方面。这些体系差异很大，它反映了这些国家不同的政治和经济体系：一些南美洲国家的平均收入比其他国家高 10 倍以上。然而，从总体上看，南美洲的环境影响评价的发展受到了政治不稳定、政府办事效率低、经济停滞以及外债的制约（Brito 和 Verocai，1999）。最近这种现象可能有所改变，秘鲁就是个例子，不过南美洲的环境影响评价经常仅由中央执行，很少或者没有公众参与，如果有，往往也是在项目已经被批准之后才进行的（Glasson 和 Salvador，2000）。

亚洲的环境影响评价的差距也很大，如柬埔寨没有相应的法律法规，而中国香港则已经在广泛经验的基础上建立了包括战略影响评价在内的强大的环境影响评价制度。20 世纪 80 年代末，许多亚洲国家都建立了环境影响评价制度，应投资机构的要求，所有国家在投资地区都开展了环境影响评价。另一方面，Briffett（1999）认为亚洲许多国家的环境影响评价质量差，评价范围确定和影响预测水平低，公众参与也很有限。部分原因是人们认为 EIA 可能会阻碍经济增长，因为一些国家会以展示其大型建筑和基础设施项目来显示国家的财富（Briffett，1999）。很多国家，正在修订其早期的 EIA 体系以试图解决这些问题。

自 20 世纪 80 年代以来，中欧和东欧国家（“转型国家”）经历了从计划体系向市场体系的巨大转变，其中包括国有企业向私营企业的转变。很多这些国家的经济主要依托于能源和资源高消耗的重工业，很多这些国家也经历了 20 世纪 90 年代的经济危机。在这些转型国家中，中欧和东欧国家已经加入（如波兰）或意在加入（如土耳其）欧盟，并使它们的 EIA 法规与《欧洲环境影响评价指令》相协调。从苏联中新独立出来的国家拥有相似的体系，均以在苏联领导下发展的“国家生态专家鉴定/评价体系”为基础。欧洲东南部国家，比如阿尔巴尼亚、波斯尼亚-黑塞哥维纳（波黑）、克罗地亚、南斯拉夫和马其顿，有着相对不成熟的环境影响评价体系。为了使其环境影响评价法规与《欧洲环境影响评价指令》相协调，这个地区最近颁布了很多新的环境影响评价法规。

专栏 10.2 总结了一些影响发展中国家环境影响评价实施的因素。

专栏 10.2　一些影响发展中国家环境影响评价实施的因素

- 影响发展中国家环境影响评价实施的因素在热带地区或其附近区域的许多国家，其所用的环境模型、数据要求和标准均不同于温带地区。
- 社会文化背景、传统、等级制度和社会关系可能很不一样。
- 采用的技术可能会存在适用范围、时期和使用标准等方面的不同，而且可能带来更大的事故风险和更高的废弃物系数。
- 对各种影响的显著性认识可能存在很大差别。
- 环境影响评价实施的体制结构可能是疲软的和脱节的，可能存在人员不足、实践和技能不充分、水平低下以及机构之间不协调等问题。
- 环境影响评价介入规划程序较晚，因此可能对项目规划仅有有限的影响，或可能被用来证明该项目是合理的。
- 开发援助机构可以为许多项目筹集资金，但它们的环境影响评价可能要求评估相当多的影响。
- 环境影响评价报告可能是保密的，几乎没有人意识到它们的存在。
- 由于政府（过去）等因素，公众参与可能会很薄弱，公众在环境影响评价中的作用可能定义不清。
- 决策可能是不公开、不透明的，援助机构的参与可能使决策变得相当复杂。
- 环境影响评价可能与发展计划不协调。
- 环境影响评价制度的实施和法律合规性可能很薄弱，环境监测非常有限或不存在。

西欧国家的环境影响评价体系被修改过至少两次，构成了每个成员国中 EIA 法规的多样性：85/337 号指令被修正为 97/11 号指令和 2003/35/EC 号指令，而且许多国家在 1985 年原始指令之前已经有自己的环境影响评价体系。一份最近对欧盟成员国环境影响评价现状的回顾（GHK，2010）表明环境影响评价正在全部 27 个欧盟成员国中实施，其中包括 12 个新成员国，每年开展的环境影响评价总数达到了 16000 左右，且这个数字还在每年递增。然而，即使经过了三轮的“协调”，欧洲各国的环境影响评价实践仍千差万别。

北美洲、澳大利亚和新西兰的环境影响评价程序仍是世界上最好的，它们对公众参与、替代方案和累积影响有着良好的规定。有些国家针对联邦和州/省级项目有着单独的程序。但为了应对经济衰退和程序过度复杂及繁重的问题，有些体系正经历被精简或者可以说是弱化的过程。

针对上述讨论内容，以下各节举例介绍了世界各国的环境影响评价体系。

10.3 贝宁

贝宁位于非洲西部，曾经被海岸线后面浓密的热带雨林所覆盖。如今大部分热带雨林已被砍伐，由棕榈树取代。在较干旱的北部主要是草地。贝宁主要出口棉花、原油、棕榈产品和可可粉。与其他邻国相比，贝宁虽然人均 GDP 相对较低，但自 1995 年起，贝宁就拥有一个功能完善的环境影响评价体系，被认为是有最先进的环境影响评价体系的三个说法语的非洲国家之一（d'Almeida，2001）。其环境影响评价体系的特点是透明度高、公众参与性强，并且综合考虑了环境问题、国家规划和强硬的行政制度手段（Baglo，2003）。

与其他非洲国家一样，贝宁早期的环境影响评价是在世界银行、非洲发展银行等援助机构的要求下实施的。1990 年 12 月的贝宁《宪法》特别重视环境保护，1993 年批准了环境行动规划。1995 年贝宁环境局成立，并负责国家环保政策的落实、管理和影响评估方面的研究，准备国家环境报告及监控环境规章制度的实施情况等。虽然该机构要向环境、生物栖息地和城镇规划部门报告，但它拥有法人地位和独立的财政。该机构在 2001 年制定了环境影响评价法规以及一系列的环境影响评价指南（如需要环境影响评价的项目、燃气管道和灌溉项目的环境影响评价；Baglo，2003）。贝宁也有一个环境影响评价专业人员国家协会。

贝宁的环境影响评价程序由贝宁环境局开展。虽然只有两名负责环境影响评价的官员，但可以从公共和私人机构邀请专家来准备或评审具体的环境影响评价。这种做法让环保局可以以最少的人员配置来运作，确保与其他机构的持续合作，并履行决策的广泛参与原则（Baglo，2003）。

图 10.3 总结了贝宁的环境影响评价程序，包括如下：

① 筛选：这项工作是分散的，将责任分配给负责拟建项目的各个部门；

② 项目注册，确定评价范围和职权范围；

③ 准备起草环境影响报告书，主要通过广播和非政府组织（NGOs）来“公布”报告；

④ 若需要，环境部门成立专案公开审理，由委员会来安排一次公开听证会，讨论环境影响评价的内容；

⑤ 环境机构对环境影响报告书进行分析，如果环境影响报告书是令人满意的，环境机构发布可行性公告；

⑥ 根据环境可行性公告，相关部门做出项目的决定，开展技术可行性研究和经济可行性研究；

⑦ 跟踪评价和审计。

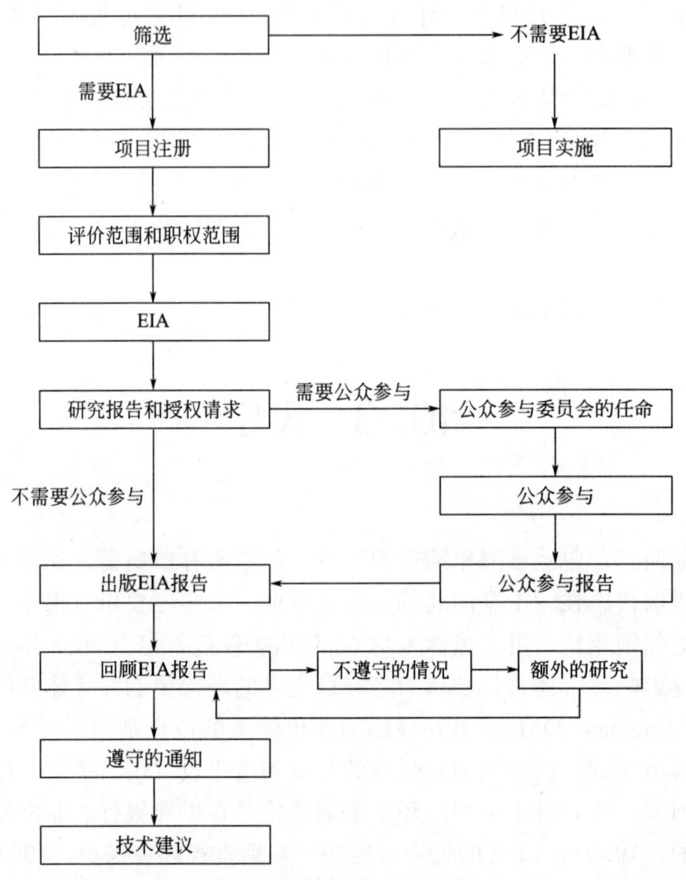

图 10.3 贝宁的 EIA 程序

1997～2002 年，有 78 份环境影响评价被评审，61 份环境影响报告书通过可行性公告验证（Yaha，2007）。贝宁执行环境影响评价的项目包括家畜发展项目、公路改进项目、2010 年洪水后的 Cotonou 排水系统改善项目、Quesse 垃圾填埋场扩展项目、城市和工业发展项目以及西非燃气管道项目。这些项目费用由捐助（如世界银行）、私有财产和国家政府资金来提供。

贝宁的环境影响评价实践同其他非洲国家一样，因为一些关键部门之间合作效率低、公众环境意识水平低、文化水平低和贫穷等而受到限制。通过贝宁环境局组织的一系列环境影响评价训练课程、环境影响评价专业人员组成的国家协会活动以及参与泛非能力建设活动（pan-African capacity building activities），早期缺少本土专家的困难有所缓解（见 10.2 节）。

10.4 秘鲁

秘鲁，南美洲第三大国家，包括了沿海岸狭窄而干旱的地带、安第斯山脉的小山和高地的肥沃山脊以及亚马逊盆地。渔业、农业和采矿业是秘鲁的主要行业。几年的经济困难之后，1990 年秘鲁政府的转变导致了国家的重组并提出一个广泛的私有化计划，包括许多国有矿山的私营化，以鼓励外国投资的大量增加。

1990 年 9 月，秘鲁政府颁布了 613-90 号法令，即《环境与自然资源法》，其规定任何主要开发项目均要强制进行环境影响评价，但并没有指定环境影响评价的内容或法定程序。能源和矿业部门通过最高法令 014-92-EM，成为第一个将 613-90 号法令付诸实施的部门。此后，渔业、农业、交通与通信、住房与建筑部门在 1994 年、1995 年都提出了相似但又彼此独立的要求。早期的迹象表明，当时环境影响报告书的质量很高，但对减缓措施的讨论较弱(Iglesias，1996)，公众参与也有限。

最高法令 019-2009-MINAM 在 2009 年 10 月重申了 2001 年的 27446 法，用一个同样适用于政策、计划和方案的全国统一系统取代了环境影响评价的独立体系。本节以下部分将集中说明这个新的体系。

最高法令 019-2009-MINAM 的附件Ⅱ广泛列举了要求做评价的项目和规划，并按有关主管机关分类（如农业部、能源与矿业部等）。被列入附件Ⅱ的任何开发项目，包括对已有项目的修改，只要其会产生显著影响，就必须编写一个初步评价。初步评价包括项目详情、市民参与计划、对项目可能产生的环境影响和减缓措施的简要描述。主管机关须在 30 天之内做出决定，判别项目属于Ⅰ类（轻微环境影响）、Ⅱ类（中度影响）或是Ⅲ类（显著影响)。最高法令的一个附件列举了主管机关做此裁决必须采用的标准：包括与人体健康、自然资源、指定区域、生物多样性及其组成以及人们的生活方式等相关的因素。

如果项目属于Ⅰ类，那初步评价就变成了一份“对环境影响的声明”，不需要采取进一步行动。如果项目属于Ⅱ类或Ⅲ类，就分别需要一份较详细或详细的环境影响报告书。最高法令详述了环境影响报告书必须包含的信息，如执行摘要、项目介绍、本底数据、对公众和法定咨询者如何参与的信息以及公开会议上做出的任何评论，包括与环保标准比较、环境管理、应急和关闭计划在内的影响评价等。只有国家环保部门管理的登记机构才有资格编写环境影响报告书。所有环境影响报告书都是公开文献。最高法令支持环境影响评价程序中的公众参与，但并未强制规定相关的具体技术。

环境影响报告书编制完成后，开发商将其提交给主管机关，主管机关有 40 天（Ⅱ类项目）或 70 天（Ⅲ类项目）的时间来评审并按需要开展公众咨询。如有必要，另有 20 天的决议过程，开发商在这有限的时间内需对评审进行评论并提供进一步的证据。环境影响评价评审过程总时间不应超过 30 天（针对Ⅰ类项目)、90 天（针对Ⅱ类项目）和 120 天（针对Ⅲ类项目)。完成之时，主管机关可批准或拒绝批准环境影响报告书。开发商可能只有在取得环境影响报告书批准之后才能申请其他牌照或许可证以及开始项目建设。环境影响报告书的批准也提出了具体的义务，以预防或减轻潜在的环境影响。

一旦项目被批准，开发商必须执行管理程序，对整个运营阶段实施控制和监测，以确保环境管理计划的实施。项目开始运作后，环境管理计划、应急计划、社区关系计划以及/或退役计划（如适用）必须每五年更新一次，以确保其仍然适用和及时更新。

新体系的优势包括：

① 批准系统，旨在解决秘鲁以前经常有项目在建设开始之后才编制环境影响报告书的问题（Brito 和 Verocai，1999)。

② Ⅱ类和Ⅲ类项目环境影响报告书的具体要求，可帮助确保环境影响报告书的高质量并提供主管机关用于明智决策的必要信息。

③ 明确而相对有限的环境影响报告书评审期限，可防止环境影响报告书的编制拖延必要的发展。

④ 对影响管理，应急计划和监测的强调，可帮助确保项目影响被有效管理。

⑤ 弱点包括：负责推进某类开发活动的主管机关也是批准此类开发活动环境影响报告书的主管机关，导致可能出现“偷猎者与狩猎人”的情况；环境影响评价程序中缺乏明确的、强有力的公众参与要求。

10.5 中国

中国自1978年起一直处于经济增长、工业化、民营企业和城市化的迅速转变过程中，也催生了很多对环境有显著影响的开发项目。中国的环境影响评价制度一直在努力跟上(Moorman和Ge，2007)。该制度仍在快速演变，并在2009年加强在项目规划中的应用。

中国的环境影响评价是1979年通过的《环境保护法》正式引入的，该法在1989年被修正。在该法引入后不久，就为环境影响评价的实施准备了一些导则，最重要的一个是1981年的《城市建设项目环境保护管理条例》。这些导则在1986年被修正并正式实施，包含了EIS的时间、资金提供、准备、检查和批准细节。此外，1990年和1998年12月政府还颁布了更有效的法令。1998年《建设项目环境管理法》要求对区域发展规划以及单个项目进行环境影响评价，增加了公众参与的要求，加强了执法力度和对违反环境影响评价要求的惩罚。

在这个阶段，EIA仍旧扮演的是自上而下的行政手段的角色（Wang等，2003)。

尽管每年中国有成千上万份环境影响报告书或报告表被编制（Ortolano，1996)，但是一些应该开展环境影响评价的项目并没有进行评价，个别项目未批先建，环境影响报告书的质量也参差不齐（Mao和Hills，2002；China，1999)。

2002年10月制定了《环境影响评价法》，也带来了环境影响评价制度的重大变革。表10.2列举了中国现有的环境影响评价法规和指南，图10.4总结了这些新的环境影响评价程序。

表10.2 中国环境影响评价法规、政府文件及导则

名　　称	年份
建设项目环境保护管理条例(国务院令第253号)	1998
关于执行建设项目环境影响评价制度有关问题的通知(国家环境保护总局 环发〔1999〕107号)	1999
建设项目环境影响报告表和环境影响登记表(国家环境保护总局 环发〔1999〕178号)	1999
建设项目竣工环境保护验收管理办法(国家环境保护总局令第13号)	2001
中华人民共和国环境影响评价法(中华人民共和国主席令第77号)	2002
国家环境保护总局建设项目环境影响评价文件审批程序规定(国家环境保护总局令第29号)	2005
关于印发《环境影响评价公众参与暂行办法》的通知(环发〔2006〕28号)	2006
建设项目环境保护分类管理名录(环境保护部令第2号)	2008
关于发布《环境保护部直接审批环境影响评价文件的建设项目目录》及《环境保护部委托省级环境保护部门审批环境影响评价文件的建设项目目录》的公告(环境保护部公告第7号)	2009
建设项目环境影响评价文件分级审批规定(环境保护部令第5号)	2009

资料来源：杨，个人交流。

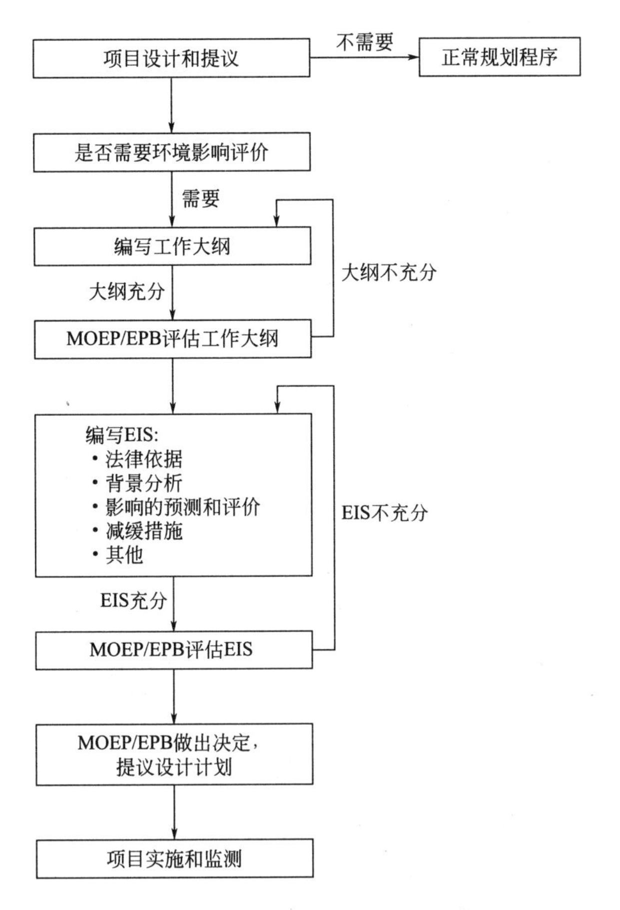

图 10.4　中国环境影响评价程序

[资料来源：Wang 等（2003）]

生态环境部（MOEP，原国家环境保护总局）掌管有关国家经济或具有重要战略意义的项目 EIA 的审批。MOEP 第 5 号法令以及 2009 年的指南列举了这些重点项目，包括核项目及跨省项目。省级环境保护局（EPB）负责具有重要区域意义的项目 EIA 的审批，如垃圾焚烧厂、冶炼及化工厂；对于其他类型的项目，省级环境保护局可以决定其由省级、市级或县/区级环保局审批。通常区或县级环保局只能审批环境影响报告表。

中国的环境影响评价程序由开发商咨询主管部门开始，以决定拟议的项目是否要进行全面的环境影响评价。跟秘鲁相同，中国的 EIA 也分三个不同层次。具有重大潜在环境影响的项目要求编制完整的环境影响报告书，影响相对有限的项目要求编制环境影响报告表，影响很小的项目只要求编制环境影响登记表，无须再提供数据。MOEP 第 2 号法令明确了不同项目需要进行的 EIA 类型，依据是项目的影响和受体环境的敏感性。

主管部门职员进行初步研究然后做出判断，有时需要外界专家的帮助。如果需要环境影响评价，主管部门将识别出那些很可能影响环境的因子并简单说明该项目的环境影响情况，然后由开发商委托有资质的环境影响评价机构专家编写环境影响报告。专家更详细地分析相关的影响，提议减缓措施，之后根据导则编写环境影响报告书，报告书需要探讨如下内容：

① 一般的法律依据；

② 拟议的项目，包括原材料的消费和生产；

③ 环境背景状况和周边环境；

④ 工程产生的短期环境影响和长期环境影响；

⑤ 影响显著性和可接受性；

⑥ 环境影响的成本-效益分析；

⑦ 监测计划；

⑧ 结论。

在环境影响报告书初稿完成前，公众有机会对其发表评论，虽然公众参与的形式并没有被规定。环境影响报告书的编制机构必须正式考虑利益相关者、专家和公众的意见，并在报告书中说明接受或拒绝这些意见的理由。

环境影响报告书或环境影响报告表提交给主管部门，主管部门检查拟议项目是否符合相关的环保法规和规划，确定项目所在区域是否有足够的环境承载力，考虑建议的减缓措施是否可能有效，并咨询相关专家的意见，最后做出决策。对于有争议的项目以及跨区域的项目，需将相关文件提交给更高一级的政府机构进行审批。如果项目被批准，环境保护计划可被包括进去，如监测和核查程序。主管部门必须提交一份报告说明项目将如何执行，所需的环境保护措施将如何贯彻。该项目一旦被省级主管部门批准，批准文件将被发布。

2002 年后的变化旨在纠正中国早期环境影响评价制度的许多问题。开发商在开展环境影响评价之前不准开始项目建设。环保部运用法律掀起了几场“环境风暴”，叫停了未批先建的项目。环保部也在 2007 年宣布对几个环境纳污能力很低的城市的新建项目不予批准（Yang，2008）。

然而，有时以发展为重心的地方政府与环保部门之间，人们对改善环境与改善生活水平之间的优先级仍然存在利益冲突。因此，个别地方环保主管部门有时不愿对抗强烈支持拟建项目的地方政府，或者以环境影响评价的形式对其施加约束（Mao 和 Hills，2002；Lindhjem 等，2007）。中国的环境影响评价程序有时会被批评过于复杂化，对历史的大气、水、土壤污染情况关注尚需加强，对公众参与要求需进一步提高，需增加可替代方案的考虑等（Lindhjem 等，2007；Wang 等，2003；Yang，2008）。

10.6 波兰

波兰与其他转型国家一样经历了环境政策的快速发展，并体现在其环境影响评价法规和制度上。

在共产主义后期，波兰政府只提供了最基本的服务，环境问题一直被忽视（Fisher，1992）。波兰的许多地区受到严重的污染，引起了广泛关注。为解决这个问题，政府于 1980 年颁布了《环境保护法》(EPA)。该法案之后也通过各种修改和 1994 年的土地使用规划法案以及 1990 年一项针对对环境非常有害的开发活动的行政命令得以加强。在该体系下，环境影响评价开始于开发商询问当地环境部门是否需要环境影响评价。如果需要，环境部门将草拟一个适宜开展这项工作的咨询服务机构名单。选定的机构一旦完成了环境影响报告书，就可以进行公众咨询，但这并不是强制的。如果该环境影响报告书被环境部门所接受，那么当地规划部门可以发布“地方指示”，列出项目的各个替代位置。在此基础上，开发商选择一个位置，继续工程设计，并由咨询机构编制最终的环境影响报告书。之后，开发商向当地规划部门递交环境影响报告书和规划申请，在对项目做决定之前规划部门再次征询当地环境部门的意见。项目建设的批准要求在项目的技术设计以外编制第三份环境影响报告书。

该体系被批评缺乏筛选标准，为每个项目准备的多种环境影响报告书既麻烦又多余，公众参与的程序太少以及评审环境影响评价的委员会的资源受限制（Jendroska 和 Sommer，1994）。它也不能应对波兰在经历了 1989 年的社会变革之后所产生的巨大的社会和经济变化：

不再有经济计划和中央策划者，货币可自由兑换，最佳技术可得，整个经济私有化。此外，为了成立一个环境监察的强大中央机构，行政安排被重新设计……但是旧工业仍在运行。从 1989 年后环境报告中看到的改进只是衰退的副作用……波兰的《环境影响评价法》仍旧反映了原来的两个特征：不愿在决策中引入公众参与以及不愿制定程序规则来解决争端。这意味着《环境影响评价法》不仅从环境的角度来看不够有效率，也与向更开放和民主社会以及自由市场的转型不相匹配。(Jendroska 和 Sommer，1994)

1994 年，波兰与欧盟形成伙伴关系，2004 年，波兰正式加入欧盟。这个转变要求波兰快速地改变其法律和法令，使之与欧盟相衔接。此过程开始于 1995 年和 1998 年对 1990 年行政命令的修订，之后又出台了三个关键的环境影响评价法规：即 2000 年关于获取环境及其保护、环境影响评价相关信息的法案，2001 年《环境保护法》(EPL) 及 2008 年取代 EPL 的环境及环境保护、环境保护和环境影响评价的公众参与法案。

随时间变化，波兰原环境影响评价制度的变化包括 EIA 应用于更广泛的项目；筛选的新要求，包括两份项目清单要求其环境影响评价体现环境影响评价指令附件Ⅰ和Ⅱ；对范围确定更加重视；对公众咨询要求更高；对跨边界影响评价的要求；努力将环境影响评价的重心从编制报告书转向整体程序（Wiszniewska 等，2002；Woloszyn，2004）。

新的体系代表了欧盟比较典型的环境影响评价体系，但包括一些有趣的差异（PIFIA，2008；EC，2009）：

① 将《栖息地指令》对适宜性评价的要求纳入了环境影响评价程序。不仅是与环境影响评价指令附件Ⅰ和附件Ⅱ项目相当的可能造成显著环境影响的项目要求进行环境影响评价，《栖息地指令》要求适宜性评价的项目也需要进行环境影响评价。这些 EIAs 必须评价欧盟自然 2000 保护地的项目影响，讨论其对自然 2000 保护地的影响的替代方案和减缓措施。

② 对于可能造成显著环境影响的项目（与附件Ⅱ项目相当），开发商必须在 EIA 筛选阶段提供“环境信息卡”。若主管机关决定不需要进行 EIA，环境信息卡就将作为正式“退出筛选”公告的基础。如需要进行 EIA，则环境信息卡就是强制要求的范围确定的基础。

③ 环境影响报告书必须包括对分析方案的描述，包括申请者倾向的方案、合理替代方案以及对环境最有利的方案；还须包括对项目相关的可能的社会冲突的分析。

④ 任何附件Ⅰ或附件Ⅱ相当项目必须取得“环境状况决议”才能得到规划许可。“环境状况决议”是基于环境影响评价或“环境信息卡”，来自法定咨询者、公众、其他国家的对 EIA 的评论以及其他一些如地籍图等的信息。

⑤ 如果 EIA 程序显示项目不应按照申请者倾向的方案实施，主管机关将发布“环境状况决议”来明确应该实施的方案。如果申请者有异议，则主管机关将不予批准该项目。

据不同数据来源（EU，2009；GHK，2010），波兰每年开展的环境影响评价大致为2200～4000。其中，只有一小部分涉及有跨边界影响的项目。然而，因为项目基数很大，GHK（2010）表示波兰的环境影响评价可能占了欧盟27成员国环境影响评价总数的1/3。波兰开展了很多跨边界环境影响评价，包括道路、铁路和输电线路项目。波兰与欧盟几个成员国签署了将环境影响评价应用到跨边界项目领域的双边协议，规定了EIA文件的翻译、时间表、公众参与原则和形式等内容（EU，2009）。

10.7 加拿大

加拿大是一个有着长久和强大的环境影响评价体系的国家，这个体系目前正处于不太稳定的状态。加拿大有着丰富的自然资源，起初被煤炭、钢铁、石油和铁路巨头信托肆意掠夺，缺乏强有力的规划和土地使用法规，强势的省级政府需求的矛盾，都催生了一个广泛的预防环境危害的机制的发展。加拿大的环境影响评价体系的特点是：国家和省级的评价程序有所不同；通过不同类型的环境影响评价程序来评价不同类型项目的程序非常复杂；在环境影响评价中采用了更新颖的调解以及公众参与方法。

在加拿大，环境影响评价由联邦和省级政府共同负责。联邦程序适用于政府有决策权的项目。早期的联邦环境影响评价导则在20世纪70～80年代有了进一步加强，并在1989年立法通过。然而，由于“环境评价及评审程序”的局限性，导则于1995年被加拿大《环境评价法》取代。2001年和2010年又对法案做了修订。从1993年开始要求进行政策的战略影响评价，1999年该要求被进一步加强。Gibson（2002）对2002年以前加拿大联邦政府的环境影响评价体系进行了很好的回顾。

加拿大环境评价机构（CEAA）管理加拿大《环境评价法》。按照该法案，由拟议行动的负责机构最初的自我评价来决定是否需要做环境影响评价。如：

① “项目”是否符合法案定义要求；

② 按照法案的排除列表规定，项目是否不被排除在外；

③ 是否涉及联邦机构；

④ 依照法案，是否需要开展环境影响评价。

排除项目列表给出了那些不需要环境影响评价的项目，因为它们的负面环境影响不是很严重（如简单的更新工程）。

一旦决定需要进行环境影响评价，可以采用以下四种途径进行决策：筛选、综合性研究、调解或评审小组。大多数工程需要对与项目环境影响以及建议的减缓措施有关的文件进行“筛选”。“等级筛选”可以用于那些已知影响易于减缓的评价项目。“模型等级筛选”为在一个等级所有的项目提供一般性评价：负责机构用模型报告作为模板，考虑特定区域和特殊项目信息。“取代等级筛选”适用于不需要特定区域或特殊项目信息的项目。

少数项目，通常少于1%，将需要更完全的“综合性研究”，综合性研究列表规则中包括核电厂、大型石油和天然气开发以及工业工厂等。

如果通过筛选需要进一步评价，那么要涉及调解或评审小组。相似地，在综合性研究的早期，环境部长必须决定项目是否应该继续作为综合性研究进行评价或者是否应该涉及调解或评审小组。通常，对于涉及调解或评审小组的项目（实际上是将其从由负责机构的自我评价转变为独立的外界评价），它们影响的显著性往往是不确定的，很可能导致重大有害的环境影响，这些影响识别是否公正是不确定的，或者通过公众考虑的方式来保证（CEAA，2003）。评审小组是由环境部长指定的，去调查和评审那些很有可能产生显著环境影响的项目。从 1995 年起，调解方法是自愿过程，由部长指派一位独立的调解员，通过无敌对的、合作解决问题的方法帮助有关团体解决他们的问题。很少有项目是采用评审小组或调解的途径。

任何需要综合性研究、调解或评审小组的项目必须考虑执行项目的替代方法、项目的目的以及可再生资源持续利用的影响；必须包括追踪调查计划。在项目决策时，负责机构必须考虑综合性研究的结果或者是调解员或评价小组的建议（CEAA，2003）。在环境影响评价程序的每个阶段都必须要考虑公众建议，尤其是筛选阶段的要求更为严格。参与者筹资的提案允许参与者参与综合性研究以及小组的评审和调解，土著居民还有一套特别的咨询程序（Sinclair 和 Fitzpatrick，2002）。

加拿大《环境评价法》（CEAA）在其网站上公布了许多报告。2006～2010 年开展了超过 22000 个联邦环境影响评价：其中 99.6%经过了筛选，但是只有 97 个进行了综合性研究，还有 19 个进行了小组报告。综合性研究项目包括管道、海洋油气、矿山、废物处理中心以及一系列早前项目的退役（CEAA，2011）。

加拿大的多数省份在自己的权限下，对项目的环境影响评价制度进行了相当大的改进。包括 1976 年的安大略省的环评法，在当时非常先进，但在 1977 年却衰退了；还有 2001 年马尼托巴省的持续发展行为守则，要求政府官员推进在环境影响评价中对可持续影响的考虑。1988 年初，联邦及省级环境部长签署了环境影响评价统一协定，以促进合作使用已有程序，减少重复和无效（Gibson，2002）。

考虑到联邦和省级政府之间功能的重复、几个政府机构参与后引起环境影响评价的领导权不清晰以及冗长的政府程序，加拿大《环境评价法》在 2010 年做出了调整，新的限制综合研究时长的法规也被提出。法案的修改被广泛认为是对加拿大环境影响评价程序的削弱（如 Green Budget Coaliation，2010），修改内容包括 CEAA 授权国家能源局和加拿大核安全委员会受理重大能源项目，许多基础设施项目不再需要进行环境影响评价，CEAA 被赋予更强大的协调功能，环境部长被赋予更高的筛选 EIA 的权力（对显著影响或部分项目可能被错误筛选存在担心）。始于 2010 年秋季的对加拿大《环境评价法》的议会检讨，很可能会引起进一步的调整。

10.8 澳大利亚及西澳大利亚

像加拿大一样，澳大利亚也有一个由强大的各州组成的联邦（共同财富）体系。其环境政策包括环境影响评价政策，有一些有趣的特征，但总的来说，没有加拿大那样强有力。联邦环境影响评价体系最早在 1974 年依据（拟建项目影响的）《环境保护法案》建立，仅适用于联邦行为。在法案有效期间内（1974～2000 年）大约评价了 4000 个提案，但平均每年不到 10 个项目开展了正式的评价（Wood，2003）。同样，各州依据它们自己的法规或程序来确定自身开发活动需要环境影响评价的范围。而且许多州的环境评价体系要比国家体系更强大、更有效。

随着时间的推移，人们逐渐认识到各州环境影响评价程序及实施之间的差异，于是试图在澳大利亚各州之间进行协调和统一（澳大利亚和新西兰环境与保护委员会——ANZECC，1991，1996，1997；也可见 Harvey，1998；Thomas，1998）。此外，在 1994 年对联邦环境影响评价程序进行了重要评审，并出台了一系列有用的报告，具体涉及了累积影响和战略评价、社会影响评价、公众参与、公众质询过程、澳大利亚环境影响评价实践及国外相应的环境影响评价实践对比（CEPA，1994）。除了这些问题以外，评审还突出了改革联邦环境影响评价的必要——包括更好地考虑累积影响、社会及健康影响、战略环境评价、公众参与和监测等。

在政府改革和联邦/国家在环境保护中的作用被进一步评审之后，澳大利亚撤销了联邦环境影响评价法规和许多其他环境法令，并于 1999 年颁布了《环境与生物多样性保护法案》（EPBCA）。与最初的环境影响评价法案相比，EPBCA 提供了更多的程序细节以及解释程序的一系列文件（澳大利亚环境，2000），要求对世界遗产、Ramsar 湿地、受威胁和迁移的物种、联邦海洋环境以及核活动等国家环境重要性事件进行环境影响评价。法案促进了生态持续发展，也有助于战略影响评价（IEMA，2002）。

Padgett 和 Kriwoken（2001）、Scanlon 和 Dyson（2001）、Marsden 和 Dovers（2002）评论了法案及澳大利亚环境影响评价和战略环境影响评价的发展。在 EPBCA 颁布的第一年，联邦环境影响评价活动并未得到很快增长，2009～2010 年有了实质性的增长，每年大概有 420 个提案。这些提案集中在主要资源开发的昆士兰州和西澳大利亚，还有维多利亚州，尤其是采矿和矿产资源勘查以及住宅开发（澳大利亚政府环境部，2010）。澳大利亚政府对 EPBCA 实施的第一个 10 年做了回顾（澳大利亚政府，环境部，2009），指出了很多正面的特征：对有着全国环境重要性事项的明确说明；环境部长担任决策者的角色；公众参与规定；对社会经济事项的明确考虑；法定咨询机制以及一个强大的合规和执行制度。重要的改进建议包括：将法案重新命名为《澳大利亚环境法案》；成立一个独立的环境委员会来为政府提供项目审批和战略批评家等的咨询，为环境绩效审计提供支持；建立一个环境修复基金以及增加决策透明度等。

西澳大利亚（WA）的环境影响评价如下。

西澳大利亚（WA）环境影响评价体系提供了一个州体系的有趣例子，该体系包括了许多创新的特征。西澳大利亚体系的成功主要归功于环保局（EPA）的作用（Wood 和 Bailey，1994）。环保局（EPA）是由西澳大利亚国会出于保护国家环境的远大目标而建立的，它也是独立的环境顾问，向西澳大利亚政府建议项目是否可接受，具有政治指导的独立性。环保局决定评价的形式、内容、时间和程序，并能征集所有的相关信息，为环境部长提供的建议必须予以公布。事实上，环境评价优先于所有的其他法规，环境决策在新提案的批准中起支配作用。其他的批复必须在基于环境影响评价的环境批文之后。

任何决策机构、建议者、环境部长、环保局或是任何公众成员都可将建议提交到环保局，并由环保局决定评价的等级。直到 2010 年年底，共有 5 个评价等级，最为全面的是公众环境评审（PER）和环境评审与管理计划（ERMP）。在新的程序之下（WAEPA，2010），这些等级被减少到了 2 个：无公众评审的建议者信息评价；为期 4～12 周的公众环境评审。实际上，这并未在很大程度上改变环保局的 EIA 程序，因为之前的 5 个评价等级广义上本来就可以被分为无公众评审的评价或是要求公众评审的评价。表 10.3（a）给出了决定评价等级的标准。

范围确定和 PER 内容等方面可参考评价导则，见表 10.3（b）和（c）。建议者制定环境评审文件并开展公众评审。PER 或 ERMP 确定范围的导则包含一些有趣的特征，尤其是与同行评审和公众咨询有关的部分。PER 评价特别关注区域位置，寻找最可能的“致命缺陷”。Waldeck 等（2003）发现这样的环境影响评价导则增加了咨询者的参与，可以有效地增强环境影响评价程序的结果——包括增加环境影响评价程序结果的确定性；更好地对拟建项目进行设计；从初期阶段就满足其环境目标。

表 10.3　西澳大利亚环境影响评价体系的一些特征

(a)决定评价等级的标准
提案若要以 API 等级来评价，必须满足以下所有标准：
• 提案引起了有限的易被管理的重要环境因素，并且已建立条件设定框架；
• 提案与已有的环境政策框架、指南和标准一致；
• 建议者能证明其已开展适宜和有效的利益相关者参与；
• 提案中只涉及有限的或本地的兴趣。
如果根据推荐信息，EPA 否定提案的环境可行性，EPA 主席将鼓励建议者撤回推荐或提交新的显著修改过的提案。附带的行政程序中的“B”类列出了决定一个提案是否不可行的标准（如：与环境政策框架不一致，可能产生显著影响，提案不能被很容易修改等）。
如果提案满足以下任何标准，EPA 将开展 PER 等级的评价：
• 提案具有区域或西澳大利亚州级重要性；
• 提案具有几个显著的环境问题或因素，有些被认为很复杂或具有战略本质；
• 要求对提案进行大量和具体的评价来决定环境问题是否能被管理，如果是，如何管理；
• 公众对提案的感兴趣水平保证了需要一个公开审查小组。
(b)确定 PER 的环境评价范围的正式步骤
一份确定环境评价范围的文件(ESD)被设计用来指导建议者需要关注的关键问题，并识别影响预测和环境信息，应包括：
• 拟议项目及周边环境的简要描述；
• 识别关键环境因素及其他与拟议项目相关的环境因子；
• 识别与每个环境因子有关的现有政策；
• 初步识别潜在环境影响；
• 工程范围，列出提议的环境研究，识别或预测拟议项目的直接和间接环境影响，包括完成时间表（研究和调查应该与已识别环境影响和因子明确关联）；
• 要求识别环境管理方案；
• 要求识别空间数据集、产品信息及数据库；
• 如必要，拟一份提供评价范围、方法、勘查和调查的发现与/或结论的同行评审的人员清单；
• 利益相关者咨询要求。
(c)PER 评价材料的形式及内容导则
建议者应该保证 PER 文件重点描述关键的环境问题/因素，应该包括以下方面：
• 描述提案及考虑的替代方案，包括使环境影响最小的替代选址；
• 受体环境及其保护价值，关键生态系统程序的描述以及在区域环境内其重要性的讨论（应集中在可能影响或被提案影响的环境成分上）；
• 识别关键问题（并列出与这些问题有关的环境因素）及其潜在的“致命缺陷”；
• 在本地和区域背景之下讨论和分析提案的直接与间接影响，包括累积影响；
• 开展勘查和调查的结果（并在附录提供技术报告）；
• 识别减缓显著不利影响的建议措施；
• 在减缓程序里其他所有步骤都耗尽之后，如适宜，识别任何弥补措施；
• 环境管理方案；
• 证明环境影响评价程序中其他预期工作都已被实施；
• 利益相关者和政府机构咨询的细节、收到的评论如何被回复及提案的任何后续修改

资料来源：WA EPA，2010。

然后，环保局基于评审文件、公众建议、倡导者反馈、专家意见及其自身的调查来评估提案的环境可行性。最后，EPA 就提案的环境及其他方面的可行性向部长汇报，并提供用于部长最终批准的参考依据。图 10.5 提供了 PER 评价的完整的程序框架。用于为部长决策提供相关环境信息的 EPA 评审核心在 WA 体系中尤显重要，它有其自身的优势，而且该体系强调环境影响评价结果对决策的影响。WA 体系拥有高水平的公众参与，尤其是在有争议的环境影响评价中。环保局的核心职能也确保了体系的一致性。

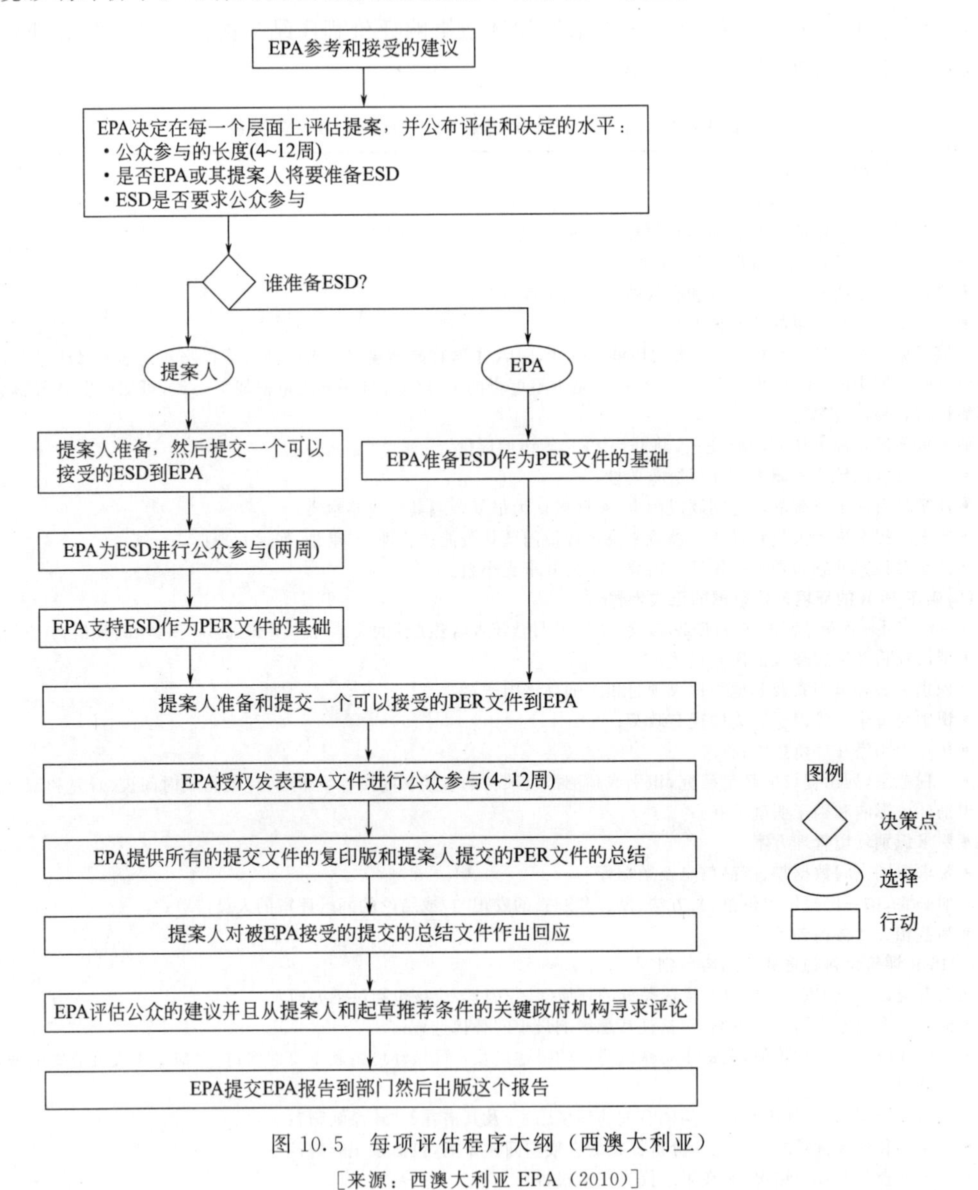

图 10.5　每项评估程序大纲（西澳大利亚）

［来源：西澳大利亚 EPA (2010)］

然而，环境影响评价和规划程序的脱节是西澳大利亚程序较薄弱的一点。为了更好地结合，20 世纪 90 年代中期对其进行了修改，改进了土地利用规划的环境影响评价，修改也反映了控制重心从环保局转至规划部门。这是一个有效体系面临挑战的表现。但有趣的是，现在又规定由 EPA 来评价战略提案，包括政策、规划、计划和开发活动。也规定在已评价的战略提案（已同意实施，环境问题都已充分考虑）中鉴别出的衍生提案，除了检查是否任何与提案相关的实施情况需要改变外，无须再通过 EPA 进一步评价。关于社会-经济影响，在 1993 年，西澳大利亚失去了其首创的社会影响单元，该单元曾经就社会影响提供专家意见，与此同时存在一个强大的发展游说势力，在这个高度依赖能源和矿产项目的州，来进一步“弱化绿色法律”。但是，对于一些最近在州北部进行的高调和大型的能源与矿产开发活动，建议者和他们的顾问已开始采取在环境-社会-经济方面更为创新的方法来开展环境影响评价。

10.9 国际机构

很多主要的国际基金组织和其他国际组织都已建立了环境影响评价程序。其中一些程序已经随着时间而演变，一些手册或导则现在已发行第二版或第三版或已更新多次。

在贷款机构中，欧洲重建和发展银行的环境影响评价要求很典型。银行 1996 年的环境程序在 2003 年更新，并于最近被拓宽至环境和社会程序（EBRD，2010）。程序的目的是确保潜在的银行财政投资和技术合作项目的环境与社会影响能在银行计划和决策过程初期就被识别和评价，确保环境与社会方面需要考虑的事项——包括潜在效益与项目的准备、批复和实施相结合。

EBRD 的评价系统作为一个尽职调查的过程被很好地融入了其项目开发过程。主要的评价步骤包括：

① 初步信息收集；

② 项目分类，决定所需环境影响评价等级；

③ 影响评价以告知详细调查时考虑的内容；

④ 信息发布；

⑤ 环境风险和机会描述，风险管理。

接下来是董事会批准、最终谈判和签署、实施和监测（EBRD，2010）。

初步信息包括项目选址、项目所在地过去与现在的土地使用情况、拟议的建设活动、重新安置或经济迁移是否有可能发生、当地人口（包括弱势群体）的特征、是否有重大环境和/或社会关注问题、主要的项目利益相关者是哪些以及客户的环境和社会名声。这些信息被用于银行和项目发起人之间的筛选讨论来将拟议项目归于以下几个类别之一：

① A 类：项目可能导致重大不利影响，而且在筛选过程中不易被识别或评价。此类项目需要完整的环境影响评价。一些类型的项目自动归为 A 类；其他的则在逐案分析的基础上划为 A 类。

② B 类：与 A 类项目相比，B 类项目可能导致的不利影响要轻一些，需要的环境分析也不是很严格，环境分析的范围在逐案分析的基础上决定。

③ C 类：C 类项目可能导致的不利影响最小或无不利影响，此类项目不需要环境分析。

④ FI 类：项目将通过一个金融中介开发，需要以下讨论的变通的环境影响评价程序。

之后，EBRD 的环境和可持续部门的官员建立环境和社会评价的职权范围，包括利益相关者的参与以及如何评价合规性。客户在 EBRD 官员的支持下开展这些研究。管理和减缓拟议项目影响的措施通常在环境和社会行动规划中列出。公众信息披露和咨询要求贯穿 A

类和B类项目的规划和评价/分析过程。EBRD为每个项目建立环境标准。它也可以指定一个由发起者执行的环境行动计划或监测，作为投资的一个条件。在最终决定是否为该项目提供贷款之前，银行职员需要评审环境信息的有效性，保证拟议的减缓措施得到项目发起人的认可，强调环境改进的机会，确定环境监测要求，识别应当采取的技术/环境合作主动性以及项目是否符合银行环境政策和程序。

亚洲发展银行的《环境评价导则》（ADB，2003）有着相似的筛选和评价要求。导则就项目与计划、不同部门、公平投资等方面的评价给出了详细的信息，它们也强调公众咨询。

世界银行将环境影响评价看作是环境与社会安全的关键政策之一。其1991年的政策/银行程序4.01（最近在2007年更新）要求对相关的贷款业务开展环境影响评价。其环境评价资料读物（世界银行，1991）及各种更新版本都介绍了环境影响评价程序：银行职员对项目的环境影响评价筛选与分类；倡导者准备环境评价报告；银行职员对该报告进行评审；评估任务，即银行职员与倡导者进行磋商并解决环境问题；结论文件；在项目实施期间进行监督和评价（世界银行，1999）。1989～1995年，银行筛选了上千个有潜在环境影响的项目：其中10%属于A类（主要是能源、农业及交通项目），41%属于B类，49%属于C类（世界银行，1997）。A类项目被认为有“敏感的、不可逆的和多种多样的有害影响”（世界银行，1999），它们需要完整的环境影响评价。对于B类项目，其影响“没有那么显著——没那么敏感、多、重要或复杂多样”，不需要完整的环境影响评价，但某些环境分析是必须的。C类项目对实际环境有可忽略的或最小的直接影响，既不需要环境影响评价也不需要环境分析。典型的C类项目主要为教育、计划生育、健康和人力资源开发。

世界银行程序的显著特征是包含全面的环境定义，包括自然、生物和社会-经济各个方面，公众咨询的高透明度以及对项目实施的高度关注。世界银行的一份报道（世界银行，1995）确定了未来面临的5个主要挑战：推动环境影响评价往“上游”发展（进入项目设计阶段，针对部门与地区层面）；更有效的公众咨询；环境影响评价与项目工作计划更好地协调（包括减缓措施、监测和管理计划）；从项目实施中获取经验（“反馈回路”）；使私人部门参与（特别是金融家和项目发起人）以确保项目服从质量合格的环境影响评价。Mercier（2003）强调当前EIA的重点应放在减缓措施、预防与补偿措施的执行上。另外，由于当前许多客户国都有自己的环境影响评价要求、评价队伍和评价机制，世界银行越来越多地参与到项目准备阶段的前期能力提升当中。

非洲发展银行（ADB，2003）和欧洲投资银行（EIB，2007）的环境影响评价导则没有那么全面，但都要求实施环境影响评价，推进环境影响评价过程的公众参与，并在决定是否给某一项目提供贷款时考虑这些内容。非洲发展银行的导则针对的是环境和社会综合影响。

其他的组织也颁布了环境影响评价导则。例如，UNEP非常实用的环境影响评价培训资料手册，现在已出到第二版（UNEP，2002），包括了案例研究（主要来自发展中国家）、透明度及环境影响评价不同阶段的细则。英国国际发展部和丹麦外交事务部也制定了一个类似的环境影响评价导则（DFID，2003；Danida，2009）。其中，赤道原则的重要性越来越突出，该原则是基于世界银行导则和国际金融公司（IFC）关于社会和环境可持续性的执行标准。赤道原则金融机构目前已包括70多家主要国内和国际银行，致力于不提供贷款给某些

（超过1000万美元限值的）项目，因为这些项目的借款人不会或不能遵守国内或国际环境影响评价标准，无论哪种标准最高（IFC，2006）。

10.10 总结

2002年，Gibson认为世界范围的环境影响评价正向以下方面发展：

① 介入规划的时期更早（以目标和广泛的替代方案开始）；

② 更开放和更好的公众参与（不仅仅是倡导者、政府官员和技术专家）；

③ 更全面（不仅是生态环境、本地影响、资本项目、单一任务）；

④ 更具效力（以政策为基础逐渐转向以法律为基础）；

⑤ 更密切的监测（由法院、知情民间社会团体和政府审计员执行）；

⑥ 应用更广泛（通过不同级别的法律、土地使用规划、自愿合作精神等）；

⑦ 更一体化（考虑系统的影响，而不仅仅是个体的影响）；

⑧ 更具挑战性（整体的可持续性，而不仅是保证个体“可接受性”）；

⑨ 更谦逊（识别和关注不确定性、应用预防措施）。

差不多10年之后，本章内容显示10年前的趋势仍在延续。但在世界范围内，环境影响评价仍然受制于政治意愿的缺乏、实施建议减缓措施的经费的不足以及机构能力的缺乏，如Goodland和Mercier（1999）之前所提到的一样。加拿大、澳大利亚、荷兰（GHK，2010）及其他地方出现的新趋势表明，即使在发达国家，环境影响评价也趋于被调整以适应其他需求——有些可以说是因为与经济发展相冲突而被冲淡了。

第11章和第12章吸收了这里讨论的及其他方面的一些观点来识别未来的可能性，主要集中于英国体系，但也考虑了欧盟和全球背景。

问　题

以下问题旨在帮助读者关注本章的重点问题。

1. 图10.1显示发展更成熟EIA体系中开展的EIA数量要大于发展不太成熟的EIA体系中开展的EIA数量。这反过来是否意味着我们可以从一个国家EIA开展数量来判断这个国家的EIA体系发展有多成熟呢？请说明你的理由。

2. 将本章中描述的两个或三个EIA体系与第3章介绍的英国体系或2.2节介绍的美国体系相比较，比较从以下方面进行：

① 哪些项目要求进行EIA，筛选如何开展；

② 环境影响报告书必须包含哪些内容；

③ 谁来开展EIA；

④ 公众参与；

⑤ 决策，包括与其他执照和许可的联系。

3. 图10.2显示更发达的国家的EIA体系一般比不太发达国家的EIA体系更为强大。但贝宁、秘鲁、中国和波兰的EIA体系每一个都有比英国或美国体系更强或更创新的方面。

这些方面具体有哪些？为什么会形成？

4.本章所讨论的国家都有广泛不同的方法来考虑EIA中的替代方案。哪一个国家的方法最强？哪一个最弱？

5.秘鲁和中国只允许注册环境影响评价专家开展EIA。这种做法有什么优势和劣势？

6.秘鲁的体系可以被批判为存在类似“偷猎者与狩猎人”的方法，即负责推进某些项目的部门也同样负责管理这些项目的EIA程序。为什么“偷猎者与狩猎人”一次被用于此种情况？本章中有没有“偷猎者与狩猎人”的其他例子？

7.西澳大利亚EIA体系被认为有着很多创新的特点，是一个良好体系的表现。这些特点都有哪些？是否也存在弱点？

8.几家国际基金组织要求综合环境和社会评价。但个别国际的法规关注的是环境评价。这种差异的原因何在？

9.世界银行和其他基金组织对EIA在很多发展中国家的应用起了关键作用。这些组织对EIA感兴趣的原因可能有哪些？

10 Comparative practice

10.1 Introduction

Most countries in the world have EIA regulations and have had projects subject to EIA. However the regulations vary widely, as do the details of how they are implemented in practice. This is due to a range of political, economic and social factors.

Environmental impact assessment is also evolving rapidly worldwide. When the second edition of this book was being written in 1999, for instance, many African countries and countries in transition had only recently enacted EIA regulations; by now, some of these countries have had considerable experience with EIA, and are developing more detailed guidelines and regulations. Just since the last edition of 2005, the EIA systems of three of the seven countries discussed in that edition—Peru, Poland and Canada—have undergone major changes, and half of the international organizations discussed at Section 10.9 have updated their EIA guidance.

This chapter aims to illustrate the range of existing EIA systems and act as comparisons with the UK and EC systems discussed earlier. It starts with an overview of EIA practice in the various continents of the world, and some of the factors that influence the development of EIA worldwide. It then discusses the EIA systems of countries in six different continents, focusing in each case on specific aspects of the system:

- Benin has one of the most advanced EIA systems in Africa, with good transparency, considerable public participation, integration of environmental concerns with national planning and robust administrative and institutional tools.

• Peru's EIA system is typical of many South American countries in its sector-specific orientation, relative lack of public participation and transparency, and late timing in project planning.

• Poland resembles several other countries in transition in that its EIA system has changed dramatically since the early post-communist days. In 2000, Poland enacted radical new regulations, which brings its EIA system in line with EU requirements.

• China's EIA system is discussed because of the worldwide effect that any Chinese environmental policy is likely to have in the future. China's environmental policies are restricted by the need to harmonize them with plans for economic development.

• Canada is known for its progressive environmental policies. Its federal EIA system has good procedures for mediation and public participation. Significant changes are currently being made to this system.

• Australia's EIA system, like Canada's, is split between the federal and state governments. The state of Western Australia provides a particularly interesting example of a good state system with many innovative features.

The chapter concludes with a discussion of the role of international institutions in developing and spreading good EIA practice for the projects and programmes they fund and support.

10.2 EIA status worldwide

Table 10.1 and Figure 10.2 show, to the authors' best knowledge, the status of EIA regulations worldwide in mid-2011. Box 10.1 gives an initial list of sources of information about EIA worldwide. More than 140 countries have some form of EIA regulation.

Table 10.1 Existing EIA systems worldwide (with date of original implementation)

Country	Implementation	Country	Implementation	Country	Implementation
Western Europe		Portugal	1987	Czech Rep.	1991,2001
Austria	1993,2000	Spain	1986,2008	Estonia	1993,2005
Belgium	1985—92	Sweden	1987—91	Georgia	2002
Denmark	1989,1999	Switzerland	1985	Hungary	2005
Finland	1994,2004	United Kingdom	1988	Kazakhstan	1997
France	1976	**Central and Eastern Europe**		Kosovo	2009
Germany	1990,2005—06	Albania	2004	Kyrgyzstan	1997
Greece	1986,2002	Armenia	1995	Latvia	2004
Iceland	2000,2005	Azerbaijan	1996	Lithuania	1996—2005
Ireland	1989-2000	Belarus	1992	Macedonia	2005
Italy	1986-96	Bosnia and Herzeg	2003	Moldova	1996
Luxembourg	1994	Bulgaria	2002,2009	Montenegro	2005
Netherlands	1987,2002	Croatia	2006	Poland	1990—2008
Norway	1990				

Continued

Country	Implementation
Central and Eastern Europe	
Romania	1995
Russian Fed.	2000(partial)
Serbia	2004
Slovakia	1994
Slovenia	1996
Tajikistan	2006
Turkey	1983,1997
Turkmenistan	2001
Ukraine	1995
Uzbekistan	2000
Africa and Middle East	
Algeria	1983,1990
Angola	2004
Bahrain	?
Benin	2001
Botswana	2005
Burkina Faso	1997
Cameroon	2005
Comoros	1994
Cote d'Ivoire	1996
Dem. Rep. of Congo	2002(partial)
Egypt	1995
Ethiopia	2002
Gabon	1979
Gambia	2005?
Ghana	1999
Guinea	1987
Iran	1994—2005
Iraq	1997(partial)
Israel	1982,2003
Jordan	1995
Kenya	2002—03
Kuwait	1990,1995—96
Lebanon	in development
Lesotho	2003
Liberia	2003
Madagascar	1997,2004
Malawi	1997(guidance)

Country	Implementation
Mali	2008
Mauritania	2004(partial)
Mauritius	2002
Morocco	2003
Mozambique	1998
Namibia	2011
Niger	1998
Nigeria	1992
Oman	1982,2001
Palestinian Auth.	2000
Rwanda	2005
Qatar	2002
Senegal	2001
Seychelles	1994
Sierra Leone	2008
South Africa	1997
Swaziland	2002
Syria	2002,2008
Tanzania	2004—05
Togo	1988
Tunisia	1988,1991
Uganda	1998
United Arab Em.	1993,2009
Yemen	1995
Zambia	1997
Zimbabwe	1997(partial)
Americas	
Argentina	1994—96
Belize	1995
Bolivia	1995
Brazil	1986,1997
Canada	1992
Chile	1997
Colombia	1997—2005
Costa Rica	1998(partial)
Cuba	1999
Dominican Rep.	2002
Ecuador	1999

Country	Implementation
El Salvador	1998
Guatemala	2003
Guyana	1996
Honduras	1993
Mexico	1988
Nicaragua	1994
Panama	2000
Paraguay	1994
Peru	2009
Uruguay	1994
USA	1969
Venezuela	1976
Asia	
Afghanistan	2007
Bangladesh	1995
Bhutan	2000
Cambodia	1999(partial)
China	2002
India	1994
Indonesia	1999
Japan	1997
Korea	1977—2000
Lao PDR	2000
Malaysia	1987
Mongolia	1998
Nepal	1997
Pakistan	1997,2000
Papua New Guinea	1978(partial)
Philippines	2003
Sri Lanka	1988
Thailand	1992
Vietnam	1994
Australia et al.	
Antarctica	1991
Australia	1999
New Zealand	1991

Note: This list represents, to the best of our knowledge, the current status of original EIA legislation worldwide, but we cannot confirm its accuracy. Many EIA regulations have been updated or fine-tuned since the last edition of this book, and in other cases specific EIA regulations have supplemented more general framework regulations. This accounts for many of the changes in this list since 2005.

Sources: Badr 2009; CISDL 2009; Economic Commission for Africa 2005; World Bank 2006; and many other Internet sites and journal articles.

Box 10.1 Some references of EIA systems worldwide

Although some of these references are country-specific, they discuss wider issues related to the region in question.

General: Lee and George (2000), Wood (2003); websites of affiliates and branches of the International Association for Impact Assessment (www.iaia.org/affiliates-branches), EIA Wiki (eia.unu.edu/wiki/index.php/EIA_Systems) publications by the OECD, UNEP and World Bank; and articles in e.g. *Environmental Impact Assessment Review*, *Impact Assessment and Project Appraisal* and the *Journal of Environmental Assessment Policy and Management*.

Western Europe: EC (2009), GHK (2010), Wood (2003).

Central and Eastern Europe: Kovalev *et al*. (2009), Rzeszot (1999), Unalan and Cowell (2009), World Bank (2002).

Australia, *New Zealand*: Elliott and Thomas (2009), Wood (2003).

Asia: Briffett (1999), Vidyaratne (2006), World Bank (2006).

North America: Wood (2003).

South America: Brito and Verocai (1999), Chico (1995), CISDL (2006), Kirchhoff (2006).

Africa and Middle East: Almagi *et al*. (2007), Appiah-Opoku (2005), Bekhechi and Mercier (2002), Economic Commission for Africa (2005), El-Fadl and El-Fadel (2004), Kakonge (1999), Marara *et al*. (2011), Southern African Institute for Environmental Assessment (2003).

However, EIA practice is not even across different countries worldwide. Figure 10.1 summarizes the evolution of EIA in a typical country: it begins with an initial limited number of EIAs carried out on an ad hoc basis in response to public concerns, donor requirements or industries based in a country with EIA requirements carrying out EIAs of their activities in the country without EIA requirements. Over time, the country institutes EIA guidelines or regulations. These may prompt a rapid surge in the number of EIAs carried out in that country, as was the case in most EU countries. However, the regulations may apply to only a limited number of projects, or may be widely ignored, leading to only a small increase. Over time, the regulations may be fine-tuned or added to, the number of EIAs carried out annually levels off or may even shrink, and the EIA system is effectively 'mature'. The current status of EIA in different countries can be roughly charted on this continuum. Figure 10.2 broadly shows this status by continent, though individual countries may vary from this.

The situation in *Africa* is changing rapidly, with many countries having recently instituted EIA-specific regulations to complement earlier framework regulations. This development has been brought about by a range of initiatives including the 1995 African Ministerial Conference on Environment that committed African environment ministers to formalize the

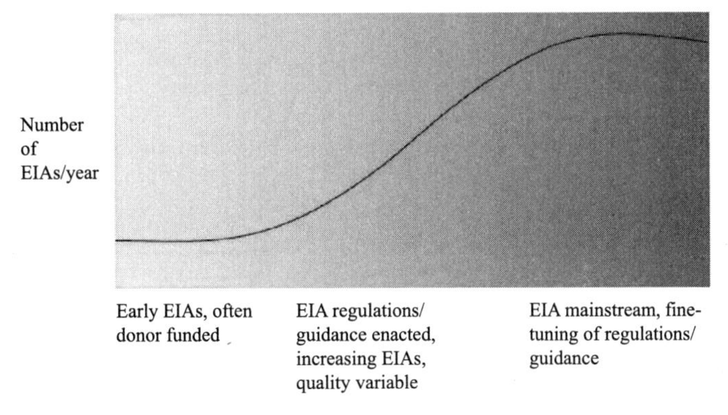

Figure 10. 1 Evolution of EIA systems

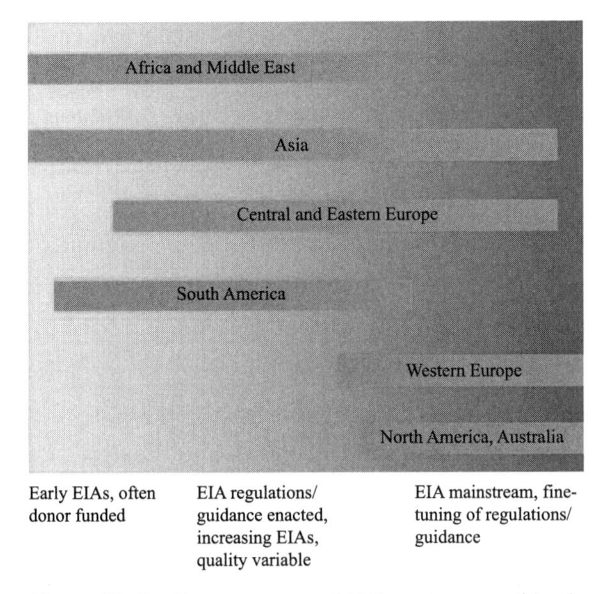

Figure 10. 2 Current status of EIA systems worldwide

use of EIA; and the establishment of several organizations that aimed to improve EIA capacity and collaboration between African countries (e. g. Capacity Development and Linkages for Environmental Impact Assessment in Africa, Partnership for Environmental Assessment in Africa, and several African branches of the International Association for Impact Assessment). On the other hand, EIA in Africa is still beset by a lack of trained personnel, cost, concern that EIA might hold back economic development and lack of political will (e. g. Appiah-Opoku 2005; Kakonge 1999). Even where EIA legislation exists, this does not mean that it is put into practice, is carried out well, involves the public or is enforced (Okaru and Barannik 1996). As such, while some African countries such as Benin, the Seychelles and Tunisia have good regulations and considerable EIA practice, others such as Burundi and Guinea-Bissau remain on the far left of Figure 10. 2.

Almost all countries in *South and Central America* have some form of legal system for environmental protection, including at least aspects of EIA. These systems vary widely, reflecting the countries' diverse political and economic systems: average income in some

South American countries is more than ten times that in others. In general, the development of EIA in South America has been hampered by political instability, inefficient bureaucracy, economic stagnation and external debt (Brito and Verocai 1999). This may be changing recently—our example of Peru is an example of this—but EIA in South America is often still carried out centrally, with little or no public participation, and often after a project has already been authorized (Glasson and Salvador 2000).

Environmental impact assessment in *Asia* also varies widely, from very limited legislation (e. g. Cambodia) through to extensive experience with robust EIA regulation set within the context of SEA (e. g. Hong Kong of China). EIA regulations were established in many Asian countries in the late 1980s, and EIA is practised in all countries of the region through the requirements of donor institutions. On the other hand, Briffett (1999) suggests that many Asian EIAs are of poor quality, with poor scoping and impact prediction, and limited public participation. This is due in part to the perception that EIA may retard economic growth—symbolized by the wish, in some countries, to expose large buildings and infrastructure projects to show off the country's wealth (Briffett 1999). Many countries, like our case study of China, are in the process of revising their early EIA systems in an attempt to deal with these problems.

Since the late 1980s, the *Central and Eastern European* countries ('countries in transition') have been going through the enormous change from centrally planned to market systems. This has included, in many cases, a move from publicly to privately owned enterprises. Many of these countries' economies were based on heavy industry, with concomitant high use of energy and resources, and many went through an economic crisis in the 1990s. Of the countries in transition, the Central and Eastern European countries have achieved (e. g. Poland) or are aiming towards (e. g. Turkey) EU accession, and are harmonizing their EIA legislation with the European EIA Directives. The Newly Independent States of the former Soviet Union all had similar systems, based on the 'state ecological expertise/review system' developed under the former Soviet Union. The countries of southeast Europe-Albania, Bosnia Herzegovina, Croatia, Yugoslavia and Macedonia—had relatively undeveloped EIA systems. The move from these systems to one aligned with the EIA Directives accounts for many of the recent EIA regulations in that region.

Box 10. 2 summarizes some of the factors that affect the application of EIA in developing countries.

The EIA systems of *Western European* countries have already been amended at least twice, accounting for a multiplicity of EIA regulations in each Member State: Directive 85/337 was amended by Directives 97/11 and 2003/35/EC, and some countries already had EIA systems in place before the original Directive of 1985. A recent review of the status of EIA in EU Member States (GHK 2010) showed that EIAs are being carried out in all 27 European Member States, including the 12 'new' Member States, with a total of about 16,000 EIAs being carried out each year by the EU-27; and there has been a general increase in the number of EIAs carried out each year. However, even after three rounds of harmonization, EIA practice still varies widely across Europe.

Box 10.2 Factors affecting the implementation of EIA in developing countries

- In countries in or near tropical areas, environmental models, data requirements and standards from temperate regions may not apply.
- Socio-cultural conditions, traditions, hierarchies and social networks may be very different.
- The technologies used may be of a different scale, vintage and standard of maintenance, bringing greater risks of accidents and higher waste coefficients.
- Perceptions of the significance of various impacts may differ significantly.
- The institutional structures within which EIA is carried out may be weak and disjointed, and there may be problems of understaffing, insufficient training and know-how, low status and a poor co-ordination between agencies.
- EIA may take place late in the planning process and may thus have limited influence on project planning, or it may be used to justify a project.
- Development and aid agencies may finance many projects, and their EIA requirements may exert considerable influence.
- EIA reports may be confidential, and few people may be aware of their existence.
- Public participation may be weak, perhaps as a result of the government's (past) authoritarian character, and the public's role in EIA may be poorly defined.
- Decision-making may be even less open and transparent, and the involvement of funding agencies may make it quite complex.
- EIAs may be poorly integrated with the development plan.
- Implementation and regulatory compliance may be poor, and environmental monitoring limited or non-existent.

Environmental impact assessment procedures in *North America*, *Australia* and *New Zealand* are still among the strongest in the world, with good provisions for public participation, consideration of alternatives and consideration of cumulative impacts. Several have separate procedures for federal and state/provincial projects. However, several of these systems are in the process of being streamlined—some would say weakened—in response to the economic recession and perceived problems of over-complexity and ponderousness.

The following sections discuss the EIA systems of different countries worldwide as examples of the concepts discussed above.

10.3 Benin

Benin, in West Africa, was once covered by dense tropical rainforest behind a coastal strip. This has largely been cleared and replaced by palm trees. Grasslands predominate in the drier north. Benin's main exports are cotton, crude oil, palm products and cocoa. Even compared to other nearby countries, Benin has a relatively low GDP per capita. However, it has had a fully functioning EIA system since 1995, and was identified as one of three francophone African countries that is most advanced in EIA terms (d'Almeida 2001). Its EIA system is characterized by transparency, public participation, integration of environmental con-cerns with national planning and robust administrative and institutional tools (Baglo 2003).

Like many other African countries, Benin's early EIAs were carried out at the behest of funding institutions such as the World Bank and African Development Bank. Benin's constitution of December 1990 placed particular emphasis on environmental protection, and in 1993 the Environmental Action Plan was adopted. In 1995 the Benin Environmental Agency was created and made responsible for, *inter alia*, implementing national environmental policy, conducting and evaluating impact studies, preparing State of the Environment reports and monitoring compliance with environmental regulations. Although the Agency reports to the Ministry of the Environment, Habitat and Town Planning, it has corporate status and financial independence. The Agency subsequently developed regulations on EIA in 2001, as well as a range of EIA guides (e. g. on projects that require EIA, EIA for gas pipelines and irrigation projects; Baglo 2003). Benin also has a national association of EIA professionals.

Benin's EIA process is led by the Benin Environmental Agency. Although this has only two officers responsible for EIA, it can also draw on a forum of experts from public and private institutions to prepare or review specific EIAs. This approach allows the Agency to operate with minimal staff, ensures ongoing cooperation with other institutions and puts into practice the principle of broad participation in decision-making (Baglo 2003).

Figure 10. 3 summarizes Benin's EIA process, which consists of:

- screening: this work is decentralized, with responsibility allocated to the ministries responsible for the proposed project's sector;
- project registration, scoping, and development of terms of reference for the EIA;
- preparation of a draft EIS, and 'publication' of the report primarily through the radio and non-government organizations (NGOs);
- if necessary, a public hearing to discuss the EIS, arranged by the Minister of Environment through the creation of ad hoc Public Hearing Commissions;
- review of the EIS by the Agency: where the EIS is adequate, the Agency issues a certificate of environmental conformity;
- decision about the project by the ministry responsible for the sector concerned, taking into account the notice of environmental compliance, technical feasibility study and economic feasibility study; and
- follow-up assessment and audit.

Between 1997 and 2002, 78 EIAs were reviewed and 61 EISs were validated with certificates of environmental conformity (Yaha 2007). Projects that have been subject to EIA in Benin include livestock development projects, road improvements, drainage improvement in Cotonou following the 2010 floods, expansion of the Ouesse landfill site, urban and indus-

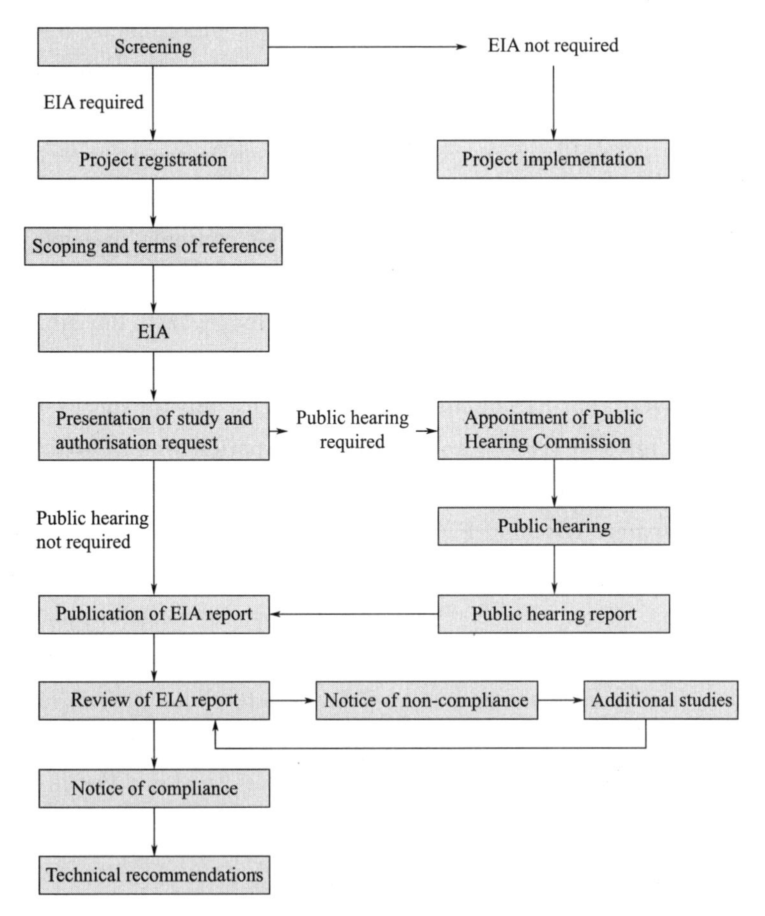

Figure 10.3 Benin's EIA procedures

Source: Based on Sutherland *et al*. 2005

trial development projects, and the West African Gas Pipeline. These have been paid for through a mixture of donor (e. g. World Bank), private and national government funds.

Environmental impact assessment practice in Benin, as in many other African countries, is limited by poor collaboration between some key ministries, a low level of public environmental awareness, illiteracy and poverty. Early problems with lack of indigenous expertise have been reduced through a range of EIA training courses run by the Benin Environmental Agency, the activities of its national association of EIA professionals and its participation in pan-African capacity-building activities (see Section 10.2).

10.4 Peru

Peru, the third largest country in South America, includes a thin dry strip of land along the coast, the fertile sierra of the Andean foothills and uplands, and the Amazon basin. Fishing, agriculture and mining are the main industries. A change in government in

1990 led to the reconstruction of the country after years of economic difficulties, and an extensive privatization programme—including the privatization of many of the state-owned mines—encouraged dramatic increases in foreign investment.

In September 1990, the Peruvian government enacted Decree 613—90, the Code of Environment and Natural Resources. This established EIA as a mandatory requirement for any major development project, but did not specify the EIA contents or legal procedures. The Ministry of Energy and Mining was the first ministry to put this decree into practice through Supreme Decree 016-93-EM, followed in 1994 and 1995 by separate but similar requirements by the ministries for fishing, agriculture, transport and communication, housing and building. Early indications were that EIS quality was quite high, but the discussions of mitigation measures were weak (Iglesias 1996), and public participation was limited.

Law 27, 446 of 2001—reaffirmed in October 2009 through Supreme Decree 019—2009-MINAM—replaced this disparate system of EIA with a unified national system that also applies to policies, plans and programmes. The remainder of this section focuses on this new system.

Annex Ⅱ of Supreme Decree 019—2009-MINAM lists a wide range of projects and plans that require assessment, under headings that represent the relevant competent authority (e. g. agriculture ministry, energy and mining ministry). Any developer whose project is listed in Annex Ⅱ, including modifications to pre-existing projects if these will have significant impacts, must prepare a preliminary evaluation. This includes details about the project, a plan for citizen participation, a brief description of the project's possible environmental impacts and mitigation measures. The competent authority must decide within 30 days whether the project is Category Ⅰ (minor environmental impacts), Ⅱ (moderate impacts), or Ⅲ (significant impacts). An annex in the Supreme Decree lists the criteria that competent authorities must use in making this judgement: they include factors relating to human health, natural resources, designated areas, biodiversity and its components, and people's lifestyles.

If the project is Category I, then the preliminary evaluation becomes a 'declaration of environmental impact' and no further action is required. If, instead, the project is Category Ⅱ or Ⅲ, it requires, respectively, a semi-detailed or detailed EIS. The Supreme Decree details what information these EISs must contain, including an executive summary, project description, baseline data, information on how the public and statutory consultees were involved and any comments made at public meetings, impact assessment including comparison with environmental standards, and environmental management, contingency and closure plans. Only organizations on a register of institutions managed by the national environmental agency are allowed to prepare EISs. All EISs are public documents. The Supreme Decree supports public involvement in the EIA process but does not mandate specific techniques for this.

Once the EIS has been prepared, the developer submits it to the competent authority, who then has 40 (Category Ⅱ) or 70 (Category Ⅲ) days in which to review it and, if appropriate, consult the public on it. If necessary, the developer then has a limited time to comment on the review and provide further evidence, with a further 20 day resolution process. The EIA review process should take no longer than 30 days in total for Category Ⅰ, 90 days for Category Ⅱ and 120 days for Category Ⅲ projects. At the end of the process, the competent authority can either provide or refuse to provide a certificate of EIS. Developers may only apply for other licenses or permits, and may only start project construction if they have received a certificate of EIS. The certificate also sets out specific obligations to prevent or mitigate potential environmental impacts.

Once a project is approved, the developer must carry out programmes of management, control and monitoring throughout the operations to ensure that the environmental management plan is adhered to. The environmental management plan, contingency plan, community relations plan, and/or decommissioning plan (as appropriate) must be updated five years after the project becomes operational, to ensure that they remain relevant and up to date.

Strengths of this new system include:

• The certificate system, which aims to deal with past problems of EISs in Peru being frequently prepared after the project construction has begun (Brito and Verocai 1999).

• The detailed requiremens for Category Ⅱ and Ⅲ EISs, which should help to ensure that the EISs are of good quality and provide the information necessary for the competent authorities to make informed decisions.

• The clearly stated and relatively limited time period for EIS review, which should prevent EISs from slowing down needed development.

• The emphasis on impact management, contingency planning and monitoring, which should help to ensure that project impacts are effectively managed.

• Weaknesses include the fact that the competent authorities charged with promoting certain types of development are also those that provide EIS certificates for those developments, leading to a potential 'poacher-gamekeeper' situation; and the lack of clear, forceful requirements for public participation in the EIA process.

10.5 China

Since 1978, China has been undergoing a rapid shift towards economic growth, decentralization of power, industrialization, private enterprise and urbanization. This has engendered many development projects with significant environmental impacts. China's EIA system has struggled to keep up, and to date has not managed to prevent some serious environmental harm (Moorman and Ge 2007). The system is still in a process of rapid evolution, with the strengthening in 2009 of its applications to plans and programmes.

EIA in China formally began with the enactment of the Provisional Environmental Protection Law of 1979, which was revised and finalized in 1989. Shortly after the law's introduction, several guidelines for its implementation were prepared, of which the central ones were the Management Rules on Environmental Protection of Capital Construction

Projects of 1981. These were revised and formalized in 1986, with details on timing, funding, preparation, review and approval of EISs. Further, stronger ordinances were enacted in 1990 and December 1998, which required EIA for regional development programmes as well as individual projects, and strengthened legal liabilities and punishment for violation of EIA requirements.

At this stage, EIA still acted very much as a top-down administrative instrument. (Wang *et al*. 2003)

Although tens of thousands of EISs or environmental impact forms were being prepared in China every year (Ortolano 1996), some projects that should have been subject to EIA were not, some EIAs were carried out post-construction, and the quality of EISs was variable (Mao and Hills 2002; China 1999).

Major changes to this system were brought about by the enactment of the EIA Law in October 2002. Table 10.2 lists existing EIA legislation and guidelines in China, and Figure 10.4 summarizes these new EIA procedures.

Table 10.2 Chinese EIA regulations, governmental documents and guidelines

Name	Year
Regulations on Environmental Management of Construction Projects(State Council Decree 253)	1998
Circular on Relevant Issues of Executing EIA for Construction Projects(107 SEPA 1999)	1999
Forms for Environmental Impact Form and Environmental Impact Registration Form(draft)(178 SEPA 1999)	1999
Acceptance of Construction Project Environmental Protection Management Regulation(SEPA Decree 13)	2001
Law of the People's Republic of China on EIA(Presidential Decree 77)	2002
Provisions on Examination and Approval Procedure for EIA Documents of Construction Projects by the SEPA (SEPA Decree 29)	2005
Circular of Printing and Distributing 'Provisional Regulation for Public Participation in EIA'(28 SEPA 2006)	2006
Classified Directory for EIA of Construction Projects(MOEP Decree 2)	2008
Announcement on the Catalogue of Construction Projects with EIA Documents Directly Examined and Approved by the MOEP and Catalogue of Construction Projects with EIA Documents Examined and Approved by Provincial Environmental Bureaus as Enstrusted by the MOEP(MOEP Announcement 7)	2009
Provision of Approval of EIA for Construction Projects by Categories(MOEP Decree 5)	2009

Source: Yang, personal communication

The competent authority for projects of national economic or strategic significance is the Ministry of Environmental Protection (MOEP, formerly the State Environmental Protection Agency). These projects are listed in MOEP Decree 5 and guidance of 2009, and include nuclear projects and projects crossing provincial boundaries. For projects of regional importance such as waste incinerators, smelters and chemical plants the competent authority is the provincial Environmental Protection Bureau (EPB); and for other types of project the provincial EPBs can determine whether the competent authority is the provincial, city or county/district EPBs. Usually district or county EPBs only examine EIA forms.

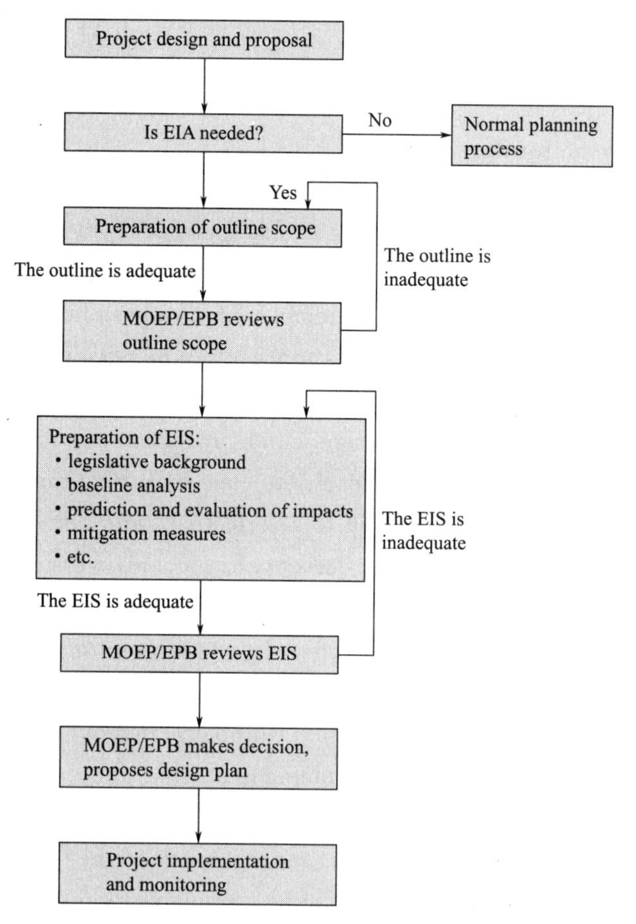

Figure 10.4 China's EIA procedures

Source: Based on Wang *et al*. (2003)

The EIA process begins when a developer asks the competent authority to determine whether or not a proposed action requires full EIA. As in Peru, three different levels of EIA apply in China. Projects with major potential environmental impacts require a full EIS, those with more limited impacts require an environmental impact form, and those with minimal impacts require only an environmental impact registration form with no further data provision. MOEP Decree 2 specifies what kind of EIA is needed for different projects, based on the project's impacts and the sensitivity of the receiving environment.

The competent authority personnel, sometimes assisted by outside experts, conduct a preliminary study and then makes a ruling. If an EIA is needed, the competent authority identifies those factors most likely to affect the environment and prepares a brief. The EIS's preparation is then entrusted to licensed, state-approved experts, who work to the brief. The expert analyses the relevant impacts and proposes mitigation measures. An EIS is then produced, which, according to the guidelines, needs to discuss:

- the general legislative background;
- the proposed project, including materials consumed and produced;
- the baseline environmental conditions and the surrounding area;

- the short-term and long-term environmental impacts of the project;
- impact significance and acceptability;
- a cost-benefit analysis of the environmental impacts;
- proposals for monitoring; and
- conclusions.

The public is to be given an opportunity to comment prior to the completion of a draft EIS, although the form of public participation is not prescribed. The organization that prepares the EIS must formally consider the opinions of relevant stakeholders, experts and the public, and include in the EIS their reasons for accepting or rejecting these opinions.

The EIS or environmental impact form is submitted to the competent authority, which checks the proposal against relevant environmental protection regulations and plans, confirms whether the area has the carrying capacity to cope with the project, considers whether the proposed mitigation measures are likely to be effective, and takes into account the comments of relevant experts before making a decision. For a controversial project, and projects that cross provincial boundaries, the document is submitted to the higher authority for examination and approval. If the project is approved, conditions for environmental protection may be included, such as monitoring and verification procedures. The competent authority must submit a report that states how the project will be carried out and how any required environmental protection measures will be implemented. Once this has been approved by the provincial authorities, a certificate of approval is issued.

The changes post-2002 certainly aimed to redress many of the problems of China's earlier EIA system. Developers are not permitted to begin construction until they have carried out EIA. The MOEP has used the law as the basis for carrying out several 'environmental storms' that have stopped projects whose construction had started before their EIA reports were approved. MOEP also announced in 2007 that no new projects would be approved in several cities with low environmental capacity to handle more pollutants (Yang 2008).

However, there is still often a conflict of interest between development-oriented local governments and the environmental protection agencies that they fund, and between people's priorities for environmental growth for improvement of their living standards. As such, local authorities are often unwilling to antagonize local leaders who strongly favour the proposed project, or impose constraints in the form of EIA (Mao and Hills 2002; Lindhjem *et al*. 2007). China's EIA process also continues to be criticized for its complexity, narrow historic focus on air, water and soil pollution, relatively low requirements for public participation, and lack of consideration of alternatives (Lindhjem *et al*. 2007; Wang *et al*. 2003; Yang 2008).

10.6 Poland

Poland, like other countries in transition, has undergone a rapid evolution in environmental policy, which is reflected in its EIA legislation and system.

In the latter days of communism, the communist regime was providing only the most basic services and environmental issues were being virtually ignored (Fisher 1992). Several areas of Poland were subject to severe pollution, causing widespread concern. In response to this, the government enacted the Environmental Protection Act (EPA) in 1980. This was subsequently strengthened through various amendments, the Land Use Planning Act of 1994, and an executive order of 1990 on developments exceptionally harmful to the environment. Under this system, an EIA began when a developer asked a local environmental authority whether an EIA was needed. If it was, the authority drew up a list of suitable consultants to carry out the work. Once the chosen consultant had completed the EIS, consultation with the public might be carried out but was not mandatory. If the EIS was accepted by the environmental authority, then the local planning authority could issue a 'location indication,' which listed alternative locations for the project. Based on this, the developer chose a site and continued to design the project, and the environmental consultants prepared a final EIS. The developer delivered the EIS along with a planning application to the local planning authority, which again consulted with the local environmental authority before making a decision about the project. Construction consent required yet a third EIS to accompany the technical design of the project.

This system was criticized for lacking screening criteria, being cumbersome and redundant with multiple EISs prepared for each project, having minimal procedures for public participation and for resource constraints on the commission that reviewed the EISs (Jendroska and Sommer 1994). It also could not deal with the huge social and economic changes that Poland went through after the overthrow of the Communist regime in late 1989:

> There are no more economic plans and central planners, the currency is convertible and the best technology accessible, and the whole economy is being privatised. Moreover, administrative arrangements have been redesigned in order to create a strong central agency as an environmental watchdog... [but] old industry is still operating. The observed improvement of environmental records since 1989 is only a side effect of the recession ... EIA law in Poland still reflects two characteristic features of the Communist regime: an aversion to getting the general public involved in decision-making, and a reluctance to developing procedural rules for dispute settlement. This means that this legislation not only is not efficient enough from the 'environmental' point of view, but also does not match the political and economic transformation towards an open and democratic society and a free market. (Jendroska and Sommer 1994)

In 1994, Poland became an associate partner of the EU, and it formally joined the EU in 2004. This process has required Poland to incrementally change its laws and statutes to progressively bring them in line with those of the EU. This process began with 1995 and 1998 amendments to the executive orders of 1990, and subsequently continued with three key EIA regulations: the 2000 Act on Access to Information on the Environment and its

Protection and on EIA; the 2001 Environmental Protection Law (EPL), and the pithily named 2008 Act on Providing Information on the Environment and Environmental Protection, Public Participation in Environmental Protection and on Environmental Impact Assessment, which replaces the EPL.

Changes made over time to Poland's original EIA system include application of EIA to a wider range of projects; new requirements for screening, including two lists of projects requiring EIA that reflect Annexes Ⅰ and Ⅱ of the EIA Directive; greater emphasis on scoping; greater requirements for public consultation; requirements for transboundary impact assessment; and a concerted attempt to shift the emphasis of EIA from the preparation of a report to the process as a whole (Wiszniewska *et al*. 2002; Woloszyn 2004).

The new system is relatively typical of EIA in the EU, but includes some interesting variants (PIFIA 2008; EC 2009):

- It integrates the Habitat Directive's requirements for appropriate assessment into the EIA process. EIA is required not only for the equivalent to EIA Directive Annex Ⅰ projects and Annex Ⅱ projects that are likely to have significant environmental impacts, but also for projects that require appropriate assessment under the Habitat Directive. These EIAs must assess the impact of the project on Natura 2000 sites, and discuss alternatives and mitigation measures considered in relation to their impact on Natura 2000 sites.
- For projects that may have sigificant environmental impacts (Annex Ⅱ equivalent), the developer must prepare an 'environmental information card' as a basis for the EIA screening stage. Where the competent authority decides that EIA is not required, the environmental information card serves as the basis for the official 'screening out' statement. Where EIA is required, the card serves as the basis for a mandatory scoping opinion.
- The EIS must include a description of analysed options, including the option favoured by the applicant, a rational alternative option, and the most advantageous option for the environment; and an analysis of probable social conflicts related to the project.
- A Decision on Environmental Conditions is required before any Annex Ⅰ or Ⅱ equivalent projects can get planning permission. The Decision on Environmental Conditions is based on the EIA or 'environmental information card', comments on the EIA from statutory consultees, the public, and other countries where relevant, and additional information such as land registry maps.
- If the EIA process shows that the project should be implemented according to a scenario other than that favoured by the applicant, the authorities issuing the Decision on Environmental Conditions are expected to specify that this scenario is the one that should be implemented. If the applicant does not agree to this, the authorities are expected to refuse permission for the project.

Depending on the source (EU 2009 or GHK 2010), approximately 2,200 to 4,000 EIAs are carried out in Poland every year. Of these, only a small proportion concern projects with transboundary impacts. However, because of the large initial number—GHK (2010) suggests that Poland might account for one third of all of the EIAs carried out in the 27 EU

Member States—Poland has been involved in a high number of transboundary EIAs, including roads, railway lines and transmission lines. Poland has bilateral agreements with several other European Member States on applying EIA in a transboundary context, which regulate such issues as the translation of EIA documentation, timeframes, principles and formats of public participation, etc. (EU 2009)

10.7 Canada

Canada is an example of a country with a long-standing and strong EIA system that is currently in a state of flux. Canada's wealth of natural resources, which were originally plundered indiscriminately by the giant 'trusts' in coal, steel, oil and railroads, its lack of strong planning and land-use legislation, and the conflicting needs of its powerful provincial governments all prompted the development of a mechanism by which widespread environmental harm could be prevented. Canada's EIA system is characterized by a split between national and provincial procedures, quite complex routeing of different types of projects through different types of EIA processes, and innovative approaches to mediation and public participation in EIA.

Responsibility for EIA in Canada is shared between the federal and the provincial governments. The *federal* procedures apply to projects for which the government of Canada has decision-making authority. Early federal EIA guidelines were progressively strengthened throughout the 1970s and 1980s, and made legally binding in 1989. However concern over the limitations of this 'Environmental Assessment and Review Process' caused it, in turn, to be replaced in 1995 by the Canadian Environmental Assessment Act. Amendments to the act were made in 2001 and 2010. SEA of policy has been required since 1993, and SEA requirements were strengthened in 1999. Gibson (2002) gives a useful review of the development of Canada's federal EIA system up to 2002.

The Canadian Environmental Assessment Agency (CEAA) administers the Canadian Environmental Assessment Act. An initial self-assessment by the responsible agency proposing the action determines whether the action requires EIA under the Act; that is, whether it:

- is a 'project' as defined by the Act;
- is not excluded by the Act's Exclusion List regulation;
- involves a federal authority; and
- triggers the need for an EIA under the Act.

The Exclusion List Regulation identifies projects for which EIAs are not required because their adverse environmental effects are not regarded as significant (e.g. simple renovation projects).

Once an EIA is determined to be required, a decision is made as to which of four EIA tracks to follow: screening, comprehensive study, mediation or review panel. Most projects require a 'screening' involving documentation of the project's environmental effects and recommended mitigation measures. 'Class screening' may be used to assess projects with known effects that can easily be mitigated. 'Model class screenings' provide a generic assessment of all projects within a class: the responsible authority uses a model report as a template, accounting for location- and project- specific information. 'Replacement class screenings' apply to projects for which no location—or project-specific information—is needed.

A small number of projects—typically less than one per cent—will require a fuller 'comprehensive study'. These projects are listed in the Comprehensive Study List Regulations and include, for instance, nuclear power plants, large oil and gas developments and industrial plants.

If a screening requires further review, it is referred to a mediator or review panel. Similarly, early in a comprehensive study, the Minister of the Environment must decide whether the project should continue to be assessed as a comprehensive study, or whether it should be referred to a mediator or review panel. Projects are normally referred to a mediator or review panel—essentially changing them from self-assessment by the respon-sible agency to independent, outside assessment—where the significance of their impacts is uncertain, where the project is likely to cause significant adverse environmental effects and there is uncertainty about whether these are justified, or where public concern warrants it (CEAA 2003). A review panel is a group of experts approved by the Minister of the Environment, which reviews and assesses a project with likely adverse environmental impacts. The mediation option, new since 1995, is a voluntary process in which an independent mediator appointed by the Minister helps the interested parties to resolve their issues through a non-adversarial, collaborative approach to problem-solving. Very few projects go through the review panel or mediation route.

Any project requiring comprehensive study, mediation or a review panel must consider alternative means of carrying out the project, the project's purpose and its effects on the sustainability of renewable resources; and must include a follow-up programme. The responsible authority must take the results of the comprehensive study, or the mediator's or review panel's recommendations into account when making a decision on the project (CEAA 2003). Public comments must be considered at various stages of the EIA process, though it is more restricted for screenings. A participant-funding programme allows stakeholders to participate in comprehensive studies, panel reviews and mediation, and Aboriginal people have a specific consultation process (Sinclair and Fitzpatrick 2002).

The CEAA publishes many of its reports on the Web. Between 2006 and 2010, more than 22,000 federal EIAs were carried out: 99.6 per cent were screenings, but 97 comprehensive studies and 19 panel reports were also initiated during that time. The comprehensive studies included studies for pipelines, offshore oil and gas projects, mines, waste treatment centres, and decommissioning of a range of former projects (CEAA 2011).

Most of Canada's *provinces* have quite widely varying EIA regulations for projects under their own jurisdictions. These include Ontario's EA Act of 1976, very advanced at the time, but subsequently weakened in 1997; and Manitoba's sustainable development code of practice of 2001, which requires public officials to promote consideration of sustainability impacts in EIA. In early 1998, federal and provincial environment ministers signed an accord on EIA harmonization, which promotes cooperative use of existing processes to reduce duplication and inefficiency (Gibson 2002).

Concern over the duplication of functions between the federal and provincial levels, unclear leadership of EIAs where several government agencies are involved, and lengthy government processing led to changes to the Canadian Environmental Assessment Act in 2010, and to proposed regulations that would restrict the time that comprehensive studies can take. The changes to the act, which are widely perceived as weakening Canada's EIA process (e. g. Green Budget Coalition 2010), include delegation of the authority for EIA of major energy projects from the CEAA to the National Energy Board and Canadian Nuclear Safety Commission, exemption of many infrastructure projects from EIA, greater coordination functions being given to the CEAA, and powers being given to the Minister of the Environment to scope EIAs (with concerns that important impacts, or parts of projects, could be scoped out). A Parliamentary review of the Canadian Environmental Assessment Act, begun in autumn 2010, may well lead to further changes.

10.8 Australia and Western Australia

Like Canada, Australia also has a federal (Common-wealth) system with powerful individual states. Its environmental policies, including those on EIA, have some interesting features but are generally not as powerful as those of Canada. The Common-wealth EIA system was established as early as 1974 under the Environmental Protection (Impact of Proposals) Act. It applied only to federal activities. During the life of the Act (1974—2000) about 4,000 proposals were referred for consideration, but on average less than 10 formal assessments were carried out each year (Wood 2003). As such, the states put in place their own legislation or procedures to extend the scope of EIA to their own activities, and many of these state systems have become stronger and more effective than the national system.

Over time there has been concern about the variation in EIA procedures, and their implementation, between states in Australia and there have been attempts to increase harmonization (Australian and New Zealand Environment and Conservation Council—ANZECC 1991, 1996, 1997; see also Harvey 1998; Thomas 1998). In addition, a major review of Commonwealth EIA processes was undertaken in 1994, producing a set of very useful reports on cumulative impact and strategic assessment, social impact assessment, public participation, the public inquiry process, EIA practices in Australia and overseas comparative EIA practice (CEPA 1994). The review highlighted, among other issues, the need to reform EIA at the Common-wealth level—including a better consideration of cumulative impacts, social and health impacts, SEA, public participation and monitoring.

Following government changes and a further review of federal/state roles in environmental protection, Australia repealed its Commonwealth EIA legislation, and several other environmental statutes, to create the Environmental Protection and Biodiversity Conservation Act (EPBCA) in 1999. The EPBCA provides a lot more procedural detail than the original EIA legislation, and a range of documents has been produced to explain the processes (Environment Australia 2000). EIA is undertaken for matters of national environmental significance, defined as World Heritage properties, Ramsar wetlands, threatened and migratory species, the Commonwealth marine environment and nuclear actions. The Act promotes ecologically sustainable development; it also provides for SEA (IEMA 2002).

Padgett and Kriwoken (2001), Scanlon and Dyson (2001), and Marsden and Dovers (2002) provide early commentary on the Act and some developments in EIA and SEA in Australia. In its first year, the EPBCA did not appear to increase much the rate of Commonwealth EIA activity (Wood 2003), but by 2009—10 there had been a substantial increase to about 420 referrals in the year. The referrals were concentrated on proposals in the main resource development states of Queensland and Western Australia, and also Victoria, especially for mining and mineral exploration, and for residential development (Australian Government Department of Environment 2010). A review was undertaken of the first 10 years of operation of the EPBCA (Australian Government, Department of Environment 2009). Many positive features were noted, including: clear specification of matters of national environmental significance, the Environment Minister's role as decision-maker, public participation provisions, the explicit consideration of socio-economic issues, statutory advisory mechanisms and a strong compliance and enforcement regime. Important recommendations for change included: renaming it as the Australian Environment Act, establishing an independent Environment Commission to advise the government of project approvals, strategic assessments etc, provide for environmental performance audits, set up an Environmental Reparation Fund, and improve transparency in decision-making.

10.8.1 *EIA in Western Australia*

The Western Australia (WA) EIA system provides an interesting example of a good state system that includes many innovative features. Central to the success of the Western Australian system is the role of the EPA (Wood and Bailey 1994). The Environmental Protection Authority (EPA) was established by the WA Parliament as an Authority with the broad objective of protecting the State's environment and it is the independent environmental adviser that recommends to the WA government whether projects are acceptable. It is independent of political direction. The EPA determines the form, content, timing and procedures of assessment and can call for all relevant information; the advice it provides to the Minister for the Environment must be published. The EPA overrides virtually all other legislation, and the environmental decision (with conditions) is central to the authorization of new proposals. Other permits must await the environmental approval, based on the EIA.

Proposals may be referred to the EPA by any decision-making authority, the proponent, the Minister for the Environment, the EPA or any member of the public. The EPA determines the level of assessment. Until late 2010 there were five levels of assessment, the most comprehensive being the Public Environmental Review (PER) and the Environmental Review and Management Programme (ERMP). Under new procedures (WA EPA 2010), these have now been reduced to two: Assessment on Proponent Information with no public review, and Public Environmental Review with a public review period of generally 4—12 weeks. In practice this does not significantly alter the EPA EIA procedures as the previous five levels of assessment could broadly be divided into either assessment without public review or assessment requiring a public review. Criteria for deciding the levels of assessment are set out in Table 10. 3 (a).

Guidance is provided on scoping and on the content of the PER, as set out in Table 10. 3 (b) and (c). The PER document is produced by the proponent, and it is subject to public review. The guidance on scoping for a PER contains interesting features, especially in relation to peer review and public consultation. The PER assessment pays particular attention to the regional setting, and seeks to highlight potential 'fatal flaws'. Waldeck *et al.* (2003) found that such EIA guidance influenced the practice of consultants and was perceived as effective in enhancing the outcomes of the EIA process—including increased certainty of outcome of the EIA process, and better design of proposals to meet environmental objectives from the outset.

Table 10. 3 Some features of the Western Australian EIA system

(a) Criteria for deciding levels of assessment

For a proposal to be assessed at an API level, it must meet all of the following criteria:

- the proposal raises a limited number of significant environmental factors that can be readily managed, and for which there is an established condition-setting framework;
- the proposal is consistent with established environmental policy frameworks, guidelines and standards;
- the proponent can demonstrate that it has conducted appropriate and effective stakeholder participation; and
- there is limited, or local, interest only in the proposal.

If, based on the referral information, the EPA considers that the proposal is environmentally unacceptable, the chairman of the EPA will encourage the proponent to withdraw the referral or submit a new significantly modified proposal. The criteria for determining whether a proposal is unacceptable are set out in the Category 'B' of the accompanying Administrative Procedures (e. g. inconsistent with environmental poicy framework, likely to have significant impacts, proposal cannot be easily modified etc).

If a proposal meets *any* of the following criteria the EPA will apply a PER level of assessment:

- the proposal is of regional or WA state-wide significance;
- the proposal has several significant environmental issues or factors, some of which are considered to be complex or of a strategic nature;
- substantial and detailed assessment of the proposal is required to determine whether, and if so, how the environmental issues could be managed; or
- the level of interest in the proposal warrants a public review panel.

(b) A formal Environmental Scoping stage for PER

An Environmental Scoping Document (ESD), designed to direct the proponent on key issues to address, and to identify impact predictions and information on the environmental setting, shall include:

- a concise description of the proposal and its environmental setting;
- the identification of the key environment factors and other environmental factors relevant to the proposal;
- the identification of the existing policy context relevant to each factor;

Continued

- the preliminary identification of the potential environmental impacts;
- a Scope of Works, setting out the proposed environmental studies and designed to identify or predict the direct and indirect environmental impacts of the proposal, including timeline for completion (the studies and investigations should be clearly linked to the identified environmental impacts and factors);
- the identification of an environmental management programme required;
- the identification of the spatial datasets, information products and databases required;
- a list of people, if necessary, proposed to provide peer review of the scope, methodologies, findings and/or conclusions of the surveys and investigations; and
- stakeholder consultation requirements.

(c) Guidance on the form and content of the assessment document for PER

The proponents should ensure that the PER document focuses on the environmental issues/factors of key significance. The document should include the following:

- a description of the proposal and alternatives considered, including alternative locations, with a view to minimizing environmental impacts;
- a description of the receiving environment, its conservation values and key ecosystem processes, and discussion of their significance in a regional setting-this should focus on those elements of the environment that may affect or be affected by the proposal;
- identification of the key issues (and list the environmental factors associated with these issues) and their potential 'fatal flaws';
- discussion and analysis of the direct and indirect impacts of the proposal, in a local and regional context, including cumulative impacts;
- findings of the surveys and investigations undertaken (and technical reports provided as appendices);
- identification of the measures proposed to mitigate significant adverse impacts;
- identification of any offsets, where appropriate, after all other steps in the mitigation sequence have been exhausted;
- environmental management programme;
- demonstration that the expectations for EIA identified elsewhere in the procedures have been carried out; and
- details of stakeholder and government agency consultation, how comments received have been responded to, and any subsequent modifications of the proposals.

Source: WA EPA 2010

The EPA then assesses the environmental acceptability of the proposals on the basis of the review document, public submissions, proponents' response, expert advice and its own investigations. The resulting EPA report to the Minister for the Environment pronounces on the environmental acceptability or otherwise of the proposal and on any recommended conditions to be applied to ministerial approval. Figure 10. 5 provides an outline of the full procedure for PER assessment. The centrality of the EPA's review of the relevant environmental information to the Minister's decision, which itself has predominance, is the most remarkable aspect of the WA system, and one which highlights the significance of the EIA impact on decisions. The WA system also has a high level of public participation, especially in controversial EIAs. The central role of the EPA also ensures consistency.

However, the limited integration of the EIA and planning procedures and a biophysical focus to assessment have been weaker features of the WA procedures. Amendments in the mid-1990s were designed to secure better integration, improving the EIA of land-use schemes, but they did also reflect a shift of control away from the EPA to the Ministry of Planning. This was symptomatic of challenges faced by an effective system. Interestingly though, there is now provision for the EPA to assess strategic proposals, including poli-

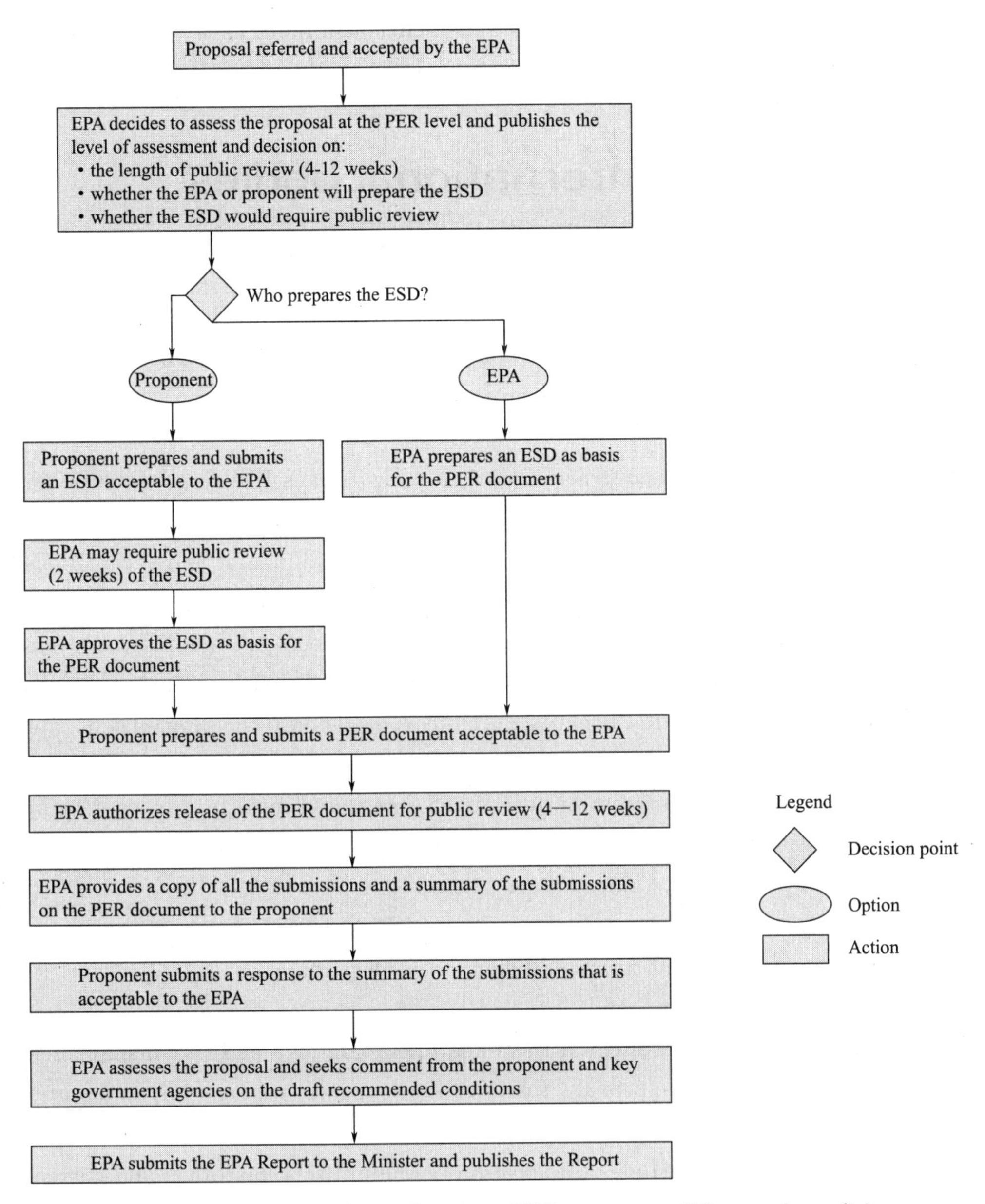

Figure 10.5 Outline of procedure for a PER assessment (Western Australia)

Source: Western Australia EPA (2010)

cies, plans, programmes and developments. There is also provision for 'derived proposals' identified in assessed strategic proposals (which themselves have been agreed for implementation, and where environmental issues have been adequately addressed); in such cases the derived proposal would not require further assessment by the EPA except for checking on whether any implementation conditions relating to the proposal should change. With regard to socio-economic impacts, in 1993, WA lost its pioneering Social Impact Unit, which had provided expert advice on social impacts, and there is a continuing strong development lobby, in a state highly dependent on major energy and mineral projects, to further 'soften green laws'. However, for some recent high-profile and major energy and mineral developments in the north of the state, proponents and their consultants have been adopting more innovative environmental-social-economic approaches to assessment.

10.9 International bodies

Many of the major international funding institutions and other international organizations have established EIA procedures. In several cases these have evolved over time, with some handbooks or guidance manuals now on their second or third edition, or with multiple updates.

The *European Bank for Reconstruction and Development*'s EIA requirements are typical of those of other lending institutions. The bank's environmental procedures of 1996 were updated in 2003 and have recently been widened out to environmental and social procedures (EBRD 2010). These aim to ensure that the environmental and social implications of potential bank-financed investment and technical co-operation projects are identified and assessed early in the bank's planning and decision-making process, and that environmental and social considerations—including potential benefits—are incorporated into the preparation, approval and implementation of projects.

The EBRD's assessment system is very much integrated into its project development process, as a process of due diligence. Main assessment steps involve:

- collection of preliminary information;
- project categorization to determine the level of EIA needed;
- impact assessment to inform due diligence considerations;
- disclosure of information; and
- characterization of environmental risks and opportunities, and risk management.

This is followed by Board approval, final negotiations and signing, implementation and monitoring (EBRD 2010).

Preliminary information includes the location of the project, historical and current land uses at the site, proposed construction activities, whether resettlement or economic displacement is likely to occur, characteristics of the local population including vulnerable groups, whether there are significant environmental and/or social issues of concern, who the main project stakeholders are, and the environmental and social reputation of the client. This information is used in screening discussions between the bank and the project sponsor to sort the proposed project into one of several categories:

- *Category A*: projects likely to cause significant adverse impacts, which, at the time of screening cannot readily be identified or assessed. A full EIA is required for these projects. Some types of projects automatically come under Category A; others are put into Category A on a case-by-case basis.
- *Category B*: projects likely to have less severe impacts than Category A pro-

jects. These require a less stringent environmental analysis. The scope of the environmental analysis is determined on a case-by-case basis.

- *Category C*: projects likely to result in minimal or no adverse impacts and that do not need analysis.
- *Category FI*: projects that will be developed through a financial intermediary. These require a variant on the EIA process discussed below.

Staff from the EBRD's Environment and Sustainability Department then establish terms of reference for the environmental and social assessment, including stakeholder engagement, and how compliance will be assessed. The client carries out these studies with support of EBRD staff. Measures to manage and mitigate a proposed project's impacts are typically set out in an Environmental and Social Action Plan. Public disclosure and consultation is required throughout the planning and assessment/analysis processes of Categories A and B projects. The EBRD sets environmental standards for each project. It may also specify an environmental and social action plan and/or monitoring to be carried out by the sponsor as a condition of investment. Prior to making a final decision about whether to lend money for the project, bank officials review the environmental due diligence information available, ensure that proposed mitigation measures are agreed with the project sponsor, highlight opportunities for environmental improvements, identify environmental monitoring requirements and any technical/environmental cooperation initiatives that should be undertaken and advise on whether the project complies with the Bank's environmental policy and procedures.

The *Asian Development Bank*'s *Environmental Assessment Guidelines* (ADB 2003) have similar screening and assessment requirements. The guidelines give more detailed information about assessment of projects vs. programmes, a range of different sectors, equity investments, etc.; they also stress consultation.

The *World Bank* perceives EIA as one of its key environmental and social safeguard policies. Its Operational Policy/Bank Procedures 4.01 (most recently updated in 2007) require EIA for relevant lending operations, and its *Environmental assessment sourcebook* (World Bank 1991) and various updates explain the EIA process. EIA involves screening of the project into assessment categories by World Bank staff; preparation of an environmental assessment report by the proponent; review of the report by World Bank staff; an appraisal mission in which World Bank staff discuss and resolve environmental issues with the proponent; documentation of the findings; and supervision and evaluation during project implementation (World Bank 1999). Between 1989 and 1995, the World Bank screened over a thousand projects for their potential environmental impacts: of these 10 per cent were in Category A (primarily energy, agriculture and transport projects), 41 per cent in Category B, and 49 per cent in Category C (World Bank 1997). Category A projects are those expected to have 'adverse impacts that may be sensitive, irreversible and diverse' (World Bank 1999), and they require a full EIA. For Category B projects, where impacts are 'less significant—not as sensitive, numerous, major or diverse', a full EIA is not required, but some environmental analysis is necessary. Category C projects have negligible or minimal direct disturbance on the physical setting, and neither EIA nor environmental analysis is required. Typical category C projects focus on education, family planning, health and human resource development.

Notable features of the World Bank process include a holistic environment definition, including physical, biological and socio-economic aspects, a high profile for public consultation and considerable focus on project implementation. A report (World Bank 1995) identified five main challenges ahead: moving EIA 'upstream' (into project design stages and at sectoral and regional levels); more effective public consultation; better integration of EIA into the project work programme (including mitigation, monitoring and management plans); learning from implementation (the 'feedback loop'); and engaging the private sector (especially financiers and project sponsors) to ensure that projects are subject to EIA of acceptable quality. Mercier (2003) reinforces the emphasis now placed on implementation of the mitigation, prevention and compensation measures contained in the EIA. Also, because many of the client countries now have their own EIA requirements and their own EIA staff and review mechanisms, the World Bank is increasingly involved in enhancing that capacity upfront during project preparation.

The *African Development Bank* (ADB 2003) and *European Investment Bank* (EIB 2007) have less comprehensive EIA guidance, but both require EIA to be carried out, promote public participation in the EIA process, and take account of these when deciding on whether to fund a project. The ADB guidance is for integrated environmental and social impacts.

Other organizations have also published EIA guidance. For instance UNEP's very useful *Environmental impact assessment training resources manual*, now in its second edition (UNEP 2002), includes case studies (primarily from developing countries), transparencies and detailed chapters on various stages of EIA. The UK Department for International Development and the Ministry of Foreign Affairs of Denmark have produced a similar guides (DFID 2003; Danida 2009). Of particularly increasing importance are the *Equator Principles*, based on both World Bank guidance, and on the International Finance Corporation (IFC) Performance Standards on social and environmental sustainability. Equator Principle Financial Institutions, now including over 70 major national and international banks, commit to not providing loans to projects (>US $10m value threshold) where the borrower will not, or is not able, to comply with domestic standards for EIA or international standards, whichever is the highest (IFC, 2006).

10.10 Summary

In 2002, Gibson suggested that EIA worldwide has been moving towards being:

- earlier in planning (beginning with purposes and broad alternatives);

• more open and participative (not just proponents, government officials and technical experts);

• more comprehensive (not just biophysical environment, local effects, capital projects, single undertakings);

• more mandatory (gradual conversion of policy-based to law-based processes);

• more closely monitored (by the courts, informed civil society bodies and government auditors);

• more widely applied (through law at various levels, but also in land-use planning, through voluntary corporate initiatives, and so on);

• more integrative (considering systemic effects rather than just individual impacts);

• more ambitious (overall sustainability rather than just individually 'acceptable' undertakings); and

• more humble (recognizing and addressing uncertainties, applying precaution).

Almost 10 years later, this chapter shows that these trends are still continuing a decade later. However, worldwide, EIA is still constrained by lack of political will, insufficient budget to implement proposed mitigation measures and lack of institutional capacity, as noted earlier by Goodland and Mercier (1999). Emerging trends in Canada, Australia, the Netherlands (GHK 2010) and elsewhere suggest that, even in developed countries, EIA also tends to be adapted—some might say watered down—where it is perceived to conflict with economic development.

Chapters 11 and 12 draw on some of the ideas discussed here, and elsewhere, to identify possibilities for the future, focusing primarily on the UK system, but set in the wider EU and global context.

SOME QUESTIONS

The following questions are intended to help the reader focus on the key issues of this chapter.

1 Figure 10.1 suggests that a greater number of EIAs are carried out in more evolved EIA systems than in less evolved systems. Does that, in turn, mean that one can tell how evolved a country's EIA system is on the basis of how many EIAs are carried out in that country? Explain your reasoning.

2 Compare two or three of the EIA systems described in this chapter with the UK system described at Chapter 3, or the US system of Section 2.2, in terms of:

• which projects require EIA, and how screening is carried out;

• what an EIS must contain;

• who carries out the EIA;

• public involvement; and

• decision-making, including links to other licenses and permits.

3 Figure 10.2 suggests that more developed countries tend to have stronger EIA sys-

tems than less developed countries. However the EIA systems of Benin, Peru, China and Poland each have strong or innovative aspects that go beyond the UK or US systems. What are these aspects, and why might they have been instituted?

4 The countries discussed in this chapter have widely varying approaches to the consideration of alternatives in EIA. Which has the strongest approach? The weakest?

5 Peru and China allow only registered EIA experts to carry out EIA. What are the advantages and disadvantages of such an approach?

6 Peru's system can be criticised for having a 'poacher-gamekeeper' approach, where the ministry in charge of promoting certain projects is also responsible for the EIA process for those projects. Why is the term 'poacher-gamekeeper' used for such a scenario? Are there any other examples of 'poacher-gamekeeper' in this chapter?

7 The Western Australian EIA system is seen as one with many innovative features, which contribute to a good state system. What are these features? Are there also weaknesses?

8 Several of the international funding organisations require integrated environmental and social assessment. However individual countries' legislation focuses on environmental assessment. What might account for this discrepancy?

9 The World Bank and other funding organizations have played a key role in the application of EIA in many developing countries. What are possible reasons for these organizations' interest in EIA?

References

ADB (African Development Bank) 2003. *Integrated environmental and social impact assessment guidelines*.

ADB (Asian Development Bank) 2003. *Integrated environmental and social impact assessment guidelines*. Available at: www. adb. org.

Almagi, D., Sondo, V. A. and Ertel, J. 2007. Constraints to environmental impact assessment practice: a case study of Cameroon. *Journal of Environmental Assessment Policy and Management* 9 (3), 357-80.

ANZECC (Australia and New Zealand Environment and Conservation Council) 1991. *A national approach to EIA in Australia*. Canberra: ANZECC.

ANZECC (Australia and New Zealand Environment and Conservation Council) 1996. *Guidelines and criteria for determining the need for and level of EIA in Australia*. Canberra: ANZECC.

ANZECC (Australia and New Zealand Environment and Conservation Council) 1997. *Basis for a national agreement on EIA*. Canberra: ANZECC.

Appiah-Opoku, S. 2005. *The need for indigenous knowledge in environmental impact assessment: the case of Ghana*. New York: Edwin Mellen Press.

Australian Government Department of Environment 2009. *Independent review of the EPBC (1999) Act*. Canberra: Government of Australia.

Australian Government Department of Environment 2010. *Annual report 2009-2010*. Government of Australia.

Badr, E-S. A. 2009. Evaluation of the environmental impact asessment system in Egypt. *Impact Assessment and Project Appraisal* 27 (3), 193-203.

Baglo, M. A. 2003. *Benin's experience with national and international EIA processes*. Available at:

www. ceaa. gc. ca/default. asp? lang=En&n=B4993348-1&offset=4.

Bekhechi, M. A. and J. R. Mercier 2002. *The legal and regulatory framework for environmental impact assessments*. Available at: www. scribd. com/doc/ 16060372/The-Legal-and-Regulatory-Framework-for-Environmental-Impact-Assessments-A-Study-of-Selected-Countries-in-SubSaharan-Africa.

Briffett, C. 1999. Environmental impact assessment in East Asia. In *Handbook of environmental impact assessment*, J. Petts (ed), vol. 2, 143-67. Oxford: Blackwell Science.

Brito, E. and Verocai, I. 1999. Environmental impact assessment in South and Central America. In *Handbook of environmental impact assessment*, J. Petts (ed), vol. 2, Chapter 10, 183-202. Oxford: Blackwell Science.

CEAA (Canadian Environmental Assessment Agency) 2003. *Canadian environmental assessment act: an overview*. Ottawa: CEAA. www. ceaa. gc. ca.

CEAA (Canadian Environmental Assessment Agency) 2011. *Canadian environmental assessment agency*. Available at: www. ceaa. gc. ca.

CEPA 1994. *Review of Commonwealth environmental impact assessment*. Canberra: CEPA.

Chico, I. 1995. EIA in Latin America. *Environmental Assessment* 3 (2), 69-71.

China 1999 (in Chinese). *A summary of the environmental protection and management work of construction projects*. Available at: www. China-eia. com/chegxu/ hpcx _ main0. htm.

CISDL (Centre for International Sustainable Development Law) 2009. *Eco-health Americas law project*. Available at: www. cisdl. org/ecohealth/impact _ assessment001. htm.

d'Almeida, K. 2001. *Cadre institutional législatifet réglementaire de l'évaluation environnementale dansles pays francophones d'Afrique et de l'Océan Indien*. Montréal, Canada: EIPF et Secrétariat francophone de l'AiEi/IAIA.

Danida 2009. *Danida environment guide: Environmental assessment for sustainable development*, 3rd edn. Copenhagen: Danida. Available at: www. danida devforum. um. dk/NR/rdonlyres/3409F0D0-D7BA-4815- BF49-C7420D8DC9CF/0/DanidaGuideto EnvironmentalAssessmentefternyAMG. pdf.

DFID (Department for International Development) 2003. *Environment guide*. London: DFID. Available at: www. eldis. org/vfile/upload/1/document/0708/ DOC12943. pdf.

EBRD (European Bank for Reconstruction and Development) 2010. *Environmental and social procedures*. Available at: www. ebrd. com/downloads/ about/sustainability/esprocs10. pdf.

EC (European Commission) DG ENV 2009. *Study concerning the report on the application and effectiveness of the EIA Directive*. Available at: ec. europa. eu/environment/eia/pdf/eia _ study _ june _ 09. pdf.

Economic Commission for Africa 2005. *Review of the application of environmental impact assessment in selected African countries*. Addis Ababa: ECA. www. uneca. org/eca _ programmes/sdd/documents/ eia _ book _ final _ sm. pdf.

EIB (European Investment Bank) 2007. *EIB environmental assessment*. Luxembourg: European Investment Bank. Available at: www. eib. org/attachments/ thematic/environmental-assessment. pdf.

El-Fadl, K. and El-Fadel, M. 2004. Comparative assessment of EIA systems in MENA countries: challenges and prospects. *Environmental Impact Assessment Review* 24 (6), 553-93.

Elliott, M. and Thomas, I. 2009. *Environmental impact assessment in Australia*, 5th edn. Annandale: The Federation Press.

Environment Australia 2000. *EPBC Act: various documents, including environmental assessment processes; administrative guidelines on significance; and frequently asked questions*. Canberra: Department of Environment and Heritage.

Fisher, D. 1992. *Paradise deferred: environmental policymaking in Central and Eastern Europe*. London: Royal Institute of International Affairs.

GHK 2010. *Collection of information and data to support the impact assessment study of the review of the EIA Directive*. Available at: www. ec. europa. eu/ environment/eia/pdf/collection _ data. pdf.

Gibson, R. 2002. From Wreck Cove to Voisey's Bay: the evolution of federal environmental assessment in Canada. *Impact Assessment and Project Appraisal* 20 (3), 151-60.

Glasson, J. and Salvador, N. N. B. 2000. EIA in Brazil: a procedures-practice gap. A comparative study with reference to EU, and especially the UK. *Environmental Impact Assessment Review* 20, 191-225.

Goodland, R. and Mercier, J. R. 1999. The evolution of environmental assessment in the World Bank: from 'Approval' to results, *World Bank environment department paper* no. 67, Washington, DC: World Bank.

Green Budget Coalition 2010. *Budget* 2010: *Environmental impact summary and analysis*. Available at: www. greenbudget. ca/pdf/Green%20 Budget%20Coalition%27s%20Environmental%20 Impact%20Sum-mary%20and%20Analysis%20of%20Budget%202010%20%28July%202010%20. pdf.

Harvey, N. 1998. *EIA: procedures and prospects in Australia*. Melbourne: Oxford University Press.

IEMA (Institute of Environmental Management and Assessment) with EIA Centre (University of Manchester) 2002. *Environmental assessment yearbook* 2002. Manchester: EIA Centre, University of Manchester.

Iglesias, S. 1996. *The role of EIA in mining activities: the Peruvian case*. MSc dissertation, Oxford Brookes University, Oxford.

International Finance Corporation (IFC) 2006. *Equator Principles*. Available at: www. equator-principles. com.

Jendroska, J. and Sommer, J. 1994. Environmental impact assessment in Polish law: the concept, development, and perspectives. *Environmental Impact Assessment* Review 14 (2/3), 169-94.

Kakonge, J. O. 1999. Environmental impact assessment in Africa. In *Handbook of environmental impact assessment*, J. Petts (ed), vol. 2, 168-82. Oxford: Blackwell Science.

Kirchhoff, D. 2006. Capacity building for EIA in Brazil: preliminary considerations and problems to be overcome. *Journal of Environmental Assessment Policy and Management* 8 (1), 1-18.

Kovalev, N., Köppel, J., Drozdov, A. and Dittrich, E. 2009. Democracy and the environment in Russia. *Journal of Environmental Assessment Policy and Management* 11 (2), 161-73.

Lee, N. and George, C. 2000. *Environmental assessment in developing and transitional countries: principles, methods and practice*. Chichester: Wiley.

Lindhjem, H., Hu, T., Ma, Z., Skjelvik, J. M., Song, G., Vennemo, H., Wu, J. and Zhang, S. 2007. Environmental economic impact assessment in China: problems and prospects. *Environmental Impact Assessment Review* 27 (1), 1-25.

Mao, W. and Hills, P. 2002. Impacts of the economic-political reform on environmental impact assessment implementation in China. *Impact Assessment and Project Appraisal* 29 (2), 101-11.

Marara, M., Okello, N., Kuhanwa, Z., Douven, W., Beevers, L. and Leentvaar, J. 2011. The importance of context in delivering effective EIA: case studies from East Africa, *Environmental Impact Assessment Review* 31 (3), 286-96.

Marsden, S. and Dovers, S. 2002. *Strategic Environmental Assessment in Australasia*. Annadale, New South Wales: Federation Press.

Mercier, J. R. 2003. Environmental assessment in a changing world at a changing world bank. In IEMA/EIA Centre 2003, *Environmental assessment outlook*. Manchester: EIA Centre, University of Manchester.

Moorman, J. L. and Ge, Z. 2007. Promoting and strengthening public particiaption in China's environmental impact assessment process: comparing China's EIA law and U. S. NEPA. *Vermont Journal of Environmental Law* 8, 281-335.

Okaru, V. and Barannik, A. 1996. Harmonization of environmental assessment procedures between the World Bank and borrower nations. In *Environmental assessment (EA) in Africa*, R. Goodland, J. R. Mercier and S. Muntemba (eds), 35-63. Washington, DC: World Bank.

Ortolano, L. 1996. Influence of institutional arrangements on EIA effectiveness in China. *Proceedings of the 16th annual conference of the international association for impact assessment*, 901-05. Estoril: IAIA.

Padgett, R. and Kriwoken, L. K. 2001. The Australian Environmental Protection and Biodiversity Conservation Act 1999: what role for the Common-wealth in environmental impact assessment? *Australian Journal of Environmental Management* 8, 25-36.

PIFIA (Polish Information and Foreign Investment Agency) 2008. *Providing information on the environment and environmental protection, public participation in environmental protection and on environmental impact assessment*. Available at: www. paiz. gov. pl/ polish _ law/environmental _ impact _ assessment.

Rzeszot, U. A. 1999. Environmental impact assessment in Central and Eastern Europe. In *Handbook of environmental impact assessment*, J. Petts (ed), vol. 2, Chapter 7, 123-42. Oxford: Blackwell Science.

Scanlon, J. and Dyson, M. 2001. Will practice hinder principle? Implementing the EPBC Act. *Environment and Planning Law Journal* 18, 14-22.

Sinclair, A. J. and Fitzpatrick, P. 2002. Provisions for more meaningful public participation still elusive in proposed Canadian EA Bill. *Impact Assessment and Project Appraisal* 20 (3), 161-76.

Southern African Institute for Environmental Assessment 2003. *Environmental impact assessment in southern Africa*. Available at: www. saiea. com/saiea-book.

Sutherland, J. W., Agadzi, K. O. and Amekor, E. M. K. 2005. *Rationalising the environmental impact assessment procedures in ECOWAS Member Countries, Union of Producers, Transporters and Distributors of Electric Power in Africa*. Available at: www. updeaafrica. org/updea/archiv/15CongresUPDEA%20 EIA%20Presentation. pdf.

Thomas, I. 1998. *EIA in Australia*, 2nd edn. New South Wales: Federation Press.

Unalan, D. and Cowell, R. 2009. Adoption of the EU SEA Directive in Turkey, *Environmental Impact Assessment Review* 29 (4), 243-51.

UNEP (United Nations Environment Programme) 2002. *UNEP Environmental Impact Assessment Training Resource Manual*, 2nd edn. Available at: www. unep. ch/etu/publications/EIAMan _ 2edition _ toc. htm.

Vidyaratne, H. 2006. EIA theories and practice: balancing conservation and development in Sri Lanka. *Journal of Environmental Assessment Policy and Management* 8 (2), 205-22.

WAEPA (Western Australian Environmental Protection Authority) 2010. *Environmental impact assessment administrative procedures*. Perth: EPA.

Waldeck, S., Morrison-Saunders, A. and Annadale, D. 2003. Effectiveness of non-legal EIA guidance from the perspective of consultants in Western Australia. *Impact Assessment and Project Appraisal* 21 (3), 251-56.

Wang, Y., Morgan, R. and Cashmore, M. 2003. Environmental impact assessment of projects in the People's Republic of China: new law, old problem. *Environmental Impact Assessment Review* 23, 543-79.

Wiszniewska, B., Farr, J. and Jendroska, J. 2002. *Handbook on environmental impact assessment procedures in Poland*. Warsaw: Ministry of Environment.

Woloszyn, W. 2004. Evolution of environmental impact assessment in Poland: problems and prospects, *Impact Assessment and Project Appraisal* 22 (2), 109-19.

Wood, C. 2003. *Environmental impact assessment: a comparative review*, 2nd edn, Prentice Hall.

Wood, C. and Bailey, J. 1994. Predominance and independence in EIA: the Western Australian model. *Environmental Impact Assessment Review* 14 (1), 37-59.

World Bank 1991. *Environmental assessment sourcebook*. Washington, DC: World Bank. Available at: www. worldbank. org.

World Bank 1995. *Environmental assessment: challenges and good practice*. Washington, DC: World Bank.

World Bank 1997. *The impact of environmental assessment: a review of World Bank experience*. World

Bank Technical Paper, no. 363, Washington, DC: World Bank.

World Bank 1999. *Environmental assessment*, BP 4. 01, Washington, DC: World Bank.

World Bank 2002. *Environmental impact assessment systems in Europe and Central Asia Countries*. Available at: www. worldbank. org/eca/environment.

World Bank 2006. *Environmental impact assessment regulations and strategic environmental assessment requirements: practices and lessons learned in East and Southeast Asia. Environment and social development safeguard dissemination note no*. 2. Available at: www. vle. worldbank. org/bnpp/files/ TF055249EnvironmentalImpact. pdf.

Yaha, P. Z. 2007. Benin: experience with results based management, in *Managing for development results*, Sourcebook 2nd edn, 131-42. Available at: www. mfdr. org/sourcebook/2ndEdition/4-5BeninRBM. pdf.

Yang, S. 2008. Public participation in the Chinese environmental impact assessment (EIA) system. *Journal of Environmental Assessment Policy and Management*, 10 (1), 91-113.

第四部分

展望

11 范围拓展：战略环境评价

11.1 导言

近年来，EIA 被越来越广泛地应用在政策、规划和计划（PPP）层面上。在美国，所谓的战略环境评价（SEA）从 20 世纪 70 年代以来作为项目 EIA 的延伸悄然展开。在其他国家，SEA 的进展比较缓慢，但是引起了较大的反响。欧盟 2004/SEA 指令的实施使其部分成员国的规划体系发生了重大改变。中国在 2009 年通过了具体的 SEA 法律，SEA 在全球其他很多地方的发展势头也很强劲。

本章主要讨论 SEA 的必要性和局限性。首先回顾 SEA 在美国、欧盟、联合国欧洲经济委员会（UNECE）以及中国的进展情况。接着详细地讨论了欧洲 SEA 指令在英国的实施情况。本章总结了近期关于 SEA 指令有效性的研究成果。根据需要，这一章简化了 SEA 许多方面的内容，读者可参考 Sadler 等（2010）以及 Therivel（2010）来深入理解。同时也可参考第 9 章有关 SEA 的两个案例。

11.2 战略环境评价

11.2.1 定义

战略环境评价可定义为：

一个对拟议政策、规划和计划进行环境结果评价的系统过程，其目的是保证在决策制定的尽可能的最早阶段全面考虑社会和经济因素。(Sadler 和 Verheem，1996)

换言之，SEA 是 EIA 在政策、规划和计划层面（PPP）上的一种应用形式。但是需要谨记的是，战略层面上的环境影响评价与项目层面上的环境影响评价不尽相同。

在 Sadler 和 Verheem 对 SEA 的定义中有几点非常重要：首先，SEA 是一个过程，而非过程末端的瞬间行为或者附加行为。它应与规划的制定过程同步进行，在各个相关阶段为规划提供相关环境信息。定义也强调了将 SEA 融入决策中的重要性。图 11.1 说明了政策、规划和计划（PPP）的制定与 SEA 之间的联系。

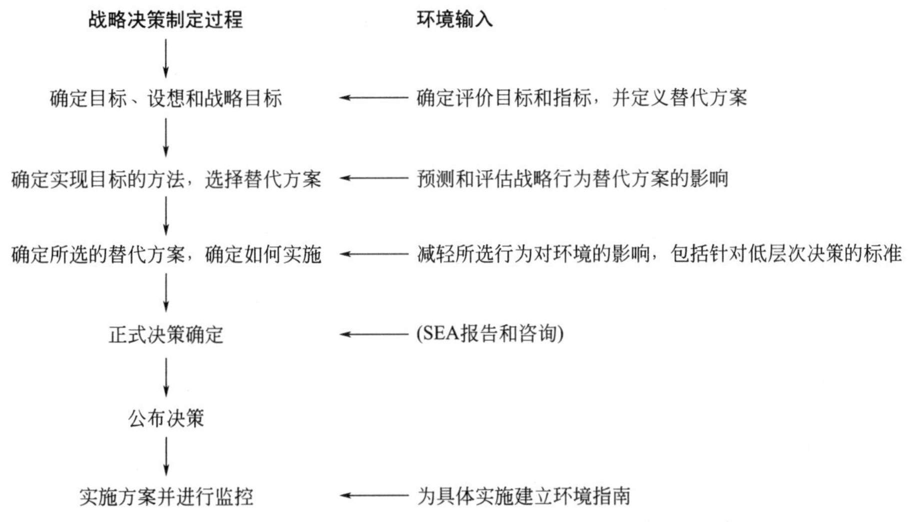

图 11.1　SEA 和 PPP 制定过程之间的关系

政策、规划和计划（PPP）定义的区别。尽管在有关 SEA 的文献中，它们经常被放到一起，但三者彼此并不相同，需要的 SEA 形式也不同。政策通常定义为一种行为的激励和指导（如“供电满足国家的需要”），规划是政策实施的一系列有序的、有时间限制的目标（如“到 2020 年为止具备 X 百万瓦发电能力”），计划是某一特定区域的一系列项目（如“到 2020 年在 Y 地区建成 4 个新的组合循环燃气涡轮发电站”）(Wood，1991)。PPP 与具体部门（如运输、采矿）或某特定区域的所有活动有关（如土地利用、发展或区域规划）。

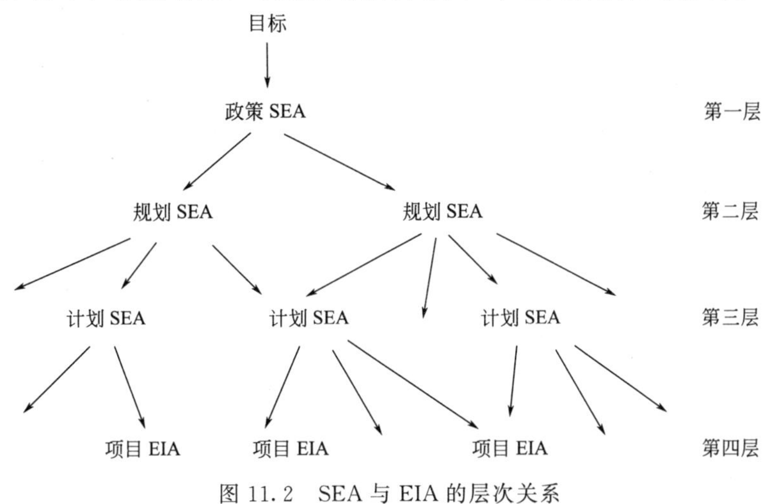

图 11.2　SEA 与 EIA 的层次关系

理论上 PPP 是分层次的：政策为规划的制定提供框架，规划为计划的制订提供框架，计划产生项目。在实践中，这种层次关系是无形且不稳定的，而且没有清晰的界限。针对不同层次 PPP 的 SEA 也是分层次的，如图 11.2 所示，所以在较高层次中考虑的问题就不需要在较低层次中再考虑。不仅仅是开发项目，PPP 也可能会有对环境产生影响的行为，如私有化、不同形式的土地管理或者撤销规划。表 11.1 总结了 EIA 与 SEA 之间的一些主要不同点。

表 11.1　EIA 与 SEA 的主要不同点

比较方面	SEA	EIA
行动的性质	战略、愿景、计划	建设/操作行动
影响的规模	宏观：全球、国家、区域	微观：局部、场地
时间尺度	长期或中期	中期或短期
数据	主要是描述性的，也有一些定量的/图表的	主要是定量的/图表的
替代方案	财政措施，经济、社会或物理策略，技术、地点的空间均衡	明确的替代地点、设计、时机
评价标准	可持续的标准及目标	法律限制及好的实例
精确/不确定性	不精确，不确定性多	较精确，不确定性小
输出结果	粗略	详细

资料来源：Partidario（2003）。

11.2.2　SEA 的必要性

为了使 EIA 更具有战略性，出现了各种各样的争论，这些争论大多都与项目 EIA 体系中存在的问题有关。项目 EIA 只能对开发提案做出回应，而不能提前参与其中。所以，项目 EIA 不能使开发项目选择环境承载力较大的地区或者远离环境敏感的区域。

项目 EIA 没有充分考虑由几个项目或单个项目下的子项目或补充项目所造成的累积影响。举例来说，单个的小型矿产开采可能不需要进行环境影响评价，但是多个此类项目的总影响却是显著的。9.5 节给出了另一个例子。目前，对于此类项目，大多数国家还没有立法要求进行综合累积影响评价。

项目 EIA 没有考虑到一些潜在的破坏性行为所产生的影响，这些行为在项目的审批过程中也没有得以控制[1]。此类行为有：农场的管理实践、私有化和新技术如转基因生物。项目 EIA 不会考虑某种特定类型的项目大体上需要发展到什么样的程度，当然，也完全不会考虑是否需要某个特定的项目。项目 EIA 也不能完全为开发行为提供替代方案类型和选址或针对整个影响范围的减缓措施，因为这些替代方案通常会受到早期的特别是战略层次上选择的限制。很多情况下，在项目已经计划得十分具体且在战略层次上已经不能再更改时才开展 EIA。

由于资金和项目规划申请时间的限制，项目 EIA 通常进行得很快。这样就限制了基础数据的收集数量和分析质量。例如，很多项目的规划期都要求在冬天进行生态影响评价，而在冬天植物都很难识别，动物也大多迁徙或进入冬眠状态。类似地，项目 EIA 中所采用的公众咨询类型和数量也会受到限制。

通过在决策制定过程早期引入 SEA，考虑特定区域或某一类型的所有项目或开发行为，SEA 可以对替代方案进行更好的评估，考虑累积影响，对公众进行更好的咨询，并且可以

[1] 参考 12.3 节有关累积影响的讨论与 9.8 节和 9.9 节两个英国案例的研究。

主动地而非被动地对单个项目做出决策。

SEA有助于促进可持续发展。比如，在英国，SEA通常拓展或结合到可持续评价/评估中去。这不仅将评价范围扩大到考虑社会和经济问题，而且也潜在地确立了可持续性目标并检验PPP是否有助于目标的实现。换句话说，可持续性评价能检验PPP是否有助于促进可持续性发展。

11.2.3 SEA存在的问题

在SEA的早期，由于缺乏经验和技术，限制了SEA的质量。随着SEA实践的发展，这些问题减少了，但是又出现了其他问题。

第一，很多PPP都模糊不清，且其数量增加，发展方式不明，所以不能确定对其进行环境影响评价的最佳时间："政策制定的过程具有动态性，这就意味着在这个过程中对一些问题可能会重复讨论，这个过程很可能需要一系列的行动，即使没得到决策和法规政策的支持"（Therivel等，1992）。在实践中，SEA通常是在规划制定完成之后开展，这时大部分的决策已经制定了。

第二，当仅仅是计划和（或）规划而非政策需要进行SEA时，比如欧盟SEA指令，一项环境不友好的政策能导致环境不友好的规划。在这种情况下，规划层次上的SEA最多能减轻规划的负面影响，而不能为政策考虑更可持续性的替代方案。

第三，正如在第1章中所讲，计划与规划战略层面上的评价应该要为大多场地具体项目的评价提供有用的工作框架，希望减少工作压力，形成更加简明有效的EIA。但是事实证明，对于各种评价间预期的层次关系，其理论要远多于实践，这就导致了一些不必要的活动以及活动的重复。

第四，多项PPP会共同影响同一地区或资源。例如，能源和交通以及很多其他的PPP都会影响气候变化。正如土地利用和交通PPP一样，废物和矿产PPP通常也相互联系。正因为这样，PPP通常很难独立评价。

对于SEA是否应扩展到社会和经济问题上也有相当大的不确定性。在考虑环境问题时，剥离社会和经济问题可能会在决策制定过程中给决策者们带来额外的负担，有助于保持环境评价的完整性。另一方面，可持续发展评估（SA）更能真实地反映决策，而且对于英国的多数PPP来说这是法律上的要求，所以将上述两方面相结合对提高效率是很有意义的。

最后也是最重要的是，政策制定是一个政治过程。决策制定者们会在他们及其政治组织的较大利益基础上权衡PPP中暗含的环境影响。SEA并不能做出最后的决定，仅仅是（有时很让人恼火）给出建议。

11.3 全球战略环境评价

尽管存在这些问题，SEA仍很快在全球范围内得以实施。例如，美国、欧盟成员国和中国都建立了SEA法规；加拿大内阁认为他们有SEA的需求；南非制定了SEA指南；SEA在中国香港特别行政区和其他地方也正在执行。在这一节中，我们讨论美国、欧盟和联合国欧洲经济委员会（UNECE）的SEA系统，因为这些系统发展较好，而且已验证了许多评价方法。这些系统的不同之处在于：是真的需要SEA还是仅仅鼓励这种做法；需要进行 SEA 的战略行为的类型；SEA 是仅考虑了环境问题还是全面可持续性问题以及 SEA 执行的详细程度。

11.3.1 美国

美国没有独立的 SEA 法规。1969 年的《国家环境政策法案》(NEPA) 对此有相关要求：

包括所有的联邦政府机构在内，在对人类环境质量有很大影响的立法和其他主要的联邦行为的建议或报告中，都必须由相关责任的职员做出详细报告。

- 拟议行动的环境影响；
- 如果有无法避免的不利环境影响，提案是否还要实施；
- 拟议行动的替代方案；
- 人类环境的短期使用与生产力的长期保持和提高之间的关系；
- 如果存在不可逆的且不可挽回的资源利用问题，提议行动是否还要实施。(42USC § 4332)

环境质量法庭 (CEQ，1978) 的相关条文对“联邦行为”的解释是“新的以及一些继续的行动，包括由联邦机构全部或部分赞助、支持、实施、管理或者赞同的项目或规划；新的或修订的规则、条例、计划、政策及程序以及立法的提案”。诸如此类 PPP，联邦机构必须准备多达三个阶段的详细评价，直到他们表示不需要下一阶段的评价：①初始分析包括了“绝对的排斥”测验 (此活动不会引起重大环境影响的早期决定)；②环境评价；③规划环境影响声明 (PEIS)。如果环境评价阶段决定这个活动不需要一个全面的 PEIS，它就会将“无重大环境影响的发现”(FONSI) 或者“缓和的 FONSI”作为结论。这表明，采取一定的减缓措施这个行动就不会产生重大的环境影响。

到目前为止，在 NEPA 要求下已经完成了几百个 PEIS，尽管这仅占美国开展的所有评价的很小比例。最近开展的 PEISs 包括西部公共土地的风能开发 (土地管理局)、西南 6 个州的太阳能开发 (能源效率和能源再利用办公室及其他)、放射性废物管理 (能源部) 以及深水区域石油泄漏影响的修复 (国家海洋与航空管理部门及其他)。

美国 50 个州中只有少部分有 SEA 法规。其中，1986 年《加利福尼亚环境质量法案》(CEQA，California，1986) 建立的 SEA 系统开展得最好。与地理学相关的大型项目“规划环境影响报告”(PEIR) 作为预期行为链的逻辑部分，保持与相关的法律法规相一致，在总体上具有相似环境影响的多个单独行为的实施也要求进行规划环境影响报告 (CEQA，15168)。同项目 EIA 一样，PEIRs 必须包括对行为的描述、对本底环境的调查、行为影响的评估、替代方案筛选、对未进行评估的影响的原因说明、咨询的组织以及这些组织关于 EIS 的反应和有关机构针对这些反应所做出的回应。

适逢 NEPA 40 周年纪念日，环境质量委员会准备了关于 SEA 各方面的指导草案，他们觉得 SEA 应该要加强实行力度且更加符合现状：联邦机构要考虑何时、如何将温室气体排放以及气候变化纳入拟议行动中；FONSI 的适宜性以及何时对环境减缓措施的承诺进行监督；绝对排斥的使用；加强运用公共手段监督 NEPA 的活动 (CEQ，2010)。

11.3.2 欧盟与 UNECE

最初希望一个欧洲指令就能覆盖所有的项目和 PPP，但是直到 1985 年 85/337 指令通过时，它的应用还是仅限于项目方面。欧盟成员国经过了25年的谈判和讨论，欧盟委员会最后于2001年7月21日通过了2001/42/EC指令“关于特定规划和计划的环境影响评价”(EC，2001)。指令的完整内容见附录3。该指令于2004年7月21日开始施行。同EIA指令一样，SEA指令并不会直接对某个欧盟成员国产生影响，但却需要进入到每个成员国的法规当中。11.4节讨论了SEA在英格兰的进展情况。

2001/42/EC号指令要求进行规划和计划的（不包括政策）SEA，这些规划和计划是：

① 遵从主管部门的意见。

② 在法律的、规章的或管理条款有要求的。

③ 可能产生重大环境影响的。

④ a. 为农业、林业、渔业、能源、工业、交通、废物管理、水资源管理、电信、旅游、TC&P或土地利用做准备，并且确定需要EIA项目的开发批复框架；

b. 或考虑到可能对选址的影响，在栖息地指令下进行适宜性评价；

c. 或者其他规划和计划，并由成员国建立未来开发项目的批复框架。

专栏11.1总结了SEA的指令要求。起草的规划和计划必须附有讨论当前基本情况的“环境报告”、规划和计划的可能影响和替代方案、如何使负面影响最小化以及提出监控方案。拟议规划和计划连同环境报告必须进行公众咨询，而且准备规划及计划的部门要将报告的信息及讨论结果公布。欧洲指南（EC，2003）关于指令的一些方面做了详细说明。

专栏11.1 欧盟SEA指令的要求

准备一份环境报告，识别、描述并评估该规划实施后可能产生的显著环境影响以及结合目标和计划实施的地理范围考虑合理的替代方案。初始信息包括（第5条和附件Ⅰ）：

(a) 该规划的内容概述、主要目标以及它和其他有关规划、计划之间的关系。

(b) 环境现状的有关情况和不实行此规划可能的发展。

(c) 可能受到显著影响的区域的环境特征。

(d) 有关此规划存在的任何环境问题，特别是具有特殊环境重要性的区域，如79/409/EEC指令和92/43/EEC指令中所列区域。

(e) 在准备期间应该重视建立在国际、联盟或国家层次上的环境保护目标，它与规划、实现目标的方法和一切对环境的考虑都有关。

(f) 可能对环境产生显著影响，包括生物多样性、人口、身体健康、动物、土壤、水、空气、气候因素、物质资源、包含建筑学和建筑学遗产在内的文化遗产、景观以及上面各种因素相互关系等问题（这些影响包括二次的、累积的、协同的、短期的、中间的、长期永久的、暂时的、积极的和消极的影响）。

(g) 防止、减少和尽可能抵消规划实施后对环境产生的任何负面影响的方法。

(h) 简要说明选择替代方案的原因，描述怎样进行评价，包括对待收集所需信息时遇到的困难（如技术缺陷或知识缺乏）。

(i) 按照第10条监控方法进行说明。

(j) 对上述标题下所提供的信息进行非技术性总结。

报告必须包含以下信息：现有评价的知识和方法；规划内容和详细程度；在决策制定过程中所处的位置；为避免重复评价在不同水平上特定事物评价的适度（5.2条）。

咨询

• 有环境责任的专家，当要决定在环境报告中必须包括的内容和信息的详细水平时（第 5.4 条）。

• 公众和有环境责任的专家，在采纳规划之前，在适当时候给他们早期和有效的机会表达他们对规划草案和环境报告的意见（第 6.1 和 6.2 条）。

• 其他欧盟成员国，在规划实施后可能对这些国家产生显著影响时（第 7 条）。

在决策制定中考虑环境报告书和咨询结果（第 8 条）。

为决策提供信息

规划实施后，依照第 7 条，必须将如下信息通知给参与磋商的公众和各个国家：

• 规划被采纳；

• 总结陈述：如何将环境结合到规划中进行考虑、第 5 条的环境报告、依照第 6 条的观点表达和有关咨询结果依照第 7 条，同时将第 7 条纳入与第 8 条的一致性考虑、选择采纳规划的原因以及其他的合理的替代方案；

• 决定监控的方法（第 9 条）。

规划实行后的显著影响监控（第 10 条）。

资料来源：EC，2001

2003 年 5 月，联合国欧洲经济委员会（UNECE）采纳了类似欧洲 SEA 指令的《SEA 协定书》（*SEA Protocol*）作为对它 1991 年《跨界环境影响评估公约》（Espoo 公约）的补充。协定书的要求大部分与欧盟指令类似且相通。协定书要求执行 SEA 的计划和规划类型与欧盟指令大体相同；协定书所要求的环境报告书和指令所要求的报告书非常相似。尽管它仅要求计划和规划进行 SEA，但协定书更关注健康影响，更多涉及公众参与，更多遵从政策和法规要求。虽然谈判是在 UNECE（它覆盖欧洲国家、美国、加拿大、高加索地区和亚洲主要国家）进行的，但协定书对所有联合国成员开放。该协议在 2010 年 7 月正式施行，到目前为止（2011 年 5 月）已有 38 个签署国及 22 个党派。

11.3.3 中国

与美国和欧洲国家相比，中国 SEA 的实践仍然处在一个早期阶段。自从 2003 年 9 月《环境影响评价法》（既适用于规划也适用于项目）施行以来，中国就要求进行 SEA。2009 年 8 月，中国政府基于《环境影响评价法》的基础出台了新的规章，但此规章仅适用于规划。

在中国，有两类规划需要进行 SEA。比较简短的 A 类型评价程序适用于土地利用规划、区域发展规划、水域及海洋开发规划、建设和利用规划以及一些高层次概念规划。对于此类规划，规划局在规划草案中必须要备有环境章节或者注释，而且必须是公开的，并将其与规划草案一起提交到授权中心。

按照要求，许多行业（例如，工业、农业、能源行业及交通运输行业）规划需要进行比较复杂的 B 类型 SEA 程序。这些规划的草案必须包含一份完整的环境影响报告（EIR）；关于规划草案以及环境影响报告，规划局必须要听取相关机构、专家以及公众的意见，如果他们的观点分歧很大，规划局必须组织不同的团体参加后续会议，在最终的环境影响报告中必须包含意见是否被采纳这一细节。相关环保部门必须组成一个审查小组，对环境影响报告书进行审查并提出自己的观点，最后，审批机构必须将该意见作为其决定规划通过与否的依据。

中国的SEA体系有其独到之处：B类规划环境影响报告书的质量检查，正式要求授权机构给予SEA结论足够的重视，强调累积影响与承载力。监督与跟踪程序可以将B类型规划实施的真正影响与EIR的预测影响进行对比。迄今为止，中国SEA中存在的一些缺点包括：个别SEA没有在相关规划中实施，或者有时在规划制定过程中实施得太晚，难以对规划产生影响；事实上，在中国进行SEA的人有些是项目EIA方面的专家，所以有时最终的SEA常常让人觉得像是修改后的EIA（Therivel，2010）。

11.4 英国的战略环境评价

SEA指令对欧洲的SEA实践产生了巨大的影响，同时间接通过UNECE协定产生了世界范围的影响。这一部分把英国SEA实践作为这种影响的一个例子：SEA的历史与立法以及由此产生的问题；SEA中涉及的典型步骤；在英国SEA的有效性。

11.4.1 历史与立法

在英国，为了应对早期的政府指导，从1990年开始广泛开展了一种简略形式的SEA——“环境评价”。环境评价的重点在于根据环境目标的框架预测规划草案的影响。它没有要求收集基本数据或政策内容，不要求考虑替代方案，不要求进行监督。1999年，新的政府指导建议规划局在范围比较大的“可持续性评价中”将规划的社会经济影响及环境影响一起考虑。到2001年10月，超过90%的英格兰和威尔士的地方政府及区域政府已经有了评价的相关经验。大约一半的评价是“环境的”，另一半是“可持续的”（Therivel和Minas，2002）。

2004年，SEA指令的实施产生了更多正式的、严格的、详细的SEA。在英国，SEA指令在英格兰、威尔士、苏格兰以及北爱尔兰通过不同的规章来施行，这些规章都是受共同商定的SEA指令实用指南支持的（ODPM等，2006年）。专栏11.2是实用指南推荐的SEA步骤，这符合SEA指令的要求，但也包含了一些附加步骤（A4、B1、D2、E2），这些附加步骤反映了英国规划制定的过程。其他政府机构的进一步指导阐述了如何在EIA中考虑一些如气候变化与生物多样性等的特殊问题（例如，Environment Agency，2011；CCW，2009）以及对于特殊类型的规划如何开展SEA（例如，PAS，2010；DfT，2009）。

专栏11.2 发展规划的SEA步骤

(A) 确定内容和目标，确立评价基础及确定评价范围

(A1) 确定其他相关的计划、规划及环境保护目标

(A2) 收集基本信息

(A3) 识别环境问题

(A4) 提出 SEA 的目标
(A5) 就有关 SEA 范围的问题进行咨询
(B) 确定和改善 SEA 的替代方案及其评价影响
(B1) 根据 SEA 的目标检查计划及规划的目标
(B2) 确定战略替代方案
(B3) 预测规划草案或计划草案（包括替代方案）的影响
(B4) 评价规划草案或计划草案（包括替代方案）的影响
(B5) 考虑减缓负面影响的措施
(B6) 提出监控规划或计划实施带来的环境影响的措施
(C) 准备环境影响报告
(D) 针对计划或规划草案及环境报告进行咨询
(D1) 针对计划或规划草案及环境报告进行咨询
(D2) 对重大变化进行评价
(D3) 决策制定及信息提供
(E) 监控计划或规划实施
(E1) 确定监控目标及方法
(E2) 采取措施应对负面影响
资料来源：ODPM 等，2006 年

许多关于如何在英国施行 SEA 指令的讨论仍在进行，目前这些讨论的关注点在于应如何将 SEA 与可持续评价（SA）联系起来。《规划与强制购买法案》(2004）是在英国立法实行 SEA 指令的前两个月制定的，要求英格兰及威尔士地区区域及地方层面上的规划要进行 SA，而且没有详述这些 SA 应该包含什么以及应该怎样与 SEA 联系起来。之后的指南（ODPM，2006）建议应该为诸如此类的规划准备联合的 SA/SEA（本质上，SEA 涉及的东西是非常多的，当然也应该包括社会及经济问题)，而不能将 SEA 报告与 SA 报告分离开来，也不能以 SEA 报告为中心而把 SA 仅看作是“SA 附录”。由于在英格兰及威尔士地区，区域及地方层次的空间规划在所有的规划中占了很大的比例，英国机构过去关于 SA 有丰富的经验以及政府对于可持续发展的要求，英格兰及威尔士地区大部分其他规划 SEA 也拓展到了 SA/SEA。然而在苏格兰及北爱尔兰地区这点就没有实现，在这里只要求进行 SEA。

11.4.2 实践中的 SEA 程序

在 SEA 指令施行之前，在英国，规划的 SA 大多是由规划者自己在内部进行。发布的指令，有些 SA/SEA 完全是在内部进行，有些完全由顾问进行，有些是由顾问和内部规划者共同进行。目前没有明确的趋势表明 SA/SEA 到底应该由谁执行，也没有可以使规划及大多的替代性规划发生变化的方法（Therivel 和 Walsh，2005；Sherston 2008)。大体算来，开展一次 SA/SEA，每天需要 6～100 人。这一节剩下的部分将要讨论对于空间规划开展传统的 SA/SEA 会涉及什么。

11.4.2.1 确定内容和目标，建立基础以及确定范围

对其他相关规划、计划以及环境目标进行分析是一项全面且枯燥的事情。通常是通过一个长表格的形式呈现出来，这份表格里会列出其他一些规划、这些规划的内容及其为正在进行的规划带来的其所需的启示。

在 SEA 指令施行后的最初几年里，基本信息大多也是通过表格的形式呈现的。这些表格会向当地政府展示基本数据，作为进行比较的区域或国家层面上相似的数据、相关的目标以及数据来源。这些表格很快会被整理编译且允许将政府制定基线基准，但这些表格很难读懂而且没有提供空间信息。最近，对于 SA/SEA 基础的描述越来越空间化，越来越描述性，例如，展示自然保护区或景观设计的地图，在有些案例中提供了限制或机会的覆盖地图。

部分基线描述也包括预测规划缺失的情形下未来可能的情形。例如，由于 Large Combustion Plant 指令的实施，英国汽车排放标准会更加严格，而且会关闭一些发电站，这些措施又会使英国空气质量得到预期的改善；由于政府海上能源开发的政策，英国周围的海域预期将会受到更多的影响。这些信息使我们可以对规划的累积影响（进行中的规划加上其他规划、项目及基线趋势）进行评价。

已存在的环境或可持续问题通常是作为一个整体来识别。问题包括，哪里的环境目标没有实现，哪里超过了环境标准，哪里的规划区域比相似区域差，整个过程情况持续恶化、当地居民不满意的地方。

对于以目标为导向的 SA/SEA，确立有关环境、经济及社会目标与指标的 SA/SEA 框架。这些目标是作为独立的“量尺”或者一系列规划的影响可能涉及的问题。这些指标对于描述或检测基础环境也有用。表 11.2 展示了典型的 SA/SEA 框架的一部分。对于由基准指导的 SA/SEAs，不需要太多的框架。

表 11.2　典型 SA/SEA 框架的部分示例

SA/SEA 目标	子目标：帮助计划实施	指标
1. 帮助实现机会平等以得到机会	(a)强调机会不平等、被剥夺、被排斥等不平衡现象； (b)增加接受教育、终生学习以及接受培训的机会； (c)增加得到住房及就业的机会，尤其是社会的弱势群体	1.1　在受剥削最严重的10%的区域中所占比例； 1.2　房屋均价与年收入均价的比值； 1.3　每年所提供的保障性住宅单元的数量级比例； 1.4　每1000个家庭中没有住房的家庭的数量
2. 保持及改善空气质量	(a)减少经过该地区的机会，设计新的开发方向，提供公共交通工具，提倡步行及自行车； (b)避免在空气质量可能会危及人体健康的地区开发新的项目	2.1　空气质量管理区域的数量
3. 保护及增加生物多样性及动植物数量	(a)保持和实现国内外重要自然保护区的舒适条件； (b)保持当地特定的地点及优先保护的栖息地范围不变，并改善其质量； (c)保持与改善半自然保护区走廊的连通性	3.1　重要特定地址的数量和范围(以公顷表示)； 3.2　原始森林覆盖的区域； 3.3　优先保护的栖息地总的范围(用公顷表示)； 3.4　国内外条件舒适的特定地点的比例

所有这些信息都会被整理进审查报告中，这个报告会被送至法律顾问［在英格兰，是环境机构、英格兰自然（资源）局及英国遗产局］处 5 周，以便他们对其进行评论。

11.4.2.2 发展与完善替代方案及评价影响

SEA指令需要环境报告以评估规划的影响及替代方案（考虑规划的目标及地理范围）的合理性。尽管存在指导方针（PAS，2008）进行替代方案的识别，SA/SEA中涉及的替代方案的质量也有所改善。但就历史来看，这一阶段还是不尽完善，而且有些SA/SEAs对于替代方案的考虑依旧是限制在存在拟议规划与没有规划之间的比较。在英国，与SEA有关的法律挑战大多都涉及替代方案的进行与评价。例如，考虑到东英格兰地区空间战略的绿化带房屋建设提案时，Justice Mitting得出了这样的结论：

SEA指令要求这一具有挑战的政策要有合理的替代方案，且要在做决定之前对其替代方案进行识别、描述以及评价。而由ERM对此提案做的环境报告并未涉及这一点。但它应当是要涉及这一点的，因而在这项工作进行之前，国务大臣是不能决定是否要采纳这一政策的。忽略法规、不遵守其要求的后果就如结果所示。如果替代方案中没有考虑到侵蚀性这一点，那么就等于是做了一个侵蚀城市绿化带的决策。

相似地，柯林斯司法在2011年规定[1]：

咨询者无法从《森林健康核心战略的SEA》中知悉否决城市开发替代方案的理由或者增加住宅开发不会产生影响的理由。先前的报告中都未能恰当地给出必要的解释及理由，而且不论基于什么样的情况，都未进行充分的总结。在最终的报告中，也没有相关的章节段落。[2]

两个计划中的相关部分都被撤销了。

以目标为导向的影响评价涉及了对于每一个规划替代方案的测试以及对SA/SEA框架中每一个目标的子目标的测试。表11.3展示了一个对规划替代方案进行评价及比较的示例。在每一个案例中，表格的每一栏里，都是一个一个的替代方案或者一个一个的子目标，这些替代方案、子目标：

表11.3 以目标为导向的评价的典型示例

SA/SEA目标	替代方案		
	开发×处，促进就业	开发×处，改善居住条件	在计划阶段保护×处免于开发
1. 帮助实现机会平等以得到机会	— 占用可用于房屋建造的土地	+	0(无变化)
2. 保持及改善空气质量	—— 因修路可能会增加就业	— 会导致交通堵塞，但可以减少某些旅程的距离及时间	
3. 保护及增加生物多样性及动植物数量	? 生物多样性状况还不清楚，需进一步研究		0(无变化)

① 是否清晰地被表述：如果没有，可能要重新编写以使其清晰；

② 是否有负面影响（—）：如果有，这些影响可能需要提出缓解措施，比如重新编写子部分或者增加一个不同的子部分等；

③ 是否有正面影响（+）：如果有，可能需要重新编写，增强其正面效果；

---

[1] 奥尔本城镇委员会与社区和地方政府国务大臣间的对抗，[2009] EWHC 1280 (Admin)。

[2] Save Historic Newmarket有限责任公司及其他公司与森林健康区议会及其他委员会的对抗，[2011] EWHC 606 (Admin)。

④ 是否有不确定影响（?）：如果有，可能有必要在评价完成之前以及规划结束之前收集更多的数据；

⑤ 是否存在决定规划施行方式的影响（I）：如果有，可能要重新编写规划以保证其施行带来的是正面积极的影响，不会有严重的影响（0）。

相反，以基线为导向的影响评价包括对预期的“规划”情形与预期的“无规划”情形的比较，然后做出规划对于事情的改变是会使其更好还是更坏的决定。在“规划”情形比“无规划”情形要糟得多的情况下，要考虑采取减缓措施。

评价阶段的重点不应该集中在变化的标志及其精确数量，而是应该集中于对规划做适当的改变：有按照 SEA 指令要求制定的减缓措施。这一指令暗示了减缓措施的层级性。一般认为，通过采取一些措施（例如，将提议的开发项目从敏感地区搬离或者不允许进行一些特定类型的活动等）来避免或者阻止影响要比削减影响或者将影响最小化（如要求开发商采用特定的工艺或者达到一定的指标）更加合适。补偿措施（允许影响的存在，通过提供一些补偿利益来弥补）是最不合适的措施。

在规划的发展期间，某些特定的规划可能需要进行好几次不同层次水平上的影响评价与减缓措施：

① 较多的战略替代方案（是否要在已存在的城镇的边缘进行房屋建设，还是通过部门干涉分散开来或者建一个新的大型城镇）。这些方案需要在优先替代方案达成一致之前，在计划制定过程中进行评估、比较。

② 关于该规划更多详细的子内容（例如，房屋建设密度及设计的规划政策）。这些需要在规划快要完成之时进行评价与微调。

③ 开发的建议地点（例如，具体的建房地址）。这可能需要在与项目 EIA 接近的水平上进行评价与微调。

11.4.2.3 准备环境报告书

A、B 两阶段的发现都会在规划草案附属的 SA/SEA（或环境）报告中发布，而且公众与法律顾问能够得到这些资料。环境报告书中也包括了 SEA 指令的一些其他要求，也就是在报告的信息整理编辑的过程中会遇到的问题以及提议的监督安排。

11.4.2.4 对规划草案或者项目及其环境报告的咨询

在咨询回复被接受后，必须将其纳入最终规划的考虑范围内。一旦最终规划达成一致，它在公布的同时必须相应地公布一份声明，这份声明要就政府是如何把 SA/SEAs 的发现结果及咨询回复纳入考虑范围做出解释，还要解释鉴于其他合理的替代方案，选择采纳这个规划的原因。这份声明还要确认即将实施的监督措施。

11.4.2.5 监督规划的实施

最后，相关部门必须监督规划实施引起的重大环境影响。

11.4.3 SA/SEA 在英国的实施效果

英国规划者（Therivel 和 Minas，2002 年；Therival 和 Walsh，2005 年；Sherston，

2008年；Yamane，2008年）的一系列调查已经向我们展示了英国SA/SEA体系的效果以及由于SEA指令该体系发生的变化。受其直接影响，超过80%的规划者在报告中提及SA/SEAs程序，这使他们的规划发生了一些变化，而且SEA指令的实施增加了这些改变。然而，对于个别规划政策，其变化数量是有限的。只有小部分规划会因为SA/SEA发生实质性的改变（图11.3和图11.4）。最近的DCLG研究也证实了这一点，DCLG得出结论认为SA/SEA的作用一般是对规划进行微调而不是塑造规划（DCLG，2010年）。

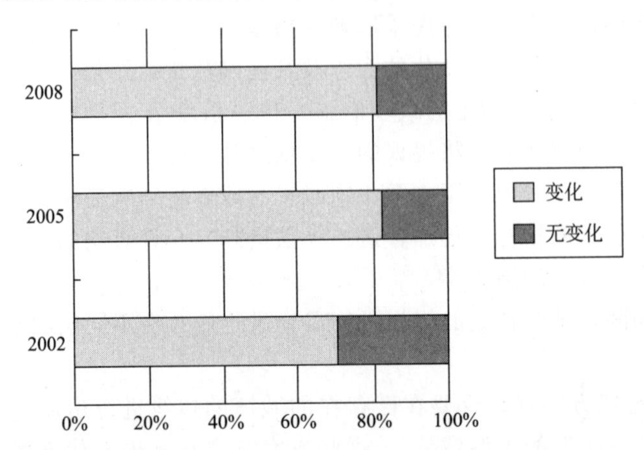

图 11.3　2002、2005及2008年由于SA/SEA规划发生的变化

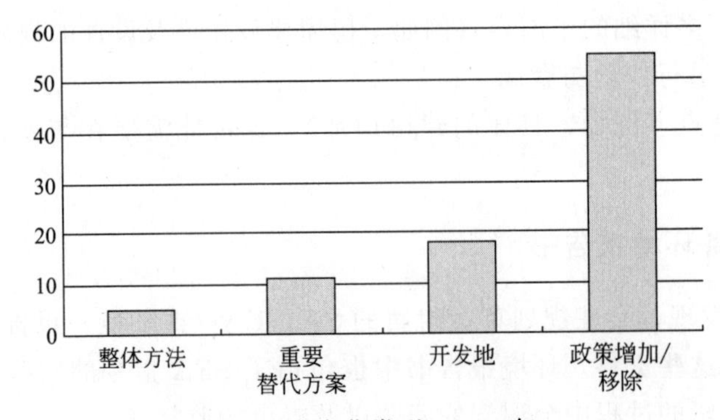

图 11.4　变化类型，2008年

规划者在报告中指出，无论如何，SA/SEA都带来了可观的额外附加效益，这些效益包括增加了对规划及可持续问题的理解，规划制定更加透明化而且启发激励了下一轮的规划制定，见图11.5。规划者也察觉到尽管SA/SEA过程略微偏向于环境，但它可以平衡规划制定的过程。规划制定偏向于支持社会与经济问题（Sherston，2008；Yamane，2008）。

DCLG（2010）的研究得出了一系列改善SA/SEA的建议，其中许多都与项目层次的EIA有关：

① 规划主体应该将规划早期的证据获取阶段与SA/SEA程序整合在一起，以制订一个更加高效且有效的方法。

② SA/SEA基本证据应该包括大量的空间信息，且要反映规划的空间属性。

③ 评价范围应该对所考虑到的替代方案有所反映。

④ 进行评价的人员不应该忽略范围的问题，尽管这一问题经常得不到重视，因而这一问题的处理应当透明化，并给予合理的解释。

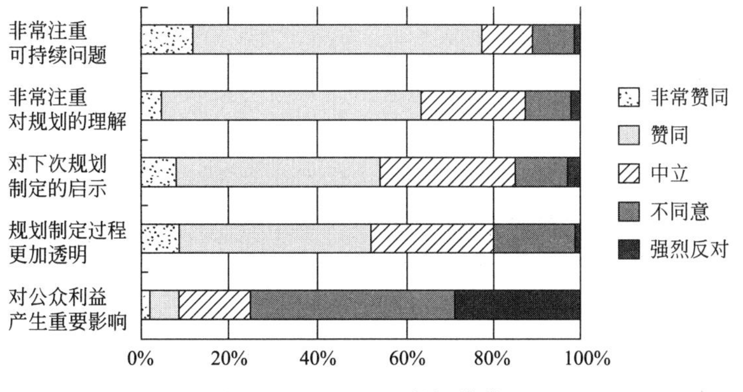

图 11.5 SA/SEA 的间接效益

⑤ 规划制定应该经过深思熟虑并且要将替代方案清晰地表达出来。

⑥ 以基本情境作为参考对规划的影响进行识别和评价。

⑦ 评价具体的细节水平应当要反映出规划的细节水平。

⑧ 评价应该考虑选择和政策可以得到有效落实的程度，要避免不切实际的评价结果。

⑨ 另外，为方便公众参与的进行，应当准备非技术性的易于理解的 SA/SEA 报告总结。

⑩ 有更大的范围让公众参与 SA/SEA，特别是关注选择的涉众事件的使用。

⑪ 一些特定主题的评价可以整合进 SA/SEA 进程中，但是栖息地法规评价要基于大范围独立的基础进行。

⑫ 理想情况下，进行评价的人员应当做出回应，为规划制定者提供明确的建议。

⑬ 加强 SA/SEA 和年度监测报告之间的联系，因为评价中识别的重大影响也是年度监测报告指标的监测对象。见 Hanusch 和 Glasson（2008）对 SAE/SA 中监测的重要性的讨论。

11.5 小结

在过去的 10 年间，SEA 快速地普及应用，而且在可预见的未来也可能会继续如此发展。对 EIA 来说，它最大的限制是它的结果仅仅只能作为参考。在实际中，这会导致经济与社会问题优先于环境问题（Therivel 等，2009）。然而，战略层面上而非项目层面上的影响评价，考虑的问题更广泛（如气候变化与贫困），考虑更多的战略替代方案（如需要发展到什么样的程度，发展的方向是什么）以及考虑更好的分析累积影响。这些也都为项目开发及项目水平上的 EIA 建立了一个有用的框架。这些问题在最后一章 EIA 的前景中将详细论述。

问　题

以下几个问题可有助于读者抓住本章的重点，增加对 SEA 及战略层面上评价的理解。

1. 在英国，典型的 SEA 范围也考虑到了社会及经济问题。这种方法有什么样的优缺点？

2. 提出 SEA 的一个根据是 SEA 无法充分地考虑累积影响。在 11.3 节中所描述的 SEA 体系是否清楚地考虑了累积影响?

3. 图 11.3 是一个城市扩建提议的限制地图。与项目层次上的限制地图相比,有何不同之处?造成这些不同的原因可能有哪些?

4. 表 11.2 列举了一个战略层次上的影响评价。这与第 5 章的项目层次上评价的技术方法有何区别?造成这些不同的原因可能有哪些?

5. 从图 11.5 可以看出,规划者发现 SEA 可以加深他们对于自己所制定的规划的理解。为什么会出现这样的情况?

6. 在 11.2.2 节中所列举的 EIA 问题中,能否根据你自身的 EIA 实践经验列举具体的示例,或者是从本书其他部分找一下示例?你是否认为 SEA 正如本章中所描述的一样,可以解决这些问题?

Part 4

Prospects

11

Widening the scope: strategic environmental assessment

11.1 Introduction

EIA has increasingly been applied at the level of policies, plans and programmes (PPPs) as well as projects. In the USA, this so-called strategic environmental assessment (SEA) has been carried out in a relatively low-key manner since the 1970s as an extension of project EIA. In other countries, SEA roll-out has been slower but has caused more of a splash. The European SEA Directive of 2004 has led to significant changes in the planning system in some Member States, China has passed specific SEA legislation in 2009, and SEA is also a strong growth area in other parts of the world.

This penultimate chapter discusses the need for SEA and some of its limitations. It reviews the status of SEA in the USA, European Union, UNECE, and China. It then discusses in more detail how the European SEA Directive is being implemented in the UK. It concludes with the results of recent research into the effectiveness of the SEA Directive. By necessity this chapter must radically simplify many aspects of SEA. The reader is referred to Sadler *et al*. (2010) and Therivel (2010) for a more in-depth discussion. Chapter 9 presents two SEA case studies.

11.2 Strategic environmental assessment (SEA)

11.2.1 *Definitions*

Strategic environmental assessment can be defined as:

> a systematic process for evaluating the environmental consequences of proposed policy, plan or programme initiatives in order to ensure they are fully included and appropriately addressed at the earliest appropriate stage of decision making on par with economic and social considerations. (Sadler and Verheem 1996)

In other words, SEA is a form of EIA for PPPs, keeping in mind that evaluating environmental impacts at a strategic level is not necessarily the same as evaluating them at a project level.

Several things are important in Sadler and Verheem's definition. First, SEA is a process, not a snapshot or stapled-on addition at the end of a process. It should take place in parallel with the plan-making process, providing environmental information at all relevant stages. The definition also emphasizes the importance of integrating SEA in decision-making. Figure 11.1 shows the links between PPP-making and SEA.

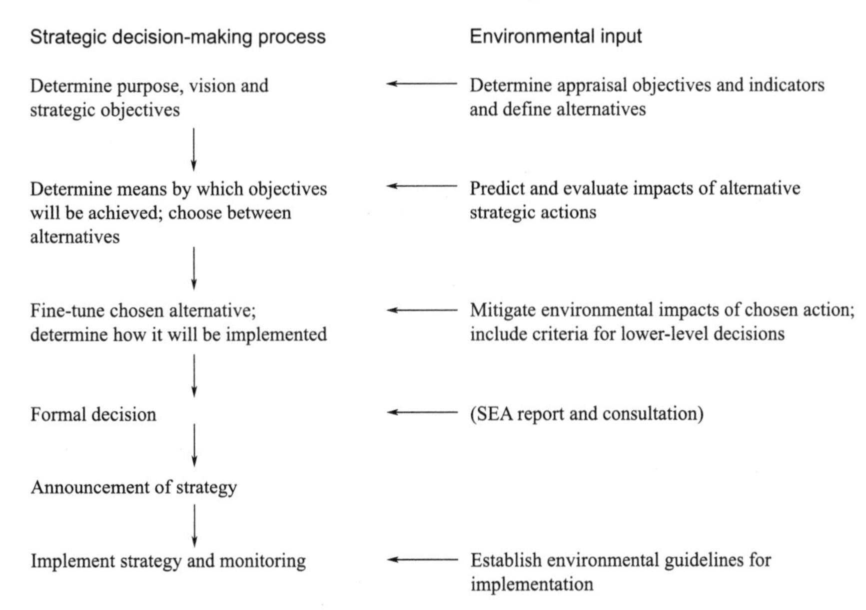

Figure 11.1 Links between SEA and the PPP-making process

The definition distinguishes between policies, plans and programmes (PPPs). Although they are often lumped together in the SEA literature, PPPs are not the same things, and may require quite different forms of SEA. A policy is generally defined as an inspiration and guidance for action (e.g. 'to supply electricity to meet the nation's demands'), a plan as a set of co-ordinated and timed objectives for the implementation of the policy (e. g. 'to build X megawatts of new electricity generating capacity by 2020') and a programme as a set of projects in a particular area (e. g. 'to build four new combined cycle gas turbine power stations in region Y by 2020') (Wood 1991). PPPs can relate to specific sectors (e. g. transport, mineral extraction) or to all activities in a given area (e. g. land use, development or territorial plans).

In theory PPPs are tiered: a policy provides a framework for the establishment of plans, plans provide frameworks for programmes, programmes lead to projects. In practice, these tiers are amorphous and fluid, without clear boundaries. SEAs for these different PPP tiers can themselves be tiered, as shown in Figure 11. 2, so that issues considered at higher tiers need not be reconsidered at the lower tiers. PPPs can also result in activities that have environmental impacts but are not development projects, such as privatization, different forms of land management, or indeed the revocation of a plan. Table 11. 1 summarizes some of the major differences between EIA and SEA.

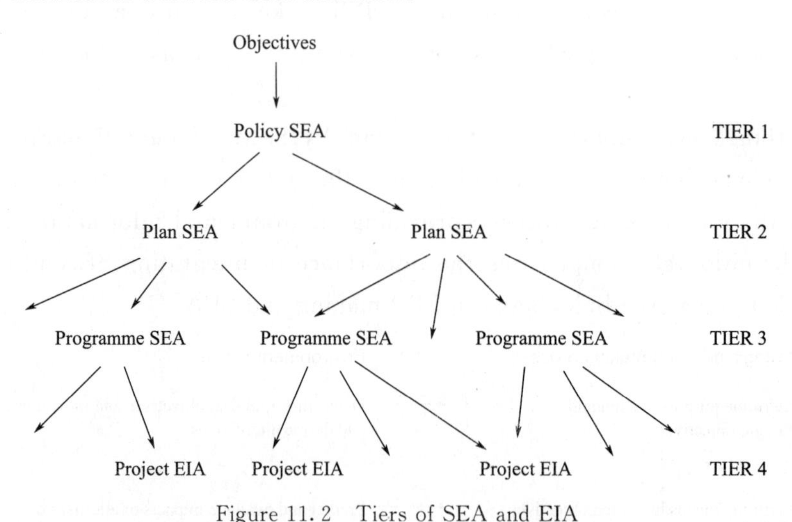

Figure 11. 2 Tiers of SEA and EIA

Table 11. 1 Main differences between SEA and EIA

	SEA	EIA
Nature of the action	Strategy, visions, plans	Construction/operation actions
Scale of impacts	Macro: global, national, regional	Micro: local, site
Timescale	Long to medium term	Medium to short term
Data	Mainly descriptive but mixed with quantifiable/mappable	Mainly quantifiable/mappable
Alternatives	Fiscal measures, economic, social or physical strategies, technologies, spatial balance of location	Specific alternative locations, design, timing
Assessment benchmarks	Sustainability criteria and objectives	Legal restrictions and best practice
Rigour/uncertainty	Less rigour, more uncertainty	More rigour, less uncertainty
Outputs	Broad brush	Detailed

Source: Based on Partidario (2003)

11.2.2 The need for SEA

Various arguments have been put forward for a more strategic form of EIA, most of which relate to *problems with the existing system of project EIA*. Project EIAs react to development proposals rather than anticipating them, so they cannot steer development towards environmentally robust areas or away from environmentally sensitive sites.

Project EIAs do not adequately consider the cumulative impacts caused by several projects, or even by one project's subcomponents or ancillary developments.[1] For instance, small individual mineral extraction operations may not need an EIA, but the total impact of several of these projects may well be significant. Section 9.5 provides another example. At present in most countries there is no legal requirement to prepare comprehensive cumulative impact statements for projects of these types.

Project EIAs cannot consider the impacts of potentially damaging actions that are not regulated through the approval of specific projects. Examples of such actions can include farm management practices, privatization and new technologies such as genetically modified organisms. Project EIAs do not consider how much total development of a particular type is needed, and so they do not consider whether a given project is required at all. They also cannot fully address alternative types/modes and locations for developments, or the full range of possible mitigation measures, because these alternatives will often be limited by choices made at an earlier, more strategic level. In many cases a project will already have been planned quite specifically, and irreversible decisions taken at the strategic level, by the time an EIA is carried out.

Project EIAs often have to be carried out very quickly because of financial constraints and the timing of planning applications. This limits the amount of baseline data that can be collected and the quality of analysis that can be undertaken. For instance, the planning periods of many projects may require their ecological impact assessments to be carried out in the winter months, when it is difficult to identify plants and when many animals either are dormant or have migrated. The amount and type of public consultation undertaken in project EIA may be similarly limited.

By being carried out early in the decision-making process and encompassing all the projects or actions of a certain type or in a certain area, SEA can ensure that alternatives are better assessed, cumulative impacts are considered, the public is better consulted, and decisions about individual projects are made in a proactive rather than reactive manner.

Strategic environmental assessment can also help to *promote sustainable development*. In the UK, for instance, SEA is often expanded or integrated into sustainability assessment/appraisal. This not only involves broadening the scope of assessment to also consider social and economic issues, but also potentially setting sustainability objectives and testing whether the PPP will help to achieve them. In other words, sustainability assessment can test whether the PPP helps to promote a sustainability vision.

11.2.3 *Problems with SEA*

In the early days of SEA, lack of experience and appropriate techniques limited the quality of SEAs. As SEA practice has evolved, these problems have eased but others have emerged.

First, many PPPs are nebulous, and they evolve in an incremental and unclear fashion, so there is no clear time when their environmental impacts can be best assessed: 'the dynamic nature of the policy process means issues are likely to be redefined throughout the process, and it may be that a series of actions, even if not formally sanctioned by a decision, constitute policy' (Therivel *et al*. 1992). In practice, SEAs are often started late in the plan-making process, when major decisions have already been made.

Second, where SEA is required only for programmes and/or plans but not policies, as is the case with the European SEA Directive, an environmentally unfriendly policy can lead to environmentally unfriendly plans: in such a case, the plan-level SEA can at best mitigate the plan's negative impacts, not consider more sustainable policy level alternatives.

Third, and as noted in Chapter 1, strategic levels of assessment of plans and programmes should provide useful frameworks for the more site specific project assessments, hopefully reducing workload and leading to more concise and effective EIAs. But the anticipated tiered relationship has proved to be more in theory than practice, leading to unnecessary and wasteful duplication of activity.

Fourth, multiple PPPs often affect a single area or resource. For instance, energy and transport PPPs—and many others—affect climate change. Waste and minerals PPPs are often integrally interconnected, as are land-use and transport PPPs. As such, it is often difficult to assess a PPP on its own.

There has also been considerable uncertainty about whether SEA should be broadened out to also cover social and economic issues. Considering environmental issues separately from social and economic issues may give them an additional 'weight' in decision-making and helps to keep the integrity of the environmental assessment. On the other hand, sustainability appraisal (SA) more closely reflects actual decision-making, and is legally required for many UK PPPs anyway, so integrating the two procedures makes sense in terms of efficiency.

Finally, and most importantly, policy making is a political process. Decision-makers will weigh up the implications of a PPP's environmental impacts in the wider context of their own interests and those of their constituents. SEA does not make the final decision: it merely (sometimes maddeningly so) informs it.

11.3 SEA worldwide

Despite these problems, SEA has been increasingly carried out worldwide. For instance, the USA, European Union Member States, and China have all established SEA regulations; Canada requires SEA by cabinet decision; South Africa has guidance on SEA; and SEAs are regularly carried out in Hong Kong of China and elsewhere. This section discusses the SEA systems of the USA, the EU and UNECE because they are well developed and demonstrate a range of possible approaches. They differ in terms of whether they require or just encourage the preparation of SEAs; the types of strategic actions that require SEA; whether the SEAs consider only environmental issues or the full range of sustainability considerations; and the level of detail that they go into.

11.3.1 The USA

The USA has no separate SEA regulations. Instead, the National Environmental Policy Act of 1969 requires that

> all agencies of the Federal Government shall include in every recommendation or report on proposals for legislation and other major Federal *actions* significantly affecting the quality of the human environment, a detailed statement by the responsible official on
>
> - the environmental impact of the proposed action;
> - any adverse environmental effects that cannot be avoided should the proposal be implemented;
> - alternatives to the proposed action;
> - the relationship between local short-term uses of man's environment and the maintenance and enhancement of long-term productivity; and
> - any irreversible and irretrievable commitments of resources that would be involved in the proposed action should it be implemented. (42 USC § 4332)

The term 'federal actions' has been interpreted through Council on Environmental Quality regulations (CEQ 1978) as meaning 'new and continuing activities, including projects and programs entirely or partly financed, assisted, conducted, regulated, or approved by federal agencies; new or revised agency rules, regulations, plans, policies, or procedures; and legislative proposals'. For such PPPs, federal agencies must prepare up to three stages of progressively more detailed assessment, until they can show that the next stage is not needed: (1) an intial analysis that includes a test of 'categorical exclusion' (a previous determination that the action would not result in significant environmental impacts); (2) environmental assessment; and (3) programmatic environmental impact statement (PEIS). If the environmental assessment stage determines that the action would not require a full PEIS, it instead concludes with a finding of no significant impact (FONSI) or a 'mitigated FONSI,' which shows that, with mitigation, the action would not have significant environmental impacts.

Hundreds of PEISs have been prepared to date under the NEPA, although these form only a small percentage of all the assessments carried out in the USA. Recent PEISs include those for wind energy development on western public lands (Bureau of Land Management), solar energy development in six southwestern states (Office of Energy Efficiency and Renewable Energy and others), radioactive waste management (Department of Energy), and restoration of the impacts of the Deepwater Horizon oil spill (National Oceanic and Atmospheric Administration and others).

Only a few of the USA's 50 states have SEA regulations. Of these, the SEA system established by the California Environmental Quality Act of 1986 (State of California 1986) is the most well developed. 'Program environmental impact reports' (PEIRs) are required for series of actions that can be characterized as one large project and are related geographically, as logical parts in a chain of contemplated actions, in connection with the issuance of rules or regulations, or as individual activities carried out under the same authority and having generally similar environmental effects (CEQA 15168). Like project EIAs, PEIRs must include a description of the action, a description of the baseline environment, an evaluation of the action's impacts, a reference to alternatives, an indication of why some impacts were not evaluated, the organizations consulted, the responses of these organizations to the EIS and the agency's response to the responses.

In conjunction with the 40th anniversary of NEPA, the Council on Environmental Quality prepared draft guidance on aspects of SEA that they felt needed to be modernised and strengthened: when and how Federal agencies must consider greenhouse gas emissions and climate change in their proposed actions; the appropriateness of FONSIs and when environmental mitigation commitments need to be monitored; the use of categorical exclusions; and enhanced public tools for reporting on NEPA activities (CEQ 2010).

11.3.2 *European Union and UNECE*

It was initially intended that one European Directive would cover projects and PPPs, but by the time that Directive 85/337 was approved in 1985, its application was restricted to projects only. After 25 years of discussion and negotiations between the European Member States, the European Commission finally agreed on Directive 2001/42/EC 'on the assessment of the effects of certain plans and programmes on the environment' (EC 2001) on 21 July 2001. The full text of the Directive is given in Appendix 3. The Directive became operational on 21 July 2004. Like the EIA Directive, the SEA Directive does not have a direct effect in individual European Member States, but instead needs to be interpreted into regulations in each Member State. Section 11.4 discusses how this has been done in England.

Directive 2001/42/EC requires SEA for plans and programmes (not policies) that:

1 are subject to preparation and/or adoption by an authority *and*

2 are required by legislative, regulatory or administrative provisions *and*

3 are likely to have significant environmental effects *and*

4 (a) are prepared for agriculture, forestry, fisheries, energy, industry, transport,

waste management, water management, telecommunications, tourism, TC&P or land use *and* set the framework for development consent of projects listed in the EIA *or*

(b) in view of the likely effect on sites, require an appropriate assessment under the Habitats Directive *or*

(c) are other plans and programmes determined by Member States to set the framework for future development consent of projects.

Box 11.1 summarizes the SEA Directive's requirements. Draft plans and programmes must be accompanied by an 'environmental report' that discusses the current baseline, the likely effects of the plan or programme and reasonable alternatives, how the negative effects have been minimized and proposed monitoring arrangements. The public must be consulted on the proposed plan or programme together with the environmental report, and the authority preparing the plan or programme has to show how the information in the report and the comments of consultees have been taken on board. European guidance (EC 2003) gives more details on some aspects of the Directive.

Box 11.1 Requirements of the EU SEA Directive

Preparing an environmental report in which the likely significant effects on the environment of implementing the plan, and reasonable alternatives taking into account the objectives and geographical scope of the plan, are identified, described and evaluated. The information to be given is (Article 5 and Annex I):

(a) An outline of the contents, main objectives of the plan, and relationship with other relevant plans and programmes.

(b) The relevant aspects of the current state of the environment and the likely evolution thereof without implementation of the plan.

(c) The environmental characteristics of areas likely to be significantly affected.

(d) Any existing environmental problems that are relevant to the plan including, in particular, those relating to any areas of a particular environmental importance, such as areas designated pursuant to Directives 79/409/EEC and 92/43/EEC.

(e) The environmental protection objectives, established at international, community or national level, which are relevant to the plan and the way those objectives and any environmental considerations have been taken into account during its preparation.

(f) The likely significant effects on the environment, including on issues such as biodiversity, population, human health, fauna, flora, soil, water, air, climatic factors, material assets, cultural heritage including architectural and archaeological heritage, landscape and the interrelationship between the above factors. (These effects should include secondary, cumulative, synergistic, short, medium and long-term permanent and temporary, positive and negative effects).

(g) The measures envisaged to prevent, reduce and as fully as possible offset any significant adverse effects on the environment of implementing the plan.

(h) An outline of the reasons for selecting the alternatives dealt with, and a description of how the assessment was undertaken including any difficulties (such as technical deficiencies or lack of know-how) encountered in compiling the required information.

(i) A description of measures envisaged concerning monitoring in accordance with Article 10.

(j) A non-technical summary of the information provided under the above headings.

The report must include the information that may reasonably be required taking into account current knowledge and methods of assessment, the contents and level of detail in the plan, its stage in the decision-making process and the extent to which certain matters are more appropriately assessed at different levels in that process to avoid duplication of the assessment (Article 5.2).

Consulting

- Authorities with environmental responsibilities, when deciding on the scope and level of detail of the information that must be included in the environmental report (Article 5.4).
- Authorities with environmental responsibilities and the public, to give them an early and effective opportunity within appropriate time frames to express their opinion on the draft plan and the accompanying environmental report before the adoption of the plan (Articles 6.1, 6.2).
- Other EU Member States, where the implementation of the plan is likely to have significant effects on the environment in these countries (Article 7).

Taking the environmental report and the results of the consultations into account in decision-making (*Article* 8)

Providing information on the decision:

When the plan is adopted, the public and any countries consulted under Article 7 must be informed and the following made available to those so informed:

- the plan as adopted;
- a statement summarising how environmental considerations have been integrated into the plan and how the environmental report of Article 5, the opinions expressed pursuant to Article 6 and the results of consultations entered into pursuant to Article 7 have been taken into account in accordance with Article 8, and the reasons for choosing the plan as adopted, in the light of the other reasonable alternatives dealt with; and
- the measures decided concerning monitoring (Article 9).

Monitoring the significant environmental effects of the plan's implementation (Article 10)

Source: EC 2001

In May 2003, the United Nations Economic Commission for Europe (UNECE) adopted an SEA Protocol similar to the European SEA Directive as a supplement to its 1991 Convention on EIA in a Transboundary Context (the Espoo Convention). The Protocol's requirements are broadly similar to, and compatible with, those of the EU Directive. Broadly, the same types of plans and programmes require SEA under the Protocol; the environmental report required by the Protocol is similar to that required by the Directive, and the consultation requirements are similar. The Protocol is more focused on health impacts, makes more references to public participation, and addresses policies and legislation, although it only requires SEA of plans and programmes. Although negotiated under the UNECE (which covers Europe, the USA, Canada, the Caucasus and Central Asia), the Protocol is open to all UN members. It entered into force in July 2010, and currently (May 2011) has 38 signatories and 22 parties.

11.3.3 *China*

Compared with the US and Europe, SEA practice in China is still in its relatively early days. SEA has been required in China since the Environmental Impact Assessment Law, which applied to both plans and projects, became operational in September 2003. In August 2009, the Chinese government published new regulations, based on the EIA Law, but which apply specifically to plans.

Two types of plans require SEA in China. The shorter Type A process relates to land use plans, regional development plans, watershed and marine development plans, construction and utilization plans, and high-level conceptual plans. For such plans, the planning authority must prepare an environmental chapter or note that must be made publicly available, and must be submitted to the authorization authority alongside the draft plan.

The more rigorous Type B SEA process is required for a range of sectoral plans, for instance for industry, agriculture, energy and transport. Drafts of these plans must be accompanied by a full environmental impact report (EIR); the planning authority must seek the opinions of relevant institutions, experts and the general public on the draft plan and its EIR; it must arrange follow-up meetings with various parties if they have strongly divergent views; and it must include details in the final EIR of whether the opinions were adopted. The relevant environmental protection authority must form a review group that examines the EIR and submits its opinion, and the authorization authority must use this opinion as the main basis for its decision on the plan.

The Chinese SEA system has particular strengths: the quality check of Type B plans' EIRs, the formal requirement for authorization authorities to give considerable weight to SEA findings, and the emphasis on cumulative impacts and carrying capacities. A monitoring and follow-up process compares the actual impacts of implementing Type B plans against those predicted in the EIR. Weaknesses to date include SEAs not being carried out for relevant plans, or being carried out too late in the plan-making process to influence the plan; and the fact that most of the people who carry out SEA in China are project EIA experts, so the resulting SEAs often feel like modified EIAs (Therivel 2010).

11.4 SEA in the UK

The SEA Directive has had a huge influence on SEA practice in Europe, and, indirectly through the UNECE Protocol, worldwide. This section considers SEA practice in the UK as an example of this influence: the history and legislation of SEA and issues raised by these; typical steps involved in SEA; and effectiveness of SEA in the UK.

11.4.1 *History and legislation*

In the UK, in response to early government guidance, an abbreviated form of SEA— 'environmental appraisal'—was widely carried out from 1990. Environmental appraisal focused on testing the impacts of a draft plan against a 'framework' of environmental objectives. It required no collection of baseline evidence or policy context, consideration of alternatives, or monitoring. In 1999, new government guidance advised planning authorities to consider their plans' social and economic as well as environmental effects in a broader 'sustainability appraisal'. By October 2001 over 90 per cent of English and Welsh local authorities and all regional authorities had had some experience with appraisal. About half of the appraisals were 'environmental' and the other half 'sustainability' (Therivel and Minas 2002).

The implementation of the SEA Directive in 2004 led to much more formal, rigorous and detailed SEAs. In the UK, the SEA Directive is being implemented through different regulations in England, Wales, Scotland and Northern Ireland, supported by a jointly agreed Practical Guide to the SEA Directive (ODPM *et al.*, 2006). Box 11.2 shows the SEA steps recommended in the Practical Guide: these clearly link to the requirements of the SEA Directive, but include some additional stages (A4, B1, D2, E2) which reflect the UK's plan-making process. Further guidance by other government bodies addresses how to consider specific topics such as climate change and biodiversity in SEA (e.g. Environment Agency 2011; CCW 2009), and how to carry out SEA for specific types of plans (e.g. PAS 2010; DfT 2009).

Box 11.2 SEA steps for development plans

(A) Setting the context and objectives, establishing the baseline and deciding on the scope

(A1) Identifying other relevant plans, programmes and environmental protection objectives

(A2) Collecting baseline information

(A3) Identifying environmental problems

(A4) Developing SEA objectives

(A5) Consulting on the scope of SEA

(B) Developing and refining alternatives and assessing effects

(B1) Testing the plan or programme objectives against the SEA objectives

(B2) Developing strategic alternatives

(B3) Predicting the effects of the draft plan or programme, including alternatives

(B4) Evaluating the effects of the draft plan or programme, including alternatives

(B5) Considering ways of mitigating adverse effects

(B6) Proposing measures to monitor the environmental effects of plan or programme implementation

(C) Preparing the Environmental Report

(D) Consulting on the draft plan or programme and the Environmental Report

(D1) Consulting on the draft plan or programme and Environmental Report

(D2) Assessing significant changes

(D3) Decision making and providing information

(E) Monitoring implementation of the plan or programme

(E1) Developing aims and methods for monitoring

(E2) Responding to adverse effects

Source: ODPM *et al.* 2006

Much of the discussion—still ongoing—about how to implement the SEA Directive in the UK has been about how SEA should relate to sustainability appraisal (SA). The Planning and Compulsory Purchase Act 2004, which was enacted only two months before the UK legislation implementing the SEA Directive, requires SA for regional and local level spatial plans in England and Wales, without specifying what these SAs should include or how they should relate to SEA. Subsequent guidance (ODPM 2006) suggested that joint SA/SEAs-essentially SEAs with a wider remit that also covers social and economic issues—should be prepared for such plans rather than, say, separate SEA and SA reports, or an 'SA addendum' to a central SEA report. Because regional and local level spatial plans account for a large proportion of all plans in England and Wales, UK authorities' past wider experience of SA, and government requirements regarding sustainable development, most other plan SEAs in England and Wales are also broadened out to SA/SEAs. The same does not hold true in Scotland and Northern Ireland, which require only SEA.

11.4.2 The SA/SEA process in practice

Prior to the SEA Directive, SAs of plans in the UK were mostly carried out in-house by the planners themselves. Post-Directive, some SA/SEAs are carried out completely in-house, some completely by consultants, and some by a mixture of consultants and in-house planners. There is no clear trend in who is carrying out SA/SEAs, nor what approach leads to the most changes to the plan or the most sustainable plan (Therivel and Walsh 2005; Sherston 2008). Very roughly, an SA/SEA will take 60—100 person days. The rest of this section discusses what carrying out a typical SA/SEA for a spatial plan would involve.

(A) Setting the context and objectives, establishing the baseline and deciding on the scope

The analysis of *other relevant plans, programmes and environmental objectives* is typically very comprehensive and seriously boring. It is usually presented as a long table that lists the other plan, what the other plan says, and what implications this has for the plan in question.

In the first few years after the SEA Directive was implemented, *baseline information* was also mostly presented in tables, which showed baseline data for the local authority, similar data at the regional and/or national level as a comparator, relevant targets, and data sources. The tables were quick to compile and allowed the authority to benchmark its baseline, but were difficult to read and provided no spatial information. More recently, SA/SEA baseline descriptions have become more descriptive and spatial, for instance showing maps of nature conservation areas or landscape designations, and in some cases providing overlay maps of constraints or opportunities.

Part of the baseline description also includes predicting the *likely future situation in the absence of the plan*. For instance, air quality is expected to improve in the UK generally due to tightening European standards for vehicle emissions and the closure of some power stations as a result of the Large Combustion Plant Directive; and the marine areas around the UK are expected to be subject to many more impacts as a result of government policies on offshore energy production. This information allows the cumulative impacts of the plan—the plan plus other plans, projects and baseline trends—to be assessed.

Existing *environmental or sustainability problems* are often identified as a group exercise. Problems include where environmental targets are not achieved, environmental standards are exceeded, the plan area is doing worse than other similar areas, the situation is worsening over time, and things that local residents are unhappy about.

For objectives-led SA/SEAs, an *SA/SEA framework* of environmental, social and economic objectives and indicators would then be set up. This will act as an independent 'measuring stick' or series of questions against which the plan's impacts can be tested. The indicators are also useful for describing and monitoring the baseline environment. Table 11. 2 shows part of a typical SA/SEA framework. For baseline-led SA/SEAs, no such framework would be prepared.

Table 11. 2 Example of part of a typical SA/SEA framework

SA/SEA objective	Sub objective: will the plan	Indicators
1 Help deliver equality of opportunity and access for all	1(a) Address existing imbalances of inequality, deprivation and exclusion 1(b) Improve access to education, lifelong learning and training opportunities 1(c) Improve accessibility to affordable housing and employment opportunities, particularly for disadvantaged sections of society	1. 1 Percentage of areas in the most deprived 10% areas 1. 2 Average house price compared to average annual salary 1. 3 Number and percentage of affordable housing units provided per year 1. 4 Number of homeless per 1000 households

Continued

SA/SEA objective	Sub objective: will the plan	Indicators
2 Maintain and improve air quality	2(a) Reduce the need to travel through the location and design of new development, provision of public transport infrastructure and promotion of cycling and walking 2(b) Avoid locating new development where air quality could negatively impact upon peoples' health	2.1 Number of air quality management areas
3 Protect and enhance biodiversity, flora and fauna	3(a) Maintain and achieve favourable condition of international and national sites of nature conservation importance 3(b) Maintain the extent and enhance the quality of locally designated sites and priority habitats 3(c) Maintain and enhance connectivity of corridors of semi-natural habitats	3.1 Number and extent (in hectares) of enhance designated sites of importance 3.2 Area of ancient woodland cover 3.3 Total extent (in hectares) of priority habitats 3.4 Percentage of features of internationally and nationally designated sites in favourable condition

All of this information is collated into a scoping report, which is sent to the statutory consultees (in England these are the Environment Agency, Natural England and English Heritage) for five weeks, to allow them to comment on it.

(B) Developing and refining alternatives and assessing effects

The SEA Directive requires the environmental report to evaluate the effects of the plan 'and reasonable alternatives taking into account the objectives and the geographical scope of the plan'. Although guidance exists on *alternatives identification* (PAS 2008) and the quality of the alternatives being considered in SA/SEAs is generally improving, historically this stage has not been done well, and some SA/SEAs continue to limit their consideration of alternatives to a comparison of the proposed plan vs. no plan. Most of the successful SEA-related legal challenges in the UK have been around the development and assessment of alternatives. For instance, concerning a proposal in the East of England Regional Spatial Strategy to build housing on the Green Belt, Justice Mitting concluded that

> [The SEA Directive] required that reasonable alternatives to the challenged policies be identified, described and evaluated before the choice was made. The environmental report produced by ERM did not attempt that task. It should have done so and the Secretary of State should not have decided to adopt the challenged policies until that had been done. The consequence of omitting to comply with the statutory requirement is demonstrated by the outcome. A decision has been made to erode the metropolitan green belt in a sensitive area without alternatives to that erosion being considered. [2]

Similarly, Justice Collins ruled in 2011 that:

> It was not possible for the consultees to know from [the SEA for the Forest Heath Core Strategy] what were the reasons for rejecting any alternatives to the urban devel-

opment where it was proposed or to know why the increase in the residential development made no difference. The previous reports did not properly give the necessary explanations and reasons and in any event were not sufficiently summarized nor were the relevant passages identified in the final report. [3]

The relevant parts of both plans were quashed.

Objectives-led *impact assessment* involves testing how well each plan alternative or sub-component fulfils each SA/SEA objective in the SA/SEA framework. Table 11. 3 shows an example of how plan alternatives can be assessed and compared. In each case, the table cells are filled in, alternative by alternative or sub-component by sub-component, noting whether the alternative/sub-component:

Table 11. 3 Example of part of a typical objectives-led assessment

SA/SEA objective	Alternative		
	Develop site X for employment	Develop site X for housing	Protect site X from development for the plan period
1 Help deliver equality of opportunity and access for all	— Takes up land that could potentially be used for housing	+	0(no change)
2 Maintain and improve air quality	— — Would support employment that requires road access	— Would add to congestion, but could shorten the length and duration of some journeys	
3 Protect and enhance biodiversity, flora and fauna	? Status of biodiversity is unclear-requires further study		0(no change)

- is clearly written: if not, it might be possible to rewrite it to make it clearer;
- has a negative impact (—): if so, this impact might be mitigated, for instance by rewriting the sub-component, adding a different subcomponent, etc. ;
- has a positive impact (+): if so, it might be possible to rewrite it to make it even more positive;
- has an uncertain impact (?): if so, it may be necessary to collect further information before the assessment can be completed, and the plan finalized;
- has an impact that depends on how the plan is implemented (I): if so, it may be possible to rewrite the plan to ensure that it is implemented positively; has no significant impact (0).

Baseline-led impact assessment, instead, involves comparing the expected 'with plan' situation against the expected 'without plan' situation, and determining whether the plan would change things for better or worse. Where the 'with plan' situation would be significantly worse than the 'without plan' situation, mitigation measures would be considered.

The focus of the assessment stage should not be on the symbol or the precise quantity of change, but rather on making appropriate changes to the plan: these are the *mitigation measures* required by the SEA Directive. The Directive implies a hierarchy of mitigation. Avoidance or prevention of impacts, for instance by moving proposed development away from a sensitive site or not allowing certain types of activities, is generally considered preferable to reduction or minimisation of impacts, for instance requiring developments to use certain technologies or achieve certain standards. Compensatory measures or offsets—allowing the impacts to happen to providing some kind of counterbalancing benefit—is the least preferable measure.

A given plan may require several rounds of impact assessment and mitigation, at different levels of detail, during the development of the plan:

• Broad strategic alternatives (e. g. whether housing should be at the edge of existing towns, scattered throughout an authority or in one large new town). These may need to be evaluated and compared early in the plan-making process before preferred alternative (s) can be agreed on.

• More detailed sub-components of the plan (e. g. plan policies on housing density and design). These may need to be evaluated and fine-tuned once the plan is closer to completion.

• Proposed locations for development (e. g. specific housing sites). These may need to be evaluated and fine-tuned at a level of detail close to that of project EIA.

(C) Preparing the Environmental Report

The findings of Stages A and B are published in an SA/SEA (or Environmental) Report alongside the draft plan, and made available to the public and statutory consultees. The Environmental Report also covers the remaining requirements of the SEA Directive, namely any problems faced in compiling the information in the report, and proposed monitoring arrangements.

(D) Consulting on the draft plan or programme and the Environmental Report

After the consultation responses have been received, they must be 'taken into account' in the final plan. Once the final plan has been agreed, it must be published alongside a statement that explains how the authority has taken the findings of the SA/SEA and the consultation responses into account, and 'the reasons for choosing the plan ... as adopted, in the light of the other reasonable alternatives dealt with'. The statement must also confirm the monitoring measures that will be carried out.

(E) Monitoring the implementation of the plan

Finally, the authority must monitor the significant environmental impacts of the plan's implementation.

11.4.3 *SA/SEA effectiveness in the UK*

A sequence of surveys of UK planners (Therivel and Minas 2002; Therivel and Walsh

2005; Sherston 2008; Yamane 2008) has given an indication of the effectiveness of the UK SA/SEA system, and changes to that system triggered by the SEA Directive. In terms of *direct effects*, more than 80 per cent of planners report that the SA/SEA process has led to some changes being made to their plan, with the SEA Directive leading to a noticeable increase in this. However most of these changes are limited to additions, deletions or rewording of individual plan policies, with only a limited number of plans being substantially changed as a result of SA/SEA (see Figures 11.3 and 11.4). This is confirmed by recent DCLG research which concludes that 'SA/SEA generally plays a "fine-tuning" rather than a "plan-shaping" role' (DCLG 2010).

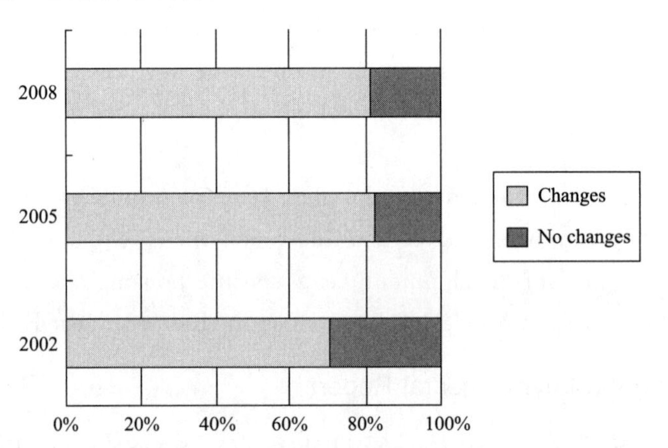

Figure 11.3 Proportion of plans changed as a result of SA/SEA in 2002, 2005 and 2008

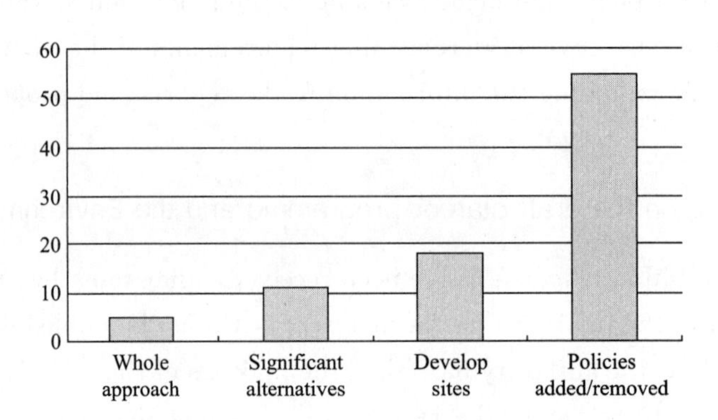

Figure 11.4 Type of changes, 2008

Planners reported, however, that SA/SEA has considerable additional *indirect benefits*, including greater understanding of their plan and of sustainability issues, more transparent plan-making, and inspiration for the next round of plan-making: see Figure 11.5. Planners also feel that the SA/SEA process, although itself biased slightly towards the environment, balances out the plan-making process that itself is biased in favour of social and economic concerns (Sherston 2008; Yamane 2008).

The DCLG (2010) research concluded with a range of recommendations for improving SA/SEA, many of which are also relevant for project-level EIA:

- Planning bodies should integrate the early, evidence gathering stages of the plan-

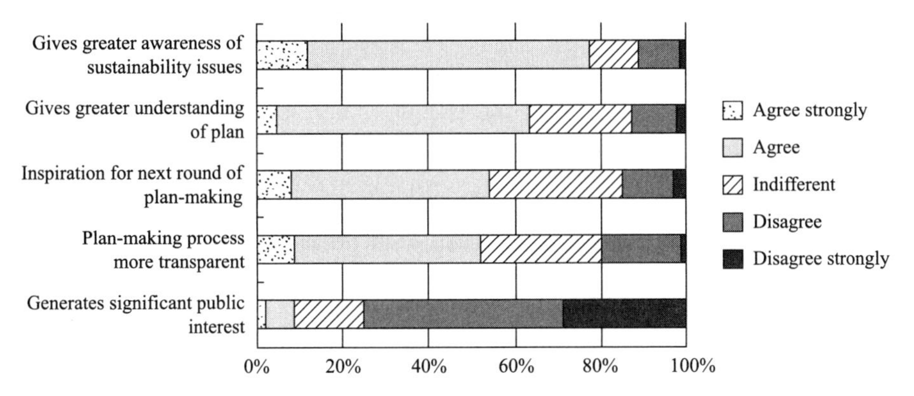

Figure 11.5 Indirect benefits of SA/SEA

making and SA/SEA processes in order to foster a more efficient and effective approach.

- The evidence base for SA/SEA should include a greater focus on sptaial information and reflect the spatial nature of the plan.
- The scope of the appraisal should reflect the alternatives being considered.
- Those undertaking the appraisal should not be afraid to omit from its scope issues that are not likely to be significant; however, this should be done transparently with a clear explanation.
- Plan-making should generate well thought out and clearly articulated alternatives.
- Plan impacts should be identified and evaluated with reference to the baseline situation.
- The level of detail the appraisal enters into should reflect the level of detail in the plan.
- The appraisal should consider the extent to which options and policies will be effectively delivered on the ground to help avoid unrealistic assessment results.
- Separate, understandable non-technical summaries of SA/SEA reports should be prepared to facilitate public engagement.
- There is further scope to engage the public in SA/SEA, particularly through the use of stakeholder events focused on options.
- Some topic-specific assessments can be integrated into the SA/SEA process, but Habitats Regulations Assessment should be undertaken on a largely separate basis.
- Those undertaking the appraisal should ideally provide plan-makers with explicit recommendations to which they can respond.
- Links between SA/SEA and Annual Monitoring Reports should be strengthened with significant effects identified by the appraisal monitored through indicators included in the Annual Monitoring Reports. See Hanusch and Glasson (2008) for discussion of importance of monitoring in SEA/SA.

11.5 Summary

SEA has spread and evolved rapidly over the last decade, and is likely to continue to do so for the foreseeable future. Its main limitation is, like that of EIA, that its findings only have to be 'taken into account'. In practice, this still leads to economic and social issues frequently being prioritized over environmental issues (Therivel *et al*. 2009).

However, impact assessment at the strategic rather than just the project level allows for an improved consideration of wider issues (such as climate change and deprivation), consideration of more strategic alternatives (such as how much development is needed and broadly where it should go) and better analysis of cumulative impacts. These all set a useful framework for project development and project-level EIA. These themes are taken further in the final chapter on prospects in EIA.

SOME QUESTIONS

The following questions are intended to help the reader focus on the key issues of this chapter, and to start building some understanding about SEA and the strategic level of assessment.

1 In the UK, SEAs are typically widened out to also consider social and economic topics. What are the advantages and disadvantages of such an approach?

2 One argument put forward for needing SEA is that EIAs do not adequately address cumulative impacts. Of the three SEA systems described in Section 11. 3, do any clearly consider cumulative impacts?

3 Figure 11. 3 shows a constraints map for an urban extension proposal. How is it different from the project-level constraints map? What might account for the differences?

4 Table 11. 2 shows an example of a strategic level of impact assessment. How does this differ from the project-level techniques of Chapter 5? What might account for the differences?

5 Figure 11. 5 shows that planners find that SEA helps them to have greater understanding of their plan. Why might that be the case?

6 Of the problems with EIA listed in Section 11. 2. 2, can you provide any specific examples from your own EIA practice, or find any in the rest of this book? Do you think that SEA, as described in this chapter, solves these problems?

Notes

1 See Section 12. 3 for a discussion of cumulative impacts and Sections 9. 8 and 9. 9 for two UK case studies.

2 City and District Council of St. Albans vs. Secretary of State for Communities and Local Government, [2009] EWHC 1280 (Admin).

3 Save Historic Newmarket Ltd and others vs. Forest Heath District Council and others, [2011] EWHC 606 (Admin).

References

CCW (Countryside Council for Wales) 2009. SEA topic guidance for practitioners. Available at: www. ccw. gov. uk/landscape—wildlife/managing-land-and-sea/environmental-assessment/sea. aspx.

CEQ (Council on Environmental Quality) 1978. *Regulations for implementing the procedural provisions of the National Environmental Policy Act*, 40 Code of Federal Regulations 1500-1508.

CEQ 2010. New proposed NEPA guidance and steps to modernize and reinvigorate NEPA. Available at: www. whitehouse. gov/administration/eop/ceq/ initiatives/nepa.

DCLG (Department for Communities and Local Government) 2010. Towards a more efficient and effective use of strategic environmental assessment and sustainability appraisal in spatial planning: summary. Available at: www. communities. gov. uk/ documents/planningandbuilding/pdf/15130101. pdf.

DfT (Department for Transport) 2009. Strategic environmental assessment for transport plans and programmes, TAG Unit 2. 11, 'in draft' guidance. Available at: www. dft. gov. uk/webtag/documents/ project-manager/pdf/unit2. 11d. pdf.

Dover District Council 2010. Whitfield Urban Extension Masterplan SPD Sustainability Appraisal. Available at: www. dover. gov. uk/pdf/WUE%20Masterplan%20SPD%20Draft%20SA%20Scoping% 20Report%20REVISED. pdf.

EA (Environment Agency) 2011. *Strategic environmental assessment and climate change: guidance for practitioners*. Reading: Environment Agency.

EC (European Commission) 2001. *Directive 2001/42/ec on the assessment of the effects of certain plans and programmes on the environment*. Brussels: European Commission. Available at: www. europa. eu. int/comm/ environment/eia/full-legal-text/0142 _ en. pdf.

EC 2003. *Implementation of Directive 2001/42 on the assessment of the effects of certain plans and programmes on the environment*. Brussels: European Commission. Available at: europa. eu. int/comm/environment/eia/030923 _ sea _ guidance. pdf.

Hanusch, M. and Glasson, J. 2008. Much ado about SEA/SA monitoring: the performance of English Regional Spatial Strategies, and some German comparisons. *Environmental Impact Assessment Review*. 28, 601-617.

ODPM (Office of the Deputy Prime Minister), Scottish Executive, Welsh Assembly Government, and Department of the Environment, Northern Ireland 2006. A practical guide to the strategic environmental assessment directive. Available at: www. communities. gov. uk/documents/planning and building/pdf/practicalguidesea. pdf.

Partidario. M. R. 2003. Strategic environmental assessment (SEA) IAIA '03 premeeting training course. Available at: www. iaia. org/publicdocuments/ EIA/SEA/SEAManual. pdf.

PAS (Planning Advisory Service) 2008. *Local development frameworks: options generation and appraisal*. London: PAS.

PAS 2010. Sustainability appraisal: advice note. Available at: www. pas. gov. uk/pas/aio/627078.

Sadler, B. and Verheem, R. 1996. *SEA: status, challenges and future directions*, Report 53. The Hague, The Netherlands: Ministry of Housing, Spatial Planning and the Environment.

Sadler, B., Aschemann, R., Dusik, J., Fischer, T. and Partidario, M. (eds) 2010. *Handbook of strategic environmental assessment*. London: Earthscan.

Sherston, T. 2008. The effectiveness of strategic environmental assessment as a helpful development plan making tool. MSc dissertation, Oxford Brookes University.

State of California 1986. *The California environmental quality act*. Sacramento, CA: Office of Planning and Research.

Therivel, R. 2010. *Strategic environmental assessment in action*, 2nd edn. London: Earthscan.

Therivel, R. and Minas, P. 2002. Ensuring effective sustainability appraisal. *Impact Assessment and Pro-*

ject Appraisal 19 (2), 81-91.

Therivel, R. and Walsh, F. 2006. The strategic environmental assessment directive in the UK: one year on. *Environmental Impact Assessment Review* 26 (7), 663-75.

Therivel, R., Wilson, E., Thompson, S., Heaney, D. and Pritchard, D. 1992. *Strategic environmental assessment*. London: RSPB/Earthscan.

Therivel, R., Christian, G. Craig, C. Grinham, R., Mackins, D., Smith, J., Sneller, T., Turner, R., Walker, D. and Yamane, M. 2009. Sustainability-focused impact assessment: English experiences. *Impact Assessment and Project Appraisal* 27 (2), 155-68.

Wood, C. 1991. EIA of policies, plans and programmes. *EIA Newsletter* 5, 2-3.

Yamane, M. 2008. Achieving sustainability of local plan through SEA/SA. MSc dissertation, Oxford Brookes University.

12

项目环境影响评价效力的提高

12.1 导言

总体来说，环境影响评价目前取得的经验可以用一句谚语来总结：副牧师的鸡蛋——好坏参半。本书1.6节里已经简单阐述过环境影响评价程序目前存在的问题，包括评估方法、环境影响评价程序的质量和效率、监控和后决策、管理更加广义和复杂的环境影响评价活动、整体的有效性等。本书接下来介绍了环境影响评价程序的各步骤和最佳实践，第8章讨论了英国迄今为止的环评实践的数量和质量。对成功的实践进行深入的案例研究和对国际经验进行比较研究可以为环境影响评价的进一步发展提供思路。主要参与者——包括顾问、地方主管部门、中央政府、开发商和受影响群体，在环境影响评价程序中所积累的经验可以解释上述部分问题，但在某些情况下仍显得不足。

在欧共体85/337号指令实施将近25年后，人们普遍认同了环境影响评价的价值，对许多方面的疑虑也有所降低，但环境影响评价仍然存在一些无法避免的不足，其质量仍然有很大的提升空间。尽管如此，环境影响评价的实践、基础知识和人们对其的理解都在不断发展，环境影响评价仍将继续处在学习曲线的陡峭阶段。毫无疑问，正如对其他国家的比较研究已经证明的，环境影响评价的流程、程序和实践仍将进一步发展。欧盟国家可以从1988年以后的英国和他们本国的实践中吸取经验。

本章将重点讨论项目EIA的发展前景。接下来的部分将简要讨论环境影响评价程序中各参与者对EIA改进的看法。然后讨论环境影响评价程序的一些重要领域和环境影响报告书的本质等方面的可能的发展。本章最后将讨论环境管理体系和审计的互补和发展。通过以上话题的讨论，给出环境影响评价未来发展的行动清单。

12.2 对环境影响评价改进的看法

任何关于环境影响评价的讨论都要涉及的一个基本主题是改进（change）。本书的不同章节中已经多次体现了这个问题。许多国家的环境影响评价体系和流程都在不断改变。正如O'Riordan（1990）所指出的（1.4节），面对环境价值、政策和管理能力的不断变化，我们应该期望环境影响评价也做出改进。这并不是要贬低环境影响评价所取得的成就。世界银行（1995）也指出，“在过去的10年里，环境影响评价已经从开发计划的边缘部分变成一个广受认可的完善项目决策的工具”。

欧盟成员国在现有体系下开展的环境影响评价实践也取得了长足的进步（第2章、第8章和第10章）。随着欧盟环境影响评价指令修订的完成以及进一步的修正案的提出，这种发展有望在将来继续保持。环境影响评价流程的改进，就像最初引入环境影响评价法规一样，很可能会在政府层面引发大量冲突：如在联邦和各州之间、在国家和地方之间，甚至在欧洲、在欧盟和其成员国之间。也可能在环境影响评价过程中的其他参与者之间引发冲突：如开发商、受影响群体和调解方等（图3.1）。

欧洲共同体委员会（the Commission of the European Communities，CEC）对环境影响评价一直非常积极主动。欧洲共同体委员会在85/337号指令中希望各国制定普通法，对项目应该提供的信息和成功案例共享进行了规定，但又担心不同国家的环境影响评价体系缺乏兼容性、程序实施过程不透明、公众参与被限制、程序缺乏连续性等问题。这给指令的修订带来了很大压力，之后的各次修订中实现了一些目标，包括2004年战略环境影响评价的实施（第11章）等。然而，正如2.8节所提到过的，目前仍然存在一些难以解决的问题，包括：筛选标准不同、跨界问题、质量控制、没有强制要求考虑替代方案、缺乏监控、环境影响评价和战略环境影响评价的关系问题等（CEC，2009）。委员会负责对环境影响评价流程进行评审和更新，这毫无疑问能促进环境影响评价的进一步改进。其他领域包括累积评价、公众参与、经济价值评估、发展援助项目中的环境影响评价流程等也需要予以关注。和欧洲共同体委员会相比，成员国对EIA改进的态度则比较消极和被动。它们在欧盟的动作面前倾向于坚持“辅助性原则”（subsidiarity），这也一直是环境影响评价中的一个问题（CEC，2009）。在全球经济不景气和竞争日益激烈的大背景下，各国政府也对加强经济发展活动的控制非常敏感。

例如，在英国政府内部，社区与地方政府部（DCLG）(原副首相办公室；环境、交通及区域部；环境事务部）一直致力于整理项目环境影响评价流程中含糊不清的地方，对指南和非正式流程等进行完善，但对出台新条例却十分谨慎。不过，社区与地方政府部委托有关机构编制了一些研究报告，如环境影响评价成功案例指南、对环境影响评价中的环境信息和减缓措施的评估和审查以及最近出台的指南和规定等，都反映出政府对环境影响评价的价值的认同。英国的地方政府也开始接受环境影响评价，有迹象表明，这些积累了丰富经验的主管部门（如Essex，Kent，Cheshire）成长非常迅速，它们在实施有关规定和指南时采取用户友好的“定制”形式来帮助辖区内的开发商和受影响群体，并按项目支持者的期望提高标准。例如，最近版的埃塞克斯环境影响评价指南（Essex Planning Officers' Association，2007）非常实用，并且可以在网上免费下载。

压力团体（pressure groups）——在英国指英格兰乡村保护运动（the Campaign to Protect Rural England，CPRE）、英国皇家鸟类保护学会（the Royal Society for the Protection of Birds，RSPB）和地球之友（Friends of the Earth，FoE）等，以及可能受拟议开发行动影响的群体将项目环境影响评价视为一个非常有用的工具，它能增加人们获取项目信息的机会，特别有利于加强对自然环境的保护。这些压力团体一直热衷于改进环境影响评价的程序和流程，例子可参见英国英格兰乡村保护运动和英国皇家鸟类保护学会的报告（1991，1992）。许多开发商对强化环境影响评价流程的热情较小，但仍然希望政府尽快将模糊之处解释清楚（社区与地方政府部，2011），特别是涉及他们的某个项目是否需要进行环境影响评价的时候。调解方（如顾问、律师等）一直对环境影响评价持欢迎态度，特别是对增加和拓展环境影响评价流程很感兴趣，因为这能带给他们更多的业务机会。

英国环境影响评价程序的其他参与者，例如环境管理和评价机构（IEMA)(12.3.2 节)、学术机构和一些环境咨询公司等，正在对最佳实践指南、环境影响评价中的货币评价法和生态系统服务评估、丰富影响研究的类型等课题开展开拓性的研究。另外，英国和许多国家每年要编写几百份环境影响报告书，这催生了大量的专业技术、创新方法和比较研究。为了发现和减少不良实践，对环境影响报告书的评审也越来越多。环境影响评价技能培训也在不断发展。

12.3 环境影响评价程序可能的改进：未来议程概述

12.3.1 环境影响评价有效性的国际研究

最近关于环境影响评价有效性的一些重大国际研究包括上文和第 2 章提到的由欧盟委员会开展的研究（CEC，2009）、由 Sadler 及其同事在 2011 年对国际影响评价协会（IAIA）1996 年发表的环境评价有效性国际研究（International Study of the Effectiveness of Environmental Assessment）进行的跟踪研究（1996，2011）等。正如第 8 章所提到的，在英国的一个关于不断发展的国际有效性研究的研讨会上，Sadler（2010）提出一个“有效性分级(Effectiveness Triage)”理论，为环境影响评价设置了三个筛选项目：什么是必须或应该做的，包括法律、制度框架和方法现状；什么是做了的，包括成功实践案例；其成果是什么，特别是对政策制定和环境效益的贡献。他谈到了目前环境影响评价实践的本质，并为上面提到的第二项列举了很多问题：咨询是环评程序的基础还是被高估而且表现不佳？筛选和确定范围是否聚焦于关键的环境影响？对环境影响的重大程度的评估是否有充足的证据？是否充分制定了减缓措施？监控是否仍是薄弱环节？这些问题与早期国际环境影响评价有效性研究中对国际最佳和最差表现做的实用总结有异曲同工之妙（Sadler，1996；专栏 12.1）。

专栏 12.1 国际上环境评价的“最佳”和“最差”表现

最佳表现

环境评价过程：

- 通过对拟议活动的后果和影响进行清晰的、结构严谨的、公正的分析，使决策更加合理；

• 协助对替代方案做出选择，包括可行性最佳的方案和最能体现环境友好的方案；

• 通过筛选对环境不利的方案和对可行方案进行修正，影响项目选择和政策制定；

• 涵盖所有相关问题和因素，包括累积效应、社会影响和健康风险；

• 对正式批复进行指导（不是指示），包括条款的建立、实施和跟踪的条件等；

• 用常规和定制的技术手段对拟议行动的不利影响及其减缓效果进行准确预测；

• 是一个有适应能力的、有组织的学习过程，取得的经验教训可以被反馈到政策、制度和项目设计中。

最差表现

环境评价过程：

• 应用于拟议开发活动时不一致，一些部门和活动被忽略；

• 作为一个“独立”的过程而存在，与项目周期和批准过程的关联性差，只能产生很小的影响；

• 后续跟踪很弱或不存在，有关条款、条件和效果监控缺乏实施和监督；

• 不考虑累积影响、社会、健康和风险因素；

• 很少或根本没有考虑公众，咨询只是敷衍了事，没有或没有充分考虑受影响群体的特殊要求；

• 环境评价报告冗长、结构散乱、只是描述性的技术文件；

• 提供的信息对政策制定没有太大帮助甚至毫无关联；

• 效率低下，浪费时间，相对于取得的收益而言成本过高；

• 低估且并未减轻环境影响，没有可信度。

来源：Sadier 1996。

2010 年的研讨会特别讨论了环境影响评价背景正在发生快速变化——正如 1.5.1 节里已经提到的，尽管气候变化可被视为加强了环境影响评价的重要性，但经济衰退和严重的财政挑战在“经济重振”之前可能会使对环保问题的支持面临被削弱的风险。英国的地球之友（2011）提供了一个这样的例子。当然，应该通过如进行绿色基础设施投资等手段增加互相的协同，但这需要政府具有开明且有远见的管理思维。

12.3.2 英国环境管理与评价协会对环境影响评价有效性的研究

英国环境管理与评价协会（Institute of Environmental Management and Assessment，IEMA）很及时地对英国环境影响评价的实践状态进行了研究。环境影响评价在英国虽然有 25 年以上的良好发展，但该研究的结论是“（英国的）环境影响评价实践并不完美，如果它要在 21 世纪里继续提供价值，必须要进一步改善才行”。这项研究指出了英国环境影响评价程序中的许多重要问题，涉及诸如筛选方法、确定范围、收集基准数据、评估环境影响的重大程度、减缓和改善、累积影响、监控和适应性环境影响评价（adaptive EIA）等，其中的不少问题已经在本书的前面的章节中讨论过。IEMA 报告为英国的环境影响评价总结了这样一个愿景：“环境影响评价必须不断发展，为开发商、社区和环境服务，帮助他们获得节约资源同时产生改善环境的效果的权利。”表 12.1 提供了实现该愿景的六个关键领域。

表 12.1 实现 IEMA 环境影响评价愿景和促进英国环境影响评价实践在未来取得成功的关键领域

1. 强调交流环境影响评价产生的附加价值：加强交流环境影响评价产生的积极效益（例如环境影响评价引导项目设计的改善）；
2. 意识到一个有效的 EIA 协调（EIA co-ordination）的功效：认识到一个好的 EIA 协调对环境影响评价的有效运行和有效应用所具有的价值；
3. 发展新的伙伴关系，促进环境影响评价程序：衡量与规划者、法律顾问、设计团队、建筑承包商等的伙伴关系，特别是参与管理环境影响评价产出的相关方的伙伴关系；
4. 有效倾听社区的呼声，与社区进行交流互动；
5. 从业者积极协作，共同解决环境影响评价中的难题，为棘手的问题寻求实用的解决方案；
6. 提供现在和将来都有效的环境产出：意识到设计措施的重要性，使其被有效实施后所产生的影响最大化；实施有效的监控

资料来源：摘自 IEMA（2011）。

12.4 环境影响评价程序可能的改进：一些更具体的例子

我们可能做什么？一个比较实际的方法是把未来议程细分成多个改进环境影响评价程序的提案，这通常比那些拓展环境影响评价流程的提案更快且更容易实施，而后者往往提出过程很长且实施起来比较困难。对项目环境影响评价的改进应当包含欧盟指令修正案中已经提到的一些改进，包括改进筛选方法、强制要求考虑替代方案、鼓励在项目开发周期的早期划定评价范围等。另外也应更多地支持流程透明化、鼓励咨询、鼓励对决策进行解释和发布、鼓励给累积影响和风险评估更多的权重等。

评估方法也应得到更多的关注。对未知事物的不确定性可能导致环境影响评价程序启动过晚，无法整合进项目的生命周期管理中。环境影响评价程序和得到的环境影响报告书可能会缺乏平衡，过于侧重对项目及其基准环境的直接描述，而对影响识别、预测和评价关注不足。大多数情况下环境影响评价中用到的预测方法都缺乏解释（8.4 节）。另外，也应该对预测影响等级和确定影响的重大程度（包括采用多标准和货币评价技术）进行更实用的改善。一份好的“方法说明书”应该从技术、咨询、专家和其他相关角色的角度解释研究是如何进行的，它应该是任何环境影响报告书的基本构成要素。

拓展环境影响评价的应用范围，特别是通过引入战略环境影响评价发展多层次的环境评价（正如第 11 章所讨论的）。环境影响评价应用范围的另一个重要拓展是通过更多地利用监控和审计来“完善评价周期”。不幸的是，虽然欧盟指令经过了若干次修订，这一至关重要的步骤却仍然是非强制的。其他更宽泛的改进包括尝试进行“全环境”（whole of environment）评价、更均衡地考虑生物物理影响和社会-经济影响。以上这些拓展都将使环境影响评价更完整。现在还有一个发展趋势叫“环境影响设计”（environmental impact design），即用环境影响评价来识别项目的环境约束和环境限制，而不仅仅是关注项目产生的环境影响。衡量一个项目对未来改变和冲击的抵御能力也可能会成为环境影响评价的内容之一。

下面各小节将讨论一些短期和长期提案的可能性：更多地考虑累积影响，将社会-经济影响纳入评价范围；考虑健康影响评价（health impact assessment）、公平影响评价（equality impacts assessment）、适宜性评价（appropriate assessment）和弹性思维（resilience thinking）等新兴内容；将气候变化纳入环境影响评价；发展综合影响评价（integrated impact assessment)，探讨环境影响设计（environmental impact design）的发展趋势等。

12.4.1 累积影响

许多项目的单独影响很小，但集合起来却会对环境造成重大影响。诸如住宅开发、农业和家庭行为（household behaviour）等活动通常都不在传统环境影响评价的关注范围内。这些活动的集体影响所造成的生态响应可能直到突破阈值前都不会被注意到，直到其以突然和剧烈的形式爆发出来（如洪水）。Odum（1982）提出“小决策效应”（tyranny of small decisions），描述了小型开发活动的持续增长所引发的后果；累积影响也可以用“千刀万剐”（death by a thousand cuts）这个词来形容。虽然目前人们对于累积影响的构成还没有形成共识，但一个被较广泛引用的分类是加拿大环境评价研究委员会（the Canadian Environmental Assessment Research Council，CEARC）提出的分类标准（Peterson 等，1987），它包括：

① 时间跨度小：开发行为在时间上非常接近，以至于上一个影响还未消散，下一个影响又接着发生了；

② 空间拥挤：开发行为在空间上非常接近，以至于其影响相互重叠；

③ 协同效应：当不同类型的影响在同一区域内同时发生时，其影响可能会相互作用，对受纳环境在质量和数量上产生不同的响应；

④ 间接影响：那些在最初扰动的某个时间或距离处产生的，或者是通过一个复杂的途径产生的；

⑤ 蚕食效应：包括对资源的逐步消耗直至其发生显著改变或被用尽。

“累积影响评价就是预测和评价由于时间跨度小、空间拥挤、协同效应、间接影响或蚕食效应引起的，或其他可能在现在、过去以及可以合理预见的环境影响。”（CEPA，1994）

人们很早就认识到环境影响评价应该包含累积影响评价。1970 年的《加利福尼亚环境政策法案》中提出“如果一个项目单独产生的影响很有限，但累积起来却很大”的话，就要被视为具有重大影响。下一个涉及这一问题的立法是 1991 年的新西兰《资源管理法案》，其中明确提到了累积效应；欧盟指令修正案也提到需要考虑那些“会和其他项目一起产生累积影响”的项目特征。加拿大的“累积效应评价”（cumulative effects assessment，CEA）处于世界领先地位，联邦政府和一些省的法律中均对累积效应有明确和强制的要求。英国的环境影响评价指南对累积效应的考虑则比较有限：

总的来说，考虑那些尚未获得许可的开发活动的累积效应并不是明智的选择。不过，在一些情况下，两个或两个以上的开发活动的实施应该被一起考虑。例如，当拟议开发活动彼此之间没有直接竞争关系，其中的两个或全部均可能被批准时；当拟议活动对环境产生的整体影响可能大于单个部分产生的影响的总和时。（社区与地方政府部，2011）

然而，在实际实施累积影响评价时，还是暴露出一些问题和缺陷，成功案例和实用方法直到最近都还非常有限。在澳大利亚，大部分评价都是由主管部门而不是项目倡议人来进行的，并且主要关注的是区域空气质量和流域下游的水质和盐度（CEPA，1994）。图 12.1 是 Lane 及其同事提出的一个简单扰动影响模型（1988）。它本质上是一个“影响树”，将开发活动的诱因、所导致的主要扰动、主要生物物理影响和社会-经济影响以及二次影响联系在了一起。该图显示了大量区域旅游开发活动可能产生的一些潜在累积影响。

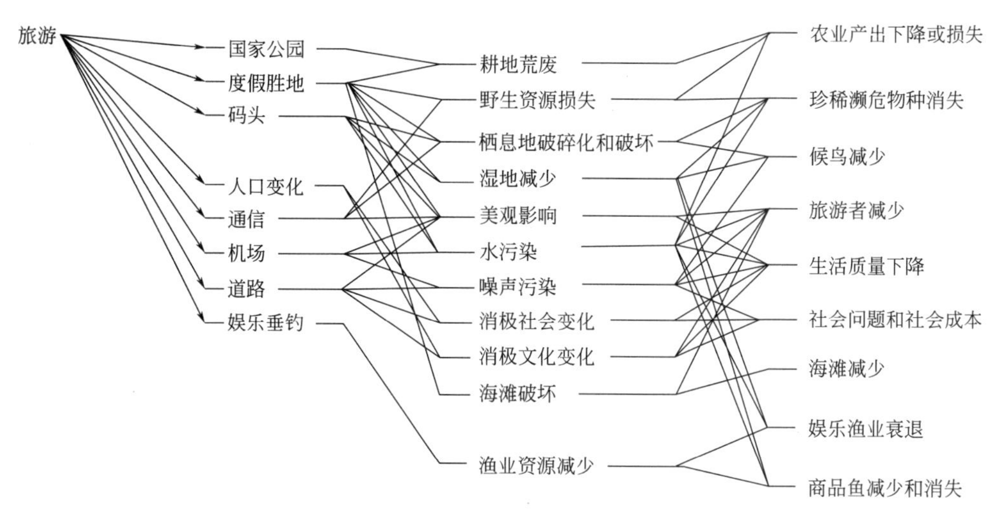

图 12.1　累积影响：扰动影响模型

（资料来源：Lane 及其同事，1988）

我们可以举出一些良好实践指南的成功例子。美国环境质量委员会在大量案例研究的基础上发布了一个实践指南“考虑累积环境效应”（considering cumulative effects）(CEQ，1997)。该指南将累积效应评价分为三个阶段、11 个步骤（参见第 2 章，2.2.4 节）。在加拿大，“累积效应评价从业指南”（cumulative effects assessment practitioners guide）(CEAA，1999）非常实用地概述和解释了相关术语和原理，提供了累积效应评价的操作方法，并对项目倡议人使用的方法做了案例研究。该指南做了一些非常清晰简单的定义——“累积效应是由于某个活动与过去、现在和未来的人类活动共同作用所导致的环境的变化。累积效应评价就是对这些影响进行评价”。加拿大后来一直致力于推进累积效应评价的实践（Baxter 等，2001）。欧盟也通过制定“间接和累积影响评价指南”（guidance for the assessment of indirect and cumulative impacts）(Hyder Consulting，1999）来试着支持相关领域的实践。Piper (2000，2001a、b）通过大量的案例研究，证明英国也进行了累积效应评价实践（见 9.5 节）。

这些指南和评价实践反映出在考虑累积影响/效应时的一些关键的过程问题和组织问题。过程问题包括：确定分析的地理范围（影响区域将有多大），建立分析的进程表（不光包括当前项目，而且也应包括那些在非即时进程表上的过去的和可合理预见的将来的项目），确定影响的重大程度。对英国来说（Piper，2001b），一个关键的组织问题是“哪些组织有责任去开展或委托他人开展累积效应评价？”这个问题在涉及多个主管部门时会非常复杂，而这是累积效应评价中经常遇到的情况（Piper，2001b；Therivel 和 Ross，2007）。

Canter 和 Ross（2010）总结了累积影响评价和管理（cumulative effects assessment and management，CEAM）的近期发展情况，其中“管理”这个词反映出人们越来越关注累积效应的管理和减缓。Canter 和 Ross 提出了一个六步框架，如表 12.2 所示，并讨论了累积影响评价和管理在实践中的“好的”“坏的”和“糟糕的”经验。“好的”经验包括：采纳那些基于已定价值的生态组成部分（valued ecosystem component，VEC）形成的观点；机构/倡议人和公众共同参与确定评价范围；当可合理预见未来的行动存在不确定性时，分情景讨论；推广成功实践。“坏的”经验包括：过于注重生态环境组成，对社会-经济造成损失；文献术语含糊不清，不利于开展研究；指南不充分；缺乏专门和复杂研究。而“糟糕的”经验包括：对累积影响评价和管理的关注太小；缺乏关键决策者的承诺；缺乏多方利益相关者的合作，甚至是态度，导致累积影响评价和管理无法完成等。在此基础上，Canter 等人（2010）总结道：“从各方面来看，分析累积效应将继续是一个持久挑战，虽然有证据表明它的实践正朝着更好的方向发展。”

表 12.2 累积影响评价和管理框架的关键步骤

1. 通过识别拟议项目（或政策、规划或计划）对项目周边的指定 VECs（已定价值的生态组成部分）施加的直接和间接效应，启动累积效应评价程序。
2. 识别时间和空间范围内的任何已经、正在或可能对 VECs 及其指标产生累积效应（压力）的过去、现在和可预见的将来的其他行动。
3. 为指定 VECs 的指标收集适当的信息，并说明和评估这些 VECs 的历史、现在甚至将来的状态。
4. 将累积影响评价和管理研究范围中的拟议项目（或其他 PPP 项目）和其他拟议行动“连接”到指定的 VECs 及其指标。
5. 评估随着时间推移对各 VEC 的累积效应的显著程度。
6. 如果拟议项目将会对 VECs 及其指标施加积极或消极的影响，并且累积效应是显著的，则应对这些影响采取适当的行动或针对性的“减缓措施”

资料来源：摘编自 Canter 和 Ross（2010）。

12.4.2 社会-经济影响

议程中非常重要的一项内容是更好地拓展环境影响评价的应用范围，使其包括社会-经济影响的内容。虽然欧盟和英国的法规里列出的环境受体绝大多数都是生物物理性质的自然环境，但将“人类”也纳入环境影响评价受体的范围表明“环境”的定义变得更宽泛了，增加了人类（如社会、经济和文化）这一维度。由于社会-经济和生物物理影响是相互关联的，将社会-经济影响纳入环境影响评价的范畴有助于更好地识别项目所有潜在的生物物理影响（Newton，1995）。尽早将社会-经济因素纳入环境影响评价可以提供修改项目设计或实施的机会，以便将负面的社会-经济影响最小化，将正面影响最大化（Chadwick，2002）。环境影响报告中包含社会-经济影响，可以通过一致的、公众可获得的文件形式详细地说明项目影响。如果没有包含社会-经济影响，主管部门可能会要求就有关事项补充资料，从而导致环境影响评价过程的延迟。

虽然对社会-经济或社会影响的评价范围有不同的解释，但不少报告都开始强调这方面的重要性（例如，CEPA，1994；国际影响评价协会，1994；Vanclay，2003；Glasson，2009）。Burdge 认为社会-经济影响评价“也许可以提前对拟议行动可能给个人和社会产生的影响进行系统分析”。大部分开发活动决策都要权衡它的生物物理影响和社会-经济影响。此外，开发项目对不同群体的影响是不同的，会不可避免地出现受益者和受害者。但在实践中，将社会-经济影响纳入评价的方式往往五花八门，而且往往很薄弱。一些国家有很好的实践经验，并有相关立法来推动社会-经济影响评价（例如美国、加拿大和澳大利亚的某些州）。国际资助机构也越来越重视这类影响，例如本书 10.9 节所述。

不过，欧洲对社会-经济影响评价的关注仍比较少（Chadwick，2002；Glasson，2009）。由于对此类影响的态度不明确，对相关评价也缺乏最佳实践指南的规范，导致实践方法尚不完善。当涉及社会-经济影响时，它关注的更多的仍是可衡量的直接就业影响。对社会-文化影响（例如分割、异化、社会两极分化、犯罪和健康）的考虑仍然十分有限。虽然现在对健康影响评价的呼声越来越高，但重要的犯罪和安全领域仍然只有一个模糊的轮廓。Glasson和Cozens（2011）重新研究了如果在环境影响评价实践中增加对这些专题的考虑要面临哪些关键问题：需要收集有意义的数据，包括“犯罪恐惧”的内容；创新指标利用的方法；使用环境犯罪学（environmental criminology）领域的证据和概念等。

然而，更全面地考虑社会-经济影响面临着许多问题和挑战，如关于影响的类型、测量、公众参与的角色和其在环境影响评价中的位置等。有一种分类将社会-经济影响分为了：①可量化的影响，如人口变化、拟议项目对就业机会或当地财政的影响；②不可量化的影响，如对社会关系、心理诉求、社会凝聚力、文化生活或社会结构的影响（CEPA，1994）。这些影响的发生范围广泛，大都不易测量，直接询问人们对社会-经济影响的看法往往是记录这些影响的唯一方法。改善公众参与方法和更全面地评价社会-经济影响往往是相辅相成的。社会-经济影响评价应该建立一套指标体系，为公众参与提供框架，以更好地识别与拟议开发活动相关的问题。和环境影响评价中包含的常规内容相比，这些问题可能会更局部、主观、非正式和武断，但却不容忽视。人们如何看待一个项目的影响及这些影响的分布，通常在很大程度上决定着不同群体在该项目中的立场和相关争论。

12.4.3 健康影响评价

健康影响评价（Health impact assessment，HIA）是影响评价中一个重要的发展领域，近年来国际影响评价协会的年度会议中都会有健康影响评价的内容，其受关注的程度可见一斑。有关健康影响评价的学术论文、评论、指南和网站也在快速增长。但什么是健康影响评价？它是如何实践的？它和环境影响评价的关联是什么？

“健康”包括社会、经济、文化和心理的福祉以及适应日常压力的能力（加拿大卫生部，1999）。许多环境因素都会对健康有正面或负面的影响，图12.2是西澳大利亚发布的一个关于健康影响评价的实用概括（西澳大利亚州卫生部，2007）。健康影响是指一个特定地域的人口在一段时期内的现有健康状态的变化。健康影响评价通过结合流程和方法，判断一项政策、规划、计划或项目可能对人群健康产生的影响。它提供了一种非常实用灵活的方式来衡量开发活动和提案对人们健康（health）和健康不平等（health inequalities）的潜在（或实际）影响，从而对提案进行改善和提高（Taylor和Quigley，2002；Taylor和Blair-Stevens，2002；WHO Regional Office for Europe，2003；Douglas，2003）。

健康影响评价在许多发达国家取得了不错的进展，特别是加拿大、荷兰、斯堪的纳维亚半岛的部分地区以及近年来的澳大利亚和英国。一些发展中国家也发现开展健康影响评价很有必要（Phoolchareon等，2003，泰国）。各级政府也提出要制定相关政策，如英国卫生部和英国公共卫生观察协会（Association of Public Health Observatories）的网站。欧盟的战略环境影响评价指令也特别明确提出要考虑规划和计划对人体健康的影响（专栏11.1）。

健康影响评价的主要步骤与环境影响评价相似，包括筛选、确定范围、现状概述（利用现有的健康指标和人口数据，识别评价范围内当前人群的健康状况）、评价（健康影响评价强调要与社区群体进行协商，以确定潜在影响）、实施和决策、监控和跟踪评审（Douglas，2003）。此外，还有很多实用的国家指南，例如前文提到的爱尔兰（IPHI，2009）和西澳大利亚。

由于健康影响评价和环境影响评价在程序和基线数据分类（图12.2表明健康影响评价

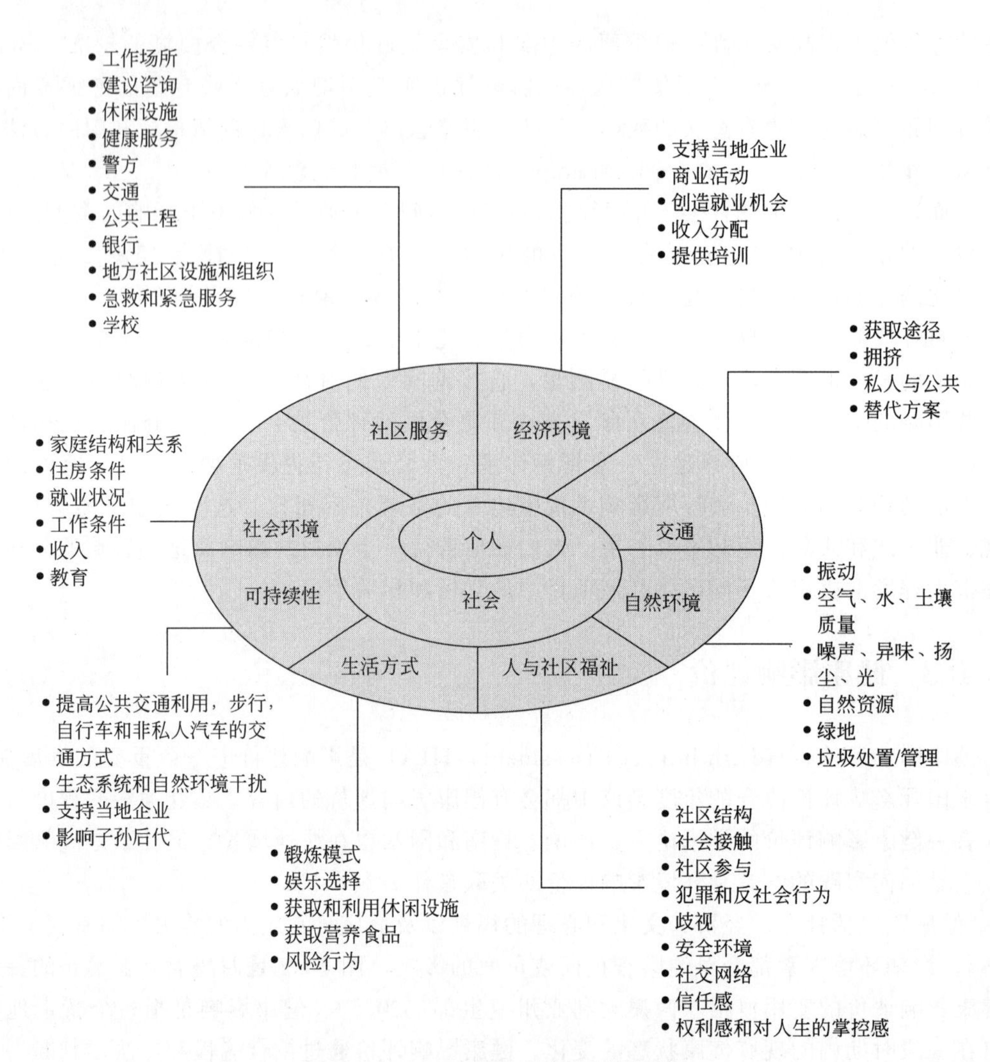

图 12.2　一些会影响健康的环境因素

（资料来源：西澳大利亚州卫生部，2007）

的潜在覆盖范围非常广）上的相似性，人们自然会提出为什么不将二者更好地整合到一起的疑问。Ahmad（2004）对这个问题给出了很多很有意义的解释，包括很难建立人口健康和多种污染物之间的因果关系；环境影响评价的时间往往很紧张，进行此类评价的资源比较有限；某些健康数据是保密的；缺乏强制要求进行健康影响评价的法律框架；进行环境影响评价的专家通常是工程和生态背景。但 Ahmad 也认为更紧密地结合两者可以在共享经验、流程、数据和价值等方面得到许多益处。特别是对于最后一点，健康影响评价可以给环境影响评价“注入平等、透明的利用证据，考虑政策或项目对不同人群的不同影响等价值”（Ahmad，2004）。在更好地整合健康影响评价和环境影响评价这个问题上，欧盟战略环境影响评价 2001/42/EC 指令是一个重要的里程碑，这一概念将在 12.4.8 节做进一步说明。

12.4.4　公平影响评价

本书里多次提到拟议开发活动的影响并非均匀地作用于受影响的各方和各地区，然而环境影响评价基本都忽略这种分布上的不均匀。例如，英国的 Walker 等人（2005）的研究发现，现有的环境影响评价，或者更广泛地说，影响评价方法没有公平地考虑过行政决策或者项目审核的影响的分布问题。一个特定的项目在整体上也许会被评定为是有益的，但可能一些团体或地区承担了大部分不利影响，主要的收益都在其他地方。现在越来越多的文献在关注环境公平问题（或者说更准确地叫作环境不公），其中某些章节提到由于某些地方的社区的环境负担日益增加，陷入社会和环境的宜居性下降、风险增加的恶性循环中（Agyeman 和 Evans，2004；Downey，2005）。

这些问题近年来已经开始得到解决，其中部分问题的解决便是得益于日益发展的公平影响评价。最早开展的是性别影响评价，1995 年在北京举行的世界妇女大会呼吁政府“将性别观点纳入所有政策和方案中，以便在做出决策前分析决策对男女的不同影响”，使性别影响评价在国际上声名鹊起。这一要求后来被写入 1997 年的《阿姆斯特丹条约》的第 2 章和第 3 章中。英国最早的公平影响评价是爱尔兰关于禁止歧视团体的宗教信仰或政治观点的立法，该法案中后来加入了残疾人平等的要求，该公平法案自 2010 年以来已经成为英国支柱性的法律框架。

公平影响评价关注的是项目、规划、方案和政策如何通过不同的方式对不同人群产生任何积极或消极的影响。人群的关键划分因子包括年龄、性别、种族/民族（包括吉卜赛人和流浪者）、宗教、残疾和性取向等，其他划分维度包括城市和农村、贫穷和富裕、有车族和无车族等。公平影响评价的程序步骤和环境影响评价非常相似，人们自然又萌生出了将其整合进环境影响评价程序中的想法。公平影响评价中非常重要的一点是识别可能受拟议开发活动影响的人群的基本公平特征和需求。例如：劳动力是否男性化和年轻化？某些民族在劳动力中的代表性是否显著不足？各种社会-经济问题是否集中在特定城镇的特定地区？在英国，一些空间不平等的具体细节可以用复合剥夺指数（Index of Multiple Deprivation，IMD，2011）确定；贫困相对水平（可以确定到区一级）可以用英国各地区的相对排名位置确定；复合剥夺指数的定期更新还提供了有价值的趋势数据。通过对拟议开发活动的潜在影响进行评估并提出减缓和改善措施可以改进公平状况。表 12.3 对一个公平影响评价中涉及的公平影响进行了总结，该公平影响评价是伦敦市内一个涉及将近 5000 户混合土地使用权住户的集卫生、社区、休闲、教育和零售设施为一体的大型复合重建工程的环境研究的一部分（Scott Wilson，2006）。

表 12.3　EqIA 分析的影响举例

项目	受影响群体	影响
(a)涉及特定公平群体的重大不利影响总结		
健康	精神健康患者，包括黑色人种和少数族裔	拆除和重建导致压力暂时增加
社区凝聚力	黑色人种和少数族裔、妇女，包括单亲、子女、男同性恋和女同性恋人群	搬迁安置/重建导致现有社会网络暂时或永久破坏、人际隔离增加
社区设施	黑色人种和少数族裔，特别是土耳其社区	社交聚会设施的永久丧失
福利	老年妇女	移居带来的临时或永久的孤独感或隔离感

续表

项目	受影响群体	影响
福利	残疾人群	移居和安置过程中个体需求被忽略所导致的临时风险
	老年人群	由于包括既定惯例的改变带来的临时/永久的压力、干扰和焦虑
	儿童	身处重大重建过程导致童年生活环境暂时/长期中断
休闲场地、空地和福利	儿童和年轻人群	大龄儿童和青少年无法使用私人庭院的永久影响，包括对他们的福利和社区凝聚力造成的影响
休闲场地、空地	儿童和年轻人群	施工导致开放空间、公共空间和游乐场地暂时无法利用
(b)涉及特定公平群体的重大有利影响(仅限住房、就业和技能)总结		
住房	黑色人种和少数族裔人群住房	提高住房质量 降低拥挤 年轻人群的新住房 提高保障性住房的置业水平
	妇女/单亲住房	住房更合适幼儿
	残疾人群	独立生活更加方便
	老年人群	保温和供暖改善，家更暖和 可以独立生活的终身住房
	儿童	降低拥挤 住房部门为儿童提供更宽敞的卧室和储藏空间
	所有人群	更大的储藏空间，家庭事故更少
就业和技能	黑色人种和少数族裔人群	使从事建造业 有针对性的技能培训
	妇女	儿童看护机构，使妇女可以去找工作
	妇女	使从事建造业 有针对性的技能培训

资料来源：Scott Wilson (2006)。

12.4.5 适宜性评价

适宜性评价（appropriate assessment）是欧洲特有的一种评价形式，用于评价某个项目或规划对国际重要自然保护地的完整性的影响：鸟类特别保护区，栖息地和物种特别保护区等。《欧盟栖息地指令》(*Habitats Directive*) 第6.3条和6.4条要求进行适宜性评价：

6.3 任何不直接与保护区有关或保护区必需但会对保护区产生重大影响的规划或项目，不论是单独还是作为其他规划或项目的一部分，都必须对其实施后对保护区的保护对象造成的影响进行适宜性评价。根据评价结论和第4段中的规定，国家主管部门只有在查明它不会对保护区的完整性造成不利影响并适当征集公众意见后，才能批准该规划或项目。

6.4 当评价认为规划或项目实施可能会造成不利影响且缺乏其他解决方法时，如果规划或项目具有能够压倒公众利益（包括社会和经济方面）的必须实施的原因，成员国应采取一切必要的补偿措施确保欧盟2000自然保护区（Natura 2000）的整体的一致性（overall coherence）得到保护。并告知委员会所采取的补偿措施。

适宜性评价的内容可以写一整本书，但出于简明扼要的考虑，本书只介绍其中与环境［主要是特别保护区（Special Protection Areas，SPAs）］和特殊保育区（Special Areas of

Conservation，SACs）有关的具体部分，考虑累积［结合（in combination）］影响，论述得十分谨慎（某个规划或项目只有在被证明不会对保护区完整性造成重大影响且所有严格的测试都通过的情况下才会被批准）。适宜性评价之所以会叫“适宜”，是因为评价的详细程度以及该评价可以视项目/规划和相关特别保护区/特殊保育区的情况而被中止。欧洲委员会（2000）发布了一个指南，解释了适宜性评价应该按照如下四个步骤开展以及各步骤得到什么样的发现时下一个步骤才需要被进行：

- 筛选：“结合”其他规划或项目，确定一项规划是否可能会对欧洲保护区产生重大不利影响。
- 适宜性评价：“结合”其他项目或规划，在考虑到保护区的结构、功能和保护目标的前提下，确定该规划对欧洲保护区的完整性的影响。当会产生不利影响时，评估这些影响的潜在减缓效果。当不会产生不利影响时，项目可以继续按计划进行。
- 替代方案评估：当项目被认为会对欧洲保护区的完整性产生不利影响（或存在这样的风险）时，应研究是否有替代途径在实现规划目标的同时避免对欧洲保护区完整性产生不利影响。
- 对没有替代方案且会产生不利影响时的评价：当规划具有能够压倒公众利益的必要理由，被认为必须进行时，评估相应的补偿措施。

适宜性评价的一个典型例子是鹿特丹港扩建项目。该项目将对大约3000公顷海洋和黄条背蟾蜍的栖息地产生重大影响，同时创造上万个长期工作机会，但没有替代方案。该项目被批准的条件是补偿一个新的海洋保护区、25000公顷保护面积和新沙丘。另一个对照案例是英格兰南部Dibden海湾的集装箱码头（港口）项目，该项目会影响到Solent和Southampton水域特别保护区的完整性。该项目未能通过适宜性评价的原因是英国的其他港口可以提供足够的集装箱吞吐能力，替代方案可行。

12.4.6 气候变化和环境影响评价

气候变化是全球所有国家面临的一个根本挑战。而环境影响评价（和战略环境影响评价）似乎是与其直接相关的且非常适合的工具：评估开发活动对气候变化的影响、气候变化对开发活动的影响以及寻求合适的减缓和适应措施。欧盟《环境影响评价指令》的附件Ⅳ给出了需要开发商提供的重要信息：

3. 对拟议项目可能产生重大影响的各个环境要素进行描述，特别包括人口、动物、植物、土壤、水、空气、气候因子、重大资产……以及这些因素之间的相互作用关系。（作者强调）

英国政府（社区与地方政府部，2007）进一步指出“当必需的信息可以通过……环境影响评价提供时，地方规划管理部门不应再进行具体和单独的（气候变化）评价”。2009年，一项由英国环境管理和评价机构（IEMA，2010a）进行的环评从业者调查发现，88%的从业者认为当有必要时，碳排放应该被考虑进环境影响评价并被写入报告书里。

然而，最近的实践表明，环境影响评价并未在气候变化问题上发挥其全部潜力。虽然《环境影响评价指令》里明确提到了气候因素，但2009年对欧盟《环境影响评价指令》的第7次评审（CEC，2009）指出，指令并没有对气候变化问题进行明确表述，各成员国也意识到他们并未在环境影响评价实践中对这个问题进行充分识别和评价。Wilson

和 Piper（2010）指出了加拿大的一个类似的例子，一份来自加拿大环境评价机构（CEEA，2003）的报告表示气候变化在大多数环评报告中没有得到良好的体现。他们举出了数个理由来解释这种限制，包括，例如与气候变化相比更短的时间范围、应对气候变化不确定性的困难、EIA 的碎片性（本书第 1 章）和联系不同因素间相互关系的困难（CEC，2009）。

通往环境影响评价和气候变化的道路究竟应该是什么样的？英国环境管理和评价机构（IEMA，2010a、b）提出了一些与减缓和适应气候变化有关的评估原则。表 12.4 列举了一些最高原则和具体的评价原则。

表 12.4　环境管理和评价机构提出的减缓温室气体排放的环评原则

最高原则

• 所有项目的温室气体排放都会贡献于气候变化；是最重要的相互关联的累积环境影响。

• 气候变化的后果可能会对《环境影响评价指令》中的所有因素产生重大环境影响，如人口、动物、土壤等。

• 英国已经设立了具有法律约束力的温室气体减排目标，因此，环境影响评价必须考虑一个项目将如何贡献于实现这些目标。

• 温室气体排放造成的综合环境效应正逼近科学界定的环境极限，因此，一个项目的任何温室气体排放和减排都应被认为具有显著影响。

• 环境影响评价过程应该在早期影响项目的设计和选址，优化温室气体排放水平，限制可能的温室气体排放贡献。

更具体的评价原则

• 在筛选阶段，应综合考虑气候变化问题及减缓的可能性，确保这些问题被整合进项目设计中。

• 在界定温室气体排放的范围时，必须考虑到相关（从地方到全球的）政策框架，并回顾所有可持续性评价/战略环境影响评价得到的有关成果。

• 在评估替代方案时，必须结合一系列环境标准对各选项的相关温室气体排放水平进行考虑。

• 划定与温室气体排放有关的基线时，必须考虑到相关政策框架和当前情景，在可能的情况下，还应考虑到可能的未来基准情景。

• 对温室气体排放进行定量分析（如碳计算器）不是环境影响评价的必需内容；但如果用到定量分析（如建设实践中的排放趋势），就必须做到完整、透明和合理。

• 评价必须考虑全生命影响（如隐含能耗，相关施工期、运行期和退役期的排放等）。

• 应当基于一个项目的净温室气体影响（net GHG effects）判断其排放是否显著，该值有可能是正的（减排）也有可能是负的（增排）。

• 如果温室气体排放无法避免，环境影响评价应努力减轻项目在各阶段——设计期、建设期、运营期等的排放造成的环境影响。

• 如果温室气体排放量仍然显著，且不可被进一步削减，应考虑采取措施对项目剩余排放量进行弥补

资料来源：英国环境管理和评价机构（2010a）。

12.4.7　弹性思维

项目规划通常假定在未来发生的变化是渐进的和可预测的，然而现实中这些变化通常发生得十分急剧且不可预测：洪水、火山爆发、流行疾病、经济危机、停电等。弹性思维并不是要建立一个能保护人群和抵抗所有消极的未来变化（如风险评估）的系统，而是要使人们在突发情况发生时有能力应对这些打击。相当于教给孩子们安全骑车的技巧和自信，而不是害怕发生意外和被人欺负就不让他们去上学。

弹性思维背后是一些微妙而复杂的原理，本书只能对其做简要概括。首先是变化发生的必然性以及适应周期（adaptive cycles）的概念。所有社会-经济系统都必然要经历低增速和积累的初始阶段，不论最终形成的是一片森林还是一个社区团体。初始阶段之后往往接着一

个短而陡峭的下降，它就是突发情况带来的打击，如森林火灾或社会团体关键成员的死亡等。根据系统的抵御能力，打击的结果既有可能是重组为一个和原来同样“好”的新状态——如一片新的森林，也有可能是比原来要差的状态，如森林成为不毛之地、社会团体解散等。一些举措如扑灭森林火灾等可以延缓这一下降阶段，但也只能简单延迟和升高这一变化发生造成的影响。

其次是阈值或临界点的重要程度。社会-环境系统有一定的自我恢复能力，但如果超过了一定的阈值，系统就会进入到一个新的状态，且往往极难再恢复。例如，将一个已经富营养化的湖泊恢复到健康状态要比防止一个健康的湖泊发生富营养化难得多。

最后是“慢变量”(slow variables) 的重要程度——如气候、土壤、全球经济系统、社交网络等。这些系统在功能上相当于一些更小、更快的系统（如生境、物种、社区和个人等）的缓冲器。然而，当这些慢变量本身被消耗时（如排放温室气体或土壤侵蚀），其缓冲功能也将被削弱。

以上三点对重大项目的规划意味着什么呢？一些类型的开发活动，包括项目类型和它们的设计和实施，将比其他开发活动更具弹性。弹性项目将：

(1) 拥抱变化，而不是试图控制变化　与其增加防洪投入、无库存生产和使用空调，不如设计应对洪水的项目、使工业生产过程不会因迟到的部件而受阻、在办公室和公共交通上打开窗户保持通风等。

(2) 在构建时保留一定的冗余或重复　这样的话当其中一个方面失效时另一个可以及时接替。如果BP公司的石油平台有备用的石油防喷器，2010年的墨西哥湾漏油事故将只会是一个很小的事件。修建公路和铁路时可以在主要干道的中央分车带设置应急门以方便车辆掉头，确保当事故、洪水或堵塞发生时能灵活处理。房地产开发时的“备用”土地可以在需要时转换为粮食产地或临时避难所等。

(3) 保留一定的模块化或不连贯　这是考虑到过度连接的系统可能容易被冲击并且传递冲击的速度很快。中世纪的村庄在瘟疫时期会切断其同外部的联系。这在现代就相当于堤坝、堤岸、安全屏障/安全门和其他通道控制等。

(4) 认识到慢变量的重要性　这意味着在项目规划和决策时要对一些问题分配更大的权重，如高品质农业用地的丧失、水循环、温室气体排放、习惯和语言的丧失、社区安置等。

(5) 在人类活动和环境结果之间建立更紧密的反馈回路　我们造成的许多影响都不会发生在我们身边：我们的粮食和衣服在其他国家生产，我们的能源生产和废物处置在英国的其他地区进行。建立反馈回路包括更强调地区供应而不是集中生产（如能源、水、粮食），在垃圾处理项目选址、社区发展和基础设施所有权问题上强调就近原则等。

(6) 以各种形式（环境的、社会的和经济的）促进和维持多样性　这一原则看上去与通常的政策制定思路相悖，后者强调提高效率、目标和保障，但多样性使得人们在面临冲击时能做出不同的反应，并提供未来选择的渠道。促进和维持多样性的例子包括保护生态多样性地区、为新技术（如各种形式的潮汐能或碳储存）提供项目试点、保护土著居民的生活方式等（Walker和Salt，2006）。

表12.5展示了弹性评价（resilience assessment）的结果和与传统项目规划的结果有何不同。

表 12.5　弹性思维和传统项目规划结果的对比

例子	传统项目规划的结果	弹性思维的结果	例子来源
巴巴多斯的一个非常小规模（三人）的生物柴油生产商有助于减少进入填埋场的厨余垃圾数量，并能提供"本地"燃料，但和其他竞争者相比，能用作燃料的废弃食用油很有限，且生产规模小使得生产成本很高	扩大生产规模，机械化，提高生产效率	促进小规模生物柴油生产商之间相互合作；将生物柴油开发为旅游项目，获取旅游收入和销售收入	Gadreau 和 Gibson(2010)
由于基础设施老化、人口和牲口增加，塔吉克斯坦的两个小村庄正面临水资源短缺问题	新建水库	培训当地人修理和维护现有的水利基础设施，改进计划生育，通过规则和配额来规范用水	Fabricius 等(2009)
农业径流中越来越频繁的"高泥沙洪水"每年给英国和比利时造成几百万欧元的损失	改善防洪措施	改进"档案记忆"：梳理对引发洪水的因素和如何管理洪水的历史认识，衡量这些历史认识的价值并予以维持	Boardman 和 Vandaele(2010)

弹性联盟（Resilience Alliance）还开发了一个弹性能力评估框架（RA，2010）。这个评估框架和环境影响评价的不同之处在于它更强调不确定性和干扰（包括过去的和未来的），考虑更高规模的行动或事件如何对项目规模产生影响，识别变化发生的阈值以及这些阈值被突破时可能导致的变化结果，并明确关注管理这些变化的治理机制。我们并不清楚目前是否有弹性评价和环境影响评价结合使用的例子，但很明显，弹性评价具有强化环境影响评价的潜力，并可使环境影响评价项目更好地应对未来变化，因此，我们认为弹性思维在未来可能会被整合进环境影响评价中。

12.4.8　综合影响评价

正如本书在开头时（1.5.3 节）提到的，环境评价的相关术语出现了爆炸式的增长。其中一个术语就是综合环境评估，它既涉及环境主题也涉及技术。其中，主题（themes）指的是将前面讨论过的社会-经济、健康、公平和弹性等内容更明确地纳入环境影响评价的范围，这就是综合影响评价（integrated impact assessment，IIA），综合影响评价的决策部分取决于各种生物物理、社会和经济影响的可交易程度（图 12.3）。例如，决策者可能不太愿意交易那些对就业或生活方式很重要的生物物理资产（如主要河流系统和供水水质），但愿意交易那些不太重要的生物物理资产。综合影响评价和传统环境影响评价的不同之处在于它有意识地使用跨学科的工作方法，认为公众参与或环境影响评价的最终用户不是一成不变的，并且认为大多数环境决策复杂且充满不确定性（Bailey 等，1996；Davis，1996）。因此，综合影响评价在评估和判断替代方案时使用的方法和观点（定量和定性的，经济学和社会学的，计算机建模和口头陈述的）更加宽泛。然而，这种综合本身也有一些问题，如评价方法的可转移性就存在一定局限（参见项目评估，1996）。在 2002 年的国际影响评价协会年会上，综合评价研讨分会重点讨论了综合评价中的社会过程（social processes）等一系列问题（国际影响评价协会，2002）。

另一个同样重要的观点是对有关规划、环境保护和污染过程进行综合。一个极端例子是

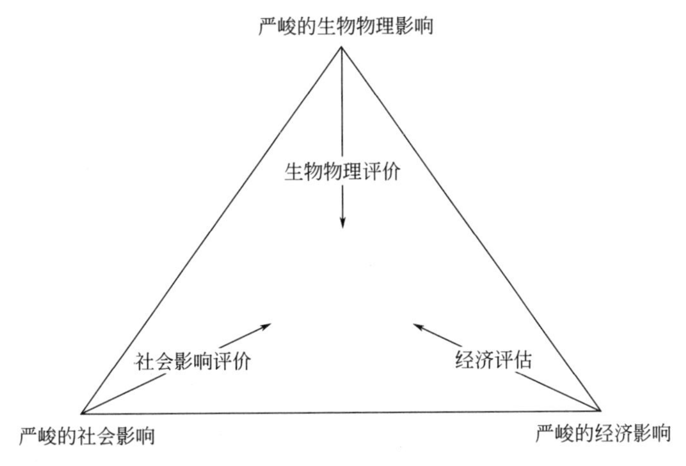

图 12.3 综合影响评价

英国仍有多个环境影响评价法规，环境影响评价体系的流程分散于一系列有关规划和其他法律中；新西兰的“一站式”资源管理法案则是另一个极端。可以说，如何更好地综合相关程序是大多数环境影响评价体系面临的另一个挑战。

12.4.9 将环境影响评价扩展到项目设计：面向环境影响的设计

环境影响评价的一个重要而积极的趋势就是它越来越早地应用于项目规划阶段。例如，美国交通部 1983 年的《环境评估手册》仅适用于具体的路线选择，而其后的《道路和桥梁设计手册》已经广泛涵盖了路线走廊、路线选择和选定线路三个阶段（公路管理局，2011)。美国国家电网对输电线路进行多层次环境分析，从大致的可行性研究阶段到详细设计阶段都要进行分析（国家电网，20××)。将环境影响评价应用于项目设计的早期阶段有助于改善项目设计，避免设计完成后才识别环境约束，造成时间延误和成本增加。

McDonald 和 Brown (1995) 建议环境影响评价小组中应该包含项目设计人员：

当前，环境影响评价的大多数正式的管理和报告要求都基于其初始的角色设定，报告成了一个单独开展的不同但平行于项目设计的东西……我们可以通过将现有环境评价中采用的思想、视角和技术应用到项目规划和设计中，来克服这一环境影响评价的局限。

这一理念的进一步发展是在项目设计开始之前用环境影响评价来识别基本环节约束，然后才允许设计者在满足这些约束条件的前提下自由设计出那些创新和具有吸引力的结构。图 12.4 就是这种理念的一个例子：一座位于维也纳的宏伟的 Hundertwasser 垃圾焚烧厂，一座人们真正可能接受的市内垃圾焚烧厂。

Holstein (1996) 将这种后现代的理念称为“环境影响设计” (environmental impact design，EID)，以区别于传统的建立在保护之上的环境影响评价。以下是 Holstein 对环境影响设计的观点。

现在的环境影响评价是对一个地点进行拆解：它将环境单独拿出来对其中互相联系的组成要素进行研究（例如土壤、水、植物)。环境影响评价总是假设这个地方除了开发之外还有别的（环境）功能。然而由于环境影响评价是建立在保护之上的，这种被拆解的关系实际上非常肤浅，它很难撼动现代主义里牢固的层次结构，而正是这些结构支撑着这样的关系，

图 12.4　艺术与开发项目的结合：维也纳 Hundertwasser/施比特劳垃圾焚烧厂
（资料来源：Wikimedia Commons）

如开发带来经济增长和技术革新。环境影响评价中的环境设计往往仅是评价的一个副产品，甚至被交给开发商成为他们设计时的另一个负担。它对用艺术性的象征手法再现或恢复人文景观价值作用很小，没有试图尝试将环境的功能和形式分开，也没有制造相对个人主义的需求来让“不激发就不感兴趣”的人群去思考文化的可持续性（后现代主义的全部特点）。在这种消极的环境影响评价下，动态设计的所有关键要素——时间、空间、信息沟通和领导能力都丢失了。

虽然如此，最初也有人争辩过真正的后现代主义是一个完全超越了环境影响评价的存在，环境影响评价只追求客观地进行评价，而不是要实现什么审美目的。以上描述就可被称为环境影响设计。环境影响设计强调让环境影响评价服务于审美，它要求的是和纯粹的环境影响评价完全不同的实现方法（可能是个人的）、创意和文化视角。实际上，环境影响设计中的某些原则在某种程度上已经被应用到了环境影响评价中，如环境影响报告书的减缓措施的章节就多少涉及了环境影响设计的内容，而那些将正式的环境影响评价过程视为其设计策略的一个外延的大型开发商（如公用事业）的环保部门则应用得更多。尽管如此，人们很少认为环境影响设计是一种艺术行为。

环境影响评价和环境影响设计的关键差异在于“不可修改的设计”（unmodifiable design）理念。传统意义上的环境影响评价是在项目的大部分结构都已经敲定后进行的。而如果采取环境影响设计的方式，“不可修改的设计”成分会更少，更多地让环境无害设计成为减缓措施的一部分。后现代环境影响评价（postmodern EIA）是一种更激进的方法，它几乎颠覆了“不可修改的设计”的理念，直接让环境影响评价成为设计的领导角色（Holstein，1996）。

12.4.10 补充性改进：提高技能和知识

前面的讨论已经表明，随着环境的内涵越来越丰富，环评从业者需要掌握更多的实质性知识。“环境状况报告”和建立“承载力和可持续指标”也非常重要（如果它没有被理解得过于狭隘的话），也占据着非常重要的角色。承载力是多维度多角度的（图 12.5，以旅游影响评价为例），同时也是一个富有弹性的概念，良好的管理可以提高承载能力。

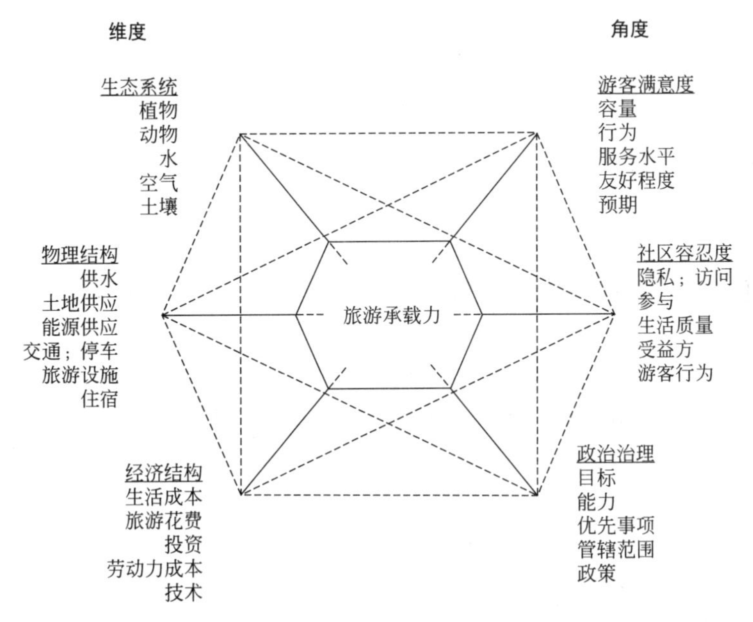

图 12.5 承载力：以旅游影响评价为例

（资料来源：Glasson 等，1995）

从业者还需要开发“技术”和“参与”方法，比如使用焦点小组访谈法（focus group）、德尔菲法（Delphi）和协商调解法。长久以来环境影响评价都是靠“专家经验”进行的，缺乏定量分析。然而，一些新兴技术仍然有应用空间——包括专家系统和 GIS（Rodriguez-Bachiller with Glasson，2004）等。环境影响评价的专业知识领域和相关研究领域也需要更多的能力建设，如需要开展更多的比较研究和纵向研究（即全生命周期影响评价——向适应性环境影响评价转变）。

12.5 将环境影响评价扩展到项目实施阶段：环境管理体系和环境审计

和环境影响评价一样，环境管理体系（environmental management system，EMS）是

一个管理工具，它通过制订目标、实施和监控目标是否实现来促使组织（organizations）对其行为更加负责。然而，环境管理体系与针对未来开发活动的环境影响评价不同，它主要是评审、评价和不断改进组织现有的环境影响。环境管理体系可以被视为环境影响评价原则在项目实施阶段的延续。从本质上说，环境管理体系和环境影响评价是目标互补的环保工具，环境影响评价是在新建项目的规划和设计阶段预测并减轻/增强其环境影响，而环境管理体系是帮助组织在项目全生命周期内有效管理日常的环境影响（Palframan，2010）。

环境管理体系源于环境审计，环境审计最初是20世纪70年代的美国的私人公司出于金融和法律原因开展起来的，是财务审计的一个延伸。欧洲的私人公司之后也开始采用环境审计，到了20世纪80年代后期，地方政府为了回应来自公众的环保压力也开始开展环境审计。20世纪90年代早期，环境审计又进一步被强化和扩展成为组织实现全面质量管理的手段之一，这就是环境管理体系的由来。环境管理体系现在被认为是一个比较成功的实践，它基本上包含了环境审计。

本小节将回顾现有的环境管理体系标准，简要介绍环境管理体系和环境审计在私人企业和地方政府的应用，最后讨论环境管理体系和环境影响评价的关系，介绍环境管理计划（environmental management plans，EMPs）等其他工具。

12.5.1 环境管理体系的相关标准和法规

英国有三套环境管理体系标准：一套是欧盟委员会1993年发布并于2001年修订的生态管理和审核计划（Eco-Management and Audit Scheme，EMAS）；另一套是国际标准化组织的ISO 14000系列标准以及最近的英国标准（British Standard，BS）8555。这三套标准是相互兼容的，但在具体的要求上略有不同。

1993年7月发布的1836/93号法令正式采纳了欧盟EMAS的内容，并于1995年4月开始执行。EMAS最初仅是用来规制工业部门的企业，但在2001年欧盟761/2001号法令（EC，2001a）取代了1993年的1836/93号法令后，EMAS可适用于包括公共和私有服务在内的所有经济部门。2010年1月生效的2009/10号法令对1221/2009号法令进行了升级。EMAS是一个自愿性的计划，对它的应用也因地而异。一个组织如果要获得EMAS认证，必须：

① 制定高层管理者认可的环境方针（environmental policy），包括遵守环境法律法规的条款以及持续改进环境绩效的承诺。

② 开展初始环境评审，评审包括：组织的活动、产品和服务的环境影响；组织的环境法规框架；现有的环境管理实践。

③ 根据环境评审的结果建立环境管理体系，以实现环境方针的要求。环境管理体系必须包括对职责、目标、方法、操作流程、培训需求、监控和信息交流的系统的解释。

④ 开展环境审计以评估环境管理体系是否适当，是否符合组织的方针和方案，是否与相关环境法规冲突。

⑤ 提供一份环境绩效报告，详细说明环境目标的实现情况，提出组织持续改进环境绩效的步骤（EC，2001b）。

环境评审、环境管理体系、审计流程和环境报告（environmental statement，ES）必须由有资质的生态管理和审核计划（EMAS）认证员进行验证。验证后的报告书必须送到EMAS主管机构登记，并在该组织被允许使用EMAS标志前进行公示。英国的EMAS主管机构是环境管理和评价机构（IEMA）。虽然生态管理和审核计划最初面向的是规模较大的私人组织，但它也可以适用于地方政府和小型公司。

国际标准化组织的ISO 14000系列标准在1991年被首次讨论，1996年9月发布了一套完整的环境管理体系标准。这个系列标准包括ISO 14001环境管理体系认证（ISO，1996a）、ISO 14004环境管理体系通用指南（ISO，1996b）和用于指导环境审计和评审的ISO 14010～14014。EMAS和ISO是相互兼容的，但也存在一些不同。见表12.6。

表12.6　EMAS和ISO 14001的不同

比较项目	EMAS	ISO 14001
初始环境评审	需验证初始评审	无评审
外部沟通和验证	公开环境方针、目标、环境管理体系和具体的组织绩效情况	公开环境方针
审计	环境管理体系审计和环境绩效审计的频率和方法	环境管理体系审计(未对频率和方法做明确要求)
承包商和供应商	需要了解对承包商和供应商的影响	提供了与承包商和供应商进行沟通的相关流程
承诺和要求	要求雇员参与环境绩效的持续改进和满足环境法规	要求对持续改进环境管理体系做出承诺，而不仅仅是持续改进环境绩效

资料来源：EC，2001b。

2003年，英国提出了一个新的环境管理体系倡议——BS 8555（包括环境绩效评估的环境管理系统实施指南，*Guide to the implementation of an environmental management system including environmental performance evaluation*），来帮助机构特别是中小企业实施环境管理体系和完成EMAS注册（DEFRA，2003）。该标准包含对如何建立指标的指南，因此机构从一开始就可能知道是否成功减小了环境影响。“IEMA Acorn计划”提供了一个官方认可的环境管理体系标准，用英国标准BS 8555将环境管理体系的实施路线拆分成了一系列符合逻辑的、便捷的和便于管理的阶段，并给出了最终实现ISO 14001认证或EMAS注册的清晰的路线图。

12.5.2　环境管理体系和环境审计的实施

截至2010年，世界范围内已经有超过13000家、英国有超过10000家组织获得了ISO 14001认证。此外，英国有超过150家组织获得了BS 8555和IEMA Acorn计划认证，另外有500多家组织正在认证中。欧洲有4500家组织参加了生态管理和审核计划，其中70多家都在英国。尽管这些组织普遍将环境管理体系视作一种通过良好的管理实践如减少废弃物和提高能效等来降低企业成本的手段、一种良好的宣传手段和间接提高员工士气的手段，但环境管理体系在私人企业的实施仍然面临着商业机密、法律责任、成本和缺乏承诺的问题。小企业尤其会受到成本因素的制约，建立环境管理体系的速度要比大公司慢。地方主管部门在实施环境管理体系时受到的限制因素包括中央财政缩减、政府重组和公众对经济的关注比对环境问题的关注高等。

12.5.3 环境管理体系和环境影响评价的联系

12.5.3.1 环境信息

环境管理体系的发展对环境影响评价非常重要，有如下几个原因。公共部门和私人组织开展环境管理体系认证都能积累更多的环境信息，这些信息对开展环境影响评价也非常有用。例如，地方政府发布的环境状况报告中的环境质量数据可以用于环境影响评价的环境现状研究。环境状况报告中提供了包括诸如当地空气和水质量、噪声、土地利用、景观、野生动物栖息地和交通等方面的信息。

相比之下，私人企业的环境审计结果通常是保密的。上面的 12.5.2 节里提到了一个值得注意的现象，即许多企业都愿意选择 ISO 14001 认证，因为它和 EMAS 认证相比只需要披露很有限的信息。这使得环境管理体系提供的信息只有在私人企业打算设立另一个相似的设施时才会对环境影响评价有用。然而，环境审计提供的信息，如不同工艺过程的污染物排放水平、使副产品最小化的污染治理设施的类型和操作流程、上述设施和措施的有效性等，这些信息在未来建设类似项目时对确定项目影响和减缓措施效果也会非常有用。其中一些环境审计还能提供其他公司在现在和将来的运行管理中都十分需要的“最佳实践”的样本。不过，最有趣的一点是，项目环境影响评价越来越被用作该项目环境管理体系认证的起点。如环境影响评价中给出的排放限值，只要是可行的，就可被用于企业环境管理体系的管理目标。环境管理体系也可以验证环境影响评价提出的减缓措施是否被实施以及是否有效运行。总的来说，环境管理体系可以提高环境监测水平、环境意识和环境数据的可得性，这些对环境影响评价都非常有帮助。

12.5.3.2 环境管理计划

环境影响评价和环境管理体系之间的另一个非常实用的联系是环境管理计划（environmental management plans，EMPs)。环境管理计划的目的是确保在环境影响评价程序的预申请（pre-application）和同意阶段（consent）所做的努力被有效传递到同意后的环节(post-consent)。这需要克服环境影响评价和环境管理体系之间的一系列障碍，包括审批制度、时段、人员以及缺乏资源等（Palframan，2010)。近期的环境管理计划正在向非正式、简单化、去官僚化以及“精简版环境管理体系”的方向发展（Marshall，2004)。环境管理和评价机构是英国地区环境管理计划的强力倡导者，它在《环境管理计划从业者指南》(IEMA，2008）中对环境管理计划进行了定位。环境管理计划是这样一个文件：它规定了对与开发活动的生命周期相关的环境和社会风险进行管理所需的行动，明确了采取行动所需的资源、何时需要采取行动以及行动负责人等问题。它是一座桥梁，将预同意阶段（pre-consent）的环境影响评价程序和同意后阶段的由各利益相关者（如项目建设承包商、项目运营经理）操作的环境管理体系连接起来。图 12.6 对这种桥梁角色进行了简单说明；表 12.7 归纳了采用环境管理计划的进一步理由及环境管理计划的关键构成。

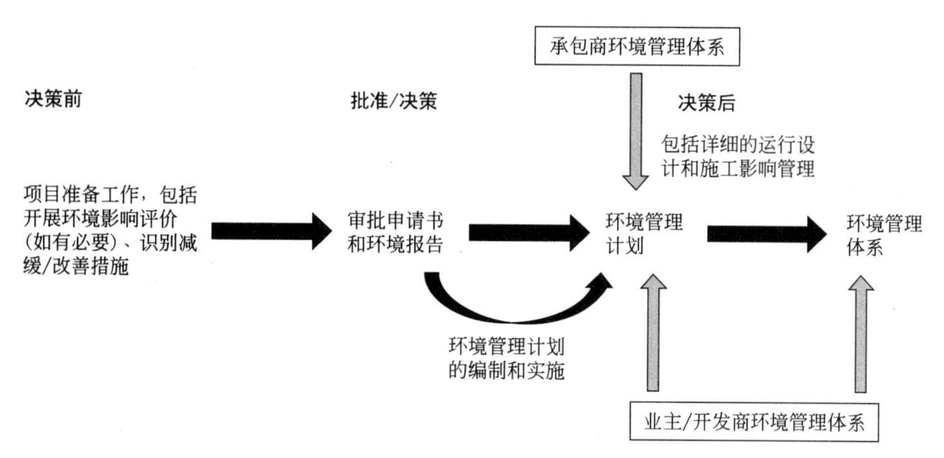

图 12.6　环境影响评价、环境管理计划和环境管理体系之间的联系

（资料来源：环境管理和评价机构，2008）

表 12.7　开展环境管理计划的好处

序号	好　处
1	构建一个框架，确保项目与有关法律要求和环境报告中给出的条件和减缓措施相符合
2	提供一个持续的连接或“桥梁”，将项目的设计阶段和建造阶段（可能还包括运行阶段）连接起来
3	确保运营商、承包商、环境经理（或环境顾问）以及终端的管理者之间有有效的沟通或反馈机制
4	在项目初期阶段编制环境管理计划初稿，说明即将采取的减缓措施，并对咨询顾问和同意部门（consenting authority）提出的疑虑进行解释，减少项目在同意后的延误
5	通过改善环境风险管理来削减成本

资料来源：摘自 IEMA 2008，2011。

环境管理和评价机构相信英国将会开展越来越多的环境管理计划。一项针对从业者的调查发现，80%的受访者认为环境影响报告书中应当包括一个环境管理计划初稿（IEMA，2011）。尽管其形式和名称可能不尽相同。例如，所有由英格兰和威尔士环保署（Environment Agency for England and Wales）提供的环境影响报告书都包括环境行动计划（表 12.8）的内容（environmental action plan，EAP）。环境行动计划相当于环境影响评价和环境管理体系之间的界面，确保环境影响评价在整个项目周期内都是一份“活的”文件。其他开发商的项目必须有施工环境管理计划（Construction Environmental Management Plans，CEMPs），CEMPs 主要是在项目施工阶段实施有效的环境管理，虽然很局限，但考虑到项目建设阶段是整个项目周期中容易受到干扰的阶段，CEMPs 仍然非常有意义（Palframan，2010）。

表 12.8　一个成功的环境管理计划的构成模块

序号	具体说明
1	在项目规划阶段制定减缓措施时，应包括项目发起方、施工方和承包方，确保： —减缓措施是可交付使用的； —减缓行动的成本在项目详细设计阶段被记入，在此基础上形成项目的建设预算； —环评阶段制定的减缓措施获得相应的认同和承诺，以增加实施的有效性
2	在设计和规范减缓行动时有环境专家参与，确保： —明确相关要求； —提高要求被成功传达的概率

续表

序号	具体说明
3	确保在项目规划阶段识别减缓措施时有利益相关者的参与，或明确给出同意后进一步协商的时间表和步骤
4	在同意前阶段明确识别和有效细化减缓措施(可在准备环境报告的同时准备一份环境管理计划框架或草案，以在编写完整的环境管理计划时提供详细的细节)
5	确保申请开发同意的资料中有关于拟议减缓措施的明确描述
6	确保减缓措施作为开发商愿意实施的拟建项目的一个组成要素被包含在最终的同意文件(consenting documentation)中

资料来源：摘自 IEMA 2008，2011。

12.6 总结

和第 10 章中讨论的一些其他国家一样，英国的项目环境影响评价实践在欧盟的大背景下已经进入了学习曲线的快速上升阶段。当然，该实践在取得成功的同时也遇到了许多问题。这些问题的解决和未来的进一步发展取决于各相关方的相互协作。欧盟对《环境影响评价指令》的各修正案改进了环境影响评价程序中的某些步骤，包括筛选、确定范围、考虑替代方案和公众参与等。但一些关键问题仍有待解决，如缺乏对强制性监控的支持等。本章讨论了其他一些可能的改进，包括累积影响、社会-经济影响、健康影响和公平影响评价的有关领域、适宜性评价、弹性思维、气候变化的关键领域、综合影响评价和环境影响设计等。其中的有些改进相对容易实现，而且这个充满活力的领域肯定还会出现其他的问题，取得新的发展成果，随着环境议程和人们管理学和方法学能力的提高，环境影响评价体系和流程将会持续不断发展。

目前也迫切需要"闭环（close the loop）"以总结经验。虽然关于强制性监控的实践进展很不顺利，但在环境管理和审计制度方面已经取得了长足的发展。现有组织（不管是私人部门还是地方主管部门）对环境管理体系的发展有助于环境影响评价的发展。环境管理和环境审计所积累的信息，加上环境管理计划（作为环境影响评价、环境管理体系以及同意前和同意后阶段的桥梁）在最近取得的重要进展，都可以显著改善环境影响评价的数量和开展质量。

随着环境影响评价活动的发展和传播，越来越多的团体将会参与其中。能力建设和培训对环境影响评价程序（相当于不同国家间的通则）及环境影响评价流程（相当于针对某个特定国家的国情需求的细则）都非常重要。随着环境这一概念不断拓展，环境影响评价从业者也需要切实积累知识，不断提高他们在环境影响评价程序中的技术能力和参与方式。

问 题

下列问题有助于读者理解本章重点，促使读者开始思索应如何提高项目评价的有效性。

1. 改进是环境影响评价体系的特征，改进的内在动力是什么？

2. 什么利益相关者对环境影响评价体系的未来发展表现积极？什么利益相关者表现消

极？为什么？

3. 回顾 Sadler 在 1996 年提出的环境评价的“最佳”表现的特征，这些特征在今天仍然适用吗？是否有其他新的特征？

4. 为什么累积影响评价对环境影响评价而言是一项充满挑战的任务？请思考一些可能的解释。

5. 如何理解“环境公平”？

6. 请举一个你熟悉的新建重大项目，识别它有哪些潜在的关键社会-经济影响。

7. 这些影响（问题 6）在项目的建设和运营阶段会如何变化？

8. 使用问题 6 中列举的项目，考虑其潜在的健康影响，并考虑这些影响在项目的建设和运营阶段会如何变化？

9. 使用问题 6 中列举的项目，考虑其潜在的公平影响。

10. 作为一个更综合的包含了生物物理和社会-经济影响的环境影响评价的一部分，请举出一个综合了健康和公平影响的社会-经济影响评价的例子。

11. 欧盟何时开展了适宜性评价？如何理解“能够压倒公众利益的必须实施的原因”这一表述？为什么该表述可用于鹿特丹港扩建项目，对 Southampton 水域的 Dibden 海湾项目就不行？

12. 为什么气候变化没有被很好地纳入大多数环境影响评价中？

13. 回顾表 12.4 提到的适用于环境影响评价的气候变化具体原则，思考哪一条最难实现。

14. 什么样的弹性“特征”可以被融入城市房地产开发、农村垃圾填埋场的设计中？

15. 如何理解“环境影响设计”这一术语？

16. 什么是环境管理体系？简要概述生态管理和审核计划和 ISO 14001 环境管理体系的不同。

17. 环境管理计划可以使环境影响评价更高效，讨论环境管理计划的潜在价值。

12

Improving the effectiveness of project assessment

12.1 Introduction

Overall, the experience of EIA to date can be summed up as being like the proverbial curate's egg: good in parts. Current issues in the EIA process were briefly noted in Section 1.6: they include EIA methods of assessment, the quality and efficiency of the EIA process, the relative roles of the participants in the process, EIS quality, monitoring and post-decision, managing the widening scope and complexity of impact assessment activity, plus concern about its overall effectiveness. The various chapters on steps in the process have sought to identify best practice, and Chapter 8 provides an overview of the quantity and quality of UK practice to date. Detailed case studies of good practice and comparative international experience provide further ideas for possible future developments. The evolving, but still in some cases limited, experience in EIA among the main participants in the process—consultants, local authorities, central government, developers and affected parties—explains some of the current issues.

However, almost 25 years after the implementation of EC Directive 85/337, there is less scepticism in most quarters and a general acceptance of the value of EIA. There are still some substantial shortcomings, and there is considerable scope for improving quality, but practice and the underpinning knowledge and understanding have developed and EIA continues on its steep learning curve. The procedures, process and practice of EIA will undoubtedly evolve further, as evidenced by the comparative studies of other countries. The EU countries can learn from such experience and from their own experience since 1988.

This chapter focuses on the prospects for project-based EIA. The following section briefly considers the array of perspectives on change from the various participants in the EIA process. This is followed by a consideration of possible developments in some important areas of the EIA process and in the nature of EISS. The chapter concludes with a discussion of the parallel and complementary development of environmental management systems and audits. Together, these topics act as a kind of action list for future improvements to EIA.

12.2 Perspectives on change

An underlying theme in any discussion of EIA is change. This has surfaced several times in the various chapters of this book. EIA systems and procedures are changing in many countries. Indeed, as O'Riordan (1990) noted (see Section 1.4), we should expect EIA to change in the face of shifting environmental values, politics and managerial capabilities. This is not to devalue the achievements of EIA; as the World Bank (1995) noted, 'Over the past decade, EIA has moved from the fringes of development planning to become a widely recognized tool for sound project decision making.'

The practice of EIA under the existing systems established in the EU Member States has also improved rapidly (see Chapters 2, 8 and 10). This change can be expected to continue in the future, as the provisions of the regularly amended EU EIA Directive work through, and even further amendments are introduced. Changes in EIA procedures, like the initial introduction of EIA regulations, can of course generate considerable conflict between levels of government: between federal and state levels, between national and local levels and, in the case of Europe, between the EU and its Member States. They also generate conflict between the other participants in the process: the developers, the affected parties and the facilitators (see Figure 3.1).

The *Commission of the European Communities* (CEC) is generally seen as positive and proactive with regard to EIA. The CEC welcomed the introduction of common legislation as reflected in Directive 85/337, the provision of information on projects and the general spread of good practice, but was concerned about the lack of compatibility of EIA systems across frontiers, the opaque processes employed, the limited access to the public and lack of continuity in the process. It pressed hard for amendments to the Directive, and has achieved some of its objectives in the various subsequent amendments. In addition the SEA Directive was implemented from 2004 (see Chapter 11). However, as noted in Section 2.8, there are some continuing and stubborn issues, including: variations in screening, transboundary issues, quality control, the absence of mandatory consideration of alternatives, lack of monitoring, and tiering issues between EIA and SEA (CEC 2009). The Commission is committed to reviewing and updating EIA procedures and there will no doubt be further changes. Other areas of attention include, for example, cumulative assessment, public participation, economic valuation and EIA procedures for development aid projects. In contrast with the CEC, Member States tend to be more defensive and reactive. They are generally concerned about maintaining 'subsidiarity' with regard to activities involving the EU; this has been an ongoing issue with EIA (CEC 2009). Governments are also sensitive to increasing controls on economic development in an increasingly difficult, competitive and global economy.

For example, within the UK *government*, the DCLG (formerly ODPM; DETR; DoE) has been concerned to tidy up ambiguities in the project-based procedures, and to improve guidance and informal procedures for example, but is wary of new regulations. However, it has commissioned and produced research reports, for example on an EIA good practice guide, on the evaluation and review of environmental information and on mitigation in EIA, and its recent guidance and regulations reflect an acceptance of the value of EIA. *Local government* in the UK has begun to come to terms with EIA, and there is evidence that those authorities with considerable experience (e. g. Essex, Kent, Cheshire) learn fast, apply the regulations and guidance in user-friendly 'customized' formats to help developers and affected parties in their areas, and are pushing up the standards expected from project proponents. For example see the latest version of the very useful Essex Guide to EIA (Essex Planning Officers' Association, 2007) which can be freely downloaded.

Pressure groups—exemplified in particular in the UK by the Campaign to Protect Rural England (CPRE), the Royal Society for the Protection of Birds (RSPB) and Friends of the Earth (FoE)—and those parties affected by development proposals view project EIA as a very useful tool for increasing access to information on projects, and for advancing the protection of the physical environment in particular. They have been keen to develop EIA processes and procedures; see, for example, the reports by CPRE (1991, 1992) and RSPB (2000). Many *developers* are less enthusiastic about strengthening EIA procedures, but will welcome the government's recent clarification on ambiguities (DCLG 2011)—especially on whether EIAs are needed in the first place for their particular projects. For *facilitators* (consultants, lawyers, etc.), EIA has been a welcome boon; their interest in longer and wider procedures, involving more of their services, is clear.

Other participants in the process in the UK, such as the IEMA (see 12. 3. 2 below), academics and some environmental consultancies, are carrying out groundbreaking studies into topics such as best-practice guidelines, the use of monetary valuation and ecosystem services approaches in EIA and approaches to widening types of impact study. In addition, the production of several hundred EISs a year in the UK and in many other countries worldwide is generating a considerable body of expertise, innovative approaches and comparative studies. EISs are also becoming increasingly reviewed, and hopefully bad practice will be exposed and reduced. Training in EIA skills is also developing.

12.3 Possible changes in the EIA process: overviews of the future agenda

12.3.1 International studies of EIA effectiveness

Examples of recent key international studies of EIA effectiveness include those by the European Commission, as already noted above and in Chapter 2 (CEC 2009), and the 2011 update of the 1996 *International study of the effectiveness of environmental assessment* for the IAIA, by Sadler and colleagues (1996, 2011). As noted in Chapter 8, in a UK workshop discussion for the evolving international effectiveness study, Sadler (2010) identified an 'effectiveness triage' involving three clearance bars for EIA: *what must or should be done*, including legal and institutional framework and methodological realities; *what is done*, including cases of good practice; and *what is the outcome*, in particular the contribution to decision-making, and environmental benefits. He raises, for example, many questions about the second of the above, the nature of current practice. Is consultation a procedural cornerstone or overrated and under-performing? Are screening and scoping focusing on the impacts that matter? Is the evaluation of significance based on adequate evidence? Are mitigation measures sufficiently tailored? Is monitoring still the weak link? Such points resonate with much of the content in the very useful summary of international best and worst case EIA performance contained in the earlier international effectiveness study (Sadler 1996; see Box 12.1).

Box 12.1 Summary of international best-and worst-case EA performances

Best-case performance

The EA process:

- facilitates informed decision-making by providing clear, well-structured, dispassionate analysis of the effects and consequences of proposed actions;
- assists the selection of alternatives, including the selection of the best practicable or most environmentally friendly option;
- influences both project selection and policy design by screening out environmentally unsound proposals, as well as modifying feasible action;
- encompasses all relevant issues and factors, including cumulative effects, social impacts and health risks;
- directs (not dictates) formal approvals, including the establishment of terms and conditions of implementation and follow-up;

- results in the satisfactory prediction of the adverse effects of proposed actions and their mitigation using conventional and customized techniques; and
- preserves as an adaptive, organizational learning process in which the lessons experienced are fed back into policy, institutional and project designs.

Worst-case performance

The EA process:

- is inconsistently applied to development proposals with many sectors and classes of activity omitted;
- operates as a 'stand alone' process, poorly related to the project cycle and approval process and consequently is of marginal influence;
- has a non-existent or weak follow-up process, lacking surveillance and enforcement of terms and conditions, effects monitoring, etc.;
- does not consider cumulative effects or social, health and risk factors;
- makes little or no reference to the public, or consultation is perfunctory, substandard and takes no account of the specific requirements of affected groups;
- results in EA reports that are voluminous, poorly organized and descriptive technical documents;
- provides information that is unhelpful or irrelevant to decision-making;
- is inefficient, time consuming and costly in relation to the benefits delivered; and
- understates and insufficiently mitigates environmental impacts and loses credibility.

Source: Sadler 1996

But in particular, the 2010 workshop raised the issue of the rapidly changing context of EIA—as noted in Section 1.5.1. While climate change can be seen as reinforcing the importance of EIA, economic recession and severe financial challenges may be more counter-productive with the risk of weakening of support for green issues 'until we get the economy back on its feet!' FoE (2011) provides a UK example of this. Of course, there should be much mutual synergy through: for example, investment in green infrastructure, but this needs enlightened and far-seeing governance.

12.3.2 UK IEMA study of EIA effectiveness

The UK IEMA study (IEMA 2011) provides a very timely study on the state of UK EIA practice. While good progress has been made over 25 years, the study concludes that '(UK) EIA practice is not perfect and further improvements must be made if it is to continue to offer value in the 21st Century.' The study raises many significant issues about the UK EIA process, relating for example to issues in the approach to screening, scoping, assembling baseline data, assessment of significance, mitigation and enhancement, cumulative impacts, monitoring and adaptive EIA-many of which are also covered in the earlier chapters of this book. The IEMA study concludes with a vision for UK EIA: 'EIA must enable developments that work for the developer, community and environment and that getting this right will save resources at the same time as generating improved environmental outcomes.' Table 12.1 provides a set of six key areas for delivering that vision.

Table 12.1 Key areas to deliver IEMA vision for EIA and to facilitate future success of UK EIA practice

1 **A focus on communicating the added value generated by EIA**: enhance the communication of the positive effects of EIA (e.g. EIA leading to improvements in project design)
2 **Realizing the efficiencies of effective EIA co-ordination**: recognize the value that a good EIA co-ordinator brings to the efficient running and effective application of the assessment
3 **Developing new partnerships to enhance the EIA process**: value effective partnerships with planners; legal advisers; design teams; and construction contractors, etc., especially those involved in managing EIA outcomes
4 **Listening, communicating and engaging effectively with communities**
5 **Practitioners actively working together to tackle the difficult issues in EIA** to generate pragmatic solutions to difficult EIA issues
6 **Delivering environmental outcomes that work now and in the future**: recognize the importance of designing measures in a way that maximizes the chance of their being implemented effectively; plus effective monitoring

Source: Adapted from IEMA (2011)

12.4 Possible changes in the EIA process: more specific examples

So what might be done? A pragmatic approach to change could subdivide the future agenda into proposals to *improve* EIA procedures, usually sooner and maybe more easily than proposals to *widen* the scope of EIA, which are likely to come later and will probably be more difficult to implement. *Improvements to project EIA* cover some of the changes introduced by the various amendments to the EC Directive, including developments in approaches to screening, the mandatory consideration of alternatives and a strong encouragement to undertake scoping at an early stage in the project development cycle. There could also be more support for more transparent procedures, and encouragement for consultation, for the explanation and publication of decisions and for greater weight to be given to cumulative impacts and risk assessment.

The methods of assessment could also benefit from further attention. Uncertainty about the unknown may mean the EIA process starts too late and results in a lack of integration with the management of a project's life cycle. The EIA process and the resulting EISs may lack balance, focus on the more straightforward process of describing the project and its baseline environment and consider much less the identification, prediction and evaluation of impacts. The forecasting methods used in EIA are not explained in most cases (see Section 8.4). Practical advances in predicting the magnitude of impacts and determining their importance (including the array of multi-criteria and monetary evaluation techniques) would be beneficial. A good 'method statement', explaining how a study has been conducted—in terms of techniques, consultation, the relative roles of experts and others—should be a basic element of any EIS.

Widening the scope of eia includes, in particular, the development of tiered assessment through the introduction of SEA (as discussed in Chapter 11). Another important extension of the scope of EIA includes 'completing the circle' through the more widespread use of monitoring and auditing. Unfortunately, this vital step in the EIA process is still not mandatory after several amendments to the EC Directive. More wide-ranging changes include the move to a 'whole of environment' approach, with a more balanced consideration of both biophysical and socio-economic impacts. Such widening of scope should lead to more integrated EIA. There may also be a trend towards using EIA to identify environmental limits and environmental constraints on the project, rather than focusing only on identifying the project's impacts on the environment—through what might be termed environmental impact design. Testing a project's resilience to future changes and shocks could also become a component of EIA.

The following sections discuss possibilities for some of these short-and longer-term proposals: better consideration of cumulative impacts, widening the scope to include socio-economic impacts; embracing the growing areas of health impact assessment, equality impacts assessment, appropriate assessment and resilience thinking; building climate change centrally into EIA; developing integrated impact assessment and moving towards environmental impact design.

12.4.1 Cumulative impacts

Many projects are individually minor, but collectively may impose a significant impact on the environment. Activities such as residential development, farming and household behaviour normally fall outside the scope of conventional EIA. The ecological response to the collective impact of such activities may be delayed until a threshold is crossed, when the impact may come to light in sudden and dramatic form (e.g. flooding). Odum (1982) refers to the 'tyranny of small decisions' and the consequences arising from the continual growth of small developments; cumulative impacts can also be described as 'death by a thousand cuts'. While there is no particular consensus on what constitutes cumulative impacts, the categorization by the Canadian Environmental Assessment Research Council (CEARC) (Peterson *et al.* 1987) is widely quoted, and includes:

- time-crowded perturbations: which occur because perturbations are so close in time that the effects of one are not dissipated before the next one occurs;
- space-crowded perturbations: when perturbations are so close in space that their effects overlap;
- synergisms: where different types of perturbation occurring in the same area may interact to produce qualitatively and quantitatively different responses by the receiving ecological communities;

- indirect effects: those produced at some time or distance from the initial perturbation, or by a complex pathway; and
- nibbling: which can include the incremental erosion of a resource until there is a significant change/it is all used up.

> Cumulative impact assessment is predicting and assessing all other likely existing, past and reasonably foreseeable future effects on the environment arising from perturbations which are time-crowded; space-crowded; synergisms; indirect; or, constitute nibbling. (CEPA 1994)

The need to include cumulative impact assessment in EIA has been long recognized. In the CEQA of 1970, significant impacts are considered to exist if 'the possible effects of a project are individually limited but cumulatively considerable'. Subsequent legislative reference is found in the 1991 Resource Management Act of New Zealand, which makes explicit reference to cumulative effects, and now also in the amended EU Directive, which refers to the need to consider the characteristics of projects having regard to 'the cumulation with other projects'. In Canada, which has been at the forefront in the development of 'cumulative effects assessment' (CEA), the consideration of cumulative effects is explicit and mandatory in legislation both federally and in several provinces. The UK guidance is for rather more limited consideration of cumulative effects:

> Generally, it would not be sensible to consider the cumulative effects with other applications which have yet to be determined, since there can be no certainty that they will receive planning permission. However, there could be circumstances where two or more applications for development should be considered together. For example, where the applications in question are not directly in competition with one another, so that both or all of them might be approved, and where the overall combined environmental impact of the proposals might be greater than the sum of the separate parts. (DCLG 2011)

However, it is in the practical implementation of the consideration of cumulative impacts that the problems and deficiencies become clear, and cases of good practice and, until recently, useful methodologies have been limited. In Australia, assessments have largely been carried out by regulatory authorities rather than by project proponents, and have focused on regional air quality and the quality and salinity of water in catchment areas (CEPA 1994). Figure 12.1 provides an example of a simple perturbation impact model developed by Lane and associates (1988). It is basically an 'impact tree' that links (a) the principal causes driving a development with, (b) the main perturbations induced with, (c) the primary biophysical and socio-economic impacts and (d) the secondary impacts. The figure shows some of the potential cumulative impacts associated with a number of area-related tourism developments.

There are some significant examples of good practice guidance. In the US, the CEQ produced a practice guide *Considering cumulative effects* (CEQ 1997), based on numerous case studies. The guide consists of 11 steps for CEA, in three main stages (see also Chapter 2, Section 2.2.4). In Canada, a *Cumulative effects assessment practitioners guide* (CEAA

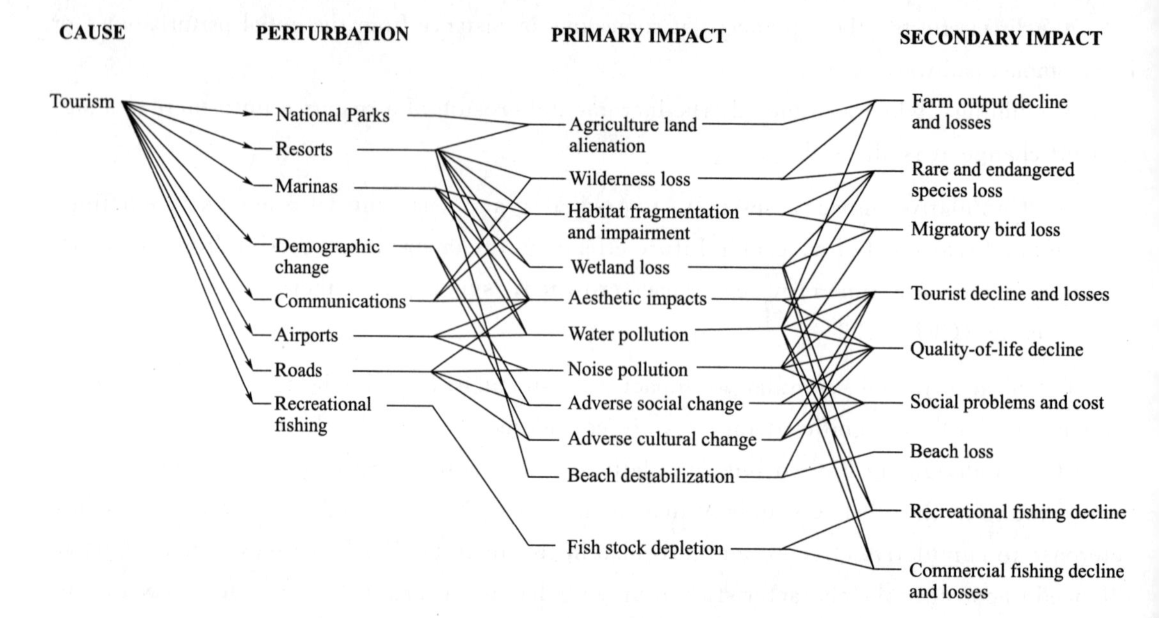

Figure 12.1 Cumulative impacts: perturbation impact model

Source: Lane and associates 1988

1999) provides a very useful overview and clarification of terms and fundamentals, of practical approaches to completing CEAs, and case studies of approaches used by project proponents. The guide provides some clear and simple definitions- 'Cumulative effects are changes to the environment that are caused by an action in combination with other past, present and future human actions. A CEA is an assessment of those effects.' Further Canadian work has sought to improve the practice of CEA (see Baxter *et al*. 2001). In the EU, there has also been an attempt to support practice in the area through the development of *Guidance for the assessment of indirect and cumulative impacts* (Hyder Consulting 1999). Piper (2000, 2001a, b) provides valuable evidence on the state of UK practice in CEA, drawing on research on a number of case studies (see Section 9.5).

Between them, these guides and assessments of practice highlight some of the key process and organizational issues in considering cumulative impacts/effects. Process issues include, for example: establishing the geographic scope of the analysis (how wide should the impacts region be), establishing the time frame for the analysis (including not only present projects, but also those in the non-immediate time frame—past and reasonably foreseeable future) and determining the magnitude and significance of the effects. A key organizational question in the UK (see Piper 2001b) is 'Which organization has the responsibility to require or commission the CEA work?' This is complicated when, as is often the case in CEA, there is more than one competent authority involved (Piper 2001b; Therivel and Ross 2007).

Canter and Ross (2010) provide a recent state of play of CEAM (cumulative effects assessment and management); the inclusion of 'M' in the term reflects the increasing attention being given to the management and mitigation of cumulative effects. They identify a six-step framework, as set out in Table 12.2. They also discuss the good, bad and ugly lessons of the practice of CEAM. 'Good', for example, includes: adoption of a valued ecosystem component (VEC)-based perspective; agency/proponent and public context scoping; use of scenarios where reasonably forseeable future actions are uncertain; and dissemination of good practice. Examples of 'bad' lessons include: over focus on biophysical environmental components at the expense of socio-economic; vague terms of reference for studies; inadequate guidance; lack of expertise and overcomplex studies. Downright 'ugly' include for example: minimal attention to CEAM; lack of commitment by key decision makers; lack of multi-stakeholder collaboration and—on occasion—an attitude that CEAM cannot be done. In this context, Canter *et al.* (2010) conclude: 'By all accounts, cumulative effects continue to be a persistent analytical challenge, although there is evidence of progress towards better practice.'

Table 12.2 Key steps in a CEAM framework

1	Initiate the CEA process by identifying the incremental direct and indirect effects of the proposed project(or policy, plan or programme)on selected VECs(valued ecosystem components)within the environs of the project location.
2	Identify other past, present, and reasonably foreseeable future actions within the space and time boundaries that have been, are, or could contribute to cumulative effects(stresses)on the VECs and their indicators.
3	For the selected VECs, assemble appropriate information on their indicators, and describe and assess their historical to current and even projected conditions.
4	'Connect' the proposed project(or other PPP)and other actions in the CEAM study study area to the selected VECs and their indicators.
5	Assess the significance of the cumulative effects on each VEC over the time horizon for the study.
6	For VECS or their indicators that are expected to be subject to negative incremental impacts from the proposed project and for which the cumulative effects are significant, develop appropriate action or activity-specific 'mitigation measures' for such impacts.

Source: Adapted (substantially simplified) from Canter and Ross (2010)

12.4.2 Socio-economic impacts

Widening the scope of EIA to include socio-economic impacts in a much better way is seen as a particularly important item for the agenda. Although most of the environmental receptors listed in EC and UK regulations are still biophysical in nature, the inclusion of 'human beings' as one of the receptors to be considered in EIA does imply a wider definition of 'the environment', encompassing its human (i.e. social, economic and cultural) dimensions. The inclusion of socio-economic impacts can help to better identify all of the potential biophysical impacts of a project, because socio-economic and biophysical impacts are interrelated (Newton 1995). Early inclusion of socio-economic considerations in the EIA can provide an opportunity to modify project design or implementation to minimize adverse socio-economic effects and to maximize beneficial effects (Chadwick 2002). It also allows a more complete picture of a project's impacts, in a consistent format, in a publicly available document. Failure to include such impacts can lead to delays in the EIA process, since the competent authority may request further information on such matters.

While there are varying interpretations of the scope of socio-economic or social impacts, over time a number of reports have highlighted the importance of this area (see, for example, CEPA 1994; IAIA 1994; Vanclay 2003; Glasson 2009). SIA has been defined by Burdge (1999) as 'the systematic analysis, in advance, of the likely impacts a proposed action will have on the life of individuals and communities'. Most development decisions involve trade-offs between biophysical and socio-economic impacts. Also, development projects affect various groups differently; there are invariably winners and losers. Yet the consideration of socio-economic impacts is very variable in practice, and often very weak. Some countries have useful practice and associated legislative impetus for SIA (for example, the USA, Canada and some states of Australia). International funding institutions are also increasingly giving a high profile to such impacts, as shown at Section 10. 9.

However, in Europe the profile is lower, and the consideration of socio-economic impacts has continued to be the poor relation (Chadwick 2002; Glasson 2009). The uncertain status of such impacts, plus the lack of best-practice guidance on their assessment, has resulted in a partial approach in practice. When socio-economic impacts are included, there tends to still be a focus on the more measurable direct employment impacts. The consideration of the social—cultural impacts (such as severance, alienation, social polarization, crime and health) is often very marginal. Although there is now increasing momentum behind the assessment of health impacts, the important area of crime and safety has had a much lower profile. Glasson and Cozens (2011) provide an update on some key issues for advancing the better consideration of these topics in EIA practice including: the need to employ meaningful data, including 'fear of crime' considerations; the consideration of innovative approaches to the use of indicators; and use of evidence and concepts from the field of environmental criminology.

However, the fuller and better consideration of socio-economic impacts does raise issues and challenges, for example about the types of impact, their measurement, the role of public participation and their position in EIA. One categorization of socio-economic impacts is into: (a) quantitatively measurable impacts, such as population changes, and the effects on employment opportunities or on local financial implications of a proposed project, and (b) non-quantitatively measurable impacts, such as effects on social relationships, psychological attitudes, community cohesion, cultural life or social structures (CEPA 1994). Such impacts are wide-ranging; many are not easily measured, and direct communication with people about their perceptions of socio-economic impacts is often the only method of documenting such impacts. There is an important symbiotic relationship between developing public participation approaches and the fuller inclusion of socio-economic impacts. SIA can establish the baseline of groups that can provide the framework for public participation to further identify issues associated with a development proposal. Such issues may be more local, subjective, informal and judgemental than those normally covered in EIA, but they cannot be ignored. Perceptions of the impacts of a project and the distribution of those impacts often largely determine the positions taken by various groups on a given project and any associated controversy.

12. 4. 3 *Health impact assessment*

Health impact assessment (HIA) is a major growth area in the field of impact assessment, as evidenced by the popularity in recent years of the HIA 'track' in the annual conference of the influential International Association for Impact Assessment. There has been a surge of academic papers, reviews, guidelines and websites relating to HIA (Ahmad 2004), but what is HIA, where is it best practised, how is it practised and how does it relate to EIA as discussed in this book?

'Health' includes social, economic, cultural and psychological well-being—and the ability to adapt to the stress of daily life (Health Canada 1999). Many environmental factors give rise to positive and negative health outcomes, as exemplified in Figure 12. 2 from a very useful Western Australian publication on HIA (Western Australia Department of Health 2007). Health impact refers to a change in the existing health status of a population within a defined geographical area over a specified period of time. HIA is a combination of procedures and methods by which a policy, plan, programme or project may be judged as to the effects it may have on the health of a population. It provides a useful, flexible approach to helping those developing and delivering proposals to consider their potential (and actual) impacts on people's health, and on health inequalities, and to improve and enhance a proposal (Taylor and Quigley 2002; Taylor and Blair-Stevens 2002; WHO Regional Office for Europe 2003; Douglas 2003).

HIA is well advanced in a number of developed countries, particularly Canada, the Netherlands, in parts of Scandinavia, and more recently in Australia and the UK. Some developing countries are also finding it very relevant to their needs (see Phoolchareon *et al*. 2003, for Thailand). Policy drivers can be found at various levels of government. In the UK, for example, see the Department of Health and the Association of Public Health Observatories websites. In the EU, the Directive on SEA specifically and very usefully refers to the impact of plans and programmes on human health (see Box 11. 1).

The main stages in the HIA process are very similar to those used in EIA, including: screening, scoping, profiling (identifying the current health status of people within the defined spatial boundaries of the project using existing health indicators and population data), assessment (HIA stresses the importance of consultation with community groups to identify potential impacts), implementation and decision-making, and monitoring and continual review (Douglas 2003). There are now also many useful national guides; for example, for Ireland (IPHI 2009) and for Western Australia as previously noted.

The overlap between HIA and EIA in terms of process, and in terms of many categories of baseline data (Figure 12. 2 indicates the potential wide coverage of HIA) does raise questions as to why HIA and EIA are not better integrated. Ahmad (2004) suggests an interesting list of reasons for this, including the difficulty of establishing causality between population health and multiple pollutants; limitations on resources to carry out such assessments within the often tight timeframes of EIA; confidentiality of some health data; lack of mandatory legal framework requiring HIA; and bias among EIA professionals towards engineering and ecology backgrounds. However, he also concludes that there are many benefits to be gained from closer integration, in terms of shared experience, procedures, data and values. With regard to the last, HIA can bring to EIA 'values such as equity, transparent use

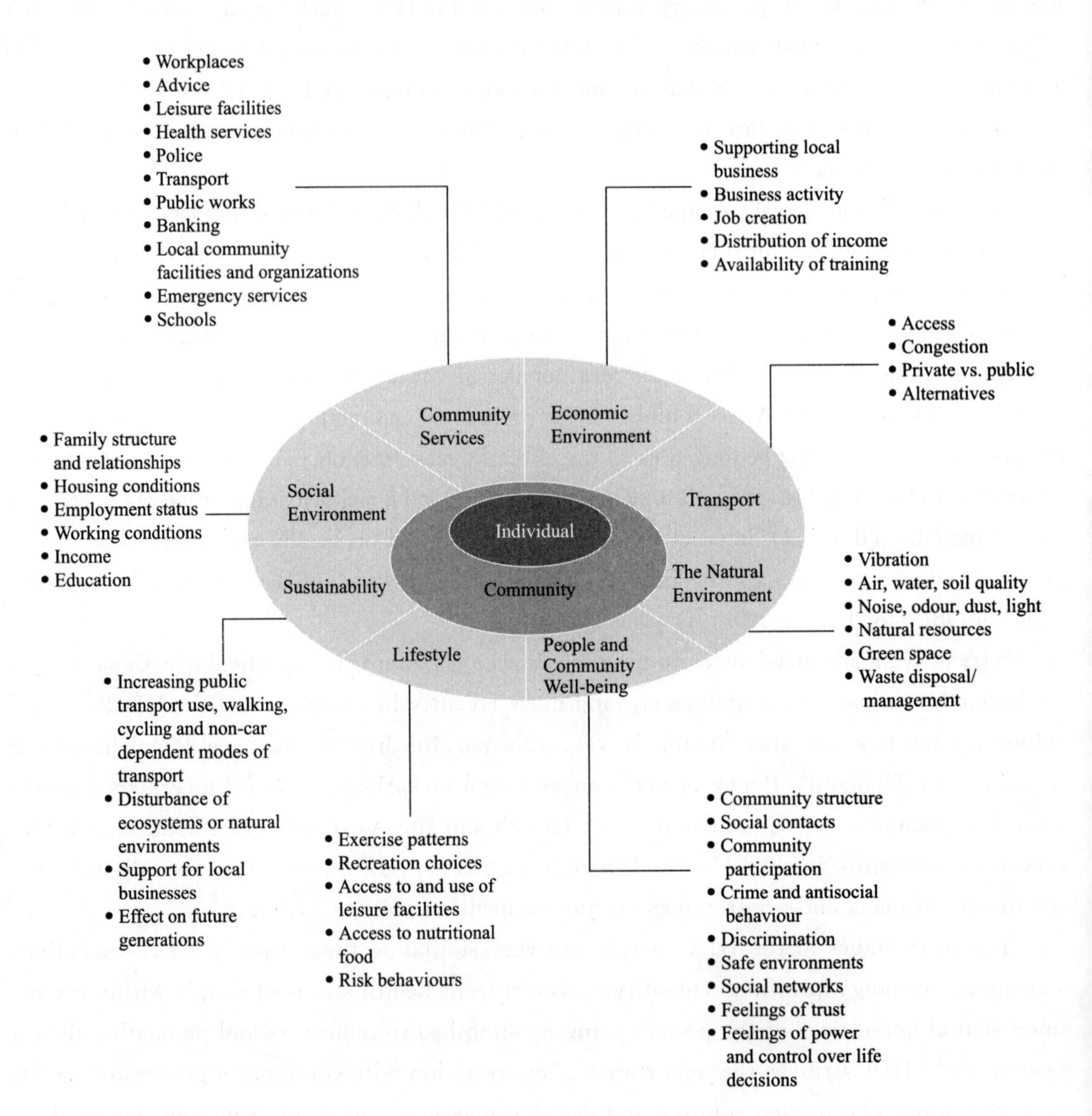

Figure 12.2 Some of the many environmental determinants of health outcomes

Source: Western Australia Department of Health 2007

of evidence and the consideration of differential impacts of the policy or project on various population subgroups' (Ahmad 2004). The SEA Directive 2001/42/EC provides an important milestone on the desirable path to a more integrated approach—a concept that is developed a little further in Section 12.4.8.

12.4.4 Equality impact assessment

It has been noted several times in this book that the distributional impacts of proposed developments rarely fall evenly on affected parties and areas and are mostly ignored in EIA. For example, a UK study by Walker *et al.* (2005), which examined the extent to which EIA, and impact assessment methodologies more generally, involved an assessment of the distribution of impacts likely to result from policy-making or project approval, found that the methodologies used provided no effective consideration of distributional or environmental justice concerns. Though a particular project may be assessed as bringing a general benefit, some groups and/or geographical areas may be receiving most of any adverse effects, the main benefits going to others elsewhere. This raises important equality issues, and there is a growing literature on issues of environmental justice (or perhaps it should be environmental injustice), where certain sections/areas of the community receive increasing environmental burdens, become socially and environmentally more unacceptable places to live, and risk becoming trapped in a vicious circle of decline (Agyeman and Evans 2004; Downey 2005).

In recent years such issues have begun to be addressed, partly through the growth of equality impact assessment (EqIA). An early step was gender impact assessment; this received international prominence through the World Conference on Women at Beijing, which in 1995 called on governments to 'mainstream a gender perspective into all policies and programmes so that, before decisions are taken an analysis is made of the effects on women and men respectively.' This requirement was then built into the Treaty of Amsterdam, Articles 2 and 3, 1997. In the UK EqIAs were introduced first in Northern Ireland where legislation had made it unlawful to discriminate on the grounds of religious belief or political opinion. Disability equality requirements followed and since 2010, the Equality Act has provided the underpinning legal framework in the UK.

EqIA is about considering how projects, plans, programmes and policies may impact, either positively or negatively, on different sectors of the population in different ways. The key sectors typically considered include age, gender, race/ethnicity (including gypsies and travellers), religion, disability and sexual orientation, although other dimensions could include rural vs. urban, poor vs. rich, or people with vs. without access to cars. The steps in the process are also very similar to those used in EIA, and again there might be merit in the integration of the approach into the wider EIA process. It is important to identify the baseline equality characteristics and needs of the population likely to be affected by the development proposal. For example: is the workforce skewed towards male and young employees? Are some ethnic groups substantially under-represented in the workforce? Are various socio-economic issues concentrated in certain wards of particular towns? In England some of the spatial inequalities can be identified in some detail by using information from the Index of Multiple Deprivation (IMD 2011); relative levels of deprivation (for local authority areas, down to ward level) can be assessed by their rank position relative to all other English local authority areas; and regular updating of the IMD provides valuable trend data. The potential impacts of a proposed development on the baseline can be assessed and mitigation and enhancement measures can be introduced to hopefully improve the equality outcomes. Table 12. 3 provides a summary example of equalities impacts in an EqIA that accompanied an ES for a large mixed-use redevelopment of nearly 5000 homes of mixed tenure, along with associated health, community, leisure, education and retail facilities in inner London (Scott Wilson 2006).

Table 12.3 Illustration of EqIA impacts

Issue	Affected group	Impact
(a) Summary of significant adverse impacts affecting specific equalities groups		
Health	Mental health sufferers, including among BME population	Temporary increase in stress as result of demolition and redevelopment
Community cohesion	BME groups, women, including lone parents, children, gay and lesbian people	Temporary or permanent disruption of existing social networks, increased isolation as result of rehousing/redevelopment
Community facilities	BME groups, particularly Turkish community	Permanent loss of facilities for social gatherings
Well-being	Older women	Temporary or permanent loneliness, isolation as result of decant process
Well-being	Disabled people	Temporary risk of individual needs being overlooked during decant and redevelopment process
Well-being	Older people	Temporary/permanent stress, disruption, anxiety increased as result of change, including change to established routine
Well-being	Children	Temporary/long-term disruption to living environment during childhood as a result of living on major redevelopment
Leisure and open spaces and well-being	Children and young people	Possible permanent impact of private courtyards actively excluding casual use by older children including teenagers-both on their well-being and courtyard cohesion
Leisure and open spaces	Children and young people	Temporary loss of access to open spaces, hang-outs in public spaces, play areas during construction
(b) Summary of significant beneficial impacts affecting specific equalities groups (housing, employment and skills only)		
Housing	BME households	Improved housing quality Reduced overcrowding New homes for young people Increased home ownership levels in affordable housing
	Women/single parent households	More appropriate housing for young children
	Disabled people	More accessible homes to enable independent living
	Older people	Improved insulation and heating for warmer homes Lifetime homes support independent living
	Children	Reduced overcrowding More generous bedroom and storage provision for children in social rented sector
	All groups	More storage, less accidents around home
Employment and skills	BME	Target group for construction employment opportunities Targeted skills training
	Women	Children's centre facilitates women to seek employment
	Women	Target group for construction employment Targeted skills training

Source: Scott Wilson (2006)

12.4.5 Appropriate assessment

Appropriate assessment is a Europe-specific form of assessment that tests the impacts of a project or plan on the integrity of internationally important nature conservation sites: Special Protection Areas for birds, and Special Areas of Conservation for habitats and species.[1] Appropriate assessment is required through Articles 6.3 and 6.4 of the Habitats Directive:

> 6.3 Any plan or project not directly connected with or necessary to the management of the site but likely to have a significant effect thereon, either individually or in combination with other plans or projects, shall be subject to appropriate assessment of its implications for the site in view of the site's conservation objectives. In the light of the conclusions of the assessment of the implications for the site and subject to the provisions of paragraph 4, the competent national authorities shall agree to the plan or project only after having ascertained that it will not adversely affect the integrity of the site concerned and, if appropriate, after having obtained the opinion of the general public.
>
> 6.4 If, in spite of a negative assessment of the implications for the site and in the absence of alternative solutions, a plan or project must nevertheless be carried out for imperative reasons of overriding public interest, including those of social or economic nature, the Member State shall take all compensatory measures necessary to ensure that the overall coherence of Natura 2000 is protected. It shall inform the Commission of the compensatory measures adopted.

An entire book could be written just about appropriate assessment but, in short, it focuses on a very specific part of the environment (the integrity of SPAs and SACs), considers cumulative ('in combination') impacts, and is very precautionary (a plan or project may only be permitted if it will have no significant impact on site integrity or if other very tough tests are passed). Appropriate assessment is called 'appropriate' because the level of detail of the assessment, and when the assessment can stop, depends on the project/plan and relevant SPA/SACs. The European Commission (2000) has published guidance that explains how appropriate assessment can be carried out in up to four steps, with the findings of each step determining whether the next step is needed:

- *Screening*: Determine whether the plan, 'in combination' with other plans and projects, is likely to have a significant adverse impact on a European site.
- *Appropriate assessment*: Determine the impact on the integrity of the European site of the plan, 'in combination' with other projects or plans, with respect to the site's structure, function and conservation objectives. Where there are adverse impacts, assess the potential mitigation of those impacts. Where there aren't, then the plan can proceed as it is.
- *Assessment of alternatives solutions*: Where the plan is assessed as having an adverse effect (or risk of this) on the integrity of a European site, examine alternative ways of achieving the plan objectives that avoid adverse impacts on the integrity of the European site.

• *Assessment where no alternative solutions* remain and where adverse impacts remain: Assess compensatory measures where, in the light of an assessment of imperative reasons of overriding public interest, it is deemed that the plan should proceed.

A high-profile example of appropriate assessment was the extension to Rotterdam Harbour. The project would significantly affect about 3,000 hectares of marine and natterjack toad habitats, but there were no alternatives and it would create roughly 10,000 long-term jobs. The project was given permission on condition that compensation was provided in the form of a new marine reserve, 25,000 hectares of protected area, and new dunes. A contrasting case was a proposed container terminal (port) at Dibden Bay in southern England, which would have affected the integrity of the Solent and Southampton Water SPA. This proposal was refused on appropriate assessment grounds because alternatives were available in the form of other UK ports that could provide enough container capacity.

12.4.6 Climate change and EIA

Climate change presents a fundamental challenge for all countries worldwide. EIA (and SEA) would seem directly relevant and very appropriate as tools: to assess the impacts of development actions on climate change, and climate change impacts on those development actions, and to advance appropriate mitigation and adaptation measures. Annex Ⅳ of the EU EIA Directive does identify the following in the important information to be supplied by the developer:

> 3. A description of the aspects of the environment likely to be significantly affected by the propsed project, including in particular, population, fauna, flora, soil, water, air, *climatic factors*, material assets ... and the inter-relationships between the above factors. (author emphasis)

Further, the UK government (DCLG 2007) noted that 'LPAs should not require specific and standalone assessments (of climate change) where the requisite information can be provided through...environmental impact assessments'. In 2009, a survey of EIA practitioners by the UK Institute of Environmental Assessment and Management (IEMA 2010a) found that 88 per cent felt that, where relevant, carbon emissions should be considered in the EIA and reported in the ES.

Yet recent practice suggests that EIA is not fulfilling its potential with regard to climate change. While the EIA Directive does mention climatic factors, the 2009 review of the EU EIA Directive (CEC 2009) notes that climate change issues are not expressly addressed in the Directive and that Member States recognize that they are not adequately identified and assessed within EIA practice. Wilson and Piper (2010) note similar experience from Canada, where a report by the Canadian Environmental Assessment Agency (CEEA 2003) found that climate change had not been well covered in most EAs. They suggest a number of reasons for this limited take-up including, for example, the often shorter term time horizons of

EIA compared with climate change, difficulties in dealing with climate change uncertainty, some fragmentation of EIA (as noted in Chapter 1 of this book) and the difficulty of addressing interrelationships of factors (CEC 2009).

So what might be the way forward for EIA and climate change? The IEMA (2010a, b) has produced assessment principles relating to both climate change mitigation and adaptation; Table 12.4 sets out some over-arching principles, and more specific EIA assessment principles.

Table 12.4 IEMA principles for EIA mitigation of GHG emissions

Overarching principles
• The GHG emissions from all projects will contribute to climate change; the largest inter-related cumulative environmental effect.
• The consequences of a changing climate have the potential to lead to significant environmental effects on all topics in the EIA Directive—e. g. population, fauna, soil, etc.
• The UK has legally binding GHG reduction targets; EIA must therefore give due consideration to how a project will contribute to the achievement of these targets.
• GHG emissions have a combined environmental effect that is approaching a scientifically defined environmental limit; as such any GHG emissions or reductions from a project might be considered to be significant.
• The EIA process should, at an early stage, influence the design and location of projects to optimise GHG performance and limit likely contributions to GHG emissions.
More specific assessment principles
• During scoping, climate change and mitigation issues and opportunities should be considered alongside each other to ensure integration in project design.
• The scope of GHG emissions must consider the relevant policy framework (local to global) and should also review the relevant findings in any associated SA/SEA.
• When assessing alternatives, consideration of the relative GHG emissions performance of each option should be considered alongside a range of environmental criteria.
• Baseline considerations related to GHG emissions should refer to the policy framework and also include the current situation and, where possible, take account of the likely future baseline situation.
• Quantification of GHG emissions (e. g. carbon calculators) will not always be necessary within EIA; however where qualitative assessment is used (e. g. emissions trends related to construction practices) it must be robust, transparent and justifiable.
• The assessment should aim to consider whole life effects (e. g. embodied energy, and emissions related to construction, operation and decommissioning—as relevant).
• The significance of a projects's emissions should be based on its net GHG effects, which may be positive (reduced) or negative (additional).
• Where GHG emissions cannot be avoided, the EIA should aim to reduce the residual significance of a project's emissions at all stages—design, construction, operation, etc.
• Where GHG emissions remain significant, but cannot be reduced further, approaches to compensate the project's remaining emissions should be considered.

Source: IEMA (2010a)

12.4.7 *Resilience thinking*

Project planning typically assumes that future changes will be gradual and predictable, whereas in reality they often come as sharp, unforeseen shocks: floods, volcanic eruptions, pandemics, economic crises, power outages etc. Resilience thinking is about how to deal with such shocks not by setting up systems to protect people and developments against all negative future change (as in risk assessment), but rather by making them able to cope with the shocks when they do come. It is the equivalent of teaching a child safe cycling and assertiveness rather than keeping them home from school for fear of accidents and bullying.

Some quite subtle and complex principles underlie resilience thinking, which can only be briefly summarized here. First is the *inevitability of change*, and the concept of adaptive cycles. All socio-economic systems go through an initial period of slow growth and accumulation, be it the formation of a woodland or a community group. This is typically followed by a short sharp period of decline, precipitated by a shock, say a woodland fire or the death of a key member of the community group. Depending on the system's resilience, the end result can be a reorganization into an equally 'good' new state—say, a new young woodland—or a worse state like a charred unproductive field or the disbanding of the community group. The phase of decline might be delayed, for instance by putting out forest fires, but this simply delays and escalates the impacts of the change when it does occur.

The second is the *importance of thresholds or tipping points*. Socio-environmental systems have a certain ability to recover from impacts, but if they are tipped over a threshold, then they plunge into a new state, which is normally disproportionately hard to recover from. For instance, it is much harder to return a eutrophic lake to a healthy state than to prevent a healthy lake from becoming eutrophic.

Third is *the importance of 'slow variables'* like climate, soil, global economic systems, or social networks. These systems act as buffers and reservoirs for the regeneration of smaller, faster systems such as habitats, species, communities and individuals. However, when the slow variables themselves are worn away (e. g. through a drop-feed of greenhouse gases or soil erosion) then this buffer and regeneration function is also worn down.

What does this mean for the planning of major projects? Some types of development—both the types of projects and how they are designed and implemented—are more resilient than others. Resilient projects would:

> *Embrace variability rather than control it*. Instead of increased flood defences, 'just in time' production and air conditioning, this would involve designing projects to cope with floods, having industrial processes that can cope with delayed parts, and having windows that open in offices and on public transport.
>
> *Build in redundancy or duplication*, so that if one aspect fails the other one can take over. The Deepwater Horizon oil spill would have been a minor blip for BP if the oil platform had had a back-up blow-out preventer. Providing access to a development by both road and rail, and putting emergency gates in the central reservation of major roads to allow cars to turn around, allows flexibility in case of accidents, flooding or congestion. Housing developments with 'spare' land can convert this to food production or temporary shelter if necessary.
>
> *Maintain some modularity or disconnectedness*, since over-connected systems are susceptible to shocks and transmit them rapidly. In the Middle Ages, villages would shut them-

selves off during times of plague. Modern equivalents are dykes, bunds, security barriers/gates and other access controls.

Recognize the importance of slow variables. This would mean giving greater weight in project planning and planning decisions to things like loss of high-quality agricultural land, water cycles, emissions of greenhouse gases, loss of customs and languages, and resettlement of communities.

Create tighter feedback loops between human actions and environmental outcomes. Many of our impacts occur away from us: in other countries where our food is grown and our clothes are manufactured, or in other parts of the UK where our energy is produced and waste disposed. Examples of this approach include greater emphasis on local rather than centralized production (e. g. energy, water, food), the proximity principle in siting waste disposal projects, and community development and ownership of infrastructure projects.

Promote and sustain diversity in all forms (environmental, social and economic). This principle runs counter to the common approach to decision-making, which promotes efficiency, targets and guarantees, but it is diversity that allows different responses to shocks and provides a source of future options. Examples include protecting areas of ecological diversity, promoting pilot projects for new technologies (e. g. various forms of tidal energy or carbon storage), and protecting indigenous people's lifestyles (adapted from Walker and Salt 2006).

Table 12. 5 shows how the results of resilience assessment might differ from the results of traditional project planning.

Table 12. 5 Contrasting resilence and traditional project planning approaches

Scenario	Results of traditional project planning	Results of resilience thinking	Source on which this example was based
A very small-scale (three persons) biodiesel producer in Barbados helps to reduce waste going to landfill and provides 'indigenous' fuel, but competes with others for limited waste cooking oil as fuel, and production costs are high due to the small scale nature of the operation	Increase the size of the operation, mechanize it, improve its efficiency	Develop a co-operative of small-scale biodiesel producers; develop biodiesel as a tourism project that provides tourism income as well as money from selling the diesel	Gadreau and Gibson(2010)
Two small villages in Tajikistan are facing water shortages due to ageing infrastructure and increasing population of humans and livestock	Build a new reservoir	Train local people to repair and maintain the existing water infrastructure, improve family planning, regulate water use through quotas and rules	Fabricius *et al.* (2009)
Increasing number of 'muddy floods' from agricultural run-off in England and Belgium cause millions of euros of damage each year	Improve flood defences	Improve 'institutional memory': value and maintain historical understanding of factors leading to these floods and how to manage the floods	Boardman and Vandaele(2010)

The Resilience Alliance has also developed a resilience assessment framework (RA 2010). This differs from EIA in that, *inter alia*, it places greater emphasis on uncertainty and disruptions (both past and future), considers higher scale actions and events that could affect the project scale, identifies thresholds of change and alternate states that could result from exceeding these thresholds, and focuses explicitly on the governance systems that manage changes. We are not aware of any cases where resilience assessment and EIA have been integrated, but clearly resilience assessment has the potential to strengthen EIA and make EIA projects better able to cope with future change, and we expect resilience thinking to be increasingly integrated into EIA.

12.4.8 Integrated impact assessment

As noted at the beginning of the book (Section 1.5.3), there has been an explosion of terms in relation to environmental assessment. One of these is that of integrated environmental assessment, which can relate to both environmental themes and techniques. In terms of themes, the preceding discussion of widening of scope to include more clearly socio-economic, health, equality and resilience content, can lead to a more integrated impact assessment (IIA), with decisions based partly on the extent to which various biophysical, social and economic impacts can be traded (Figure 12.3). For example, decision-makers might be unwilling to trade critical biophysical assets (e.g. a main river system and the quality of water supply) for jobs or lifestyle, but willing to trade less critical biophysical assets. Integrated impact assessment differs from traditional EIA in that it is consciously multi-disciplinary, does not take citizens' participation or the ultimate users of EIA for granted and recognizes the critical role of complexity and uncertainty in most decisions about the environment (Bailey *et al.* 1996; Davis 1996). Hence it tolerates a much broader array of methods and perspectives (quantitative and qualitative, economic and sociological, computer modelling and oral testimony) for evaluating and judging alternative courses of action. However, integration is not without its problems, including limitations on the transferability of assessment

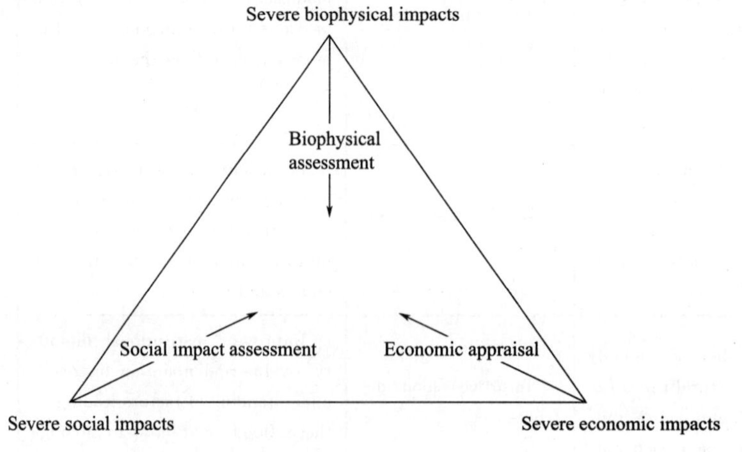

Figure 12.3 Integrated impact assessment

methods (see *Project Appraisal* 1996). The Integrated Assessment Workshop at the IAIA 2002 Conference highlighted the continuing problems of including social processes in integrated assessment (IAIA 2002).

Another equally important perspective is of the integration of relevant planning, environmental protection and pollution procedures. At the one extreme the UK still has multiple regulations for EIA, grafting the procedures into an array of relevant planning and other legislation; there is also parallel environmental protection and pollution legislation. At the other, there is the New Zealand 'one-stop shop' *Resource Management Act*. A better integration of relevant procedures represents another challenge for most EIA systems.

12.4.9 Extending EIA to project design: towards environmental impact design

An important and positive trend in EIA has been its application at increasingly early stages of project planning. For instance, while the DoT's 1983 *Manual of environmental appraisal* applied only to detailed route options, its later *Design manual for roads and bridges* requires a staged approach covering, in turn, broadly defined route corridors, route options and the chosen route (Highways Agency 2011). National Grid also uses multiple levels of environmental analysis for its transmission lines, from broad feasibility studies to detailed design (National Grid 20xx). This application of EIA to the early stages of project planning helps to improve project design and to avoid the delayed and costly identification of environmental constraints that comes from carrying out EIA once the project design is completed.

McDonald and Brown (1995) suggest that the project designer must be made part of the EIA team:

> Currently, most formal administrative and reporting requirements for EIA are based on its original role as a stand alone report carried out distinct from, but in parallel with the project design ...We can redress [EIA limitations] by transferring much of the philosophy, the insights and techniques which we currently use in environmental assessments, directly into planning and design activities.

A further evolution of this concept is to use EIA to identify basic environmental constraints before the design process is begun, but then allow designers freedom to design innovative and attractive structures as long as they meet those constraints. Figure 12.4 is an example of this approach: the magnificent Hundertwasser incinerator in Vienna, an incinerator that people might actually *want* to have in their city.

Holstein (1996) calls this postmodern approach 'environmental impact design' (EID), and distinguishes it from EIA's traditionally conservative, conservation-based focus.[2] The following paragraphs explain Holstein's view of EID.

> EIA as presently practised deconstructs a site: it takes an environment apart to highlight the different interacting components within it (e.g. soil, water, flora). EIA

Figure 12.4 Art and development project combined: the Hundertwasser/Spittelau incinerator in Vienna
Source: Wikimedia Commons

suggests that the site has another (environmental) function other than that for which it is being developed. Yet this relationship to deconstruction is only superficial because EIA is conservation based; it makes little challenge to the fixed hierarchies of modernism that underpin it, such as development-induced growth and technological subservience. Environmental design within EIA is too often merely a byproduct of assessment or is even handed back to the developer to have another shot at the design themselves. It makes little use of artistic-based metaphors to provide any re-enchantment or return to human landscape values, it makes no attempt to rip apart environmental function and form, and creates no demand for the kind of relative individualism needed to reflect cultural sustainability to an uninterested-unless-aroused population (all characteristics of postmodernism). Through this passivity of EIA, time, space, communication, leadership—all the key elements of good flowing design are lost.

This said, initially it might be argued that true postmodernism is simply beyond the remit of an EIA that exists for objective assessment rather than artistic purposes. The above description should be called EID. EID emphasizes the artistic contribution to EIA; it requires a different set of approaches (and probably personnel) than pure EIA, as well as creativity and elements of cultural vision. To an extent, some of the principles of EID are already being undertaken in EIA, in the mitigation sections of EISs, and especially within environmental divisions of the larger developers (e. g. the utilities) who often seem to see the formal EIA process as merely a lateral extension to their own design policies. Even so, rarely is it recognized as an artistic activity.

The key difference between EIA and EID lies in the concept of 'unmodifiable design'. Traditionally, EIAs are carried out on projects in which most of the structural elements have already been finalized. In more EID-oriented approaches, there is less unmodifiable design, and thus more scope for introducing environmentally sound design as mitigation measures. An even more radical path would be a postmodern EIA which aims to begin with so few unmodifiable design ideas that the EIA essentially becomes the leading player in design (adapted from Holstein 1996).

12. 4. 10 Complementary changes: enhancing skills and knowledge

The previous discussions indicate that EIA practitioners need to develop further their substantive knowledge of the wider environment. There is also an important role for 'State of the environment reports' and the development of 'carrying capacity and sustainability indicators'—if not interpreted too narrowly. Carrying capacity is multi-dimensional and multi-perspective (see Figure 12. 5 for an example for tourism impact assessment). Carrying capacity is also an elastic concept, and the capacity can be increased through good management.

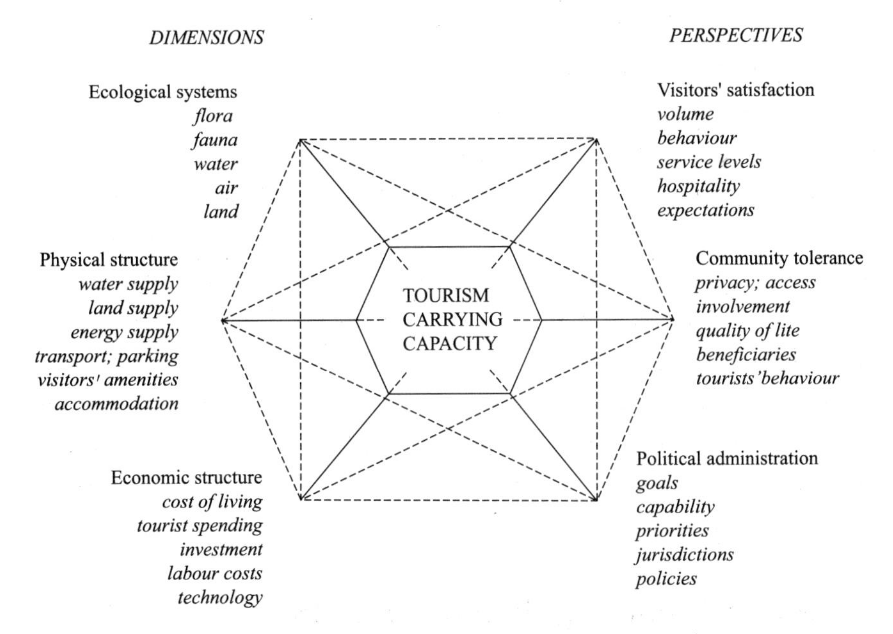

Figure 12. 5 Carrying capacity: a tourism example

Source: Glasson *et al*. 1995

Practitioners also need to develop both 'technical' and 'participatory' approaches, using, for example, focus group, Delphi and mediation approaches. EIA has been too long dominated by the 'clinical expert' with the detached quantitative analysis. However, there is still a place for the sensible use of the rapidly developing technology—including expert systems and GIS (Rodriguez-Bachiller with Glasson 2004). There is also a need for more capacity building of EIA expertise, plus relevant research, including, for example, more comparative studies and longitudinal studies (following impacts over a longer life cycle—moving towards adaptive EIA).

12.5 Extending EIA to project implementation: environmental management systems, audits and plans

An environmental management system (EMS), like EIA, is a tool that helps organizations to take more responsibility for their actions, by determining their aims, putting them into practice and monitoring whether they are being achieved. However, in contrast with the orientation of EIA to future development actions, EMS involves the review, assessment and incremental improvement of an *existing organization*'s environmental effects. EMS can thus be seen as a continuation of EIA principles into the implementation stage of a project. In essence, EMS and EIA can be seen as environmental protection tools with complementary purposes, with EIA seeking to anticipate and mitigate/enhance impacts of proposed new projects at the planning and design stage, and EMS helping organizations to effectively manage the day to day impacts during the full life cycle of such projects (Palframan 2010).

EMS has evolved from environmental audits, which were first carried out in the 1970s by private firms in the USA for financial and legal reasons as an extension of financial audits. Auditing later spread to private firms in Europe as well and, in the late 1980s, to local authorities in response to public pressure to be 'green'. In the early 1990s environmental auditing was strengthened and expanded to encompass a total quality approach to organizations' operations through EMS. EMS is now seen as good practice and has mostly subsumed environmental auditing.

This section reviews existing standards on EMS, briefly discusses the application of EMS and environmental auditing by both private companies and local authorities, and concludes by considering the links between EMS and EIA, using environmental management plans (EMPs) and other vehicles.

12.5.1 *Standards and regulations on EMS*

Three EMS standards apply in the UK: the EC's Eco-Management and Audit Scheme (EMAS) of 1993, which was revised in 2001; the International Organization for Standardization's (ISO) series 14000; and the more recent British Standard (BS) 8555. The schemes are compatible with each another, but differ slightly in their requirements.

The EC's EMAS scheme was adopted by EC Regulation 1836/93 in July 1993 (EC 1993), and became operational in April 1995. It was originally restricted to companies in industrial sectors, but since the 1993 regulations were replaced in 2001 by Regulation 761/2001 (EC 2001a) it has been open to all economic sectors, including public and private services. It was most recently updated in 2009/10 by Regulation 1221/2009, which came into force in January 2010. It is a voluntary scheme and can apply on a site-by-site basis. To receive EMAS registration, an organization must:

- Establish an environmental policy agreed by top management, which includes provisions for compliance with environmental regulation, and a commitment to continual improvement of environmental performance.
- Conduct an environmental review that considers the environmental impacts of the organization's activities, products and services; its framework of environmental legislation; and its existing environmental management practices.
- Establish an EMS in the light of the results of the environmental review that aims to achieve the environmental policy. This must include an explanation of responsibilities, objectives, means, operational procedures, training needs, monitoring and communication systems.
- Carry out an environmental audit that assesses the EMS in place, conformity with the organization's policy and programme and compliance with relevant environmental legislation.
- Provide a statement of its environmental performance that details the results achieved against the environmental objectives, and steps proposed to continuously improve the organization's environmental performance. (EC 2001b)

The environmental review, EMS, audit procedure and environmental statement (ES) must be approved by an accredited eco management and audit scheme (EMAS) verifier. The validated statement must be sent to the EMAS competent body for registration and made publicly available before an organization can use the EMAS logo. In the UK the competent body is the IEMA. Although EMAS was originally oriented towards larger private organizations, it can also apply to local authorities and smaller companies.

The *International Organization for Standardization's ISO 14000* series was first discussed in 1991, and a comprehensive set of EMS standards was published in September 1996. These include ISO 14001 on EMS specifications (ISO 1996a), ISO 14004 on general EMS guidance (ISO 1996b) and ISO 14010—14014, which give guidance on environmental auditing and review. EMAS and ISO 14001 are compatible, but have some differences. These are shown in Table 12.6.

Table 12.6 Differences between EMAS and ISO 14001

	EMAS	ISO 14001
Preliminary environmental review	Verified initial review	No review
External communication and verification	Environmental policy, objectives, EMS and details of organization's performance made public	Environmental policy made public

Continued

	EMAS	ISO 14001
Audits	Frequency and methodology of audits of the EMS and of environmental performance	Audits of the EMS(frequency of methodology not specified)
Contractors and suppliers	Required influence over contractors and suppliers	Relevant procedures are communicated to contractors and suppliers
Commitments and requirements	Employee involvement, continuous improvement of environmental performance and compliance with environmental legislation	Commitment of continual improvement of the EMS rather than a demonstration of continual improvement of environmental performance

Source: EC 2001b

In 2003 the UK government introduced a new EMS initiative—*BS* 8555 (*Guide to the implementation of an environmental management system including environmental performance evaluation*)—to assist organizations, in particular small and medium sized enterprises, to implement an environmental management system and subsequently achieve EMAS registration (DEFRA 2003). The standard includes guidance on how to develop indicators, so right from the start it is possible to know whether environmental impacts have been successfully reduced. The IEMA Acorn Scheme provides an officially recognized EMS standard recommended by the government. Acorn provides a route to EMS implementation, broken down into a series of logical, convenient, manageable phases using the British Standard BS 8555, plus a clearly defined route plan to ISO 14001 certification and/or EMAS registration.

12.5.2 Implementation of EMS and environmental auditing

By 2010, more than 130,000 organizations worldwide had gained ISO 14001 certification, with over 10,000 in the UK. In addition in the UK over 150 organizations have achieved the BS8555 and IEMA Acorn scheme, and another 500 are working through the scheme. Europewide, 4,500 organizations have participated in EMAS, with over 70 of these in the UK. Organizations perceive EMS as a way to reduce their costs through good management practices such as waste reduction and energy efficiency. They also see EMS as good publicity and, less directly, as a way of boosting employees' morale. However, private companies still have problems implementing EMS due to commercial confidentiality, legal liability, cost and lack of commitment. Smaller companies are especially affected by the cost implications of establishing EMS systems, and have been slower than the larger companies in applying it to their operations. The use of EMS by local authorities has been limited by cutbacks in central government funding, government reorganization and growing public concerns about economic rather than environmental issues.

12.5.3 *Links between EMS and EIA*

Environmental information

The growth in EMS is important to EIA for several reasons. First, EMS of both public sector and private sector organizations will increasingly generate environmental information that will also be useful when carrying out EIAs. For example, local authorities' State of the Environment Reports provide data on environmental conditions in areas that can be used in EIA baseline studies. Generally, such reports will contain information on such topics as local air and water quality, noise, land use, landscape, wildlife habitats and transport.

In contrast, private companies' environmental audit findings have traditionally been kept confidential, and it is noticeable from Section 12.5.2 that many more companies have opted for ISO 14001 accreditation—which requires only limited disclosure of information—than EMAS accreditation. Thus a private company's EMS is likely to be useful for EIA only if that company intends to open a similar facility elsewhere. However, environmental auditing information about levels of wastes and emissions produced by different types of industrial processes, the types of pollution abatement equipment and operating procedures used to minimize these by-products, and the effectiveness of the equipment and operating procedures will be useful for determining the impact of similar future developments and mitigation measures. Some of these audits are also likely to provide models of 'best practice', which other firms can aspire to in their existing and future facilities. Most interestingly, however, project EIAs are increasingly used as a starting point for their projects' EMSs. For instance, emission limits stated in an EIA can be used as objectives in the company's EMS, once it is operational. The EMS can also test whether the mitigation measures discussed in the EIA have been installed and whether they work effectively in practice. Overall, EMS is likely to increase the level of environmental monitoring, environmental awareness and the availability of environmental data; all of this can only be of help in EIA.

Environmental Management Plans

A very practical link between EIAs and EMSs can be provided by environmental management plans (EMPs)—to add yet another acronym! The aim of EMPs is to ensure that the effort put into the EIA process pre-application and consent is effectively delivered post-consent. This may involve overcoming a variety of perceived barriers linking EIA and EMS, including different consenting regimes, time periods and personnel, and lack of resources (Palframan 2010). Recent experience of EMPs has been of a less formal, simpler, less bureaucratic, and '*EMS-lite*' approach (Marshall 2004). The IEMA has been a strong UK advocate of the EMP approach, and set out its position in its practitioner guide on *Environmental Management Plans* (IEMA 2008). An EMP is a document that: sets out the actions that are needed to manage environmental and community risks associated with the life cycle of a development, identifies what is needed, when, and who is responsible for its delivery.

It is a bridge between the EIA process pre-consent and EMS operated by various stakeholders (e. g. project construction contractors, project operation managers) post-consent. Figure 12. 6 provides a simple illustration of this bridging role; Table 12. 7 outline further reasons for using EMPs and the key building blocks for the approach.

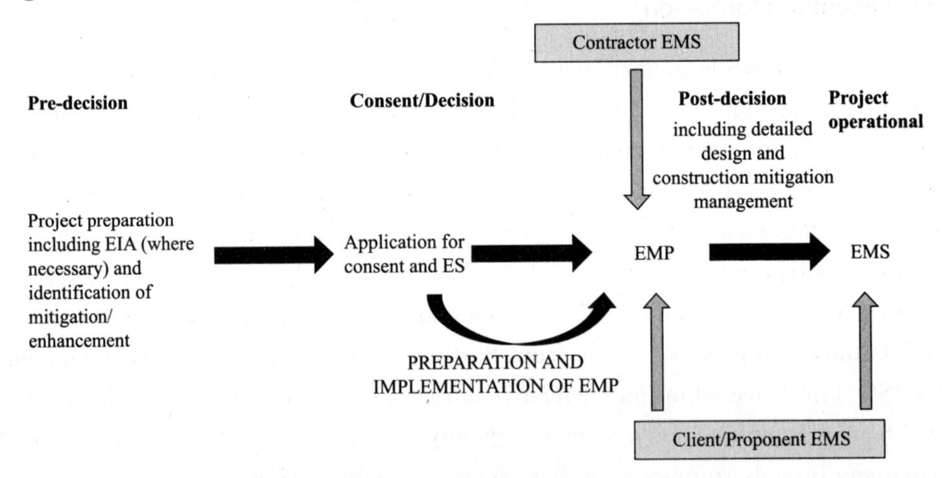

Figure 12. 6 Linkages between EIA, EMPs and EMS

Source: IEMA 2008

Table 12. 7 Benefits of preparing an EMP

1 Creates a framework for ensuring and demonstrating conformance with legislative requirements, conditions and mitigation set out in ESs.
2 Provide a continuous link or 'bridge' between the design phase of a project and the construction, and possibly operational, phase.
3 Ensures an effective communication or feedback system is in place between the operators on site, the contractor, the environmental manager(or consultant) and ultimately the regulator.
4 Preparing a draft EMP at an early stage demonstrates commitment to mitigation and can help reduce delays post-consent by showing how consultees and consenting authority concerns will be addressed.
5 Drives cost savings through improved environmental risk management.

Source: Adapted from IEMA 2008, 2011

Table 12. 8 The building blocks for developing a successful EMP

1	Involvement of the proponent, construction teams and contractors during the formulation of mitigation when planning the development to ensure: -the mitigation is deliverable; -costs of mitigation actions are factored into the detailed design stage, while the construction budget is still being developed; and -there is buy-in and commitment when developing mitigation at the EIA stage, to increase likelihood of effective implementation.
2	Involvement of competent environmental professionals in the design and specification of mitigation actions to ensure: -that the requirements are very clear; and -to improve the chances of successful delivery.
3	Ensuring mitigation measures identified while planning the development have been formulated with the involvement of relevant stakeholders, or a clear timetable and process for further consultation post-consent is set out(where mitigation requires more detailed design post-consent).
4	Clearly identified and sufficiently detailed mitigation measures in the pre-consent phase(a framework or draft EMP prepared alongside an ES is helpful as it can set out additional detail in preparation for a full EMP).
5	Ensuring mitigation proposals are identifiable in documents accompanying the application for development consent;
6	Ensuring mitigation measures are presented as elements that the proponent would be willing to have included in the final consenting documentation.

Source: Adapted from IEMA 2008, 2011

The IEMA believes that EMPs will become increasingly common in UK practice; a practitioner survey found that almost 80 per cent of respondents agreed that a draft EMP should be required within the EIS (IEMA 2011). However, the format and name may vary. For example, all EISs produced by the Environment Agency for England and Wales include an environmental action plan (EAP). This acts as an interface between the EIA and the EMS and keeps the EIA as a 'live' document through the project life. Projects for other developers have included Construction Environmental Management Plans (CEMPs) which focus on implementation during the project construction stage; these are more limited, but are still valuable for this often very disruptive stage in a project's life (Palframan 2010).

12.6 Summary

As in a number of other countries discussed in Chapter 10, the practice of EIA for projects in the UK, set in the wider context of the EU, has progressed rapidly up the learning curve. Understandably, however, practice has highlighted problems as well as successes. The resolution of problems and future prospects are determined by the interaction between the various parties involved. In the EU the various amendments to the EIA Directive have helped to improve some steps in the EIA process, including screening, scoping, consideration of alternatives and participation. However, some key issues remain unresolved, including the lack of support for mandatory monitoring. This chapter has identified an agenda for other possible changes, including cumulative impacts, socio-economic impacts and the linked areas of health impact and equalities impact assessment, appropriate assessment, resilience thinking, the key area of climate change, IIA and EID. Some of these will be easier to achieve than others, and there will no doubt be other emerging issues and developments in this dynamic area, and systems and procedures will continue to evolve in response to the environmental agenda and to our managerial and methodological capabilities.

There is also an urgent need to 'close the loop', to learn from experience. While the practice of mandatory monitoring is still patchy, notable progress has been made in the development of environmental management and auditing systems. Assessment can be aided by the development of EMSs for existing organizations, be they private-sector firms or local authorities. The information from such activities, plus the recent and important development of EMPs as a bridge between EIA and EMS activity, pre-and post-consent, could provide a significant improvement in the quality and effective implementation of EIAs.

As EIA activity spreads, more groups will become involved. Capacity building and training is vital both in the EIA process, which may have some commonality across countries, and in procedures that may be more closely tailored to particular national contexts. EIA practitioners also need to develop their substantive knowledge of the wider environment and to improve both their technical and participatory approaches in the EIA process.

SOME QUESTIONS

The following questions are intended to help the reader focus on the important issues of this chapter, and to start to consider potential approaches for improving the effectiveness of project assessment.

1 Change is a feature of EIA systems; what is driving such change?

2 Which stakeholders are more positive about further developing EIA systems and which are less so, and why?

3 Review the features of 'best-case' EA performance identified by Sadler in 1996; do those features still all apply today? Would you add any new features?

4 Consider possible explanations as to why the assessment of cumulative impacts has been such a challenging task for EIA?

5 What do you understand by 'environmental justice'?

6 Identify the potential key socio-economic impacts of a new major project with which you are familiar.

7 How might such impacts vary between the construction and operational stages of the project?

8 For the same project used in Question 6, consider the potential health impacts, again considering how such impacts might vary between the construction and operational stages of the project?

9 Similarly, for the same project again, consider the potential equality impacts.

10 Examine the case for integrating health and equality impacts within a socio-economic impact assessment as part of a more integrated approach to EIA that includes both bio-physical and socio-economic impacts.

11 When does appropriate assessment apply in the EU? What do you understand by the term 'imperative reasons of overriding public interest'? Why might they have been significant for the Rotterdam Harbour extension, but not for Dibden Bay on Southampton Water?

12 Why has climate change not been well covered in many EIAs?

13 Review the specific climate assessment principles for EIA from Table 12.4; consider which might be most difficult to achieve.

14 What kind of resilience 'features' could one integrate into the design of an urban housing development? A rural landfill site?

15 What do you understand by the term 'environmental impact design'?

16 What is an EMS? Outline some of the differences between the EMAS and ISO 14001 EMS systems.

17 Discuss the potential value of environmental management plans for more effective EIA.

Notes

1 In practice, it is also applied to Ramsar wetland sites, candidate SACs, and European Marine Sites.

2 This term was originally coined for a slightly different context by Turner (1995).

References

Agyeman, J. and Evans, B., 2004. Just sustainability: the emerging discourse of environmental justice in Britain? *The Geographical Journal*, 170, 2, 155-64.

Ahmad, B. 2004. Integrating health into impact assessment: challenges and opportunities. *Impact Assessment and Project Appraisal* 22 (1), 2-4.

Bailey, P., Gough, P., Chadwick, C. and McGranahan, G. 1996. *Methods of integrated environmental assessment: research directions for the European Union*. Stockholm: Stockholm Environmental Institute.

Baxter, W., Ross, W. and Spaling, H. 2001. Improving the practice of cumulative effects assessment in Canada. *Impact Assessment and Project Appraisal* 19 (4), 253-62.

Boardman, J. and Vandaele, K. 2010. Soil erosion, muddy floods and the need for institutional memory. *Area* 42 (2), 502-13.

Burdge, R. 1999. The practice of social impact assessment-background. *Impact Assessment and Project Appraisal* 17 (2), 84-8.

Canter, L and Ross, W. 2010. State of practice of cumulative effects assessment and management: the good, the bad and the ugly. *Impact Assessment and Project Appraisal* 28 (4), 261-68.

CEAA (Canadian Environmental Assessment Agency) 1999. *Cumulative effects assessment: practitioners' guide*. Quebec: CEAA (www. ceaa. gc. ca).

CEC 2000. *Assessment of plans and programmes significantly affecting Natura 2000 sites: Methodological guidance on the provisions of Article 6 (3) and (4) of the Habitats Directive 92/43/EEC*. Brussels: CEC.

CEC 2009. *Study concerning the report on the application and effectiveness of the EIA Directive: final report*. Brussels: DG Environment.

CEPA (Commonwealth Environmental Protection Agency) 1994. *Review of Commonwealth environmental impact assessment*. Canberra: CEPA.

CEQ (Council on Environmental Quality, USA) 1997. *Considering cumulative effects under the National Environmental Policy Act*. Washington, DC: Office of the President. Available at: www. ceq. eh. doe. gov/ nepa/nepanet/htm.

Chadwick, A. 2002. Socio-economic impacts: are they still the poor relations in UK environmental statements? *Journal of Environmental Planning and Management* 45 (1), 3-24.

CPRE (Council for the Protection of Rural England) 1991. *The environmental assessment directive-five years on*. London: Council for Protection of Rural England.

CPRE 1992. *Mock directive*. London: Council for Protection of Rural England.

Davis, S. 1996. *Public involvement in environmental decision making: some reflections on West European experience*. Washington, DC: World Bank.

DCLG (Department of Communities and Local Government) 2007. *Planning Policy Statement: Planning and Climate Change. Supplement to PPS1*. London: DCLG.

DCLG 2011. *Guidance on the Environmental Impact Assessment (EIA) Regulations 2011 for England*. London: DCLG.

DEFRA 2003. *An Introductory Guide to EMAS*. London: DEFRA.

Douglas, C. H. 2003. Developing health impact assessment for sustainable futures. *Journal of Environmental Assessment Policy and Management* 5 (4), 477-502.

Downey, L. 2005. Assessing environmental inequality: How the conclusions we draw vary according to the definitions we employ. *Sociological Spectrum*, 25, 349-69.

EC (European Commission) 1993. *Regulation no. 1836/93 allowing voluntary participation by companies in the industrial sector in an eco-management and audit scheme*. Brussels: EC.

EC 2001a. *Regulation no. 761/2001 of the European Parliament and of the Council of 19* March *2001 allowing voluntary participation by organisations in a community eco-management and audit scheme (emas)*. Brussels: EC.

EC 2001b. *EMAS and ISO/EN ESO 14001: differences and complementarities*. Available at: www. europa. eu. int/comm/environment/emas/pdf/factsheets.

Essex Planning Officers' Association 2007. *The Essex Guide to Environmental Impact Assessment*. Chelmsford: Essex County Council.

Fabricius, C., Quinlan, A. and Otambekov, A. 2009. *Resilience assessment in Roghun, Tajikistan*. Available at: www. nmmu. ac. za/documents/ SRU/Resilience%20Assessment%20in%20Roghun%2017%20May. pdf.

FoE (Friends of the Earth) 2011. *The greenest government ever: one year on*. London: FoE.

Gadreau, K. and Gibson, R. 2010. Illustrating integrated sustainability and resilience based assessments: a small-scale biodiesel project in Barbados. *Impact Assessment and Project Appraisal* 28 (3), 233-43.

Glasson, J. 2009. Socio-economic impacts. In *Methods of environmental impact assessment*. 3rd edn, P. Morris and R. Therivel (eds), Chapter 2. London: Spon.

Glasson, J. and Cozens, P. 2011. Making communities safer from crime: an undervalued element in environmental assessment. *EIA Review* 31, 25-35.

Glasson, J., Godfrey, K. and Goodey, B. 1995. *Towards visitor impact management*. Aldershot: Avebury.

Health Canada 1999. *Canadian handbook of health impact assessment*. vol. 1: *The basics*. Ottawa: Health Canada.

Highways Agency 2011. *Design manual for roads and bridges*, London: the Stationery Office.

Holstein, T. 1996. *Reflective essay: postmodern EIA or how to learn to love incinerators*. Written as part of MSc course in Environmental Assessment and Management, Oxford Brookes University, Oxford.

Hyder Consulting 1999. *Guidelines for the assessment of indirect and cumulative impacts as well as impact interactions*. Brussels: CEC-DGXI. Available at: www. europa. eu. int/comm/environment/eia/eia_support. htm.

IAIA (International Association for Impact Assessment) 1994. Guidelines and principles for social impact assessment. *Impact Assessment* 12 Summer.

IAIA 2002. Workshop on Integrated Assessment, in *IAIA Annual conference*. The Hague, The Netherlands: IAIA.

IEMA (Institute of Environment Management and Assessment) 2008. *Practitioner series No 12: environmental management plans*. Lincoln: IEMA.

IEMA 2010a. *IEMA Principles series: climate change mitigation and EIA*. Lincoln: IEMA.

IEMA 2010b. *IEMA Principles series: climate change adaptation and EIA*. Lincoln: IEMA.

IEMA 2011. *The state of environmental impact assessment practice in the UK*. Lincoln: IEMA.

IMD (Index of Multiple Deprivation) 2011. *The English indices of multiple deprivation 2010*. London: DCLG.

IPHI (Institute of Public Health in Ireland) 2009. *Health impact assessment guidance*. Dublin and Belfast: IPHI.

ISO (International Organization for Standardization) 1996a. *ISO 14001 Environmental management sys-*

tems-specification with guidance for use.

ISO 1996b. *ISO 14004 Environmental management systems-general guidelines on principles, systems and supporting techniques.*

Lane, P. and associates 1988. *A reference guide to cumulative effects assessment in Canada*, vol. 1, Halifax: P Lane and Associates/CEARC.

Marshall, R. 2004. Can industry benefit from participation in EIA follow up? in Morrison-Saunders, A. and J. Arts (eds). *Assessing impact: handbook of EIA and SEA follow-up.* London: Earthscan.

McDonald, G. T. and Brown, L. 1995. Going beyond environmental impact assessment: environmental input to planning and design. *Environmental Impact Assessment Review* 15, 483-95.

National Grid 2011. *Methodological guidence for options appraisal.* Warwick: National Grid.

Newton, J. 1995. *The integration of socio-economic impacts in EIA and project appraisal.* MSc dissertation, University of Manchester Institute of Science and Technology.

Odum, W. 1982. Environmental degradation and the tyranny of small decisions. *Bio Science* 32, 728-29.

O'Riordan, T. 1990. EIA from the environmentalist's perspective. VIA 4, March 13.

Palframan, L. 2010. *The integration of EIA and EMS systems: experiences from the UK.* 30th Annual Conference of IAIA Proceedings.

Peterson, E. *et al.* 1987. *Cumulative effects assessment in Canada: an agenda for action and research.*

Quebec: Canadian Environmental Assessment Research Council (CEARC).

Phoolchareon, W., Sukkumnoed, D. and P. Kessomboon 2003. Development of health impact assessment in Thailand: recent experiences and challenges. *Bulletin of the World Health Organization* 81 (6), 465-67.

Piper, J. 2000. Cumulative effects assessment on the Middle Humber: barriers overcome, benefits derived. *Journal of Environmental Planning and Management* 43 (3), 369-87.

Piper, J. 2001a. Assessing the cumulative effects of project clusters: a comparison of process and methods in four UK cases. *Journal of Environmental Planning and Management* 44 (3), 357-75.

Piper, J. 2001b. Barriers to implementation of cumulative effects assessment. *Journal of Environmental Assessment Policy and Management* 3 (4), 465-81.

Project Appraisal 1996. Special edition: *Environmental Assessment and Socio-economic Appraisal in Development* 11 (4).

RA (Resilience Alliance) 2010. *Assessing resilience in social-ecological systems.* Workbook for Practitioners. Version 2.0. Available at: www. resalliance. org/index. php/resilience _ assessment.

RSPB 2000. *Biodiversity and EIA: a good practice guide for road schemes.* RSPB, WWF, English Nature and Wildlife Trusts.

Rodriguez-Bachiller, A. with J. Glasson 2004. *Expert systems and geographical information systems for impact assessment.* London: Taylor and Francis.

Sadler, B. 1996 *International study in the effectiveness of environmental assessment.* Ottawa: Canadian Environmental Assessment Agency.

Sadler, B. 2012. Latest EA effectiveness study (not available at time this book went to press).

Scott Wilson (2006), *ES for Woodberry Down (London)*, London Borough of Hackney: Scott Wilson for Hackney Homes.

Taylor, L. and Blair-Stevens, C. 2002. *Introducing health impact assessment (HIA): informing the decision-making process.* London: Health Development Agency.

Taylor, L. and Quigley, R. 2002. *Health impact assessment: a review of reviews.* London: Health Development Agency.

Therivel, R. and Ross, B. 2007. Cumulative effects assessment: does scale matter? *Environmental Impact Assessment Review* 27, 365-85.

Turner, T. 1995. *City as landscape: a post-postmodernist view of design and planning*. London: E&FN Spon.

Vanclay, F. 2003. International principles for social impact assessment. *Impact Assessment and Project Appraisal* 21 (1), 5-12.

Walker, B. and Salt, D. 2006. *Resilience thinking*. Washington, DC: Island Press.

Walker, G., Fay, H. and Mitchell, G. 2005. *Environmental justice impact assessment: an evaluation of requirements and tools for distributional analysis (A report for Friends of the Earth)*. Leeds: Institute for Environment and Sustainability Research, University of Leeds.

Western Australia Department of Health 2007. *Health impact assessment in WA: summary document*. Perth: Western Australia Department of Health.

WHO Regional Office for Europe 2003. *Health impact assessment methods and strategies*. Available at: www. euro. who. int/eprise/main/WHO/progs/ HMS/home.

Wilson, E. and Piper, J. 2010. *Spatial planning and climate change*. Abingdon: Routledge.

World Bank 1995. *Environmental assessment: challenges and good practice*. Washington, DC: World Bank.

附 录

附录1　欧盟委员会环境影响评价指令全文（合并版）

1985年6月27日关于特定公共和私人项目的环境影响评价议会指令（85/337/EEC）

修订纪录：

1997年3月3日议会指令97/11/EC；

欧洲议会2003/35/EC指令和2003年5月26日议会指令；

欧洲议会2009/31/EC指令和2009年4月23日议会指令。

第1条

1. 该指令适用于评价可能对环境产生显著影响的公共和私人项目。

2. 在该指令中“项目”是指：

建筑以及其他设施或方案的执行，其他对自然环境和景观有干扰的。包括矿产资源开采。

“开发商”是指：

授权开发私人项目的申请者或者是发起项目的公共机构。

“开发许可”是指：

主管部门授权开发商开发项目。

“公众”是指：

一个或者更多自然人或者法律人以及根据国家法律规定和实际情况，他们的协会组织或者团体。

“公众关注”是指：

受到或者有可能受到影响的公众，按照2条第2款中规定的对环境决策有兴趣的公众；该定义的目的，认定非政府组织促进环境保护和满足国家法律法规均被认为是有兴趣的。

3.主管部门是成员国指派的执行本指令的责任人。

4.根据本国法律，在具体问题具体分析的基础上，如果项目会有消极影响，成员国可以决定为国防服务的项目不包含在此指令中。

5.该指令不适用于被国家法律系统的特定法案采纳的项目，因为该指令的目标以及所提供的信息是通过立法程序实现的。

第2条

1.在项目获得批准前，成员国应尽可能采用必要的措施以确保可能产生显著环境影响的项目，其属性、大小或坐落位置都要符合开发的要求，并针对影响做出相应的评价。这些项目将在第4条中定义[1]。

2.环境影响评价可能结合到成员国现有的项目批准程序中，若失败，那么结合到其他程序或者为实现指令目标所建立的程序中。

成员国可以提供一个独立的程序，以履行该指令要求和1996年9月24日通过的关于综合污染与防控的96/61/EC议会指令的要求。

3.并非因为对第7条有偏见，对于一些特殊案例，成员国可以全部或部分豁免执行指令条款。在此情况下，成员国应该：

(a) 考虑是否有其他更适合的评价形式；

(b) 使公众能获取在(a)中其他评价形式的信息以及有关豁免的信息和项目批准的原因；

(c) 在项目批准前，要将其豁免批准原因告知委员会，同时将其提交给项目所在国。

委员会应立即将递交的文件转达给其他成员国。委员会每年应向议会报告本节的实施情况。

第3条

环境影响评价应以适当的方法，根据单个案例和第4～11条的内容，在识别、描述和评价项目时，下面这些因素产生的直接和间接影响包括：

- 人、动物和植物；
- 土壤、水、空气、气候和景观；
- 矿产资源和文化遗产；
- 以上三种因素的相互作用。

第4条

1.依照第2条第3款，附件Ⅰ中所列的项目应根据第5～10条进行评价。

2.依照第2条第3款，附件Ⅱ中所列的项目，成员国应通过以下条款进行确定：

(a) 逐案分析；

(b) 成员国所设立的阈值或标准。

项目是否依照第5～10条进行评价。

成员国可以选择同时遵守(a)和(b)的两个程序。

---

[1] OJ No L 257，10.10.1996，p.26。

3. 当执行逐案分析或按照第 2 款设置阈值或标准时，应考虑附件Ⅲ中有关的备选标准。

4. 成员国应确保公众能获得主管部门在第 2 款下做出的决定。

第 5 条

1. 按照第 4 条，项目必须根据第 5～10 条进行环境影响评价。成员国应采取必要的措施确保开发商以适当形式提供附件Ⅳ中所列的信息，这是由于：

(a) 成员国认为，该信息与批准程序的指定阶段有关，与该项目的特性或项目类型以及可能被影响的环境特征有关。

(b) 成员国认为，有理由要求开发商根据现有的知识和评价方法把这些信息和其他信息汇总。

2. 在开发商提出开发许可申请之前，成员国要采取必要的措施，以确保主管部门能根据第 1 款对开发商所提交的信息提出一定的意见。在给出意见前，权力机构应参考第 6 条第 1 款与开发商和专家进行协商。在此指导下所提出的意见不排除要求开发商后期提供进一步的信息。

成员国要求主管部门给出此类意见，并不考虑开发商是否要求。

3. 根据第 1 款，开发商所提供的信息应至少包括：

• 对项目的位置、设计和规模等描述；

• 对为避免、减少或者尽可能弥补显著的不利影响而设计的措施的描述；

• 有关识别和评价项目对环境可能造成的主要影响的数据；

• 考虑到环境影响，开发商所研究的主要替代方案的简述及其做出选择的主要原因的表述；

• 非技术性总结关于在前面的省略中提到的信息。

4. 若有必要，成员国应确保所有专家掌握相关信息，特别参照第 3 条，使开发商可以获取这些信息。

第 6 条

1. 成员国应采取必要措施，确保可能由于项目或特殊的环境责任的原因而备受关注的专家们有机会表达他们对开发商所提供信息的意见和关于开发批准的要求。然后，成员国应任命提供咨询的专家，不论该咨询是总体形式的还是以具体案例为基础的。根据第 5 条，收集得到的信息应提供给专家。咨询的具体安排应由成员国拟订。

2. 公众应当被通知，无论是使用公众通知板还是其他合适的方式，如果有条件可以使用电子媒体，并根据第 2 条第 2 款所要求的尽早提供相关环境决策程序，无论如何，以下信息应该尽可能合理地提供：

(a) 发展咨询的要求；

(b) 项目进行环境影响评价程序的事实，如果相关，事实可应用第 7 条；

(c) 主管部门根据所得相关信息、提交的评论或疑问而负责做出决定的细节、公布评论或疑问时间表的细节；

(d) 潜在决定的实质或者其草案；

(e) 根据第 5 条所收集的信息可用性的迹象；

(f) 时间、地点以及相关信息可见的方法的迹象；

(g) 根据第 5 条公众参与安排的细节。

3. 成员国应确保公众在合理的时间内有机会：

(a) 按照第 5 条收集到的信息；

(b) 根据本条第 2 款，当公众有顾虑时，根据本国法律，向主管部门提交的主要报告和建议；

(c) 根据欧盟国会 2003/4/EC 指令文件和理事会 2003 年 1 月 28 日关于公众可以获取环境信息当决议[1]；除了本条第 2 款中提到的信息与第 8 条信息相关外，其信息根据本条第 2 段，只允许在公众关注以后可以获得。

4. 根据第 2 条第 2 款的要求，需要保证公共顾虑在环境决议以前尽早有效地表达，为此，公众可以在项目开发以前表达对相关主管部门的建议和观点。

5. 成员国可以决定通知公众的安排（例如，在一定范围内粘贴通告）和咨询公众顾虑（例如，通过书面咨询表或是公开询问）。

6. 确定程序不同阶段的时间限制，根据本条，保证足够的时间通知公众准备和有效参与环境决策。

第 7 条

1. 当某成员国意识到某项目可能对另一成员国的环境产生显著影响时，或当受到显著性影响的成员国提出要求时，准备在其领土上执行项目的成员国应尽快将有关信息送交到受影响的成员国，并马上通知本国的公众，其中包括：

(a) 项目描述和关于可能跨界影响的一切可获取的信息；

(b) 所采用决策信息的性质。

应给另一成员国适当的时间，以确定是否愿意参加该环境影响评价程序，同时应包括第 2 条第 2 款中提到的相关信息。

2. 如果成员国收到如第 1 条的信息，表明它们根据第 2 条第 2 款打算参加该环境影响评价程序，准备在其领土上执行项目的成员国应根据第 6 条第 2 款发送所收集的信息和考虑上述程序有关信息给受影响的成员国，并且根据第 6 条第 3 款（a）和（b）中的要求使信息可以获得。

3. 有关的成员国，在所有相关的领域内，也应：

(a) 安排第 1 款和第 2 款的信息可以获得，使第 6 节第 1 款中涉及的专家以及成员国范围内的可能受到显著影响的相关公众有机会获取该信息；

(b) 在项目开发许可批准前，确保有关专家和公众有机会和时间对主管部门提供给成员国的信息表达他们的意见。

4. 相关成员国应进行磋商，考虑项目潜在的跨界影响，并拟订减少或消除这种影响的措施，同时就咨询期间的合理时间框架达成一致。

5. 执行本条的详细安排将由相关成员国决定，并且根据第 2 条第 2 款有关项目的规定，应该确保受到影响的成员国中在该领域有顾虑的公众可以有效参与环境决策制定程序。

第 8 条

按第 5～7 条收集的信息和咨询结果应在开发许可过程中予以考虑。

第 9 条

1. 当批准或拒绝开发许可的决策实施后，主管部门或专家们将以适当的方式通知相关的

---

[1] OJ L 41，14. 2. 2003，p. 26。

公众，使公众获得以下信息：

• 决策的内容和其所有附加的情况；

• 已经审查的公众以前表达过的顾虑，包括公众参与程序在内的决策所依据的主要理由和考虑因素；

• 若有必要，对避免、减少和消除主要负面影响的措施进行描述。

2. 主管部门或专家应把第 1 节所涉及的信息通知给第 7 条中规定的成员国。

被咨询过的成员国需保证信息可以被有过顾虑的本国人民在合适的方式下获得。

第 10 条

本指令的条款不影响主管部门尊重国家法规和管理条款所加的限制和接受有关商业和工业机密的法律实践的义务，包括知识产权和公众利益保护。应用第 7 条时，成员国之间的信息传递和接收都应服从各成员国对拟议项目的限制。

成员国必须确保遵守相关国家法律系统，公众关注要点是：

(a) 有足够的兴趣或者其他；

(b) 当所在国家管理程序需要其作为先决条件的时候，维持其权利不受到侵害。

有权在法院或独立公正的法律机构前接受审查程序，对受本指令公众参与条款约束的决定、行为或不作为的实体或程序合法性提出质疑。

成员国应决定在什么时期对决议、作为或不作为提出质疑。

成员国应该决定什么是充分的利益，维持其权利不受到侵害，让公众参与可以被法律保护。为此目的，符合第 1 条第 2 款所述要求的任何非政府组织的利益应被认为足以满足本条 (a) 项的目的。相关组织也应该认为具有根据本条 (b) 项的目的而可能受到损害的权利。

这一条的条文应该不包括在行政审批之前进行预先审查程序的可能性，当在本国法律要求存在的情况下，也不应该在影响行政审批程序之前要求法律审批。

该程序应该公正、平等、实效以及控制成本。

为了确保该条款条文的有效性，成员国应该确保在行政和法律程序中的实际信息对公众可见。

第 11 条

1. 成员国和委员会应交流在指令应用中所获得的经验信息。

2. 要特别指出的是，根据第 4 条第 2 款，成员国应将拟议项目所选用的标准和/或阈值告知委员会。

3. 在指令生效 5 年后，委员会应向欧洲议会提交关于本指令作为修改指令的应用和效果报告。报告应以本条上述条款的信息交流为基础。

4. 在该信息交流的基础上，委员会应向议会提交关于确保在指令应用中的进一步协调的附加建议。

第 12 条

1. 在 3 年通知期以内，成员国应该采取必要措施遵循该指令[1]。

2. 成员国需要向委员会报告其本国法律法规与该指令有交集的地方。

---

[1] 这份指令已经于 1985 年 7 月 3 日通知了成员国。

第 13 条

该指令对欧盟成员国生效。

附件 I：隶属于第 4 条第 1 款中的项目

1. 原油冶炼（不包括以原油为原料生产润汾油）设施以及日产量 500t 以上的气化和液化煤或沥青的设施。

2. 产热量在 300MW 及其以上的热电厂和其他燃烧装置；核电站及其他核反应堆，包括已拆除或退役的核电站或反应堆❶（不包括研究装置、可引起核分裂的和能产生裂变物质的材料的转换，它的连续热量负荷不超过 1kW）。

3. (a) 非放射核燃料的再处理装置。

(b) 装置设计用于：为了生产或浓缩核燃料；为了处理非放射性核燃料或高水平的放射性废物；为了非放射性核燃料的最终处置；仅仅为了放射性废物的最终处置；仅仅为了储存（计划 10 年以上）非放射性核燃料或放射性废物，将它们储存于非生产地点的不同地方。

4. (a) 有关铸铁和钢的初炼的综合工作；

(b) 通过冶金、化学或电解过程，从矿石、浓缩物或二次加工的原材料中获得非有色金属的装置。

5. 提炼、加工、转化石棉的设施和过程以及含石棉的产品：对于含石棉的水泥产品，年产量超过 20000t；对于含石棉的摩擦材料，年产量超过 50t；其他用途石棉，每年超过 200t。

6. 一体化化学设备，例如，将那些利用化学过程生产工业物质的设备联用，同时功能上可以与各单独的设备互补，它们是：

(a) 主要有机化学物的生产设备；

(b) 主要无机化学物的生产设备；

(c) 含磷、氮或钾肥料（单一或混合肥料）的生产；

(d) 基本植物健康产品和生物杀虫剂的生产；

(e) 以化学或生物过程进行基本药品的生产；

(f) 炸药的生产。

7. (a) 远距离铁路运输线和 2100m 以上长度跑道的机场❷的基本跑道建设；

(b) 高速公路和快速路的建设❸；

(c) 四车道及以上的新公路的建设，或将已有的两车道或以下道路合并或加宽建为四车道或以上的公路，改建后的新公路，其连续长度在 10km 以上的。

8. (a) 内陆水路运输的航道和港口，能容纳 1350 吨位以上的船只通行；

(b) 连接陆地和外港（不包括摆渡码头）货物装卸的贸易港、码头，能容纳 1350 吨位以上的船只。

---

❶ 当所有核燃料以及放射性物质的基础设施被永久移除时，核电站和其他核反应堆才能停止建设基础设施。

❷ 本指令的目的是，“机场”符合 1944 年《芝加哥公约》中的《国际民用航空组织》中的定义（附录 14）。

❸ 本指令的目的是，“快速路”1975 年 10 月 15 日发布的《欧洲国际主要交通干线协定》中的定义。

9. 75/442/EEC[1] 指令附件ⅡA 标题 D9 中定义的焚烧、化学处理废物的装置，或将危险废物进行土地填埋的（例如 91/689/EEC[2] 指令）。

10. 75/442/EEC 指令附件ⅡA 标题 D9 中定义的非危险性废物，将其进行焚烧或化学处理的装置日处理能力超过 100t 的。

11. 地下水抽取或人工地下水再生计划每年抽取或再生水的容积相当于或超过 1000 万立方米的。

12.（a）水资源在流域之间调度以阻止可能的水短缺，水调度的总量超过 1 亿立方米/年的；

（b）在全部其他情况中，水资源在流域之间的调度，多年来抽取流量平均超过 20 亿立方米/年，水调度总量超过流量 5%的。

这两种情况不包括管道饮用水的调度。

13. 污水处理厂治理能力超过 15 万人的排放量，相当于 91/271/EEC[3] 指令第 2 条第 6 款指令定义的。

14. 商业目的的石油和天然气提炼，其提取石油数量超过 500t/d，天然气为每天 $500000m^3/d$ 的。

15. 为了控水或蓄水而设计的大坝和其他设施，其新建的或附属设施的蓄水量超过 1 千万立方米的。

16. 管道直径大于 800mm、长度大于 40km 的：

（a）输送天然气、石油或化学物质；

（b）输送以地质储存为目的的二氧化碳蒸气，包括附属增压站。

17. 饲养家禽或猪的养殖场，其集中程度要大于：

（a）能容纳小鸡 85000 只，母鸡 60000 只；

（b）能容纳猪 3000 头（超过 30kg）；

（c）能容纳母猪 900 头。

18. 木材加工：

（a）以木材或类似的纤维物质为原料进行生产的；

（b）纸和木板的生产，其生产能力超过 200t/d 的。

19. 矿石开采场和露天矿场占地面积超过 $25hm^2$，或泥煤冶炼场占地超过 $150hm^2$ 的。

20. 电压 220kV 以上、长度 15km 以上悬空电线的建设。

21. 储存石油、石化制品或化学产品的设备，其储存能力为 200000t 以上的。

22. 在本附件中任何项目的改变或者扩建，都需要满足其设置在附件中的阈值。

23. 储藏地需要遵守欧盟国会 2009/31/EC 指令以及 2009 年 4 月 23 日关于地质储藏二氧化碳的议会指令[4]。

24. 获取二氧化碳蒸气地质储存目的的基础设施建设需要遵守 2009/31/EC 指令包含本指令附件的内容，或者其每年获取二氧化碳等于或者多余 1.5Mt。

---

[1] OJ No L 194，25.7.1975. p.39。94/3/EC 欧盟决议对指令进行了最后的修订（OJ No L 57.1.1994，p.15）。

[2] OJ No L 337，31.12.1991，p.20。94/31/EC 指令对指令进行了最后的修订（OJ No L 168，2.7.1994，p.28）。

[3] OJ No L 135，30.5.1991，p.40。1994 年《欧盟加入法案》对指令进行了最后的修订。

[4] OJ L 140，5.6.2009，p.114。

附件Ⅱ：隶属于第4条第2款的项目

1.农业、林业和水产业

(a) 农田保留重建项目；

(b) 使用非农耕地或半自然区域的密集型农业用地项目；

(c) 农业系统的水体管理项目，包括灌溉用水和土地排水项目；

(d) 为了转化转换使用类型而进行的初期造林和伐木；

(e) 家禽集中饲养设施（不包括附件Ⅰ的项目）；

(f) 集中养鱼场；

(g) 填海造田。

2.冶炼工业

(a) 采石场、露天矿和泥煤冶炼（不包括附件Ⅰ中的项目）。

(b) 地下开采。

(c) 从海底或河流底泥中提炼矿物。

(d) 深度钻井，特别是：

- 地热井钻凿；
- 用于储存核废料的钻井；
- 用于水供应的钻井；
- 不包括用于调查土壤稳定性的钻井。

(e) 用于冶炼煤、石油、天然气、矿石和含沥青的页岩的初级工业设施。

3.能源工业

(a) 电力、蒸汽和热水生产的工业设施（不包括附件Ⅰ中的项目）；

(b) 输送气、蒸汽和热水的工业设施项目（不包括附件Ⅰ中的项目）；

(c) 天然气的地表储存；

(d) 易燃气体的地表储存；

(e) 化石燃料的地表储存；

(f) 工业煤砖和褐煤砖；

(g) 加工和储存放射性废物的设施（不包括附件Ⅰ中的项目）；

(h) 水力发电装置；

(i) 风力发电装置（风力农场）；

(j) 获取二氧化碳蒸气地质储存目的的基础设施建设需要遵守2009/31/EC指令而不包含本指令附件的内容。

4.金属的生产和加工

(a) 生产生铁或钢的设施（初级或二次熔解，包括连续铸造）。

(b) 含铁金属的加工设施：

- 热轧钢厂；
- 锻造；
- 保险丝金属涂层的应用。

(c) 含铁金属铸造品。

(d) 熔炼设施，包括炼制合金、除重金属以外的其他非含铁金属、产品（精炼、铸造等）。
(e) 金属表面处理设施和电解或化学过程中塑料材料的使用。
(f) 汽车的生产和组装，汽车发动机的生产。
(g) 造船。
(h) 制造飞机和修理飞机的设备。
(i) 铁路制造设备。
(j) 爆炸冲模。
(k) 金属矿石的焙烧和熔结设施。
5. 矿石工业
(a) 焦炉（干煤蒸馏）；
(b) 生产水泥的设施；
(c) 生产石棉及石棉制品的设施（不包括附件Ⅰ中的项目）；
(d) 生产玻璃及玻璃纤维的设施；
(e) 熔炼矿石物质以及生产矿物纤维的设施；
(f) 通过煅烧生产陶器制品，特别是屋顶用瓦、砖、耐火砖、瓷砖、陶器或瓷器。
6. 化学工业（不包括附件Ⅰ中的项目）
(a) 中间产品的生产及处理；
(b) 杀虫剂和药品、涂料和油漆、人造橡胶和过氧化物的生产；
(c) 储存石油、石化制品和化学制品的设备。
7. 食品工业
(a) 植物油、动物油和脂肪的生产；
(b) 动物和植物产品的包装；
(c) 日常用品的生产；
(d) 酿造和麦芽制造；
(e) 糖果和糖浆的生产；
(f) 牲畜屠宰设施；
(g) 工业淀粉生产设施；
(h) 鱼粉和鱼油制品工厂；
(i) 制糖工厂。
8. 纺织、皮革、木材和造纸工业
(a) 制造纸张和木板的工厂（不包括附件Ⅰ中的项目）；
(b) 预处理工厂（如清洗、漂白、丝光处理等），或纤维染色或纺织工厂；
(c) 制革工厂；
(d) 纤维素加工和生产设施。
9. 橡胶工业
生产和处理含橡胶成分的产品。
10. 基础设施项目
(a) 工业区开发项目；
(b) 城市发展项目，包括购物中心和停车场的建设；
(c) 铁路和联合运输的中转设施的建设，联合运输终端的建设（不包括附件Ⅰ中的项目）；

(d) 机场建设（不包括附件Ⅰ中的项目）；

(e) 公路、港口和港口设施，包括渔港设施的建设（不包括附件Ⅰ中的项目）；

(f) 不包括附件Ⅰ中的内陆水路建设、运河和泄洪作业；

(g) 为长期控水和蓄水而设计的大坝和其他设施（不包括附件Ⅰ中的项目）；

(h) 专门或主要用于客运的电车、轻轨和地铁、磁悬浮线路或特殊类型相似的线路；

(i) 石油和输气管道设备（不包括附件Ⅰ中的项目）；

(j) 长距离输水管设备；

(k) 防止沿海侵蚀的作业和海事作业，通过该建设能改变海岸带，例如，堤、防洪堤、码头和其他海防工作，其中不包括该工作的维护和重建；

(l) 不包括附件Ⅰ中的地下水抽取和人工地下水充注计划；

(m) 不包括附件Ⅰ的流域间水资源的调度作业。

11. 其他项目

(a) 动力化交通工具的永久轨道和测试轨道；

(b) 废物处置设施（不包括附件Ⅰ中的项目）；

(c) 污水处理厂（不包括附件Ⅰ中的项目）；

(d) 污泥储存处；

(e) 废铁储存，包括报废汽车；

(f) 引擎、涡轮机或反应器的测试台；

(g) 人工矿物纤维的生产设施；

(h) 回收或破坏易爆物质的设施；

(i) 牲畜屠宰场。

12. 旅游休闲

(a) 滑雪跑道、载送滑雪者上坡的缆车设施以及相关开发项目；

(b) 码头；

(c) 市区外的度假村和旅店以及相关开发项目；

(d) 固定的露营地和活动住房地；

(e) 主题公园。

13. (a) 附件Ⅰ或Ⅱ所列的项目中已授权的、已经或正在执行的改变或扩展，其可能对环境产生的重大影响（不包括附件Ⅰ中的改变或扩展）；

(b) 附件Ⅰ中所列的项目主要是开发和测试新方法或产品的项目以及使用期在两年以下的项目。

附件Ⅲ：第4条第3款中提到的标准选择

1. 项目的属性

必须要考虑到项目的本质，特别是以下的情况：

- 项目的规模；
- 与其他项目的累积影响；
- 对自然资源的利用；

- 废弃物的产生；
- 污染和损害；
- 事故风险，特别是考虑物质或技术使用带来的风险。

2. 项目的选址

必须考虑到可能受项目影响的环境敏感点，尤其是：

- 目前的土地利用情况；
- 该地区自然资源的相对丰裕度、质量和再生能力；
- 自然环境的自净能力，对以下区域应给予特别关注：

(1) 湿地；

(2) 沿海地区；

(3) 山地和森林地区；

(4) 自然遗址和公园；

(5) 在成员国的法律中已进行分类的地区或受保护的地区；依照 79/409/EEC 指令和 92/43/EEC 指令，明确制定受特殊保护的地区；

(6) 在所要求的环境质量标准已经超出欧盟委员会法律所规定的区域；

(7) 人口密集区域；

(8) 具有历史、文化或考古意义的景观区。

3. 潜在影响的特征

项目的潜在影响必须与一定的标准相关，该标准是在以上第 1 款和第 2 款下制定的，同时要特别考虑：

- 影响的范围（受影响的地理区域和人口多少）；
- 影响的边界性质；
- 影响的大小和复杂程度；
- 影响的可能性；
- 影响的持续性、频率和可逆性。

附件Ⅳ：有关第 5 条第 1 款的信息

1. 对项目描述，尤其要包括：

- 对整个项目自然特征和在建设和运行阶段对土地利用的描述；
- 对生产过程中主要特点的描述，例如所使用材料的性质和数量；
- 对项目运行阶段的废渣和排放物（水、空气和土壤污染、噪声、振动、光、热、辐射等）类型和数量的预测。

2. 开发商为考虑环境影响而设计的主要替换方案概述以及选择该方案的主要原因。

3. 拟议项目对环境组分可能产生的显著影响描述，特别是人口、动物、植物、土壤、水、空气、气候因素、物质资源等，包括建筑和考古遗迹、景观和以上因素之间的相互关系。

4. 拟议项目对环境可能产生的显著影响来源的描述[1]：

---

[1] 此描述包括对项目产生的直接影响，间接影响，二次影响，累积影响，短期、中期和长期影响，持久和暂时影响，消极和积极影响的描述。

- 已有的项目；
- 自然资源的使用；
- 污染物的排放、损害的产生和废物的排出以及开发商对评价环境影响所用预测方法的描述。

5. 为尽可能阻止、减少和消除显著消极环境影响而设计的措施的描述。
6. 以上条款所提供信息的非技术性总结。
7. 开发商在收集所需信息时可能遭遇的困难（技术缺陷或知识缺陷）。

附录 2 城乡规划（EIA）法规 2011 清单 2（规定 2.1）

在“清单 2”中定义的开发的描述以及阈值和标准的应用。

在下表中：

- “工作区域”包括仪器、装备、机械、材料、工厂、矸石或其他设施、建筑或基础设施的存储；
- “控制水域”在 1991 年《水资源法》中规定定义的同样内容[1]；
- “建筑面积”是在建筑中的建筑面积。

下表是在“清单 2”中定义的开发描述以及阈值和标准的应用。

专栏一：开发描述	专栏二：阈值和标准的应用
1. 农业和水产业	
(a)将未开垦的或者半自然的土地用于密集的农业生产	开发的面积超过 0.5hm²
(b)农业的水资源管理，包括灌溉和地面排水工程	工程的面积超过 1hm²
(c)密集型家畜养殖场(除非包含在清单 1 中)	新的建筑面积超过 500m²
(d)密集型渔场	每年养殖超过 10t 鱼(净重)的开发项目
(e)围海造田	所有的开发

专栏一：开发描述	专栏二：阈值和标准的应用
2. 采掘业	
(a)采石场、露天采矿和泥炭开采(除非包含在清单 1 中) (b)地下矿物开采	除了新建建筑面积建设房屋和其他辅助设施不超过 1000m² 以外的所有开发
(c)在河流中提取矿物	所有开发
(d)深层钻探，特别是： • 地热钻探； • 以核废料存储为目的的钻探； • 水供应钻探； 不包括以调查土地稳定性为目的钻探	(i)任何种类工程面积超过 1hm² 的钻探； (ii)在控制水域附近 100m 以内的地热钻探和以核废料存储为目的的钻探
(e)用于开采煤、石油、天然气以及矿石和沥青页岩的地面工业设施	开发的面积超过 0.5hm²

[1] 1991 c.57，见 104 节。

专栏一:开发描述	专栏二:阈值和标准的应用
3. 能源工业	
(a)发电,制造蒸汽和热水的工业设施(除非包含在清单1中)	开发的面积超过 0.5hm^2
(b)运输天然气、蒸汽和热水的工业设施	工程的面积超过 1hm^2
(c)天然气地上储存设施 (d)地下存储可燃气体的设施 (e)化石燃料地上储存设施	(i)新的房屋、安置处或者建筑面积超过 500m^2; (ii)在控制水域附近 100m 以内新的房屋、安置处或者建筑
(f)褐煤和煤炭的压制机	新的建筑的面积超过 1000m^2
(g)放射性废物的处理和储存设施(除非包含在清单 1 中)	(i)新的建筑的面积超过 1000m^2; (ii)在《环境许可法》(英格兰和威尔士)要求下,开发中的基础设施安装需要取得环境许可。2010 年法规[1]中关于放射性实质活动,第 5 条第 2 款(b),第 2 款(c)或者清单 23 中第 2 部分第 4 款,或者其变种需要取得类似许可
(h)水力发电站	新建水力发电站的发电能力在 0.55MW 以上
(i)以风为动力的能源生产设施(风力发电场)	(i)如果风力发电场有 2 个以上的涡轮机; (ii)或其发电设备中任何建筑物高于 15m 以上
(j)根据 2009/31/EC 指令规定的以地质储存为目的的获取二氧化碳蒸气体的设施	所有的开发

专栏一:开发描述	专栏二:阈值和标准的应用
4. 金属的生产和处理	
(a)生产生铁或钢的设施(初级或二次熔解),包括连续铸造。 (b)含铁金属的加工设施: (i)热轧钢厂; (ii)锻造; (iii)保险丝金属涂层的应用。 (c)含铁金属铸造品。 (d)熔炼设施,包括炼制合金、除重金属以外的其他非含铁金属、产品(精炼、铸造等)。 (e)金属表面处理设施和电解或化学过程中塑料材料的使用。 (f)汽车的生产和组装,汽车发动机的生产。 (g)造船。 (h)制造飞机和修理飞机的设备。 (i)铁路制造设备。 (j)爆炸冲模。 (k)金属矿石的焙烧和熔结设施	新的建筑的面积超过 1000m^2

[1] S. I. 2010/675。

专栏一:开发描述	专栏二:阈值和标准的应用
5.矿业	
(a)焦炉(干煤蒸馏); (b)生产水泥的设施; (c)生产石棉及石棉制品的设施(除非包含在清单1中); (d)生产玻璃及玻璃纤维的设施; (e)熔炼矿石物质以及生产矿物纤维的设施; (f)通过煅烧生产陶器制品,特别是屋顶用瓦、砖、耐火砖、瓷砖、陶器或瓷器	新的建筑的面积超过1000m^2

专栏一:开发描述	专栏二:阈值和标准的应用
6.化学工业	
(a)中间产品的生产及处理; (b)杀虫剂和药品、涂料和油漆、人造橡胶和过氧化物的生产	新的建筑的面积超过1000m^2
(c)储存石油、石化制品和化学制品的设备	(i)新的建筑的面积超过0.05hm^2; (ii)单次存储超过200t的石油、石化制品和化学制品

专栏一:开发描述	专栏二:阈值和标准的应用
7.食品工业	
(a)植物油、动物油和脂肪的生产; (b)动物和植物产品的包装; (c)日常用品的生产; (d)酿造和麦芽制造; (e)糖果和糖浆的生产; (f)牲畜屠宰设施; (g)工业淀粉生产设施; (h)鱼粉和鱼油制品工厂; (i)制糖工厂	新的建筑的面积超过1000m^2

专栏一:开发描述	专栏二:阈值和标准的应用
8.纺织、皮革、木材和造纸工业	
(a)制造纸张和木板的工厂(除非包含在清单1中); (b)预处理工厂(如清洗、漂白、丝光处理等),或纤维染色或纺织工厂; (c)制革工厂; (d)纤维素加工和生产设施	新的建筑的面积超过1000m^2

专栏一:开发描述	专栏二:阈值和标准的应用
9.橡胶工业	
生产和处理含橡胶成分的产品	新的建筑的面积超过1000m^2

专栏一：开发描述	专栏二：阈值和标准的应用
10. 基础设施项目	
(a)工业区开发项目； (b)城市发展项目，包括购物中心、停车场、体育馆、娱乐中心和多功能剧院的修建等； (c)联合运输的运输设备和总站(除非包含在清单1中)	开发的面积超过0.5hm^2
(d)铁路的建设(除非包含在清单1中)	工程的面积超过1hm^2
(e)机场建设(除非包含在清单1中)	(i)新跑道扩展的开发； (ii)工程的面积超过1hm^2
(f)公路的建设(除非包含在清单1中)	工程的面积超过1hm^2
(g)港口和港口设施，包括渔港设施的建设(除非包含在清单1中)	工程的面积超过1hm^2
(h)不包括在附件Ⅰ中的内陆水路建设、运河和泄洪作业； (i)为长期控水和蓄水而设计的大坝和其他设施(除非包含在清单1中)； (j)专门或主要用于客运的电车、轻轨和地铁、磁悬浮线路或特殊类型相似的线路	工程的面积超过1hm^2
(k)石油和输气管道设备(除非包含在清单1中)； (l)长距离输水管设备	(i)工程的面积超过1hm^2； (ii)在天然气管道中，绝缘设计操作压力超过7bar(1bar=100kPa)
(m)防止沿海侵蚀的作业和海事作业，通过该建设能改变海岸带，例如，堤、防洪堤、码头和其他海防工作，其中不包括该工作的维护和重建	所有开发
(n)不包括在附件Ⅰ中的地下水抽取和人工地下水充注计划； (o)不属于附件Ⅰ的流域间水资源的调度作业	工程的面积超过1hm^2
(p)高速路服务区域	开发的面积超过0.5hm^2

专栏一：开发描述	专栏二：阈值和标准的应用
11. 其他项目	
(a)动力化交通工具的永久轨道和测试轨道	开发的面积超过1hm^2
(b)废物处置设施(除非包含在清单1中)	(i)焚烧废弃物； (ii)开发的面积超过0.5hm^2； (iii)在控制水域附近100m以内的基础设施
(c)污水处理厂(除非包含在清单1中)	开发的面积超过1000m^2
(d)污泥储存处； (e)废铁储存，包括报废汽车	(i)存储或者存放地超过0.5hm^2； (ii)在控制水域附近100m以内的存储或者报废存放处
(f)引擎、涡轮机或反应器的测试台； (g)人工矿物纤维的生产设施； (h)回收或破坏易爆物质的设施； (i)牲畜屠宰场	新的建筑面积超过1000hm^2

专栏一：开发描述	专栏二：阈值和标准的应用
12.旅游休闲	
(a)滑雪跑道、载送滑雪者上坡的缆车设施以及相关开发项目	(i)开发的面积超过 $1hm^2$； (ii)高度超过15m以上的任何房屋和建筑
(b)码头	封闭水域面积超过 $1000m^2$
(c)市区外的度假村和旅店以及相关开发； (d)主题公园	开发的面积超过 $0.5hm^2$
(e)固定的露营地和活动住房地	开发的面积超过 $1hm^2$
(f)高尔夫球场	开发的面积超过 $1hm^2$

<table>
<tr><th>专栏一：开发描述</th><th colspan="2">专栏二：阈值和标准的应用</th></tr>
<tr><td colspan="3">13.改建或扩建</td></tr>
<tr><td rowspan="22">(a)附件Ⅰ所列的开发项目中已授权的、已经或正在执行的改建或扩建(不同于在本清单中21款提到的项目改建或者扩建)</td><td colspan="2">两者之一：
(i)开发项目的改建和扩建可能给环境带来严重消极影响；
(ii)在本表中专栏一“开发描述”与对应的如下专栏二“应用的阈值和标准”有所改变、扩展或者超过其标准</td></tr>
<tr><td>清单1</td><td>本表</td></tr>
<tr><td>1</td><td>6(a)</td></tr>
<tr><td>2(a)</td><td>3(a)</td></tr>
<tr><td>2(b)</td><td>3(g)</td></tr>
<tr><td>3</td><td>3(g)</td></tr>
<tr><td>4</td><td>4</td></tr>
<tr><td>5</td><td>5</td></tr>
<tr><td>6</td><td>6(a)</td></tr>
<tr><td>7(a)</td><td>10(d)铁路或者10(e)机场</td></tr>
<tr><td>7(b)(c)</td><td>10(f)</td></tr>
<tr><td>8(a)</td><td>10(h)</td></tr>
<tr><td>8(b)</td><td>10(g)</td></tr>
<tr><td>9</td><td>11(b)</td></tr>
<tr><td>10</td><td>11(b)</td></tr>
<tr><td>11</td><td>10(n)</td></tr>
<tr><td>12</td><td>10(o)</td></tr>
<tr><td>13</td><td>11(c)</td></tr>
<tr><td>14</td><td>2(e)</td></tr>
<tr><td>15</td><td>10(i)</td></tr>
<tr><td>16</td><td>10(k)</td></tr>
<tr><td>17</td><td>1(c)</td></tr>
<tr><td></td><td>18</td><td>8(a)</td></tr>
<tr><td></td><td>19</td><td>2(a)</td></tr>
<tr><td></td><td>20</td><td>6(c)</td></tr>
<tr><td>(b)在本表格中罗列的1～12部分专栏一中的开发中已授权的、已经或正在执行的改建或扩建</td><td colspan="2">两者之一：
(i)开发项目的改建和扩建可能给环境带来严重消极影响；
(ii)在本表中专栏一“开发描述”与对应的专栏二“应用的阈值和标准”有所改变、扩展或者超过其标准</td></tr>
<tr><td>(c)附件Ⅰ中所列的项目主要是开发和测试新方法或产品的项目以及使用期在两年以下的项目</td><td colspan="2">所有开发项目</td></tr>
</table>

附录3　欧盟委员会战略环境影响评价指令全文

欧盟国会2001/42/EC以及2001年6月27日议会关于评价特定的规划和项目的环境影响的指令

第1条　目标

指令的目标是能够更好地保护环境，在规划和计划的制订和实施过程中综合考虑各种环境因素，以促进社会的可持续发展。为实现此目标，依照本指令，要对可能产生重大环境影响的规划和计划实施环境评价。

第2条　定义

本指令的目的：

(a)“规划和计划”指规划、计划及其修改，包括欧洲共同体共同筹资的规划和计划。

• 依照议会或政府制定的法律程序，由国家、区域或地方层次的权力机构制定和/或实施的规划或计划。

• 法律、规章或行政规定所必需的规划或计划。

(b)“环境评价”指准备环境报告、进行咨询、决策过程中考虑环境报告和咨询结果，依照第4～9条提供决策过程中需要的信息。

(c)“环境报告”指规划或计划文件中包含第5条和附件Ⅰ所需信息的部分。

(d)“公众”指单个或多个自然人或法人以及国家法律、惯例规定所组成的协会、组织或团体。

第3条　范围

1. 依据第4～9条的规定，在第2～4款中提到的可能有重大环境影响的规划和计划应该执行环境评价。

2. 根据第3款，以下规划和计划应执行环境评价：

(a) 农业、林业、渔业、能源、工业、交通运输、废物管理、水资源管理、电信、旅游、城乡规划或土地利用领域的规划或计划，为85/337/EC号指令的附件Ⅰ和附件Ⅱ中涉及的项目未来发展树立框架的规划或计划；

(b) 考虑到可能产生的影响，依据92/43/EEC号指令第6条或第7条需要评价的规划和计划。

3. 对于第2款中决定地方性小面积土地利用的规划和计划以及对第2款涉及的规划和计划的微小修改，只有当成员国确定可能有重大环境影响时才执行环境评价。

4. 除第2款中为项目未来发展树立框架的规划和计划外，成员国应确定规划和计划是否可能有重大环境影响。

5. 各成员国应通过分析案例、细化规划和计划类型或二者相结合的方法，判断第3款、第4款涉及的规划和计划是否有重大环境影响。为实现此目的，各成员国在所有的案例中均应考虑附件Ⅱ中所列的相关标准，以确保本指令能覆盖所有可能有重大环境影响的规划和项目。

6. 在依据第 5 款逐案分析、细化规划和计划类型时，应该咨询第 6 条第 3 款涉及的权力机构。

7. 各成员国应确保依照第 5 款得到的结论向公众公开，包括依照第 4～9 条不需要执行环境评价的原因。

8. 下列规划和计划不服从本指令：

- 以提供国防和国民应急为唯一目的的规划和计划；
- 财政或预算的规划和计划。

9. 本指令不适用于那些依照欧盟理事会法规 No 1260/1999[1] 和 No 1257/1999 在各自计划周期[2]内共同筹资的规划和计划[3]。

第 4 条　一般义务

1. 第 3 条涉及的环境评价应在准备规划或计划过程中、实施规划或计划之前或进入法律程序之前实施。

2. 本指令各项要求应融入各成员国有关采用规划和计划的现有程序中，或者与遵照本指令而确定的程序合并。

3. 当对某一层次的规划和计划实施评价时，为了避免重复评价，成员国应该考虑到依照本指令，评价将会在该层次的不同层面上进行。为实现此目的，成员国应该应用第 5 条第 2 款和第 3 款的规定。

第 5 条　环境报告

1. 依据第 3 条第 1 款，需要进行环境评价时，要准备一份环境报告，报告包括识别、描述和评估执行规划、计划可能造成的环境影响，以及在考虑到规划或计划的目标和地理范围时的合理替代方案。附件Ⅰ涉及了这方面的信息。

2. 依据第 1 款，准备的环境报告应该包括：鉴于现有的评价知识和评价方法所需要的信息；规划或计划的内容和详细程度及其在决策制定过程中所处的阶段；对某些问题进行不同层次评价时所应达到的深度，以避免重复评价。

3. 那些可以获得的关于规划和计划的环境影响，以及从其他层次的决策过程或其他共同体立法过程中所获得的相关信息，可能会被用于提供附件Ⅰ中涉及的信息。

4. 在确定环境报告所需信息的深度和广度时，应该向第 6 条第 3 款涉及的权力机构咨询。

第 6 条　咨询

1. 拟订的规划或计划草案及依照第 5 条准备的环境报告应该向公众和本条第 3 款涉及的权力机构公开。

2. 在规划或计划被采纳或进入法律程序之前，应该安排适当的时间为第 3 款涉及的权力机构和第 4 款涉及的公众提供早期、有效的机会，以表达他们对拟订的规划或计划及附随的环境报告的意见。

---

[1] 1999 年 6 月 21 日通过的欧洲共同体理事会法规 No 1260/1999，该法规制定了关于结构基金的总则（OJ L 161，26.6 1999，p.1）。

[2] 欧盟法规 No 1260/1999 在 2000～2006 的规划期以及欧盟法规 No 1257/1999 在 2000～2006 和 2000～2007 的规划期。

[3] 1999 年 5 月 17 日通过的欧洲共同体理事会法规第 1257/1999 号，支持乡村的发展，并对特定规章进行了修订和撤销，该法规要求欧洲农业指导与保证基金（EAGGF）（OJ L 160，26.6 1999，p.80）。

3. 各成员国应指定咨询的权力机关，因为它们有着特殊的环境职责，从而很可能关心执行规划或计划造成的环境影响。

4. 为实现第 2 款的目的，各成员国应确定公众，包括受影响公众、可能受影响公众、服从本指令并对决策过程感兴趣的公众，包括相关非政府组织，如那些促进环境保护的组织和其他相关组织。

5. 各成员国应确定关于告知和咨询权力机构及公众的详细安排。

第 7 条　跨界咨询

1. 当某成员国认为一份涉及领土问题的规划或计划的实施可能会对其他成员国造成重大环境影响，或者该成员国可能受到重大影响时，在规划和计划被采纳或进入法律程序之前，规划或计划的发起成员国应将拟订的规划或计划草案及相关环境报告转送给其他成员国。

2. 如果某成员国转送第 1 款提到的拟订的规划或计划草案及环境报告，这预示着其他成员国在规划和计划被采纳或进入法律程序之前是否希望进入咨询，如果是这样，就应该向相关成员国咨询执行规划或计划可能造成的跨界环境影响以及减少或消除不利影响的措施。

进行上述咨询后，相关成员国应就详细安排达成协议，以确保在第 6 条第 3 款和第 6 条第 4 款涉及的可能会受到重大影响的成员国的权力机构和公众被告知，并在合理的时间安排下有机会提出他们的意见。

3. 按照本条需要进入咨询的成员国，在咨询初期应就咨询持续时间的合理安排达成协议。

第 8 条　决策

按照第 5 条准备的环境报告、按照第 6 条表达的意见和按照第 7 条进行跨界咨询的结果，在规划或计划准备过程中和采纳或进入法律程序之前均应被考虑到。

第 9 条　决策信息

1. 某规划或计划被采用时，成员国应确保公众、第 6 条第 3 款涉及的权力机构和第 7 条咨询的其他任何成员国被告知，并向其公开以下条款：

(a) 采用的规划或计划；

(b) 一份综述，包括：如何将环境因素纳入规划中，如何使第 5 条提到的环境报告、第 6 条中的各方意见以及第 7 条所咨询的意见跟第 8 条保持一致，与其他合理的替代方案相比，采用这个规划的原因；

(c) 根据第 10 条决定相关的监控措施。

2. 第 1 款涉及的相关信息的详细安排应由成员国决定。

第 10 条　监测

1. 为了识别早期无法预料的不利影响从而采取适当的补救措施，各成员国应监测实施规划和计划后造成的重大环境影响。

2. 为了服从第 1 款，如果合适可以使用现有的监测安排，从而避免重复监测。

第 11 条　与其他共同体立法的关系

1. 按照本指令要求执行的环境评价对 85/337/EEC 号指令和其他任何共同体法律不造成侵害。

2. 对于同时要遵循本指令和其他共同体立法要求进行环境影响评价的规划或计划，为避免重复评价，各成员国可以提供同等的或共同的程序来完成相关共同体立法的要求。

3. 对于欧洲共同体共同筹资的规划和计划，依据本指令执行的环境评价应与相关共同体立法的具体规定相一致。

第 12 条　信息、报告和审查

1. 成员国和委员会应该就应用本指令获得的经验交流信息。

2. 成员国应确保环境报告的质量足以满足本指令的要求，同时就与这些报告质量相关的拟采取措施与委员会交换意见。

3. 委员会应在 2006 年 7 月 21 日之前向欧洲议会和理事会提交第一份关于指令应用和效果的报告。

为了更好地保护环境，依照条约第 6 条的规定，同时考虑成员国应用本指令获得的经验，如果适当，第一份报告应附有关于修改本指令的建议，尤其是委员会应考虑扩展本指令应用范围到其他区域/部门和其他类型的规划和计划的可能性。

以后每 7 年提交一份新的评估报告。

4. 为确保本指令和后续的共同体法规相一致，委员会应在法规提供的计划期满之前报告本指令和欧盟法规（第 1260/1999 号和第 1257/1999 号）的关系。

第 13 条　指令的执行

1. 成员国应遵照本指令，在 2004 年 7 月 21 日之前实施必要的法律、规章和行政规定，同时他们应该立即将此告知委员会。

2. 当成员国实施（指令中）这些规则时，应该包括一份关于本指令的参考，或者在官方公布时随附一份这样的参考。提供参考的方法应由成员国确定。

3. 第 4 条第 1 款涉及的义务应该应用于第一次正式预案在第 1 款提到的日期之后的规划和计划，第一次正式预案在第 1 款提到的日期之前的规划和计划以及 2 年以后才被采用或服从法律程序的规划和计划，服从第 4 条第 1 款涉及的义务，除非成员国在分析案例的基础上确定不可行，并将他们的决定告知公众。

4. 成员国应在 2004 年 7 月 21 日以前与委员会沟通，除了第 1 款涉及的措施之外，第 3 款规定的另外一些有关规划和计划类型的分散信息也应该服从本指令实施环境评价。委员会应该向成员国公开这些信息，同时还要对信息定期更新。

第 14 条　生效

本指令应在《欧洲共同体公报》出版之日起生效。

第 15 条　接受人

本指令呈送给各成员国。2001 年 6 月 27 日订于卢森堡。

欧洲议会会长 N. FONTAINE

理事会会长 B. ROSENGREN

附件 I：第 5 条第 1 款涉及的信息

第 5 条第 1 款所提供的服从于第 5 条第 2 款和第 3 款的信息如下：

(a) 规划或计划的内容、主要目标以及与其他相关规划、计划的联系；

(b) 环境现状的相关内容以及不执行该规划或计划可能发生的环境演变；

(c) 可能会受到重大影响的区域的环境特征；

(d) 任何已经存在的与规划或计划相关的环境问题，尤其是与具有特殊环境意义的区域相关的，比如依照 79/409/EEC 和 92/43/EEC 号指令指定的区域；

(e) 建立在国际、共同体或成员国基础之间的环境保护目标以及把该目标和任何环境因素纳入规划制定过程中的方法；

(f) 可能会对环境产生的重大影响[1]，包括对生物多样性、人口、人类健康、动物、植物、土壤、水、空气、气候特征、有形资产、文化遗产（包含建筑学和考古学遗产）、自然景观以及以上这些因素之间的相互关系的影响；

(g) 为了预防、减少、尽可能抵消该规划或计划的重大不利环境影响而采取的措施；

(h) 列出选择替代方案的原因，叙述评价工作是如何展开的，包括在编辑所需信息时遇到的一些困难（例如，技术缺陷或者知识的缺乏）；

(i) 按照第 10 条的要求介绍所涉及的监控措施；

(j) 将以上信息进行非技术性总结。

附件Ⅱ：第 3 条第 5 款涉及的用于确定可能造成重大环境影响的标准

1. 规划和计划的特征，尤其要考虑：

• 规划或计划对项目及其他活动指导和规范的程度，通过对选址、性质、规模和运行条件的考虑，或者是通过对资源的分配；

• 规划或计划对其他同一层次规划或计划的影响程度；

• 规划或计划综合考虑环境因素的程度，尤其是对促进可持续发展的考虑；

• 规划或计划的相关环境问题；

• 规划或计划与执行环境共同体立法的相关性（例如，与废物管理或水资源保护相关的规划和计划）。

2. 影响特点和可能受影响地区的特点，尤其是：

• 影响的可能性、持久性、频率和可逆性。

• 影响的累积性。

• 影响的跨界特点。

• 对人类健康或环境的风险（例如，由于意外事故）。

• 影响的程度和空间范围（可能受影响的地理面积和人口数量）。

• 可能受影响地区的价值和脆弱性：

  ○特殊的自然特性或文化遗产；

  ○超过环境质量标准值或极限值；

  ○集约型土地利用。

• 对国家、共同体或世界公认的保护区或景观的影响。

---

[1] 此描述包括对项目产生的直接影响，间接影响，二次影响，积累影响，短期、中期和长期影响，持久和暂时影响，消极和积极影响的描述。

附录4 The Lee和Colley评审标准

Lee 和 Colley 方法从四个方面评审 EIS，每个方面均有自己独立的指标：

1.对开发项目、当地环境和基准环境条件的描述

• 开发的描述；
• 选址的描述；
• 残留物；
• 环境基准情况。

2.关键影响的识别和评价

• 影响的识别；
• 影响大小的预测；
• 影响显著性的评价。

3.替代方案和减缓措施

• 替换方案；
• 减缓措施；
• 承诺的减缓措施。

4.结果交流讨论

• 表达；
• 平衡；
• 非技术性总结。

总之，环境评价的内容和质量根据以上指标进行评审，可分为以下级别 A～F：

• 等级 A 总体上很好地完成工作，没有重大的遗漏。
• 等级 B 总体上满意，只有较少的遗漏和不足。
• 等级 C 基本上满意，但是有一些遗漏或不足。
• 等级 D 一部分工作完成得不错，但是总的来说因为遗漏和不足不能让人满意。
• 等级 E 有重大的遗漏或不足，令人不满意。
• 等级 F 由于重要的工作完成得很差或没有进行，令人非常不满意。

根据内容分析，给出四个核查指标的累加分数，同时最后的综合等级附在整个评价中。

附录5 环境影响报告评审系统牛津布鲁克斯大学影响评价小组（IAU）

评审系统的使用

IAU 评审系统是为了研究 EIS 对项目质量的改进，由能源部、苏格兰和威尔士办公室 1995/1996 筹资开发的。评审系统是一个系统地评审 EIS 的强大机制。完整的评审系统现在

已进行了更新，以满足2011年的EIA法规、能源部检查表、曼彻斯特大学开发的评审系统、EU检查表和IAU提出的最佳实践概念的要求。系统共分为8个部分，同时每一部分都细分为一系列评审标准。总之，评审依据92条标准评价EIS的质量，其中的一些标准并不适用于所有项目。每个标准按照提供的材料的质量基础进行分级，并由此得出每部分的等级。通过对每部分所给出的等级综合就可以得到EIS总的等级。IAU评审等级建立在由曼彻斯特大学开发的关于评审的等级系统上。这些等级是：

A＝总体上很好地完成工作，没有重大的遗漏；

B＝总体上满意，只有较少的遗漏和不足；

C＝基本上满意，但是有一些遗漏或不足；

D＝一部分工作完成得不错，但是总的来说因为遗漏和不足不能让人满意；

E＝有重大的遗漏或不足，令人不满意；

F＝由于重要的工作完成得很差或没有进行，令人非常不满意。

这些等级可以用来检验EIS是否符合相关法规，等级C与等级D是通过与不通过的分界。通过使用该等级系统，评审者可以更容易地识别EIS需要进一步完善的方面，因为该等级是由主管部门精心设定的，能准确判断进一步信息的需求。根据这些等级评价EIS质量更像一篇学术论文，单独的等级属性和单独的标准本质上是都是主观的，但行为（即评审）是独立的、客观的和系统的。一种减少评审主观性的方法是使用"双盲法"让两个独立的评审者对EIS进行评价。每个评审者用标准和等级A～F评价EIS，最后，评审者对结果和等级进行比较。

为了由所有单独等级完成整体等级，必须做出某些决定，例如某区域的A等级是否优于另一区域的D等级。这应从整体出发，单独的评审者可能认为某些方面更重要，所以把所有的ABC累加起来，并在此基础上给出总的等级，并不是一件简单的事情。有些情况下，由于某些方面特别重要，一个规则最低要求中的法定要求（如非技术性总结）可以作为整个EIS失败的依据。其他方面（如替代方案的考虑）如果不是与拟议项目特别相关的话，就可以认为不那么重要。根据该标准，在这些方面的F等级可能不会妨碍EIS获得C等级。通过该过程得到的EIS总的等级，要求评审者判断单个评审方面的权重，给予规划足够的重视。

EIS评审的成功在很大程度上取决于评审者的经验和他们通过对EIS各部分系统评价后对整体质量进行判断的能力。在评审EIS时，评审者应对所提供的信息进行权衡：

- 必须包含什么；
- 能够包含什么；
- 合理地预期到什么。

牛津布鲁克斯大学影响评价小组（IAU）

环境影响报告(EIS)评审系统
项目名称：
EIS递交者：
递交日期：
评审等级：
A＝总体上很好地完成工作，没有重大的遗漏；
B＝总体上满意，只有较少的遗漏和不足；
C＝基本上满意，但是有一些遗漏或不足；
D＝一部分工作完成得不错，但是总的来说因为遗漏和不足不能让人满意；
E＝有重大的遗漏或不足，令人不满意；
F＝由于重要的工作完成得很差或没有进行，令人非常不满意。
NA＝EIS内容或项目根本不能实施。

1. 对开发的描述

标　　准	评审等级	注释
项目的基本特征		
1.1 解释开发的目的和目标		
1.2 对已准备环境信息的决策的性质和状态进行鉴定		
1.3 对建设期、运行期进行估算，并应对废止的阶段及其规划进行估算		
1.4 提供包括开发的地点、设计和规模的信息的描述①		
1.5 提供对开发带有图表、计划或地图和照片的描述		
1.6 指出已完成的开发造成的自然环境压力或表现		
1.7 描述建设方法		
1.8 描述生产的性质和方法，或描述所实施项目的其他方面		
1.9 描述所有与项目有因果关系的附加服务（水、电、能源服务等）以及开发要求		
1.10 描述项目的潜在事故、危险和紧急事件		
用地要求		
1.11 确定由于开发和/或者建筑和其他相关的布置、辅助设施和景观区域而占用的土地区域，同时在地图上标出。对于带状项目，描述其土地通道、垂直和水平图和地道与土木工程需求		
1.12 描述土地的用途，并划分不同的土地利用区域		
1.13 描述在建设期土地的恢复和后续利用情况		
项目输入		
1.14 描述在建设和运行阶段所需材料的性质和数量		
1.15 估计在建设和运行阶段参与项目的工人和来访者的数量		
1.16 描述能进入现场的路径以及使用的运输方法		
1.17 指出在建设和运行阶段运输材料和产品到项目现场的方式以及相关的移动频次		
残余和排放物		
1.18 估计在建设和运行阶段产生的废物、能量（噪声、振动、光、热、放射物等）和残余物的类型和数量以及它们所达到的程度		
1.19 指出这些废物和放射性物质如何优先处理治理，而不是排放/丢弃以及它们通过何种途径排放到环境中		
1.20 识别所有特殊和危险废物（定义中要求的）是如何产生的，描述处置方法可能带来的主要环境影响		
1.21 指出估算残余物和废物数量的方法。识别不确定因素，尽可能给出范围或者置信区间		
本节的评审等级＝ 注释		

① 清单4第二部分标准（2011 EIA法规）。

2. 环境的描述

标　　准	评审等级	注释
对项目所占用区域和周边环境的描述		
2.1 在适当的地图辅助下，指出预期项目产生显著影响的区域。解释这些影响可能发生的时间		
2.2 描述项目所在地和周边区域的土地利用情况		
2.3 确定建设和运行阶段受到的影响环境，包括潜在的显著影响。例如，可能由于污染物的分解、项目的基础设施建设、交通等引起的影响		

续表

标　　准	评审等级	注释
基本情况		
2.4 指明并描述受到项目潜在影响的环境要素		
2.5 受影响环境所使用的方法要与评价任务的规模和复杂程度相一致。指出其不确定性		
2.6 预测如果不实施项目可能的未来的环境状况，论述自然环境系统和人类利用的可变性		
2.7 现有的技术数据资源的利用，包括环保局以及特殊利益群体的执行记录和研究		
2.8 检验开发建议与地方、区域和国家的规划与政策的一致性，检验为将来的环境状况预测所收集的必要数据。当开发建议与规划和政策不一致时，修正开发建议		
2.9 研究地方、区域和国家部门所掌握的基本环境状况的信息		
本节的评审等级 = 注释		

3. 范围、咨询和影响识别

标　　准	评审等级	注释
范围和咨询		
3.1 尽力联系一般公众、相关公众机构、有关专家和特殊利益群体，对项目和相关设施进行评价。列出有关小组		
3.2 联系法定咨询机构。列出有关咨询者		
3.3 在咨询的基础上确定环境属性的价值		
3.4 确定对有价值环境属性产生显著影响的项目活动。为进一步的调查确定和选择关键影响。描述并证实所使用方法的范围		
3.5 包括法定咨询机构和公众得到的主要评论的副本或总结以及对这些评论采取的措施		
影响的识别		
3.6 提供识别开发对环境产生的主要影响所需要的数据		
3.7 考虑项目建设、运行以及退役后所造成的直接或间接/二次影响（包括积极影响和消极影响）。考虑“后期”开发是否会增加影响		
3.8 调查上述影响类型对人类、植物、动物、土壤、水、空气、气候、景观的相互作用、物质资源、文化遗产的影响		
3.9 噪声、土地利用、历史遗产、社区的影响		
3.10 如果上面的任何因素在具体项目和项目所在地没有引起关注的话，在这里要清楚地指明		
3.11 使用系统的方法来识别影响，如项目检查表、矩阵、专家小组、广泛咨询等。对使用的方法和使用它们的理由进行描述		
3.12 调查影响类型是否显著影响决策。避免不必要的信息，关注关键问题		
3.13 考虑本身并不重要但会逐渐放大的影响		
3.14 考虑由非标准的运行条件带来的影响、事故和紧急情况		
3.15 如果项目的性质如3.14所述，则事故将有可能对周围环境引起严重的损害，对事故发生的可能性和可能产生的结果要进行评价，并对主要结论加以报告		
本节的评审等级 = 注释		

4. 影响的预测及评估标准

标　　准	评审等级	注释
对影响范围的预测		
4.1 根据所产生变化的性质和大小以及影响受体的性质、位置、数量、价值、敏感性描述影响		
4.2 预测产生影响的时间跨度，以便明确影响是短期的、中期的还是长期的，短期的还是永久的，可恢复的还是不可恢复的		
4.3 若有可能，用定量的方式来表达影响预测。必要时，要尽可能定量描述		
4.4 描述影响发生的可能性以及其结果的不确定程度		
方法和数据		
4.5 提供评价开发项目对环境造成的主要影响所需要的数据		
4.6 针对项目的复杂程度和重要性、该方法是恰当，对用于预测影响性质、大小、尺度的方法加以描述		
4.7 用于评价主要影响大小和范围的数据必须充足，能清楚描述和确定资料来源，识别并指出数据缺口		
影响显著性评价		
4.8 根据对当地社区（包括影响的分配）和受保护的环境资源的影响对影响的显著性进行讨论		
4.9 对用来进行显著性评价的标准、假设和价值体系加以讨论		
4.10 当缺少被普遍接受的显著性评价标准或准则时，讨论替换方案，在这种情况下，就要在事实、假设和专业判断间做出清晰的分辨		
4.11 在适当的章节，对有关显著性影响的国家和国际标准或准则加以讨论。另外，应对影响的大小、位置和持续时间与资源的综合价位、敏感性和稀有性结合起来进行讨论		
4.12 项目造成的影响差别是在不实施项目的情况下对比中产生的		
4.13 清楚说明哪些影响比较显著，哪些影响不显著，并说明判断理由		
本节的评审等级 = 注释		

5. 替代方案

标　　准	评审等级	注释
5.1 提供主要替代方案的概述，指出考虑环境影响后，做出该选择的主要原因		
5.2 在项目规划的初期，考虑“零行为”替代方案，替代过程、比例、布局图、设计和运行环境，同时调查其主环境的优缺点		
5.3 如果在调查期间证实了有出乎意料的显著负面影响，很难减轻，则在较早阶段被否决的替代方案可以被重新评议		
5.4 替代方案是客观真实的		
5.5 对拟议项目的替换方案造成的主要环境影响与不实施该项目的环境状况，做出清晰客观的比较		
本节的评审等级 = 注释		

6. 减缓措施和监测

标　准	评审等级	注释
减缓措施的描述		
6.1 提供为了避免、减少和尽可能修复显著负面影响而拟订的措施的描述		
6.2 减缓措施要考虑到对项目设计的修正、建设和运行，设施/资源的更新，新资源的开发以及污染控制的末端治理技术		
6.3 描述选择某特定类型减缓措施的原因和其他可用的备选方案		
6.4 说明减缓措施将发生的效用。当效力不明确或减缓措施不起作用时，可以引用数据来证实这些假设的可接受性		
6.5 指明所有残留或未减缓影响的显著性，并说明这些影响为何没有被减缓		
减缓措施和监测承诺		
6.6 指出怎样执行减缓措施和在必要时间内的作用		
6.7 对所有显著影响提出监测、安排，尤其当存在不确定因素时，检验由于项目的实施带来的环境影响是否与预测影响相符合		
6.8 任何拟检测、安排范围相对的潜在预计影响的范围和重要性		
减缓措施的环境影响		
6.9 调查和描述减缓措施的消极影响		
6.10 减缓措施的优点和其造成的消极影响之间潜在的冲突		
本节的评审等级 = 注释		

7. 非技术性总结

标　准	评审等级	注释
7.1 清单4第2部分第1～4节所提供的信息的非技术性总结		
7.2 非技术总结至少要包括项目和环境的简要描述、开发商采取的主要减缓措施的说明以及所有残余影响的描述		
7.3 总结应该避免的专业术语、数据罗列和过于详细的科学原理解释		
7.4 总结要表达评价的主要见解，并涵盖信息中提到的所有问题		
7.5 总结包括整个评价方法的简历注解		
7.6 总结要简洁明了，要在结论中表现出来		
本节的评审等级 = 注释		

8. 信息的组织和表达

标　准	评审等级	注释
信息的组织		
8.1 信息的逻辑顺序		
8.2 确定信息在表格和清单中的位置		
8.3 各章节概述各阶段进行调查而得到的主要阶段		
8.4 当引用外部资源的信息时，应对该资源进行说明		

续表

标　　准	评审等级	注释
信息的表达		
8.5 有关 EIA 法律，开发商的名称，主管部门的名称，EIS 编制机构的名称以及联系人的名称、地址和电话		
8.6 包括项目简介、评价目的和所使用方法的简单介绍		
8.7 完整的报告书。附录中的数据在报告书主体中得到充分讨论		
8.8 提供信息和分析支持所有结论		
8.9 为非专业人员提供全面的信息。可以使用地图、表格、图表材料和其他适当的方式。避免出现不必要的技术或不明确的语言		
8.10 对所有重要的数据和结果加以综合讨论		
8.11 避免多余的信息(即决策中不需要的信息)		
8.12 以简洁的形式、一致的专业术语和不同章节间的逻辑联系来表达信息		
8.13 保证和强调显著的负面影响、可持续的环境利益和有争议的问题		
8.14 专业术语、缩写和词首大写字母的规定		
8.15 信息是客观的，不随任何其他特殊观点而改变。负面影响也不能通过委婉的说法和陈词滥调而掩饰		
汇编信息时遇到的困难		
8.16 标明所要求的数据缺口，并解释在评价中对它们的处理方法		
8.17 收集和分析预测影响所需数据时，在识别和解释中遇到的所有困难或信息的一切问题		
本节的评审等级 = 注释		

收集清单

第 2 部分的最低要求（2011 EIA 法规）

标准	总分	需要更多信息的方面
1. 对项目位置、设计和规模的有关信息的描述		
2. 为了避免、减少或补救(如果可能)显著环境消极影响而拟设措施的描述		
3. 识别或评价开发可能对环境造成的主要影响所需要的数据评审等级需要更多信息的领域		
4. 结合考虑环境影响，对所研究的主要替代方案和选择它们的主要原因的概述		
5. 上面 1～4 提到的信息的非技术性总结		
总的等级(A～F)：		
要求完善 EIS 的信息列表		

IAU 最佳实践必要条件

标准	总分	需要更多信息的方面
开发项目的描述		
环境的描述		
范围、咨询和影响识别		
影响的预测和评估		
替代方案		
减缓和监测		
非技术性总结 信息的组织和表达		
总的等级(A～F)：		
注释		

Appendices

Appendix 1 Full text of the European Commission's EIA Directive (the Consolidated EIA Directive)

Council Directive of 27 June 1985 on the assessment of the effects of certain public and private projects on the environment (85/337/EEC)

Amended by:

Council Directive 97/11/EC of 3 March 1997;

Directive 2003/35/EC of the European Parliament and of the Council of 26 May 2003; and

Directive 2009/31/EC of the European Parliament and of the Council of 23 April 2009.

Article 1

1 This Directive shall apply to the assessment of the environmental effects of those public and private projects that are likely to have significant effects on the environment.

2 For the purposes of this Directive: '*project*' *means*:

-the execution of construction works or of other installations or schemes;

-other interventions in the natural surroundings and landscape including those involving the extraction of mineral resources;

'*developer*' *means*:

-the applicant for authorization for a private project or the public authority that initiates a project;

'development consent' means:

-the decision of the competent authority or authorities that entitles the developer to proceed with the project;

'public' means:

-one or more natural or legal persons and, in accordance with national legislation and practice, their associations, organizations or groups;

'public concerned' means:

-the public affected or likely to be affected by, or having an interest in, the environmental decision-making procedures referred to in Article 2 (2); for the purposes of this definition, non-governmental organizations promoting environmental protection and meeting any requirements under national law shall be deemed to have an interest.

3 The competent authority or authorities shall be that or those that the Member States designate as responsible for performing the duties arising from this Directive.

4 Member States may decide, on a case-by-case basis if so provided under national law, not to apply this Directive to projects serving national defence purposes, if they deem that such application would have an adverse effect on these purposes.

5 This Directive shall not apply to projects the details of which are adopted by a specific act of national legislation, since the objectives of this Directive, including that of supplying information, are achieved through the legislative process.

Article 2

1 Member States shall adopt all measures necessary to ensure that, before consent is given, projects likely to have significant effects on the environment by virtue, *inter alia*, of their nature, size or location are made subject to a requirement for development consent and an assessment with regard to their effects. These projects are defined in Article 4.

2 The environmental impact assessment may be integrated into the existing procedures for consent to projects in the Member States, or, failing this, into other procedures or into procedures to be established to comply with the aims of this Directive.

2 (a) Member States may provide for a single procedure in order to fulfil the requirements of this Directive and the requirements of Council Directive 96/61/EC of 24 September 1996 on integrated pollution prevention and control. [1]

3 Without prejudice to Article 7, Member States may, in exceptional cases, exempt a specific project in whole or in part from the provisions laid down in this Directive.

In this event, the Member States shall:

(a) consider whether another form of assessment would be appropriate;

(b) make available to the public concerned the information obtained under other forms of assessment referred to in point (a), the information relating to the exemption decision and the reasons for granting it;

(c) inform the Commission, prior to granting consent, of the reasons justifying the exemption granted, and provide it with the information made available, where applicable, to their own nationals.

The Commission shall immediately forward the documents received to the other Member States.

The Commission shall report annually to the Council on the application of this paragraph.

Article 3

The environmental impact assessment shall identify, describe and assess in an appropriate manner, in the light of each individual case and in accordance with Articles 4 to 11, the direct and indirect effects of a project on the following factors:

- human beings, fauna and flora;
- soil, water, air, climate and the landscape;
- material assets and the cultural heritage;
- the interaction between the factors mentioned in the first, second and third indents.

Article 4

1 Subject to Article 2 (3), projects listed in Annex Ⅰ shall be made subject to an assessment in accordance with Articles 5 to 10.

2 Subject to Article 2 (3), for projects listed in Annex Ⅱ, the Member States shall determine through:

(a) a case-by-case examination, or

(b) thresholds or criteria set by the Member State

Whether the project shall be made subject to an assessment in accordance with Articles 5 to 10.

Member States may decide to apply both procedures referred to in (a) and (b).

3 When a case-by-case examination is carried out or thresholds or criteria are set for the purpose of paragraph 2, the relevant selection criteria set out in Annex Ⅲ shall be taken into account.

4 Member States shall ensure that the determination made by the competent authorities under paragraph 2 is made available to the public.

Article 5

1 In the case of projects that, pursuant to Article 4, must be subjected to an environmental impact assessment in accordance with Articles 5 to 10, Member States shall adopt the necessary measures to ensure that the developer supplies in an appropriate form the information specified in Annex Ⅳ inasmuch as:

(a) the Member States consider that the information is relevant to a given stage of the consent procedure and to the specific characteristics of a particular project or type of project and of the environmental features likely to be affected;

(b) the Member States consider that a developer may reasonably be required to compile this information having regard *inter alia* to current knowledge and methods of assessment.

2 Member States shall take the necessary measures to ensure that, if the developer so requests before submitting an application for development consent, the competent authority

shall give an opinion on the information to be supplied by the developer in accordance with paragraph 1. The competent authority shall consult the developer and authorities referred to in Article 6 (1) before it gives its opinion. The fact that the authority has given an opinion under this paragraph shall not preclude it from subsequently requiring the developer to submit further information.

Member States may require the competent authorities to give such an opinion, irrespective of whether the developer so requests.

3 The information to be provided by the developer in accordance with paragraph 1 shall include at least:

(a) a description of the project comprising information on the site, design and size of the project;

(b) a description of the measures envisaged in order to avoid, reduce and, if possible, remedy significant adverse effects;

(c) the data required to identify and assess the main effects that the project is likely to have on the environment;

(d) an outline of the main alternatives studied by the developer and an indication of the main reasons for his choice, taking into account the environmental effects;

(e) a non-technical summary of the information mentioned in the previous indents.

4 Member States shall, if necessary, ensure that any authorities holding relevant information, with particular reference to Article 3, shall make this information available to the developer.

Article 6

1 Member States shall take the measures necessary to ensure that the authorities likely to be concerned by the project by reason of their specific environmental responsibilities are given an opportunity to express their opinion on the information supplied by the developer and on the request for development consent. To this end, Member States shall designate the authorities to be consulted, either in general terms or on a case-by-case basis. The information gathered pursuant to Article 5 shall be forwarded to those authorities. Detailed arrangements for consultation shall be laid down by the Member States.

2 The public shall be informed, whether by public notices or other appropriate means such as electronic media where available, of the following matters early in the environmental decision-making procedures referred to in Article 2 (2) and, at the latest, as soon as information can reasonably be provided:

(a) the request for development consent;

(b) the fact that the project is subject to an environmental impact assessment procedure and, where relevant, the fact that Article 7 applies;

(c) details of the competent authorities responsible for taking the decision, those from which relevant information can be obtained, those to which comments or questions can be submitted, and details of the time schedule for transmitting comments or questions;

(d) the nature of possible decisions or, where there is one, the draft decision;

(e) an indication of the availability of the information gathered pursuant to Article 5;

(f) an indication of the times and places where and means by which the relevant information will be made available;

(g) details of the arrangements for public participation made pursuant to paragraph 5 of this Article.

3 Member States shall ensure that, within reasonable time-frames, the following is made available to the public concerned:

(a) any information gathered pursuant to Article 5;

(b) in accordance with national legislation, the main reports and advice issued to the competent authority or authorities at the time when the public concerned is informed in accordance with paragraph 2 of this Article;

(c) in accordance with the provisions of Directive 2003/4/EC of the European Parliament and of the Council of 28 January 2003 on public access to environmental information,[2] information other than that referred to in paragraph 2 of this Article that is relevant for the decision in accordance with Article 8 and that only becomes available after the time the public concerned was informed in accordance with paragraph 2 of this Article.

4 The public concerned shall be given early and effective opportunities to participate in the environmental decision-making procedures referred to in Article 2 (2) and shall, for that purpose, be entitled to express comments and opinions when all options are open to the competent authority or authorities before the decision on the request for development consent is taken.

5 The detailed arrangements for informing the public (for example by bill posting within a certain radius or publication in local newspapers) and for consulting the public concerned (for example by written submissions or by way of a public inquiry) shall be determined by the Member States.

6 Reasonable time-frames for the different phases shall be provided, allowing sufficient time for informing the public and for the public concerned to prepare and participate effectively in environmental decision-making subject to the provisions of this Article.

Article 7

1 Where a Member State is aware that a project is likely to have significant effects on the environment in another Member State or where a Member State likely to be significantly affected so requests, the Member State in whose territory the project is intended to be carried out shall send to the affected Member State as soon as possible and no later than when informing its own public, *inter alia*:

(a) a description of the project, together with any available information on its possible transboundary impact;

(b) information on the nature of the decision that may be taken.

And shall give the other Member State a reasonable time in which to indicate whether it wishes to participate in the environmental decision-making procedures referred to in Article 2 (2), and may include the information referred to in paragraph 2 of this Article.

2 If a Member State that receives information pursuant to paragraph 1 indicates that it intends to participate in the environmental decision-making procedures referred to in Article 2 (2), the Member State in whose territory the project is intended to be carried out shall, if it has not already done so, send to the affected Member State the information required to be given pursuant to Article 6 (2) and made available pursuant to Article 6 (3) (a) and (b).

3 The Member States concerned, each insofar as it is concerned, shall also:

(a) arrange for the information referred to in paragraphs 1 and 2 to be made available, within a reasonable time, to the authorities referred to in Article 6 (1) and the public concerned in the territory of the Member State likely to be significantly affected; and

(b) ensure that those authorities and the public concerned are given an opportunity, before development consent for the project is granted, to forward their opinion within a reasonable time on the information supplied to the competent authority in the Member State in whose territory the project is intended to be carried out.

4 The Member States concerned shall enter into consultations regarding, *inter alia*, the potential transboundary effects of the project and the measures envisaged to reduce or eliminate such effects and shall agree on a reasonable time frame for the duration of the consultation period.

5 The detailed arrangements for implementing this Article may be determined by the Member States concerned and shall be such as to enable the public concerned in the territory of the affected Member State to participate effectively in the environmental decision-making procedures referred to in Article 2 (2) for the project.

Article 8

The results of consultations and the information gathered pursuant to Articles 5, 6 and 7 must be taken into consideration in the development consent procedure.

Article 9

1 When a decision to grant or refuse development consent has been taken, the competent authority or authorities shall inform the public thereof in accordance with the appropriate procedures and shall make available to the public the following information:

(a) the content of the decision and any conditions attached thereto,

(b) having examined the concerns and opinions expressed by the public concerned, the main reasons and considerations on which the decision is based, including information about the public participation process,

(c) a description, where necessary, of the main measures to avoid, reduce and, if possible, offset the major adverse effects.

2 The competent authority or authorities shall inform any Member State that has been consulted pursuant to Article 7, forwarding to it the information referred to in paragraph 1 of this Article.

The consulted Member States shall ensure that information is made available in an appropriate manner to the public concerned in their own territory.

Article 10

The provisions of this Directive shall not affect the obligation on the competent authorities to respect the limitations imposed by national regulations and administrative provisions and accepted legal practices with regard to commercial and industrial confidentiality, including intellectual property, and the safeguarding of the public interest.

Where Article 7 applies, the transmission of information to another Member State and the receipt of information by another Member State shall be subject to the limitations in force in the Member State in which the project is proposed.

Article 10 (a)

Member States shall ensure that, in accordance with the relevant national legal system, members of the public concerned:

(a) having a sufficient interest, or alternatively;

(b) maintaining the impairment of a right, where administrative procedural law of a Member State requires this as a precondition.

Have access to a review procedure before a court of law or another independent and impartial body established by law to challenge the substantive or procedural legality of decisions, acts or omissions subject to the public participation provisions of this Directive.

Member States shall determine at what stage the decisions, acts or omissions may be challenged.

What constitutes a sufficient interest and impairment of a right shall be determined by the Member States, consistently with the objective of giving the public concerned wide access to justice. To this end, the interest of any non-governmental organisation meeting the requirements referred to in Article 1 (2), shall be deemed sufficient for the purpose of subparagraph (a) of this Article. Such organisations shall also be deemed to have rights capable of being impaired for the purpose of subparagraph (b) of this Article.

The provisions of this Article shall not exclude the possibility of a preliminary review procedure before an administrative authority and shall not affect the requirement of exhaustion of administrative review procedures prior to recourse to judicial review procedures, where such a requirement exists under national law.

Any such procedure shall be fair, equitable, timely and not prohibitively expensive.

In order to further the effectiveness of the provisions of this article, Member States shall ensure that practical information is made available to the public on access to administrative and judicial review procedures.

Article 11

1 The Member States and the Commission shall exchange information on the experience gained in applying this Directive.

2 In particular, Member States shall inform the Commission of any criteria and/or thresholds adopted for the selection of the projects in question, in accordance with Article 4 (2).

3 Five years after notification of this Directive, the Commission shall send the Euro-

pean Parliament and the Council a report on its application and effectiveness. The report shall be based on the aforementioned exchange of information.

4 On the basis of this exchange of information, the Commission shall submit to the Council additional proposals, should this be necessary, with a view to this Directive's being applied in a sufficiently coordinated manner.

Article 12

1 Member States shall take the measures necessary to comply with this Directive within three years of its notification.[3]

2 Member States shall communicate to the Commission the texts of the provisions of national law that they adopt in the field covered by this Directive.

Article 13

This Directive is addressed to the Member States.

Notes

1 OJ No L 257, 10. 10. 1996, p. 26.

2 OJ L 41, 14. 2. 2003, p. 26.

3 This Directive was notified to the Member States on 3 July 1985.

Annex Ⅰ Projects subject to article 4 (1)

1 Crude-oil refineries (excluding undertakings manufacturing only lubricants from crude oil) and installations for the gasification and liquefaction of 500 tonnes or more of coal or bituminous shale per day.

2 -Thermal power stations and other combustion installations with a heat output of 300 megawatts or more, and

-Nuclear power stations and other nuclear reactors including the dismantling or decommissioning of such power stations or reactors[1] (except research installations for the production and conversion of fissionable and fertile materials, whose maximum power does not exceed 1 kilowatt continuous thermal load).

3 (a) Installations for the reprocessing of irradiated nuclear fuel;

(b) Installations designed:

-for the production or enrichment of nuclear fuel;

-for the processing of irradiated nuclear fuel or high-level radioactive waste;

-for the final disposal of irradiated nuclear fuel;

-solely for the final disposal of radioactive waste;

-solely for the storage (planned for more than 10 years) of irradiated nuclear fuels or radioactive waste in a different site than the production site.

4 -Integrated works for the initial smelting of cast-iron and steel;

-Installations for the production of nonferrous crude metals from ore, concentrates or secondary raw materials by metallurgical, chemical or electrolytic processes.

5 Installations for the extraction of asbestos and for the processing and transformation of asbestos and products containing asbestos: for asbestoscement products, with an annual production of more than 20,000 tonnes of finished products, for friction material, with an annual production of more than 50 tonnes of finished products, and for other uses of asbestos, utilization of more than 200 tonnes per year.

6 Integrated chemical installations, i. e. those installations for the manufacture on an industrial scale of substances using chemical conversion processes, in which several units are juxtaposed and are functionally linked to one another and that are:

(i) for the production of basic organic chemicals;

(ii) for the production of basic inorganic chemicals;

(iii) for the production of phosphorous-, nitrogen- or potassium-based fertilizers (simple or compound fertilizers);

(iv) for the production of basic plant health products and of biocides;

(v) for the production of basic pharmaceutical products using a chemical or biological process;

(vi) for the production of explosives.

7 (a) Construction of lines for long-distance railway traffic and of airports[2] with a basic runway length of 2100 m or more;

(b) Construction of motorways and express roads;[3]

(c) Construction of a new road of four or more lanes, or realignment and/or widening of an existing road of two lanes or less so as to provide four or more lanes, where such new road, or realigned and/or widened section of road would be 10 km or more in a continuous length.

8 (a) Inland waterways and ports for inland waterway traffic that permit the passage of vessels of over 1350 tonnes;

(b) Trading ports, piers for loading and unloading connected to land and outside ports (excluding ferry piers) that can take vessels of over 1350 tonnes.

9 Waste disposal installations for the incineration, chemical treatment as defined in Annex Ⅱ A to Directive 75/442/EEC[4] under heading D9, or landfill of hazardous waste (i. e. waste to which Directive 91/689/EEC[5] applies).

10 Waste disposal installations for the incineration or chemical treatment as defined in Annex Ⅱ A to Directive 75/442/EEC under heading D9 of non-hazardous waste with a capacity exceeding 100 tonnes per day.

11 Groundwater abstraction or artificial groundwater recharge schemes where the annual volume of water abstracted or recharged is equivalent to or exceeds 10 million cubic metres.

12 (a) Works for the transfer of water resources between river basins where this transfer aims at preventing possible shortages of water and where the amount of water transferred exceeds 100 million cubic metres/year;

(b) In all other cases, works for the transfer of water resources between river basins where the multi-annual average flow of the basin of abstraction exceeds 2000 million cubic metres/year and where the amount of water transferred exceeds 5 per cent of this flow.

In both cases transfers of piped drinking water are excluded.

13 Waste water treatment plants with a capacity exceeding 150,000 population equivalent as defined in Article 2 point (6) of Directive 91/271/EEC. [6]

14 Extraction of petroleum and natural gas for commercial purposes where the amount extracted exceeds 500 tonnes/day in the case of petroleum and 500,000 m^3/day in the case of gas.

15 Dams and other installations designed for the holding back or permanent storage of water, where a new or additional amount of water held back or stored exceeds 10 million cubic metres.

16 Pipelines with a diameter of more than 800 mm and a length of more than 40 km:

-for the transport of gas, oil, chemicals; and,

-for the transport of carbon dioxide (CO_2) streams for the purposes of geological storage, including associated booster stations.

17 Installations for the intensive rearing of poultry or pigs with more than:

(a) 85,000 places for broilers, 60,000 places for hens;

(b) 3,000 places for production pigs (over 30 kg); or

(c) 900 places for sows.

18 Industrial plants for the

(a) production of pulp from timber or similar fibrous materials;

(b) production of paper and board with a production capacity exceeding 200 tonnes per day.

19 Quarries and open-cast mining where the surface of the site exceeds 25 hectares, or peat extraction, where the surface of the site exceeds 150 hectares.

20 Construction of overhead electrical power lines with a voltage of 220 kV or more and a length of more than 15 km.

21 Installations for storage of petroleum, petrochemical, or chemical products with a capacity of 200,000 tonnes or more.

22 Any change to or extension of projects listed in this Annex where such a change or extension in itself meets the thresholds, if any, set out in this Annex.

23 Storage sites pursuant to Directive 2009/31/EC of the European Parliament and of the Council of 23 April 2009 on the geological storage of carbon dioxide. [7]

24 Installations for the capture of CO_2 streams for the purposes of geological storage pursuant to Directive 2009/31/EC from installations covered by this Annex, or where the total yearly capture of CO_2 is 1.5 megatonnes or more.

Notes

1 Nuclear power stations and other nuclear reactors cease to be such an installation when all nuclear fuel and other radioactively contaminated elements have been removed permanently from the installation site.

2 For the purposes of this Directive, 'airport' means airports that comply with the definition in the 1944 Chicago Convention setting up the International Civil Aviation Organization (Annex 14).

3 For the purposes of the Directive, 'express road' means a road that complies with the definition in the

European Agreement on Main International Traffic Arteries of 15 November 1975.

4 OJ No L 194, 25. 7. 1975, p. 39. Directive as last amended by Commission Decision 94/3/EC (OJ No L 5, 7. 1. 1994, p. 15).

5 OJ No L 377, 31. 12. 1991, p. 20. Directive as last amended by Directive 94/31/EC (OJ No L 168, 2. 7. 1994, p. 28).

6 OJ No L 135, 30. 5. 1991, p. 40. Directive as last amended by the 1994 Act of Accession.

7 OJ L 140, 5. 6. 2009, p. 114.

Annex Ⅱ Projects subject to article 4 (2)

1 Agriculture, silviculture and aquaculture

(a) projects for the restructuring of rural land holdings;

(b) projects for the use of uncultivated land or seminatural areas for intensive agricultural purposes;

(c) water management projects for agriculture, including irrigation and land drainage projects;

(d) initial afforestation and deforestation for the purposes of conversion to another type of land use;

(e) intensive livestock installations (projects not included in Annex Ⅰ);

(f) intensive fish farming;

(g) reclamation of land from the sea.

2 Extractive industry

(a) quarries, open-cast mining and peat extraction (projects not included in Annex Ⅰ);

(b) underground mining;

(c) extraction of minerals by marine or fluvial dredging;

(d) deep drillings, in particular:

-geothermal drilling,

-drilling for the storage of nuclear waste material,

-drilling for water supplies,

with the exception of drillings for investigating the stability of the soil;

(e) surface industrial installations for the extraction of coal, petroleum, natural gas and ores, as well as bituminous shale.

3 Energy industry

(a) industrial installations for the production of electricity, steam and hot water (projects not included in Annex Ⅰ);

(b) industrial installations for carrying gas, steam and hot water; transmission of electrical energy by overhead cables (projects not included in Annex Ⅰ);

(c) surface storage of natural gas;

(d) underground storage of combustible gases;

(e) surface storage of fossil fuels;

(f) industrial briquetting of coal and lignite;

(g) installations for the processing and storage of radioactive waste (unless included in Annex Ⅰ);

(h) installations for hydroelectric energy production;

(i) installations for the harnessing of wind power for energy production (wind farms);

(j) installations for the capture of CO_2 streams for the purposes of geological storage pursuant to Directive 2009/31/EC from installations not covered by Annex Ⅰ to this Directive.

4 Production and processing of metals

(a) installations for the production of pig iron or steel (primary or secondary fusion) including continuous casting;

(b) installations for the processing of ferrous metals:

(i) hot-rolling mills,

(ii) smitheries with hammers,

(iii) application of protective fused metal coats;

(c) ferrous metal foundries;

(d) installations for the smelting, including the alloyage, of non-ferrous metals, excluding precious metals, including recovered products (refining, foundry casting, etc.);

(e) installations for surface treatment of metals and plastic materials using an electrolytic or chemical process;

(f) manufacture and assembly of motor vehicles and manufacture of motor-vehicle engines;

(g) shipyards;

(h) installations for the construction and repair of aircraft;

(i) manufacture of railway equipment;

(j) swaging by explosives;

(k) installations for the roasting and sintering of metallic ores.

5 Mineral industry

(a) coke ovens (dry coal distillation);

(b) installations for the manufacture of cement;

(c) installations for the production of asbestos and the manufacture of asbestos-products (projects not included in Annex Ⅰ);

(d) installations for the manufacture of glass including glass fibre;

(e) installations for smelting mineral substances including the production of mineral fibres;

(f) manufacture of ceramic products by burning, in particular roofing tiles, bricks, refractory bricks, tiles, stoneware or porcelain.

6 Chemical industry (Projects not included in Annex Ⅰ)

(a) treatment of intermediate products and production of chemicals;

(b) production of pesticides and pharmaceutical products, paint and varnishes, elastomers and peroxides;

(c) storage facilities for petroleum, petrochemical and chemical products.

7 Food industry

(a) manufacture of vegetable and animal oils and fats;

(b) packing and canning of animal and vegetable products;

(c) manufacture of dairy products;

(d) brewing and malting;

(e) confectionery and syrup manufacture;

(f) installations for the slaughter of animals;

(g) industrial starch manufacturing installations;

(h) fish-meal and fish-oil factories;

(i) sugar factories.

8 Textile, leather, wood and paper industries

(a) industrial plants for the production of paper and board (projects not included in Annex Ⅰ);

(b) plants for the pretreatment (operations such as washing, bleaching, mercerization) or dyeing of fibres or textiles;

(c) plants for the tanning of hides and skins;

(d) cellulose-processing and production installations.

9 Rubber industry

Manufacture and treatment of elastomer-based products.

10 Infrastructure projects

(a) industrial estate development projects;

(b) urban development projects, including the construction of shopping centres and car parks;

(c) construction of railways and intermodal transshipment facilities, and of intermodal terminals (projects not included in Annex Ⅰ);

(d) construction of airfields (projects not included in Annex Ⅰ);

(e) construction of roads, harbours and port installations, including fishing harbours (projects not included in Annex Ⅰ);

(f) inland-waterway construction not included in Annex Ⅰ, canalization and flood-relief works;

(g) dams and other installations designed to hold water or store it on a long-term basis (projects not included in Annex Ⅰ);

(h) tramways, elevated and underground railways, suspended lines or similar lines of a particular type, used exclusively or mainly for passenger transport;

(i) oil and gas pipeline installations and pipelines for the transport of CO_2 streams for the purposes of geological storage (projects not included in Annex Ⅰ);

(j) installations of long-distance aqueducts;

(k) coastal work to combat erosion and maritime works capable of altering the coast through the construction, for example, of dykes, moles, jetties and other sea defence works, excluding the maintenance and reconstruction of such works;

(l) groundwater abstraction and artificial groundwater recharge schemes not included in

Annex Ⅰ;

(m) works for the transfer of water resources between river basins not included in Annex Ⅰ.

11 Other projects

(a) permanent racing and test tracks for motorized vehicles;

(b) installations for the disposal of waste (projects not included in Annex Ⅰ);

(c) waste-water treatment plants (projects not included in Annex Ⅰ);

(d) sludge-deposition sites;

(e) storage of scrap iron, including scrap vehicles;

(f) test benches for engines, turbines or reactors;

(g) installations for the manufacture of artificial mineral fibres;

(h) installations for the recovery or destruction of explosive substances;

(i) knackers' yards.

12 Tourism and leisure

(a) ski-runs, ski-lifts and cable-cars and associ-ated developments;

(b) marinas;

(c) holiday villages and hotel complexes outside urban areas and associated developments;

(d) permanent camp sites and caravan sites;

(e) theme parks.

13 -Any change or extension of projects listed in Annex Ⅰ or Annex Ⅱ, already authorized, executed or in the process of being executed, which may have significant adverse effects on the environment (change or extension not included in Annex Ⅰ);

-Projects in Annex Ⅰ, undertaken exclusively or mainly for the development and testing of new methods or products and not used for more than two years.

Annex Ⅲ Selection criteria referred to in article 4 (3)

1 Characteristics of projects

The characteristics of projects must be considered having regard, in particular, to:

-the size of the project;

-the cumulation with other projects;

-the use of natural resources;

-the production of waste;

-pollution and nuisances;

-the risk of accidents, having regard in particular to substances or technologies used.

2 Location of projects

The environmental sensitivity of geographical areas likely to be affected by projects must be considered, having regard, in particular, to:

-the existing land use;

-the relative abundance, quality and regenerative capacity of natural resources in the area;

-the absorption capacity of the natural environment, paying particular attention to the following areas:

(a) wetlands;

(b) coastal zones;

(c) mountain and forest areas;

(d) nature reserves and parks;

(e) areas classified or protected under Member States' legislation; special protection areas designated by Member States pursuant to Directive 79/409/EEC and 92/43/EEC;

(f) areas in which the environmental quality standards laid down in Community legislation have already been exceeded;

(g) densely populated areas;

(h) landscapes of historical, cultural or archaeological significance.

3 Characteristics of the potential impact

The potential significant effects of projects must be considered in relation to criteria set out under 1 and 2 above, and having regard in particular to:

-the extent of the impact (geographical area and size of the affected population);

-the transfrontier nature of the impact;

-the magnitude and complexity of the impact;

-the probability of the impact;

-the duration, frequency and reversibility of the impact.

Annex Ⅳ Information referred to in article 5 (1)

1 Description of the project, including in particular:

-a description of the physical characteristics of the whole project and the land-use requirements during the construction and operational phases;

-a description of the main characteristics of the production processes, for instance, nature and quantity of the materials used;

-an estimate, by type and quantity, of expected residues and emissions (water, air and soil pollution, noise, vibration, light, heat, radiation, etc.) resulting from the operation of the proposed project.

2 An outline of the main alternatives studied by the developer and an indication of the main reasons for this choice, taking into account the environmental effects.

3 A description of the aspects of the environment likely to be significantly affected by the proposed project, including, in particular, population, fauna, flora, soil, water, air, climatic factors, material assets, including the architectural and archaeological heritage, landscape and the inter-relationship between the above factors.

4 A description[1] of the likely significant effects of the proposed project on the environment resulting from:

-the existence of the project;

-the use of natural resources;

-the emission of pollutants, the creation of nuisances and the elimination of waste; and

-the description by the developer of the forecasting methods used to assess the effects on the environment.

5 A description of the measures envisaged to prevent, reduce and where possible off-set any significant adverse effects on the environment.

6 A non-technical summary of the information provided under the above headings.

7 An indication of any difficulties (technical deficiencies or lack of know-how) encountered by the developer in compiling the required information.

Notes

1 This description should cover the direct effects and any indirect, secondary, cumulative, short, medium and long-term, permanent and temporary, positive and negative effects of the project.

Appendix 2 Town and Country Planning (EIA) Regulations 2011—Schedule 2 (Regulation 2.1)

Descriptions of development and applicable thresholds and criteria for the purposes of the definition of 'Schedule 2 development'

In the table below:

- 'area of the works' includes any area occupied by apparatus, equipment, machinery, materials, plant, spoil heaps or other facilities or stores required for construction or installation;
- 'controlled waters' has the same meaning as in the Water Resources Act 1991;[1]
- 'floorspace' means the floorspace in a building or buildings.

The table below sets out the descriptions of development and applicable thresholds and criteria for the purpose of classifying development as Schedule 2 development.

Column 1: Description of development	Column 2: Applicable thresholds and criteria

The carrying out of development to provide any of the following:

1 Agriculture and aquaculture	
(a) Projects for the use of uncultivated land or semi-natural areas for intensive agricultural purposes;	The area of the development exceeds 0.5 hectares.
(b) Water management projects for agriculture, including irrigation and land drainage projects;	The area of the works exceeds 1 hectare.
(c) Intensive livestock installations (unless included in Schedule 1);	The area of new floorspace exceeds 500 square metres.
(d) Intensive fish farming;	The installation resulting from the development is designed to produce more than 10 tonnes of dead weight fish per year.
(e) Reclamation of land from the sea.	All development.

Column 1: Description of development	Column 2: Applicable thresholds and criteria
2 Extractive industry	
(a) Quarries, open-cast mining and peat extraction (unless included in Schedule 1); (b) Underground mining;	All development except the construction of buildings or other ancillary structures where the new floorspace does not exceed 1,000 square metres.
(c) Extraction of minerals by fluvial or marine dredging;	All development.
(d) Deep drillings, in particular: (i) geothermal drilling; (ii) drilling for the storage of nuclear waste material; (iii) drilling for water supplies; with the exception of drillings for investigating the stability of the soil;	(i) In relation to any type of drilling, the area of the works exceeds 1 hectare; or (ii) in relation to geothermal drilling and drilling for the storage of nuclear waste material, the drilling is within 100 metres of any controlled waters.
(e) Surface industrial installations for the extraction of coal, petroleum, natural gas and ores, as well as bituminous shale.	The area of the development exceeds 0.5 hectares.
3 Energy industry	
(a) Industrial installations for the production of electricity, steam and hot water (unless included in Schedule 1);	The area of the development exceeds 0.5 hectares.
(b) Industrial installations for carrying gas, steam and hot water;	The area of the works exceeds 1 hectare.
(c) Surface storage of natural gas; (d) Underground storage of combustible gases; (e) Surface storage of fossil fuels;	(i) The area of any new building, deposit or structure exceeds 500 square metres; or (ii) a new building, deposit or structure is to be sited within 100 metres of any controlled waters.
(f) Industrial briquetting of coal and lignite;	The area of new floorspace exceeds 1,000 square metres.
(g) Installations for the processing and storage of radioactive waste (unless included in Schedule 1);	(i) The area of new floorspace exceeds 1,000 square metres; or (ii) the installation resulting from the development will require the grant of an environmental permit under the Environmental Permitting (England and Wales). Regulations 2010(2) in relation to a radioactive substances activity described in paragraphs 5(2)(b), (2)(c) or (4) of Part 2 of Schedule 23 to those Regulations, or the variation of such a permit.
(h) Installations for hydroelectric energy production;	The installation is designed to produce more than 0.5 megawatts.
(i) Installations for the harnessing of wind power for energy production (wind farms);	(i) The development involves the installation of more than 2 turbines; or (ii) the hub height of any turbine or height of any other structure exceeds 15 metres.
(j) Installations for the capture of carbon dioxide streams for the purposes of geological storage pursuant to Directive 2009/31/EC from installations not included in Schedule 1.	All development.

Column 1: Description of development	Column 2: Applicable thresholds and criteria

4 Production and processing of metals	
(a) Installations for the production of pig iron or steel (primary or secondary fusion) including continuous casting; (b) Installations for the processing of ferrous metals: (i) hot-rolling mills; (ii) smitheries with hammers; (iii) application of protective fused metal coats. (c) Ferrous metal foundries; (d) Installations for the smelting, including the alloyage, of nonferrous metals, excluding precious metals, including recovered products (refining, foundry casting, etc.); (e) Installations for surface treatment of metals and plastic materials using an electrolytic or chemical process; (f) Manufacture and assembly of motor vehicles and manufacture of motor-vehicle engines; (g) Shipyards; (h) Installations for the construction and repair of aircraft; (i) Manufacture of railway equipment; (j) Swaging by explosives; (k) Installations for the roasting and sintering of metallic ores.	The area of new floorspace exceeds 1,000 square metres.

5 Mineral industry	
(a) Coke ovens (dry coal distillation); (b) Installations for the manufacture of cement; (c) Installations for the production of asbestos and the manufacture of asbestos-based products (unless included in Schedule 1); (d) Installations for the manufacture of glass including glass fibre; (e) Installations for smelting mineral substances including the production of mineral fibres; (f) Manufacture of ceramic products by burning, in particular roofing tiles, bricks, refractory bricks, tiles, stoneware or porcelain.	The area of new floorspace exceeds 1,000 square metres.

6 Chemical industry (unless included in Schedule 1)	
(a) Treatment of intermediate products and production of chemicals; (b) Production of pesticides and pharmaceutical products, paint and varnishes, elastomers and peroxides;	The area of new floorspace exceeds 1,000 square metres.
(c) Storage facilities for petroleum, petrochemical and chemical products.	(i) The area of any new building or structure exceeds 0.05 hectares; or (ii) more than 200 tonnes of petroleum, petrochemical or chemical products is to be stored at any one time.

Column 1: Description of development	Column 2: Applicable thresholds and criteria
7 Food industry	
(a) Manufacture of vegetable and animal oils and fats; (b) Packing and canning of animal and vegetable products; (c) Manufacture of dairy products; (d) Brewing and malting; (e) Confectionery and syrup manufacture; (f) Installations for the slaughter of animals; (g) Industrial starch manufacturing installations; (h) Fish-meal and fish-oil factories; (i) Sugar factories.	The area of new floorspace exceeds 1,000 square metres.
8 Textile, leather, wood and paper industries	
(a) Industrial plants for the production of paper and board (unless included in Schedule 1); (b) Plants for the pre-treatment (operations such as washing, bleaching, mercerisation) or dyeing of fibres or textiles; (c) Plants for the tanning of hides and skins; (d) Cellulose-processing and production installations.	The area of new floorspace exceeds 1,000 square metres.
9 Rubber industry	
Manufacture and treatment of elastomer-based products.	The area of new floorspace exceeds 1,000 square metres.
10 Infrastructure projects	
(a) Industrial estate development projects; (b) Urban development projects, including the construction of shopping centres and car parks, sports stadiums, leisure centres and multiplex cinemas; (c) Construction of intermodal transshipment facilities and of intermodal terminals (unless included in Schedule 1);	The area of the development exceeds 0.5 hectares.
(d) Construction of railways (unless included in Schedule 1);	The area of the works exceeds 1 hectare.
(e) Construction of airfields (unless included in Schedule 1);	(i) The development involves an extension to a runway; or (ii) the area of the works exceeds 1 hectare.
(f) Construction of roads (unless included in Schedule 1);	The area of the works exceeds 1 hectare.
(g) Construction of harbours and port installations including fishing harbours (unless included in Schedule 1);	The area of the works exceeds 1 hectare.
(h) Inland-waterway construction not included in Schedule 1, canalisation and flood-relief works; (i) Dams and other installations designed to hold water or store it on a long-term basis (unless included in Schedule 1); (j) Tramways, elevated and underground railways, suspended lines or similar lines of a particular type, used exclusively or mainly for passenger transport;	The area of the works exceeds 1 hectare.

Column 1: Description of development	Column 2: Applicable thresholds and criteria

Continued

10 Infrastructure projects	
(k) Oil and gas pipeline installations and pipelines for the transport of carbon dioxide streams for the purposes of geological storage (unless included in Schedule 1); (l) Installations of long-distance aqueducts;	(i) The area of the works exceeds 1 hectare; or (ii) in the case of a gas pipeline, the installation has a design operating pressure exceeding 7 bar gauge.
(m) Coastal work to combat erosion and maritime works capable of altering the coast through the construction, for example, of dykes, moles, jetties and other sea defence works, excluding the maintenance and reconstruction of such works;	All development.
(n) Groundwater abstraction and artificial groundwater recharge schemes not included in Schedule 1; (o) Works for the transfer of water resources between river basins not included in Schedule 1;	The area of the works exceeds 1 hectare.
(p) Motorway service areas.	The area of the development exceeds 0.5 hectares.

11 Other projects	
(a) Permanent racing and test tracks for motorised vehicles;	The area of the development exceeds 1 hectare.
(b) Installations for the disposal of waste (unless included in Schedule 1);	(i) The disposal is by incineration; or (ii) the area of the development exceeds 0.5 hectare; or (iii) the installation is to be sited within 100 metres of any controlled waters.
(c) Waste-water treatment plants (unless included in Schedule 1);	The area of the development exceeds 1,000 square metres.
(d) Sludge-deposition sites; (e) Storage of scrap iron, including scrap vehicles;	(i) The area of deposit or storage exceeds 0.5 hectare; or (ii) a deposit is to be made or scrap stored within 100 metres of any controlled waters.
(f) Test benches for engines, turbines or reactors; (g) Installations for the manufacture of artificial mineral fibres; (h) Installations for the recovery or destruction of explosive substances; (i) Knackers' yards.	The area of new floorspace exceeds 1,000 square metres.

12 Tourism and leisure	
(a) Ski-runs, ski-lifts and cable-cars and associated developments;	(i) The area of the works exceeds 1 hectare; or (ii) the height of any building or other structure exceeds 15 metres.
(b) Marinas;	The area of the enclosed water surface exceeds 1,000 square metres.
(c) Holiday villages and hotel complexes outside urban areas and associated developments; (d) Theme parks;	The area of the development exceeds 0.5 hectares.
(e) Permanent camp sites and caravan sites;	The area of the development exceeds 1 hectare.
(f) Golf courses and associated developments.	The area of the development exceeds 1 hectare.

Column 1: Description of development	Column 2: Applicable thresholds and criteria
13 Changes and extensions	
(a) Any change to or extension of development of a description listed in Schedule 1 (other than a change or extension falling within paragraph 21 of that Schedule) where that development is already authorised, executed or in the process of being executed.	Either: (i) The development as changed or extended may have significant adverse effects on the environment; or (ii) in relation to development of a description mentioned in a paragraph in Schedule 1 indicated below, the thresholds and criteria in column 2 of the paragraph of this table indicated below applied to the change or extension are met or exceeded. *Paragraph in Schedule 1* — *Paragraph of this table* 1 — 6(a) 2(a) — 3(a) 2(b) — 3(g) 3 — 3(g) 4 — 4 5 — 5 6 — 6(a) 7(a) — 10(d) (in relation to railways) or 10(e) (in relation to airports) 7(b) and (c) — 10(f) 8(a) — 10(h) 8(b) — 10(g) 9 — 11(b) 10 — 11(b) 11 — 10(n) 12 — 10(o) 13 — 11(c) 14 — 2(e) 15 — 10(i) 16 — 10(k) 17 — 1(c) 18 — 8(a) 19 — 2(a) 20 — 6(c)
(b) Any change to or extension of development of a description listed in paragraphs 1 to 12 of column 1 of this table, where that development is already authorised, executed or in the process of being executed.	Either— (i) The development as changed or extended may have significant adverse effects on the environment; or (ii) in relation to development of a description mentioned in column 1 of this table, the thresholds and criteria in the corresponding part of column 2 of this table applied to the change or extension are met or exceeded.
(c) Development of a description mentioned in Schedule 1 undertaken exclusively or mainly for the development and testing of new methods or products and not used for more than two years.	All development.

Notes

1 1991 c. 57. See section 104.

2 S. I. 2010/675.

Appendix 3 Full text of the European Commission's SEA Directive

Directive 2001/42/EC of the European Parliament and of the Council of 27 June 2001 on the assessment of the effects of certain plans and programmes on the environment

Article 1: Objectives

The objective of this Directive is to provide for a high level of protection of the environment and to contribute to the integration of environmental considerations into the preparation and adoption of plans and programmes with a view to promoting sustainable development, by ensuring that, in accordance with this Directive, an environmental assessment is carried out of certain plans and programmes that are likely to have significant effects on the environment.

Article 2: Definitions

For the purposes of this Directive:

(a) 'plans and programmes' shall mean plans and programmes, including those co-financed by the European Community, as well as any modifications to them:

-which are subject to preparation and/or adoption by an authority at national, regional or local level or which are prepared by an authority for adoption, through a legislative procedure by Parliament or Government, and

-which are required by legislative, regulatory or administrative provisions;

(b) 'environmental assessment' shall mean the preparation of an environmental report, the carrying out of consultations, the taking into account of the environmental report and the results of the consultations in decision-making and the provision of information on the decision in accordance with Articles 4 to 9;

(c) 'environmental report' shall mean the part of the plan or programme documentation containing the information required in Article 5 and Annex Ⅰ;

(d) 'The public' shall mean one or more natural or legal persons and, in accordance with national legislation or practice, their associations, organisations or groups.

Article 3: Scope

1 An environmental assessment, in accordance with Articles 4 to 9, shall be carried out for plans and programmes referred to in paragraphs 2 to 4, which are likely to have significant environmental effects.

2 Subject to paragraph 3, an environmental assessment shall be carried out for all plans and programmes,

(a) which are prepared for agriculture, forestry, fisheries, energy, industry, transport, waste management, water management, telecommunications, tourism, town and country planning or land use and which set the framework for future development consent of projects listed in Annexes I and II to Directive 85/337/EEC, or

(b) which, in view of the likely effect on sites, have been determined to require an assessment pursuant to Article 6 or 7 of Directive 92/43/EEC.

3 Plans and programmes referred to in paragraph 2 that determine the use of small areas at local level and minor modifications to plans and programmes referred to in paragraph 2 shall require an environmental assessment only where the Member States determine that they are likely to have significant environmental effects.

4 Member States shall determine whether plans and programmes, other than those referred to in paragraph 2, which set the framework for future development consent of projects, are likely to have significant environmental effects.

5 Member States shall determine whether plans or programmes referred to in paragraphs 3 and 4 are likely to have significant environmental effects either through case-by-case examination or by specifying types of plans and programmes or by combining both approaches. For this purpose Member States shall in all cases take into account relevant criteria set out in Annex Ⅱ, in order to ensure that plans and programmes with likely significant effects on the environment are covered by this Directive.

6 In the case-by-case examination and in specifying types of plans and programmes in accordance with paragraph 5, the authorities referred to in Article 6 (3) shall be consulted.

7 Member States shall ensure that their conclusions pursuant to paragraph 5, including the reasons for not requiring an environmental assessment pursuant to Articles 4 to 9, are made available to the public.

8 The following plans and programmes are not subject to this Directive:

-plans and programmes the sole purpose of which is to serve national defence or civil emergency;

-financial or budget plans and programmes.

9 This Directive does not apply to plans and programmes co-financed under the current respective programming periods[1] for Council Regulations (EC) No 1260/1999[2] and (EC) No 1257/1999.[3]

Article 4: General obligations

1 The environmental assessment referred to in Article 3 shall be carried out during the preparation of a plan or programme and before its adoption or submission to the legislative procedure.

2 The requirements of this Directive shall either be integrated into existing procedures in Member States for the adoption of plans and programmes or incorporated in procedures established to comply with this Directive.

3 Where plans and programmes form part of a hierarchy, Member States shall, with a view to avoiding duplication of the assessment, take into account the fact that the assessment will be carried out, in accordance with this Directive, at different levels of the hierarchy. For the purpose of, *inter alia*, avoiding duplication of assessment, Member States

shall apply Article 5 (2) and (3).

Article 5: Environmental report

1 Where an environmental assessment is required under Article 3 (1), an environmental report shall be prepared in which the likely significant effects on the environment of implementing the plan or programme, and reasonable alternatives taking into account the objectives and the geographical scope of the plan or programme, are identified, described and evaluated. The information to be given for this purpose is referred to in Annex Ⅰ.

2 The environmental report prepared pursuant to paragraph 1 shall include the information that may reasonably be required taking into account current knowledge and methods of assessment, the contents and level of detail in the plan or programme, its stage in the decision-making process and the extent to which certain matters are more appropriately assessed at different levels in that process in order to avoid duplication of the assessment.

3 Relevant information available on environmental effects of the plans and programmes and obtained at other levels of decisionmaking or through other Community legis-lation may be used for providing the information referred to in Annex Ⅰ.

4 The authorities referred to in Article 6 (3) shall be consulted when deciding on the scope and level of detail of the information that must be included in the environmental report.

Article 6: Consultations

1 The draft plan or programme and the environmental report prepared in accordance with Article 5 shall be made available to the authorities referred to in paragraph 3 of this Article and the public.

2 The authorities referred to in paragraph 3 and the public referred to in paragraph 4 shall be given an early and effective opportunity within appropriate time frames to express their opinion on the draft plan or programme and the accompanying environmental report before the adoption of the plan or programme or its submission to the legislative procedure.

3 Member States shall designate the authorities to be consulted that, by reason of their specific environmental responsibilities, are likely to be concerned by the environmental effects of implementing plans and programmes.

4 Member States shall identify the public for the purposes of paragraph 2, including the public affected or likely to be affected by, or having an interest in, the decision-making subject to this Directive, including relevant non-governmental organisations, such as those promoting environmental protection and other organisations concerned.

5 The detailed arrangements for the information and consultation of the authorities and the public shall be determined by the Member States.

Article 7: Transboundary consultations

1 Where a Member State considers that the implementation of a plan or programme being prepared in relation to its territory is likely to have significant effects on the environment in another Member State, or where a Member State likely to be significantly affected so requests, the Member State in whose territory the plan or programme is being prepared

shall, before its adoption or submission to the legislative procedure, forward a copy of the draft plan or programme and the relevant environmental report to the other Member State.

2 Where a Member State is sent a copy of a draft plan or programme and an environmental report under paragraph 1, it shall indicate to the other Member State whether it wishes to enter into consultations before the adoption of the plan or programme or its submission to the legislative procedure and, if it so indicates, the Member States concerned shall enter into consultations concerning the likely transboundary environmental effects of implementing the plan or programme and the measures envisaged to reduce or eliminate such effects.

Where such consultations take place, the Member States concerned shall agree on detailed arrangements to ensure that the authorities referred to in Article 6 (3) and the public referred to in Article 6 (4) in the Member State likely to be significantly affected are informed and given an opportunity to forward their opinion within a reasonable time-frame.

3 Where Member States are required under this Article to enter into consultations, they shall agree, at the beginning of such consultations, on a reasonable timeframe for the duration of the consultations.

Article 8: Decision making

The environmental report prepared pursuant to Article 5, the opinions expressed pursuant to Article 6 and the results of any transboundary consultations entered into pursuant to Article 7 shall be taken into account during the preparation of the plan or programme and before its adoption or submission to the legislative procedure.

Article 9: Information on the decision

1 Member States shall ensure that, when a plan or programme is adopted, the authorities referred to in Article 6 (3), the public and any Member State consulted under Article 7 are informed and the following items are made available to those so informed:

(a) the plan or programme as adopted;

(b) a statement summarizing how environmental considerations have been integrated into the plan or programme and how the environmental report prepared pursuant to Article 5, the opinions expressed pursuant to Article 6 and the results of consultations entered into pursuant to Article 7 have been taken into account in accordance with Article 8 and the reasons for choosing the plan or programme as adopted, in the light of the other reasonable alternatives dealt with; and

(c) the measures decided concerning monitoring in accordance with Article 10.

2 The detailed arrangements concerning the information referred to in paragraph 1 shall be determined by the Member States.

Article 10: Monitoring

1 Member States shall monitor the significant environmental effects of the implementation of plans and programmes in order, *inter alia*, to identify at an early stage unforeseen adverse effects, and to be able to undertake appropriate remedial action.

2 In order to comply with paragraph 1, existing monitoring arrangements may be used if appropriate, with a view to avoiding duplication of monitoring.

Article 11: Relationship with other Community legislation

1 An environmental assessment carried out under this Directive shall be without prejudice to any requirements under Directive 85/337/EEC and to any other Community law requirements.

2 For plans and programmes for which the obligation to carry out assessments of the effects on the environment arises simultaneously from this Directive and other Community legislation, Member States may provide for coordinated or joint procedures fulfilling the requirements of the relevant Community legislation in order, *inter alia*, to avoid duplication of assessment.

3 For plans and programmes co-financed by the European Community, the environmental assessment in accordance with this Directive shall be carried out in conformity with the specific provisions in relevant Community legislation.

Article 12: Information, reporting and review

1 Member States and the Commission shall exchange information on the experience gained in applying this Directive.

2 Member States shall ensure that environmental reports are of a sufficient quality to meet the requirements of this Directive and shall communicate to the Commission any measures they take concerning the quality of these reports.

3 Before 21 July 2006 the Commission shall send a first report on the application and effectiveness of this Directive to the European Parliament and to the Council.

With a view further to integrating environmental protection requirements, in accordance with Article 6 of the Treaty, and taking into account the experience acquired in the application of this Directive in the Member States, such a report will be accompanied by proposals for amendment of this Directive, if appropriate. In particular, the Commission will consider the possibility of extending the scope of this Directive to other areas/sectors and other types of plans and programmes.

A new evaluation report shall follow at seven-year intervals.

4 The Commission shall report on the relationship between this Directive and Regulations (EC) No 1260/1999 and (EC) No 1257/1999 well ahead of the expiry of the programming periods provided for in those Regulations, with a view to ensuring a coherent approach with regard to this Directive and subsequent Community Regulations.

Article 13: Implementation of the Directive

1 Member States shall bring into force the laws, regulations and administrative provisions necessary to comply with this Directive before 21 July 2004. They shall forthwith inform the Commission thereof.

2 When Member States adopt the measures, they shall contain a reference to this Directive or shall be accompanied by such reference on the occasion of their official publica-

tion. The methods of making such reference shall be laid down by Member States.

3 The obligation referred to in Article 4 (1) shall apply to the plans and programmes of which the first formal preparatory act is subsequent to the date referred to in paragraph 1. Plans and programmes of which the first formal preparatory act is before that date and which are adopted or submitted to the legislative procedure more than 24 months thereafter, shall be made subject to the obligation referred to in Article 4 (1) unless Member States decide on a case by case basis that this is not feasible and inform the public of their decision.

4 Before 21 July 2004, Member States shall communicate to the Commission, in addition to the measures referred to in paragraph 1, separate information on the types of plans and programmes that, in accordance with Article 3, would be subject to an environmental assessment pursuant to this Directive. The Commission shall make this information available to the Member States. The information will be updated on a regular basis.

Article 14: Entry into force

This Directive shall enter into force on the day of its publication in the Official Journal of the European Communities.

Article 15: Addressees

This Directive is addressed to the Member States. Done at Luxembourg, 27 June 2001.

For the European Parliament
The President
N. FONTAINE
For the Council
The President
B. ROSENGREN

Notes

1 The 2000-06 programming period for Council Regulation (EC) No 1260/1999 and the 2000-06 and 2000-07 programming periods for Council Regulation (EC) No 1257/1999.

2 Council Regulation (EC) No 1260/1999 of 21 June 1999 laying down general provisions on the Structural Funds (OJ L 161, 26. 6. 1999, p. 1).

3 Council Regulation (EC) No 1257/1999 of 17 May 1999 on support for rural development from the European Agricultural Guidance and Guarantee Fund (EAGGF) and amending and repealing certain regulations (OJ L 160, 26. 6. 1999, p. 80).

Annex I Information referred to in Article 5 (1)

The information to be provided under Article 5 (1), subject to Article 5 (2) and (3), is the following:

(a) an outline of the contents, main objectives of the plan or programme and relationship with other relevant plans and programmes;

(b) the relevant aspects of the current state of the environment and the likely evolution thereof without implementation of the plan or programme;

(c) the environmental characteristics of areas likely to be significantly affected;

(d) any existing environmental problems that are relevant to the plan or programme including, in particular, those relating to any areas of a particular environmental importance, such as areas designated pursuant to Directives 79/409/EEC and 92/43/EEC;

(e) the environmental protection objectives, established at international, Community or Member State level, which are relevant to the plan or programme and the way those objectives and any environmental considerations have been taken into account during its preparation;

(f) the likely significant effects[1] on the environment, including on issues such as biodiversity, population, human health, fauna, flora, soil, water, air, climatic factors, material assets, cultural heritage including architectural and archaeological heritage, landscape and the interrelationship between the above factors;

(g) the measures envisaged to prevent, reduce and as fully as possible offset any significant adverse effects on the environment of implementing the plan or programme;

(h) an outline of the reasons for selecting the alternatives dealt with, and a description of how the assessment was undertaken including any difficulties (such as technical deficiencies or lack of know-how) encountered in compiling the required information;

(i) a description of the measures envisaged concerning monitoring in accordance with Article 10;

(j) a non-technical summary of the information provided under the above headings.

Notes

1 These effects should include secondary, cumulative, synergistic, short, medium and long-term, permanent and temporary, positive and negative effects.

Annex Ⅱ Criteria for determining the likely significance of effects referred to in Article 3 (5)

1 The characteristics of plans and programmes, having regard, in particular, to

-the degree to which the plan or programme sets a framework for projects and other activities, either with regard to the location, nature, size and operating conditions or by allocating resources;

-the degree to which the plan or programme influences other plans and programmes including those in a hierarchy;

-the relevance of the plan or programme for the integration of environmental considerations in particular with a view to promoting sustainable development;

-environmental problems relevant to the plan or programme;

-the relevance of the plan or programme for the implementation of Community legislation on the environment (e. g. plans and programmes linked to waste-management or water protection).

2 Characteristics of the effects and of the area likely to be affected, having regard, in particular, to

-the probability, duration, frequency and reversibility of the effects;

-the cumulative nature of the effects;

-the transboundary nature of the effects;

-the risks to human health or the environment (e. g. due to accidents);

-the magnitude and spatial extent of the effects (geographical area and size of the population likely to be affected);

-the value and vulnerability of the area likely to be affected due to:

(i) special natural characteristics or cultural heritage;

(ii) exceeded environmental quality standards or limit values;

(iii) intensive land-use;

-the effects on areas or landscapes that have a recognized national, Community or international protection status.

Appendix 4 The Lee and Colley review package

The Lee and Colley method reviews EISs under four main topics, each of which is examined under a number of sub-headings:

1 Description of the development, the local environment and the baseline conditions:

- description of the development
- site description
- residuals
- baseline conditions

2 Identification and evaluation of key impacts:

- identification of impacts
- prediction of impact magnitudes
- assessment of impact significance

3 Alternatives and mitigation:

- alternatives
- mitigation
- commitment to mitigation

4 Communication of results:

- presentation
- balance
- non-technical summary

In outline, the content and quality of the environmental statement is reviewed under each of the subheads, using a sliding scale of assessment symbols A-F:

Grade A indicates that the work has generally been well performed with no important omissions.

Grade B is generally satisfactory and complete with only minor omissions and inadequacies.

Grade C is regarded as just satisfactory despite some omissions or inadequacies.

Grade D indicates that parts are well attempted but, on the whole, just unsatisfactory because of omissions or inadequacies.

Grade E is not satisfactory, revealing significant omissions or inadequacies.

Grade F is very unsatisfactory with important task (s) poorly done or not attempted.

Having analysed each sub-head, aggregated scores are given to the four review areas, and a final summary grade is attached to the whole statement.

Appendix 5 Environmental impact statement review package (IAU, Oxford Brookes University)

Using the review packages

The IAU review package was developed for a research project into the changing quality of EISs that was funded by the DoE, the Scottish and Welsh Offices in 1995/96. The package is a robust mechanism for systematically reviewing EISs. The full review package has been updated to combine the requirements of the 2011 EIA Regulations, the DoE checklist, a review package developed by Manchester University, an EU review checklist as well as notions of best practice developed by the IAU. The package is divided into 8 sections and within each section are a number of individual review criterion. In all, the package assesses the quality of an EIS against 92 criteria, some of which are not necessarily relevant to all projects. Each criterion is graded on the basis of the quality of the material provided and each section is then awarded an overall grade. From the grades given to each section an overall grade for the EIS is arrived at. The IAU review grades are based upon the grading system developed by Manchester University for their review package. These grades are:

A = indicates that the work has generally been well performed with no important omissions;

B = is generally satisfactory and complete with only minor omissions and inadequacies;

C = is regarded as just satisfactory despite some omissions or inadequacies;

D = indicates that parts are well attempted but, on the whole, just unsatisfactory because of omissions or inadequacies;

E = is not satisfactory, revealing significant omissions or inadequacies;

F = is very unsatisfactory with important task (s) poorly done or not attempted.

These grades can be used to test an EIS's compliance with the relevant Regulations, with the pass/fail mark lying between grades 'C' and 'D'. By using this grading system the reviewer can more readily identify the aspects of the EIS that need completing and because the grades are well established the competent authority can confidently justify any requests for further information. The assessment of EIS quality against these grades is rather like the marking of an academic essay in that while the activity—i. e. review—is carried out independently, objectively and systematically, the attributing of individual grades to individual criterion is inherently subjective. One way of reducing the subjectivity of the review is for the EIS to be assessed by two independent reviewers on the basis of a 'double blind' approach. Here each reviewer assesses the EIS against the criteria and grades the EIS on the basis of 'A' to 'F' for each criterion and for the ES as a whole. The reviewers then compare results and agree grades.

In arriving at overall grades, from all of the individual grades, a decision must be made over whether, for example, an 'A' grade for one area outweighs a 'D' grade for another area. This will depend entirely on perspective, as an individual reviewer may consider some aspects to be more important than others and so it is not a simple matter of counting up all of the 'A', 'B' and 'Cs' and giving an overall grade based on the most common or average grade. In some cases a clear 'F' grade for one of the minimum regulatory requirements (e. g. non-technical summary) could be seen as resulting in an overall fail for the EIS because of the importance of that particular aspect. Other areas (e. g. consideration of alternatives) may be seen as less crucial where that aspect is not of particular relevance to the project in question. An 'F' grade for one such criteria, may not, in such cases, prevent an EIS being attributed a 'C' grade, or above, overall. Attributing the overall grade for an EIS through this process requires the reviewer to come to a judgement on the weight to be given to the individual review areas and is rather like attributing weight to planning considerations.

The success of EIS review relies a great deal on the experience of the reviewer and their ability to make a judgement on the quality of the EIS as a whole, based upon the systematic assessment of its parts. In reviewing the EIS a reviewer should come to a view on the information provided based upon a balance between:

- what it 'must' contain;
- what it could contain; and
- what it can be reasonably expected to contain.

Oxford Brookes University

Impacts Assessment Unit

Environmental Impact Statement Review Package

Name of Project:

EIS Submitted by:

Date Submitted:

Review Grades

A = Relevant tasks well performed, no important tasks left incomplete.

B = Generally satisfactory and complete, only minor omissions and inadequacies.

C = Can be considered just satisfactory despite omissions and/or inadequacies.

D = Parts are well attempted but must, as a whole, be considered just unsatisfactory because of omissions and/or inadequacies.

E = Not satisfactory, significant omissions or inadequacies.

F = Very unsatisfactory, important task(s) poorly done or not attempted.

NA = Not applicable in the context of the EIS or the project.

1 DESCRIPTION OF THE DEVELOPMENT

Criterion	Review grade	Comments
Principal features of the project		
1. 1 Explains the purpose(s) and objectives of the development.		
1. 2 Indicates the nature and status of the decision(s) for which the environmental information has been prepared.		
1. 3 Gives the estimated duration of the construction, operational and, where appropriate, decommissioning phase, and the programme within these phases.		
1. 4 Provides a description of the development comprising information on the site, design and size of the development. [1]		
1. 5 Provides diagrams, plans or maps and photographs to aid the description of the development.		
1. 6 Indicates the physical presence or appearance of the completed development within the receiving environment.		
1. 7 Describes the methods of construction.		
1. 8 Describes the nature and methods of production or other types of activity involved in the operation of the project.		
1. 9 Describes any additional services (water, electricity, emergency services etc.) and developments required as a consequence of the project.		
1. 10 Describes the project's potential for accidents, hazards and emergencies.		
Land requirements		
1. 11 Defines the land area taken up by the development and/or construction site and any associated arrangements, auxiliary facilities and landscaping areas, and shows their location clearly on a map. For a linear project, describes the land corridor, vertical and horizontal alignment and need for tunnelling and earthworks.		
1. 12 Describes the uses to which this land will be put, and demarcates the different land use areas.		
1. 13 Describes the reinstatement and after-use of landtake during construction.		
Project inputs		
1. 14 Describes the nature and quantities of materials needed during the construction and operational phases.		
1. 15 Estimates the number of workers and visitors entering the project site during both construction and operation.		
1. 16 Describes their access to the site and likely means of transport.		
1. 17 Indicates the means of transporting materials and products to and from the site during construction and operation, and the number of movements involved.		
Residues and emissions		
1. 18 Estimates the types and quantities of waste matter, energy (noise, vibration, light, heat, radiation etc.) and residual materials generated during construction and operation of the project, and rate at which these will be produced.		
1. 19 Indicates how these wastes and residual materials are expected to be handled/treated prior to release/disposal, and the routes by which they will eventually be disposed of to the environment.		
1. 20 Identifies any special or hazardous wastes (defined as ...) which will be produced, and describes the methods for their disposal as regards their likely main environmental impacts.		
1. 21 Indicates the methods by which the quantities of residuals and wastes were estimated. Acknowledges any uncertainty, and gives ranges or confidence limits where appropriate.		
Overall Grade for Section 1 = Comments		

1 Schedule 4 Part 2 Criteria (2011 EIA Regulations)

2 DESCRIPTION OF THE ENVIRONMENT

Criterion	Review grade	Comments
Description of the area occupied by and surrounding the project		
2. 1 Indicates the area expected to be significantly affected by the various aspects of the project with the aid of suitable maps. Explains the time over which these impacts are likely to occur.		
2. 2 Describes the land uses on the site(s)and in surrounding areas.		
2. 3 Defines the affected environment broadly enough to include any potentially significant effects occurring away from the immediate areas of construction and operation. These may be caused by, for example, the dispersion of pollutants, infrastructural requirements of the project, traffic etc.		
Baseline conditions		
2. 4 Identifies and describes the components of the affected environment potentially affected by the project.		
2. 5 The methods used to investigate the affected environment are appropriate to the size and complexity of the assessment task. Uncertainty is indicated.		
2. 6 Predicts the likely future environmental conditions in the absence of the project. Identifies variability in natural systems and human use.		
2. 7 Uses existing technical data sources, including records and studies carried out for environmental agencies and for special interest groups.		
2. 8 Reviews local, regional and national plans and policies, and other data collected as necessary to predict future environmental conditions. Where the proposal does not conform to these plans and policies, the departure is justified.		
2. 9 Local, regional and national agencies holding information on baseline environmental conditions have been approached.		
Overall Grade for Section 1 = Comments		

3 SCOPING, CONSULTATION AND IMPACT IDENTIFICATION

Criterion	Review grade	Comments
Scoping and consultation		
3. 1 There has been a genuine attempt to contact the general public, relevant public agencies, relevant experts and special interest groups to appraise them of the project and its implication. Lists the groups approached.		
3. 2 Statutory consultees have been contacted. Lists the consultees approached.		
3. 3 Identifies valued environmental attributes on the basis of this consultation.		
3. 4 Identifies all project activities with significant impacts on valued environmental attributes. Identifies and selects key impacts for more intense investigation. Describes and justifies the scoping methods used.		
3. 5 Includes a copy or summary of the main comments from consultees and the public, and measures taken to respond to these comments.		
Impact identification		
3. 6 Provides the data required to identify the main effects that the development is likely to have on the environment. [1]		

Continued

Criterion	Review grade	Comments
3.7 Considers direct and indirect/secondary effects of constructing, operating and, where relevant, after-use or decommissioning of the project (including positive and negative effects). Considers whether effects will arise as a result of 'consequential' development.		
3.8 Investigates the above types of impacts in so far as they affect: human beings, flora, fauna, soil, water, air, climate, landscape, interactions between the above, material assets, cultural heritage.		
3.9 Also noise, land use, historic heritage, communities.		
3.10 If any of the above are not of concern in relation to the specific project and its location, this is clearly stated.		
3.11 Identifies impacts using a systematic methodology such as project specific checklists, matrices, panels of experts, extensive consultations, etc. Describes the methods/approaches used and the rationale for using them.		
3.12 The investigation of each type of impact is appropriate to its importance for the decision, avoiding unnecessary information and concentrating on the key issues.		
3.13 Considers impacts that may not themselves be significant but that may contribute incrementally to a significant effect.		
3.14 Considers impacts that might arise from non-standard operating conditions, accidents and emergencies.		
3.15 If the nature of the project is such that accidents are possible that might cause severe damage within the surrounding environment, an assessment of the probability and likely consequences of such events is carried out and the main findings reported.		
Overall Grade for Section 1 = Comments		

4 PREDICTION AND EVALUATION OF IMPACTS

Criterion	Review grade	Comments
Prediction of magnitude of impacts		
4.1 Describes impacts in terms of the nature and magnitude of the change occurring and the nature, location, number, value, sensitivity of the affected receptors.		
4.2 Predicts the timescale over which the effects will occur, so that it is clear whether impacts are short, medium or long term, temporary or permanent, reversible or irreversible.		
4.3 Where possible, expresses impact predictions in quantitative terms. Qualitative descriptions, where necessary, are as fully defined as possible.		
4.4 Describes the likelihood of impacts occurring, and the level of uncertainty attached to the results.		
Methods and data		
4.5 Provides the data required to assess the main effects that the development is likely to have on the environment.[1]		
4.6 The methods used to predict the nature, size and scale of impacts are described, and are appropriate to the size and importance of the projected disturbance.		

Continued

Criterion	Review grade	Comments
4.7 The data used to estimate the size and scale of the main impacts are sufficient for the task, clearly described, and their sources clearly identified. Any gaps in the data are indicated and accounted for.		
Evaluation of impact significance		
4.8 Discusses the significance of effects in terms of the impact on the local community (including distribution of impacts) and on the protection of environmental resources.		
4.9 Discusses the available standards, assumptions and value systems that can be used to assess significance.		
4.10 Where there are no generally accepted standards or criteria for the evaluation of significance, alternative approaches are discussed and, if so, a clear distinction is made between fact, assumption and professional judgement.		
4.11 Discusses the significance of effects taking into account the appropriate national and international standards or norms, where these are available. Otherwise the magnitude, location and duration of the effects are discussed in conjunction with the value, sensitivity and rarity of the resource.		
4.12 Differentiates project-generated impacts from other changes resulting from non-project activities and variables.		
4.13 Includes a clear indication of which impacts may be significant and which may not and provides justification for this distinction.		
Overall Grade for Section 4 = Comments		

5 ALTERNATIVES

Criterion	Review grade	Comments
5.1 Provides an outline of the main alternatives studied and gives an indication of the main reasons for their choice, taking into account the environmental effects. [1]		
5.2 Considers the 'no action' alternative, alternative processes, scales, layouts, designs and operating conditions where available at an early stage of project planning, and investigates their main environmental advantages and disadvantages.		
5.3 If unexpectedly severe adverse impacts are identified during the course of the investigation, which are difficult to mitigate, alternatives rejected in the earlier planning phases are re-appraised.		
5.4 The alternatives are realistic and genuine.		
5.5 Compares the alternatives' main environmental impacts clearly and objectively with those of the proposed project and with the likely future environmental conditions without the project.		
Overall Grade for Section 5 = Comments		

6 MITIGATION AND MONITORING

Criterion	Review grade	Comments
Description of mitigation measure		
6. 1 Provides a description of the measures envisaged in order to avoid, reduce and, if possible, remedy significant adverse effects.		
6. 2 Mitigation measures considered include modification of project design, construction and operation, the replacement of facilities/resources, and the creation of new resources, as well as 'end-of-pipe' technologies for pollution control.		
6. 3 Describes the reasons for choosing the particular type of mitigation, and the other options available.		
6. 4 Explains the extent to which the mitigation methods will be effective. Where the effectiveness is uncertain, or where mitigation may not work, this is made clear and data are introduced to justify the acceptance of these assumptions.		
6. 5 Indicates the significance of any residual or unmitigated impacts remaining after mitigation, and justifies why these impacts should not be mitigated.		
Commitment to mitigation and monitoring		
6. 6 Gives details of how the mitigation measures will be implemented and function over the time span for which they are necessary.		
6. 7 Proposes monitoring arrangements for all significant impacts, especially where uncertainty exists, to check the environmental impact resulting from the implementation of the project and its conformity with the predictions made.		
6. 8 The scale of any proposed monitoring arrangements corresponds to the potential scale and significance of deviations from expected impacts.		
Environmental effects of mitigation		
6. 9 Investigates and describes any adverse environmental effects of mitigation measures.		
6. 10 Considers the potential for conflict between the benefits of mitigation measures and their adverse impacts.		
Overall Grade for Section 6 =Comments		

7 NON-TECHNICAL SUMMARY

Criterion	Review grade	Comments
7. 1 There is a non-technical summary of the information provided under paragraphs 1 to 4 of Part 2 of Schedule 4. 1.		
7. 2 The non-technical summary contains at least a brief description of the project and the environment, an account of the main mitigation measures to be undertaken by the developer, and a description of any remaining or residual impacts.		
7. 3 The summary avoids technical terms, lists of data and detailed explanations of scientific reasoning.		
7. 4 The summary presents the main findings of the assessment and covers all the main issues raised in the information.		
7. 5 The summary includes a brief explanation of the overall approach to the assessment.		
7. 6 The summary indicates the confidence that can be placed in the results.		
Overall Grade for Section 7 = Comments		

8 ORGANISATION AND PRESENTATION OF INFORMATION

Criterion	Review grade	Comments
Organisation of the information		
8.1 Logically arranges the information in sections.		
8.2 Identifies the location of information in a table or list of contents.		
8.3 There are chapter or section summaries outlining the main findings of each phase of the investigation.		
8.4 When information from external sources has been introduced, a full reference to the source is included.		
Presentation of information		
8.5 Mentions the relevant EIA legislation, name of the developer, name of competent authority(ies), name of organisation preparing the EIS, and name, address and contact number of a contact person.		
8.6 Includes an introduction briefly describing the project, the aims of the assessment, and the methods used.		
8.7 The statement is presented as an integrated whole. Data presented in appendices are fully discussed in the main body of the text.		
8.8 Offers information and analysis to support all conclusions drawn.		
8.9 Presents information so as to be comprehensible to the non-specialist. Uses maps, tables, graphical material and other devices as appropriate. Avoids unnecessarily technical or obscure language.		
8.10 Discusses all the important data and results in an integrated fashion.		
8.11 Avoids superfluous information (i.e. information not needed for the decision).		
8.12 Presents the information in a concise form with a consistent terminology and logical links between different sections.		
8.13 Gives prominence and emphasis to severe adverse impacts, substantial environmental benefits, and controversial issues.		
8.14 Defines technical terms, acronyms and initials.		
8.15 The information is objective, and does not lobby for any particular point of view. Adverse impacts are not disguised by euphemisms or platitudes.		
Difficulties compiling the information		
8.16 Indicates any gaps in the required data and explains the means used to deal with them in the assessment.		
8.17 Acknowledges and explains any difficulties in assembling or analysing the data needed to predict impacts, and any basis for questioning assumptions, data or information.		
Overall Grade for Section 8 = Comments		

COLLATION SHEET

Minimum requirements of Schedule 4 Part 2 (2011 EIA Regulations)

Criterion	Overall Grade	Areas where more information required
(1)A description of the development comprising information on the site, design and size of the development.		
(2)A description of the measures envisaged in order to avoid, reduce and, if possible, remedy significant adverse effects.		
(3)The data required to identify and assess the main effects that the development is likely to have on the environment.		
(4)An outline of the main alternatives studied and an indication of the main reasons for their choice, taking into account the environmental effects.		
(5)A non-technical summary of the information provided under 1 to 4 above.		
Overall Grade(A-F):		
List of Information that is required to complete the EIS		

IAU Best Practice Requirements

Criterion	Overall Grade	Areas where more information required
Description of the development		
Description of the environment		
Scoping, consultation, and impact identification		
Prediction and evaluation of impacts		
Alternatives		
Mitigation and monitoring		
Non-technical summary		
Organisation and presentation of information		
Overall Grade(A-F):		
Comments